T0207004

Observing handbook and
Catalogue of deep-sky objects

Observing Handbook and Catalogue of Deep-Sky Objects

CHRISTIAN B. LUGINBUHL
BRIAN A. SKIFF

PUBLISHED BY THE PRESS SYNDICATE OF THE UNIVERSITY OF CAMBRIDGE
The Pitt Building, Trumpington Street, Cambridge CB2 1RP, United Kingdom

CAMBRIDGE UNIVERSITY PRESS
The Edinburgh Building, Cambridge CB2 2RU, UK http://www.cup.cam.ac.uk
40 West 20th Street, New York, NY 10011-4211, USA http://www.cup.org
10 Stamford Road, Oakleigh, Melbourne 3166, Australia

First published 1990
First paperback edition 1998

A catalogue record for this book is available from the British Library

Library of Congress Cataloguing in Publication data

Luginbuhl, Christian B.
Observing handbook and catalogue of deep-sky objects/Christian
B. Luginbuhl, Brian A. Skiff.
p. cm.
Bibliography: p.
ISBN 0-521-25665-8
1. Astronomy–Handbooks, manuals, etc. 2. Astronomy—Observers' manuals. I. Skiff, Brian A. II. Title.
QB64.L84 1990
522—dc20 89–7318 CIP

ISBN 0 521 25665 8 hardback
ISBN 0 521 62556 4 paperback

Transferred to digital printing 2003

CS

For David Philip Forsyth

Contents

Charts, tables and figures

Charts

Tables

Figures

Prolegomenon

This book grew out of the lack of a comprehensive modern manual to aid in observing deep-sky objects. All earlier guides are either outdated, lacking in descriptive detail, or limited in their scope.

We began by undertaking a systematic program of visual observations, the aim being to observe each of approximately 1500 galaxies, star clusters, and nebulae through three telescopic apertures commonly used by amateur astronomers. As observing proceeded, we collected data on each object from a wide variety of sources, mostly reference catalogues and scientific papers in the professional literature. Accurate total magnitudes were newly derived for open clusters and planetary nebulae. Further, each object was examined after observation on at least one of three photographic sky atlases or on large-scale telescopic photographs to ensure the authenticity of details we recorded visually.

Consequently, what we present is not merely a list of visual observations, but a compendium of information relevant to viewing the brightest deep-sky objects. Included are dimensions, magnitudes, orientations, and angular distances, verified in nearly all cases through photography, photometry, or astrometry. Of the 88 constellations in the sky, 68 north of Declination $-50°$ are encompassed by the present survey. The objects included range from those best viewed in binoculars and small telescopes, such as the Pleiades, to fifteenth magnitude galaxies. Hundreds of companions to bright galaxies and lesser-known open clusters likely to be of interest in modest instruments are described, many for the first time. In all, about 2050 objects are described. For a number of crowded fields where ambiguity may arise in identifying objects, finder charts that include faint telescopic stars are provided. Six of the brighter clusters of galaxies are also mapped. To aid in determining approximate visual magnitudes at the telescope, we provide photographic charts of ten star clusters with accurate V magnitudes conveniently labelled. This collection of observations and relevant data, though destined in many instances to become outdated as astronomical research progresses, is the most comprehensive and up-to-date available.

Many people have been of considerable aid in the realization of this book, and deserve grateful acknowledgement. Thanks go above all to Hank Messick for his extensive contributions to many phases of the project; also to Richard C. Hall for material help in many ways; Robert L. Wildey for providing the 30 cm telescope, lending a copy of the Lick Sky Atlas, and for puzzled looks and encouragement; William G. Delinger and Randy Johnson for early computer assistance in compiling the catalogue; Harold G. Corwin (University of Texas) and Gösta Lyngå (Lund Observatory), who helped with and confirmed identities of galaxies and open clusters; James B. Kaler (University of Illinois) and Jack B. Marling (Lumicon) for providing data for planetary nebulae prior to publication; Lowell Observatory for the use of their extensive library facilities, and for permission to reproduce photographic plates from their collection; the staff of the U.S. Naval Observatory Flagstaff Station for aluminizing the 25 cm and 30 cm mirrors, use of the Palomar Sky Survey prints, information on double stars, photographs and references from their library, and for friendly help and interest all along the way. William R. Willis, past chairman of the Northern Arizona University Astronomy and Physics department, deserves our apologies for our late-night "theft" of University optical property.

Some of the material in this book has appeared in a different form in *Deep Sky Monthly* and its quarterly successor *Deep Sky*, published by Astromedia Corporation; we are grateful to editor Dave Eicher for permission to use it here. Carol Vireday magically transformed a very rough draft into a crisply typed manuscript, a considerable undertaking by any measure. Helen Horstman converted the final draft into electronic form.

And thanks to Hank Messick, Sr., who thought of it in the first place.

Christian B. Luginbuhl
Brian Skiff

Flagstaff, Arizona

Amateur observing

Telescopes

There is no ideal telescope for the amateur observer, no telescope that will provide the best views of all the myriad objects in the night sky one may wish to look at. However, almost any telescope from a simple 6 cm refractor to a large observatory instrument will give good views of some deep-sky objects; even the unaided eye will show many objects in a clear dark sky! A single telescope has its limitations following from its aperture and focal length: the 6 cm or 8 cm refractor with a 2°.5 field provides wonderful views of the Pleiades or the North America nebula, but cannot be expected to resolve faint clusters or show fifteenth magnitude galaxies; a 40 cm Cassegrainian will show only a fraction of the star cluster and almost none of the nebula, but reveals thousands of faint galaxies.

Deep-sky observing can profitably begin even with common hand-held binoculars. From a dark site these will show stars to tenth magnitude; scanning the rich star fields of the Milky Way is a joy best appreciated with such low-magnification, wide-field instruments. Dozens of open clusters are clearly resolved into stars, and many galactic nebulae can be seen with ease. Many globular clusters and galaxies can be spotted, though most will appear as little more than tiny smudges of light.

Small telescopes in the 6 cm to 8 cm range will show stars to magnitude 12 and hundreds of star clusters, nebulae, and galaxies, as the descriptive notes in this book demonstrate. Under even mediocre conditions all objects in Messier's list are visible, and that's just a start. Our experience indicates that most galaxies brighter than V magnitude 11 can be identified at 20× to 50× in such instruments.

Larger telescopes give better views of fainter and smaller objects, but suffer on objects like the Pleiades or the North America nebula. With a common 15 cm telescope, either a large refractor or a more economical reflector, many globular clusters begin to resolve into stars, and upwards of one thousand galaxies are visible to about V magnitude 12.5. At high powers the brighter planetary nebulae become interesting, detailed objects rather than mere points of light or diffuse amorphous blobs. 25 cm aperture will reach about a magnitude fainter than 15 cm, encompassing something like 10,000 deep-sky objects, most of them galaxies. Details in a few galaxies can be seen with 15 cm, but with 25 cm many begin to show features hinting at spiral arms, dark lanes and star clouds. Telescopes of 30 cm aperture can show every object in the NGC and more: over 20,000 galaxies are within reach in dark skies. Visible details in the brighter objects become increasingly common and varied. With apertures larger than 30 cm the number of accessible galaxies increases exponentially until the desire to see them all is akin to that of reckoning sand (apologies to Archimedes).

Of the many types of optical systems, the refractor is clearly to be preferred for instruments of 10 cm aperture or less. Larger than this they become, lamentably, both expensive and unwieldy. Compromise is thus sought among the various types of reflecting telescopes. For all larger apertures the choice is often a simple, robust Newtonian of moderate to short focal ratio, typically f/6. In circumstances where portability is an overriding consideration, a more expensive Cassegrainian or catadioptric telescope can be chosen. Among the largest amateur telescopes, a Newtonian on an altazimuth mount is perhaps the most common, a design that is nevertheless quite bulky. A large number of variations involving ingenious compromises optimizing cost, weight, bulk, and optical performance have appeared in the popular literature, and the observer is directed there for further discussion of telescope designs, materials, and construction.

The telescope's mount is an important consideration for the observer, and indeed can account for a substantial portion of the cost of many commercially made products. An equatorial mount has a definite advantage over an altazimuth mount for finding faint, isolated objects, but it need not be complicated to serve quite well. Slow-motions, quartz-controlled variable-rate clock drives, and other fancy accoutrements add more to the cost than to the usefulness of an amateur's telescope. Most commercially manufactured instruments are mounted equatorially either on German or fork mounts. Many amateurs place larger instruments on so-called Dobson mounts, which are altazimuth mounts where the telescope rotates in azimuth on a large baseplate or ball-bearing turntable (among many variations). With equatorial mounts cardinal directions are directly established by motion of the mount, and this makes finding easy.

Altazimuth mounts cannot easily be swung only in right-ascension or declination. Even when finding is accomplished with the aid of computers, directions in the eyepiece are difficult to determine accurately. Altazimuth mounts are also difficult to maneuver near the zenith, exactly as equatorials are at the celestial pole, but many more objects pass near the zenith than are within a small distance of either pole.

For visual amateur observing a clock drive is not really a necessity. Even at 200×, the diurnal motion of the field is not objectionable, though at such high powers it is a nuisance to have to re-center the object often. One advantage the motion offers, since the field drifts exactly east to west, is that these directions will rarely be confused in the telescope.

A significant part of the amateur deep-sky observer's experience is learning the sky well enough to find many objects without having always to refer to information in catalogues and atlases: to know these objects as parts of the familiar naked-eye constellations. Many professional astronomers can tell you in wonderful and exciting detail about what they have learned or deduced about this galaxy or that star-forming region, but haven't more than a vague notion of where the object lies in the panorama of the night sky. This is an understandable if unfortunate consequence of the efficient pursuit of scientific results, and the necessary expedience of setting circles and computer control. For the amateur there is time for more than just the objects themselves, time to learn their environs and appreciate the spectacle overhead. On large, permanently mounted instruments setting circles may indeed be a worthwhile and time-saving convenience, especially if the view of the sky is obstructed by a dome or other housing. For the typical portable telescope that most amateurs use, however, setting circles are superfluous. Unless you have a permanent pier, or spend a lot of time carefully positioning the mount each time you observe, setting circles become mostly an expensive cosmetic affectation that provide little real benefit. Likewise, the increase in computer control among amateurs is welcome and commendable, making the process of finding an object fast and sure. But it limits your perspective and can take the fun out of observing.

Eyepieces

With eyepieces, it appears now that generally you get what you pay for: more expensive designs will provide a higher degree of correction and sharper images over wider fields of view than cheaper ones. Bear in mind, however, that optical systems slower than about f/10 require less well-corrected oculars to give sharp images than, say, an f/4 Newtonian. For all eyepieces, anti-reflection coatings or "multicoatings" are strongly recommended on all glass to air surfaces. Many manufacturers advertise "coated" lenses, but unless it says "fully coated," make sure it is before you buy.

Table I. *An eyepiece set for a 25 cm telescope.*

Focal Length	Optical Design	Magnification		True Field	
		1×	2×	1×	2×
45 mm	Plössl	50×	100×	60′	30′
24 mm	König	95×	190×	40′	20′
16 mm	König	145×	290×	27′	13′

A large array of eyepieces is extravagant and usually unnecessary for deep-sky viewing. A good basic set would include two eyepieces chosen to give low and medium power, plus a 2× Barlow lens that yields high power with the second eyepiece. A third eyepiece may be added that gives a medium-high power alone, and a very high power in combination with the Barlow. The relative terms used here to describe magnifications are difficult to make specific. The actual magnifications that are "best" for any particular telescope depend on obvious factors like focal length and focal ratio, but also on less obvious and less quantifiable things like the particular object being viewed and the observer doing the viewing. Our preferences are illustrated by the set of eyepieces listed in Table I, which is the set used with a 25 cm f/9 Ritchey-Chrétien.

With this combination, a target (or its expected position) is located and centered in the 50× eyepiece, then examined using 95× or 190×, depending on the size and nature of the object. Some small galaxies of low surface brightness are better viewed at 145× instead of 190×, while globular clusters and planetary nebulae often profit from the use of 290×. Objects within a few degrees of one spot are easily found by sweeping with 95× rather than switching all the way back to 50×, where faint objects are often invisible.

Our experience has been that each aperture has a "highest working power" at which image brightness, contrast, and resolution are optimized for objects near the limit of faintness for the aperture. With telescopes of 15 cm aperture and less, this magnification is about ten times the aperture in centimeters. For larger instruments (up to 40 cm aperture), the magnification is seven to eight times the aperture in centimeters. However, these values are by no means the highest useful powers on each telescope. During periods of exceptionally steady seeing, and with sharp optics, magnifications as high as 20 times the aperture can be used on planets, double stars, and the brightest planetary nebulae.

Once you have selected your set of eyepieces, there are

two items of information you will need for each: the magnification and true field of view they provide when used with your telescope. The magnification is easily found by dividing the eyepiece's focal length into the effective focal length of the telescope. The true field of view can be determined in two ways. The most direct method is to time the transit of a star across the field at the telescope. The true field in arcminutes is then:

$$\text{true field} = 15\ t \cos(\delta)$$

where

t = transit time (in minutes of time)
δ = declination of star observed

Without going to the telescope you can determine the true field if you know the magnification and the apparent field of the eyepiece:

$$\text{true field} = \text{apparent field}/\text{magnification}$$

where the apparent and true fields are expressed in arcminutes.

The apparent field is sometimes given by the manufacturer, but it can be quite easily determined as follows. Holding the eyepiece to one eye, position yourself such that the circle of light you can see through the eyepiece is the same size as something viewed simultaneously with your other eye: a window frame for example, or a standard 11-inch piece of paper on a table. After measuring the size of the object, and the distance from it to your eye, the apparent field (in arcminutes) can be computed as:

$$\text{apparent field} = 120 \tan^{-1} [d/(2r)]$$

where

d = size of object
r = distance from object to eye (in same units as used for d).

Finderscopes and finding

The majority of deep-sky objects visible in any telescope will be small and faint, a simple consequence of the dramatic increase in numbers of objects, particularly galaxies, with magnitude. The finderscope cannot be expected to grasp these objects directly; it need only reach stars of magnitude nine or ten. These stars are sufficiently numerous to guide the observer very close to most objects. If some care is given to the selection of a good finderscope, it can be used not only as an accessory to the main telescope, but also as a small telescope in its own right. An ideal finderscope is a small refractor of 5 or 6 cm aperture that permits the use of standard 1.25-inch eyepieces. At 20×, using a wide-field eyepiece, such an instrument typically gives a sharp 2° or 3° field, which will provide fine views of the most extended deep-sky objects. At 60× to 100× worthwhile observations can be made of many other objects. To be useful at these higher magnifications, the finder has to be of reasonably good optical quality. All too often, the finder is a neglected part of amateur instruments, of such inferior quality that it barely provides for finding, let alone observing.

The main concern when choosing a finder is its field of view on the sky. A true field of 2° to 3° at low power (10× to 25×) is a good size to allow pointing with dispatch. With practice, or with the aid of setting circles, one can do with a smaller field, especially in familiar regions of the sky. Many objects in our survey were located with a 15 cm refractor using only its 1° low-power field.

The qualities of the finder must reflect the use to which it is put, and different methods of finding may require different finderscopes. Computer-controlled setting or manual setting with accurately aligned mounts with setting circles may require no finderscope. The preferences outlined above reflect a method of finding commonly referred to as "star hopping," which we use almost exclusively with our portable instruments, and even most of the time with permanently mounted telescopes. This method begins with finding a naked-eye star in the finder, and then proceeding with the finder to fainter stars found on an atlas that lead closer and closer to the sought object. For this method, if the finder's field of view is too small, sighting the first star will be difficult, while if the field of view is too large, fainter stars and patterns of stars will be hard to see.

Star atlases

After you are outfitted with a telescope and its necessary accessories, the next concern should be your references, particularly atlases. What you are likely to find with your telescope will depend on what you know about, and that will depend largely on what references you have at your disposal. When building an observer's library, a few books stand out from the sea of those written for the amateur astronomer.

Among innumerable all-sky star cards, charts, finderwheels, and guides to the constellations for the beginner, the most sensible book remains H. A. Rey's *The Stars: A New Way to See Them*. Besides outlining the constellations with recognizable stick figures, the text provides an excellent introduction to naked-eye astronomy.

For the beginning telescopic observer there are a number of good guides, among them Menzel and Pasachoff's *Field Guide to the Stars and Planets*, Ridpath and Tirion's *Universe Guide to Stars and Planets*, and Muirden's *Astronomy with Binoculars*. Some of these works include small scale charts of the entire sky with a few hundred of the brightest deep-sky objects labelled.

Medium-scale maps are more convenient for detailed

deep-sky observation. The best map of this sort is Tirion's *Sky Atlas 2000.0*. These charts show a wealth of deep-sky objects of all kinds, and about 45,000 stars to visual magnitude 8. The "Deluxe" edition color-codes the objects and delineates the Milky Way in striking detail. The two-volume companion *Sky Catalogue 2000.0* brings together data on nearly everything plotted in the *Atlas*.

For more experienced observers, the Uranometria 2000.0 (Tirion, Rappaport, and Lovi) is arguably the best atlas for in-depth scrutiny of deep-sky objects. These large-scale maps (3.′2 per mm) have enough magnitude reach (complete to about magnitude 9.5) so that finding is made easy. This is the only atlas available that has a fairly complete selection of deep-sky objects plotted on it, including most of the NGC and all of the objects listed in the Sky Catalogue 2000.0.

Uranometria was published after our work was finished; we used instead the Atlas Borealis, Atlas Eclipticalis, and Atlas Australis by Antonin Becvar. These atlases had beautiful large-scale charts similar in star count and scale to Uranometria, but the stellar information coded on them was more complete, including color-coded spectral types, double and variable stars to magnitude 10 (to only mag. 8 on Uranometria), and in some zones field stars as faint as magnitude 14. Unfortunately no deep-sky objects were plotted, and so had to be added by hand, a very tedious process. Alas, these atlases have gone out of print; Uranometria forms a good substitute for practically all deep-sky observing.

Gadgets

Our experience with filters designed to reduce sky glow and even light pollution has been generally favorable. These filters are especially effective on HII regions and many planetary nebulae, but are not so useful on other types of objects when the sky is clear and dark. One filter, which passes the nebular Hβ and [OIII] lines at 486 nm and 501 nm, enhances emission nebulae to a remarkable degree, helping even when very thin clouds or scattered moonlight interfere. These filters are designed for nebulous objects and will not let one view fainter *stars* (therefore including most light from galaxies) than one would see without such devices: the nebula filter just described produces about a magnitude loss on stars. Because these filters have become generally available only in recent years, none of the observations included in this book make use of them.

A few planetary nebulae were examined by the authors with a direct-vision spectroscope. This device renders stars as rainbow-colored streaks, whereas nebulae remain as a single spot or appear as a string of spots, since most of their light is emitted at discrete wavelengths. The smallest planetaries can be visually distinguished from stars only by using this instrument. The central star, if visible, usually appears as a thin streak on which the nebular emission images are strung. Some central stars have emission lines in their spectra, but these occur usually at different wavelengths than those of the nebulae.

Looking through the telescope

Understanding the properties and function of your eyes leads to better appreciation of how they need to be treated to perform well at the telescope. The deep-sky observer gives his eyes a real workout, straining them to see just a little fainter galaxy, or a bit more detail in a complex planetary nebula. To get the best results, an observer must give attention to the relevant characteristics of the eye, the telescope, and the way these two work together.

Optically, the eye is a refractive device. Light that impinges upon the front of the cornea is refracted first through an adjustable aperture, the iris, then through a lens with an adjustable focal length, finally falling onto the retina, the light-sensitive tissue of the eye. Focusing of the light is primarily accomplished at the curved front surface of the cornea, the lens providing mostly for the ability to focus at different distances in front of the eye. In the healthy eye, the rays that fall onto the retina form a sharply focused image.

The retina is a marvel in its ability to function well under a tremendous range of light intensities. The dark-adapted eye is more sensitive to light than any photographic emulsion, being exceeded in sensitivity only by modern electronic detectors. Dark adaptation of the eye is due to two factors: the opening of the iris, a relatively fast response, accounting for a factor of about seven gain in sensitivity; and slow chemical changes in the photore-

Figure 1. Dark adaptation of the human eye.

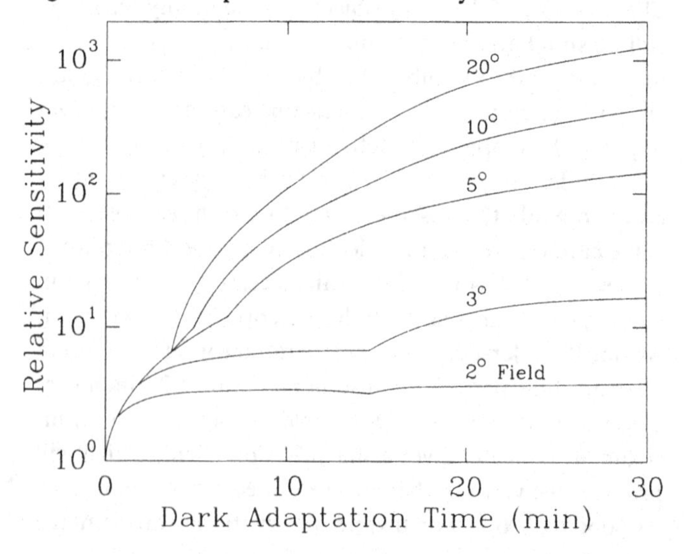

ceptive cells of the retina, requiring thirty minutes or more, which result in an overall gain of about 1000 for peripheral vision (see Figure 1).

The photoreceptive cells of the human eye are of two types, called rods and cones after their shapes. The cone cells are concentrated toward the center of vision of the retina, and are responsible for the perception of color. Away from the center of vision, the rod cells become more frequent; it is these cells that are responsible for the retina's adaptation to low light levels. Rods, however, are insensitive to color, which causes us to tend toward color blindness in dim light. The spectral sensitivity of dark-adapted (scotopic) vision peaks at about 510 nm, somewhat bluer than light-adapted (photopic) vision, and drops to one percent of maximum at 400 nm (blue-violet) and 620 nm (orange-red). The rod cells are also distributed more thinly than the cone cells, and because of this the eye's ability to distinguish fine detail is considerably diminished in faint light.

Dark adaptation must be cultivated and carefully preserved. Do not expect to be able to see your best at the telescope for *at least* half an hour after being in a lighted room, particularly one with fluorescent lights. Even after driving a car at night, dashboard illumination and reflected headlights will leave your eyes more photopic than scotopic. There are few better ways to begin a night of observing than by enjoying the twilight display as the sky darkens and your eyes adapt.

Though there is no single correct way to look into a telescope, there are certainly incorrect ways. The deep-sky observer should seek a method that blocks extraneous light, if any, but minimizes eyestrain. Both eyes should be kept open. Face the eyepiece squarely, not at an angle, and look directly into it. Get your eye as close to the eyepiece as is necessary to see the field stop, but not so close that the field is restricted: the edge of the field of view should be crisply defined. Do not hold your eye at such a distance from the eyepiece that you cannot hold the relative position fixed. Pressing the side or bridge of your nose and the edge of your eye-socket against the eyepiece will both steady your eye and help block light. If this is too close for the eyepiece, wrap your hand around the eyepiece, then press your face against your hand. If the vision from your unused (but open!) eye interferes, cup one hand over it. If nearby lights are distracting and you haven't enough hands to block it effectively, a piece of cloth or a towel attached to the telescope and draped over the head and eyepiece can help. Relax and breathe normally. This may sound obvious, but we often found that in our efforts to hold still and look into the (sometimes awkwardly positioned) telescope, we would stop breathing: the faint galaxy would slowly disappear in the dark graininess of our oxygen-starved eyes.

When using eyepieces with long eye relief (the distance back from the lens at which the exit pupil forms and at which you must hold your eye), observers who wear glasses can keep them on, though care will be needed to avoid scratching plastic eyeglass lenses. Defects of the eye are more likely to affect the image quality here, and the glasses can serve to keep the eye steady and at the right distance. At higher powers, eye relief is usually much shorter, and bespectacled observers should remove them and refocus accordingly. Since the exit pupil is smaller under these circumstances, using a smaller part of your cornea, the eye naturally forms sharper images and defects such as astigmatism are reduced. The field of view will be severely restricted if the eyeglasses are kept on.

When looking at an object in the telescope the eyes need to be used in a manner that is very different than usual: it is because the uninitiated are unfamiliar with this that they are so often disappointed by what they see in the telescope. At low light levels the eye's response is dominated by the characteristics of the rod cells. Since there are almost no rods in the center of vision or fovea of the retina, a technique called averted vision is used that utilizes the more light-sensitive area outside of the center of vision. When using this technique, you do not look directly at an object, but instead look a little off to the side. How far to the side is difficult to say, and probably varies from individual to individual. There are two factors that come into play here. First, the rod cells become more and more dominant with respect to the cone cells as the distance from the fovea increases. This would mean that the farther away from the object you looked the better you could see it. But of course this isn't true, even in faint light, because of the second factor, which is the overall *decrease* in density of photoreceptive cells as distance from the fovea increases. This second phenomenon leads to poorer and poorer ability to discern fine detail as the distance from the center of vision increases. Because of these factors, averted vision is a compromise between the ability to detect light and the ability to discern detail. For many people, the eyes almost naturally make use of averted vision when light levels are very low, but for others it takes practice. Observers should always use this technique when viewing, although details in bright objects can occasionally be better seen with direct vision.

Another useful characteristic of peripheral vision is its unusual sensitivity to a moving stimulus. Often details that are of very low contrast can be discerned more easily if the image is caused to move on the retina. This is accomplished usually by rapidly shifting your gaze from point to point about the object, but sometimes jiggling the telescope itself will bring unsuspected detail into view.

Lighting and the recording of notes

In order to keep the eyes as completely dark adapted as possible, a minimum of illumination must be used at all times while you are observing. If electrical power is available at an outdoor site, faint lights of many types can be used provided they are covered with red filters. A heavily reddened flashlight is the most useful in a powerless situation and is indeed often to be preferred in any case, since it can be held very close to what you need to see and thus made to be quite dim. Whatever light you use, remember that red light interferes with dark adaptation less than other colors, a fact well demonstrated by many studies (see Notes on references for deep-sky observers).

Lighting for the writing of notes can be eliminated by recording observations on a cassette tape recorder. Even at remote sites there is enough light from the sky to see the switches that operate the machine, thus helping to preserve dark adaptation and reducing eyestrain on the observer. We have noticed, however, that in speaking into the tape recorder we tend to become verbose and redundant, sometimes omitting important parts of descriptions (cosmic stage fright?). Since you can't easily review what you have already said as you are speaking (unlike with writing), the tape recorder requires a degree of concentration that is difficult to maintain through a long night of observing.

Whether or not notes are kept on the objects observed is an individual decision each observer must make. The number of deep-sky objects is large enough, even using small telescopes, that one cannot hope to remember a great deal about many of them. Recorded notes provide a bank of information to which the observer may refer, either when reobserving an object, or when discussing an object's appearance with other observers. Careful note-taking is the preference of the authors, and we encourage observers to make detailed notes of the things they observe: you will find that the collected notes will shortly form a valued collection of descriptions that you will often find yourself comparing to subsequent observations of your own and the notes of other observers. Writing notes forces you to look carefully at each object, noting how it is similar to other objects, and how it differs. Simply writing down or checking off a number may give you some feeling of accomplishment, but it will feel pretty thin when next week or next year you can't remember a thing about the object to which the number refers. The time spent making careful observations and notes may make the checklist shorter, but it will certainly make it richer.

When first starting to observe, the differences that make each object unique seem to many to be lost in a sea of similarities: they're all so many faint smudges. As you look at more and more, though, you'll find yourself noticing details that weren't visible before. This special kind of visual acuity will steadily improve as you continue to observe and develop your observing skills.

A quick look at some of the descriptions in this book will give an idea of the kinds of things we try to record when we observe a deep-sky object; you may note yet others. Below we briefly outline what we always try to note when we observe.

To begin with, the date, location, telescope used (if you use more than one), and general observing conditions are of interest. The observing conditions can be described by the transparency and seeing. A good way to rate the transparency is by estimating the limiting visual magnitude of stars near the zenith, either with the naked eye or with the telescope. To determine the limiting magnitude of your telescope, ten photographs of open and globular cluster fields are provided with V magnitudes labelled for a number of stars. (Following the AAVSO convention, magnitudes are given to tenths of a magnitude with the decimal point omitted.)

For seeing you might estimate the actual diameter of stellar images in arcseconds, using double stars as a gauge. Pairs may be found that are merged, or others that are resolved, but whose known separation permits an estimate of the size of the images. A list of close, equal double stars to help gauge seeing is included in the Appendix of double stars.

When taking notes on a deep-sky object, be specific. General observations such as "great," "beautiful," or "dull" may be a major part of your impression at the time, but these are least useful in comparing what you see to what others see. They should not be the major part of your written observation. We often begin by noting the relative difficulty of the object. Objects visible easily at low power may be rated "bright" or "easy;" those visible only at high-optimum powers are "difficult" or "pretty faint." Next, at the best high power, the size is estimated, either a circular diameter or the lengths of the longest and shortest dimensions. This is done from knowledge of the true angular field of the eyepiece in use and estimating what portion of that field the object covers or extends to. Doing this accurately takes practice. We find that we tend to overestimate the size of objects that are small in comparison to the size of the eyepiece field.

If the object is elongated, we note the orientation at least approximately ("elongated NE-SW"), or accurately in terms of position angle (pa). This is a compass angle measured from north counterclockwise on the sky. Thus due north is pa 0°; an object elongated exactly northeast to southwest will have a pa of 45°; elongation southeast to northwest is pa 135°. To eliminate redundancy for deep-sky objects all angles are referred to the first half of the circle: no position angles exceed 179°. Positions *relative* to a deep-sky object and double star orientations can naturally cover the full 360°. Double star position angles are always with reference to the brighter star. Using an equatorial mount enables position angles to be estimated with practice to about 15° accuracy. With equatorially mounted Cassegrainians you can do even better, since the orientation of the sky in the eyepiece (if you don't use a star diagonal) is always the same relative to the telescope: marking the cardinal directions on the tailpiece of the telescope could be useful here.

Next we note any zones of brightness in galaxies or nebulae, and the relative concentration of stars in clusters. We consistently use a number of terms to apply to specific observable phenomena in deep-sky objects. A "stellaring" is any faint star-like manifestation appearing on the surface of a nebulous object. In galaxies these may be true stars of the Milky Way superposed on it, or a bright star cloud or HII region within the galaxy that appears stellar due to the modest telescopic power in use. The terms "halo," "core," and "nucleus" refer to more-or-less well-defined zones of brightness from the edge to the center of an object.

The general terms "concentration" or "condensation" refer to any brightening in a nebula, or to the rise in apparent density of stars toward the center of a cluster. How the brightness changes with radius varies from object to object. The globular clusters, more than any other group, exhibit most clearly the idealized concentration types we recognize. In Figure 2 are brightness profiles illustrating what we call "broad," "even," and "sharp" concentrations.

The clusters Messier 55 (NGC 6809) in Sagittarius and ω Centauri (NGC 5139) show a broad concentration: the rate of increase in brightness decreases toward the center (Figure 2a). An even concentration indicates that the rate of increase in brightness toward the center is constant (Figure 2b); Messier 5 (NGC 5904) in Serpens is a good example of this type. With a sharp concentration the rate of increase in brightness increases toward the center (Figure 2c); Messier 54 (NGC 6715) and Messier 75 (NGC 6864) in Sagittarius are examples here, both having stellar centers in small apertures. Most objects, of course, exhibit a combination of these types. A hypothetical brightness profile of a galaxy is given in Figure 2d, showing an evenly concentrated halo, broadly brighter core, and a sharply concentrated nucleus.

In star clusters we try to count the stars, not merely guess at the number. If there are many stars, one may count only half or a quarter of them and multiply accordingly. In the case of open clusters, counts are restricted to some area that seems to be the natural size of the group, and this size is noted. This permits a better comparison between observations made on different dates or with other instruments.

It is often of interest to note the location of nearby stars or interesting aspects of the field around an object. Estimated magnitudes, directions, and distances of stars are often included, sometimes using cardinal point or pa and angular measure, sometimes only relative to the object ("a prominent red mag. 9 star at the west edge of the cluster;" "a mag. 13.5 star just off the NE flank of the halo"). The ten cluster charts with labelled magnitude sequences can be used to learn to estimate magnitudes at the telescope to an accuracy of perhaps half a magnitude.

Finally, if a filter is used, a comparison of the appearance of the object with and without it is noted. Such comparisons will distinguish between emission and reflection nebulae when a narrow-band "emission-line" filter is employed. When using a spectroscope on planetaries, we note the number, spacing , and relative brightnesses of the emission-line images. This tells something about the level of excitation in the nebula, while the nature of the central star may be revealed in its spectrum.

Figure 2. Brightness profiles of deep-sky objects (see text).

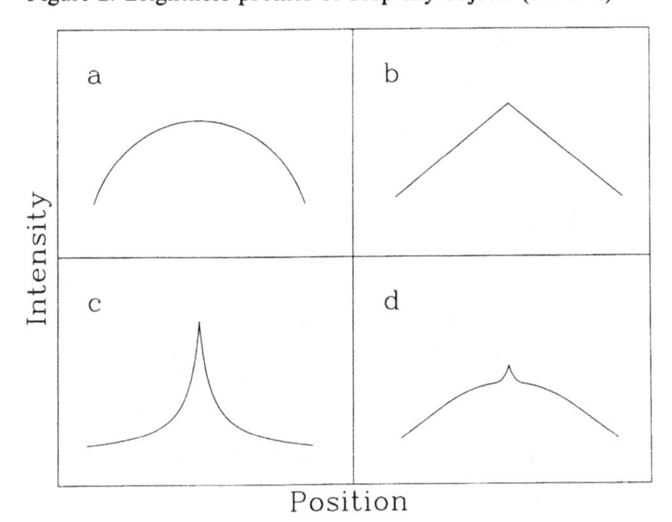

Figure 3. The authors' observing record for NGC 949 in Triangulum.

```
eg 949       dimen.  2:8x1:7        V=11.9        sfcbr=13.4
Tri
15cm- faint streak, fnt * on SSE. Roof. #2:  elong SE-NW
   1'x0:75, much brtr nearly *ar center.
   Anderson Mesa, 3SE81
25cm- inconsp, elong SE-NW, about 1:5x0:75. no features
   except brtr v sm non*ar nuc. Roof. #2:  190x, 1:5x1'
   elong pa150. lenticular w/sl bulging core. core more
   oval, sl grainy texture. mag14 * S of nuc along SW
   flank. occas *ar nuc. Anderson Mesa, 3SE81
30cm- nice, elong, mod concen @140x. 220x shows intrstng
   detail. v elong w/thin tapered tips in pa145, 2:5x
   0:75. core broad, uneven w/occas off-cntr nuc to SW
   side of bar. SW flank fades more abruptly. core has
   thin lengthwise brtning. neat. Roof, 120C77

PSS- pa140. mag13 * 38" from nuc in pa165.
UGC- pa145.
```

Sketches are a helpful supplement to and confirmation of verbal notes, though making them can be very time consuming. They are most effective with interesting, detailed objects when indicating nearby stars or features difficult to describe verbally. They can also serve to show the relationships among several objects in the same field, and provide a means of later identifying unknown objects.

We transcribe our observations verbatim onto unruled index cards that are then filed within each constellation by NGC number. Across the top are typed the identification and vital statistics; the descriptions are then listed according to aperture. A sample card is shown in Figure 3. Sketches can be put on separate cards kept with the verbal notes, or on looseleaf paper kept in a notebook.

Observing locations

The most important thing you can do to optimize the performance of your telescope is to observe from a dark site. A telescope restricted to an urban environment cannot be expected to perform nearly as well as its rural counterpart. Frequently the sole requirement for the visibility of a notoriously difficult nebula is not a large telescope or a special eyepiece, but truly dark sky. The individual in search of dark skies can do little about urban lights except travel to remote locations, something we all cannot afford to do often, especially when long distances must be covered. Filters advertised to improve viewing in light-polluted skies may help somewhat for those stranded too far from dark skies, but they also filter out some of the already scant light from most deep-sky objects. In addition, sites bright enough to need filters will also be bright enough to interfere with dark adaptation, so you lose ground on two fronts. A third factor to consider is the overall experience of observing from a dark site. Though with money and determination you may improve the viewing to a degree, it will never compare with the thrill you get from setting your telescope up in a remote meadow, where the reminders of civilization are distant and the stars seem close.

In selecting an observing site compromises must usually be made regarding transparency, darkness, seeing, and accessibility. Any site, urban or rural, should be relatively open and level, if possible. Temperature inversions, which form on clear still nights, will trap smoke and haze in low-lying areas. Thus it is worthwhile to consider a high spot on the terrain. Such sites above the local inversion will often prove to be very clear. Seeing will vary at even the best sites. There is usually nothing you can do to improve it if it is bad, unless you can determine that it is due to a local effect like a heated building or some other obstruction upwind causing turbulence near the ground. If this is the case, moving even a short distance may improve the seeing dramatically.

Instruments used in the survey of deep-sky objects

Many instruments were used in our survey. The smallest, apart from unaided eyes and an assortment of binoculars, were two 6 cm refractors. One was f/17, on an equatorial tripod, and the other, f/12, served as a finderscope for a 30 cm Cassegrainian telescope. Both instruments were fitted with 1.25-inch focusing mounts so they could accept the higher quality standard diameter eyepieces. Useful magnifications on deep-sky objects with this aperture range up to 120 ×. Despite its comparatively small size, this sort of instrument is quite useful for deep-sky observing. From a dark site all the objects in Messier's list are visible, and some of the best views of galactic nebulae are obtained with these telescopes at very low power. About one hundred observations were made with an 8 cm f/5 telescope limited by its poor optical quality to about 20 ×. Notes on objects previously observed with 6 cm showed that this instrument was roughly equivalent (even slightly inferior); observations made with it have been treated as 6 cm observations in the descriptions that follow.

The 15 cm observations were made in several ways. A majority of the notes were taken by placing an off-axis cardboard stop over a 25 cm telescope, converting it into an f/10 or f/15 system. A number of observations were made with 13 cm and 15 cm f/15 refractors and a 15 cm f/4 Newtonian Richest-Field-Telescope. The refractors had a distinct advantage when examining small galaxies and planetary nebulae thanks to their long focal lengths.

The few 20 cm observations were made through an f/10 Newtonian intended primarily for observing planets. At moderate powers it yielded bright images for the study of clusters and nebulae.

Nearly all the 25 cm observations were made with two f/6 Newtonians, one assembled from various components and without a clock drive, the other of commercial manufacture with a clock drive and a rotating tube saddle. The rest of the notes were taken with a 25 cm f/9 Ritchey-Chrétien telescope.

The majority of the 30 cm observations were made with an f/15 Cassegrainian telescope at Northern Arizona University in Flagstaff. Intended for photoelectric work, the instrument rests on a massive head and serves as a guidescope for a 35 cm Schmidt camera. Since only rather high magnifications were possible with the available eyepieces, large objects could not be viewed in their entirety, although it was still useful for examining details in them. About 400 observations were made with the 60 cm f/8 Cassegrainian telescope at the U.S. Naval Observatory Flagstaff Station fitted with a three-bladed stop that reduced the light-gathering power to that of a 30 cm telescope.

Observing sites for the survey

Much of the observing for the survey was done from the fifth-floor Rooftop Observing Platform atop the Chemistry building on the Northern Arizona University campus in Flagstaff at north latitude 35°.2 and about 2120 meters elevation. Though the sky was fully illuminated by Flagstaff, because of the transparent air, viewing was surprisingly good. On a typical observing night the limiting magnitude for a 15 cm telescope was about 13.6; a 25 cm aperture reached to magnitude 14.4. The naked-eye limit was about 6.2. The zodiacal light was usually visible in the morning or evening; the Merope nebula in the Pleiades was faintly visible in a 6 cm refractor at low power. During winter, the sky brightness was increased considerably by the lights of the city reflecting into the sky from the snow-covered ground. This effect was of course worst after the passage of snow storms, and observing was avoided on these occasions.

Large numbers of observations were made at dark sites: at the Anderson Mesa station of Lowell Observatory about 15 km southeast of Flagstaff, at the U.S. Naval Observatory Flagstaff Station about 10 km west of town, and at an ideally dark undeveloped site about 40 km north of the city. The limiting visual magnitude at these sites, all at 2200 meters elevation or higher, was about 7.0.

Deep-sky data sources

Each class of deep-sky object has its own set of basic references in which are found the best and most recent data on individual objects. In many cases, particularly for planetary nebulae and open clusters, the kinds of data relevant to amateur observing and those of interest to the professional astronomer have not been the same, and consequently there is no published set of data on a consistent system of magnitudes, populations, or sizes. In compiling this catalogue we have tried where possible to present only rigorously defined values for types, magnitudes, and so on. Sometimes this has meant deriving new data or prompting others to do so. The sources used and the meanings of the data are described below.

Galaxies

Nearly all data on galaxies are from the *Second Reference Catalogue of Bright Galaxies* (RC2) by de Vaucouleurs, de Vaucouleurs, and Corwin, and the *Southern Galaxy Catalogue* (SGC) by Corwin, de Vaucouleurs, and de Vaucouleurs. Secondary references are the six-volume *Catalogue of Galaxies and Clusters of Galaxies* (CGCG) by Zwicky and collaborators, the *Morphological Catalogue of Galaxies* (MCG) in five volumes by Vorontsov-Veljaminov and collaborators, and the *Uppsala General Catalogue of Galaxies* (UGC) by Peter Nilson with its southern counterpart *The ESO/Uppsala Survey of the ESO (B) Atlas* (ESO) by Lauberts. For some of the fainter galaxies in rich clusters data come from an assortment of papers in the literature.

Dimensions for galaxies are mostly from the RC2. The values correspond to the boundary at which the galaxy fades to a surface brightness level of B magnitude 25.0 per square arcsecond. In most cases these dimensions are significantly larger than those commonly published, but represent a rigorously defined measurement. In general these values have not been measured for every galaxy, but instead have been inferred using statistical methods from more easily determined characteristics such as integrated magnitude, galaxy type, and inclination to the line of sight. Typically these dimensions are accurate to about ten percent. Dimensions for many objects not listed in the RC2 come from the UGC, ESO or MCG (in order of preference). These data have been transformed to the standard system defined in RC2 using the statistical correction formulae given in the RC2: still their errors can be much larger than ten percent.

For most bright galaxies integrated magnitudes on the standard Johnson UBV system are available in the RC2. These data are based on photoelectric or occasionally photographic measures from many sources and have been reduced to a homogeneous system. The magnitudes represent the total integrated brightness of the galaxy. The V magnitudes are given by preference, or, in the descriptions, the B magnitude where no V data are available. These magnitudes have a standard error of about 0.1 magnitude.

For galaxies lacking detailed photometric study, magnitudes come from the CGCG (indicated by $mag._Z$) and the MCG ($mag._{VV}$). Both are photographic-blue magnitudes (not on the UBV system), but those in the CGCG are of higher accuracy. Typically, errors are less than half a magnitude, though they have systematic errors that depend on the size and morphological type among other factors. The MCG magnitudes were never intended to be photometrically precise, yet for southern galaxies especially, they are often all there is. The MCG magnitudes are given to half or whole magnitudes, but still can be in error by a magnitude or more: galaxies with $mag._{VV}15$ are often not difficult objects in moderate aperture.

The surface brightness magnitudes (sfc. br.), also from the RC2, represent the brightness (in V or B, depending on the color of the integrated magnitude) of a square arcminute patch averaged over the galaxy within the dimensions given for each. Since this value is an *average*, the central parts of a galaxy will typically have higher surface brightness and the outer parts lower. Still, we have found that this value, combined with the integrated magnitude, gives a more consistent indication of the difficulty of an object in a particular aperture than the integrated magnitude alone. As a general rule, galaxies with V surface brightnesses greater than about 13.0 will be visible in 15 cm aperture; the limits for 25 cm and 30 cm are about 14.0 and 14.5, respectively. Like the limiting stellar magnitude, these limits will vary according to sky conditions. Objects that have bright integrated magnitudes or are rather larger than average may be visible beyond these limits, but only the brightest part may be seen, *viz.* the bright core perhaps only one-fourth or one-fifth the total size of the galaxy. The faintest galaxies in terms of integrated magnitude will often be difficult to view simply because they are small, nearly stellar. That they are visible at all is due to their high surface brightness, which for quite small objects will be brighter than their integrated magnitude. Importantly, surface brightness is not a function only of a galaxy's total magnitude, but also of its size and type.

Galaxy types listed in our catalogue are nearly all from the RC2; a few are from the SGC, ESO, UGC, or our own examination of photographs. These types are on a system defined by de Vaucouleurs that is closely related to the well known Hubble system. Hubble's system comprises three basic types: "ellipticals," ranging from circular objects such as M87 (NGC 4486) to thin lens-shaped objects such as NGC 4111; "normal spirals" such as the Andromeda Galaxy (NGC 224); and "barred spirals" such as NGC 1300. The spirals are subclassed according to the degree of winding and resolution of the arms. Hubble later introduced an intermediate type, "lenticulars," for those galaxies having characteristics between elliptical and spiral objects. Chaotic or nondescript forms were classified as "peculiar" or "irregular." These included colliding systems and those with strange distortions.

The revised morphological classification system, as devised by de Vaucouleurs and adopted here, makes use of the same five classes: elliptical, lenticular, spiral (ordinary and barred), irregular, and peculiar. It also distinguishes a wide range of families, varieties, and stages (see Table II).

All elliptical galaxies are designated with an E. The degree of elongation is expressed by a number from 0 through 6 increasing with apparent elongation. Objects classed as "giant" and those with extended halos are designated E^+. Compact objects are designated cE, while dwarf ellipticals are dE.

The lenticulars occur in three families: ordinary (SA0), barred (SB0), and mixed between the two (SAB0). The degree of visibility of structure is represented by three stages: early, intermediate, and late. The normal spirals also include the ordinary, barred, and mixed families. In this class two varieties also are recognized based on the way the spiral arms manifest themselves: the S-shaped variety has typically two strong arms that give an overall S-shape to the brighter regions; other galaxies have a more or less distinct inner ring from which many arms originate and spiral outward. Stages represented by lower-case letters refer to the degree of resolution or fineness of the spiral arms, ranging from tightly wound arms (Sa) to loosely wound and barely discernable irregular spiral structure as in the Magellanic clouds (Sm).

Both lenticulars and spirals can be surrounded by an outer ring, or the spiral arms may nearly close upon themselves to form a pseudo-ring. These features are indicated by the prefixes (R) and (R′), respectively.

Table II. *Coding of revised morphological types.*

Classes	Families	Varieties	Stages	Type
Ellipticals				E
		Compact		cE
		Dwarf		dE
			Ellipticity (0–6)	E0
			(intermediate)	E1.5
		"cD"		E^+
Lenticulars				S0
	Ordinary			SA0
	Barred			SB0
	Mixed			SAB0
		Inner Ring		S(r)0
		S-shaped		S(s)0
		Mixed		S(rs)0
			Early	$S0^-$
			Intermediate	$S0^\circ$
			Late	$S0^+$
Spirals				S
	Ordinary			SA
	Barred			SB
	Mixed			SAB
		Inner Ring		S(r)
		S-shaped		S(s)
		Mixed		S(rs)
			0/a	S0/a
			a	Sa
			ab	Sab
			b	Sb
			bc	Sbc
			c	Sc
			cd	Scd
			d	Sd
			dm	Sdm
			m	Sm
Irregulars				I
	Ordinary			IA
	Barred			IB
	Mixed			IAB
		S-shaped		I(s)
			Magellanic	Im
			Non-magellanic	I0
		Compact		cI
Peculiars				P
Peculiarities			Peculiarity	P
(all types)			Uncertain	:
			Doubtful	?
			Spindle	sp
			Outer Ring	(R)
			Pseudo Outer Ring	(R')

Irregulars occur in unstructured (IA) and barred families (IB). Compact irregulars (also called "blue compact dwarfs") are prefixed by a c.

Peculiar galaxies that do not fit into the above classes are designated by a P. Peculiarities in otherwise normal galaxies (from interactions or other unexplainable causes) are suffixed by P. Edge-on lenticulars and spirals have less certain classification, and have the suffix sp (spindle). Objects that have uncertain fine structure are followed by :, while those of doubtful classification are followed by a ?.

Open clusters

Data for open clusters are from the *Catalogue of Open Cluster Data*, third edition, by Lyngå. For all the brighter objects, integrated V magnitudes were computed by B. Skiff by summing the brightnesses of individual cluster stars as published in the literature. The accuracy of these values is limited primarily by incompleteness in the photometry or uncertainties in defining the physical limits of the group. When the resultant magnitude is likely to be in error by more than about 0.2 magnitudes, the value is followed by a colon. For others lacking photometric study, Lyngå cites photographic-blue magnitudes, keyed "mag.p," come from Collinder's classic study published in 1931. A number of recently discovered clusters have no available integrated magnitudes.

The listed number of stars is either from star counts made by Lyngå on the Palomar Observatory Sky Survey (POSS) prints or from the number of stars in the magnitude sum. Most of the diameters are also from Lyngå's POSS survey; a number of others are derived from star counts (usually in papers with photometric observations) or are the area over which the magnitude sum has been strictly limited. Those less than 5′ are expressed to a precision of 0.′1. Unfortunately, diameters on a consistent system based on star counts have been determined for only a few clusters.

Cluster types in our catalogue are given where possible from Lyngå's *Catalogue* according to the Trumpler system. Roman numerals I–IV indicate decreasing central concentration or detachment from the background; arabic numerals 1–3 indicate increasing range in magnitude of cluster stars; letters r (rich), m (medium), and p (poor) indicate the relative population of the cluster. For example, Messier 11 (NGC 6705) in Scutum is classed as I 2 r: its members are well concentrated toward the center, show a moderate range of magnitudes, and are numerous. Messier 39 (NGC 7092) in Cygnus is III 2 m: here the group is poorly concentrated, again with a moderate range of magnitudes, but not as rich as Messier 11.

Globular clusters

The data for galactic globular clusters are from a 1977 paper by Alcaino, a 1985 compilation by Webbink, and a list of positions by Shawl and White in the *Astronomical Journal*. The diameters are usually from star counts, and represent the size at which the cluster members are as dense as the nonmembers of the stellar "background." These values are probably accurate to about ten percent. The magnitudes represent the total integrated brightness of the cluster. For the globulars discussed in the Andromeda Galaxy (NGC 224), we have used the identifications given in *Atlas of the Andromeda Galaxy* by Hodge. Magnitudes and diameters are from a study by Crampton *et al.*

The Shapley-Sawyer concentration classes I to XII give a measure of the concentration of stars toward the center of the cluster, where I indicates the most strongly concentrated and XII the least. Despite their subjective nature, they have proven to be entirely consistent with visual impressions.

Planetary nebulae

The most comprehensive reference for planetaries is Perek and Kohoutek's *Catalogue of Galactic Planetary Nebulae* (CGPN). Here are data listed from a multitude of sources, photographic finder charts, and bibliographic references for 1034 of these nebulae known up to 1967. Since then a number of corrections have become known, and the number of objects expanded by some 500–few of which are of visual interest except in large instruments. Positions in our catalogue come from the CGPN or from papers by Milne and Higgs. We have tried to list the most recent measures of dimensions corresponding to the brighter parts of the nebulae, but in many cases this is quite uncertain: photographically, many planetaries are surrounded by faint halos many times larger than what is seen visually even in the largest telescopes.

Integrated true visual magnitudes for all of the planetaries described here have been derived by Jack Marling, who kindly provided this information prior to formal publication. The magnitudes were computed by integrating published measurements of emission-line fluxes convolved with the standard scotopic (dark-adapted) visual response function. The zero-point was established by a similar computation using an absolute flux calibration of Vega. The values listed are probably accurate to about 0.2 magnitudes and represent a substantial improvement over commonly published estimates dating from 1930 or earlier.

Photoelectrically measured central star magnitudes for many nebulae are given in a series of papers by James B. Kaler and his colleagues, who have also kindly provided results prior to publication. The central stars of many nebulae are difficult to measure due to contamination by the nebulae they illuminate. Thus, while this information is the best available, it is by no means the last word on the subject. The remaining data, derived from historical sources, are denoted as being either photographic-blue (p) or visual (v) when photoelectric B or V is not available. Photoelectric uncertainty greater than 0.2 magnitudes is indicated by a colon.

Galactic nebulae

The galactic nebulae rarely manifest themselves as discrete bodies, but instead form a loosely structured network throughout the Milky Way. Thus the Orion Nebula (NGC 1976) is but the brightest portion of a cloud of gas and dust that covers much of the constellation of Orion. For the amateur, a useful listing of the brighter, more distinct portions can be found in the *Sky Catalogue 2000*. Our dimensional data are generally drawn from papers by Lynds, which include information for both bright and dark nebulae visible on the POSS.

Nebulae occur in two general types: e-type, or emission, characterized by emission lines in the spectrum; and c-type, or reflection, characterized by a continuous (stellar) spectrum. Emission nebulae, or HII regions, are gathered around hot stars of spectral type O and B, whose ultraviolet radiation causes the gas in the nebula to fluoresce. Reflection nebulae are centered upon cooler stars not hot enough to cause fluorescence of the gas; the dust in the nebulae merely reflects and scatters the light of the star, giving it a spectrum similar to that of the illuminating star. Some nebulae have both emission and reflection components. The magnitude listed for the illuminating star (drawn from a wide array of sources in the literature) is often for the brightest apparently associated star, not necessarily the true illuminating star, as this datum is sometimes uncertain or unknown. Photoelectric V magnitudes on the UBV system are indicated as usual by "V = ".

Double stars

Several hundred double and multiple stars are cited in the descriptions when they are near or involved in deep-sky objects. Many of them are uncatalogued pairs, being rather faint or too widely separated to be of interest to double star astronomers. For these, separations and position angles have been measured on photographs where possible and magnitudes taken either from published photoelectric measures or from the author's visual estimates confirmed on photographs.

Over one hundred fifty catalogued pairs have been identified in the U.S. Naval Observatory *Washington Visual Double Star Catalog* (WDS). The preferred designations are Aitken double star catalogue (ADS) numbers, but the discoverer's catalogue number is used for pairs not included in the ADS. Magnitudes come from a variety of sources. Photoelectric V magnitudes for some pairs are available in the *Sky Catalogue 2000.0*, in Wallenquist's *Catalogue of Photoelectric Magnitudes*, other shorter lists scattered through the literature, and from Dick Walker at the USNO Flagstaff Station by kind permission. Most of the separations, position angles, and magnitudes are from the WDS. WDS magnitudes are systematically off the standard V system, but the magnitude differences (Δm) tend to be satisfactory. In some cases, when the combined V magnitude was available, we used the WDS Δm value to calculate approximate individual V magnitudes for the components. The cited pairs are generally fixed; the few showing rapid change (usually due to orbital motion) have a date indicated. All the catalogued pairs are listed in the Appendix, which also includes a number of close, nearly equal pairs that can be used for judging seeing.

In the descriptions, each noted pair is described by four or five items that usually appear in a condensed "data block" enclosed within parentheses. First comes the pair's name, if it has one, followed by a colon. The next two numbers are the magnitudes for the two components. Photoelectric values are again indicated by "V = "; otherwise there are none available and we have had to resort to the magnitudes in the WDS or to our own visual estimates. WDS magnitudes are listed to typically tenths of a magnitude, ours to half-magnitudes. Next in the data block is the separation of the pair in arcseconds, and last is the position angle. If the pair shows substantial motion, a date appears in parentheses at the end indicating when the preceding data apply. As examples, consider (ADS 9959: V = 8.5,9.7; 7.″3; 147°), which appears in the description for I4593 in Hercules. This is an ADS pair that has been measured photoelectrically. In NGC 1662 in Orion is a pair described by (h684CD: V = 9.6,11.3; 11″; 307°). This pair is involved in an open cluster (whose stars have been studied via photometry), and was catalogued by John Herschel in the early nineteenth century as the third and fourth ("CD") components of a multiple star. Under NGC 684 in Triangulum appears (12,13,12; 11″, 32″; 243°, 246°). This is an uncatalogued triple star, showing three magnitudes, two separations, and two position angles. The separations and position angles were measured on a photograph, while the magnitudes are our visual estimates. Following convention, both separations and position angles are with respect to the brightest member of the group.

Observations

The following descriptive observations represent the majority of deep-sky objects easily observable from the northern hemisphere in telescopes up to 30 cm aperture. The objects included are essentially those on Tirion's *Sky Atlas 2000.0*. During the course of the observations many additional objects not on this atlas, including companions to brighter galaxies and recently discovered open clusters, were examined as being of interest in amateur instruments. The resulting collection of notes is arranged by constellation in alphabetical order; within each constellation objects are listed in order of NGC number. Objects without NGC numbers (*Index Catalogue* objects, those from other catalogues, or anonyma) are interpolated in order of Right Ascension. A header for each object begins with a two-letter classification: eg for galaxy, oc for open cluster, gc for globular cluster, pn for planetary nebula, gn for galactic (gaseous/dust) nebula. Following is the catalogue number, usually the NGC number, which has no special letter to distinguish it. Objects from other lists are distinguished by a letter or name, such as U for galaxies from the *Uppsala General Catalogue*. (A complete list of these and other abbreviations used throughout the book is given below.) The remainder of each heading includes data for quick comparison of objects. Closely related objects, such as interacting pairs of galaxies, are often described together under a multiple heading.

The data in the heading are the same within each class of object, but differ from one class to the next. For galaxies, the major and minor diameters, the integrated magnitude, and the surface brightness (in the same color as indicated for the integrated magnitude) are listed. Open clusters have a diameter, integrated magnitude, and estimate of the number of members. With globular clusters the number of members is replaced by the Shapley-Sawyer concentration class. Planetary nebula headers list a size, integrated visual magnitude for the nebula, and the magnitude for the central star, if known. For galactic nebulae, the approximate size of the nebula and the magnitude of the associated star is listed. Occasionally a magnitude is followed by a colon: this indicates that the value is tentative or uncertain, based either on evidently incomplete or inconsistent data. For complete descriptions of the data and their sources see the Deep-Sky Data Sources chapter.

The descriptions are all structured similarly. Messier objects and other names are announced at the opening of each descriptive block, as are naked-eye or bright telescopic stars nearby. Following are the actual visual descriptions in order of increasing aperture. Once an aperture is announced, all that follows applies to that aperture until a new aperture is given. It may be safely assumed that details noted in smaller instruments are visible in larger ones unless otherwise stated. When they are of value, magnifications are given (rounded to the nearest 25× for generality), particularly with regard to the resolution of globular clusters and details in galaxies and planetary nebulae, which can change with magnification. These observational notes are not definitive, but can be considered indicative of what one is likely to see with various apertures from a fairly dark site.

Magnitudes for stars in the field are largely based on visual estimates made at the telescope and are probably accurate to about one-half magnitude. In many star clusters and around many galaxies photoelectric photometry of stars is available, and this is cited where appropriate. For bright field stars, magnitudes come from *The Bright Star Catalogue*, *Sky Catalogue 2000.0*, and other sources.

For a number of objects sketches made at the telescope are included. Each drawing is labelled with pertinent data that describes the telescope used, magnification, and a scale bar indicating an angular length.

A number of charts have been included to help with identification and finding in particularly crowded or confused fields. These include all catalogued galaxies that are likely to be visible in even the largest amateur telescopes. Stars have been plotted from the lists given in the *Astrographic Catalogue*, otherwise from the *Bonner Durchmusterung* or *Cordoba Durchmusterung* when the *Astrographic Catalogue* does not include reduced positions for the field. All of these charts are labelled in equinox 2000 coordinates, and all objects on them not discussed in the text can be found in the Catalogue. The RA axes are labelled every minute of time with tick marks at every 10 or 30 second interval; the Dec axes are labelled every whole degree, but are always marked at 10 arcminute intervals.

Abbreviations

A	anonymous (eg,oc)
AA	*Atlas Australis*
AB	*Atlas Borealis*
ADS	Aitken Double Star Catalogue
AE	*Atlas Eclipticalis*
B	integrated B magnitude
BD	*Bonner Durchmusterung* (pn)
Be	Berkeley (oc)
Ced	Cederblad (gn)
CGCG	*Catalogue of Galaxies and Clusters of Galaxies* (eg)
Cr	Collinder (oc)
ds	double star
eg	galaxy
E	*ESO/Upssala Survey of the ESO (B) Atlas* (eg)
gc	globular cluster
gn	galactic nebula
h	John Herschel (double star)
Ha	Haffner (oc)
I	Index Catalogue
J	Jonckheere (pn)
M	Messier
M	MCG (eg)
Ma	Markarian (oc)
$mag._H$	integrated Harvard (photographic-blue) magnitude
mag.p	integrated photographic-blue magnitude
mag.v	integrated visual magnitude
$mag._{VV}$	magnitude from the MCG (eg)
$mag._Z$	magnitude from the CGCG (eg)
MCG	*Morphological Catalogue of Galaxies* (eg)
Mk	Markarian (eg)
NGC	*New General Catalogue*
oc	open cluster
OΣ(Σ)	Otto Struve double star catalogue (Supplement)
pa	position angle
pn	planetary nebula
St	Stock (oc)
To	Tombaugh (oc)
U	*Uppsala General Catalogue* (eg)
V	integrated V magnitude
Z	CGCG (eg)
Zw	Zwicky lists of compact galaxies (eg)
Σ	Wilhelm Struve double star catalogue

Andromeda

Lying along the edge of the Milky Way, this constellation contains a variety of galactic and extragalactic objects, featuring the brilliant Andromeda Galaxy group. The center of the constellation culminates at midnight about 18 September.

eg 205 dimen. 17′ × 9.′8 V = 8.0 sfc. br. 13.6
Messier 110 is a large galaxy of comparatively low surface brightness. It lies 37′ NW of the center of its brighter companion, Messier 31 (eg 224, *q.v.*). In 6 cm this galaxy appears as a large, diffuse haze weakly concentrated toward the center. It is larger than Messier 32 (eg 221, *cf.*), and elongated SE-NW. Two stars are visible about 4′ and 6′ S of the center. 15 cm shows it about 10′ × 3′ in extent. In 25 cm it is best viewed at medium power and is about 9′ × 3′, elongated in pa 165°. The core is about 3′ × 2′, but no sharp nucleus is visible. 30 cm shows a slight broad concentration and reveals a faint stellaring at the center. Many stars are associated with the galaxy, including one of mag. 14 in the SSE end.

eg 214 dimen. 2.′1 × 1.′6 V = 12.2 sfc. br. 13.4
This galaxy is a difficult object for 15 cm, appearing simply as a circular spot 1′ diameter with a stellar nucleus. It is fairly faint for 25 cm, about 1.′5 diameter, and with irregular edges. In 30 cm the very faint halo is elongated slightly NE-SW. The 50″ core is moderately concentrated and seems to have an irregular boundary. A faint stellaring is visible just S of center.

eg 221 dimen. 7.′6 × 5.′8 V = 8.2 sfc. br. 12.3
Messier 32 is the most obvious companion to the Andromeda Galaxy, Messier 31. 6 cm shows it easily at the edge of M31's halo, but it appears tiny next to the parent galaxy. The edges are irregular, and it brightens suddenly toward a very conspicuous stellar nucleus. In 15 cm it is about 3′ × 2′, elongated SE-NW, with a stellar nucleus. The core has the character of a planetary nebula. Viewed in 25 cm the halo is elongated nearly toward the nucleus of M31; the brightness rises smoothly in the outer parts, then suddenly to a nearly stellar nucleus, but there is no distinct core. 30 cm shows the core in pa 160°, about 1.′75 × 1′ in extent. The faint halo merges with that of M31, so the total size is difficult to estimate. The nucleus is less than 10″ across and clearly nonstellar. A faint star is visible 2′ E of center.

eg 224 dimen. 180′ × 63′ V = 3.5 sfc. br. 13.4
Messier 31 is the large, very bright galaxy of the Local Group. It is visible to the naked eye even on relatively poor nights or in moonlight; in a dark sky its elongation is striking in contrast to the starry foreground. A very faint star (V = 6.9) on its SW tip is a good test for sensitive eyes and dark skies: in the best conditions, the galaxy extends past this star, which implies an observed length of about 3½°. 6 cm shows the halo up to 135′ long in pa 40°, but the outer halo is faint and has no definite edge. The halo contains some weak, broad condensations, but the core appears smooth and well concentrated. As the eye considers brighter and brighter areas toward the center, the core grows progressively more eccentric in relation to the halo; the NW side of the core fades more abruptly than the SE side, thus placing the nucleus off-center to the NW. Some stars are superposed on the galaxy, including one of mag. 12 on the SSW side of the core.

15 cm reveals a 120′ × 20′ halo with a pronounced core 10′ across. The dark lane NW of center can be seen without difficulty with some faint haze beyond it. In 25 cm the halo extends to 120′ × 40′. The core is about 7′ wide, but its length is uncertain since it fades evenly along the major axis. The inner region of the core is quite circular and appears "opaque" like other unresolved objects of very high surface brightness, as in the center of Messier 87 in Virgo (eg 4486, *cf.*). The brightness rises evenly to a central pip about 10″ diameter. The dark lane on the NW side is most distinct as it passes the core, but is clearly visible extending southwest past two stars aligned NE-SW about 15′ from the center. Two more indistinct lanes are discernable, one farther NW and one along the SE flank. On the W side of two brighter stars lying about 15′ NE of center, a spike of darker material intrudes toward the nucleus. In 30 cm the galaxy is well concentrated, showing an extremely bright center in comparison to the halo. The dark lane NW appears as a sharply defined strip about 1′ wide that passes 5′ from the bright center. The nucleus has a bright, nearly stellar center, as would be expected, but there also appear to be some odd brightenings within 15″ radius. The overall size is difficult to estimate, but the brighter part of the galaxy exceeds 80′ length.

Chart I. NGC 224 — The Andromeda Galaxy. A 30-minute exposure on 103a-O with the Lowell 33 cm "Pluto" Camera taken by B. Skiff on 2 November 1981. Object names and magnitude sequence from: "Atlas of the Andromeda Galaxy," Paul Hodge; Crampton *et al.* 1985, Ap. J. 288: 494; Arp 1956, A. J. 61: 15; de Vaucouleurs and Corwin 1985, Ap. J. 295: 287. Lowell Observatory photograph.

Table III. *Globular clusters in the Andromeda Galaxy.*

Globular clusters in NGC 224

Name	V	B–V	FWQM
G 35	15.6	0.89	2″.9
G 52	15.7	1.03	2.3
G 64	15.1	0.77	2.3
G 72	15.0	0.96	2.2
G 73	14.9	0.85	—
G 76	14.2	0.82	3.6
G 78	14.2	1.12	3.2
G 87	15.6	0.83	2.9
G 96	15.5	0.87	2.7
G119	15.0	0.83	2.7
G156	15.6	0.87	2.5
G172	15.3	0.87	2.4
G205	14.8	1.31	2.9
G213	14.7	0.90	2.5
G222	15.2	0.98	3.2
G226	15.5	0.97	3.8
G229	15.0	0.76	3.4
G230	15.2	0.82	2.9
G233	15.4	0.95	2.6
G244	15.3	0.99	2.6
G254	15.7	1.00	—
G257	15.0	0.87	3.2
G272	14.7	0.84	3.4
G279	15.4	0.73	4.9
G280	14.2	0.92	2.7
G286	15.7	0.77	2.5
G287	15.8	0.87	2.2
G302	15.2	0.75	2.5
G305	15.6	0.97	2.2
G315	15.7	0.77	—

Source: Data from Crampton *et al.* 1985, Ap. J. 288: 494.
Note: The diameters represent the size in arcseconds at which the light has fallen to one-quarter of the maximum brightness (FWQM = full-width quarter-maximum).

With apertures of about 25 cm and larger, a number of faint objects within M31 are accessible in good observing conditions. Most of these lie around the periphery of the halo and include many globular clusters and a few luminous open clusters and associations.

The most conspicuous feature in M31 is the large stellar association A78, which also bears the designation NGC 206. Embedded in the southwestern arms, it is conspicuous in the field with 25 cm. The 2′ × 1′ haze has a sharply defined eastern flank, and a few faint stellarings scattered over its surface. Along its major axis, it merges smoothly into the spiral arms of the galaxy. The globular cluster G76, one of the brightest in the galaxy, lies not far SE, just S of an oblique triangle of faint stars. The cluster is just separable from a star of similar brightness immediately SE of it. Heading northeast toward M32, the open clusters C202 and C203 (separated by 16″) are visible with 25 cm and seem nonstellar at 200x when compared to faint stars about 2′ SSE.

Midway along the southeastern rim of M31 is C410, just visible as a starlike spot. G229, set against the outer reaches of the core, is difficult to distinguish but is just visible at 150 × in 25 cm. Moving northeast, A40 is barely visible as a faint haziness associated with a small triangle of unequally bright stars (the southwestern member is visible only to averted vision). Just visible 5′ NNE is A42. Although C311-C312-C313 are not visible in 25 cm, when the bright stars NE and SE are centered in a 30′ field, the brightness gradient from the disk to the sky is very strong, perhaps more conspicuous here than anywhere else around the rim of the galaxy. Far out in the northeastern extremity, A102 is discernable as a faint stellar spot of about mag. 14 with some very faint haze around it. Turning back southwest, A54 is visible in 25 cm as a broad, mottled area best viewed at 100 ×. G78 is visible only at 200 ×, as is A67.

eg 404 ***dimen. 4′.4 × 4′.2*** ***V = 10.1*** ***sfc. br. 13.2***
This galaxy lies 6′ NW of β Andromedae (V = 2.1), which interferes with viewing, especially in larger apertures. The object is just visible in 6 cm, forming a triangle with β And and another star SE. In 25 cm it is bright and circular, about 45″ diameter. A faint stellar nucleus is visible, but otherwise there is no central concentration. Viewed in 30 cm it is 2′.25 diameter with a slight concentration to a faint stellar nucleus. The halo has a smooth texture. Many stars are visible in the vicinity: one lies 2′ N, another 1′.8 SW.

oc 752 ***diam. 45′*** ***V = 5.7*** ***stars 77***
This large naked-eye cluster can be found by sweeping N from λ Arietis. It is a nice sight in 6 cm, with about 60 stars visible in a 50′ field. A bright yellow and red pair lies to the SSW [V = 5.7,5.9; 3′.6; 296°; the brighter component (56 Andromedae) is a more difficult pair (ADS 1534AP: V = 5.7,11.9; 18″; 79°)]. Many other pairs are visible, including a bright star on the E with two colorful companions. In 15 cm about 75 stars can be seen within 1° diameter. The cluster is irregularly circular in 25 cm, filling a 65′ field with long strings of stars.

eg 753 ***dimen. 2′.9 × 2′.1*** ***V = 12.4*** ***sfc. br. 14.2***
This is a faint galaxy for 25 cm, appearing less than 1′ diameter and showing no concentration. In 30 cm it has a fairly low surface brightness. The 20″ core is moderately concentrated but is without a distinct nucleus; a very faint stellaring is occasionally visible in, or just N of the center. The halo is irregularly bounded and elongated a bit SE-NW. Several faint stars lie scattered about the field.

eg 891 ***dimen. 13′ × 2′.8*** ***V = 10.0*** ***sfc. br. 13.8***
A difficult object in 6 cm, this edge-on spiral is visible as a tiny streak of low surface brightness. 25 cm shows it

faintly, about 10′ × 1′ in extent with a broad and moderately brighter core. The galaxy appears very faint and thin in 30 cm, reaching 10′ × 2.′5 in pa 25°. The broad, weak central brightening is about 4′ × 2′ with two brighter condensations, the brighter one located on the E side of the core. The halo is most extensive to the SSW, extending all the way to a mag. 13 star; before reaching this star, there is a small, slightly brighter patch in the halo 30″ across. A mag. 12.5 star lies just N of the core, and a mag. 14.5 pair (30″; 85°) is 1.′5 SSW of the core.

eg 7640 ***dimen. 11′ × 2.′5*** ***V = 10.8*** ***sfc. br. 14.3***
Visible in 6 cm, this diffuse, unconcentrated galaxy appears elongated N-S with some faint stars nearby. In 15 cm it lies along the western side of a triangle of mag. 12 stars. It is 7′ long in 25 cm with a 3′ × 1′ core mottled with dark spots. Viewed in 30 cm the galaxy is about 7′ × 2′ in pa 170°. The inner halo is mottled and concentrated only slightly to a 1.′5 core. A distracting mag. 13.5 star is visible at the SE edge of the core.

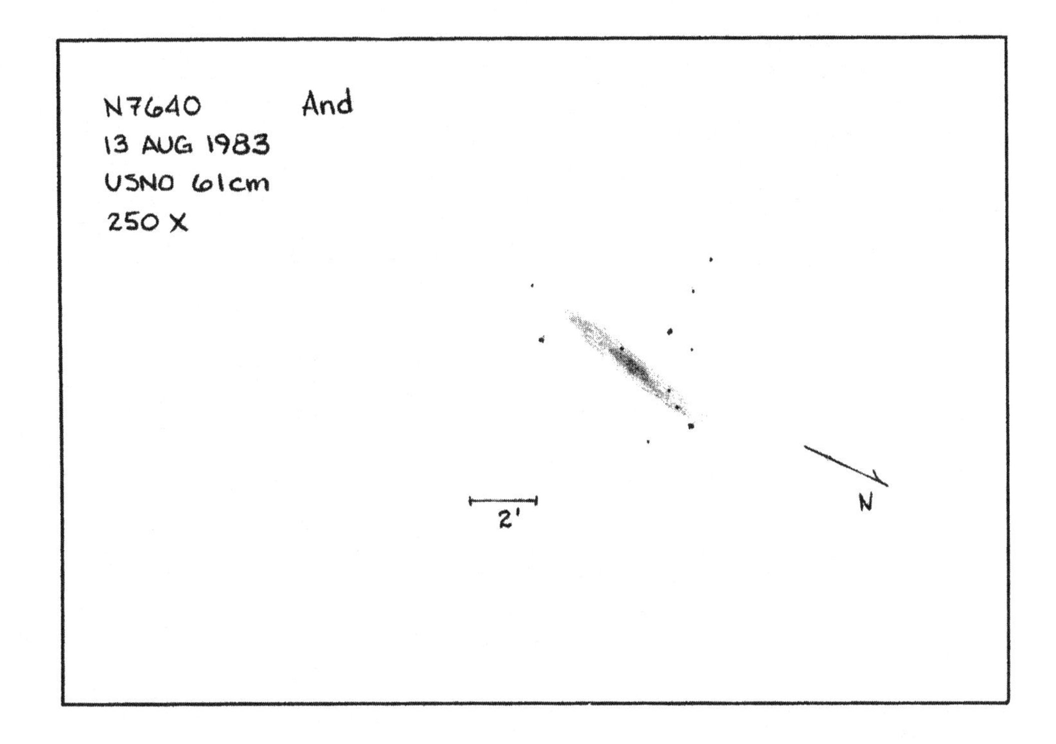

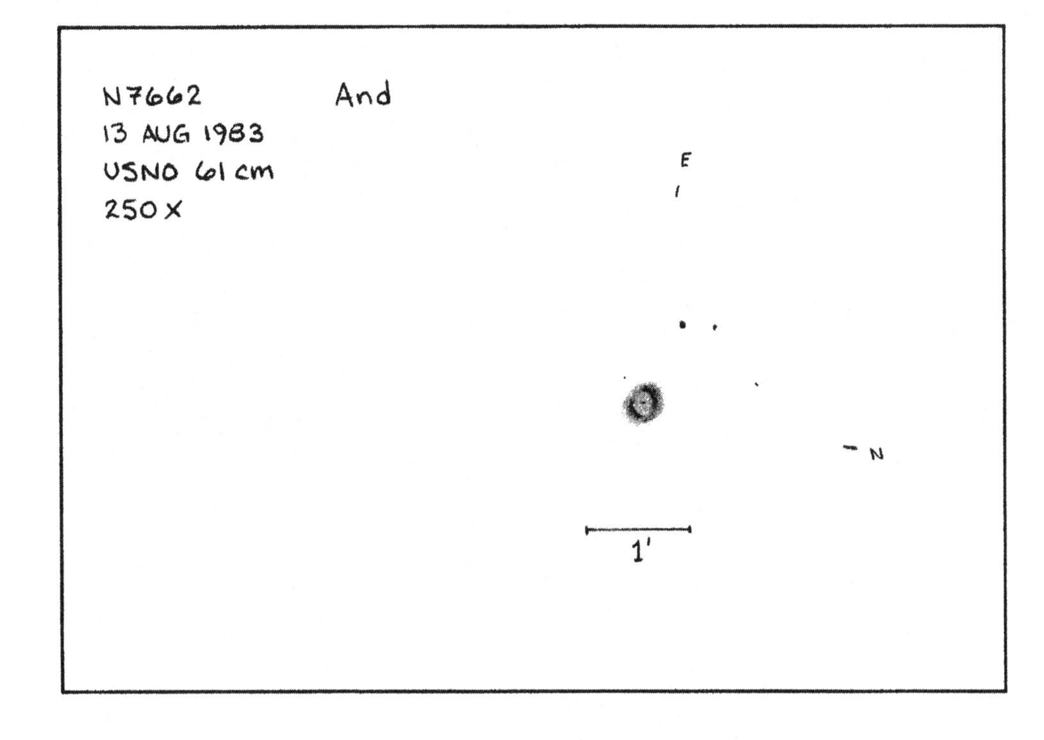

pn 7662 dimen. 32″ × 28″ nebula 8.3v star V = 13.2: This bright planetary is clearly nonstellar even to 6 cm: it appears circular and faintly bluish with a smooth texture and diffuse edges. 15 cm shows a mag. 13 star on the E edge, while the NE side of the nebula has a bright spot that appears stellar at low power. In 25 cm a small, dark patch is visible at the center. The nebula is annular in 30 cm, but the center is only slightly fainter. Slightly elongated SE-NW, the obtuse sides (on the NE and SW) are brighter, the northeastern one particularly so. No central star is detected.

oc 7686 diam. 13′ V = 5.6 stars 80
6 cm shows ten stars here scattered mostly N and W of a red mag. 6 star. In 25 cm the cluster is about 25′ diameter with 45 stars loosely spread around the reddish star. The south side is relatively devoid of stars. 30 cm reveals 50 stars in a 15′ area with stragglers extending to 25′ diameter.

Antlia

One of the three or four dullest sections of the celestial sphere to the naked eye lies in the constellation Antlia. Within its boundary there is no star brighter than magnitude 4.3. This is the astral Empty Quarter. However, the situation is not so bleak for deep-sky observers, as long as one does not expect too much. The region culminates at midnight about 24 February.

eg 2997 dimen. 8′.1 × 6′.5 B = 10.6 sfc. br. 14.7
In 15 cm this galaxy appears as a large diaphanous glow with a small, very slight central brightening. The overall size is about 4′ × 2′.5, elongated in pa 75°. A mag. 12 star is visible on the SW edge. 25 cm shows a knotty 45″ × 15″ core embedded in a 5′ × 3′ halo of comparatively low surface brightness.

eg 3001 dimen. 3′.1 × 2′.1 B = 12.8 sfc. br. 14.6
Located about 45′ N of eg 2997, *q.v.*, this is a faint object for 15 cm. The galaxy appears a bit less than 1′ across and is well concentrated toward the center. A mag. 11 star lies on the NNW edge. With 25 cm the halo is about 1′.5 diameter with only a slight concentration to an inconspicuous 20″ core. With averted vision some stellarings can be seen near the center.

eg 3038 dimen. 2′.6 × 1′.7 B = 12.7 sfc. br. 14.2
In 15 cm this galaxy appears as a small, faint glow. The outline is circular and the brightness rises to a bright nonstellar center. With 25 cm an occasionally stellar nucleus is visible inside a small, moderately conspicuous core. The halo is 1′.2 diameter. A mag. 13.5 star lies 2′ SW.

eg 3056 dimen. 2′.0 × 1′.4 B = 12.5 sfc. br. 13.5
This galaxy is very faint in 15 cm. A mag. 12.5 star is visible on the N edge. It is small and moderately faint in 25 cm. The star lies on the edge of the 40″ × 30″ halo at pa 15°; the halo is elongated in pa 10°, a little off from the star/nucleus axis. The core is moderately conspicuous, about 10″ across, and seems about 1½ magnitudes fainter than the star.

eg I2522 dimen. 2′.8 × 1′.7 B = 12.7 sfc. br. 14.2
Barely visible with 15 cm, this galaxy can be seen more easily at 100 × in 25 cm. It is larger than eg I2523, *cf.* 4′.4 S, but only slightly brighter. The halo is 1′.5 × 1′ in pa 120° and, besides some mottling, without concentration.

eg I2523 dimen. 1′.4 × 0′.8 mag.$_{\mathrm{vv}}$14 sfc. br.
Though invisible in 15 cm, 25 cm will show this object at 100 × . The galaxy is slightly smaller and fainter than eg I2522, *cf.* 4′.4 N, appearing as a smoothly textured lens 40″ × 20″ in pa 30°.

eg 3087 dimen. 1′.9 × 1′.3 V = 11.7 sfc. br. 12.6
Lying amongst a group of mag. 12 and 13 stars, this galaxy appears as a small faint spot in 15 cm. 25 cm shows it 35″ diameter with a faint stellar nucleus. Two stars of the surrounding group lie close by ENE and WSW, the first being closer and fainter. A faint companion galaxy, eg I2532 (mag.$_{\mathrm{vv}}$15), lies about 8′ E.

eg 3089 dimen. 1′.9 × 1′.3 B = 13.1 sfc. br. 13.9
Not visible in 15 cm, 25 cm will show this galaxy at medium powers about 2′ W of a mag. 8 star. The 1′ halo is unconcentrated toward the center but mottled with stellarings. A mag. 13 star is visible on the E side that has a mag. 14.5 companion to its N.

eg 3095 dimen. 3′.2 × 2′.2 B = 12.4 sfc. br. 14.3
This galaxy is barely detectable with 15 cm. 25 cm shows it to have the lowest surface brightness of three galaxies in the area (*cf.* eg I2533 and eg 3100). It appears elongated SE-NW, about 3′ × 1′ in extent, with some stellarings visible in the center. A mag. 11.5 star is on the W side, and just N of this is what appears to be a bright patch in the galaxy, which photographs show to be a faint pair of stars.

eg I2533 dimen. 1′.5 × 1′.2 mag.$_{\mathrm{vv}}$13.5 sfc. br.
With 15 cm this galaxy looks similar to eg 3100, *cf.*, but is somewhat more sharply concentrated. 25 cm gives a like impression: the object looks like the core of eg 3100, but stands out sharply from the sky. From a nearly stellar nucleus, the halo fades to an overall diameter of 45″.

eg 3100 dimen. 2′.8 × 1′.7 mag.$_{\mathrm{vv}}$12.5 sfc. br.
This galaxy is the brightest of three in the same field (*cf.* eg 3095 and eg I2533) and is visible to 15 cm with two stars of unequal magnitude (9, 10.5) about 1′.5 E. The small circular halo is 1′.5 across and has a bright center.

In 25 cm the very faint halo is 2′ × 0′.9, elongated N-S. The bright core is 45″ across with some stellarings in it. A mag. 10.5 star is visible between the galaxy and the two stars E; another is visible 30″ SSW of the core.

eg 3108 ***dimen. 2′.7 × 1′.8*** ***mag.$_{VV}$13.5*** ***sfc. br.***
Lying about 25′ E of eg 3100, *q.v.*, this galaxy is barely visible to 15 cm 2′ N of a mag. 10 star. In 25 cm the halo is 45″ across with moderate concentration toward the center, though the nucleus is not conspicuous. Three faint stars and the mag. 10 star form a box around the galaxy.

eg I2537 ***dimen. 2′.7 × 1′.9*** ***V = 12.2*** ***sfc. br. 13.8***
Barely visible to 15 cm, this galaxy is shown by 25 cm as a low surface brightness 3′ × 1′ oval, elongated nearly N-S. The surface has a smooth texture and is hardly brighter at the center.

eg 3125 ***dimen. 1′.5 × 0′.8*** ***B = 13.1*** ***sfc. br. 13.3***
While invisible to 15 cm, 25 cm shows this galaxy as a small spot at 100 ×. Higher powers reveal a stellar nucleus at the threshold of visibility in a small, nearly circular core about 40″ across. Little halo extends beyond this. Photographs show that the impression of a stellar nucleus may be due to a star (V = 15.2) superposed next to it.

eg 3175 ***dimen. 4′.8 × 1′.7*** ***V = 11.3*** ***sfc. br. 13.4***
This is a fairly faint galaxy for 15 cm. It is elongated NE-SW with an overall size of 2′ × 0′.5. Many faint stars fill a 40′ field around the object. With 25 cm the lenticular halo is 3′ × 0′.7 in pa 40°. The inner parts are brighter and mottled, but there is no well-defined core. A star or stellaring is visible off center to the W side of the core area.

eg 3223 ***dimen. 4′.1 × 2′.6*** ***B = 11.8*** ***sfc. br. 14.2***
This galaxy has a very low surface brightness and is a pretty faint object for 15 cm. A faint star is visible on the E side. The halo has moderately low surface brightness in 25 cm. Though no conspicuous nucleus is discernable, a line of three or four stellarings can be seen in the core aligned in pa 120°. The core is cigar-shaped, 1′ × 0′.3, while the halo extends to 2′ × 1′.5 overall.

eg 3241 ***dimen. 1′.5 × 1′.1*** ***V = 12.7*** ***sfc. br. 13.1***
15 cm shows this galaxy as a faint starlike object. At 100 × in 25 cm it also appears stellar. A very bright stellar nucleus is visible at 200 × in a faint halo 30″ across. Two stars to the E form a triangle with the object.

eg 3244 ***dimen. 1′.9 × 1′.7*** ***mag.$_{VV}$14*** ***sfc. br.***
This galaxy is not visible to 15 cm. With 25 cm it appears as a low surface brightness spot about 1′ diameter with little central concentration. A mag. 13.5 star is involved on the S side 30″ from the nucleus.

eg 3250 ***dimen. 3′.2 × 2′.1*** ***V = 11.0*** ***sfc. br. 13.1***
This is a difficult galaxy for 15 cm. The overall diameter is only 30″ with an unsharp central brightening. With 25 cm the halo is moderately faint and circular, extending to 1′.25 diameter. The brightness rises evenly to a small core and stellar nucleus. A bright patch can be seen in the NE side, which photographs show to be a faint pair of stars lying about 30″ ENE of the nucleus.

eg A1026–35 ***dimen.*** ***mag.*** ***sfc. br.***
eg 3257 ***dimen. 1′.1 × 1′.1*** ***mag.$_{VV}$15*** ***sfc. br.***
eg 3258 ***dimen. 1′.8 × 1′.7*** ***V = 11.7*** ***sfc. br. 12.9***
This group is not visible with 15 cm. In 25 cm eg 3258 is clearly larger and brighter than eg 3260, *cf.* 2′.6 E. The outline is roughly circular, about 1′.2 across, with moderate concentration to a starry center. At 200 ×, eg 3257 is faintly visible 4′.5 SSW, and eg A1026–35 is just discernable about 9′ SW.

eg 3260 ***dimen. 1′.3 × 1′.1*** ***V = 13.2*** ***sfc. br. 13.6***
Lying 2′.6 E of eg 3258, *q.v.*, this galaxy is invisible to 15 cm. 25 cm shows a nice, smoothly textured oval with a mag. 12.5 star embedded at pa 190°, 24″ from the center. The halo is elongated roughly toward this star (i.e., pa 10°) and has a total size of 40″ × 20″.

eg E375-G41 ***dimen. 0′.7 × 0′.1*** ***mag.$_{VV}$14.5*** ***sfc. br.***
A member of the eg 3267–81 group, this galaxy is not visible to 15 cm. It can be seen faintly with 25 cm, lying about 5′ NW of eg 3267, *q.v.* The small spindle is 1′.2 × 0′.3, elongated SE-NW.

eg 3267 ***dimen. 2′.4 × 1′.4*** ***mag.$_{VV}$13*** ***sfc. br.***
This member of the eg 3267–81 group lies 2′.5 W of eg 3268, *q.v.*, and is not visible in 15 cm. With 25 cm the object is pretty faint, having a low surface brightness halo about 1′ diameter. The core is slightly brighter. A mag. 13 star is visible about 1′.25 E, between the galaxy and −68.

eg A1027–35B ***dimen.*** ***mag.*** ***sfc. br.***
Not visible to 15 cm, this member of the eg 3267–81 group can be seen faintly with 25 cm. Lying about 3′ SSW of eg 3268, *q.v.*, it appears about 20″ diameter and is moderately concentrated to a nonstellar nucleus.

eg 3268 ***dimen. 2′.0 × 1′.7*** ***V = 11.7*** ***sfc. br. 13.1***
The brightest member of the eg 3267–81 group, 15 cm will show this galaxy as a faint circular spot 1′ diameter with a brighter center. With 25 cm it is visible at 50 ×. The halo seems elongated roughly N-S with a moder-

Chart II. NGC 3267 group.

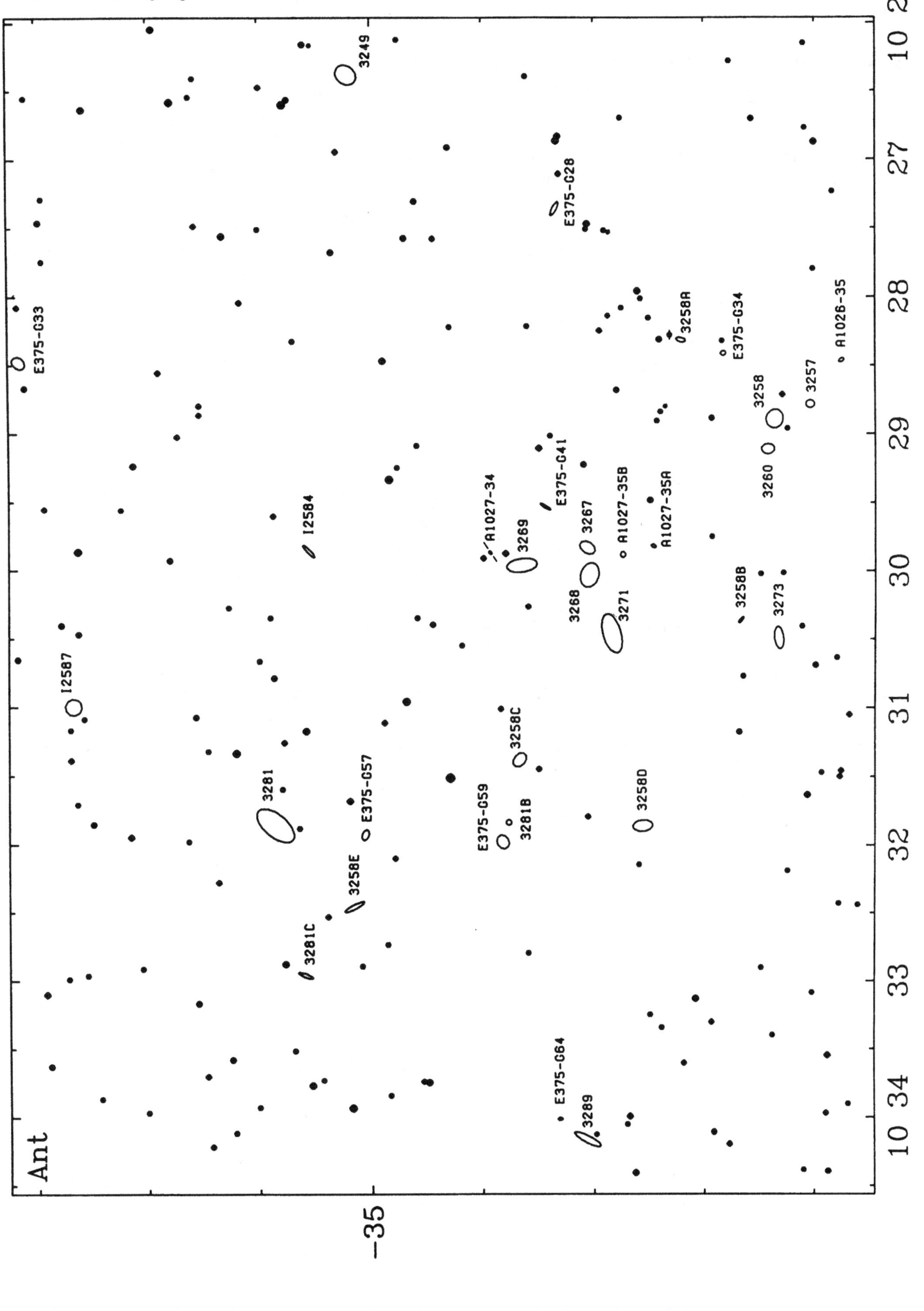

Ant
−35
10 26
27
28
29
30
31
32
33
10 34
3249
E375-G28
E375-G33
3258A
E375-G34
A1026-35
3258
3257
3260
E375-G41
A1027-34
3269
3267
A1027-35B
A1027-35A
I2584
3268
3271
3258B
3273
I2587
3258C
3281
E375-G57
E375-G59
3281B
3258D
3258E
3281C
E375-G64
3289

ately conspicuous stellar nucleus. The core is faintly detailed, and the E edge fades more abruptly than the W.

eg 3269 dimen. 3′.1 × 1′.5 mag.$_{VV}$13 sfc. br.
Not visible in 15 cm, this eg 3267–81 group member lies 6′ N of eg 3268, *q.v.* 25 cm shows it with a mag. 13 star on the SE edge. The outline is roughly circular, 1′ across, with a small, nearly stellar nucleus. About 2′.5 SW is a faint pair of stars that may appear nebulous. eg M-06-23-39 (mag.$_{VV}$15) lies about 4′ N.

eg 3271 dimen. 2′.3 × 1′.1 V = 11.7 sfc. br. 12.6
A member of the eg 3267–81 group, this galaxy is visible in 15 cm. It appears slightly fainter but larger than eg 3268, *cf.* With 25 cm the halo is elongated in pa 120°, 1′.5 × 0′.75 in extent. The core appears as a well-defined 30″ × 15″ ellipse and contains a faint stellar nucleus.

eg 3273 dimen. 2′.3 × 1′.1 V = 12.5 sfc. br. 13.4
While not visible with 15 cm, this member of the eg 3267–81 group is easy with 25 cm, lying about 15′ S of eg 3271, *q.v.* The moderately faint halo is nearly circular, 1′ diameter, with moderate concentration to a stellar nucleus at the threshold of visibility. eg 3258B (no mag. available) lies 3′.0 NNW.

eg 3275 dimen. 2′.8 × 2′.3 B = 12.5 sfc. br. 14.4
Barely visible in 15 cm, this galaxy appears moderately faint with 25 cm. The halo seems elongated slightly E-W, 1′.25 × 1′, with an inconspicuous core and faint stellar nucleus visible at high power.

eg 3281 dimen. 3′.3 × 1′.8 V = 11.7 sfc. br. 13.4
A member of the eg 3267–81 group, this galaxy is not visible to 15 cm. With 25 cm it has low surface brightness and is best viewed at medium powers. The halo is 2′ × 0′.4 in pa 135° and brightens to an elongated core that contains some stellarings. eg 3281C, *q.v.*, lies about 12′ E.

eg 3281C dimen. 1′.9 × 0′.5 mag.$_{VV}$14.5 sfc. br.
This faint member of the eg 3267–81 group lies 12′ E of eg 3281, *q.v.*, and is not visible in 15 cm. With 25 cm it is barely detectable, appearing starlike at 200 ×.

eg 3347 dimen. 4′.4 × 2′.6 B = 12.6 sfc. br. 14.4
This galaxy is visible in 15 cm as a very faint and small spot. 25 cm shows it 45″ across with a very bright stellar nucleus like a mag. 13 star. eg 3354, *q.v.*, lies 3′.5 E.

eg 3354 dimen. 0′.6 × 0′.6 V = 13.2 sfc. br. 11.8
This faint companion to eg 3347, *q.v.* 3′.5 W, is not visible with 15 cm. 25 cm reveals a smooth, well-defined 35″ × 20″ oval elongated in pa 0°. A mag. 13.5 star is embedded in the N side.

eg 3358 dimen. 3′.8 × 2′.3 B = 12.7 sfc. br. 14.8
This galaxy is barely visible to 15 cm. 25 cm shows a low surface brightness glow about 1′ × 0′.75 in size elongated in pa 165°. A stellar nucleus is clear, with some other indefinite detail visible near the center. A faint pair of stars that can be mistaken for a companion galaxy lies 2′.5 SW.

eg 3449 dimen. 2′.6 × 1′.2 B = 13.1 sfc. br. 14.2
Not visible with 15 cm, this galaxy appears moderately faint in 25 cm. The lenticular 1′.25 × 0′.6 halo is elongated in pa 135°, and the core is brighter but not well defined. A stellar nucleus is visible at high powers.

Aquarius

Aquarius is an extended constellation whose boundary extends well south of the Celestial Equator. The Helix Nebula, pn 7293, is a fine sight in a dark sky. Three globular clusters and many galaxies are also visible. The center of the constellation culminates at midnight on 26 August.

gc 6981 diam. 5′.9 V=9.3 class IX
Messier 72 can be seen faintly in 6 cm with a mag. 9.5 star lying about 5′ ESE. 15 cm reveals many faint field stars around the cluster. The center of the cluster is broadly concentrated and has a granular texture at 100×. 25 cm shows it fairly well resolved at 250×, and up to 2′.5 diameter. Two widely spaced stars lie S of the cluster. It is 2′ diameter with a 1′.25 core in 30 cm. The outliers are well resolved, particularly on the NE, and an overlying layer of faint stars is visible on the core. A conspicuous star or stellaring appears on the NE edge of the core. The concentration is broad and uneven, with dark "shadows" visible on the S and E sides.

oc 6994 diam. 2′.8 mag. stars 4
Messier saw M73 as a cluster of three or four "small stars . . . containing a little nebulosity." 25 cm shows it as a nice Y-shaped asterism of four stars: two fainter stars are on the W, leading to a mag. 10 star on the S. The fourth star, mag. 11, lies NE.

pn 7009 dimen. 44″×23″ nebula 8.0v star 11.9p
The Saturn Nebula is a large, bright planetary, which 15 cm will show as a smooth, pale-green rectangular oval elongated E-W, about 25″×20″. A central brightening is visible. With 25 cm at 300×, spike-like extensions to the oval nebula are visible at each end, each seemingly as long as the disk. A mag. 13 star lies 1′.6 WNW. 30 cm shows it to 40″×35″ in pa 80°, with hazy extensions to 1′ diameter. The core is unevenly bright with bright curds on the N side and a darker area on the E. The ansae appear only 10″ long at each end of the brighter core. A very faint star is visible 45″ E of center.

gc 7089 diam. 13′ V=6.4 class II
Messier 2 is about 5′ diameter with 6 cm. It appears well concentrated but is without a bright nucleus. A mag. 10 star is visible 4′.5 NE of center. In 15 cm the cluster has a distinct core with an almost star-like nucleus centered in a faint haze. At 50× in 25 cm it looks similar to Messier 15 in Pegasus (gc 7078, *cf.*). A well-resolved mass of stars is visible at 200×, rising to a broad, bright center. The brightest part is about 8′×6′, with outliers extending to 12′ diameter. The cluster is well resolved in 30 cm, with a smooth texture due to the uniform magnitude of the stars. The cluster is 7′ diameter with a N-S elongation expressed mostly in the halo.

eg 7171 dimen. 2′.8×1′.7 V=12.3 sfc. br. 13.8
This galaxy is faintly visible in 25 cm, lying not far SW of a line of evenly spaced mag. 10.5 stars. The center of the smooth ellipse is slightly brighter, and the halo is elongated in pa 120°. It has a low surface brightness in 30 cm, which reveals a 40″ core within the 2′×1′ halo. Several faint stars lie in the same field with the galaxy, including one of mag. 12.5 on the SE tip 50″ from center. eg I1417 (mag.$_{VV}$14.5) is located 12′ NW.

eg 7184 dimen. 5′.8×1′.8 V=11.2 sfc. br. 13.6
In 25 cm this large, fairly bright galaxy has a circular 1′ core with a stellaring involved on its S side. The halo extends to 4′.5×1′.5 in pa 60° with a mag. 11.5 star off the NE tip. It grows to 5′×1′.5 in 30 cm; the unevenly bright core is 1′.2×0′.8 and seems more definite on the NW side.

eg 7218 dimen. 2′.5×1′.3 V=12.1 sfc. br. 13.2
A difficult 2′×1′ streak in 20 cm, this galaxy is located just W of a mag. 12.5 star. The nebula seems speckled with brighter areas, suggesting detail. 25 cm shows it 1′.5×1′ with a mag. 13.5 star on the NE end. The core is slightly brighter and has a sharply defined edge. The halo has a few stellarings in it. 30 cm shows the faint star NE lying just within the halo, which is elongated in pa 30°, 2′.5×0′.7, with a 1′×0′.5 unevenly bright core.

eg 7252 dimen. 2′.2×1′.8 V=12.1 sfc. br. 13.5
This is a faint object, invisible in 15 cm, which 25 cm reveals as a small circular spot with a stellar nucleus. The curious halo is faint but well defined. In 30 cm the fairly bright nucleus seems stellar, but may have a star closely associated with it on the NW. Overall it appears 1′ diameter with a brighter center half that size. A mag. 13 star is visible 2′.2 NE.

pn 7293 dimen. 12′ × 10′ nebula 7.3:v star V = 13.4
The Helix Nebula is a large, dim planetary that can be seen with 7 × 35-mm binoculars in dark sky. It is faintly visible in 15 cm, appearing about 15′ diameter with several stars embedded. The center of the nebula is slightly darker. In 25 cm it is oblong with a fairly distinct hole in the center. Some details are visible, and a close pair on the S side is resolved. 30 cm reveals ten stars in the nebula including the central star and a companion. A brighter band of nebulosity aligned N-S lies on the E side between two stars. Bright spots are also visible on the NW and SW portions. The annularity is indistinct.

eg 7300 dimen. 2′.2 × 1′.2 V = 12.9 sfc. br. 13.8
With 25 cm this galaxy can be viewed only with averted vision. At 100 × it appears as an amorphous, ill-defined glow perhaps 1′.5 diameter without central concentration. 30 cm shows the galaxy in the N side of a 12′ trapezoid of mag. 12 stars. With very low surface brightness, the halo is about 1′.5 × 1′, elongated ESE-WNW. eg 7298 ($mag._{VV}$14) lies 11′ WSW.

eg 7302 dimen. 1′.9 × 1′.3 V = 12.1 sfc. br. 13.0
In 25 cm this galaxy is about 45″ diameter with a strongly condensed center. The halo has irregular indentations around the periphery. In 30 cm it lies 4′ N of a mag. 9 star. The halo is 1′.2 diameter here, and slightly elongated E-W. The core is broadly concentrated toward a stellar nucleus.

eg 7309 dimen. 2′.1 × 2′.0 V = 12.5 sfc. br. 14.0
30 cm will show this galaxy just off the southern point of a 10′ triangle of stars. The halo appears about 1′ diameter with a broad, weak concentration. A mag. 13 star is visible 1′.5 E.

eg 7371 dimen. 2′.1 × 2′.0 V = 12.1 sfc. br. 13.5
This galaxy appears fairly faint in 25 cm, about 1′ diameter overall. The halo is without central brightening except for a stellar nucleus that seems off-center to the NW. A star is visible 2′ SE, which 30 cm shows to be double (11.5,12.5; 21″; 55°). 30 cm reveals a broad nebula with a weak concentration and no definite nucleus, though the center has a few bright markings in it.

eg 7377 dimen. 2′.2 × 1′.8 V = 11.6 sfc. br. 13.0
Appearing fairly bright in 25 cm, this galaxy is located NE of a group of several faint stars. The halo is circular, about 45″ across, and brightens irregularly to a moderately concentrated core, but no nucleus is evident. 30 cm gives a similar view: the halo is about 50″ diameter, moderately concentrated, and without a distinct nucleus.

eg 7392 dimen. 2′.0 × 1′.3 V = 11.9 sfc. br. 12.7
In 25 cm this galaxy is elongated in pa 120°, about 1′ × 0′.6, with rounded ends and irregular edges. The core is faintly mottled. A few stars lie about 3′ N; 2′.5 ESE is a mag. 11.5 star. 30 cm reveals a poorly concentrated glow about 1′.2 × 0′.7 in extent. The acute ends have indefinite edges.

gc 7492 diam. 6′.2 V = 11.4 class XII
This faint globular cluster is not visible in 25 cm. 30 cm shows it very faintly 15′ E of two wide pairs of stars; it is poorly condensed and best viewed at high power. Several threshold magnitude stars are visible on the unevenly concentrated background. The center seems slightly darker, giving the impression of a ring-like shape.

eg 7585 dimen. 2′.3 × 1′.9 V = 11.7 sfc. br. 13.2
With 25 cm this galaxy is well concentrated and about 1′ diameter. A faint stellar nucleus lies inside an irregularly circular core which, with averted vision, seems to extend a bit more to the E than W. 30 cm shows the halo 2′.2 × 1′.5 in pa 105°, with indefinite edges. The core is small and contains a brighter nucleus. Some mottling is visible around the core at 225 × . eg 7576 (V = 13.0) lies 10′ SW; eg 7592AB ($mag._{VV}$14), a colliding pair of galaxies, is 15′ NW.

eg 7600 dimen. 2′.4 × 1′.1 V = 11.8 sfc. br. 12.9
In 15 cm this galaxy is visible as a small, faint circular spot. 25 cm shows it in pa 70°, 1′ × 0′.6, with a higher surface brightness than eg 7606, *cf.* The galaxy is moderately faint with a very small sparkling core. 30 cm gives it 1′ × 0′.5 with weak concentration to a faint stellar nucleus.

eg 7606 dimen. 5′.8 × 2′.6 V = 10.8 sfc. br. 13.5
This galaxy is only just visible in 6 cm. 15 cm shows a faint oval granular haze about 3′ × 1′ in size. 25 cm shows it 3′.5 × 1′ in pa 150°. There seems to be a dark lane near the core; the nucleus is stellar. The galaxy is bright in 30 cm, which shows the core in pa 135° while the halo stays in pa 150°. The halo is 3′.5 × 1′.3 with fair concentration overall to a bright, nonstellar nucleus.

eg 7721 dimen. 3′.4 × 1′.5 V = 11.8 sfc. br. 13.4
While only barely visible in 15 cm, 25 cm shows this galaxy as a faint lenticular glow with a weak, mottled central concentration. The halo extends to 2′ × 0′.75 in pa 20°. It grows to 3′ × 1′ in 30 cm, again with little concentration toward the elongated, mottled core. Some stellarings are visible toward the center of the galaxy at low magnifications.

eg 7723 dimen. 3′.6 × 2′.6 V = 11.1 sfc. br. 13.4
This galaxy is just visible in 6 cm, which also shows a faint star to the N. 25 cm shows a well-defined core in

which a stellar nucleus is occasionally visible. The halo seems to have a notched border. 30 cm shows a weakly concentrated oval reaching to 3′ × 2′ in pa 50°. A faint stellar nucleus is visible within the mottled, oval 1′ core.

eg 7727 dimen. 4.′2 × 3.′4 V = 10.7 sfc. br. 13.5
This is a faint object for 15 cm, appearing about 2.′5 diameter, and seemingly composed of a bunch of faint stars. 25 cm shows it elongated and smaller, but a bit brighter than eg 7723, *cf.* It appears circular in 30 cm, about 2.′8 diameter. At 150 × it is smoothly concentrated to a bright nonstellar nucleus. At 225 × it seems smaller and more sharply concentrated, the nucleus nearly stellar. A dark patch is suspected just NE of the nucleus at both magnifications. eg 7724 (mag.$_{VV}$13) lies 12′ WNW.

Aquila

Located in the thick of the Milky Way, this region boasts several planetary nebulae ranging from star-like spots to ghostly blobs. Low-power sweeping reveals many small, dark nebulae. The center of the constellation culminates at midnight on 17 July.

oc Berkeley 79 ***diam. 10′*** ***mag.*** ***stars 60***
Viewed in 25 cm, this cluster is fairly conspicuous at 75×. At 200× a total of 18 stars is visible in a 3′ group slightly elongated E-W. The four brightest stars are about mag. 11.5–12, the remainder about mag. 13.5. Lying 10′ NW is a bright, unequal pair of stars (h5501: 9.5,10.8; 23″; 7°).

oc 6709 ***diam. 13′*** ***V=6.7*** ***stars 111***
Conspicuous as a hazy spot in an 8×50 mm finder, this cluster is well resolved even with 6 cm at 25×. The overall shape is triangular, with a fairly bright pair aligned E-W on the eastern point. In 15 cm a star to the northeast gives the 25′ diameter cluster an overall square outline containing about 45 stars, mostly on the E and W sides. There is a roughly triangular hole in the middle, and a string of stars runs N-S on the W side. The E-W pair on the E side is composed of two wide pairs (22″; 205°, and 10″; 220°). The brightest stars are about mag. 9, but most are mag. 10–11.5. In 25 cm about 90 stars can be counted, including several mag. 12–13 pairs. Viewed in 30 cm, the cluster fills an 18′ field with a loose, scattered array of 75 stars, most of them brighter than mag. 11.5.

oc Berkeley 80 ***diam. 4.′0*** ***mag.*** ***stars 20***
Though hardly visible in 15 cm, 25 cm shows this cluster at 75× on the N side of an isosceles triangle of stars (mag. 9.5,10.5,10.5) with the long base on the S (plotted on the AE). The field surrounding the cluster is dark, containing only a few faint stars; 10′ W, however, is the edge of a star cloud. The hazy 2′×1′ cluster is elongated E-W. About half a dozen stars of mag. 13 and fainter are resolved at 125×.

oc 6738 ***diam. 15′*** ***mag.*** ***stars***
This cluster spreads widely over a rich Milky Way field in 15 cm. The overall size is about 30′, and a string of brighter stars runs N-S through its center. At 100× 45 stars are visible. The stars in the brightest part have a wide range of magnitudes. On the E side lies an elongated strip of a dozen mag. 12.5–13 stars. Including this group, 25 cm shows 75 stars, but the cluster does not stand out clearly against the Milky Way.

pn 6741 ***dimen. 9″×7″*** ***nebula 11.5v*** ***star V=14.7:***
This small planetary is visible with 25 cm at 200×. A mag. 11 star on the W edge is barely separable from the nebula at 250×, the latter appearing no brighter, but larger than the star. Viewed in 30 cm the greenish-blue nebula is discernable as the southeastern corner of a box of mag. 11–12 stars. At 425× it occasionally appears annular with a bright spot on the E, which looks like the central star at 225×. A mag. 14 star lies 20″ NW.

gc 6749 ***diam. 6.′3*** ***V=12.4*** ***class***
This globular cluster is difficult to view in 25 cm, lying about 10′ WSW of two mag. 9 stars 1′ apart. Barely distinguishable from the background at 200×, the cluster is unconcentrated and unresolved, about 2′ diameter, and has a very low surface brightness. A mag. 12 star lies on the E edge.

pn 6751 ***dimen. 21″×21″*** ***nebula 11.9v*** ***star V=13.9***
This object is discernable in 15 cm at 50× ESE of 12 Aquilae (V=4.0). At 200× the mag. 13.5 central star is barely visible, and another mag. 13.5 star lies 20″ E. It is 20″ diameter in 25 cm. The nebula brightens gradually toward the center, where the central star is clearly visible. Two stars lie on the E edge, and one on the western edge; the inner one on the E is the brightest. In 30 cm one or two dark spots can be seen on the S side of the broad core that otherwise fades evenly to the edge. A mag. 12 star is visible 55″ E.

oc 6755 ***diam. 15′*** ***V=7.5*** ***stars 157***
In 15 cm this cluster is fairly conspicuous. It is distinctly split into two parts by a dark lane passing through it from NE to SW; the southeastern half contains brighter stars than the NW half. Overall the cluster contains about 40 stars of mag. 10 and fainter. About 75 stars are visible in 25 cm, more than half of them in the southeastern section. A clustering of field stars lies about 15′ SE. In 30 cm the southeastern section is about 7′ diameter and

contains more than 30 stars. Curved strings of stars extend to 10′ diameter.

oc 6756 ***diam. 4.′0*** ***mag. 10.6p*** ***stars 40***
Faintly visible in 15 cm, this cluster appears about 3′ diameter. At 125 × it is partially resolved, showing about ten stars of mag. 12. The brightest star, mag. 11.5, lies next to a compact central clot of unresolved stars. About 20 stars are visible in 25 cm, but the compact group, which lies E of the geometric center of the cluster, is still unresolved. A small chain of stars extends away from the cluster to the south and west. 30 cm shows 15 stars in a 7′ area elongated ESE-WNW. The condensation is just resolved, containing about ten mag. 13.5–14 stars in a pointed form about 45″ across pointing N.

gc 6760 ***diam. 6.′6*** ***V = 9.1*** ***class XI***
Found easily in 15 cm, this globular is situated on the N side of a 25′ circlet of mag. 8–9 stars. A mag. 10.5 star lies to the NE. At 120 × it is circular, about 2′ diameter. Some granularity is visible and a few stars are resolved on the N side. In 25 cm the N edge of the cluster is more well defined, giving the impression of a slight E-W elongation. Overall it is 3.′5 diameter with a broadly concentrated 2′ core. Only partial resolution is obtained at 200 ×, with a sprinkling of mag. 13 stars over a hazy background. 30 cm reveals a circular glow 2.′5 diameter with a broad core and a few resolved outliers. A conspicuous star lies 1′ E of center.

oc Berkeley 82 ***diam. 4.′0*** ***mag.*** ***stars 20***
In 25 cm this cluster is fairly conspicuous because of an arc of three mag. 10.5 stars on its W side. The rest of the stars are mag. 13.5 and fainter. About 15 stars are resolved within 2′ diameter, including a faint pair on the S side. A rich star cloud lies 20′ W: it is elongated NE-SW, 20′ × 12′, and includes many bright and faint stars.

pn 6772 ***dimen. 70″ × 56″*** ***nebula 12.7v*** ***star V = 18.9:***
This fairly large nebula is visible at 125 × in 15 cm. The unconcentrated glow is 1′ diameter with a low surface brightness and fuzzy edges. In 25 cm the nebula is still fairly faint and without features or a central star. At 200 × it is circular, about 1.′2 diameter, but it seems to have an indefinite extension on the SW perimeter. 30 cm shows it NE of a Y-shaped asterism. Still without detail, the low surface brightness nebula appears 1′ diameter.

oc Berkeley 43 ***diam. 6′*** ***mag.*** ***stars 36***
25 cm will just show this cluster. It appears only as a large, grainy haze of very low surface brightness about 20′ diameter. A few threshold magnitude stars are resolved at 100 ×.

pn 6778 ***dimen. 25″ × 19″*** ***nebula 12.3v*** ***star 14.8p***
Lying 5′ W and a little S of a mag. 9 star, this planetary appears as a featureless grey circular puff in 15 cm. In 25 cm it is elongated E-W, about 25″ × 15″. At 200 × the central star is occasionally visible in the slightly darker center. 30 cm reveals a football-shaped object elongated E-W with a faint star associated about 40″ E. At 225 × the nebula is about 20″ × 18″ with a "hairy" edge. The central star is occasionally visible against an unevenly bright background.

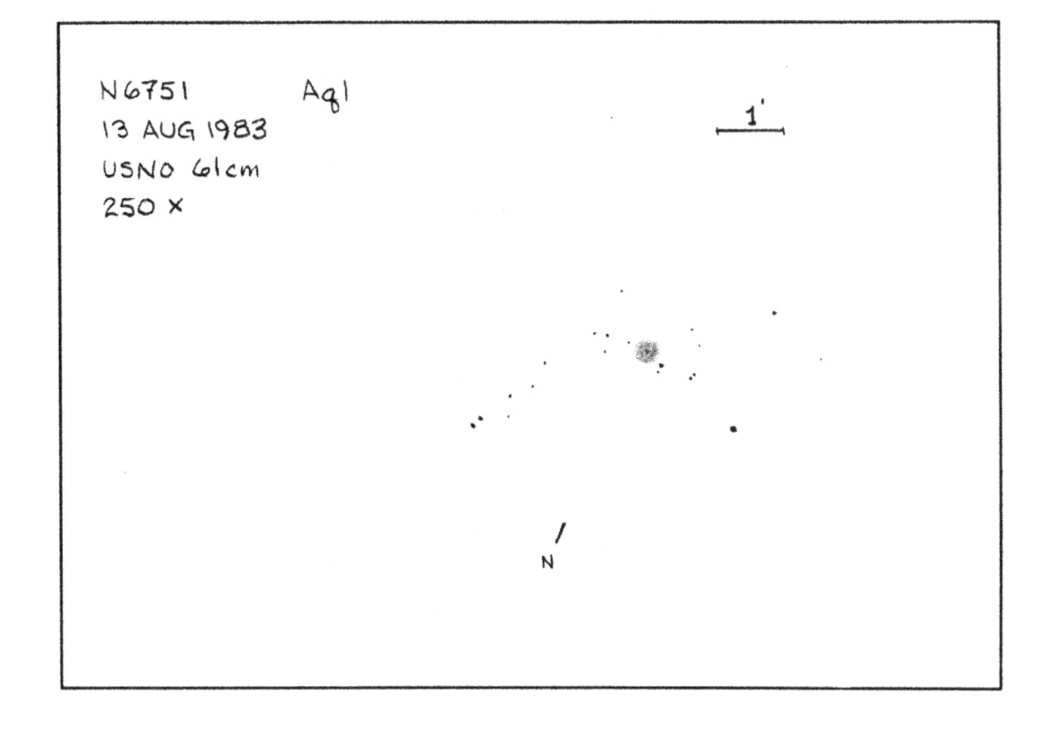

pn 6781 dimen. 1.′9 × 1.′8 nebula 11.4v star V=15.0:
In 15 cm this nebula is visible as a large faint circular spot with a mag. 12.5 star involved on its NE edge. With 25 cm the well-defined disk is about 90″ diameter. The southern half is generally brighter, and a dark patch lies just N of the geometric center. Fairly bright in 30 cm, it appears about 1.′7 diameter with a 40″ central hole.

oc Berkeley 45 diam. 4.′0 mag. stars 20
This cluster is visible but unremarkable in 25 cm. The small, 1.′5 diameter haze is unresolved at 125×.

pn 6803 dimen. 5″ × 5″ nebula 11.4v star V=15.2:
With apertures of 25 cm and less this planetary appears as a mag. 11.5 star and needs the spectroscope to reveal its nonstellar character. 30 cm shows a nearly stellar spot 5″ diameter lying just S of a mag. 11 star. The nebula has a stellar center with a very slight outer halo. A mag. 13.5 star lies NNW.

pn 6804 dimen. 62″ × 49″ nebula 12.0v star V=13.9
Faintly visible in 15 cm, this nebula appears roughly 1′ diameter with a grey color and a mag. 12.5 star on the NE side. With larger instruments the nebula provides a good contrast in the character of planetaries with pn 6803, *cf.*, which lies 50′ N. The mag. 13 central star is visible in 25 cm and seems off-center toward the E side of the nebula. A mag. 14 star lies about 1′ W. In 30 cm the star NE is just inside the border of the object. The nebula is a little less than 1′ diameter and fades gradually from the central star to the sky.

eg 6814 dimen. 3.′2 × 3.′0 V=11.2 sfc. br. 13.5
Faint for 15 cm, this Seyfert galaxy is roughly 1′ diameter with a brighter core, but without a characteristically conspicuous nucleus. With 25 cm the circular halo is about 1.′25 diameter and a faint stellar nucleus is visible. Viewed in 30 cm, the halo is slightly elongated toward a mag. 13.5 pair (22″; 87°) that lies 1.′5 ENE. Overall, it extends to 2′ × 1.′5, with a 1.′25 core and faint stellar nucleus. The galaxy is of moderate surface brightness and exhibits little concentration. A mag. 12.5 star lies 1.′2 NW, and a very faint, threshold magnitude star lies only 25″ from the nucleus in pa 265°.

Ara

Northern observers must not neglect summer nights of good definition to view the fine objects in this area. The center of the constellation culminates at midnight about 11 June.

oc 6193 diam. 16′ V=5.2 stars 14
This cluster appears very low in the sky for northern observers. In 6 cm it is small and poorly populated, containing only four or five stars, including two bright ones. In larger instruments, the brightest star is a close pair (Melbourne 8AB: V=5.7,9.0; 1″.6; 14°). Some nebulosity is involved with the cluster, but none is discernable with 6 cm.

oc 6204 diam. 6′ V=8.2 stars 13
oc Hogg 22 diam. 1′.5 V=6.7 stars 10
In 6 cm, oc 6204 is about 5′ diameter with ten stars evenly sprinkled over an unconcentrated haze. The cluster shows about 20 stars with 25 cm, irregularly scattered in a 4′ area. Hogg 22 is visible to 6 cm as an arc of four stars lying 8′ SE of 6204. With 25 cm the second star from the SE end is a pair (V=9.2,10.2; 16″; 190°).

oc Harvard 13 diam. 15′ mag. stars 15
Viewed in 25 cm, this cluster is about 30′ across. Many faint stars form a general brightening of the background, overlain by two dozen mag. 10–11 stars.

oc I4651 diam. 12′ V=6.9 stars 102
Reaching a maximum altitude of only 5° for the authors, observation of this cluster requires a clear southern horizon. In 25 cm it covers 30′ diameter with 50 stars at 100×. The stars are evenly scattered, exhibiting little concentration to the center. The brightest member is about mag. 9.

gc 6352 diam. 7′.1 V=8.1 class XI
This globular cluster appears large, bright, and broadly concentrated in 25 cm. At high powers it is granular, while a few brighter field stars lie against the SW side. The overall diameter is about 5′.

Aries

Except for the trio of stars α, β, and γ Arietis, this area is devoid of conspicuous offset stars. Though deep-sky viewing is restricted to galaxies, these abound in interesting detail. The center of the constellation culminates at midnight on 31 October.

eg 697 dimen. 4′.7 × 1′.8 mag.$_z$12.7 sfc. br.
Visible only with averted vision in 15 cm, this faint galaxy has a broad core that fades rapidly to a 1′ diameter halo. In 25 cm it appears moderately faint, set in the NW side of a field rich in mag. 10–12 stars. It is elongated E-W, about 1′.5 long, with an elongated core and stellar nucleus. 30 cm reveals a 2′ × 1′ halo in pa 115°. The texture is very smooth and the object has a broad, weak central brightening. eg 697 is the brightest of a group including eg 678 (mag.$_z$13.3), eg 680 (mag.$_z$13.0), eg 691 (mag.$_z$13.5), eg 694 (mag.$_z$13.9), and I 167 (mag.$_z$14.0), all located about 30′ SW.

eg 772 dimen. 7′.1 × 4′.5 V = 10.3 sfc. br. 13.9
This galaxy is visible in 15 cm; it has a small intense core and a stellar nucleus. In 25 cm the core is granular, elongated SE-NW, and about 45″ long. Viewed in 30 cm the halo is 3′.5 × 1′.5 in pa 135°, with fairly good concentration to a stellar nucleus. eg 770 (mag.$_z$14.2), which is possibly interacting with eg 772, lies 3′.5 SW.

eg 821 dimen. 3′.5 × 2′.2 V = 10.8 sfc. br. 13.0
Located only 1′ SE of a mag. 9 star, this galaxy is moderately faint in 25 cm. The halo is 1′.5 × 0′.6, elongated NE-SW. The broadly condensed core is surrounded by a faint halo. 30 cm shows it 1′.3 × 0′.6, elongated in pa 35°. The light is very well concentrated to a bright, nearly stellar nucleus and fades rapidly toward the halo. High magnifications reduce the apparent size and emphasize the stellar nature of the nucleus.

eg 871 dimen. 1′.3 × 0′.5 V = 13.5 sfc. br. 13.0
eg 876 dimen. 2′.1 × 0′.5 mag.$_V$ v14.5 sfc. br.
eg 877 dimen. 2′.3 × 1′.8 V = 11.8 sfc. br. 13.2
The brightest of this group is eg 877, which is a fairly faint object for 25 cm. There seems to be a faint extension to the halo on the NW, with an overall size of 1′.7 × 0′.75 in pa 135°. 30 cm shows a broad, moderately concentrated glow about 2′ × 1′ that is elongated in the direction of a mag. 13 star 1′ SE. The edges are well defined. A thin streak of brightenings, one or two of which appear stellar, run along the major axis. A mag. 14 pair (20″; 55°) is visible about 3′ SSW. Just N of these is a faint nebulous spot, which is eg 876, 2′.0 SW of eg 877. It is visible only with averted vision in 25 cm and is not much easier to view in 30 cm. About 12′ W of eg 877 is eg 871. It appears quite faint in 25 cm, about 45″ diameter with a slightly brighter center. In 30 cm it is long and thin, 1′ × 0′.3 in pa 0°, and is only a little fainter than −77. A very thin core runs almost the length of the object, and is brighter toward the center, but knotted as in −77. Another pair of stars lies 2′.5 SSW (12,12; 33″; 8°). eg 870 (mag.$_z$16.0), a compact elliptical galaxy only 0′.1 diameter, lies 1′.6 W of eg 871.

eg 972 dimen. 3′.6 × 2′.0 V = 11.3 sfc. br. 13.3
This nice galaxy is located 1′.5 NE of a wide pair of bright stars (40″; 120°). 15 cm shows a broadly brighter middle region with a distinct halo extending to 1′.5 × 0′.75. A bright object in 25 cm, it appears elongated SE-NW, about the same size as in 15 cm, but the bright, elongated core is off-center to the NE and faint extensions to the halo spread SE. Viewed in 30 cm this galaxy is 2′.2 × 1′ in pa 150°. There is good concentration toward the center of the large broad core; a faint stellar nucleus is occasionally visible E of center at 225 ×. The inner areas of the core appear granular. Three stars (including the pair) SW of the galaxy are lined up in pa 120°, each about 40″ apart, and decreasing in magnitude going to the SE.

eg 976 dimen. 1′.7 × 1′.5 V = 12.4 sfc. br. 13.3
This is a small, faint galaxy that can be barely glimpsed in 15 cm. In 25 cm it appears moderately faint with a small, well-defined core. The halo seems slightly elongated SE-NW, about 1′ × 0′.6 in size. 30 cm shows a well-concentrated object; a faint stellar nucleus is visible at 225 ×.

eg 1156 dimen. 3′.1 × 2′.3 V = 11.7 sfc. br. 13.6
In 15 cm this galaxy is well concentrated, with a few stars involved. In 25 cm it is 2′.5 × 1′.5, elongated NE-SW. A slightly darker area about 1′ long intrudes at an angle into the halo from the S. Two very faint stellarings are visible within the object, and a mag. 11.5 star is on the N side. In 30 cm it is about 2′.6 long in pa 35°, varying from 1′ to 1′.5 in width: the NE end is wider and curves a little to the E; the S end is fainter with a few stellarings embedded.

Auriga

The galactic equator passes through the southern part of this constellation, and many fine clusters are visible. The northernmost reaches of this area contain some very faint galaxies visible with large telescopes. The center of the constellation culminates at midnight about 23 December.

oc 1664 diam. 13′ V=7.6 stars 101
6 cm shows this cluster as a small hazy spot in a very rich field. About eight mag. 10 stars are resolved over the haze with a mag. 7.5 star on the SE side. 15 cm shows about 35 stars, many of them forming a 10′ dipper-shaped asterism. In 25 cm the cluster merges smoothly into the background, but is nevertheless conspicuous. About 60 stars are visible in a 12′ area. The brighter stars are grouped in short chains of three or four stars. Many fainter stars spread to the W. Viewed in 30 cm, about 50 stars are loosely grouped in a 12′ area. Several strings of brighter stars make the outline appear irregular.

oc 1778 diam. 10′ V=7.7 stars 112
Located 4° S of η Aurigae, this cluster is an easy object in 15 cm. About 20 stars are visible running SE-NW in two rows that are about 12′ long and 4′ apart. 25 cm shows 30 stars near the two lines, which are aligned in pa 150°. The brightest stars lie on the N and E sides: on the N end is a triple star (h3265: V=10.2,10.2,13.0; 15″,15″; 137°,22°); on the S is a pair.

oc King 17 diam. 1.′5 mag. stars 25
This cluster is too faint to be visible in 15 cm. In 25 cm, however, it is visible on the N side of an oblique isosceles triangle of mag. 11 stars. At 200×, about ten stars, all fainter than mag. 13.5, are resolved in a 2.′5 area.

oc 1798 diam. 6′ mag. stars 50
Lying 50′ N of a mag. 5.5 star, this faint cluster is difficult to see in 15 cm. It is just visible at 100× in 25 cm as a hazy spot about 7′ diameter. At 200× it seems elongated E-W because of a mag. 12 star on the E end. About 15 stars of mag. 13.5 and fainter are resolved. Lying 50′ SW of the cluster is a mag. 8.5 star, and 10′ NW of this is a conspicuous asterism of mag. 11 and 12 stars in the shape of a small, italic y. This little group may appear nebulous at low powers.

oc 1857 diam. 6′ V=7.0 stars 23
In 6 cm this cluster shows only as an unresolved cloud surrounding an orange mag. 7.5 star. 15 cm will show about 25 stars, but the group hardly stands out. In 25 cm, 40 stars of mag. 12 and fainter are visible within a 10′ diameter. At low power the bright central star interferes with the fainter cluster members. 30 cm resolves the cluster well, showing 46 stars in a 9′ area. It is slightly concentrated, with four brighter stars including the "central" star.

oc 1883 diam. 2.′5 mag. 12.0p stars 30
Located 3′ NNE of a mag. 10.5 star, this faint cluster cannot be seen with 15 cm. 25 cm resolves about ten mag. 13 and fainter stars in a 3′ area with little background haze. With averted vision, 30 cm shows about a dozen faint stars.

oc 1893 diam. 12′ V=7.5 stars 270
This cluster illuminates gn I 410, which is difficult to view with amateur instruments. Lying 1.°5 SW of φ Aurigae, 6 cm shows it to be elongated N-S with ten stars visible. Roughly 30 stars in a 20′×5′ area are visible to 15 cm, with some field stars trailing to the NE. It is an obvious cluster in 25 cm despite the rich fields nearby. Most of the stars form a 15′×3′ group containing about 50 stars. In the center is a nice pair (V=10.3,11.4; 14″; 40°) and a mag. 9.5 star. A well-defined group of stragglers extends N and E including another 30 stars. An equal pair of mag. 9 stars lies isolated in a field 18′ W of the cluster (ADS 3928). 30 cm shows 25 stars including ten brighter ones in a 15′×8′ area.

oc Berkeley 70 diam. 12′ mag. stars 40
At low power 25 cm shows a small clot of stars here involved with a bright nebulosity. Only a few stars of mag. 12.5 and fainter are visible in a 4′ area at 200×.

oc Stock 8 diam. 5′ mag. stars 40
Surrounding φ Aurigae, this cluster is not very conspicuous in 25 cm. At low power a slight concentration of stars around φ Aur is discernable against the background. The associated nebula, I 417, is not easily discerned visually, though photographs will show it.

Chart III. NGC 1778. A 30-minute exposure on 103a-O with the USNO 1 meter telescope taken by A. Hoag on 17 September 1958. Magnitude sequence from: Hoag *et al.* 1961, Pubs. of the U.S. Naval Observatory, second series, volume 17, part 7. U.S. Naval Observatory photograph.

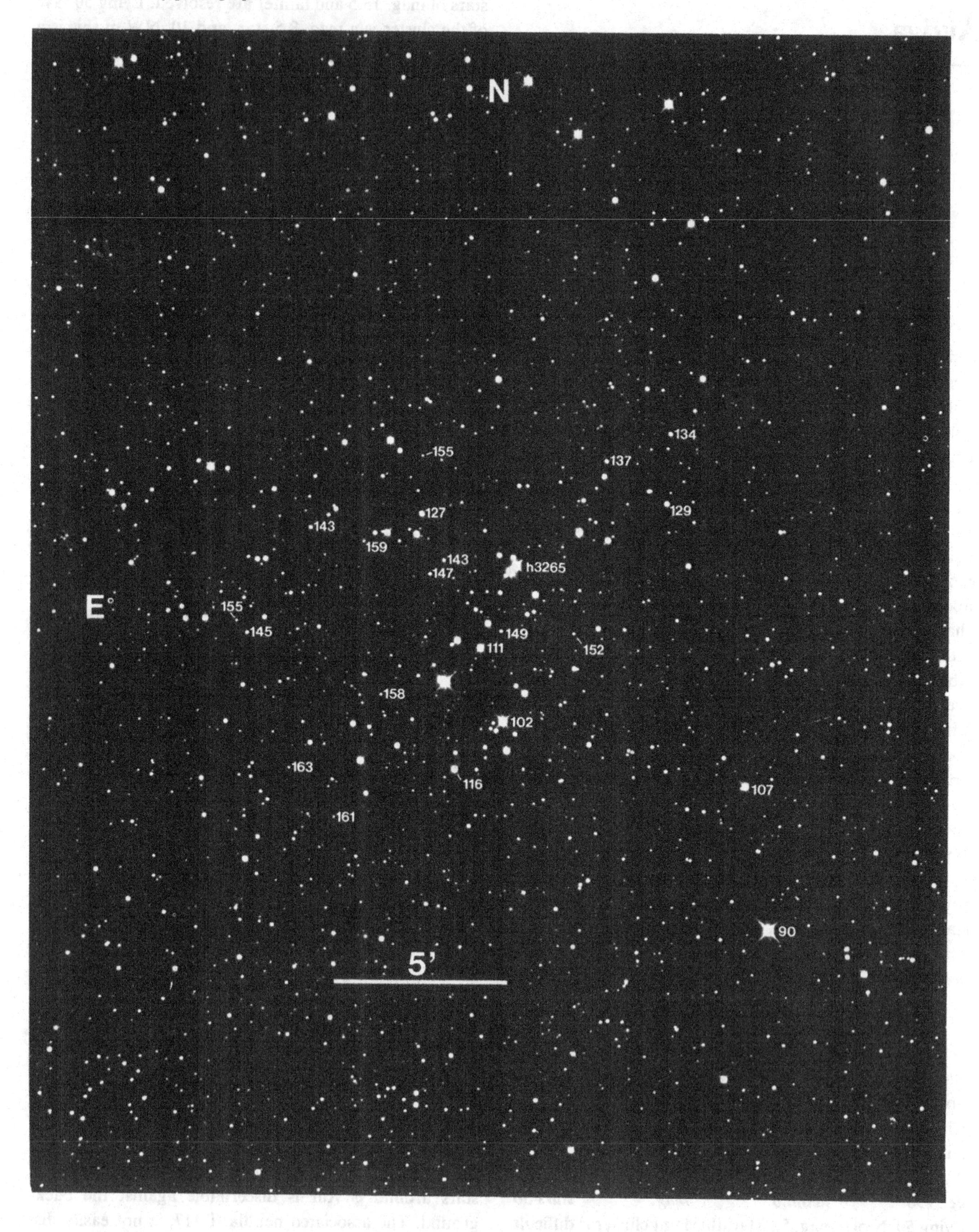

oc 1907 ***diam. 7′*** ***V=8.2*** ***stars 113***

Lying just 30′ S of Messier 38 (oc 1912, *q.v.*), this rich cluster often goes unnoticed next to its brighter companion. 6 cm will show a few stars in the compact group with a wide pair (V=9.6,9.9; 50″; 100°) 3′ SSE. At high power 15 cm shows perhaps 30 stars in a 5′ area. About 4.′5 to the NE is another, closer pair (h699: V=10.4,11.6; 9.″7; 216°). With 25 cm the main part is about 5′ across, but there are many outliers of mag. 10.5–12 spreading to 10′ diameter, past the wide pair SSE. A total of 40 stars is visible. A pretty cluster in 30 cm, it is circular with 40 stars in a 6′ area. The brightest star (V=11.3) lies in the center within a 1.′5 hole. A concentration of stars lies just NW of this hole. The central condensation looks like a miniature Messier 11.

oc 1912 ***diam. 21′*** ***V=6.4*** ***stars 160***

Messier 38 is easy to see in binoculars, and 6 cm will show at least 30 stars. The brighter stars form a distorted "X"; a brighter star (V=9.3) lies about 6′ NE. Many long strings and triangles stand out in 15 cm. With 25 cm the cluster is as much as 30′ across with about 150 stars visible within that circle. The stars are arranged in an annulus, with a conspicuous void about 10′ across surrounding a mag. 11,12,13 central triplet. Outliers extend to 25′ radius in the SE quadrant, reaching almost to the outliers of oc 1907, *q.v.* 30 cm shows some pairs, and two arms of stars extending to the N and NE.

oc-gn 1931 ***diam. 1.′0*** ***V=11.3:*** ***stars 3***
dimen. 4′×4′

This inconspicuous dust nebula surrounds a small group of faint stars (h367). At high power, 15 cm shows three stars with a bit of nebulosity. The nebula is about 1.′5×1′ in 25 cm with a mag. 11.5 central star and four fainter companions. It is smooth, featureless, and elongated NE-SW. Viewed in 30 cm, this object appears as a nebula surrounding a star on the NW side of a box-shaped asterism 4′ across. The haze is 1.′5×0.′75 in pa 45°. With careful scrutiny, four fainter companion stars can be seen in the tight asterism.

oc 1960 ***diam. 12′*** ***V=6.0*** ***stars 50***

Messier 36 is easily visible to 6 cm, and about 20 stars, including five brighter ones on the NE side, can be resolved on a hazy background. In 15 cm it appears smaller and not as rich as Messier 38 (oc 1912, *cf.*). About 60 stars are visible, including several bright, colored stars and a close pair in the center (ADS 4194: V=9.1,9.4; 11″; 305°). 25 cm shows 75 stars with a wide range in brightness in a 25′ area. The stars are more concentrated toward the center than in M38. On the N side are some faint, close pairs. In 30 cm, an even sprinkling of mag. 8.5–9.5 stars is visible amid many fainter ones. A total of 70 stars is visible in an 18′ area.

oc Stock 10 ***diam. 25′*** ***mag.*** ***stars 15***

In 6 cm this cluster is a scattered field of about 15 stars with a wide range of magnitudes. About 25 stars can be counted in 30′ diameter with 25 cm. The loose, unconcentrated cluster stands out from the background at low power because of its bright members, but no faint members are evident.

oc King 8 ***diam. 8′*** ***V=11.2*** ***stars 198***

Lying in the SW side of a 3′ triangle composed of mag. 10.5, 11.5, and 12 stars, this cluster is faintly visible in 25 cm. A few stars, mag. 13.5 and fainter, are resolved over a slight haze about 1.′5 across.

oc 2099 ***diam. 20′*** ***V=5.6*** ***stars 1842***

Messier 37 appears as a richly sparkled cloud in 6 cm. The cloud is elongated in pa 40° and is partially resolved into perhaps 30 stars, but their density precludes an accurate count. A red star (V=9.2) shines out from the center. The impression with 15 cm at high power is that of a broken-down view of Messier 22 or ω Centauri. In 25 cm the cluster is bright and well resolved at any power. The bright central star is more conspicuous at lower power. The core of the group is about 8′ across and is distinctly flattened on the SE side, where there is an area empty of bright stars. Outliers extend to 50′ diameter, with a total of around 200 stars visible. There are too many stars to count easily in 30 cm, and many of them are of a similar brightness. The dark void just SE of the central star is about 5′ across and contains a single mag. 12 star.

pn I2149 ***dimen. 15″×10″*** ***nebula 10.7v*** ***star V=11.6***

This small, bright planetary forms the center star of an M-shaped asterism. At 200× in 15 cm it is barely distinguishable from a star. 25 cm will show a faint blue halo surrounding the sharply concentrated, nearly stellar center. The halo is elongated in pa 75° in 30 cm, about 15″ long. The central star is barely visible above the bright central condensation.

oc 2126 ***diam. 6′*** ***mag. 10.2p*** ***stars 40***

This dim cluster is just visible to 6 cm as a few stray threshold stars around a mag. 6 star. 15 cm shows about 15 stars in three rows 6′ apart aligned roughly E-W next to the bright star. Viewed in 25 cm, the cluster contains 20–25 stars, mag. 11.5 and fainter, in a 7′ area; several pairs are visible. It appears comet-shaped in 30 cm with the bright star as the nucleus. The cluster stars lie to the SW of this star, with about 25 members visible in a 6′–7′ area.

oc 2192 ***diam. 6′*** ***mag. 10.9p*** ***stars 45***
This cluster is composed of stars that are mostly fainter than mag. 13.5. 15 cm will just reveal, but not resolve it lying immediately S of a reddish star (V = 7.9). It is visible in 25 cm as a faint granular spot. High power will begin to resolve the irregularly round 5′ cluster into about 15 stars. In 30 cm the brighter stars lie in the northern half.

oc 2281 ***diam. 15′*** ***V = 5.4*** ***stars 119***
The core of this cluster is noticeably elongated in 6 cm. About 25 stars are visible, including a four-star diamond-shaped asterism in the center. 15 cm shows 30 stars in a 20′ area. It is slightly concentrated to the center and contains several close pairs. In 25 cm it is a fairly sparse cluster with 35 stars in a 20′ area. Outliers scatter to about 40′ diameter. The northeastern star in the diamond is an unequal pair (ADS 5482AB: V = 9.0,10.6; 7″.7; 49°), while the southern one is a more equal pair (V = 10.1,10.6; 8″; 108°). A mag. 13.5 star lies within the diamond. 30 cm shows 40 stars in a 20′ area. The close, northeastern pair in the diamond is contrastingly colored red and blue.

Bootes

T. W. Webb characterized this region as "rich in pairs, poor in clusters and nebulae." The galaxies here are only moderately bright, and the only cluster is the feeble globular NGC 5466. The center of the constellation culminates at midnight on 2 May.

eg 5248 ***dimen. 6′.5 × 4′.9*** ***V = 10.2*** ***sfc. br. 13.8***
15 cm will show this galaxy as a broad, even glow without elongation. In 25 cm it lies on the E side of an isosceles triangle of stars 8′ long whose acute apex points S. Slightly elongated E-W, it is about 1′.5 diameter with a sharply concentrated core. A mag. 13 star is visible 1′.7 SSW. 30 cm shows a bright, broadly oval halo 3′.5 × 3′ in pa 130°. The stellar nucleus is off-center to the N in the prominent elongate 25″ core. A dark patch in the halo just S of center gives a "black-eye" appearance (*cf.* eg 4826 in Coma Berenices). Bright spots are visible in the halo on the E and W sides, the eastern one the most obvious.

gc 5466 ***diam. 11′*** ***V = 9.0*** ***class XII***
This is a very loose and poorly concentrated globular cluster, visible in 6 cm as a very faint 5′ patch WNW of a mag. 7 star. 25 cm shows it as a hazy, low surface brightness glow 6′–7′ across and slightly elongated E-W. At 200 × about two dozen faint stars are resolved. In 30 cm it is broadly and slightly concentrated with a 5′ diameter core. Faint outliers extend to 7′ diameter, and a few brighter stars are superposed on the core. The core is unevenly bright, occasionally appearing annular.

eg 5523 ***dimen. 4′.5 × 1′.4*** ***B = 12.5*** ***sfc. br. 14.3***
25 cm will show this galaxy in pa 100° with an inconspicuous circular core. The tapered halo has a low surface brightness and indefinite extent. In 30 cm the brighter part is about 1′.5 across with occasional extensions to 2′.5 × 1′.5. The light is broadly and slightly concentrated toward the center; the edges are indefinite.

eg 5533 ***dimen. 3′.2 × 2′.1*** ***V = 11.8*** ***sfc. br. 13.7***
25 cm shows a faint diffuse area here about 1′ diameter that brightens gradually toward the center. It grows to 1′.5 × 0′.8 in pa 30° with 30 cm. The nucleus is nearly stellar, while the core and halo are mottled.

eg 5548 ***dimen. 1′.9 × 1′.7*** ***V = 12.5*** ***sfc. br. 13.5***
This Seyfert galaxy is difficult to view in 15 cm, showing only a stellar nucleus within a faint halo. It appears less than 1′ diameter with a stellar nucleus in 25 cm. At high magnifications the core seems elongated SE-NW. The nearly stellar nucleus is prominent in 30 cm. Overall it is poorly concentrated, 1′ diameter, with some faint stars associated.

eg 5557 ***dimen. 2′.4 × 2′.2*** ***V = 11.1*** ***sfc. br. 12.9***
Faintly visible in 15 cm, this galaxy has a small, relatively bright nucleus. 25 cm shows a broad core that fades evenly to the 1′ halo. It is well concentrated in 30 cm, the brighter core only 30″ across, the halo extending to 1′.1. The bright inner region makes the stellar nucleus occasionally seem double.

eg 5600 ***dimen. 1′.4 × 1′.4*** ***V = 12.4*** ***sfc. br. 13.3***
25 cm will show this galaxy easily: it is quite small and has a stellar nucleus. A nice object in 30 cm, it is broadly concentrated to a 1′ core, which is surrounded by only a little halo to 1′.25 total diameter. No distinct nucleus is visible, but there are some faint stellarings in the inner regions. A mag. 11 star lies 8′ NE, and mag. 12.5 stars are about 5′ N and SE; together with the galaxy, these form a symmetric kite-shaped asterism pointing NE.

eg 5614 ***dimen. 2′.7 × 2′.3*** ***V = 11.7*** ***sfc. br. 13.5***
25 cm shows only a faint circular smudge here less than 1′ across. From diffuse edges it rises to a broadly brighter core. In 30 cm the halo is fairly well concentrated to a nearly stellar nucleus. The overall size is 1′.25, possibly with a slight elongation NE-SW. eg 5614 forms an interacting pair with eg 5613 (mag.$_{VV}$15) 2′ NNW. NGC 5615 is a knot on the NW edge of eg 5614.

eg 5633 ***dimen. 2′.3 × 1′.4*** ***V = 12.3*** ***sfc. br. 13.5***
With care, 25 cm shows this galaxy as a faint featureless spot less than 1′ across. It has diffuse edges and is slightly brighter across the center. 30 cm shows a small oval about 40″ long, elongated approximately N-S. It is unevenly bright with some central concentration, but no distinct nucleus.

eg 5641 ***dimen. 2′.7 × 1′.6*** ***B = 12.9*** ***sfc. br. 14.3***
25 cm shows this galaxy 2′.7 E of a mag. 13 star. The very faint glow is about 1′ diameter. It is still pretty faint in 30

cm, reaching 1′.8 × 0′.6 in pa 160°. The thinly tapered tips brighten broadly to an oval core and very faint nucleus.

eg 5653 dimen. 1′.8 × 1′.5 V = 12.2 sfc. br. 13.2
This galaxy is discernable in 25 cm at 50 ×. It is moderately faint and diffuse, about 1′ diameter, and uniformly bright. 30 cm also shows it about 1′ across with a faint stellar nucleus in a moderately concentrated core that seems elongated in pa 120°. This object does not bear magnification well.

eg 5660 dimen. 2′.8 × 2′.6 V = 11.8 sfc. br. 13.8
This is a faint galaxy barely observable in 15 cm. Viewed in 25 cm, it is round or slightly elongated, about 1′.25 diameter, with a broadly brighter core. 30 cm shows a weakly concentrated 1′.5 glow with well-defined edges. A very slightly brighter 5″ nucleus is occasionally visible in the center. A tiny Magellanic-type companion, M + 08–26–38 (mag.$_{VV}$16), lies 2′.6 NW.

eg 5665 dimen. 2′.1 × 1′.5 B = 12.8 sfc. br. 13.9
In 25 cm this galaxy appears less well concentrated than nearby eg 5600, *cf.* The halo is about 1′ diameter with a circular outline. 30 cm shows the halo up to 2′ × 1′.5, elongated N-S. The core is 1′ across with a mottled inner region containing a faint nucleus.

eg 5669 dimen. 4′.1 × 3′.2 B = 12.0 sfc. br. 14.6
In 25 cm this galaxy has a moderate surface brightness. Appearing slightly elongated in pa 15°, it is 2′.5 diameter and brightens gradually from diffuse edges to a broad core and stellar nucleus. 30 cm shows a broad 2′.5 × 2′ oval. The inner regions are only slightly brighter yet show the elongation most strongly. Except for a few faint spots near the center, the object has a smooth texture. Two stars 6′ NW form a triangle with the galaxy.

eg 5676 dimen. 3′.9 × 2′.0 V = 10.9 sfc. br. 13.0
In 15 cm, this is a faint, unconcentrated object about 1′ diameter. 25 cm shows it 1′.5 × 1′ in pa 50°. A stellaring is visible one-third the distance from the NE end along the major axis, although the narrow elongated core is centered. The galaxy appears much larger in 30 cm, about 3′.75 × 2′. The halo has a broad, even concentration to an oval core which, however, does not show a nucleus. eg I1029 (mag.$_Z$13.7) lies 27′ N, and eg 5673 (mag.$_Z$14.0) is 33′ NNW.

eg 5687 dimen. 2′.6 × 1′.9 V = 11.8 sfc. br. 13.4
Lying only 2′.1 N of a mag. 9 star, this galaxy is just visible to 15 cm. In 25 cm it makes a triangle with this star and a mag. 12.5 star 1′.2 SW. The galaxy is round, about 40″ across, with a stellar nucleus. A faint star is visible on the W edge 30″ from the center. It is an interestingly mottled object for 30 cm due to the associated stars. The overall size is 1′.8 × 1′ in pa 105°. No sharp nucleus is visible. Another faint star is involved on the SW edge only 36″ from center.

eg 5689 dimen. 3′.7 × 1′.2 V = 11.9 sfc. br. 13.4
In 15 cm this galaxy is faintly visible as a small concentrated brightening. In 25 cm the halo is elongated in pa 85°, about 1′ × 0′.6. The knotty core is irregularly oval, and a stellar nucleus is occasionally visible at 200 ×.

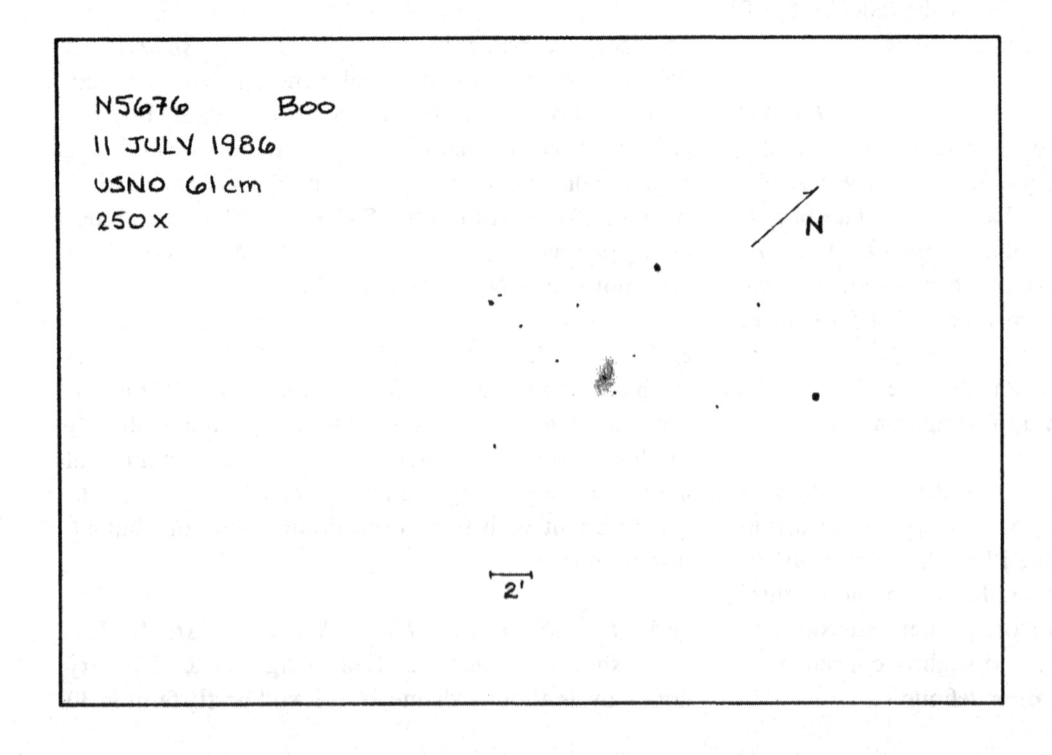

30 cm shows a bright 2′ × 0′.5 "dash," the ends visible only with averted vision. It is well concentrated to a nucleus that occasionally appears stellar. This galaxy is the brightest of a group that includes eg 5693 ($mag._Z$14.5) 12′ SE.

eg 5739 dimen. 2′.2 × 2′.1 B = 12.7 sfc. br. 14.3
In 25 cm this galaxy is moderately faint, about 1′.25 × 0′.6, elongated in pa 30°. The bright round core has a stellar nucleus. On the NE tip is a mag. 14 star. 30 cm shows a well-concentrated galaxy with a nearly stellar nucleus. Several faint stars lie E of the object.

eg 5820 dimen. 2′.5 × 2′.3 V = 11.9 sfc. br. 13.6
Situated 8′ W of a wide pair (ADS 9474: V = 6.8,7.4; 40″; 342°), this galaxy is faintly visible in 15 cm. 25 cm shows it 40″ × 20″, elongated E-W. The halo is well-defined with a nearly stellar nucleus. In 30 cm the mottled halo is slightly less than 1′ across and contains a 20″ diameter core. eg 5821 ($mag._Z$14.9) lies 3′.6 N, and photographs show several other very small, faint galaxies nearby.

eg 5874 dimen. 2′.5 × 1′.8 $mag._Z$14.1 sfc. br.
Though not visible in 15 cm, 25 cm will show this galaxy as a moderately faint glow with a fairly low surface brightness. The weakly mottled halo is broadly brighter toward the center and elongated in pa 50°.

eg 5876 dimen. 2′.6 × 1′.4 $mag._Z$13.9 sfc. br.
Although this galaxy is brighter than eg 5874, *q.v.* 20′ NW, it is still not visible in 15 cm. In 25 cm the halo is at most 45″ × 20″ in pa 50°. A conspicuous stellar nucleus is visible in the center.

eg 5899 dimen. 3′.0 × 1′.3 V = 11.8 sfc. br. 13.1
This galaxy is 2′ × 1′ in 25 cm, elongated approximately N-S, with a very gradual brightening toward the core. In 30 cm it is elongated in pa 20° and is as large as 3′ × 1′. The core is more circular, roughly 50″ across. A mag. 13.5 stellaring is visible on the SE side of the core. The object has a smooth texture and is without distinct zones of brightness. eg 5895 and eg 5896 (combined $mag._Z$15.5), a close pair, lie 13′.5 W; eg 5893 ($mag._Z$14.1) is 17′.5 WSW; and eg 5900 ($mag._Z$15.0) lies 9′.5 N.

Camelopardalis

Though its western edge dips into the Milky Way, this circumpolar constellation contains a fine selection of galaxies. eg 2403 is a large, detailed object. The center of the constellation culminates at midnight in the first week of February, but parts of it may be viewed most of the year in northern latitudes.

oc Stock 23 diam. 18′ mag. stars 25
This loose cluster is easily visible with binoculars and low-power telescopes. It is about 10′ across with 20 stars visible in 25 cm. Four brighter stars form a quadrilateral that has a few faint stars within it. The western star of the four is a pair (ADS 2426AB: 7.7,8.0; 7″.2; 142°). In 30 cm the eastern star is associated with a group of about ten stars, seven or eight of which lie in a string aligned NE-SW. The two bright stars on the W and S are in a group of about 20 stars including a string of five extending to the W. The star on the N is by itself.

eg I 342 dimen. 18′ × 17′ B=9.1 sfc. br. 15.2
This large galaxy is relatively easy to see in small apertures at low powers. In 6 cm it is a faint blob north of a loose clustering of stars. It is at least 12′ diameter, apparently elongated N-S, with a few brighter spots on a smooth background. In 25 cm a string of six stars runs SE-NW through where the object is seen in 6 cm. Northeast of this string are four mag. 11 stars, three of them in an acute triangle. A larger, brighter triangle lies farther northeast and is composed of mag. 10 stars about 8′ apart. The southern "star" of the first triangle is a circular fuzzy spot about 30″ across that brightens toward the center. A 15′ area centered on the string of stars is faintly luminous with many faint stars superposed. The bright spot looks like an elliptical galaxy in 30 cm. It is 1′.5 diameter and well concentrated to a stellar center. A faint patch of low surface brightness appears off the N end of the string of stars; it is about 2′ across and is without concentration.

oc Tombaugh 5 diam. 17′ V=8.4 stars 102
Though it is excluded from most catalogues, this cluster is visible even in 6 cm, which shows a hazy area about 5′ diameter with a half-dozen faint stars sprinkled over it. The cluster is moderately faint in 25 cm, composed of stars mag. 12 and fainter. About 30 stars are visible within a rectangular 15′ × 12′ area elongated SE-NW. The stars are uniformly distributed over a hazy background. 30 cm shows 40 stars in 12′ diameter, mostly of mag. 12 to 13. The group is loose and unconcentrated, but distinct from the background due to the similarity in brightness of the stars. A bright star lies off the W side.

eg I 356 dimen. 5′.2 × 4′.1 V=10.5 sfc. br. 13.7
In 15 cm this galaxy is not difficult to see at 75 ×. It forms a smooth glow with moderate, broad central concentration 3′.5 S of a mag. 8 star. With 30 cm the halo extends to 2′.5 × 1′.5 in pa 95°. The circular core is about 20″ across, much brighter than the halo, and shows a steady stellar nucleus at 250 ×. A mag. 13.5 star is on the NE edge of the halo, 40″ from center, and a mag. 14.5 star lies 1′.2 NW. At times a faint threshold sparkle is visible on the SW edge of the halo.

pn 1501 dimen. 56″ × 48″ nebula 11.5v star V=14.5
This planetary is a bright, even round dot in 25 cm, about 1′ diameter and without a central star. It is a beautiful and well-defined object in 30 cm. Though the nebula has a high surface brightness, the central star sparkles clearly and seems "small." At 300 ×, the annularity is visible, but the hole is only a little darker than the ring. The nebula is uniformly mottled, has unsharp edges, and is slightly elongated NE-SW.

oc 1502 diam. 8′ V=5.7 stars 63
This bright group includes a pair near the center (ADS 2984AB: V=7.0,7.0; 18″; 304°; the B component is the eclipsing binary SZ Camelopardalis, but the A component may also be variable). 25 cm shows about fifteen stars in a 5′ area. 30 cm shows 35 stars in an 8′ area, including a lacing of bright stars and a dozen mag. 9.5 stars.

oc I 361 diam. 8′ V=11.7: stars 42
A faint cluster in 25 cm, it is at best an irregular, roundish, and irresolved blur about 8′ diameter. 30 cm shows four stars leading SE from the NE side with increasing separation and brightness. The cluster is 4′.5 diameter with some concentration of light. About eight stars are resolved on a lumpy background including two brighter mag. 12 stars on the NW side.

eg 1530 ***dimen. 4′.9 × 2′.9*** ***mag.$_{z}$13.4*** ***sfc. br.***
This is a faint galaxy for 15 cm, appearing as a low surface brightness patch with a tiny substellar concentration at its center. A mag. 12.5 star lies 2′.5 N. In 30 cm the galaxy is clearly elongated, the halo extending to 1′.5 × 0′.8 in pa 120°. It shows no overall concentration toward the center except for a small 10″ × 5″ core, which rises suddenly but only moderately in brightness from the halo, and a stellar nucleus. Between the galaxy and the mag. 12.5 star to the N are two mag. 13.5 stars aligned approximately parallel to the halo (23″; 110°).

eg 1569 ***dimen. 2′.9 × 1′.5*** ***V = 11.2*** ***sfc. br. 12.6***
This galaxy is not difficult for 6 cm. It is visible as a fuzzy patch just S of a mag. 9 star. 25 cm shows a smooth, fairly bright lens in pa 120° with an oval core. The halo is bright and well concentrated in 30 cm, reaching 2′ × 0′.7. The core is oval, whereas the halo is lenticular with thinly pointed tips. At 225 ×, stellarings are visible near the center. The halo extends ESE to a mag. 12 star and seems more extensive on the south, perhaps due to the proximity of the bright star to the N.

eg I 381 ***dimen. 2′.8 × 1′.7*** ***mag.$_{z}$14.5*** ***sfc. br.***
This galaxy is only just visible to 15 cm as a small faint spot close on the S side of a mag. 12.5 star. 30 cm shows a diffuse 1′.5 × 1′ halo elongated N-S, reaching to or a little beyond the mag. 12.5 star, which lies about 35″ N of center. The 30″ × 15″ core is much brighter than the halo and contains a stellar nucleus. Off the edge of the halo, 50″ from center in pa 210°, is a 15″ pair of mag. 14 stars aligned in pa 110°.

eg 1961 ***dimen. 4′.3 × 3′.0*** ***V = 11.2*** ***sfc. br. 13.8***
A very faint object for 15 cm, this galaxy is just visible at 100 ×. 25 cm shows a moderately bright 2′ × 1′ oval in pa 90°. The halo is diffuse and without concentration, but has a few mottlings. It is 3′.2 × 1′.5 in 30 cm, with a star involved 40″ SE of the center. The core is mottled with darker spots and streaks; the halo is more sharply defined on the N side. Photographs show many faint galaxies within 30′ to the southeast.

eg 2146 ***dimen. 6′.0 × 3′.8*** ***V = 10.5*** ***sfc. br. 13.7***
This galaxy is faint in 15 cm. The halo is 1′.5 × 1′ and rises little in brightness to the center. In 25 cm the object is fairly bright but diffuse and without much concentration. With averted vision a very small faint core pops out occasionally within a 3′ × 1′.5 halo in pa 120°. It is a large, bright object for 30 cm, extending to 5′ × 1′.8. The broad core is not as elongated as the halo and is mottled, particularly around the faint stellar nucleus. A bright group of six stars lies E of the object, including a close pair near the galaxy. eg 2146A (mag.$_{z}$14.2) lies 19′ ENE.

eg 2268 ***dimen. 3′.4 × 2′.2*** ***V = 11.5*** ***sfc. br. 13.5***
25 cm shows this galaxy as a moderately faint haze, 1′.5 × 0′.5, elongated roughly NE-SW. The slightly bulging middle is weakly concentrated and contains a stellar nucleus. 30 cm shows a small, broadly concentrated oval 1′.5 × 1′ in pa 65°. A mag. 13 star is visible off the SW end, 1′.1 from the center.

eg 2314 ***dimen. 2′.1 × 1′.8*** ***V = 12.1*** ***sfc. br. 13.3***
Located E of a bright pair (ADS 5669: V = 7.1,8.2; 13″; 31°), this very small galaxy is barely visible with 15 cm. 25 cm shows a round concentrated spot 20″ diameter. A faint star is visible 40″ NE of center. It is still quite small in 30 cm. The halo is only 35″ across and is well concentrated to a fairly bright, nearly stellar nucleus. eg I2174 (mag.$_{z}$15.0) is located 5′.6 NNW.

eg 2336 ***dimen. 6′.9 × 4′.0*** ***V = 10.3*** ***sfc. br. 13.9***
This galaxy is moderately bright overall, but of fairly low surface brightness in 25 cm. It seems elongated E-W, about 3′ × 1′, but the boundaries are indefinite. The slightly brighter center has a mottled texture. The mottlings are seen clearly with 30 cm. The halo is circular or slightly elongated E-W and up to 3′ diameter. A faint, occasionally stellar nucleus is visible within the 1′ diameter core. eg I 467 (mag.$_{z}$12.7) lies 21′ SE.

eg 2347 ***dimen. 2′.0 × 1′.5*** ***V = 12.5*** ***sfc. br. 14.4***
Located 4′ S of a mag. 7.5 star, in 25 cm this galaxy is a diffuse, weakly concentrated glow about 1′.25 across. It is about 1′ across with 30 cm, brightening to a core and nonstellar nucleus. A mag. 13.5 star lies 1′.9 SW.

eg 2366 ***dimen. 7′.6 × 3′.5*** ***V = 10.7*** ***sfc. br. 14.3***
This galaxy is not visible in 15 cm, and all that can be seen with 25 cm is a spot less than 1′ diameter in a field of many faint stars. 30 cm shows this bright blob just E of what appears to be a mag. 13.5 star. Photographs show that this "star" is a bright HII region. A faint halo leads NE away from here, brightening again to a broad area of very low surface brightness before disappearing. The blob at the SW end is 50″ across and has moderately high surface brightness, so that the HII region is often difficult to distinguish. The overall size is 4′ × 1′.5.

eg 2403 ***dimen. 18′ × 11′*** ***V = 8.4*** ***sfc. br. 13.9***
This bright galaxy is a good sight even in small apertures. In 6 cm it is about 15′ × 5′, elongated SE-NW, with a mag. 11 star on either side of the core. In 25 cm at least three stars can be seen in the 10′ × 3′ nebula. The edges

are uneven, and there is no marked central condensation. Five fainter stars are associated with the core in 30 cm, the brightest of which lies just S of center. More faint stars can be seen in the low surface brightness halo. Some tenuous spiral structure is suspected: a dark area appears between the center and the mag. 11 star on the ESE, and a 2′.5 × 0′.75 dark lane runs across the N side of the core. The overall size is at least 15′ × 4′ in pa 115°.

eg 2441 dimen. 2′.2 × 2′.0 V = 12.2 sfc. br. 13.7
25 cm shows this galaxy only faintly, with a stellar nucleus suspected in the weak spot. In 30 cm it is a faint unconcentrated blob about 1′ across. At high power it is elongated in pa 135°, and a little mottling can be seen.

eg I2209 dimen. 1′.1 × 1′.0 V = 13.8 sfc. br. 13.8
This companion to eg 2460, *q.v.*, is located 5′.5 SW of the parent galaxy and 1′.5 N of a mag. 13 star. It is barely visible in 25 cm as an indefinite patch with a star-like nucleus. In 30 cm it is about 1′ diameter and without much concentration.

eg 2460 dimen. 2′.9 × 2′.2 V = 11.7 sfc. br. 13.6
At high power in 25 cm, this galaxy is moderately faint and amorphous with a brighter center and stellar nucleus. A 10″ mag. 12 pair, located 4′.5 SE, can be mistaken for the nebula. In 30 cm the core has a high surface brightness and contains an almost stellar nucleus. The overall size is 1′ × 0′.8 in pa 110° with little halo. eg I2209, *q.v.*, lies 5′.5 SW.

eg 2523 dimen. 3′.0 × 2′.0 V = 12.0 sfc. br. 13.7
Not visible in 15 cm, in 25 cm this galaxy is a diffuse and unconcentrated spot 1′ diameter. A mag. 11 star lies 1′.5 SW. With 30 cm some structure is discernable in the irregularly shaped core, notably a lump on the SW, giving a slight elongation in pa 60°. eg 2523B (mag.$_Z$14.8) lies 8′.6 E.

eg 2551 dimen. 1′.9 × 1′.3 V = 12.0 sfc. br. 12.9
Located 2′ SW of a mag. 12 star, 25 cm shows a round glow here about 45″ across that is brighter toward the center and has a granular texture. In 30 cm it is weakly condensed except for a small, bright core. At 225 × a thin form is visible, 1′ × 0′.5 in pa 90°. A faint stellar nucleus is occasionally visible within the core.

eg 2633 dimen. 2′.6 × 1′.7 V = 11.9 sfc. br. 13.3
This galaxy forms a pair with eg 2634, *q.v.* 8′.2 S. 25 cm shows a faint, round object less than 1′ diameter with a stellar nucleus. In 30 cm the halo is 2′ across, but the outer extensions do not bear magnification. A conspicuous, nearly stellar nucleus is visible with several faint stellarings around it. The inner core seems elongated N-S.

eg 2634 dimen. 2′.2 × 2′.1 mag.$_Z$12.6 sfc. br.
In 25 cm this galaxy is nearly the same size and brightness as eg 2633, *cf.* 8′.2 N. The halo is less than 1′ diameter and contains a stellar nucleus. With 30 cm it appears much smaller, though of similar magnitude to −33. The center is bright with a nucleus that appears stellar at times. At high power the nucleus seems off-center to the NW. The halo is elongated in pa 80°, 1′ × 0′.6. eg 2634A (mag.$_Z$14.3), an edge-on system, is located 2′ SSE.

eg 2646 dimen. 1′.7 × 1′.6 V = 12.0 sfc. br. 12.9
In 25 cm this galaxy is visible 2′ NNW of a 40″ unequal pair of stars. The halo is less than 1′ across and contains a stellar nucleus that seems off-center to the N. 30 cm shows it 30″ × 20″, elongated E-W, with a moderate concentration to a faint stellar nucleus. eg I 520 (B = 12.7) lies 15′ E, while eg I2389 (mag.$_Z$13.2) is 12′ NW.

eg 2655 dimen. 5′.1 × 4′.4 V = 10.1 sfc. br. 13.3
6 cm will show this galaxy as a small round spot with a bright center forming an equilateral triangle with a mag. 7.5 star SE and a mag. 9 star NE. In 15 cm, it is a well-concentrated spot about 1′ across with a stellar nucleus. With 25 cm, the stellar nucleus lies in a bright core elongated E-W. Little halo is visible, extending to only 1′.5 × 1′.25. In 30 cm, the galaxy is evenly concentrated to a very bright inner core and an 8″ nucleus. The object is elongated, particularly in the inner regions, in pa 90°. The inner core shows bright mottlings along the major axis at high power. The northern side of the 4′.2 × 3′ halo darkens abruptly.

eg 2715 dimen. 5′.0 × 1′.9 V = 11.4 sfc. br. 13.7
This distended object is barely discernable in 15 cm. 25 cm shows only a low surface brightness blob 1′.5 across without concentration or detail. In 30 cm the halo is very weakly and broadly concentrated, reaching 3′ × 2′ in pa 20°. The core is more elongated than the halo, 2′.5 × 1′, with a very faint nucleus. High power shows mottlings over most of the surface.

eg 2732 dimen. 2′.3 × 0′.9 V = 11.9 sfc. br. 12.5
An edge-on lenticular, this galaxy can be seen only very faintly in 15 cm. It is elongated in pa 70°, 1′.5 × 0′.2, in 25 cm. The tiny oval core seems displaced WSW of the center of the halo, which extends ENE just past a mag. 12 star 45″ from the center. With 30 cm the core is less elongated than the halo and well concentrated to a stellar nucleus. The halo is 1′.1 × 0′.4 and tapers thinly to pointed tips. Photographs show eg U4832 (mag.$_Z$15.0)

4.′1 E, and eg U4847 ($mag._{vv}$15) 9.′4 ENE, near a mag. 9 star.

eg 2748 dimen. 3.′1 × 1.′3 V=11.7 sfc. br. 13.1
In 25 cm this galaxy is elongated in pa 40°, 1.′25 × 0.′6, an oval glow with a weak central brightening. Several stellarings with different brightnesses appear scattered over the surface. 30 cm shows the 3′ × 1′ halo with tapered ends and a smooth texture. A faint stellar nucleus is displaced to the SW of center of the bulging core. The halo bears magnification well and at high power appears more abrupt on the NW side.

pn I3568 dimen. 18″ × 18″ nebula 10.6v star V=12.3
Viewed in 30 cm, this bright planetary forms the northern "star" in the base of a narrow isosceles triangle that points to the W. About 20″ diameter overall, it is well concentrated from diffuse edges to a bright, 10″ diameter core. The nebula is white and has a stellar center at low magnifications.

Cancer

Marked by the popular Praesepe, or Beehive cluster (oc 2632), this region contains many galaxies. Of special interest is the rich open cluster Messier 67 (oc 2682), which is one of the oldest known. The center of the constellation culminates at midnight about 30 January.

eg 2507 ***dimen. 2′.5 × 2′.1*** ***mag.$_z$14.0*** ***sfc. br.***
In 15 cm this galaxy is a faint, moderately concentrated spot lying 1′.25 NE of a mag. 12.5 star. 30 cm shows the halo to about 40″ diameter, slightly elongated NE-SW. The brightness grows toward the center in fairly distinct zones, forming a circular core and a faint substellar nucleus. A threshold magnitude star lies on the N edge. eg 2514, *q.v.*, lies 18′ ENE.

eg 2514 ***dimen. 1′.4 × 1′.4*** ***mag.$_z$14.4*** ***sfc. br.***
Though not visible to 15 cm, this object is faintly visible in 30 cm. It forms an almost unconcentrated, low surface brightness patch about 40″ diameter. The texture is uneven at 250 ×, particularly in the N side. eg 2507, *q.v.*, lies 18′ WSW.

eg 2535 ***dimen. 3′.0 × 1′.7*** ***V = 12.6*** ***sfc. br. 14.2***
eg 2536 ***dimen. 0′.9 × 0′.6*** ***V = 14.1*** ***sfc. br. 13.4***
Of this pair, only eg 2535 is visible in 15 cm. It appears as a small spot of low surface brightness with three mag. 11.5–12.5 stars nearly in contact on its NW side. In 30 cm the halo is about 30″ across and shows some weak broad concentration. The interacting companion, eg 2536, is visible in 30 cm as a tiny circular spot 1′.8 SSE. The overall size is only about 15″, but it shows relatively good concentration to a faint substellar nucleus.

eg 2545 ***dimen. 2′.2 × 1′.3*** ***V = 12.4*** ***sfc. br. 13.4***
This galaxy appears only a little brighter than other galaxies in the vicinity in 25 cm. The narrow oval halo is elongated N-S, 1′.25 × 0′.4. It brightens only gradually across the center; thus the core is ill-defined. The faint stellar nucleus is more obvious than a mag. 14.5 star on the NW tip at pa 340°. To 30 cm the halo has a low surface brightness and exhibits no overall concentration. A faint stellaring is occasionally visible on the SE edge of the core. The overall size is about 50″.

eg 2557 ***dimen. 1′.4 × 1′.3*** ***mag.$_z$14.6*** ***sfc. br.***
Though barely visible to 15 cm, this galaxy needs more aperture to be seen well. 25 cm shows the low surface brightness halo extending to 1′.25 diameter. It is gradually concentrated toward the center and contains a stellar nucleus. Some mag. 13–14 stars lie in the field nearby that are of comparable brightness to the nucleus. Photographs show eg I2293 (mag.$_z$15.2) 5′.5 SE.

eg 2560 ***dimen. 1′.7 × 0′.5*** ***mag.$_z$14.9*** ***sfc. br.***
Lying just a few arcminutes from a mag. 8 star, this galaxy is visible in 25 cm just E of a mag. 11 star. The thin, spindle-like halo is 35″ × 20″, elongated E-W. A conspicuous stellar nucleus is visible within what is otherwise a poorly concentrated object. Overall it is a little harder to see than either eg 2562 or eg 2563, *cf.*

eg 2562 ***dimen. 1′.4 × 1′.0*** ***V = 12.9*** ***sfc. br. 13.0***
15 cm will not reveal this galaxy, but 25 cm shows it 4′.7 NW of eg 2563, *q.v.* It has a higher surface brightness than −63 but is a little fainter overall. The halo is about 20″ diameter and brightens a little toward the center and a stellar nucleus.

eg 2563 ***dimen. 2′.3 × 1′.8*** ***V = 12.3*** ***sfc. br. 13.8***
Only just visible to 15 cm, this galaxy is moderately faint even in 25 cm. High powers show a stellar nucleus within an ill-defined core that fades evenly to the halo. Overall the galaxy is 45″ × 30″, elongated E-W. Lying 2′.2 W is a mag. 13 star, and a threshold magnitude star lies 1′.7 SE. eg 2562, *q.v.*, lies 4′.7 NW.

eg 2608 ***dimen. 2′.5 × 1′.6*** ***V = 12.1*** ***sfc. br. 13.5***
This galaxy is barely visible in 15 cm about 5′ N of a mag. 11.5 pair. 25 cm shows a 1′.5 diameter glow with a faint stellar nucleus occasionally visible. Viewed in 30 cm, the halo is elongated in pa 60°, 1′ × 0′.8. It is slightly concentrated, and the S edge is more definite.

oc 2632 ***diam. 95′*** ***V = 3.1*** ***stars 161***
Messier 44, or the Beehive cluster, is easily visible to the naked eye and is a fine sight in small telescopes. 6 cm shows 55 stars in 80′ diameter. The stars are associated in groups rather than strings, and many appear yellow and blue. 15 cm shows 75 stars down to mag. 12 in a 90′ low-power field, including two multiple stars. Larger instru-

ments reveal several fainter companions among the associated groups of stars. The cluster is not well viewed with longer focal lengths and small fields due to its large size.

eg 2672 dimen. 2'.6 × 2'.4 V = 11.6 sfc. br. 13.6
eg 2673 dimen. 1'.4 × 1'.4 V = 12.9 sfc. br. 13.6
This is an interesting pair of galaxies only 40″ apart. 15 cm will show −72 faintly, but its companion is too faint. With 25 cm, −72 is easy to see. The halo appears circular, about 45″ diameter, and contains a faint, nearly stellar nucleus. A mag. 13 star lies 1'.2 distant in pa 20°, and another mag. 14 star lies at the same distance in pa 135°. Only 40″ away in pa 100° lies −73. High power shows it with a 15″ diameter well-defined core and a substellar nucleus. In 30 cm, −72 has an average surface brightness halo about 1'.5 × 1′ in size, elongated in pa 120°. This elongation may, however, be due to the presence of −73, which appears only as a stellaring on the ESE side of the core of −72. These galaxies are the brightest of a group, the others being mag. 15 and fainter.

oc 2682 diam. 30′ V = 6.9 stars 324
Messier 67 is a very rich, old cluster visible faintly to the naked eye in the best conditions. 6 cm shows a bright, partially resolved haze in the shape of a cornucopia. About 20 stars can be counted. In 15 cm, the cluster appears of average size and brightness, but much richer than usual. About 50 stars brighter than mag. 12 are visible in a 15′ area. In 25 cm the view is reminiscent of Messier 11 (oc 6705, Scutum, *cf.*), but not as bright. About twenty mag. 9 stars are distributed over a host of fainter stars down to mag. 14. Outliers extend to 25′ diameter. With 30 cm, the impression is of a fiber-optic "tree" with the trunk pointing to the SW, and a mag. 8 star at its crown. Approximately 90 stars are visible in a 16′ area.

eg 2749 dimen. 2'.0 × 1'.7 V = 12.0 sfc. br. 13.3
25 cm shows this galaxy not far S of five mag. 11 stars. It is 1′ diameter with a stellar nucleus. Some threshold magnitude stars are visible to the S. In 30 cm it is small and faint, rising suddenly to a faint stellar nucleus that is off-center to the NE. A very faint star is visible immediately SW of the center. Photographs show that the galaxy is the brightest of a group including eg 2744 (V = 13.4), which lies 14′ NW, eg 2751 (mag.$_Z$15.1) 4′ SE, and eg 2752 (mag.$_Z$14.8) 5'.3 NE.

eg 2764 dimen. 1'.7 × 1'.0 V = 12.7 sfc. br. 13.2
In 25 cm this galaxy is no more than 1′ diameter and has little central concentration. Magnitude 10 stars lie 2'.2 N and SE, and a threshold magnitude star is visible on the N edge of the galaxy. Viewed in 30 cm, this fairly faint object has a broadly brighter core and little surrounding halo. A mag. 14 stellaring is occasionally visible at or near the center.

eg 2775 dimen. 4'.4 × 3'.5 V = 10.3 sfc. br. 13.1
This galaxy is easy to see in 15 cm as a uniform 2′ × 1′ glow with a stellar nucleus occasionally visible. 25 cm shows a 2′ × 1'.5 halo with a bright stellar nucleus. At low power there is an impression of a star next to the nucleus, but this disappears at 200 ×. In 30 cm, the halo is elongated in pa 155°, about 3′ × 1'.5, and rises evenly to a 40″ core and a stellar nucleus. eg 2777 (mag.$_Z$13.9) lies 11′ N, and eg 2773 (mag.$_Z$14.5) lies 13′ NW.

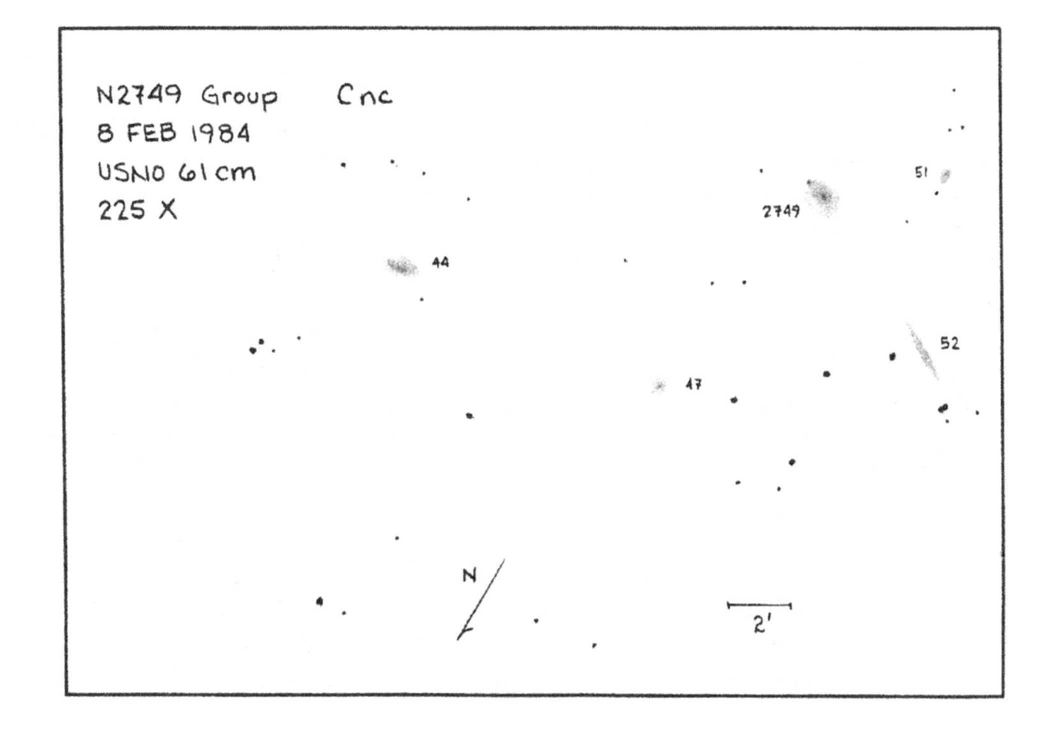

Canes Venatici

This constellation lies between Ursa Major and Coma Berenices and contains a rich string of bright galaxies. Also included is Messier 3 (gc 5272), a bright, well-studied globular cluster. The center of this constellation culminates at midnight on 5 April.

eg 4111 dimen. 4'.8 × 1'.1 V = 10.6 sfc. br. 12.3
This galaxy is visible as a nearly stellar spot in 6 cm. In 15 cm at 100×, it appears 1'.5 × 0'.5 with a conspicuous stellar nucleus. The nucleus is very bright in 25 cm, and the halo extends to 2'.25 × 0'.3 in pa 150°. The core has extremely high surface brightness. A mag. 8 star 3'.7 NE has a faint companion on its SW side (h2596: 8.2,10.7; 34″; 239°). Several bright markings can be seen with 30 cm along the major axis. The elongated core is about 1'.5 long and contains a bright nucleus; the S half has a large brighter lobe in it. The brightest part of the galaxy is 2′ × 0'.4, but a faint rounded halo to 2'.8 × 0'.75 can be discerned. eg 4109 (mag.$_Z$15.1) lies 4'.8 S, and eg 4117, *q.v.*, lies 9′ NE.

eg 4117 dimen. 2'.8 × 1'.1 mag.$_Z$14.3 sfc. br.
This faint companion 9′ NE of eg 4111, *q.v.*, is visible in 30 cm. The nebula is small and unconcentrated, about 40″ × 30″, elongated NE-SW. Photographs show eg 4118 (mag.$_Z$15.7) 1'.6 SE.

eg 4138 dimen. 2'.9 × 1'.9 B = 12.3 sfc. br. 14.0
This galaxy is moderately faint in 15 cm, located at the S end of a triangle of moderately bright stars. The halo is about 1′ diameter, weakly elongated SE-NW, with a circular core and stellar nucleus. In 25 cm there is definite elongation in pa 150°, with an overall size of 1'.25 × 0'.5. The core is circular, or possibly oval in pa 90°; an inconspicuous stellar nucleus is visible. A star or bright spot is suspected N of the core. With 30 cm the halo is 2'.25 × 1'.25, concentrated to a 1'.75 × 1′ broad, oval core and faint stellar nucleus. Thin, slightly brighter spikes extend through the halo along the major axis.

eg 4143 dimen. 2'.9 × 1'.8 B = 12.1 sfc. br. 13.8
15 cm shows this galaxy as a bright circular spot about 45″ diameter. The core is bright and diffuse with little surrounding halo. In 25 cm the object grows to 3′ × 0'.5 in pa 145°, with a sharply tapered halo. The oval core is 45″ × 20″ with a stellar nucleus. The galaxy has a peculiar twisted form in 30 cm. The nucleus is 8″ × 4″ in pa 105° with a bright nearly stellar center. A dark patch is suspected on its E edge. The core is 1'.5 × 0'.8 in pa 170°. The halo extends to 2'.8 × 1'.25 in pa 145°, with faint tapering tips.

eg 4145 dimen. 5'.8 × 4'.4 V = 11.0 sfc. br. 14.3
This low surface brightness galaxy is not visible with 15 cm. In 25 cm it appears about 2'.5 × 1′, elongated E-W. No central concentration is evident. The galaxy is a difficult object for 30 cm, located 8'.6 W of a bright star [ADS 8446: V = 7.3,8.0; 0″.4; 265° (1980)]. It has a slight, broad concentration to a mottled center, but there is no outstanding nucleus. The very faint halo extends to 2'.5 × 1'.5 in pa 90°. A mag. 14.5 star is visible about 1'.5 SSW.

eg 4151 dimen. 5'.9 × 4'.4 V = 10.4 sfc. br. 13.8
eg 4156 dimen. 1'.5 × 1'.3 V = 13.0 sfc. br. 13.8
eg 4151 is a Seyfert galaxy with a variable nucleus ($12.4 \leq B \leq 13.4$). In 15 cm it is star-like with a very faint halo. 25 cm reveals a stellar nucleus off-centered to the NE. The unconcentrated halo is about 1′ diameter. The nucleus appeared about mag. 12 in 30 cm at the time of observation. The 2′ × 1′ halo is in pa 130° and rises evenly in brightness to the center. Two faint stars lie about 1'.4 N and W; a mag. 12 star (of similar brightness to the nucleus) is 2'.3 N. eg 4156, 5'.1 NE of −51, is pretty easy to see in 30 cm. It seems slightly elongated E-W, 50″ diameter, without much concentration.

eg 4183 dimen. 5'.0 × 0'.9 B = 12.4 sfc. br. 13.9
25 cm shows this galaxy as little more than a 1′ diameter circular spot. 30 cm reveals a faint, weakly concentrated spindle, 3′ × 0'.4 in pa 165°. The elongated core is without a nucleus. A mag. 13.5 star lies 2'.7 S; a mag. 14.5 star lies only 45″ SSE of center on the E side of the spindle.

eg 4190 dimen. 1'.7 × 1'.6 B = 13.0 sfc. br. 14.0
Located about 6′ S of a mag 8.5 star, this galaxy is not visible in 15 cm and is only barely discernable with 25 cm. The object is difficult to view even with 30 cm. The halo appears about 1'.5 diameter with a mottled surface and a faint stellaring near the center.

eg 4214 dimen. 7′.9 × 6′.3 V = 9.7 sfc. br. 13.8
This bright, irregular galaxy is visible 4′.5 NW of a mag. 11 star in 6 cm. Easily seen even at low power, it is about 3′ across with a broad, slight concentration. In 15 cm it is a roughly circular glow about 1′.5 diameter. The core is bright but without a nucleus. 25 cm shows a bright object 2′ × 1′.25 in pa 130°. The edges are indefinite, and a diffuse central pip is visible. It grows to 3′.75 × 2′.25 in 30 cm. The broad, unevenly concentrated core is 2′.5 × 0′.75, tapering at both ends, particularly on the ESE. A stellar spot lies in the bar E of center, and a small, nonstellar spot lies 20″ E of this. The WNW end has a more extensive halo. A mag. 14 star is visible 3′.3 NW of center.

eg 4217 dimen. 5′.5 × 1′.8 B = 11.3 sfc. br. 13.6
This galaxy is visible as an indefinite patch in 15 cm. 25 cm reveals a fairly low surface brightness object located just S of two mag. 9 and 11 stars that interfere with viewing. It appears elongated in pa 45°, 3′.5 × 1′.5 in size. Most of the light is concentrated in a 1′.5 × 1′ blob. The faint halo extends to 4′ × 1′ in 30 cm, reaching to a mag. 13.5 star 1′.9 SW. No prominent nucleus is evident. eg 4226 (mag.$_z$14.4) lies 7′.4 SE.

eg 4218 dimen. 1′.2 × 0′.8 mag.$_z$13.2 sfc. br.
eg 4220 dimen. 4′.1 × 1′.5 B = 12.2 sfc. br. 14.0
eg 4220 is visible with 15 cm as a dull patch of uniform light. In 25 cm it is elongated SE-NW, about 2′.5 × 0′.75, and contains a stellar nucleus. Curious extensions from the core along the minor axis are suspected. The galaxy has a similar size in 30 cm. With a lenticular outline, the halo is broadly concentrated to elongated inner regions. A faint, circular and nonstellar nucleus can be seen with averted vision. The SW flank of the halo fades more abruptly than the NE, placing the concentrated inner regions off-center. eg 4218, 15′ N of −20, is a faint 1′.5 × 0′.5 spot in 25 cm, elongated SE-NW, that has a circular core.

eg 4242 dimen. 4′.8 × 3′.8 V = 11.1 sfc. br. 14.1
This galaxy has a low surface brightness and is located about 3′ W of a mag. 12 star. In 25 cm it is faint and indistinct, roughly circular, about 2′.5 diameter. The center is only a little brighter. The nebula is 3′ diameter in 30 cm, with broad, weak concentration. The brighter central parts are elongated approximately N-S, and the halo is broadly mottled. A mag. 14 star is visible on the S edge 1′.7 from the center.

eg 4244 dimen. 16′ × 2′.5 V = 10.1 sfc. br. 14.0
Viewed in 15 cm, this galaxy is a 10′ × 1′ spindle in pa 50°. The slightly brighter core is 2′ × 0′.75. 25 cm shows it 15′ × 1′, extending 1′ past a mag. 12 star that lies just N of the NE tip, and right to a mag. 11 star that has two fainter companions on the SW. With 30 cm the halo extends to 15′ × 1′.5. The core is 4′ long and bulges slightly in the center. On each end of the core, about 1′.6 from the center, are slight broad brightenings; thus the core seems in three parts, with the brightest in the center. The NE half of the halo is brighter, and on both tips are a few small, faint splotches.

eg 4248 dimen. 3′.0 × 1′.2 V = 12.6 sfc. br. 13.8
eg 4258 dimen. 18′ × 7′.9 V = 8.3 sfc. br. 13.5
Messier 106 is one of the brightest galaxies in the sky and is easily visible in 6 cm. It is elongated in pa 140° toward a mag. 11 star, with good concentration to a faint, nonstellar nucleus. 15 cm shows a large, bright core that rises abruptly in brightness to a small, well-defined nucleus. It has a granular texture, with an overall size of 10′ × 7′. In 25 cm it is 15′ × 4′.5 in pa 150°, bounded on the N by a mag. 11 star. The nucleus is stellar and surrounded by a bright oval core less than 1′ across in pa 140°. The brightest part of the halo is 5′ × 2′ and has a smooth texture. The galaxy is at least 12′ × 4′ in 30 cm. The bright inner core is 1′ × 0′.7 with a prominent but nonstellar nucleus on its SW edge: a small shadow falls just beyond it to the SW. The arms are visible as two broadly brighter areas: one extends NNW from the SE end of the 1′.5 outer core along the NE flank of the halo for about 5′, where it gets slightly brighter before fading; the other arm extends SSE from the NW end of the core along the SW flank of the halo; the arm on the NE flank is brighter. eg 4248, 13′ NW, is visible in 30 cm. Elongated in pa 110°, the halo is poorly concentrated, 1′.2 × 0′.6, with granular inner regions. A faint star or stellaring stands out on the N edge to the W of center.

eg 4346 dimen. 3′.5 × 1′.4 B = 12.2 sfc. br. 13.8
This object appears as a brighter version of eg 4218, *cf.*, in 25 cm. It is elongated E-W, about 1′.5 × 0′.5, and lies in a blank field. The core is circular and highly concentrated. In 30 cm the halo is in pa 100°, 1′.8 × 0′.75, with faint extensions to 2′.1 × 1′. The lenticular form is well concentrated to a bright nearly stellar nucleus; the bright inner parts are elongated.

eg 4369 dimen. 2′.5 × 2′.4 B = 12.3 sfc. br. 14.1
In 15 cm this galaxy appears nearly stellar. 25 cm, however, reveals a circular halo 1′ diameter, with a bright and nearly stellar nucleus. A faint pair of stars is resolved about 5′ S (12,12; 23″; 65°). In 30 cm the best view of the low surface brightness 2′ halo is had at low power, where it takes on a slight E-W elongation. The distinct, round core is 30″ across, with a nonstellar nucleus just discernable over it.

eg 4389 dimen. 2′.7 × 1′.5 B = 12.7 sfc. br. 14.1
In 25 cm this galaxy is moderately bright but of low surface brightness. The uniform circular glow is 1′.5 diameter. 30 cm shows it elongated in pa 100°, 1′.5 × 0′.8. The weakly concentrated halo has a rounded rectangular form, like a television screen, and a granular texture. A mag. 12 star is visible 2′.1 NNW; a mag. 13.5 star lies about 1′ SE, just off the E flank.

eg 4395 dimen. 13′ × 11′ V = 10.2 sfc. br. 15.4
This faint galaxy is invisible to 15 cm. 25 cm reveals a diffuse area about 8′ diameter. There is an extremely weak brightening to a small, barely noticeable core. 30 cm shows two brighter spots, like a pair of mag. 13 galaxies, on a mottled haze 8′ diameter. The two patches are in pa 125°, about 1′.25 across each, the southeastern one being brighter. A very faint star or stellaring lies NE of the northwestern patch. The brighter patch to the SE has been designated NGC 4401 and is a bright HII region.

eg 4449 dimen. 5′.1 × 3′.7 V = 9.4 sfc. br. 12.5
This bright irregular galaxy is visible in 6 cm as a broad 2′ brightening. 15 cm shows a fairly even gradation of light. Viewed in 25 cm, the galaxy appears elongated NE-SW with interesting bright spots near the center of the elongated core. Averted vision shows an arm curving NW from the 3′ × 1′ oval halo. The halo extends to 6′ × 2′.5 in 30 cm. The arm NW is seen well.

eg 4460 dimen. 4′.4 × 1′.4 B = 12.1 sfc. br. 13.9
Moderately faint in 25 cm, this galaxy appears elongated NE-SW, about 3′ × 0′.7. The core is composed of two tiny, faint blotches. A fine pair (ADS 8561: V = 7.5,8.1; 9″.9; 158°) lies nearby SW. The galaxy has a lenticular form in 30 cm, with broad concentration and no prominent nucleus. The central spots are about 30″ across each. The overall size is 2′ × 0′.9 in pa 40°.

eg 4485 dimen. 2′.4 × 1′.7 V = 12.0 sfc. br. 13.3
eg 4490 dimen. 5′.9 × 3′.1 V = 9.8 sfc. br. 12.8
These interacting galaxies lie 3′.5 apart. They can be seen as separate objects in 6 cm: −85 lies NNW of −90 and appears circular; −90 is elongated, about 1′.5 long. In 15 cm, they appear as a small dense patch (−85) next to a larger, more diffuse patch (−90). 25 cm shows −85 attached to −90. The smaller object is about 1′.5 diameter and roughly circular. The core is mildly concentrated. eg 4490 is elongated SE-NW about 5′ × 1′.5. There is a small central bulge with bright dots along the major axis near the center. eg 4485 remains circular in 30 cm, while −90 grows to 6′ × 4′ with a 45″ core. Passing on the N side of the nucleus is a dark band parallel to the major axis.

eg 4618 dimen. 4′.4 × 3′.8 V = 10.8 sfc. br. 13.8
This galaxy is easily visible about 6′ N of a mag. 10.5 star in 6 cm. It appears 1′.5 diameter with a stellar nucleus. A bright object for 25 cm, it is 3′ × 2′, elongated NNE-SSW. The nucleus is moderately bright, and a few mottlings are visible on the S. 30 cm reveals a broad nebula 4′ diameter with a knot in the arms on the S. eg 4625, *q.v.*, is a possibly interacting companion 8′.3 NNE.

eg 4625 dimen. 2′.4 × 2′.0 V = 12.3 sfc. br. 13.9
This faint galaxy can be glimpsed in 25 cm. 30 cm shows a broadly condensed patch about 2′ diameter, lying just E of a faint star. No elongation is discernable.

eg 4627 dimen. 2′.7 × 2′.0 V = 12.3 sfc. br. 14.1
eg 4631 dimen. 15′ × 3′.3 V = 9.2 sfc. br. 13.3
eg 4631 is visible as a smooth, weakly concentrated bar in 6 cm, elongated in pa 85°. Only −31 is visible in 15 cm. It is a 10′ × 1′.5 streak with a mag. 12 star on the N edge of the brightest part of the mottled spindle. 25 cm shows −31 as 13′ × 1′.5 overall. eg 4627 is faintly visible 2′.7 NW of the brightest part. The brighter galaxy is magnificent in 30 cm, extending to 17′ × 1′.25, and is as interesting as Messier 82 (eg 3034, Ursa Major, *q.v.*). The W end of the halo tapers to a thin, long point; the E end fades out more abruptly. The mag. 12 star lies on the N edge near the broadest part; a mag. 13.5 star lies 8′ W of this star on the N flank. A small knot is visible 20″ S of the fainter star; the halo extends 2′.5 W of this star. Moving E, the bar grows wider, and several bright and dark markings are visible with stellarings among them. About 1′ SE of the mag. 12 star is the brightest blob, which exhibits some concentration. A beautiful galaxy! Next to this exceptional object, −27 is small and faint. The poorly concentrated halo extends only to 50″ × 30″, elongated NNE-SSW.

eg 4656 dimen. 14′ × 3′.3 V = 10.3 sfc. br. 14.3
Lying 10′ SW of a mag. 10 star, this galaxy is barely visible in 6 cm. In 15 cm the mottled halo is 8′ × 1′ in pa 35°, with a slightly brighter core that is more pronounced than in nearby eg 4631, *cf.* 30 cm will show the halo to about 9′ × 1′.2. The SW end is exceedingly faint but has a slightly brighter patch at the tip. The NE end is smooth and of high surface brightness. The core is patchy and has two distinct parts. Curving E off the NE end is a brighter appendage that is elongated in pa 90°, designated NGC 4657. It has two brighter patches in it, the western one being larger and brighter. Evidently this is not an interacting companion, but just a brighter part of eg 4656.

eg 4736 dimen. 11′ × 9′.1 V = 8.1 sfc. br. 13.0
Messier 94 is visible in 6 cm as a bright circular spot. In 15 cm it has a bright, blazing core, rising much in bright-

ness to the center, much like an unresolved globular cluster. 25 cm shows it elongated E-W. The bright core is 45″ across and contains a faint stellar nucleus. The halo is a hazy 2.′5 × 1′ envelope. 30 cm reveals some stars on the nebula. On the S and E are knots, and the faint parts of the halo are not uniformly bright. The core grows rapidly brighter toward the center but the nucleus is nonstellar.

eg 4800 dimen. 1.′8 × 1.′4 B = 12.3 sfc. br. 13.1
A very faint object for 6 cm, this galaxy is visible using averted vision. In 15 cm it is small and circular, less than 1′ diameter. 25 cm shows a broad concentration with a stellar nucleus at the threshold of vision. About 50″ WNW is a mag. 13.5 star. 30 cm reveals a fairly well concentrated spot with a small halo. The broad core is unevenly bright and contains a nonstellar nucleus. The overall size is 1.′3 × 0.′6, elongated roughly N-S.

eg 4861 dimen. 4.′1 × 1.′6 V = 12.2 sfc. br. 14.1
A low surface brightness object, this galaxy is not visible to 15 cm. 25 cm shows an evanescent wisp aligned in pa 15° between two mag. 12 stars, the glow centered closer to the star on the S. Photographs show that this "star" is actually a bright HII region in the galaxy, designated I3961.

eg 4868 dimen. 1.′7 × 1.′6 B = 13.1 sfc. br. 14.0
With 15 cm this object appears about 1.′5 × 1′ in pa 15° with a mag. 11 star off the N edge. 25 cm reveals a circular halo 1′ diameter with a small, broadly concentrated core. A mag. 14 star is visible on the SW edge. 30 cm shows the galaxy to be fainter than nearby eg 4914, *cf.* At 150 × it appears elongated in pa 40°. Higher power shows the broadly concentrated core in pa 40°, but the halo in pa 20°, about 1.′5 × 1.′ The star SW appears inside the edge of the nebula.

eg 4914 dimen. 3.′6 × 2.′2 B = 12.3 sfc. br. 14.4
Located 18′ E of eg 4868, *q.v.*, this object is not visible with 15 cm. With 25 cm the tiny knotted core is elongated in pa 150°, about 15″ × 5.″ The halo increases the size to 1′ × 0.′3. 30 cm shows a 2′ × 1′ halo in pa 150°, with a stellar nucleus that seems multiple at low power. High power reveals a weak, broad concentration. The inner core contains stellarings that confuse the nucleus.

eg 5005 dimen. 5.′4 × 2.′7 V = 9.8 sfc. br. 12.6
This is an easy object for 6 cm. The 3′ × 1′ patch is elongated in pa 65° with a strongly concentrated center. 15 cm shows it about the same size; the nucleus appears stellar, and the halo is sharply tipped. 25 cm reveals a star-like appendage on the NE side of the core. A fairly distinct dark space lies between it and the stellar nucleus. The halo extends to 4′ × 1.′25. 30 cm provides a nice view of the 4.′5 × 1.′8 object. It is well concentrated overall, with a deeply glowing 1′ × 0.′5 core and a bright nonstellar nucleus. The halo fades more abruptly on the SE side.

eg 5033 dimen. 10′ × 5.′6 V = 10.1 sfc. br. 14.4
With 15 cm, this galaxy is 2.′5 × 0.′5 in pa 165°. The elongated core is small and contains a stellar nucleus. The halo bulges at the middle and has sharp tips. 25 cm shows a 3′ × 0.′75 halo with a mag. 14 star just W of the NW tip. 30 cm shows a well-concentrated halo rising to a mag. 13.8 stellar nucleus. The overall size is 3.′2 × 0.′7 in pa 165°, while the core seems to be elongated in pa 150°. The mag. 14 star is 1.′3 NW, just outside the nebula to the W.

eg 5055 dimen. 12′ × 7.′6 V = 8.6 sfc. br. 13.4
Messier 63 is an easy object for 6 cm, located 3.′5 E and a bit S of a mag. 8.5 star. It is elongated E-W, passing just S of the star. The broadly concentrated core has a stellar nucleus. 15 cm shows the halo 3′ × 1.′5 in pa 110°, an oval galaxy with a stellar nucleus. In 25 cm the S side appears flattened. 30 cm reveals a granular core surrounding a stellar nucleus. The halo is distinctly more abrupt on the S flank with an overall size of 3.′5 × 1.′3.

eg 5074 dimen. 1.′0 × 1.′0 V = 14.0 sfc. br. 13.8
Viewed in 30 cm, this very faint galaxy is only 30″ diameter with faint stellarings near the center. Three mag. 13 stars lie about 5′ NNE; two more lie 4′ ENE. The galaxy lies 8.′5 SSE of a mag. 8 star.

eg 5112 dimen. 3.′9 × 2.′9 B = 11.8 sfc. br. 14.3
In 30 cm this galaxy is just visible at low power NNW of a mag. 12.5 star. At 225 × it is elongated in pa 130°, about 2.′5 × 1.′5. The light is unconcentrated with a stellaring occasionally visible SE of center. The halo does not reach the star SSE. eg 5107 (mag.$_z$13.7) lies 14′ SSW.

eg 5194 dimen. 11′ × 7.′8 V = 8.4 sfc. br. 13.0
eg 5195 dimen. 5.′4 × 4.′3 V = 9.6 sfc. br. 12.9
Messier 51 is visible in 7 × 35-mm binoculars as a pair of spots in contact. In 6 cm, −94 (the spiral component) has a mottled and extensive, but well-defined halo with little brightening toward the center, then a suddenly brighter core and stellar nucleus. On the N it touches −95 (the irregular appendage), which is quite small and weakly condensed. In 15 cm it is difficult to tell which galaxy is brighter, though they are easily distinguished by size. eg 5194 has a large core that dominates the halo. No bridge is visible to −95. eg 5194 has a granular core, while −95's is diffuse. It is up to 6′ diameter, while −95 is about 3′ × 2.′ In 20 cm eg 5195 is elongated with its

major axis perpendicular to a line connecting the two objects. Both have stellar nuclei. 25 cm shows the arms clearly, brought out by dark patches N and SW of center. Bright patches are visible on the E and NE sides, and a mag. 13.5 star lies 1′.5 SW, well within the halo. A faint band reaches out to −95. M51 is a grand sight in 30 cm. The eastern arm is brighter, unwinding clockwise. Its brightest portions are on the SE and NE, after which it becomes indistinct. The western arm has one big brightening on the SW with the mag. 13.5 star on the inner side. The core is optically "deep," glowing internally like an unresolved globular cluster. eg 5194 is about 5′.5 diameter overall. eg 5195 is unevenly bright with a stellar nucleus off-centered to the SE. A dark blob intrudes on the SE, separating part of the halo. A mag. 13.5 star is visible to both 25 cm and 30 cm on the E side of −94.

eg 5198 dimen. 2′.1 × 1′.9 B = 12.6 sfc. br. 14.1
15 cm shows this galaxy as a small, moderately bright object 1′ diameter. A stellar nucleus is visible, as well as a faint star 1′ N. In 25 cm little more detail is visible, but additional stars appear N and W. 30 cm reveals an extensive halo 2′ diameter that is fairly well concentrated to a 30″ core. The star W lies in the halo 30″ distant from the center and is about mag. 14.5. eg 5173 (mag.$_Z$13.5) lies 20′ WSW; eg 5169 (mag.$_Z$14.7) lies 21′ W, or 5′.5 NW of −73; eg I4263 (mag.$_Z$15.4) lies 22′ NW.

gc 5272 diam. 16′ V = 5.9 class VI
Messier 3 is visible to 6 cm with a small core inside an extensive halo. It is granular at low power in 15 cm and is well resolved at 200 ×. The outline is oval, with the stars forming curved rays emanating from the bright central spot. 25 cm at high power resolves the cluster almost to the center of the oval core, which seems W of center. The brighter part has a rectangular form, about 10′ × 8′, elongated SE-NW. Outliers, however, fill a circular area that is up to 20′ diameter. The core is 4′–5′ across with a conspicuous star near the center. eg 5263 (mag.$_Z$14.0) lies 29′ W of the cluster.

eg 5273 dimen. 3′.1 × 2′.7 V = 11.6 sfc. br. 13.7
This object is barely visible to 15 cm. 25 cm shows a circular spot 1′ diameter. It is well concentrated to a bright core, but no sharp nucleus is visible. A faint star lies 2′.5 WNW. 30 cm shows an object of low overall surface brightness. At high power it appears well concentrated to a stellar nucleus. eg 5276 (mag.$_Z$14.6) lies 3′.3 to the SE.

eg 5297 dimen. 5′.6 × 1′.4 B = 12.4 sfc. br. 14.5
30 cm shows this galaxy 1′.9 SW of a mag. 9 star. Elongated in pa 140°, the halo is about 5′ × 0′.8, extending beyond a mag. 11.5 star 2′.4 NW. The galaxy forms a triangle with these two stars. Low power shows the core to have three or four stellarings along the major axis near the center. At 225 × the core is 2′ × 0′.7 and is broadly concentrated. eg 5296 (mag.$_Z$15.0) lies 1′.6 to the S and is possibly interacting with −97.

eg 5301 dimen. 4′.4 × 1′.1 B = 12.7 sfc. br. 14.2
This galaxy is barely visible to 15 cm. 25 cm reveals a spindle with a starry texture. The overall size is 1′.5 × 0′.5 in pa 150°. The low surface brightness makes medium power best for viewing. 30 cm shows it 1′.8 × 0′.8, lenticular in shape, with a central bulge. There is little concentration to a faint nonstellar nucleus.

eg 5311 dimen. 2′.3 × 2′.0 mag.$_Z$13.7 sfc. br.
eg 5313 dimen. 1′.9 × 1′.2 B = 13.1 sfc. br. 13.9
In 15 cm, eg 5313 is visible as a small round spot 1′ across. 25 cm shows it 1′.25 × 0′.5 in pa 40°. The core is elongated and has a starry appearance. A wide pair of stars lies about 8′ SSE (12,12; 43″; 160°) that could be mistaken for a galaxy at low power. eg 5311 lies 9′ W and appears circular, 35″ across, with a stellar nucleus. In 30 cm, −13 is broadly and weakly concentrated to a stellar nucleus occasionally visible in the smooth core. The halo extends to 1′.3 × 0′.7.

eg 5326 dimen. 2′.5 × 1′.3 B = 13.0 sfc. br. 14.2
15 cm will barely show this galaxy. 25 cm shows a 45″ × 15″ object in pa 135°. The core is small, elongated, and starry. A tantalizing object in 30 cm, it is elongated in pa 135°, about 1′.8 × 0′.7. The core is 40″ diameter and well concentrated to elongated inner regions and a steadily sparkling nucleus.

eg 5347 dimen. 1′.9 × 1′.5 V = 12.6 sfc. br. 13.6
This galaxy is not visible in 15 cm and is faint in 25 cm, appearing small and little brighter to the center. 30 cm shows a 1′ × 0′.6 halo elongated in pa 100°. It has a low surface brightness and is weakly concentrated to a faint, occasionally stellar nucleus.

eg 5351 dimen. 3′.1 × 1′.8 V = 12.1 sfc. br. 13.8
Though not visible in 15 cm, 25 cm shows this galaxy in pa 105°, about 1′.25 × 1′, with little concentration. It has a faint halo and a stellar nucleus. 30 cm gives a size of 2′.8 × 1′ with a very slight central concentration. The shape is lenticular with definite edges and an elongated core. Lying 5′ NNE are some faint stars that may look nebulous; 2′ NW is another faint star. Photographs show eg 5341 (mag.$_Z$14.1) 13′ SW and eg 5349 (mag.$_Z$15.1) 3′.4 SW.

Chart IV. NGC 5272 = Messier 3. A 30-minute exposure on 103a-O with the USNO 1 meter telescope taken by A. Hoag on 6 January 1960. Magnitude sequence from: Sandage 1953, A. J. 58: 61 and Johnson and Sandage 1956, Ap. J. 124: 379. U.S. Naval Observatory photograph.

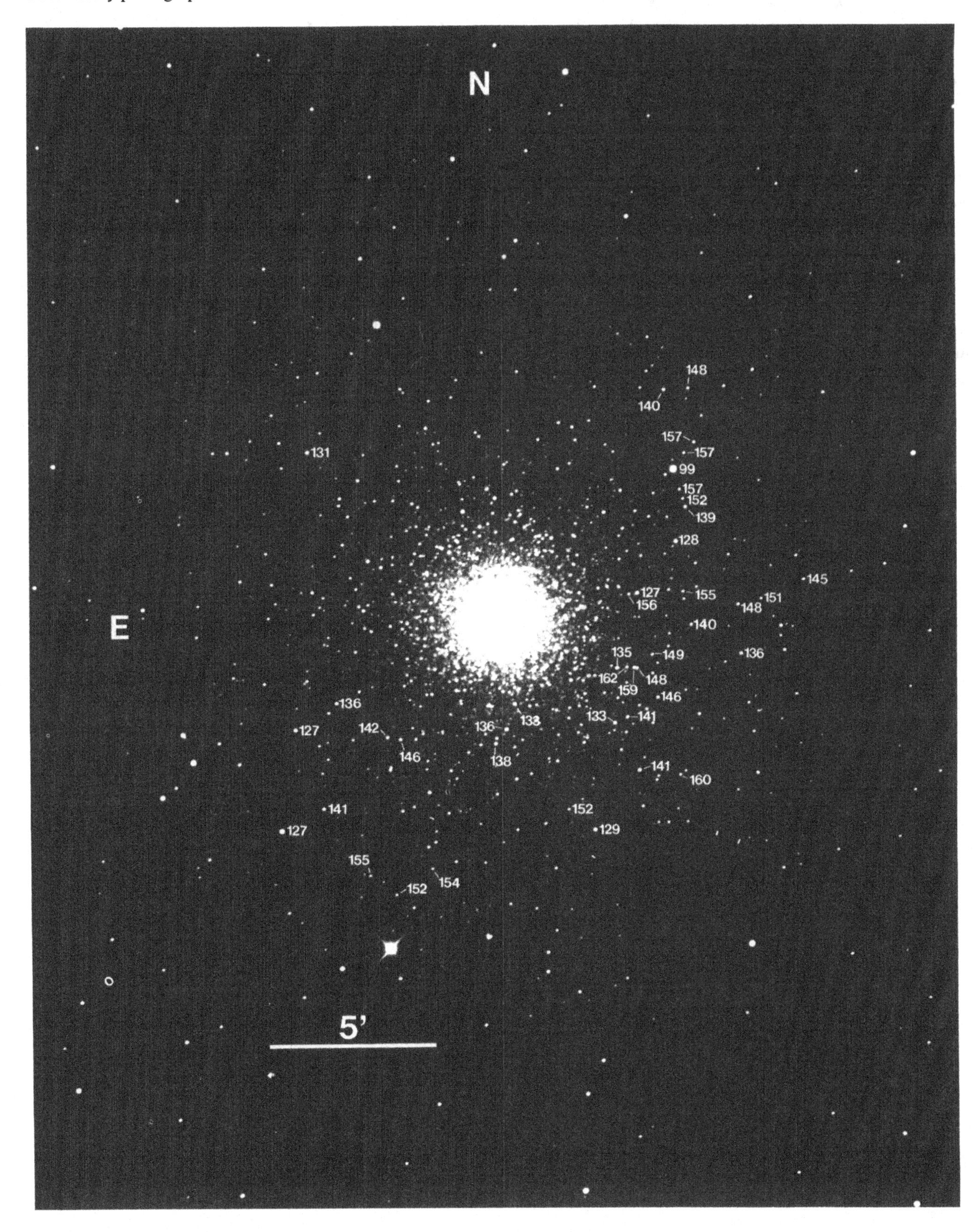

eg 5350 ***dimen. 3′.2 × 2′.6*** ***V = 11.4*** ***sfc. br. 13.5***
eg 5353 ***dimen. 2′.8 × 1′.5*** ***V = 11.1*** ***sfc. br. 12.5***
eg 5354 ***dimen. 2′.3 × 2′.0*** ***V = 11.5*** ***sfc. br. 13.0***
eg 5355 ***dimen. 1′.5 × 0′.9*** ***mag.$_z$14.0*** ***sfc. br.***
eg 5358 ***dimen. 1′.3 × 0′.4*** ***mag.$_z$14.6*** ***sfc. br.***

This is a rich and conspicuous grouping for amateur instruments. The brightest of the group, eg 5353, is distinguishable in 6 cm SE of an orange mag. 6.5 star. With 15 cm it is easily seen. In 25 cm it is 1′.5 × 0′.75 in pa 135°, with a slightly brighter core but no prominent nucleus. Well concentrated in 30 cm, it has a narrow, elongated core and a stellar nucleus. The halo extends to 1′.3 × 0′.5 and merges with the halo of eg 5354, 1′.2 N. eg 5354 is not visible to 6 cm, but is easy in 15 cm. In 25 cm it appears circular, 1′.25 diameter with a stellar nucleus. 30 cm shows it as less concentrated than −53, with a small 5″ nucleus fading evenly through the core into the halo. The halo seems elongated in pa 100°, but the proximity of −53 makes this uncertain. Barely visible to 15 cm, in 25 cm eg 5350 is diffuse and little brighter to the center. Elongated in pa 45°, the halo is 1′.5 × 1′.25. It is the largest member of the group in 30 cm, fully 2′.3 × 1′.5, but with very weak concentration and no nucleus. The halo is elongated in pa 90°, but the core is in pa 40°. eg 5355 is about 4′ E of a line joining −53 and −50. In 25 cm it is very small, less than 30″ diameter, with a stellar nucleus. Between −53 and −55 is a mag. 13,14 double star (29″;15°) that appears nebulous, and 8′ N of −50 is another faint pair (12,12.5; 29″; 55°) that can be resolved at high power. In 30 cm −55 is faint and small, about 1′.5 × 1′ in pa 30°, with some concentration to a very faint stellar nucleus. eg 5358 lies 7′ E of −53, and 1′.1 N of a mag. 13 double star of 9″ separation. In 30 cm it is 40″ diameter with moderate concentration.

eg 5362 ***dimen. 2′.4 × 1′.1*** ***B = 13.1*** ***sfc. br. 14.1***

Viewed in 30 cm, this is a very faint and unconcentrated galaxy. The halo is about 1′.5 × 0′.5 in pa 90°. A faint stellar nucleus is visible, but the galaxy exhibits no other central brightening. A mag. 14 star lies about 1′.5 ESE.

eg 5371 ***dimen. 4′.4 × 3′.6*** ***V = 10.8*** ***sfc. br. 13.6***

15 cm shows this galaxy as a 4′ × 1′.5 glow in pa 15°. The nebula has a fine-grained texture and is without any central brightening. 25 cm shows it 4′ × 2′.5. The nucleus appears stellar, but there is no brighter core. Lying W of center and the S flank are mag. 14 stars; another, brighter star lies NE. 30 cm shows a 20″ core, with an overall size of 3′ × 1′.8 in pa 15°.

eg 5377 ***dimen. 4′.6 × 2′.7*** ***V = 11.2*** ***sfc. br. 14.8***

This is a faint object for 6 cm, but visible with averted vision as a 2′ patch. 15 cm shows a 1′.5 × 0′.5 halo in pa 40°, containing a stellar nucleus. In 25 cm, the size increases to 3′ × 0′.5. The core is bright and oval, while the halo appears spindly. 30 cm shows an elongated core with pointed ends and a bright, nearly stellar nucleus. The faint halo extends to 2′ × 0′.6.

eg 5380 ***dimen. 2′.1 × 2′.1*** ***B = 12.7*** ***sfc. br. 14.3***

This galaxy appears as a threshold magnitude star in 15 cm. 25 cm shows a faint halo with a stellar nucleus. The overall size is about 40″. In 30 cm the halo is about 30″ diameter and moderately concentrated. At low power the

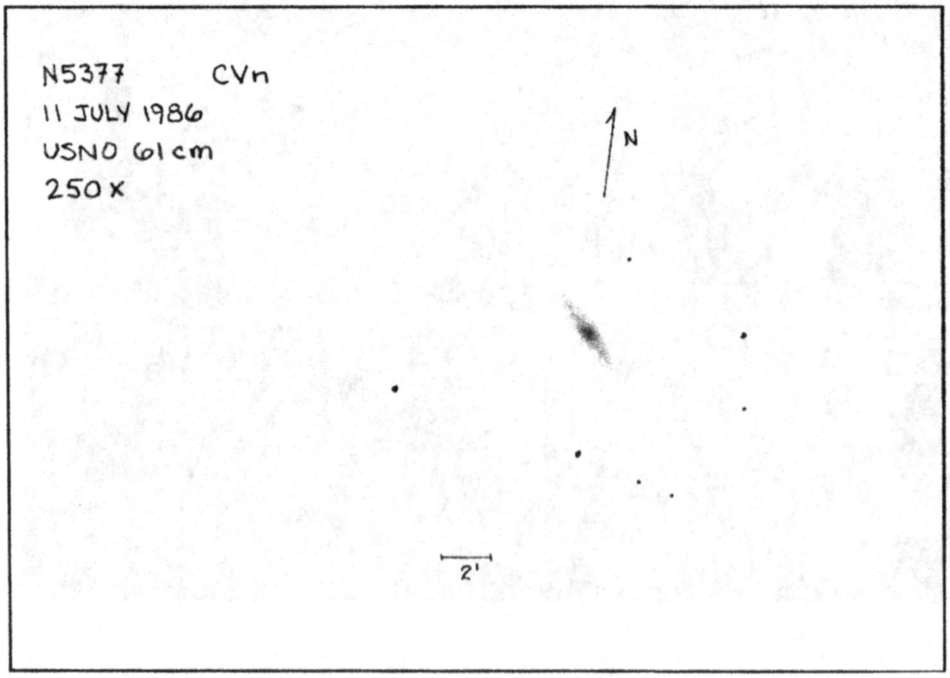

Chart V. NGC 5353 group.

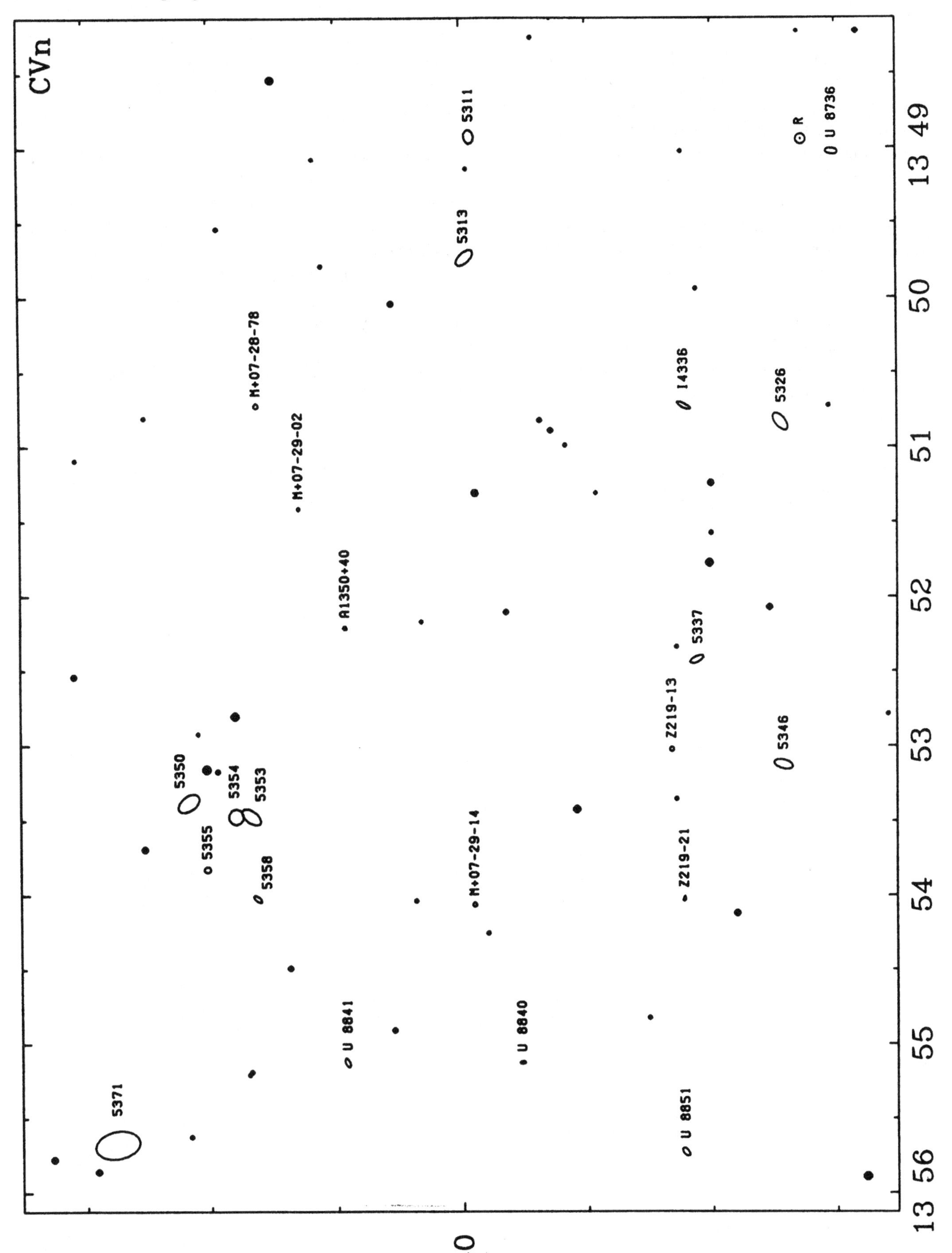

nucleus appears stellar, but not so at high power. eg 5378 (mag.$_z$13.8) lies 11′ N.

eg 5383 dimen. 3′.5 × 3′.1 V = 11.4 sfc. br. 13.9
Though not visible in 15 cm, this object can be seen in 25 cm with averted vision. It is 1′.5 × 0′.5 in pa 130°. Stellarings are seen within, but there is no central condensation. It is an interesting object for 30 cm, with many stars associated. The overall size is 2′.8 × 1′.8 in pa 130°. The core is irregularly concentrated and twisted, with its inner part elongated nearly E-W. A nonstellar nucleus is at the center. A mag. 13.5 pair (9″.3; pa 150°) is located 1′ ENE, with a wide, brighter pair to the SE.

eg 5394 dimen. 1′.9 × 1′.1 V = 13.0 sfc. br. 13.6
eg 5395 dimen. 3′.1 × 1′.7 V = 11.6 sfc. br. 13.3
These galaxies are an interacting pair. Viewed in 15 cm, eg 5395 appears to be elongated in pa 165°, but there is a threshold magnitude star on the S edge that merges with the nebula, making it appear as an elongated streak. At high power the galaxy appears round, 1′ across, and a little brighter to the center. In 25 cm, it is about 1′.5 × 1′. The mag. 13 star is visible 1′.75 S; eg 5394 appears as a mag. 14 star 1′.9 NNW. Viewed in 30 cm, −95 is brighter and larger than eg 5380, *q.v.* The smooth broad surface is about 2′ × 0′.9 in pa 165°. The inner core seems elongated in pa 30°, but there is no nucleus.

eg 5406 dimen. 2′.1 × 1′.6 B = 13.0 sfc. br. 14.1
In 15 cm this galaxy is visible faintly as an unconcentrated blob 7′ S of a mag. 7 star. In 25 cm it is round, 1′.25 diameter, and very slightly elongated E-W. There is not much concentration, but a faint stellar nucleus is discerned. In 30 cm, the star N interferes with observation. The galaxy is elongated in about pa 100°, about 1′.1 × 0′.7, and is weakly concentrated.

eg 5444 dimen. 2′.7 × 2′.3 B = 12.5 sfc. br. 14.4
Viewed in 30 cm this galaxy appears quite small, about 1′ × 0′.75, elongated E-W. It is well concentrated to a small, beady core without a nucleus. eg 5445 (mag.$_z$14.1) lies 7′ S, and eg 5440 (mag.$_z$13.4) lies 23′ SSW.

Canis Major

Led by Sirius, the brightest star in the sky, this constellation includes dozens of open clusters and reasonably bright galaxies shining through the edges of the Milky Way. Messier 41 is visible to the naked eye even on nights of fair quality. The center of the constellation culminates at midnight about 2 January.

oc 2204 diam. 13′ V=8.6 stars 353
This cluster is a very faint object for 25 cm. About 20 stars are visible over an irresolved haze 20′ diameter elongated E-W. It is barely distinguishable from the background in 30 cm. About 35 stars of mag. 12.5 and fainter, including one long string, are visible in a 13′ area. The cluster is enclosed in an east-pointing kite-shaped box of stars about 18′ long.

eg 2207 dimen. 4:3 × 1:9 V=10.7 sfc. br. 13.2
eg I2163 dimen. 2:7 × 1:2 mag.$_{vv}$13.5 sfc. br.
These interacting galaxies are 1:4 apart in pa 105°. eg 2207 is elongated in pa 70° in 25 cm, about 1:5 × 1:. The halo has a lenticular outline and a small bright core. The faint oval companion is elongated in pa 110° and has a smooth texture. The pair are 3′ × 1:5 overall. With 30 cm, eg 2207 has a bright core with a stellaring on its ENE side. The halo extends E to eg I2163, which appears as a bright lump about 30″ across. A mag. 14 star lies 1:6 NW of eg 2207. The combined size is 2:2 × 1:2.

pn I2165 dimen. 9″ × 7″ nebula 10.6v star
This tiny planetary forms the third "star" at the W end of an E-W string of stars. At 150 × in 30 cm, it appears nearly stellar. At 225 × it is elongated in pa 40°. A circular core 6″ across is revealed at 450 ×, with a little haze extending NE and SW. The core has several bright spots around the center, making it appear annular at times. Two faint stars lie 1′ NNE and E.

eg 2217 dimen. 4:8 × 4:4 V=10.4 sfc. br. 13.6
15 cm shows this object as a small, moderately faint patch with a faint star on the W. Two stars are visible on the W with 25 cm, about 1:5 from the galaxy. The galaxy lies in a rich field and appears to be elongated in pa 120°, about 50″ diameter. The broad core is much brighter, but there is no sharp nucleus. It is well concentrated in 30 cm to a 10″ core and stellar nucleus. The innermost part seems elongated NE-SW, i.e., perpendicular to the halo. The almond-shaped object extends to 1′ × 0:75. A very faint patch of nebulosity is visible about 1:5 ENE of the galaxy; a haze reaches from the galaxy to this area.

eg 2223 dimen. 3:3 × 3:0 V=11.4 sfc. br. 13.7
This is a pretty faint object for 25 cm. The core has two unequally bright stellarings in it. The oval halo is of fairly low surface brightness and extends to 2′ × 0:75 in pa 30°. 30 cm shows a mag. 13.5 star 25″ N of center. The core is 20″ across and well concentrated. The overall size is up to 2:5 at 150 ×, the brighter part being 1:5 × 1:2.

oc 2243 diam. 7′ V=9.4 stars 368
This cluster is visible in 15 cm as a faint haze 9′ SW of a mag. 7.5 star. It is confined by mag. 11.5–12 stars on the NE, NW, and SW sides. A mag. 12 star is visible in the NW side of the haze. 25 cm shows it 4′ diameter with about 20 stars resolved over the haze. 30 cm reveals a small, unevenly concentrated and poorly resolved cluster. About 25 stars are visible in a 4′ area, but at least half of these appear to be foreground stars. Three brighter stars appear on the W side; a mag. 12 pair lies to the E, with about 10″ separation in pa 55°.

eg 2263 dimen. 2:8 × 1:8 mag.$_{vv}$13 sfc. br.
This galaxy is very faintly discernable in 15 cm at 75 ×. In 30 cm the halo is 2′ × 0:5 in pa 150°, equal in length to the separation of two mag. 12.5 stars N and S of the galaxy. The light is broadly concentrated overall; the core is a weakly brighter, but well-defined streak dotted with a few spots.

eg 2280 dimen. 5:6 × 3:2 B=12.0 sfc. br. 15.0
A faint object for 15 cm, this galaxy is found within a quadrangle of mag. 12.5–13 stars about 2:5 across whose long dimension is aligned NE-SW. The halo is broadly concentrated, within which the core seems to be a thin streak in pa 165°. This streak is still visible in 30 cm: the 35″ × 10″ region is brighter in the northwestern half, and more sharply defined on the NE side; there is no conspicuous nucleus. Overall, the halo is 2′ × 1:2, with a very weak, even concentration across the center. The center of the galaxy is closer to the quadrangle stars NW, and the halo extends between them. A mag. 14 star is involved

Chart VI. NGC 2204. A 30-minute exposure on 103a-O with the USNO 1 meter telescope taken by A. Hoag on 8 November 1958. Magnitude sequence from: Hawarden 1976, M. N. R. A. S. 174: 225. U.S. Naval Observatory photograph.

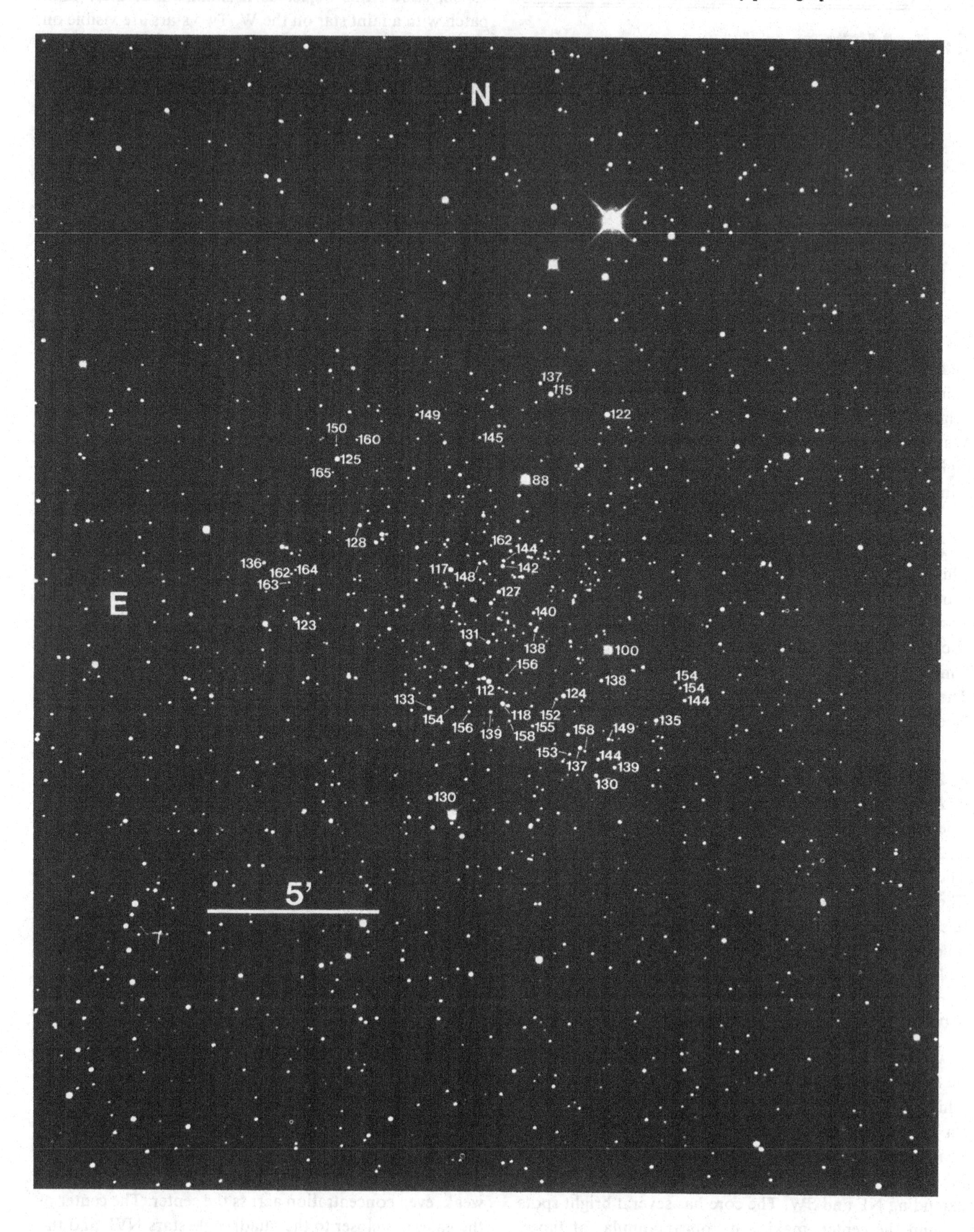

35″ SE of the center on the major axis; another slightly fainter star is 50″ due S.

eg 2283 ***dimen. 3′.4 × 2′.6*** ***mag.*** $_{vv}$***13*** ***sfc. br.***
With 15 cm this galaxy is faintly visible lying closely involved with two mag. 12.5 stars to the NW and NE. In 25 cm these two stars lie very close to the edge of the halo. The halo is nearly circular, about 2′ diameter, with only a slight central concentration. With 30 cm the galaxy appears more strongly concentrated. A mag. 13 star lies against the S edge; another, slightly fainter, lies to the N.

oc 2287 ***diam. 38′*** ***V = 4.5*** ***stars 69***
Messier 41 is a bright cluster visible to the unaided eye NW of 12 Canis Majoris (V = 6.0) and is a fine cluster for small instruments. 6 cm reveals about 30 stars including a band of stars on the NW. About 50 stars ranging in magnitude from 8 to 12 are shown by 15 cm in a 30′ area. Outliers extend to 45′ diameter in 25 cm. On the W side of the cluster are many fainter stars. Several pairs are noted in 30 cm, particularly the bright pair on the NW side (V = 7.9,8.6; 25″; 88°).

eg 2292 ***dimen.*** ***mag.*** $_{vv}$***14*** ***sfc. br.***
eg 2293 ***dimen.*** ***mag.*** $_{vv}$***13.5*** ***sfc. br.***
eg 2293 is the brightest of the trio of galaxies including eg 2295, *q.v.*, 4′ W. In 15 cm, −93 is inseparable from eg 2292, with which it interacts. The glow is somewhat larger and better concentrated than −95. Set in a rich Milky Way field, several faint stars are visible nearby. In 30 cm the two galaxies are 50″ apart, their halos in contact or merged, the common oval 1′.25 × 0′.75 envelope elongated in pa 125°. eg 2293 has a moderate, even concentration, except for a fairly conspicuous stellar nucleus. The core is not distinctly bounded, but is clearly elongated toward −92. The companion is elongated N-S, at least in the core, which contains a small substellar nucleus. Its concentration is similar to that of −93, though the galaxy is fainter overall. A tiny 25″ triangle involving four mag. 13–14.5 stars 2′.5 SW of −93 appears nebulous at 225 ×.

eg 2295 ***dimen. 2′.7 × 0′.7*** ***mag.*** $_{vv}$***14*** ***sfc. br.***
This galaxy, located about 4′ W of eg 2292/93, *q.v.*, is faintly visible in 15 cm as a substellar spot. Several stars lie close by: mag. 12.5 stars 20″ N and 1′.4 SW are matched by a mag. 13 star 20″ SW and a mag. 12 star 1′.6 N. In 30 cm the closer stars (about 40″ apart) define the limits of the galaxy's halo. The light has a weak, even concentration to a very faint substellar nucleus, which seems a bit closer to the faint star N.

oc Ruprecht 7 ***diam. 4′.0*** ***mag.*** ***stars 30***
This cluster is visible to 25 cm at low power, but needs medium power for scrutiny. At 125 × it is 6′ to 8′ diameter and partially resolved into stars of mag. 13.5 and fainter. The stars and thin background haze are broadly concentrated toward the center.

oc Tombaugh 1 ***diam. 10′*** ***V = 9.3:*** ***stars 16***
15 cm will show this cluster just NE of two mag. 7 stars that point to it. Three stars stand out: one with a fainter companion on the W side (V = 10.5,12.8; 40″; 5°), one

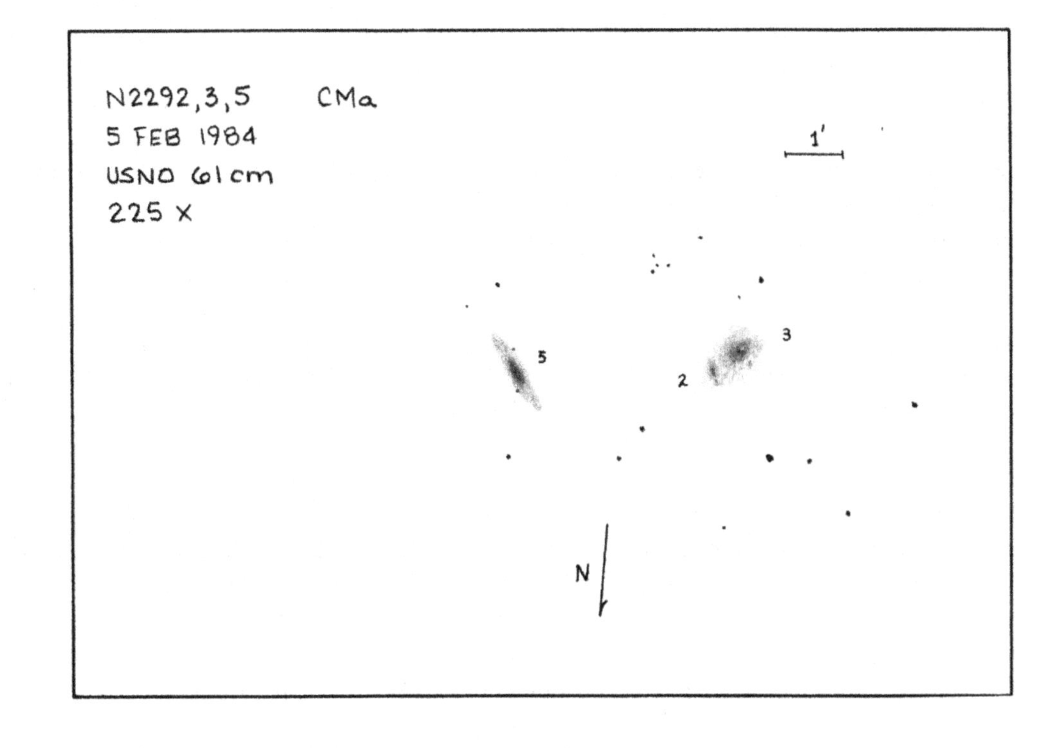

near the center (V=11.2), and on the E side (V=11.4). With averted vision about a dozen threshold magnitude stars are resolved. About 20 stars can be seen in 25 cm within 6′ diameter. With 30 cm it is a rich, pretty cluster of mostly faint stars. Forty-five stars are resolved in a 5′×3′ area elongated E-W. There is no concentration; in fact, around the mag. 11 star near the center is a 2′.5 hole containing only a few stars. Many of the other members are in groups of two to five stars, including three variously oriented wide pairs on the W side. The nearest of the bright field stars SSE is an unequal pair (ADS 5692: V=6.8,10.8; 5″.1; 143°).

oc Ruprecht 8 diam. 4′.0 mag. stars 10
Located in a rich Milky Way field, 25 cm shows this cluster at low power. At 100× the group stands out as a concentration of faint stars of mag. 12 and fainter spread across a 20′ area that is slightly elongated N-S. Around this area there appears to be a ring lacking faint stars, making it seem like the cluster is shining through a hole in obscuring matter.

eg 2325 dimen. 2′.3×1′.5 V=11.2 sfc. br. 12.5
15 cm shows a small, moderately faint spot here 1′.5 NW of a mag. 11.5 star. The 1′ halo is broadly concentrated, but has a sparkle occasionally visible in the center. In 30 cm the galaxy is 1′.2×0′.8, elongated roughly toward the star SE. The concentration remains broad; the nearly circular 15″ core is indistinct, and there is no sharp nucleus. Two mag. 13 stars are visible 25″ NW along the major axis, and 35″ SE; mag. 14 stars lie 1′ N and 1′.3 S.

oc Tombaugh 2 diam. 3′.0 mag. stars 319
Though barely visible in 15 cm, 25 cm will show this cluster more clearly about 4′ S of a mag. 8 star. At medium power it appears as a 3′ diameter smoothly hazy spot, implying that the cluster contains many very faint stars. High power will only just begin to resolve stars, showing a few of mag. 13–14, though these may be only in the foreground.

gn 2327 dimen. star V=7.2
gn I2177 dimen. 20′×20′ star
gn A0702–12 dimen. star V=8.0
This elongated nebular complex spans the Canis Major-Monoceros border, reaching from oc 2335 (*q.v.*, Monoceros) to a mag. 8 star about 2° S. In 25 cm gn 2327 is visible as a haze concentrated around an unequal double star (ADS 5761: V=7.4,9.4; 5″.3; 150°); the nebulosity is more extensive on the NW side of this pair. Faint, indefinite nebulosity is also visible nearby NNE; this is I2177. Around a mag. 8 star about a degree S is an unnumbered section (A0702–12), which 25 cm and 30 cm show as a 3′ diameter patch that fades evenly away from the illuminating star and extends mostly to the W and SW.

oc Haffner 4 diam. 2′.4 mag. stars 30
Lying between mag. 9 stars on its W and NE, this faint cluster is barely discernable against a rich field in 25 cm. About a dozen mag. 12.5 and fainter stars can be resolved in a 5′ area. The cluster exhibits only a little concentration to the center, but nevertheless looks real.

oc Ruprecht 11 diam. 2′.9 mag. stars 20
This cluster stands out from the field in 25 cm due to the equal magnitudes of its members. About 15 stars of mag. 11–12 are grouped in two clumps: overall the cluster is elongated N-S. The northern group contains about two-thirds of the stars and includes a pair (20″;45°) on its N end.

oc 2345 diam. 12′ V=7.7 stars 59
Trailing away S from a mag. 7 star, 6 cm shows 13 stars in this cluster with a little haze. With 25 cm about 30 stars can be counted in a 10′×6′ area elongated NNE-SSW. Several mag. 9–10 stars are scattered about, but the concentrated part of the cluster is comprised mostly of fainter stars. On the E edge is a tight knot of about eight faint stars including an unequal pair (V=10.4,12.4; 12″; 240°). Between this knot and the rest of the cluster is an elongated void about 1′ wide running also NNE-SSW. Another conspicuous pair lies in the N side (h3930: V=9.9,10.9; 15″; 72°). To 30 cm the knot is 30″ across and contains 10 stars, including the pair. Overall about 50 stars are visible in a 12′×8′ area.

oc 2354 diam. 20′ V=6.5 stars 297
Four groups of stars are shown by 6 cm in this area. Of the four groups, the western one is the richest, with about a dozen stars. The other groups have only three to five stars each. The cluster is also ill-defined in 15 cm. The concentrated group on the W has about 15 stars in a 7′ area. Including the extensive groups E and S, the overall size is about 40′. 30 cm shows about 50 stars in a 20′ area, including about ten brighter ones. The W side is richer, but otherwise the cluster is unconcentrated.

gn 2359 dimen. 10′×5′ star V=11.4
These merged nebulae are fairly large and bright: 25 cm will show it clearly at low power, but 15 cm shows it only weakly. The roughly rectangular unconcentrated haze spreads to 10′×5′ in 25 cm, with a broader southern base. On the S side is a mag. 11 star; on the N is an "X" of six stars, five forming a triangle.

oc 2360 diam. 13′ V = 7.2 stars 91
In 15 cm this cluster is visible lying about 10′ W of a mag. 9 star. About 75 stars are resolved in a 20′ area arranged in various geometric patterns. The cluster is bright for 25 cm. The SW side seems flattened; on the N are three parallel bands of stars aligned roughly N-S. On the SE side are other radial strings. A total of 100 stars is visible in a 12′ diameter. The cluster fills a 22′ field in 30 cm with 100 stars. The N side lacks outliers while the S side is rich in fainter stars. The interior is elongated E-W due to a group of eight brighter stars.

oc 2362 diam. 8′ V = 4.1 stars 69
This cluster surrounds the mag. 4.4 star τ Canis Majoris. With 6 cm the central star interferes with viewing the fainter cluster members, though about fifteen can be discerned. The cluster is about 5′ diameter in 15 cm; the fainter members are bluish. About 40 stars are shown by 25 cm arranged in a triangular shape. The central star has two close companions: mag. 10.5 at 85″ in pa 89°, and mag. 11.2 at 15″ in pa 79° (ADS 5977). A beautiful sight in 30 cm, 50 stars, mostly mag. 10.5, fill a south-pointing triangular shape with τ CMa north of the center.

oc 2367 diam. 3′.5 V = 7.9 stars 15
Plotted as a quadruple star on the AE, this cluster is a very nice object in 25 cm at low power. It is dominated by a fairly bright pair near the center (ADS 5997AB: V = 9.4,9.7; 4″.5; 275°). The concentrated clot of stars is 10′ × 5′ in extent, elongated N-S, and contains 18 stars. A dark area surrounds the cluster. In this area about 7′ N is a small group of four stars (two brighter) that nevertheless seems an appendage to the cluster.

oc Haffner 8 diam. 4′.2 V = 9.1 stars 71
In 25 cm this cluster is visible as a loose, 4′ diameter group of about 20 stars. On the N side is a single mag. 10.5 star; the S side of the cluster is cradled in an arc of stars mag. 11 and fainter that is concave to the N. Most of the rest of the cluster members are fainter than mag. 12.5.

oc 2374 diam. 19′ V = 8.0 stars 73
15 cm will show about 25 stars in this moderately faint group. The stars are evenly distributed in a 5′ area elongated NE-SW. Scattered brighter stars lie on the NE side, but most of the mag. 12–12.5 cluster stars are on the SW. This group alone contains about 25 stars in a 3′ area with 25 cm. On its SW side is a small arc of about six stars opening to the S.

eg 2380 dimen. 1′.2 × 1′.2 mag.$_{vv}$13.5 sfc. br.
This small elliptical galaxy is barely visible in 25 cm at low power. At high power it appears as a small, high surface brightness patch extending to at most 30″ diameter. It looks a lot like a faint, unresolved globular cluster.

oc 2383 diam. 6′ V = 8.4: stars 11
In this cluster at least 20 stars can be counted in a 5′ area with 25 cm. Three mag. 10 stars straddle the group, two lying on the NE side, the other to the W. The cluster members are a little concentrated to the center, and mostly mag. 11 and fainter. oc 2384, *q.v.*, lies just 8′ SE.

oc Ruprecht 18 diam. 4′.0 V = 9.4 stars 20
This is a faint version of the typical sparse winter Milky Way cluster. It is a pretty object in 25 cm, which shows 25 stars in an unconcentrated group about 5′ diameter. A brighter, mag. 10.5 star lies on the SW edge, but most of the stars are mag 11.5 and fainter.

oc 2384 diam. 2′.5 V = 7.4: stars 15
Lying in the same field as oc 2383, *q.v.*, this cluster is visible in 25 cm. A bright pair is conspicuous on the W side (ADS 6062: V = 9.1,9.6; 4″.5; 347°). To the E of this pair spreads a small group of about eight stars. Farther E is a mag. 8.5 star with a few companions.

oc Trumpler 6 diam. 6′ mag. 10.0p stars
This cluster appears much fainter than nearby Trumpler 7, *cf.*, to 25 cm. Loosely distributed in a 6′ area are 20 stars of mag. 11 and fainter with a little concentration to the center.

oc Trumpler 7 diam. 5′ V = 7.9 stars 16
In 25 cm this cluster is conspicuous at low power, dominated by a string of mag. 9–11.5 stars aligned SE-NW. The string includes two pairs and a faint triple star at the N end (V = 12.3,10.8,12.7). Overall the cluster is 10′–12′ long, elongated N-S, and contains about 25 stars.

oc 2396 diam. 10′ mag. 7.4p stars 30
Spilling over the border into Puppis, this widespread cluster is best viewed with smaller apertures and wide fields. Lying south of a bright star (ADS 6104AC: V = 5.8,8.5; 20″; 313°), 25 cm shows about 40 stars, mag. 9–11, scattered over a 40′ area.

Canis Minor

Marked by Procyon, this constellation holds but a few deep-sky objects. The center culminates at midnight about 15 January.

eg 2350 ***dimen. 1!6 × 0!9*** ***mag.$_z$14.1*** ***sfc. br.***
This very small object is faintly visible in 30 cm, located 10′ S of a mag. 10.5 pair of 20″ separation. The galaxy appears elongated in pa 110°, with a broad moderate concentration across the 50″ × 30″ halo. A string of four stars aligned E-W is visible about 3′ S; two faint stars lie 1′ NW; one more lies 1′ NE. Another mag. 12 pair (1!4 apart) lies 6!5 ESE.

oc 2394 ***diam.*** ***mag.*** ***stars***
In 15 cm this cluster is centered about 10′ E of η Canis Minoris (V = 5.3). About a dozen stars are resolved. 25 cm shows seventeen stars in an oddly patterned group 10′ long, elongated SE-NW. With 30 cm the group seems elongated N-S. The N side is broader and has extensions to the E and W that curve N, like a bull's horns viewed face-on.

eg 2485 ***dimen. 1!7 × 1!7*** ***mag.$_z$13.3*** ***sfc. br.***
Located 7′ N of a mag. 9.5 star, this galaxy is not visible in 15 cm. At 100 × in 25 cm, it looks like a hazy double star, but at 200 × the galaxy is revealed 30″ N of a mag. 12.5 star. The halo is 20″ across with a gradual central brightening. 30 cm shows a stellar nucleus in the evenly concentrated 20″ nebula.

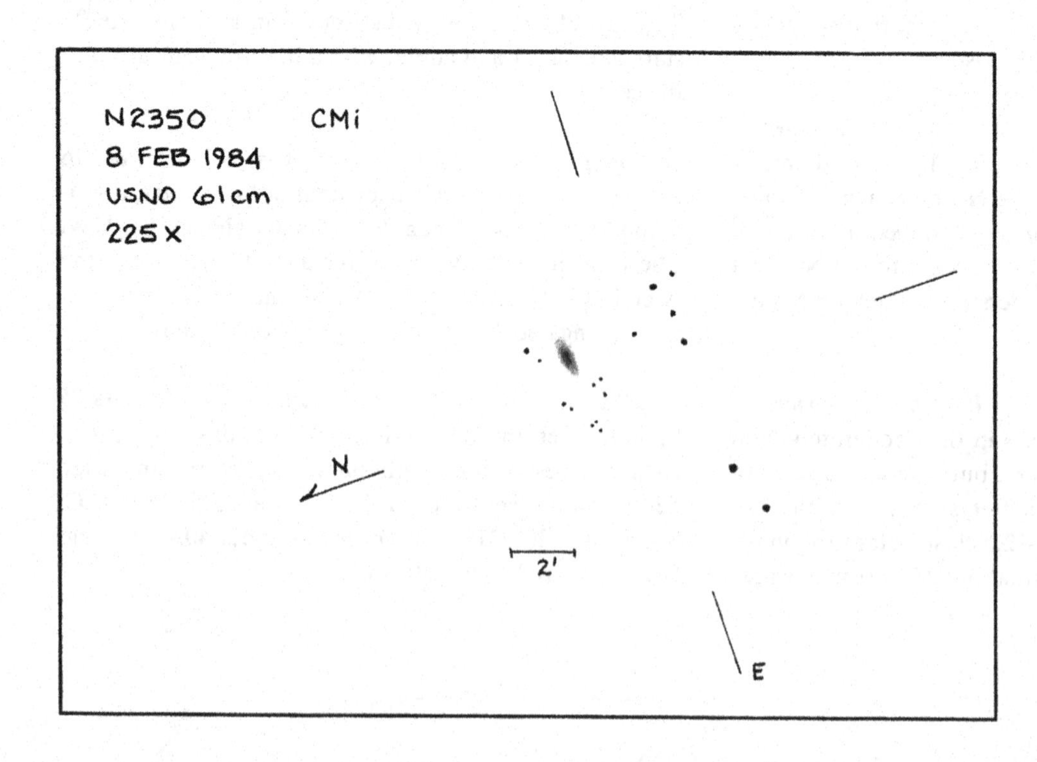

Capricornus

"Not a conspicuous constellation," wrote T. W. Webb of this area, and only two deep-sky objects easily accessible to amateur telescopes are found here. The center of the constellation culminates at midnight about 7 August.

eg 6907 ***dimen. 3.′4 × 3.′0*** ***V = 11.3*** ***sfc. br. 13.6***
This galaxy is visible to 15 cm as a small, round 1′ diameter spot. It is, however, an interesting object in 25 cm. Extending to 1.′75 × 1′ in pa 90°, it has a circular core and a tapered halo. A faint stellar nucleus is visible. On the NE flank a dark lane intrudes from the NW into the halo, separating a small patch of haze that is elongated in pa 120°. Two pairs of stars lie about 3′ SW, the farther being brighter and closer together. 30 cm shows the unevenly concentrated 1.′8 × 0.′8 halo with irregular edges and a nonstellar nucleus off-center to the NE.

gc 7099 ***diam. 11′*** ***V = 7.3*** ***class V***
Messier 30 is visible in 6 cm as a sharply concentrated 3.′5 diameter glow lying 7′ E of a mag. 8 star. 15 cm shows it about the same size, but with a broadly brighter center and irregular edges. At high power it is partially resolved. 25 cm resolves it well at 200 ×. The cluster appears elongated E-W with star chains extending N and NW, and a fainter one extending E. 30 cm shows an overall diameter of 3.′8, with an abruptly brighter 1′ core. It is well resolved except in the center. The star chain extending to the N has four stars which are, from the center of the cluster outward, mag. 13, 12, 12, 13. The skew string on the NW is of three stars, all about mag. 12.5.

Cassiopeia

This prominent group contains at least two dozen open clusters that are visible in amateur instruments; the total found photographically is over fifty. The center of the constellation culminates at midnight about 11 October.

eg I 10 ***dimen. 5!1 × 4!3*** ***V=10.3*** ***sfc. br. 13.5***
This member of the Local Group, though heavily obscured by the Milky Way, is nonetheless faintly visible in 15 cm. It appears as a diffuse unconcentrated patch elongated SE-NW, lying N of a triangle of mag. 9 stars. A faint star is visible on the W edge. With 30 cm it forms a barely concentrated granular patch in a field rich with faint stars. The brightest part lies between the mag. 13.5 star on the W and a mag. 14.5 star 45″ E. The galaxy has no distinct nucleus but is speckled with nonstellar spots. Overall the halo extends to perhaps 4′ × 2′. Photographs show that the fainter star on the E side may be an HII region.

oc 103 ***diam. 5′*** ***V=9.8*** ***stars 88***
Appearing elongated in 25 cm, this cluster is 4′ × 2′ in pa 45°. A total of 16 stars is visible, including two mag. 12 stars on the NE end. An arm of stars curves E and S from the center of the cluster. 30 cm shows a fuzzy spot 4!5 diameter containing 25 stars with a string of stars extending SW from the center. Some unresolved haze is associated, particularly with the NE side.

oc 129 ***diam. 21′*** ***V=6.5*** ***stars 193***
This cluster is visible in 6 cm as a few scattered mag. 9 stars with some faint stars associated, but it is not distinctive as a cluster. 25 cm shows an unconcentrated assemblage whose stars have a wide range of brightness. About 80 cluster members are visible in a 20′ area, including a triangle of mag. 10 stars in the center. 30 cm shows about 50 stars loosely spread in a 12′–15′ area, with the brightest ones on the S.

oc 133 ***diam. 7′*** ***mag. 9.4p*** ***stars 5***
oc King 14 ***diam. 7′*** ***V=8.5*** ***stars 186***
25 cm shows oc King 14 as a hazy patch about 10′ diameter with a few stars resolved. 30 cm reveals 25 stars in a 7′–8′ area. Five brighter mag. 10.5–12 stars are in a N-S line on the E side. There is a bright star near the center, and some fainter stars appear on the NW side. This cluster is much more conspicuous than oc 133, which is only an arc of brighter stars about 15′ NNW. oc 146, *q.v.*, lies 15′ NE.

oc 136 ***diam. 1!2*** ***mag.*** ***stars 20***
In 25 cm this cluster appears as a hazy patch at 50 ×. At 200 ×, half a dozen mag. 13 stars are visible in a 2′ diameter with some haziness. 30 cm shows the cluster 6′ NE of a reddish mag. 9.5 star with a mag. 10.5 companion 30″ S. The cluster is hazy, with ten very faint stars resolved. The stars are brighter on the NE and SW sides, and outliers trail SE to 2′ diameter.

oc 146 ***diam. 7′*** ***V=8.4*** ***stars 134***
25 cm shows this cluster to be similar in appearance to oc King 14, *cf.*, 15′ SW. It is 10′ diameter with about two dozen stars resolved over a hazy background. 30 cm shows about 20 stars in a 6′ area. Just SE of center is a close equal pair (ADS 460: 10.0,10.0; 6″8; 223°).

eg 147 ***dimen. 13′ × 8!1*** ***V=9.5*** ***sfc. br. 14.5***
This galaxy has exceptionally low surface brightness and is difficult to see in 30 cm. At 150 ×, the halo is 3′ × 2′, elongated in pa 30°. A faint star is visible on the N end of the core. At 225 ×, which shows only the core, some other faint stars are visible nearby.

eg 185 ***dimen. 11′ × 9!8*** ***V=9.2*** ***sfc. br. 14.3***
6 cm will show this galaxy as a moderately faint and weakly concentrated circular spot. Bright field stars lie W and NE. In 15 cm the halo seems elongated E-W and up to 4′ long. The core is 1′ diameter, and the E edge of the halo appears granular. In 25 cm the galaxy is circular, 5′ diameter, and looks like an unresolved globular cluster. A very faint stellar nucleus is visible, and a star lies on the NW edge. In 30 cm it is about 2!25 diameter with a broad, even concentration to a 1!25 core. Several faint stars are visible nearby.

oc 189 ***diam. 3!7*** ***V=8.8*** ***stars 90***
In 25 cm this cluster is poorly concentrated, with 20 stars within 8′ diameter. Framing the group is a 4′ rectangle of mag. 10.5–11 stars. The easternmost star in the rectangle is a pair (h1043: 11.6,12.7; 11″; 173°).

oc Stock 24 ***diam. 4′.0*** ***V = 8.8*** ***stars 180***
oc 225 ***diam. 12′*** ***V = 7.0*** ***stars 76***

Easily identifiable in 6 cm, oc 225 is nevertheless not a conspicuous object. A string of five mag. 9.5 stars aligned roughly N-S is visible on the E side. 25 cm shows a very loose group 16′ across with 20 stars down to mag. 13.5. 30 cm reveals 30 stars in a 20′ area elongated E-W. An interesting region lies 40′ W, containing a compact group of stars on its N side: this is Stock 24.

eg 278 ***dimen. 2′.2 × 2′.1*** ***V = 10.9*** ***sfc. br. 12.3***

This galaxy is barely visible in 6 cm, 3′ S of a mag. 8 star. It appears small, but nonstellar at 100×. In 15 cm it is moderately bright and best viewed at medium power. The halo is 1′ diameter, irregularly round, with a stellar nucleus. 25 cm shows a bright and concentrated object. The halo is 1′.5 diameter and brightens to a well-defined core and stellar nucleus. Viewed in 30 cm, it is 1′.2 diameter and broadly concentrated to a faint stellar nucleus.

oc-gn 281 ***diam. 4′.0*** ***mag. 7.4p*** ***stars***
dimen. 35′ × 30′

6 cm will show this nebula as a fairly low surface brightness glow, but no concentration of stars is visible in the area. In 15 cm the nebula is about 10′–12′ across and has four associated stars; the central star is a close triple at 100×. The haze extends to about 15′ in 25 cm; in 30 cm the nebula breaks into dissociated parts. Some nebulosity is visible around the central triplet and extends 1′ SW to a mag. 11.5 pair (13″; 310°). More nebulosity is visible associated with stars about 3′.5 E of the trio, and other faint patches lie SE. About 10 stars lie in a 2′ area around the central group, which is actually a quadruple (ADS 719): the two brightest stars are separated by 3″.9 in pa 134°, while a third member is 9″ S. The brightest star has a close companion 1″.5 E that is two magnitudes fainter.

oc 366 ***diam. 3′.0*** ***mag.*** ***stars 30***

In 25 cm this cluster is 2′ diameter, lying in the midst of a blank region of the Milky Way. Eleven stars are visible in pairs and triplets.

oc 381 ***diam. 6′*** ***mag. 9.3p*** ***stars 50***

This is a nice cluster in 25 cm. It is moderately concentrated, about 10′ diameter, with 30 stars. On the N side is a triple star of unequal magnitudes (the two closest form Stein 185: 10.8,12.5; 8″.1; 129°). In 30 cm this cluster is relatively indistinct from the Milky Way, appearing as a clot of background stars of mag. 11.5 and fainter. About 40 stars are visible in an 8′ area, with a few strings.

oc 436 ***diam. 6′*** ***V = 8.8*** ***stars 49***

Located 40′ NW of oc 457, *q.v.*, this cluster appears as a small, high surface brightness spot in 6 cm. Only a few stars can be resolved. In 20 cm it is 5′ diameter, with about 30 stars of mag. 10 and fainter. 25 cm shows strings of stars extending to 10′ diameter from a central condensation. 30 cm shows a 4′.5 object with a knot of four stars N of center. oc 433 lies 80′ NNW.

oc 457 ***diam. 12′*** ***V = 6.4*** ***stars 204***

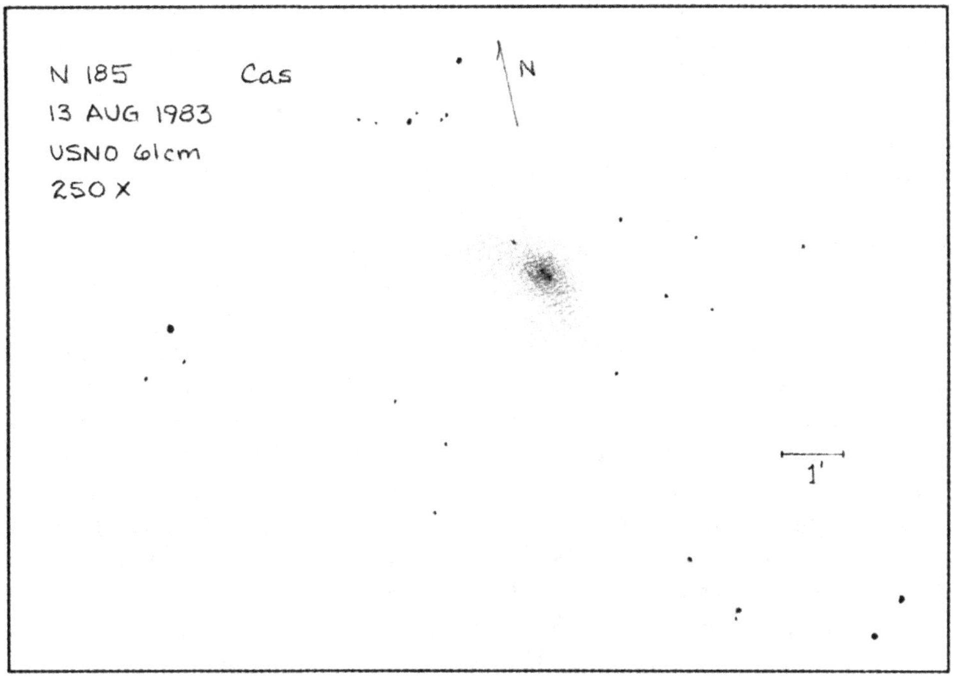

This is perhaps the most impressive of all the clusters in Cassiopeia. On the SE are $\phi^{1,2}$ Cassiopeiae (ADS 1073: V = 5.0,7.0; 134″; 231°), yellow and blue in color, which are probable cluster members. In 6 cm about 25 stars are resolved, with moderate concentration to the center. 15 cm will show 45 stars in a 15′ area. In 20 cm it is irregularly round, 20′ diameter, with about 100 stars visible. The outline appears triangular in 25 cm. South of the center is a condensation of bright stars, while N of the center is a well-defined dark patch. On the N edge is a red mag. 9 star (V466 Cas). 30 cm reveals many threshold magnitude stars interspersed among the bright members. On the NW side is a mag. 9,10 pair about 14″ apart in pa 319° (Stein 1560).

oc 559 diam. 4′.4 V=9.5 stars 120
This cluster looks similar to oc 659, *cf.*, in 6 cm. A string of four mag. 10.5 stars extends NW from the center; the rest of the cluster is granular to partially resolved. 25 cm shows the cluster as a small 3′ diameter object. Three brighter stars lie among a total of 20 stars of mag. 13 and fainter. In 30 cm the three brighter stars are in the SE sector of the cluster, which extends to 9′ × 5′ in pa 45°. It has an irregular outline, with stragglers leading off the SW side and a long string on the E. Forty stars are visible, mostly of mag. 12.

oc 581 diam. 6′ V=7.4 stars 172
Messier 103 is visible in 6 cm with a mag. 7.5 star on the NW. A mag. 8 star SE and a mag. 10.5 star SW combine with the bright star to enclose the cluster in a triangle. Overall, the group is elongated SE-NW, with 13 other stars visible including two brighter ones. 15 cm shows about 15 stars in a 7′ area. 20 cm shows 50 stars within the triangle. A similar view is obtained with 25 cm. 30 cm shows a moderately compact cluster about 8′ diameter. A very red mag. 8.5 star is SE of the center. A total of 50 stars can be counted, discounting those of threshold visibility.

oc Trumpler 1 diam. 4′.5 V=8.1 stars 112
With 25 cm this cluster is an interesting collection of stars. Most obvious is a line of four stars 1′.5 long in pa 60°. Two stars about 1′ S form a box with the central string. A total of 14 stars is scattered N and S of the bright line.

oc 637 diam. 3′.5 V=8.2 stars 55
To 6 cm this cluster is visible as a small, moderately bright spot. Four or five stars are resolved against a hazy background. 25 cm shows it as a small, uninteresting group, 4′ × 2′ in extent. A pair of stars is resolved in the center (ADS 1342: 10.0,11.2; 8″.7; 359°), and a total of ten cluster members is visible. In 30 cm, 25 stars are visible in the main part of the cluster, which is 1′.5 diameter. Another pair of 25″ separation in pa 120° lies 3′.5 E.

oc 654 diam. 8′ V=6.5 stars 83
This cluster is a small, nebulous spot in 6 cm, lying immediately NNW of a mag. 7 star. It is larger and brighter than oc 659, *cf.*, and is partially resolved at 25 ×. The cluster is small and condensed in 25 cm. About 20 stars are resolved in a roughly circular haze 3′ diameter. 30 cm reveals 30 stars fainter than mag. 10.5 in a 4′.5 area including several close groups. A mag. 9.5 star lies 4′ SW.

oc 659 diam. 5′ V=7.9 stars 186
6 cm will show this cluster 11′ NE of 44 Cassiopeiae (V=5.8). At 25 × it is granular to partially resolved. It is not very bright in 25 cm, with 22 stars in a 3′ diameter. 30 cm shows a 4′.5 object with no more than 20 stars countable at 225 ×, all fainter than mag. 10.5. The cluster lies 4′ N of a mag. 10 star.

oc 663 diam. 8′ V=7.1 stars 108
This large cluster shows about 15 stars to 6 cm, including a wide E-W pair on the N side. Many stragglers, all about the same magnitude, spread S to oc 659, *q.v.* Viewed in 15 cm, this is a typical cluster of 30 stars with some central concentration. It is about 15′ diameter, and a circlet of five mag. 7–8 stars encloses the fainter cluster members. 25 cm reveals about 60 stars in a 20′ area. Several double stars are visible, and some are faintly colored: the eastern star of the very wide pair N is ADS 1390 (9.7,10.9; 7″.6; 69°); the eastern and southern stars of a 1′ triangle on the W side are also close pairs (ADS 1384: 9.4,10.9; 9″.3; 105° and ADS 1381: 10.7,11.2; 7″.1; 38°, respectively). In 30 cm the cluster is only slightly concentrated. It appears 15′ diameter with 65 stars visible, including five of mag. 9.5. Three conspicuous concentrations of stars are visible, with the northernmost being the brightest and largest.

pn I1747 dimen. 13″ × 13″ nebula 12.1v star V=15.4
This nebula is discernable with 30 cm in the center of a 1′.5 triangle of mag. 13.5–14 stars. The object is 15″ diameter and fairly well concentrated.

oc 743 diam. 5′ mag. stars 12
Lying between Messier 103 (oc 581, *q.v.*) and Stock 2 (*q.v.*) is a large, scattered group of stars. It makes a nice low-power field for 6 cm, with many mag. 7–9 stars in a 90′ area. On the W side, seven brighter mag. 8–9 stars form a "C" opening to the E. Near the center of this extended bunch is oc 743 proper, which forms a conspicuous clustering of about ten mag. 9–11 stars. The area looks similar to oc Stock 2, but not so rich and with brighter stars. In 25 cm, oc 743 is about 7′ diameter. The cluster outline is triangular and includes 20 stars, the brightest, mag. 9, on the S point. One leg of the triangle

faces W, and on the SE leg is a wide, unequal pair (h1098: 10.3,12.7; 17″; 343°).

oc Stock 2 diam. 60′ V=4.4 stars 166
This extended cluster is easily found at low power in 6 cm by following a curving string of stars for 2° N from oc 869 (h Persei, *q.v.*). Many stars of mag. 8.5 and fainter stand out in an 80′ area. A conspicuous, bright pair lies on the NE side (V=7.0,7.2; 63″; 200°). The group is comparable to oc I4756, *cf.*, in Serpens. 25 cm shows at least 100 stars here within a 90′ field.

oc Markarian 6 diam. 4′.5 V=7.1 stars 29
In 30 cm this cluster appears as a string of six stars 8′ long trailing to the S, then W, with decreasing separation. The last and next-to-last members in the string are pairs (Σ263AB: V=8.4,11.2; 15″; 103°, and Σ264CD: V=9.9,10.9; 17″; 226°). Two other stars are visible close to the string.

oc-gn I1805 diam. 22′ V=6.5 stars 62
This cluster is involved with an extended faint nebulosity. At low power, 6 cm will show a conspicuous clustering of about 25 stars embedded in a very faint haze. 25 cm shows a loose aggregation of about ten brighter stars and as many fainter ones in a 20′ area, but no nebulosity is visible. In 30 cm, the bright mag. 8.5–9 stars are scattered in a 25′ area. Near the center, grouped about a mag. 8 star, are 25 stars within 5′ diameter.

oc 1027 diam. 20′ V=6.7 stars 152
Centered around a 7th magnitude star, 6 cm will show about 15 stars in this cluster. In 25 cm it is richer and more concentrated than oc-gn I1805, *cf.* Around the bright central star are about 40 other members. 30 cm shows about 70 stars in an 18′ diameter, including a few strings.

oc-gn I1848 diam. 12′ V=6.5 stars 74
dimen. 40′ × 10′
This cluster is more definitely associated with nebulosity than oc-gn I1805, *cf.*, in 6 cm. The stars are also fainter, less concentrated toward the center, and strung out to the E and W. No nebulosity is visible with 25 cm, but 65 stars, mostly mag. 11, can be seen in a 30′ area. On the E side of the 8′ core is a mag. 7 star. 30 cm shows the cluster stars grouped around the mag. 7 star and a mag. 8 star 2′ SSE. The fainter star is surrounded by 12 stars in a 2′ area with strings extending to the NE and SW. The brighter star is accompanied by eight stars in a 1′ area. Some nebulosity is visible around the bright stars.

pn I 289 dimen. 42″ × 28″ nebula 13.3v star
25 cm shows this planetary lying 1′.75 N of a mag. 10 star. It is a very faint circular glow about 40″ across. A mag. 13.5 star is visible 1′.25 E. The nebula is annular in 30 cm, 30″ diameter with a 15″ hole. The outer rim of the annulus is generally brighter, and a stellaring is visible on the WSW edge. A mag. 14 star is visible 50″ NE.

oc Trumpler 3 diam. 23′ mag. 7.0p stars 30
In 25 cm this cluster is a loose, poor, irregularly round group about 15′ diameter with 20 stars. A dozen stars of mag. 9–10 are visible in 30 cm out of a total of 35. The cluster is 18′ diameter, and the NW side is more concentrated.

gn 7635 dimen. 15′ × 8′ star 8.5p
This nebula is very faint and barely discernable in 15 cm. 25 cm shows it situated in a rich field 6′ NE of a mag. 7 star. The brightest part of the nebula is NW of the illuminating star in 30 cm, and another faint mag. 14 star is associated with the nebula. The overall size is 1′, with some brightening toward the middle.

oc 7654 diam. 13′ V=6.9 stars 173
Messier 52 is visible in 6 cm as a hazy, partially resolved patch with a mag. 8 star on the SW side. In 15 cm the cluster members appear faint with only 30 resolved in a 10′ area. 20 cm shows 50 stars of mag. 10 and fainter in a 10′ area. The group appears fairly large and bright in 25 cm. There is a central condensation, but elsewhere the cluster is uniformly scattered with stars. At least 80 stars are visible in a 15′ area. 30 cm shows 140–150 stars in 13′ diameter. Two clusterings 2′ across are visible on the NE side, and fainter stars trail off the SW side.

oc King 21 diam. 2′.5 V=9.6 stars 17
This faint cluster is barely visible to 25 cm at 100×; it is best viewed at 200×. In the center lie three mag. 11–12 stars, two of which form a close pair (Stein 1223: V=11.4,12.4; 9″.8; 72°). Around these, 20 stars of mag. 12.5 and fainter are visible concentrated into a 5′ × 2′.5 bar elongated NE-SW.

oc King 12 diam. 2′.0 V=9.0 stars 30
In small apertures this cluster appears as a hazy patch around a mag. 9 star. 15 cm reveals a few faint cluster members and a mag. 11 star to the NE. The cluster is elongated E-W in 25 cm, with a double star resolved on the S edge (ADS 17081: V=10.4,10.7; 6″.6; 341°). A total of 18 stars embedded in some haziness can be seen. Viewed in 30 cm, the group has about 20 members in a 2′ × 1′ area, with a thin arm extending to the W.

oc Harvard 21 diam. 4′.0 mag. 9.0p stars 6
Lying 10′ SSE of oc King 12, *q.v.*, this inconspicuous cluster is visible to 25 cm as a small group of seven stars.

Chart VII. NGC 7790. A 30-minute exposure on 103a-O with the USNO 1 meter telescope taken by A. Hoag on 31 August 1959. Magnitude sequence from: Pedreros *et al.* 1984, Ap. J. 286: 563. U.S. Naval Observatory photograph.

The five brighter members are all about mag. 10, while the other two are mag. 12.5–13.

oc 7788 ***diam. 9′*** ***mag. 9.4p*** ***stars 20***
This cluster is nested in a group of bright mag. 7–10 stars about 15′ across. Near one of these stars is the cluster. 15 cm will reveal about eight stars, lying mostly to the E of the bright star. With 25 cm, 14 stars can be counted in a 1′.5 area. The group has a triangular outline in 30 cm, with one apex pointing E. About ten stars to mag. 13 are visible in a 1′ area.

oc 7789 ***diam. 16′*** ***V=6.7*** ***stars 583***
This cluster is visible to 6 cm as a medium-bright glow without resolution. 15 cm shows it 20′ diameter, with about 60 stars resolved. There is little haziness, and a few brighter stars appear on the N side. About 100 stars are visible in 20 cm, and 25 cm will show approximately 150 stars. In 30 cm about the same number is visible. The stars are unevenly scattered over a 20′ area.

oc 7790 ***diam. 8′*** ***V=8.5*** ***stars 134***
15 cm shows this cluster about 4′ NW of a mag. 10 star, and other bright stars lie on the W and NW sides (these are the variable stars QX, CEab, and CF Cassiopeiae, respectively). About a dozen stars are visible in a rectangular area elongated E-W with a clump on the E side. In 25 cm the group is 4′ × 2′ and contains about 25 stars. Five brighter stars on the W side, including CE and CF Cas, form a circlet. On the E side, the clump of mag. 12–13 stars includes a close pair that is just resolvable at 200 ×. An arc of stars opening toward the E also lies on the E side. About 20 stars are visible in 30 cm in a 2′.5 area. An arm of faint stars extends E. The variables CE Cas are components of a close pair (Leonard 55: 10.9,11.1; 2″.4; 264°).

oc Berkeley 58 ***diam. 8′*** ***V=9.7*** ***stars 39***
At low power in 25 cm, this group is visible as a small collection of faint stars. About 35 stars of mag. 12 and fainter are revealed at 200 × in a 5′ area. The cluster is not concentrated, but includes a few pairs.

Centaurus

Though not fully visible from northern locations, this large bright constellation contains a variety of bright and curious objects. gc 5139 (ω Centauri), eg 5128, and eg 4945 are among the best objects for amateur viewing. The center of the constellation culminates at midnight about 7 April.

eg 4945 ***dimen. 20′ × 4′.4*** ***B = 9.5*** ***sfc. br. 14.2***
With 15 cm this is a large and diffuse galaxy. With averted vision the halo extends to about 11′ × 2′ in pa 40°.

eg 5011 ***dimen. 2′.0 × 2′.0*** ***B = 12.6*** ***sfc. br. 14.1***
This small object is visible to 15 cm as a nearly stellar patch 30″ diameter. No details are noticeable. 25 cm shows it less than 1′ diameter with an inconspicuous stellar nucleus. Three stars are visible to the W, one nearly touching the galaxy.

eg 5128 ***dimen. 18′ × 14′*** ***V = 7.0*** ***sfc. br. 12.9***
One of the brightest galaxies in the sky, this unusual object is also a strong radio emitter, designated Centaurus A. The strong dark lane, a striking feature on photographs, is clearly visible with 15 cm. It is aligned in pa 120° and is 1′ wide and about 6′ long; the eastern part is wider than on the west. The S half of the nebula is much larger and more concentrated, including a star near its center and one on the W just within the dark lane. In 25 cm the dark lane is easily visible: at higher powers it flares to 90″ width at the ends. Overall, the halo has a nearly circular outline.

gc 5139 ***diam. 36′*** ***V = 3.5*** ***class VIII***
ω Centauri is the brightest globular cluster in the sky, visible without difficulty to the naked eye. 6 cm will show a very large and broadly concentrated oval that is elongated E-W. At low power it has a fine, granular texture with only a few mag. 10.5 stars resolved. The cluster is very broadly concentrated in 15 cm, with a core about 8′ across, but no central pip. It is evenly resolved though somewhat hazy across the center, with outliers extending to 35′ diameter. An asterism of three or four brighter stars is visible 8′ NE of center. The cluster is huge in 25 cm; it is well resolved at 100 × with an underlying haze, and outliers extending to 45′ × 35′. Two darker patches appear on the S side of the core.

Cepheus

T. W. Webb characterized this constellation as "much more barren to the naked eye than to the telescope." The Milky Way fades here, but there remain many clusters and a few conspicuous gaseous nebulae. The bulk of the constellation culminates at midnight toward the end of August, but some areas are circumpolar.

pn 40 ***dimen. 38″ × 35″*** ***nebula 12.4:v*** ***star V = 11.6***
This is a bright nebula for 25 cm, plainly visible at 50×. The central star is conspicuous, and the nebula appears dark immediately around it; one or two other dark spots can be seen. On the SE side of the object is a bright stellaring, and a mag. 13 star is located 1.′25 to the SW. 30 cm shows an unevenly bright object with a mag. 11.5 central star. The brightening on the SE edge is conspicuous at 120° from the central star. The overall size is 55″ × 40″, elongated E-W.

oc 188 ***diam. 14′*** ***V = 8.1*** ***stars 550***
This old, faint cluster is visible in 6 cm as a large, moderately faint patch among several mag. 8–10 field stars. The unresolved glow is slightly brighter toward the center and has a smooth to slightly granular texture. In 25 cm it is distinguishable as a dim swarm of stars 20′ diameter. Two widely separated stars are visible on each side, and about 50 faint stars are resolved at high power. The large size makes it inconspicuous in 30 cm; about 50 stars of mag. 13 and fainter are visible with no unresolved haze.

eg 2276 ***dimen. 2.′6 × 2.′5*** ***V = 10.6*** ***sfc. br. 13.3***
This object is barely visible to 25 cm, 2′ ENE of a mag. 8.5 star that interferes with viewing. It appears diffuse and without central concentration. 30 cm shows a medium-sized, low surface brightness patch lying in a string of four stars aligned E-W that includes the bright star WSW. It is circular, about 1.′8 diameter, and has broad irregularities in the surface brightness.

eg 2300 ***dimen. 3.′1 × 2.′6*** ***V = 10.8*** ***sfc. br. 13.0***
This galaxy is faintly visible in 15 cm, but 25 cm will show a high surface brightness object located 6′ SE of eg 2276, *q.v.* The halo appears slightly elongated E-W and about 1.′25 diameter. The broadly brighter core has a stellar nucleus. 30 cm shows the object located off the E end of a string of stars that contains eg 2276. It is well concentrated to a bright nonstellar nucleus lying in a 10″ core. The halo is elongated in pa 120° and extends to 1.′5 × 1.′ eg I 455, *q.v.*, lies 11′ SSE.

eg I 455 ***dimen. 1.′4 × 0.′9*** ***mag.$_z$14.3*** ***sfc. br.***
Lying about 11′ SSE of eg 2300, *q.v.*, this galaxy is a small, faint object for 30 cm. The 45″ × 25″ halo is elongated in pa 80°, and fairly well concentrated to a stellar nucleus, which seems to be SW of the geometric center.

eg I 469 ***dimen. 2.′3 × 1.′3*** ***mag.$_z$13.6*** ***sfc. br.***
Located within a 6.′5 × 5′ triangle of mag. 9 stars, this galaxy is visible in 15 cm as a very faint glow. There appears to be no concentration in the 1′ halo, except for a very faint nucleus. 30 cm shows the halo to 2′ × 0.′75 in pa 90° around an indistinct 1′ × 0.′5 core and faint stellar nucleus. The light is broadly concentrated overall, with the inner regions becoming progressively more circular. At 225× the nucleus seems to lie a little S of the center.

oc 6939 ***diam. 8′*** ***V = 7.8*** ***stars 301***
In 6 cm aperture this cluster looks much like eg 6946 (*cf.*, Cygnus), which lies less than a degree SE. The unresolved haze is, however, brighter than the galaxy, and a little concentrated toward the center. A broad triangle of mag. 10 stars lies on the N side of the hazy, partially resolved patch in 15 cm. Within a 10′ circle only about ten stars can be resolved. 25 cm shows it 12′–15′ diameter, bounded on the NE, N, and W by the mag. 10 stars. Approximately 100 stars are visible with some haze. NE of center is a 45″ clump of stars that remains nebulous at 200×. 30 cm shows 50 stars, mostly mag. 12 and fainter, in a 12′ area, including a mag. 11.5 star on the SW. There is some haziness in a 3′ diameter central area, but the knot NE of center is resolved into four or five stars. Most of the outliers spread NE to 8′ radius.

eg 6951 ***dimen. 3.′8 × 3.′3*** ***V = 11.1*** ***sfc. br. 13.7***
25 cm shows this object elongated in pa 100°, about 1.′25 × 0.′75. It is fairly faint with a small oval core. On the E end is a mag. 13 star. The nucleus is stellar in 30 cm, and the object fades rapidly outward to a faint 1.′5 × 0.′9 halo. The core is indistinct and only a little brighter than the halo.

oc-gn 7129 diam. 2.′7 mag. 11.5p stars 10
dimen. 7′ × 7′
gn 7133 dimen. 3′ × 3′ star

6 cm shows oc-gn 7129 as two double stars with nebulosity surrounding and N of the northernmost pair. A few arcminutes NE is gn 7133, which appears as a small, faint patch without stars. In 25 cm the nebulous cluster has four bright stars and several fainter ones. The nebula is 4′ × 2′ and has a fairly high surface brightness. 30 cm shows the pair on the S in pa 0°, the pair N in pa 110°. The nebula is brightest around the eastern star of the northern pair, and a faint companion is suspected near this star. The nebula is mostly N of this pair, and at least two more stars are involved. It is about 2′ diameter and irregularly shaped. gn 7133 is fainter, extending to only 1′, and has a single star involved on its S side.

pn 7139 dimen. 86″ × 70″ nebula 13.3v star 17.7p

Visible at 100 × in 25 cm, this nebula appears as a faint circular patch with a mag. 13.5 star off the SE edge. With averted vision a darker center is discernable. 30 cm shows a slightly concentrated smooth object in pa 30° without detail. A threshold magnitude star lies on the S edge.

oc 7142 diam. 9′ V = 9.3 stars 186

In 25 cm this object appears as an irregularly round nebulous area 12′ in diameter. About 35 stars of mag. 12.5 and fainter are resolved. 30 cm shows an unconcentrated cluster with 35–40 stars evenly spread in an 11′ area. Three brighter stars are visible N, E, and SE of center, but the cluster members are mostly mag. 13 and fainter.

oc 7160 diam. 7′ V = 6.1 stars 61

6 cm shows a hazy cluster here elongated NE-SW and containing ten stars, including two bright stars on the NE end (ADS 15434: V = 7.0,7.9; 63″; 145°). In 25 cm the cluster looks like a group of multiple stars. About 15 stars are visible in a 5′ × 2′ area. 30 cm shows many faint stars irregularly scattered in an 8′ area around the central group of bright stars. Just SW of the brightest pair is a curved triplet of mag. 9–10 stars.

oc 7226 diam. 4.′0 V = 9.6 stars 83

This faint cluster is visible in 25 cm, though it is not well resolved even at 200 × . It appears as a 2′ hazy area with only a few mag. 14 stars resolvable. A relatively bright pair lies 1′ NE of center (Stein 2630: 11.7,12.7; 11″; 197°); a single mag. 11 star is conspicuous on the NW edge.

oc 7235 diam. 4.′0 V = 7.7 stars 98

About 15 stars are visible in this cluster with 25 cm. The stars are loosely scattered in an area about 5′ long, elongated ENE-WSW. A brighter mag. 9 star lies a little separated to the SE side, while the other members are mag. 10–12.

oc 7261 diam. 6′ V = 8.4 stars 62

This cluster contains about ten stars visible to 6 cm, with the brightest member on the E edge. 25 cm shows a 4′ × 2′ group elongated N-S. There are several brighter stars and about 20 fainter ones involved with a little haze. 30 cm shows at least 30 stars including ten brighter ones in an unconcentrated group. The N and S ends

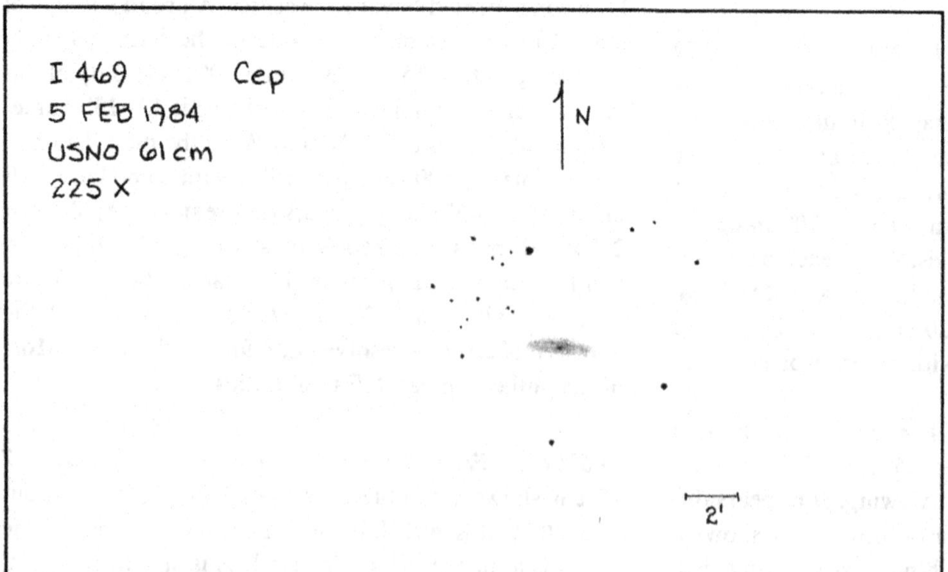

I 469 Cep
5 FEB 1984
USNO 61 cm
225 X
N
2′

curve a little westward. The cluster is 8′ × 5′, with more stars trailing off to the S.

oc 7281 ***diam. 12′*** ***mag.*** ***stars 20***
Lying about 40′ ESE of oc 7261, *q.v.*, 25 cm will show this cluster with about 35 members in a 15′ area. On the W side are three evenly spaced mag. 8.5 stars in a line aligned E-W. South of these is the greatest concentration of mag. 11 and fainter stars. To the E and N of the three-in-line are scattered stars, apparently members of the cluster. 30 cm shows 30–40 stars in a 10′ area that are not very distinct from the background.

pn 7354 ***dimen. 22″ × 18″*** ***nebula 12.2v*** ***star***
This planetary is visible at low power in 25 cm as a moderately faint and uniform circular patch about 30″ diameter with diffuse edges. A mag. 13.5 star is located on SSW, with another a little farther toward the W. 30 cm shows a grey object that has a broadly brighter center.

oc-gn 7380 ***diam. 12′*** ***V = 7.2*** ***stars 125***
Though this cluster is involved with a faint nebulosity, none is easily discernable visually in amateur instruments. 15 cm shows the cluster nestled in a circlet of bright stars. It is 10′ diameter with 20 stars resolved over haze. About 40 stars are visible in 25 cm, including a distinctive V-shaped asterism pointing N. The brightest star (V = 8.6) is on the SW at one end of the V. 30 cm reveals about 30 stars, including a bright pair about 10′ to the W (ADS 16260: V = 7.6,8.6; 31″; 117°).

oc King 18 ***diam. 4.′0*** ***mag.*** ***stars 20***
This small, fairly faint cluster is visible in 25 cm. A total of 15 stars, mag. 11 and fainter, is visible in a 4′ area.

oc 7419 ***diam. 2.′0*** ***mag. 13.0p*** ***stars 40***
Only a few stars can be resolved in this small cluster with 25 cm, and these require high power. Two bright stars lie on the NW side, and a nice pair lies 11′ WNW (ADS 16334: 7.8,9.8; 8.″3; 138°). The cluster is elongated in 30 cm due to the two bright stars on the NW side and a mag. 9.5 star on the SE side. About ten stars are visible with some haziness in a 1.′5 × 1′ area.

gn I1470 ***dimen. 70″ × 45″*** ***star V = 12.7***
This nebula is located 2′ SE of a mag. 10 star. In 25 cm the illuminating star and a very faint companion 20″ SE lie on the N edge of the object. The nebula is a uniform grey patch about 40″ diameter. 30 cm shows it broadly concentrated with "hairy" edges.

oc 7510 ***diam. 4.′0*** ***V = 7.9*** ***stars 75***
15 cm shows this cluster as a rectangular blaze about 2′ long, with one star outshining ten fainter ones. 25 cm shows bright stars at each end, between which are fifteen more in a 5′ streak. Viewed in 30 cm, 23 stars are visible in a compact group 7.′5 × 2′ in size, elongated in pa 60°.

gn 7538 ***dimen. 8′ × 7′*** ***star 10.0p***
25 cm shows two mag. 11 stars here 35″ apart in pa 30° embedded in nebulosity. The nebula extends SW from the southern star to about 1.′5 × 0.′3. A threshold magnitude star is visible 30″ W of the southern star. The nebula is unevenly bright in 30 cm, about 1.′5 diameter. The brightest parts lie W of the pair.

oc 7762 ***diam. 11′*** ***mag. 10.0p*** ***stars 40***
Lying about 15′ NE of a 5th magnitude star, 25 cm will show this cluster as a large faint glow. It appears about 20′ across with 20–30 stars visible over a granular background haze. 30 cm shows it 15′ diameter with a sparse, uneven sprinkling of mag. 10.5 stars and many fainter ones. A group of four or five stars is nebulous at 150 ×, but resolved at 225 ×.

Cetus

This widespread constellation contains several groups of bright and faint galaxies. The center of the constellation culminates at midnight about 16 October.

eg 45 dimen. 8′.1 × 5′.8 V = 10.4 sfc. br. 14.4
This large, very low surface brightness face-on spiral is invisible to 15 cm. 25 cm reveals only an indefinite patch lying 1′.5 NNW of a mag. 10 star, which in turn is about 5′ E of a mag. 7 star (both are plotted on the AE). It is very faint and difficult in 30 cm, which shows two faint patches within the object: one in the center, the other N of center.

eg 151 dimen. 3′.7 × 1′.9 V = 11.5 sfc. br. 13.6
Located immediately SW of a mag. 11 star, this is a fairly nice object in 25 cm. The halo is elongated ENE-WSW, about 1′.5 × 0′.75, with a bright core. 30 cm reveals a faint 2′ × 1′ halo in pa 50°. The core is irregular but well defined, about 45″ across, and without a prominent nucleus. The halo is much fainter.

eg 157 dimen. 4′.3 × 2′.9 V = 10.4 sfc. br. 13.0
An easy object for 15 cm, this galaxy is located between two mag. 8.5 stars. The halo extends to 3′ × 2′, sometimes looking rectangular in outline. The core region is evenly concentrated and detailed. 25 cm shows a 3′.5 × 2′ object that appears granular at high power. A mag. 12.5 star lies on the NE edge. In 30 cm it appears broadly and weakly concentrated, elongated in pa 30°, reaching just past the star on the NE. At 225 × the core is conspicuously mottled, particularly on the E. Averted vision shows very faint stellarings running along the major axis, the brightest being the nucleus and those to the SW.

eg 175 dimen. 2′.6 × 2′.4 V = 12,1 sfc. br. 13.9
This is a faint galaxy that is not visible to 15 cm. 25 cm shows it faintly, forming a triangle with two stars S and E about 5′ away. A faint stellaring is visible at the center of the 1′ diameter object. Viewed in 30 cm, it is slightly and evenly concentrated to the center with a stellaring toward the NW. The overall diameter is about 1′.5.

eg 178 dimen. 2′.0 × 1′.2 V = 12.6 sfc. br. 13.4
In 25 cm this faint object is about 1′ × 0′.5, elongated in pa 5°. The core is elongate and a stellar nucleus is occasionally visible. The halo has a smooth outline. In 30 cm the halo extends to 1′.3 × 0′.7 and is weakly concentrated with no sharp nucleus.

eg 210 dimen. 5′.4 × 3′.7 V = 10.9 sfc. br. 13.9
Lying 1′.25 ENE of a mag. 11 star, 15 cm will show this galaxy as a uniformly bright, circular patch about 1′ diameter. 25 cm shows a featureless 1′ diameter object, which is indefinitely elongated N-S. In 30 cm it is lens-shaped with an elongated core and brighter nonstellar nucleus. The overall size is 1′.5 × 0′.9 in pa 170°, with a smooth texture fading to indefinite edges.

eg 227 dimen. 2′.1 × 1′.7 B = 12.8 sfc. br. 14.1
This galaxy is not visible to 15 cm. 25 cm, however, shows it faintly with a conspicuous stellar nucleus. The small silvery core is elongated SE-NW; the halo is circular and less than 1′ diameter. In 30 cm the galaxy is dominated by its bright stellar nucleus, which seems multiple at 150 × . It is sharply concentrated from the core to the nucleus, with a total diameter of 50″.

eg 237 dimen. 1′.8 × 1′.2 B = 13.7 sfc. br. 14.4
This object is only just visible in 25 cm at 100 × , and even with 30 cm it is difficult. At high power it appears evenly and moderately concentrated to a faint nucleus. The core is 30″ across, and the overall diameter is less than 1′.

eg 245 dimen. 1′.4 × 1′.3 B = 13.2 sfc. br. 13.7
Though not visible in 15 cm, 25 cm will show this galaxy faintly in a nearly blank field. The halo is 1′.5 diameter and a little brighter toward the center. A mag. 13 star is 2′ SW. Viewed in 30 cm, it is 1′.3 diameter with a smooth texture, except near the center where brighter mottlings are visible. The halo has a slight broad concentration to a faint stellar nucleus. An additional star, mag. 14, lies toward the SW.

pn 246 dimen. 4′.6 × 4′.1 nebula 10.9v star V = 11.8
This planetary nebula is faint, but can be found easily at low power in 15 cm. Three stars are involved with the nebula. 25 cm shows the three stars in an indefinite glow 3′ diameter. 30 cm shows a total of five stars grouped in a crude "?" asterism. The three stars on the N have some

nebulosity associated with them. eg 255, *q.v.*, lies 27′ NNE.

eg 247 dimen. 20′ × 7.′4 V = 8.8 sfc. br. 14.1
This faint, extended galaxy is difficult in smaller apertures due to its low surface brightness, but even 6 cm will show it faintly. In this aperture it appears as an unconcentrated spindle elongated N-S with a mag. 8.5 star on its southern tip. It is much fainter than nearby eg 253, *cf.*, in Sculptor. It is a faint extended object in 15 cm, remaining quite featureless even in dark sky. 25 cm shows it elongated in pa 170°, about 15′ × 3.′ It has a very low surface brightness, smooth texture, and no central concentration. The mag. 8.5 star on the S is within the halo. 30 cm shows a 2′ concentrated core, with some haziness extending to a mag. 10.5 star 3′ WSW of center.

eg 255 dimen. 3.′1 × 2.′8 V = 11.8 sfc. br. 14.0
Found by moving 20′ E from the southern star of a triangle 40′ across, in 25 cm this low surface brightness object appears elongated N-S, about 1′ × 0.′6 in size. It has a granular texture and a well-defined core. 30 cm shows a 40″ core with the halo extending to 2′ diameter. A mag. 13.5 star lies 3′ ESE; a mag. 14 star is 1.′4 SSW.

eg 268 dimen. 1.′7 × 1.′3 B = 13.3 sfc. br. 14.1
This fairly faint object is located off the W side of a Delphinus-like asterism of mag. 7–8.5 stars (plotted on the AE). In 25 cm it is 1′ diameter with a slightly brighter center. 30 cm gives a similar view including some stellarings in the core. A mag. 14 star lies 2.′7 SE.

eg 273 dimen. 2.′6 × 0.′9 mag.$_{VV}$13 sfc. br.
This faint galaxy is located 11′ NNW of eg 274–75, *q.v.* It is situated near a mag. 13.5 star and appears faint and circular in 25 cm. 30 cm shows it 40″ across with a 20″ core. The mag. 13.5 star lies 40″ NW.

eg 274 dimen. 1.′7 × 1.′6 B = 12.9 sfc. br. 13.9
eg 275 dimen. 1.′5 × 1.′2 V = 12.5 sfc. br. 13.5
This is an interesting pair of galaxies separated by only 45.″ They are moderately difficult in 15 cm, which shows only a 1′ blob without separation between the objects. A faint stellaring lies on the NW side (−74). In 25 cm they are separated SE-NW. On the W side is a circular patch with a fairly bright stellar nucleus. Toward the E is a dark arc followed by a bright arm, which is the bright part of −75. Herein lies an extremely faint stellar nucleus. 30 cm shows them just touching. eg 274 has a bright 10″ core without a sharp nucleus. Its overall size is 30.″ eg 275 is bigger, but has a lower surface brightness and a faint stellar nucleus. The pair are in pa 130°. A mag. 13.5 star is 2.′4 SW.

eg 309 dimen. 3.′1 × 2.′7 V = 11.9 sfc. br. 14.0
Located 9′ SE of three mag. 9 stars, this object is not visible in 15 cm. 25 cm, however, shows a faint, circular, diffuse glow lying 1.′9 SSW of a mag. 11.5 star. No concentration is evident, but a very small central pip appears at the threshold of vision. 30 cm shows a circular, weakly concentrated halo 2′ diameter with two or three stellarings near the nucleus. eg I1602 (mag.$_{VV}$14) lies 13′ WSW.

eg 337 dimen. 2.′8 × 2.′0 V = 11.6 sfc. br. 13.4
This object is barely visible in 15 cm 9′ SSW of a mag. 9.5 star. It appears moderately bright in 25 cm, about 2′ × 0.′75 in pa 150°. The core is circular, about 30″ across, and slightly granular. The halo has thin sharp tips. 30 cm shows a nice oval galaxy, about 2′ × 1.′2 in extent. There is a slight even concentration to a 1′ × 0.′3 core. eg 337A (mag.$_{VV}$13.5) lies 25′ E.

eg 357 dimen. 2.′6 × 1.′9 V = 11.8 sfc. br. 13.4
With 25 cm this galaxy appears elongated E-W, about 1′ × 0.′75. A stellar nucleus is visible and occasionally seems like a tiny starry streak along the major axis. 30 cm shows a high surface brightness halo containing a 20″ core and a stellar nucleus. A mag. 12 star is visible 30″ ENE. eg 357 is the brightest of ten galaxies within a 12′ circle.

eg M-01-03-85 dimen. 4.′0 × 3.′5 B = 12.2 sfc. br. 14.9
In 25 cm this galaxy appears as a small faint patch with a nearly stellar center. 30 cm shows a very diffuse object, about 3.′5 × 3′, elongated roughly E-W. The surface is unevenly bright with some faint stellarings sprinkled over it. It has an extremely low surface brightness but bears magnification well. A mag. 12 pair in pa 30° lies 8′ N; a mag. 13 star is visible 4′ to the E.

eg 428 dimen. 4.′1 × 3.′2 V = 11.4 sfc. br. 14.0
A faint object for 25 cm, this galaxy appears as a circular patch 2′ diameter. It is broadly concentrated and without a distinct core. A mag. 12.5 pair (15″; 77°) is visible 2′ S; a mag. 12 star lies 1.′8 NW. 30 cm shows a weakly concentrated object, about 3′ × 2.′5 in pa 160°. The inner part of the core is most strongly elongated and without a nucleus. Some irregularities can be seen in the outer core and the halo. The galaxy forms a 6′ triangle with mag. 8.5 stars N and W.

eg 450 dimen. 3.′2 × 2.′6 V = 12.1 sfc. br. 14.3
This object is visible as a faint, circular patch about 1.′5 diameter in 25 cm. 30 cm shows a 2′ blur N of a triangular asterism. There is no concentration, but a faint stellaring is visible in the center. The overall size is 2′ × 1.′5, elongated E-W. Several faint stellarings are associated.

eg 442 (mag.$_Z$14.5) is located 16′ SW, just past 38 Ceti (V=5.7).

eg 521 ***dimen. 3.′4 × 3.′2*** ***B=12.5*** ***sfc. br. 14.9***
In 25 cm this galaxy is a faint circular glow about 45″ diameter with a slight brightening in the core. eg 533, *q.v.*, is located 15′ E and a bit N; an extremely compact anonymous galaxy lies 2.′9 WNW.

eg 533 ***dimen. 3.′7 × 2.′6*** ***B=12.5*** ***sfc. br. 15.0***
25 cm shows this galaxy as a 1′ spot with a much brighter core. Viewed in 30 cm, the small, fairly well concentrated object appears slightly elongated in pa 45°. The core is relatively bright and contains a faint stellar nucleus.

eg 578 ***dimen. 4.′8 × 3.′2*** ***V=10.9*** ***sfc. br. 13.7***
This galaxy has a low surface brightness and is only faintly visible in 15 cm. In 25 cm it appears cigar-shaped, about 2′ × 1.′2 in pa 120°, brightening a little toward the center. 30 cm shows a large object, 3′ × 2′ with faint extensions to 4′ length. The core is broad and unconcentrated with a few mottlings: when the telescope is jiggled, some of these appear stellar. Some faint field stars lie nearby.

eg 584 ***dimen. 3.′8 × 2.′4*** ***V=10.4*** ***sfc. br. 12.8***
This is a very bright object, visible in 6 cm. In 15 cm it appears elongated NE-SW, about 1.′5 diameter. The core is brighter and contains a stellar nucleus. 25 cm shows a nearly stellar nucleus within a bright 15″ × 10″ inner oval. The smooth high surface brightness core fades rapidly at its outer edges to a faint 1.′8 × 0.′9 halo in pa 65°. 30 cm also shows a stellar nucleus and a large core. With careful viewing, the halo extends to 2.′5 × 1.′ eg 586 (V=13.2) lies 4.′3 SE.

eg 596 ***dimen. 3.′5 × 2.′2*** ***V=10.9*** ***sfc. br. 13.1***
Lying 12′ W of a 6th magnitude star, 15 cm shows a small, faint circular patch here with a stellar nucleus. 25 cm shows a 1′ diameter halo that is concentrated to a tiny core and a stellar nucleus off-center to the NE. With 30 cm a nearly stellar nucleus is visible within a bright core about 25″ across. The halo extends to 1.′2 × 0.′8 in pa 35°. Several stars lie within 1.′5 of the object.

eg 615 ***dimen. 4.′0 × 1.′7*** ***V=11.5*** ***sfc. br. 13.4***
Located 6′ ENE of a mag. 8.5 star, this galaxy is easily seen in 15 cm. The round, fairly bright halo contains a stellar nucleus, but no distinct core is visible. 25 cm shows a 1.′5 × 0.′5 halo in pa 155°. The core is also elongated and seems to have a dark gap on the S side. 30 cm shows a nearly stellar nucleus off-center to the N in a moderately concentrated 2′ × 1′ halo. The object is not as concentrated as eg 596, *cf.*

eg 636 ***dimen. 2.′3 × 1.′9*** ***V=11.3*** ***sfc. br. 12.9***
A faint object for 15 cm, this galaxy looks like a hazy star located SW of three mag. 9 stars. In 25 cm the faint 1′ halo grows gradually brighter to a small core and a nearly stellar nucleus. Viewed in 30 cm, it is 1.′5 diameter with a 15″ core and a nearly stellar nucleus. Lying 3′ E is a mag. 11.5 star.

eg 681 ***dimen. 2.′8 × 1.′8*** ***V=11.8*** ***sfc. br. 13.4***
eg 681 is barely visible in 15 cm as the northernmost "star" of a 3-shaped asterism. 25 cm shows it about 1′ in diameter with a broad core; a stellar flash is occasionally visible within. The object lies in a field of faint stars, with a mag. 12.5 star 45″ NW. 30 cm shows this star to lie just within the halo. Four mag. 13 stars to the SW form a 1.′5 diamond, and a mag. 14 star lies between this and the galaxy. The object has a 1′ core with a 1.′5 diameter halo. The core is irregularly bright and without a nucleus.

eg M-02-05-53 ***dimen. 2.′8 × 2.′2*** ***mag.$_{VV}$13*** ***sfc. br.***
This galaxy lies 22′ N of eg 681, *q.v.* It is just visible in 25 cm, with some stellarings discernable on the W side of the halo. 30 cm shows a faint object about 2′ diameter that is slightly concentrated to a faint nucleus. A mag. 13 star lies 2′ to the E.

eg 701 ***dimen. 2.′5 × 1.′3*** ***V=12.2*** ***sfc. br. 13.3***
eg I1738 ***dimen. 1.′2 × 1.′0*** ***mag.$_{VV}$14*** ***sfc. br.***
eg 701 is a moderately difficult object in 15 cm, appearing as a 2′ × 1′ streak elongated NE-SW. 25 cm shows it 1.′75 × 1′ with a very narrow core 45″ long containing a stellar nucleus. In 30 cm the core is unevenly bright and has a dark bar crossing it on the SE side. The overall size is 2′ × 0.′8 in pa 45°. A mag. 14.5 star is 3′ SSW. eg I1738 lies 5.′4 to the SSE. Viewed in 30 cm it appears as an unevenly bright glow about 50″ in diameter. A mag. 14 star lies 1.′2 NNE.

eg 720 ***dimen. 4.′4 × 2.′8*** ***V=10.2*** ***sfc. br. 12.9***
In 15 cm this object looks like a faint globular cluster. The halo is about 1.′5 in diameter and is elongated SE-NW. The knotted nucleus and bright core fade quickly to the edge. 25 cm shows a 1.′5 × 1′ object with an elongated, sharply defined 30″ × 15″ core. Viewed in 30 cm, it is 1.′5 × 1′ in pa 140°. The halo brightens gradually to a nearly stellar nucleus.

eg 779 ***dimen. 4.′1 × 1.′4*** ***V=11.0*** ***sfc. br. 12.8***
This bright object is a fine sight in 25 cm. The halo is elongated in pa 160° and extends to 2.′5 × 0.′8. There is no central bulge, and occasional details appear along the major axis. On the NW side of the core is a star or knot along the major axis. The halo extends farther SE than

NW from the core. 30 cm shows a long, thin bar with a nearly stellar nucleus in a circular core. The outer regions of the core and halo are elongated. Several faint stars are visible within 3′; a mag. 11 star is located 4′ S. Photographs show eg M-01-06-23 (mag.$_{VV}$14.5) 11′ SE.

eg I 184 dimen. 1′.5 × 0′.7 V = 13.8 sfc. br. 13.6
eg 788 dimen. 1′.8 × 1′.5 V = 12.1 sfc. br. 13.0
In a blank field, 25 cm will show eg 788 elongated in pa 95°, about 1′.5 × 1′, with a bright elongated core. In 30 cm the halo is irregularly shaped and about 1′ diameter. The core is brighter, about 30″ across, with a nearly stellar nucleus. eg I 184, 18′.5 W, is located 3′ NNW of a mag. 10 star. In 30 cm it is very faint, 30″ diameter, but has some central concentration.

eg 864 dimen. 4′.6 × 3′.5 V = 11.0 sfc. br. 13.8
This is a low surface brightness object, visible in 25 cm as a faint smudge touching a mag. 10.5 star 50″ to the ESE. It appears elongated NE-SW and without central brightening. 30 cm shows the 25″ core divided into two parts. The brightest area is 2′.2 × 1′.5, but the halo has very faint extensions to 3′.2 × 1′.8. The texture of the halo is smooth except near the center.

eg 895 dimen. 3′.6 × 2′.8 V = 11.8 sfc. br. 14.1
A large, very faint object in 25 cm, this galaxy appears about 3′ diameter with a slightly brighter core and a nearly stellar nucleus. 30 cm shows it 2′ × 1′.5, elongated E-W. The concentration is uneven with a faint stellaring quite off-center to the N side. A mag. 13.5 star is visible 2′ from the center in pa 80°.

eg 908 dimen. 5′.5 × 2′.8 V = 10.2 sfc. br. 12.9
Lying in a rich field, this object is bright in 25 cm. The halo is elongated in pa 80°, about 4′ × 1′.5, occasionally extending to 5′ length. The core is circular, and a faint star is superposed only 30″ NE of the nucleus. A beautiful object for 30 cm, the halo extends to as much as 6′ × 2′. The surface brightness is fairly low, and the object is unevenly bright. Four stars lie about 4′ N; 1′.9 E of center a very faint star is visible on the S side of the bar.

eg 936 dimen. 5′.2 × 4′.4 V = 10.1 sfc. br. 13.4
Forming the period in a "?" asterism of four mag. 8.5 stars, this object appears 1′.5 diameter in 15 cm, with a bright core half the diameter. 25 cm shows the halo elongated E-W, with a concentrated core 20″ across. The brightest part of the halo is 1′ across and extends faintly to 3′ × 2′. The center of the core seems composed of three or four tiny lobes. It appears only 1′.5 diameter in 30 cm with a large brighter core but no nucleus. eg 941, *q.v.*, lies 12′.6 E.

eg 941 dimen. 2′.8 × 2′.1 V = 12.5 sfc. br. 14.3
Lying 12′.6 E of eg 936, *q.v.*, this is a faint object for 15 cm, more difficult than eg 955, *cf.* It is an even smudge in 25 cm, about 1′ in diameter, possibly elongated NE-SW. 30 cm shows a low surface brightness, broadly concentrated circular spot about 2′ diameter without a prominent nucleus.

eg 955 dimen. 3′.0 × 0′.9 V = 12.0 sfc. br. 13.0
This object is located about 25′ W of 75 Ceti (V = 5.4). It can be seen faintly in 15 cm as a small spot 2′.5 NW of a mag. 11 star. 25 cm shows a spindle-shaped halo elongated in pa 20°, about 1′ × 0′.3. The core is very thin and has a starry texture within which a stellar nucleus is occasionally visible. 30 cm shows a 1′ diameter circular object with a moderately brighter core and a faint, nearly stellar nucleus.

eg 958 dimen. 2′.8 × 1′.1 V = 12.2 sfc. br. 13.2
Though invisible to 15 cm, 25 cm shows this galaxy as a thin spindle in pa 10°, about 2′ × 0′.3 in size. The brightest part of the core is off-center toward the SE and is much more nearly round than the halo. The surface brightness is moderately low, but the object is bright overall. 30 cm shows it 1′.5 × 0′.6. The core is unevenly bright, especially on the E side.

eg 991 dimen. 2′.7 × 2′.5 B = 12.4 sfc. br. 14.3
This object is not visible in 15 cm. 25 cm shows the galaxy faintly 1′.5 NNW of a mag. 12 star. It seems elongated approximately in pa 75°, 1′.2 × 0′.5, and is without central brightening. 30 cm shows a smooth glow about 1′.2 diameter without concentration.

eg 1022 dimen. 2′.5 × 2′.1 V = 11.4 sfc. br. 13.0
A fairly faint galaxy for 15 cm, it appears circular, about 1′ diameter. The core is much brighter, but contains no distinct nucleus. 25 cm shows a broad haze with a brighter core and a stellar nucleus. There seems to be a bar of brightness in pa 120° spanning the circular 1′.5 diameter halo. A mag. 13 star is visible about 2′ NE. 30 cm shows the object situated 6′ W of a triangle of mag. 10.5 stars. The galaxy fades from an occasionally visible stellar nucleus to a broadly concentrated core about 45″ across and a 1′.2 diameter halo.

eg 1035 dimen. 2′.2 × 0′.9 B = 13.1 sfc. br. 13.6
This galaxy is only marginally visible in 15 cm, lying about 7′.5 NNE of a mag. 9 star. 25 cm shows a nicely elongated blotchy spindle in pa 150°, about 1′.5 × 0′.3. A faint star is visible on the SE tip of the halo. 30 cm shows it 2′.5 × 0′.8. The star SE is about 45″ from the center and is within the nebula. The bar-like object is unevenly

bright: the faintest part lies near the star; a brighter core-like area lies in the N and middle parts of the bar.

eg 1042 ***dimen. 4′.7 × 3′.9*** ***V = 10.8*** ***sfc. br. 14.0***
This is a low surface brightness object that is not visible in 15 cm. In 25 cm the fat, unconcentrated oval is elongated E-W, about 2′ × 1′.75. A mag. 12 star is visible 2′.5 N. 30 cm shows only an unconcentrated blob about 50″ diameter with a 30″ core.

eg 1052 ***dimen. 2′.9 × 2′.0*** ***V = 10.5*** ***sfc. br. 12.4***
15 cm shows this galaxy as a small circular spot with a sharply brighter stellar nucleus. 25 cm shows it 1′ diameter. A stellar nucleus is just discernable in the center of the bright, distinctly bounded core. A mag. 12.5 star lies 1′.2 SW. In 30 cm the halo extends to about 1′.2 diameter. The 25″ core is distinct here also, and the stellar nucleus is still just visible within it. This active radio galaxy is the brightest of a group that includes eg 1035 and eg 1042, *q.v.*

eg 1055 ***dimen. 7′.6 × 3′.0*** ***V = 10.6*** ***sfc. br. 13.8***
In 6 cm this galaxy is a moderately faint circular spot that is a little larger than nearby Messier 77 (*cf.* eg 1068). In 15 cm it is a 4′ × 1′ spindle, elongated almost E-W. 25 cm shows it a little larger. There appear to be brighter globs in the halo, and the core is only a little brighter. 30 cm shows a 1′.5 triangle of stars to the NW pointing at the object; the closest star, mag. 11, is almost touching. The halo is about 4′ × 1′.5 in pa 100° with a broadly brighter circular core 1′.2 across but no nuclear condensation. A bright spot appears on the NE side of the core. The surrounding field is rich in mag. 14 and fainter stars.

eg 1068 ***dimen. 6′.9 × 5′.9*** ***V = 8.8*** ***sfc. br. 12.7***
Messier 77 is visible in 6 cm as a very small, high surface brightness spot 1′.5 WNW of a mag. 10 star. At first appearance with 15 cm there seem to be two stars: the galaxy has a bright core and a faint, wispy halo about 1′.5 × 1′ in size. 25 cm shows an intense core about 30″ across inside a bright 3′ halo. 30 cm shows it 2′ × 1′.8 in pa 45°. There is a bright, nearly stellar nucleus inside a slightly elongated 45″ core. The halo seems to be more extensive to the NW, away from the star, giving the object a comet-like appearance. This galaxy is the prototype of the Seyfert class of galaxies.

eg 1073 ***dimen. 4′.9 × 4′.6*** ***V = 11.0*** ***sfc. br. 14.3***
This very low surface brightness object is not visible in 25 cm. In 30 cm it is situated 7′ NE of a 5′ triangle of mag. 9–10 stars. Circular, about 3′.5 diameter, the halo has a slightly brighter core 1′.5 across. Some stellarings are visible over the surface.

eg 1087 ***dimen. 3′.5 × 2′.3*** ***V = 11.0*** ***sfc. br. 13.1***
Forming a triangle with mag. 11 stars to the SE and NE, this galaxy is easily viewed in 15 cm. The slightly elongated granular glow is about 2′ × 1′ in pa 30°. With 25 cm it is moderately bright and sightly elongated. The halo is irregularly bounded, about 2′ diameter, and has a patchy texture. In 30 cm the elongation is indistinct. The middle parts are slightly mottled, and the halo is about 2′ diameter.

eg 1090 ***dimen. 3′.8 × 1′.8*** ***V = 11.9*** ***sfc. br. 13.8***
This object is located 3′ N of a mag. 10 star. 15 cm shows it fainter than and about three-fourths the size of eg 1087, *cf.* In 25 cm it appears as a circular patch 1′ in diameter. 30 cm shows it 1′.5 across with a slightly brighter core, but no nucleus. A mag. 14 star is visible 45″ S.

Columba

This southern constellation is fully visible south of 45° North latitude. It contains some nice galaxies and a globular cluster. The center of the constellation culminates at midnight on 19 December.

eg 1792 ***dimen. 4′.0 × 2′.1*** ***V = 10.2*** ***sfc. br. 12.3***
Lying in a rich field, this is a large and bright object for 25 cm. The halo is elongated SE-NW, 2′.5 × 1′, with a smooth outline and glittering spots along the major axis.

eg 1800 ***dimen. 1′.6 × 0′.9*** ***V = 12.6*** ***sfc. br. 12.9***
This peculiar object is small and faint in 30 cm, yet appears fairly well defined at low power. The broadly condensed patch is 50″ × 30″, elongated in pa 110°. A star is suspected on the W edge, and a mag. 13 star lies 1′.2 ENE.

eg 1808 ***dimen. 7′.2 × 4′.1*** ***V = 9.9*** ***sfc. br. 13.4***
In 25 cm this galaxy is elongated in nearly the same position angle as eg 1792, *cf.*, but it is narrower and longer. The faint outer halo extends to 3′.5 × 1′, encompassing a small core with a moderately bright nucleus. In 30 cm the low surface brightness halo is 5′ × 1′.5, elongated in pa 145°. A faint stellar nucleus is visible within the unevenly bright core. A mag. 12.5 star lies off the SW flank.

gc 1851 ***diam. 11′*** ***V = 7.2*** ***class II***
This cluster appears about 4′ diameter in 15 cm. It has a granular texture, and a few stars can be resolved. 25 cm reveals a sharply brighter core about 1′ across that fades suddenly to widely spaced outliers. The overall diameter is 5′, though the outline is somewhat flattened on the SE, producing a rounded triangular appearance.

eg 2090 ***dimen. 4′.5 × 2′.3*** ***B = 12.0*** ***sfc. br. 14.4***
In 25 cm this galaxy is elongated in pa 15°, about 2′.5 × 1′ in extent, with a moderately brighter core. Two mag. 12.5 stars are visible on the N side; a mag. 14 star lies W of the center.

Coma Berenices

To the naked eye this region appears as a triangular grouping of faint stars; with the telescope myriads of galaxies appear. The globular cluster Messier 53 is not to be neglected. The center of the constellation culminates at midnight on 3 April.

eg 4015A   dimen. 1'.4 × 1'.1   mag.$_z$14.2   sfc. br.
This is a moderately bright galaxy for 30 cm, lying 9′ SE of eg 4005, *q.v.*, in Leo. The well-concentrated halo is 30″ × 25″, slightly elongated NE-SW, and shows a small core and stellar nucleus. Photographs show a very faint, irregular companion (eg 4015B) attached on the NE side. eg 4021 (mag.$_z$15.3), a compact elliptical, lies 5'.3 NE, while eg 4023, *q.v.*, is 5'.8 SE.

eg 4022   dimen. 1'.6 × 1'.5   mag.$_z$14.4   sfc. br.
In 30 cm this is a small object located 12′ NE of eg 4005, *q.v.*, in Leo. At 225 × the galaxy is quite concentrated to a stellar nucleus.

eg 4023   dimen. 1'.2 × 0'.8   mag.$_z$14.6   sfc. br.
30 cm shows this galaxy faintly about 6′ SE of eg 4015, *q.v.* The diffuse halo seems elongated approximately N-S and is poorly concentrated.

eg 4032   dimen. 2'.1 × 2'.0   B = 12.8   sfc. br. 14.2
While it is not visible to 15 cm, 25 cm reveals this galaxy as a small and fairly faint object. It appears elongated in pa 45°, 50″ × 20″, with a stellar nucleus. The halo grows to an irregularly round 2'.5 diameter in 30 cm. There is a broad, moderate concentration to an elongated 1′ core, but no nucleus is evident. The halo appears to extend more to the N toward a mag. 12.5 star that lies about 3′ distant.

eg 4064   dimen. 4'.5 × 1'.9   V = 11.4   sfc. br. 13.6
This object is barely discernable in 15 cm. 25 cm reveals a 2′ × 1′ patch in pa 150°. The tips taper sharply, and the central region seems confused as though there were several bright and dark arcs curving to the center from without. The core is broadly brighter and without a stellar nucleus. A nice object for 30 cm, it grows to 3′ × 1'.25 in size. A bar mottled with bright spots runs along the major axis on either side of the round bulging core. The halo is more extensive on the W side of the core. Two mag. 13.5 stars lie 2′ E and 1'.25 SSW.

eg 4136   dimen. 4'.1 × 3'.9   B = 11.6   sfc. br. 14.4
15 cm will show this object faintly. In 25 cm the halo extends to 2'.5 × 1'.5 in pa 90° with little central concentration. 30 cm shows a large faint object about 3'.5 across. At 225 × the surface has several small faint patches on it, particularly to the SE of center. Though there is a slight central brightening, no nucleus is visible. The halo occasionally takes on a slight E-W elongation.

gc 4147   diam. 4'.0   V = 10.2   class IX
Visible at low power in 15 cm, this cluster appears as an ill-defined glow about 1′ diameter. It is about 1'.5 across in 25 cm with ragged edges and a stellar nucleus. At 250 × some resolution is indicated with an arm of stars extending N out of the cluster. Better resolution is obtained with 30 cm. A few stars stand out, particularly one on the SW edge of the core. The nucleus is bright and almost stellar. The overall size is 2'.1 × 1'.9, elongated NE-SW.

eg 4150   dimen. 2'.5 × 1'.8   V = 11.7   sfc. br. 13.1
In 15 cm this galaxy shows as a small round spot 6′ E of a mag. 9.5 star. 25 cm reveals a 1′ × 0'.75 object elongated in pa 140°. The nebula has a smooth texture, a small core, and a faint stellar nucleus. It grows to 2'.5 × 1'.5 in 30 cm with a prominent stellar nucleus but weak concentration otherwise. The core fades evenly to the halo and seems to have slight brightēnings along it for about 1'.5 length.

eg 4152   dimen. 2'.3 × 1'.9   V = 12.0   sfc. br. 13.5
Lying in an empty field, this galaxy is a moderately faint and unconcentrated circular spot in 15 cm. 25 cm reveals a round glow about 1′ across. The halo is faint, and the core is only slightly brighter. In 30 cm it is 50″ × 40″ in pa 170°. The halo shows moderate, even concentration to an ill-defined circular core about 30″ across, and a faint stellar nucleus.

eg 4158   dimen. 2'.0 × 1'.8   B = 12.8   sfc. br. 14.1
15 cm will just show this object 1'.6 NW of a mag. 10.5 star. It is 1′ × 0'.5 in 25 cm, elongated roughly E-W, and is broadly brighter across the center. 30 cm reveals a small core within a mottled inner halo. The overall diameter is about 45″, with good concentration toward the center.

eg 4162 dimen. 2.′5 × 1.′6 V = 11.5 sfc. br. 12.8
This object is barely visible with 15 cm, lying about 2.′5 ENE of a mag. 10.5 star. 25 cm shows it elongated in pa 170°, 1.′25 × 0.′75. The core is slightly brighter and contains a stellar nucleus. It grows to 2′ × 1′ in 30 cm with a rounder bulging core that is not much brighter than the halo. Averted vision reveals a star superposed on the core 12″ N of center. A mag. 12.5 star lies 2′ ENE; a wide mag. 12.5 pair (35″;30°) lies 7′ S.

eg I3044 dimen. 2.′1 × 1.′0 B = 14.0 sfc. br. 14.6
This is an exceedingly difficult object for 30 cm. The very low surface brightness spot is 1.′2 × 0.′5, elongated ENE-WSW. Further details are not certainly visible.

eg 4186 dimen. 1.′4 × 1.′1 B = 14.4 sfc. br. 14.7
Viewed in 30 cm, this galaxy is small and has a conspicuous stellar nucleus. It is elongated NE-SW, 1′ × 0.′5; the halo is faint and without central brightening except for the nucleus. This object lies 11′ SSE of Messier 98 (eg 4192, *q.v.*).

eg 4189 dimen. 2.′5 × 2.′1 V = 11.7 sfc. br. 13.4
In 15 cm this galaxy appears a bit fainter than eg 4168, *cf.* in Virgo. The halo is elongated nearly E-W and is broadly concentrated across the center. A mag. 12.5 star is visible 2.′3 ENE of center. With 25 cm the halo is 1.′5 × 0.′5 and becomes a little brighter toward the center. It grows to 2′ × 1.′5 in 30 cm, the small circular core barely discernable west of the center. A mag. 14 star lies 1.′0 NE, just within the strongly mottled halo.

eg 4192 dimen. 9.′5 × 3.′2 V = 10.1 sfc. br. 13.6
With care, Messier 98 is a fine sight even with 6 cm. It appears about 5′ × 2′ in pa 150° with very weak and irregular concentration to a mottled middle, but no nucleus. In 15 cm the halo extends to 6′ × 2′ and contains a broad, irregularly bright core and stellar nucleus. The galaxy is bright and much elongated in 25 cm. The nucleus is 30″ across and fades to a broad, somewhat patchy haze. With 30 cm it is elongated in pa 150°, 6.′5 × 1.′5, rather fainter and less extensive to the SE, and not well concentrated to the center. The E and W sides of the halo do not bulge as they pass the core: the overall outline is cigar-shaped, and the NE flank is more sharply defined. The slightly mottled core is circular, about 50″ across, with a nearly stellar nucleus that appears granular.

eg I3061 dimen. 2.′3 × 0.′6 V = 13.6 sfc. br. 13.7
This is a very faint spindle in 30 cm, but it is hard to discern many details. The halo is about 55″ × 20″, elongated in pa 120°, and shows a broad, but spotty concentration. A mag. 13.5 pair of stars (22″; 260°) is resolved 3′ to the E.

eg 4203 dimen. 3.′6 × 3.′3 V = 10.7 sfc. br. 13.2
This galaxy is visible in 6 cm as a faint patch with an indistinct central pip. With 15 cm it is about 1′ across with a stellar nucleus. 25 cm shows it 1.′25 across. The faint halo rises abruptly to a very bright sharp core. A stellar nucleus is suspected. In 30 cm the halo is slightly elongated SE-NW. The broadly concentrated core seems to have many sparkles in its central half, though none of them stand out.

eg 4212 dimen. 3.′0 × 2.′1 V = 11.7 sfc. br. 13.0
In 15 cm this galaxy has a moderately high surface brightness, but the oval halo is only broadly concentrated toward its center. A mag. 11.5 star lies 2.′3 S. 25 cm shows a fairly bright 2′ × 1′ halo elongated in pa 70°. No bright nucleus is visible, and the core fades promptly at the edges. 30 cm reveals a distinctly mottled halo up to 2.′5 × 1.′5 in extent; its southern flank has a flatter outline than the domed northern perimeter. The middle parts grow broadly but irregularly brighter toward the faint stellar nucleus; the ill-defined core seems composed of three or four very faint stellarings on the major axis near the center. eg I3061, *q.v.*, lies 13′ NW.

eg 4222 dimen. 3.′3 × 0.′6 B = 13.6 sfc. br. 14.4
Not visible in 15 cm, this galaxy is the faintest of three spindle-shaped objects in a 30′ field including eg 4206 and eg 4216, both in Virgo, *q.v.* In 30 cm the spotty, broadly concentrated halo is 1.′5 × 0.′2 in pa 55°, and without a distinct core or nucleus. A mag. 14 star is visible at 225 × just S of the NE tip, 1.′5 from the center.

eg 4237 dimen. 2.′3 × 1.′6 V = 11.7 sfc. br. 12.8
15 cm shows this galaxy as a small, concentrated spot slightly elongated ESE-WNW. In 25 cm it is an evenly bright circular patch with well-defined edges. Lying in a nearly empty 15′ field, 30 cm shows the oval halo up to 1.′1 × 0.′8 in pa 110°. The core is unevenly bright, and a very weak stellar nucleus is occasionally visible in its center.

eg 4239 dimen. 1.′9 × 1.′2 B = 13.4 sfc. br. 14.1
Lying in the NW side of a 5′ triangle of mag. 10 stars, this galaxy is a moderately faint unconcentrated spot in 15 cm. With 30 cm the halo is of moderate surface brightness, reaching only to 40″ × 30″ in pa 120°. The core consists of a slightly brighter line along the major axis, and contains a faint stellar nucleus.

eg 4245 ***dimen. 3.′3 × 2.′6*** ***V = 11.4*** ***sfc. br. 13.5***
This galaxy is visible only as a faint, dull spot in 15 cm. In 25 cm it appears at the NE end of a chain of four mag. 11 stars. The circular halo is 1′ diameter and contains a small bright core. A mottled texture is evident in 30 cm. The overall size is 1.′5 × 1′ in pa 145° with an occasional stellaring for the nucleus. eg 4253 (mag.$_z$13.7) lies 16′ NE.

eg 4251 ***dimen. 4.′2 × 1.′9*** ***B = 11.6*** ***sfc. br. 13.7***
15 cm reveals this galaxy as a bright oval glow. It is 1.′5 × 1′ in 25 cm with a brighter core and stellar nucleus. It is elongated in pa 100° in 30 cm, the halo extending to 1.′5 × 0.′6 with good concentration toward the center. The circular core is 25″ across and contains a nearly stellar nucleus.

eg 4254 ***dimen. 5.′4 × 4.′8*** ***V = 9.8*** ***sfc. br. 13.2***
Messier 99 is fairly easy to see with 6 cm. The broad 3′ × 2′ smudge is elongated E-W and shows some concentration. The galaxy looks granular in 15 cm and grows gradually brighter to a broad diffuse core. It is up to 4′ diameter in 25 cm; the brighter core, looking like an unresolved globular cluster, is about 1′ across. An arm of the galaxy extends S from the E side. A conspicuous mag. 12.5 star appears on the SE edge of the halo, 2′ from the center. In 30 cm the bright arm is clearly separated from the main body of the galaxy. It wraps close by the core on the east and south, where two brighter patches are visible, and ends directly W of the nucleus with another spot. At 225 × the southern patch is distinctly elongated. A second very short arm protrudes from the NW side of the core. The bright oval center is about 50″ across, elongated in pa 75°, and evenly concentrated to a substellar nucleus.

eg 4262 ***dimen. 2.′2 × 2.′0*** ***V = 11.5*** ***sfc. br. 13.0***
Only the nucleus of this galaxy is visible in 15 cm, where it appears like a mag. 12 star at 75 ×. 25 cm reveals a faint halo to perhaps 1′ diameter around the stellar nucleus. In 30 cm the substellar center is about 8″ across, a remarkably strong, sharp concentration dominating the very faint and otherwise featureless 45″ halo.

eg 4274 ***dimen. 6.′9 × 2.′8*** ***V = 10.3*** ***sfc. br. 13.4***
This large object can be seen in 15 cm as a broad, even glow of fairly low surface brightness. The halo is 3.′5 × 1′ with a gradual brightening toward the center. It grows to 4.′5 × 1′ in 25 cm, elongated E-W. The broad core contains a stellar nucleus. 30 cm shows it 6′ × 1.′5 in pa 100°. There is moderate central brightening and a faint nuclear condensation. The S flank fades more abruptly from the center. Many faint stars are associated with the nebula. eg 4278 and eg 4283, *q.v.*, lie 22′ S.

eg 4278 ***dimen. 3.′6 × 3.′5*** ***V = 10.2*** ***sfc. br. 13.0***
eg 4283 ***dimen. 1.′4 × 1.′4*** ***V = 12.0*** ***sfc. br. 12.8***
Both of these galaxies are visible in 15 cm, lying 3.′5 apart in pa 60°. The brighter one (−78) appears about 1′ diameter with a very bright center. The companion (−83) seems like a star with a small surrounding haze. In 25 cm, −78 is about 1.′2 diameter with some central concentration; −83 is fairly faint, no more than 30″ across, and has a stellar nucleus. In 30 cm, −78 is fairly bright, circular, and well concentrated. It is 1.′1 diameter with a

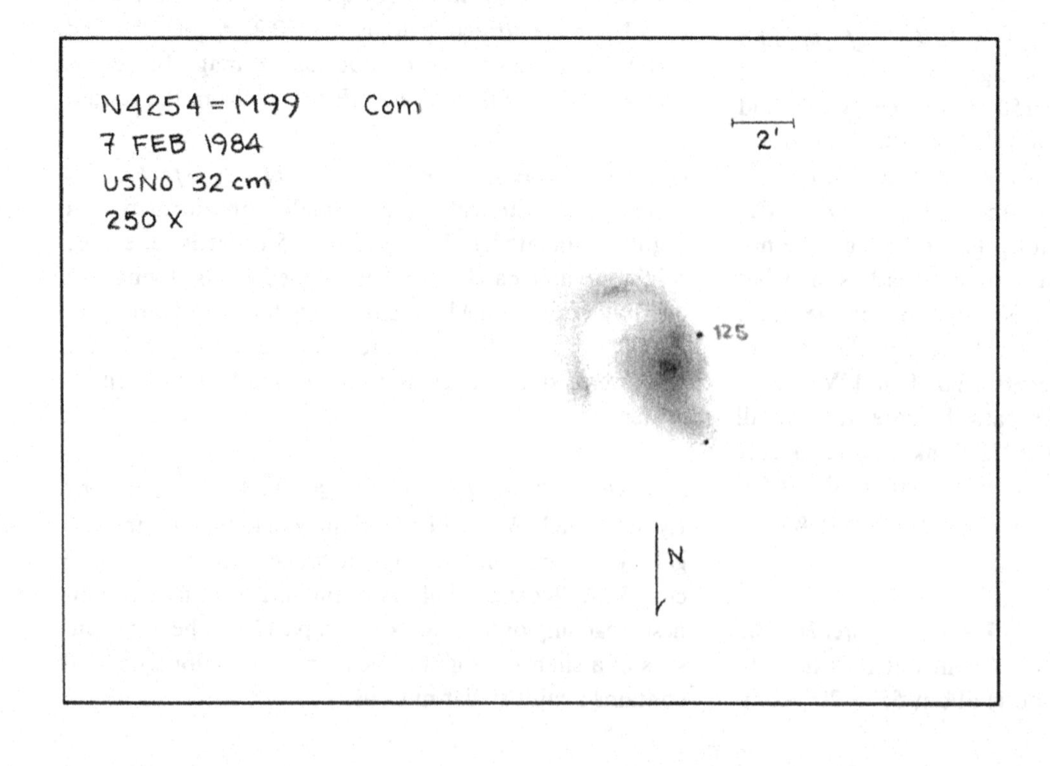

20″ core and a stellar nucleus. eg 4283 is not so well concentrated as −78. It appears 45″ across with a faint stellar nucleus occasionally visible. eg 4286 (mag.$_z$14.7) lies 8.′5 NE of −78.

eg 4293 dimen. 6.′0 × 3.′0 B = 11.2 sfc. br. 14.2
This is a large object for 15 cm, extending to as much as 3.′5 × 1′ in pa 70° with a slight central brightening. Two mag. 12.5 stars lie off the NE flank. There is no distinct core visible in 25 cm, only a slight broad brightening across the center. 30 cm shows a poorly concentrated 4′ × 1′ halo, a fat lens with somewhat rounded ends. The small oval core has a distinct dark patch on its E side and contains a very faint stellar nucleus. A mag. 13.5 star lies just S of the E tip, 2.′7 from the center; another slightly fainter one is visible 1.′5 SW, just beyond the edge of the halo.

eg 4298 dimen. 3.′2 × 1.′9 V = 11.4 sfc. br. 13.2
eg 4302 dimen. 5.′2 × 1.′1 V = 11.6 sfc. br. 13.4
15 cm shows eg 4298 as a broadly concentrated oval haze elongated SE-NW. A mag. 13 star is just visible on the

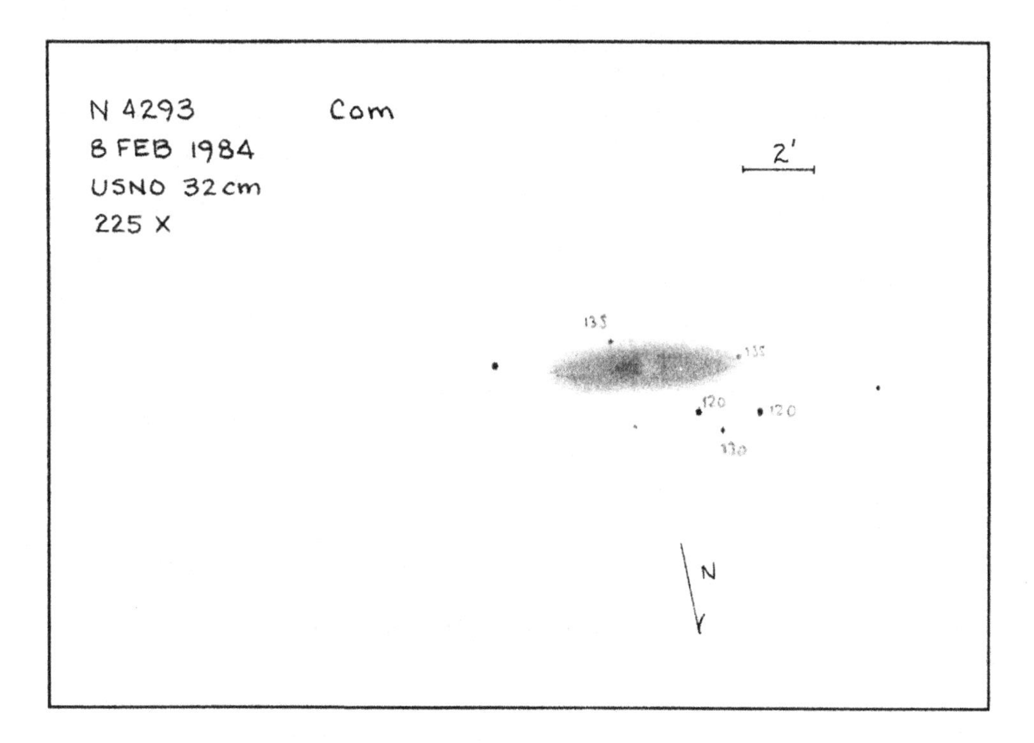

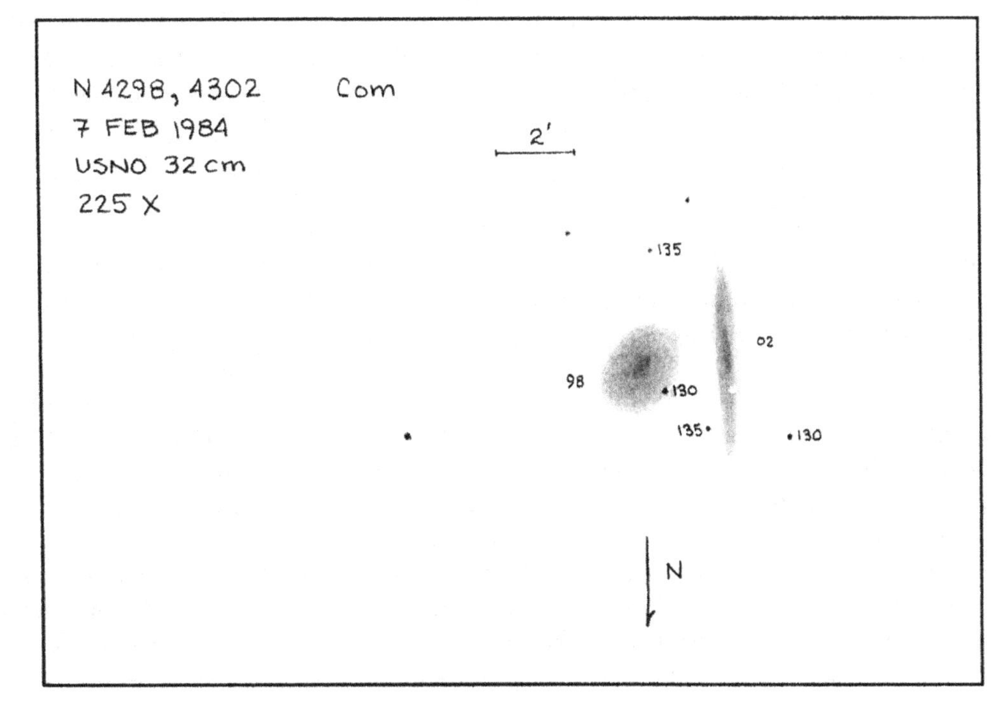

NE edge, and a conspicuous pair (11,11; 25″; 100°) is resolved about 17′ SW of the galaxy. eg 4302, 2′.4 SE, is pretty faint, but quite clearly a narrow spindle elongated in pa 0°. In 25 cm, −98 is a bright, diffuse object containing several stellarings, but there is no marked central concentration. eg 4302 seems just separated from −98 at low power, appearing as a very thin 2′.5 × 0′.5 glow without central brightening. 30 cm shows −98 as a fat 1′.5 × 1′ oval elongated in pa 140°. The central regions are slightly and broadly brighter, with a very faint nonstellar nucleus visible at times. eg 4302 is as long as 5′ × 0′.4, almost reaching a mag. 13.5 star 3′ SSW, and extending about 30″ past a star of similar brightness 2′ from the center on the NW flank. The halo is very weakly and irregularly concentrated: occasionally the ends of the spindle seem brighter than the middle.

eg I 783 dimen. 1′.5 × 1′.4 V = 13.5 sfc. br. 14.2
This galaxy lies 18′ SE of Messier 100 (eg 4321, *q.v.*). In 30 cm it is a very faint patch 45″ × 35″, elongated approximately E-W, with a very slight central brightening.

eg 4312 dimen. 4′.7 × 1′.3 V = 11.8 sfc. br. 13.6
A distant companion to Messier 100 (eg 4321, *q.v.* 18′ NNE), this galaxy is faintly visible in 15 cm. The low surface brightness lens is elongated in pa 170° and shows no central concentration. A mag. 13 star is visible 2′.5 SE, which larger apertures show to be a nearly equal pair (23″; 140°). With 30 cm the halo extends to 2′ × 0′.5 and shows a broad but very weak brightening across the center.

eg 4314 dimen. 4′.8 × 4′.3 V = 10.5 sfc. br. 13.7
This is a barred spiral with very faint, diffuse arms. 15 cm shows it 2′ × 0′.5 in pa 150° with a stellar nucleus. It is 2′.5 × 1′ in 25 cm with a bright oval 40″ × 30″ core. A very bright stellar nucleus is inconspicuously seen within the core. A star is visible off the NW tip. A fine view is obtained with 30 cm. The thin lens-shaped object is 3′ × 0′.9 with blunt ends and a circular inner core 30″ across. In good conditions a faint halo can be seen around the object: it is also elongated, but about 1′.3 wide at the middle. An additional star is visible on the E edge of the SE tip about 45″ from the center.

eg 4321 dimen. 6′.9 × 6′.2 V = 9.4 sfc. br. 13.3
Messier 100 is a bright galaxy with many fainter companions. It is visible in 6 cm as a smoothly textured glow with weak concentration to a faint nonstellar nucleus. The halo appears about 4′ × 3′ overall, elongated ESE-WNW. It is about 5′ across in 15 cm, the uniformly bright halo containing a very small and much brighter core. 25 cm shows a 5′ × 4′ halo with a curdled surface and bright core. In 30 cm the halo extends to 5′ × 4′ in pa 110°. The small bright core is 20″ across and elongated in pa 60° with little wings on the NE and SW. No nucleus is evident. Several bright and dark spots plus a few stars are scattered over the surface of the galaxy; the only conspicuous patch lies near the edge on the NW side.

eg 4322 dimen. 1′.3 × 1′.1 V = 13.9 sfc. br. 14.1
This companion 5′.2 NNE of Messier 100 (eg 4321, *q.v.*) is more difficult to see than nearby eg 4328, *cf.*, in 30 cm. At 225 × the small galaxy is only 30″ diameter, a diffuse and poorly concentrated spot.

eg 4328 dimen. 1′.5 × 1′.4 V = 13.5 sfc. br. 14.1
Lying 6′ directly E of Messier 100 (eg 4321, *q.v.*), this galaxy is a difficult diffuse spot of low surface brightness for 30 cm. The halo is perhaps 1′ across with a weak, broad concentration toward center.

eg 4340 dimen. 4′.1 × 3′.2 V = 11.0 sfc. br. 13.7
eg 4350 dimen. 3′.2 × 1′.1 V = 11.0 sfc. br. 12.3
These galaxies lie 5′.6 apart. eg 4350 is smaller and more sharply concentrated in 15 cm, and so seems the brighter of the two. The halo is elongated NNE-SSW and contains a stellar nucleus. eg 4340 has nearly the same pa, but is less definitely elongated. 25 cm shows −50 as a fairly bright silvery oval extending to about 2′ × 1′. A stellar nucleus is visible inside a bright core. The companion is more nearly circular and has a faint stellar nucleus. The impression that −50 is the brighter galaxy persists in 30 cm due to its higher surface brightness. Its total size is about 2′ × 0′.6 in pa 30°. The core is also elongated and strongly concentrated at its center toward a stellar nucleus. eg 4340 shows a 1′.6 × 1′.2 halo elongated in a pa slightly larger than its companion. The galaxy is unevenly bright and weakly concentrated except for a conspicuous 10″ core, which nearly hides an embedded stellar nucleus.

eg 4344 dimen. 1′.9 × 1′.8 B = 13.2 sfc. br. 14.4
This galaxy is visible in 15 cm as a faint unconcentrated spot 2′ NW of a mag. 12.5 star. With 30 cm this star is the brightest of three that form a 2′ box with the galaxy as the western corner. At 225 × the galaxy is 45″ diameter, a moderately low surface brightness spot with a moderate, broad concentration across the center. The central area is slightly elongated E-W.

eg 4375 dimen. 1′.6 × 1′.4 mag.$_z$13.9 sfc. br.
In 25 cm at high power this galaxy is pretty faint. The 30″ diameter circular glow is only a little concentrated to the center. A mag. 14 star lies 1′ NE.

oc Melotte 111 ***diam. 275′*** ***V=1.8*** ***stars 273***
The Coma star cluster is quite nearby and visible to the naked eye on clear nights; in city skies it takes on a fuzzy appearance. In dark sky it is a striking assemblage of perhaps two dozen stars. Wide-field binoculars give the best views of this large object.

eg 4377 ***dimen. 1′.8 × 1′.5*** ***V=11.8*** ***sfc. br. 12.7***
15 cm shows this object as a tiny concentrated round spot with a stellar nucleus. In 25 cm the overall diameter is 40″. 30 cm shows a fairly bright irregular patch 45″ across. There is little concentration except for the bright nearly stellar nucleus. eg I3327 (B=15.0) lies 14′ ENE.

eg 4379 ***dimen. 2′.1 × 1′.8*** ***V=11.5*** ***sfc. br. 12.8***
At 75× in 15 cm this galaxy appears as a small, concentrated spot with a stellar nucleus. Higher powers show the halo to be elongated slightly E-W. 25 cm will show a small, nearly circular haze 45″ diameter with a relatively large, bright core. It is a small object even for 30 cm, though its surface brightness is moderately high. The halo is 50″ × 40″ in pa 105°, brightening smoothly to a small 8″ core and stellar nucleus.

eg 4382 ***dimen. 7′.1 × 5′.2*** ***V=9.2*** ***sfc. br. 13.0***
eg I3292 ***dimen.*** ***B=15.0*** ***sfc. br.***
Messier 85 appears similar in all apertures, though the apparent brightness varies. One of the brightest Virgo cluster galaxies, it is not difficult to see in 6 cm about 5′ NW of a mag. 10 star. 15 cm shows it about 2′ across, growing gradually brighter toward the center. A mag. 12.5 star is visible embedded near the N edge of the halo. 25 cm shows the star only 50″ from the nucleus, clearly within the halo. In 30 cm the galaxy is very well concentrated, appearing like an unresolved globular cluster. The very faintest part of the halo extends to 5′–6′ length, elongated nearly N-S. The more well defined region is about 2′ × 1′.75 (the mag. 12 star lies just within this), growing much brighter to a small, intense 15″ core and substellar nucleus, which is nearly swamped by the core. A faint companion, eg I3292, is visible in 30 cm as a very faint 20″ spot of moderate concentration 8′.5 W of M85. A bright companion, eg 4394, *q.v.*, lies 7′.8 E.

eg 4383 ***dimen. 2′.2 × 1′.2*** ***B=13.0*** ***sfc. br. 13.9***
In 15 cm this galaxy appears as a substellar spot of the same brightness as a mag. 12 star 1′.7 SW. A stellar center is visible at high power. It is a moderately bright object for 25 cm, a round dot less than 1′ diameter with a stellar nucleus. 30 cm shows a well-concentrated halo only 45″ × 30″ in size, elongated in pa 30°. The 15″ core is approximately circular and contains a bright substellar nucleus.

eg 4394 ***dimen. 3′.9 × 3′.5*** ***V=10.9*** ***sfc. br. 13.6***
This galaxy looks about the same size as nearby Messier 85 (eg 4382, *cf.*) in 15 cm; though much fainter, it has a sharp stellar nucleus. The halo is lens-shaped, clearly elongated SE-NW. A mag. 12.5 star lies 3′ S. The galaxy is fairly bright in 30 cm, the halo extending to 2′ × 0′.8 in pa 140°. A faint stellar nucleus resides within the small circular core; other fainter spots appear along the bar.

eg 4396 ***dimen. 3′.5 × 1′.2*** ***B=13.0*** ***sfc. br. 14.4***
This galaxy is visible in 15 cm at 75× as a faint spindle 1′.5 S of a mag. 12 star. The unconcentrated halo is in pa 125° and seems a bit larger than eg 4379, *cf.* 9′ WSW. With 30 cm the galaxy is about 2′.5 × 0′.4 and has a mag. 13.5 star superposed 1′ from the center toward the W end. The light is barely concentrated; the core is hardly discernable, but seems nearly circular, and about 15″ across.

eg 4405 ***dimen. 2′.0 × 1′.4*** ***B=13.0*** ***sfc. br. 14.0***
A fairly faint object for 15 cm, at 75× this galaxy appears as a small spot without much concentration. With 30 cm the moderate surface brightness oval is 50″ × 40″ in extent, elongated in pa 20°. There is an ill-defined core, but the galaxy remains poorly concentrated as in 15 cm.

eg 4414 ***dimen. 3′.6 × 2′.2*** ***V=10.3*** ***sfc. br. 12.3***
This galaxy is easy to see in 6 cm. The tapered halo is clearly elongated in pa 155°. A faint prominent nucleus stands out in the otherwise only moderately concentrated bar. 15 cm shows it 1′.75 × 1′ with a sharp stellar nucleus, but no distinct core. While the halo is elongated in 25 cm, extending 2′ × 1′, the 10″ core appears circular and contains a stellar nucleus. There seem to be dark patches on the halo. It grows to 3′.2 × 1′.5 in 30 cm. Lying 50″ NW of the stellar nucleus is a bright spot; another stellaring is visible 10″ S of the nucleus. A dark lane cuts SE of the nucleus, coming in from the SE. eg 4359 (mag.$_z$13.9) lies 35′ WNW.

eg 4419 ***dimen. 3′.4 × 1′.3*** ***V=11.0*** ***sfc. br. 12.4***
This is a moderately bright galaxy for 15 cm. The sharply tipped halo is elongated SE-NW and surrounds an irregular core that seems double. A mag. 12.5 star is visible 2′ S. 25 cm shows it 2′ × 0′.5, elongated in pa 135°. The detailed core has several bright spots near the center and occasionally appears double. In 30 cm the halo is 2′.5 × 0′.6, with very smooth, even concentration. At high power the core is distinctly elongated and seems double because of a faint star or stellaring on its NW edge on the major axis.

eg 4421 ***dimen. 2′.7 × 2′.2*** ***V = 11.6*** ***sfc. br. 13.3***
Lying 2′.4 ESE of a mag. 9 star, 15 cm shows this galaxy as a circular haze about the same brightness as eg 4379, *cf.* The halo is evenly concentrated to a stellar nucleus. 30 cm shows a 1′.2 × 0′.6 oval in pa 150°. The core is very small and much brighter, exhibiting a faint stellar nucleus.

eg 4448 ***dimen. 4′.0 × 1′.6*** ***V = 11.1*** ***sfc. br. 13.0***
This galaxy appears fairly bright in 15 cm, showing a bright stellar nucleus in an elongated halo. In 25 cm the halo is 2′ × 0′.75 in pa 90° with a lenticular outline. The core is oval, but the brightest part is centered in its W side. The halo, however, extends farther and more narrowly on the E side. 30 cm shows the halo 2′.2 × 1′ in pa 90°, but the core seems elongated ESE-WNW. The nucleus is bright and distinct but nonstellar.

eg 4450 ***dimen. 4′.8 × 3′.5*** ***V = 10.1*** ***sfc. br. 13.0***
In 15 cm this large galaxy is an elongated 2′.5 glow, brightening gradually to the center except for a conspicuous stellar nucleus. A mag. 9 star lies 4′ SW. At high power, 25 cm shows a small, bright core. It is a large and pretty bright galaxy for 30 cm, reaching roughly halfway to a mag. 12.5 star 3′.8 S. The broadly concentrated, lumpy core contains a stellar nucleus slightly off-center to the E side of the high surface brightness nebula. Dark lanes are suspected on the E side. The overall size is 4′ × 1′.25 in pa 0°. eg 4489, *q.v.*, lies 48′ ESE.

eg 13392 ***dimen. 2′.3 × 1′.1*** ***B = 13.3*** ***sfc. br. 14.2***
This galaxy is visible in 15 cm as a moderately faint circular spot about 1′ diameter. Two mag. 11–12 stars lie about 3′ SW. 30 cm shows a fat lens of moderate surface brightness at 225×. The 1′.8 × 0′.8 halo is elongated in pa 40°, growing broadly brighter across the more nearly circular inner regions.

eg 4455 ***dimen. 2′.8 × 1′.0*** ***B = 12.9*** ***sfc. br. 14.0***
30 cm shows this galaxy 4′ S of a wide mag. 12 pair of stars with 1′.5 separation in pa 170°. The halo is elongated in pa 15°, 2′.25 × 1′. The only concentration is expressed by a string of faint spots along the major axis.

eg 4459 ***dimen. 3′.8 × 2′.8*** ***V = 10.4*** ***sfc. br. 12.9***
eg 4468 ***dimen. 1′.5 × 1′.1*** ***V = 13.0*** ***sfc. br. 13.5***
15 cm will show eg 4459 as a round 1′ patch with a brighter center. In 25 cm it is bright, about 1′.1 across, with an even, moderate concentration. The overall outline is approximately circular, while the inner parts seem slightly elongated E-W. The core does not stand out, but there is a stellar nucleus. eg 4468, which lies 8′.5 ENE, can be seen in 25 cm as a faint, unconcentrated circular spot about 45″ diameter with low surface brightness. eg 4474, *q.v.*, lies 14′ ENE of eg 4459.

eg 4473 ***dimen. 4′.5 × 2′.6*** ***V = 10.2*** ***sfc. br. 12.9***
Not difficult to view in 15 cm, this galaxy appears 1′ diameter with a stellar nucleus. The lenticular halo extends to 1′.5 × 1′ elongated E-W in 25 cm, and the object has a high overall surface brightness. An indistinct bar lies along the major axis, crossing a small elongated core that is much brighter than the surrounding regions. The nucleus is substellar, about 3″ across.

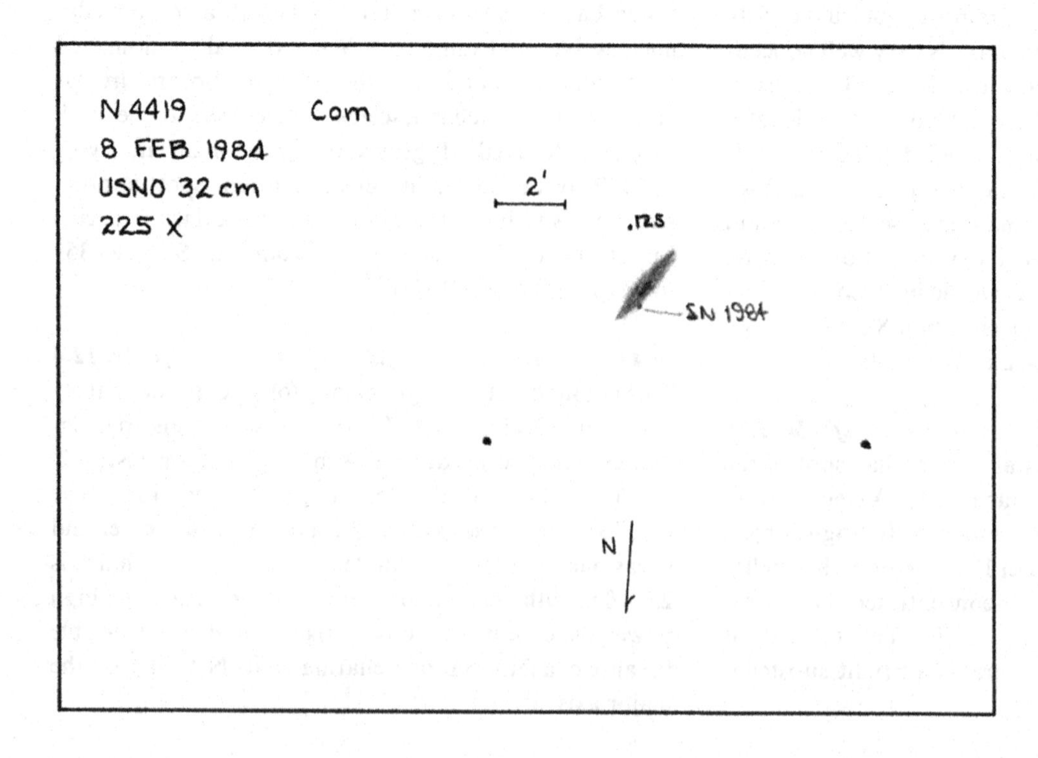

eg 4474 dimen. 2′.3 × 1′.2 V = 11.7 sfc. br. 12.7
This galaxy lies 14′ ENE of eg 4459, *q.v.* In 15 cm it is quite a bit fainter than eg 4459. The halo extends to 50″ × 35″ in 25 cm, elongated E-W. A faint stellar nucleus is visible in an elongated core. Two faint stars lie to the N with 20″ separation in pa 285°.

eg 4477 dimen. 4′.0 × 3′.5 V = 10.4 sfc. br. 13.2
This galaxy appears similar to eg 4473 (*cf.* 13′ SSW in Virgo) in 15 cm, but it is somewhat larger and has a more rounded outline. It is moderately bright in 25 cm, but fainter than −73. The halo is slightly elongated, extending to 1′.75 × 1′.5 in pa 20°. Overall the object has a strong, even concentration to a substellar nucleus, but the core is not sharply defined. eg 4479, *q.v.*, lies 5′.8 SE.

eg 4479 dimen. 1′.8 × 1′.5 V = 12.5 sfc. br. 13.5
Viewed in 25 cm, this galaxy is located 5′.3 SE of eg 4477, *q.v.* The moderately faint halo extends to only 30″ diameter and contains a sharp stellar nucleus.

eg 4489 dimen. 2′.2 × 2′.1 B = 12.8 sfc. br. 14.5
Typical of many in the region, 15 cm shows this galaxy as a small, concentrated spot with a bright nonstellar center. With 30 cm the 40″ × 30″ halo brightens moderately and evenly to an indistinct stellar nucleus. eg 4498, *q.v.*, lies 13′ NE.

eg 4494 dimen. 4′.8 × 3′.8 V = 9.9 sfc. br. 13.0
15 cm shows this object as an unimpressive circular blob about 1′ across. It is quite bright in 25 cm. At 200× a stellar nucleus is visible inside the 1′.5 × 1′ halo, which is elongated N-S.

eg 4498 dimen. 3′.2 × 1′.9 B = 12.6 sfc. br. 14.4
With 15 cm this galaxy appears to be nearly the same size and brightness as eg 4489, *cf.*, 13′ SE. The unconcentrated halo is elongated SE-NW. In 30 cm the galaxy is obviously larger, but of lower surface brightness than its companion, the halo reaching 1′.3 × 0′.5 in pa 135°. It is nearly unconcentrated, but many stellarings are visible at the threshold along the major axis. eg 4502 (B = 14.6) lies 12′ SE.

eg 4501 dimen. 6′.9 × 3′.9 V = 9.5 sfc. br. 13.0
Messier 88 is a giant spiral galaxy. The halo is distinctly elongated in 6 cm, and a faint star is visible 5′ S. With 15 cm the star S is actually a pair (11.5,12.5; 30″; 210°), and another star appears 3′.5 N. The galaxy is bright, elongated in pa 140°, but little detail is visible. 25 cm reveals a faint pair (13.8,14.2; 15″; 200°) well within the SE portion of the halo, but they hardly interfere with viewing. The halo measures 6′ × 2′.5 overall. The SE side seems triangular and of lower surface brightness than the NW, which has an oval outline. The core is a smooth oval of the same proportions as the halo with a smooth texture and a central substellar nucleus.

eg 4506 dimen. 1′.6 × 1′.3 B = 13.6 sfc. br. 14.3
In 25 cm this galaxy appears about 1′ diameter with low surface brightness and a fairly definite edge. The halo is broadly concentrated to a well-defined but only slightly brighter central pip.

eg I3476 dimen. 2′.2 × 1′.9 V = 12.8 sfc. br. 14.2
This faint galaxy has a low surface brightness in 25 cm. The slightly elongated patch is about 50″ × 35″ in pa 35°. The center is broadly and irregularly brighter and contains a few stellarings.

eg 4515 dimen. 1′.6 × 1′.3 B = 13.4 sfc. br. 14.1
A fairly difficult object for 15 cm, at 75× this galaxy appears as a tiny faint spot with a stellar center. It is only 30″ diameter in 30 cm, a circular glow sharply concentrated to a stellar nucleus.

eg 4548 dimen. 5′.4 × 4′.4 V = 10.2 sfc. br. 13.5
Messier 91 is visible in 6 cm as a small circular patch with a bright center. In 15 cm it appears slightly elongated NE-SW. The halo has a smooth texture and a small condensed core. The halo extends to 2′.5 × 1′.5 with 25 cm, elongated in pa 60°. The brightness rises smoothly to a small round core and a stellar nucleus. Identification of this galaxy with M91 is uncertain.

eg 4559 dimen. 10′ × 4′.9 V = 9.9 sfc. br. 13.9
This is an impressive object even for 15 cm, which reveals two stars nearby to the E and S. 30 cm shows it 9′ × 2′ in pa 140°. There is little brightening to a very irregular core elongated in pa 130°. The core fades more abruptly on the SE end, and the halo is not as extensive to the SE. Three stars are attending on the SE side: a mag. 12 star on the NE flank directly E of center; a mag. 13 star right on the major axis to the SE; a mag. 12.5 star on the SW flank. All three lie between 1′.5 and 2′ from the center.

eg 4565 dimen. 16′ × 2′.8 V = 9.5 sfc. br. 13.5
This edge-on galaxy presents a striking contrast to the typical galaxy in all apertures. In 15 cm the halo reaches to 7′ × 1′.5, elongated SE-NW. Close attention in good conditions will give a considerably larger length. Some faint mottlings along the major axis give the impression of a dark lane. A mag. 13.5 star is visible 1′.5 NE of the center, and another is visible farther SW. With averted vision in 25 cm it seems 20′ × 1′ in size. The bright central bulge is 4′ × 1′. A stellaring can be seen on the SE side of

the core. The galaxy appears 16′ long in 30 cm. The stellaring SE lies just off the dark lane, which is prominent at this aperture against the NE edge of the core. The lane is traceable for about 3′ beside the brighter parts of the core. About 5′ NW of center along the major axis is a bright patch about 1.′2 long, and several smaller patches are visible along the SE arm. eg 4562 (mag.$_Z$14.6) lies 13′ SW.

eg 4571 ***dimen. 3.′8 × 3.′4*** ***V = 11.3*** ***sfc. br. 14.0***
This object is faint for 25 cm, appearing about 1.′5 diameter, perhaps slightly elongated N-S. A few stellarings can be seen on the otherwise undetailed nebula.

eg 4651 ***dimen. 3.′8 × 2.′7*** ***V = 10.7*** ***sfc. br. 13.1***
In 15 cm this object is a 2.′5 × 1′ patch with a slightly brighter core. 25 cm shows the 2.′2 × 2′ halo elongated almost E-W. Three stars are visible near the nebula.

eg 4659 ***dimen. 1.′8 × 1.′3*** ***B = 13.1*** ***sfc. br. 13.9***
15 cm will barely show this faint galaxy lying 1.′4 N of a mag. 10 star. In 30 cm it appears well and evenly concentrated and about 1.′1 diameter. A stellar nucleus just stands out at the center. The halo seems elongated a little N-S, but the bright star may be the culprit. eg I3742, *q.v.*, lies 18′ SE in Virgo.

eg 4670 ***dimen. 1.′8 × 1.′4*** ***V = 12.7*** ***sfc. br. 13.5***
eg 4673 ***dimen. 1.′2 × 1.′0*** ***mag.$_Z$13.7*** ***sfc. br.***
Viewed in 30 cm, eg 4670 is sharply concentrated to a bright beady nucleus. The core fades smoothly to the halo with a total size of 1′ × 0.′75, elongated E-W. A mag. 13 star lies 2′ SW. eg 4673 lies 5.′6 SE and forms an equilateral triangle with −70 and a mag. 8 star. It is a small and fainter version of −70. The 10″ core has a nearly stellar nucleus; the overall diameter is about 30″.

eg 4689 ***dimen. 4.′0 × 3.′5*** ***V = 10.9*** ***sfc. br. 13.6***
15 cm will show this galaxy as an unevenly bright and unconcentrated patch about 3′ diameter. The galaxy forms the northern point of an almost equilateral triangle with mag. 6.5 and 8 stars about 15′ SE and SW. A faint pair of mag. 12 stars lies 3.′4 NNE with 1.′2 separation in pa 300°. The halo has a low surface brightness in 25 cm. The core is about 1.′5 across. In 30 cm the halo extends to 3.′75 × 3′ in pa 160°. The middle two-thirds of the nebula is mottled, and a faint nonstellar nucleus is occasionally visible within.

eg 4710 ***dimen. 5.′1 × 1.′4*** ***V = 11.0*** ***sfc. br. 13.0***
This galaxy is not hard to see in 15 cm. The mottled nebula is 2.′5 × 0.′5. With averted vision, 25 cm shows it up to 3′ × 0.′5 in pa 30°. The nebula is not uniformly bright, but composed of three or four large patches.

eg 4712 ***dimen. 2.′6 × 1.′3*** ***V = 13.0*** ***sfc. br. 14.1***
A very faint object for 25 cm, it appears elongated N-S with a small, faint circular core. 30 cm shows it 1.′2 × 0.′6, elongated in pa 160°. The galaxy lies 4.′5 S of a mag. 12.5 star. Two faint stars that can be mistaken for the galaxy lie SE of this star. eg 4725, *q.v.*, lies 12′ E.

eg 4725 ***dimen. 11′ × 7.′9*** ***V = 9.2*** ***sfc. br. 13.9***
Visible in 6 cm, this large galaxy is elongated NE-SW, about 8′ × 6′, and very little concentrated except for a small brightening in the center. It is about 7′ × 5′ in 15 cm with a small brighter core. The galaxy is quite bright in 25 cm, appearing as a 5′ × 3′ haze with a central stellar nucleus but no core. A mag. 12.5 star lies 2.′5 N; a threshold magnitude star is on the E edge. 30 cm shows a 1′ core and stellar nucleus with the faint halo extending to 4.′5 × 3′ in pa 45°. The star 1.′5 E of center is about mag. 14.5. Broad brightenings are visible at the acute ends of the halo 2′ from the nucleus, each about 1′ diameter. A slight brightening along the major axis connects these with the core. The halo is oval and has extremely faint extensions to about 7′ length. eg 4712, *q.v.*, lies 12′ W; eg 4747, *q.v.*, 24′ NE.

eg 4747 ***dimen. 3.′6 × 1.′4*** ***V = 12.4*** ***sfc. br. 14.1***
This galaxy lies 7′ N of a wide unequal pair, the brightest being an orange mag. 7.5 star. It is quite faint for 25 cm, elongated NE-SW, about 3′ × 1.′ The nebula is without central concentration, though some sparkles are visible near the center. 30 cm shows it 1.′5 × 0.′5 in pa 40°. Here also it exhibits no overall concentration and a few faint spots in the middle. eg 4725, *q.v.*, lies 24′ SW.

eg 4793 ***dimen. 2.′9 × 1.′7*** ***V = 11.7*** ***sfc. br. 13.2***
In 30 cm this object is located 2′ S of the western tip of an equilateral triangle of mag. 9 stars 10′ across. The halo is elongated in pa 45°, about 2′ × 1′, with a slight broad concentration to a faint substellar nucleus. The middle two-thirds is mottled with small bright and dark spots.

eg 4826 ***dimen. 9.′3 × 5.′4*** ***V = 8.5*** ***sfc. br. 12.6***
Messier 64, the Blackeye Galaxy, is easily visible in 6 cm. It is elongated in pa 115°, about 6′ × 3′, with a prominent nucleus. The inner regions seem more elongated, and the nucleus appears closer to the S side. A mag. 11 star lies 4.′25 NE. At 125 × in 15 cm the dark patch can be seen with averted vision. 25 cm, however, shows the dark patch best at 200 × using direct vision. The patch is kidney-shaped and is nestled on the NNE side of the nucleus. The halo grows suddenly brighter to the 8″ nucleus, and fades smoothly to the sky. The dark lane is obvious with 30 cm using either direct or averted vision. The nonstellar nucleus is right up against the dark lane,

which is widest on the NE side. A smooth rim of brightness curves around the outside of the dark lane. The smoothly textured halo extends to 6′ × 3′.

eg 4839 dimen. 4′.2 × 2′.1 V = 12.4 sfc. br. 14.6
This Coma cluster member is not difficult to view in 15 cm, lying between a mag. 12 star 2′.6 NE and a mag. 13.5 star 2′.3 SW. The small galaxy is diffuse and weakly brighter across the center. It is one of the larger galaxies in the cluster region with 30 cm, reaching 1′.1 × 0′.7 in pa 60°. The halo has a flat brightness profile in the outer regions, then becomes suddenly and sharply brighter across the core, which contains a faint, but distinct stellar nucleus. Photographs show that the observed elongation may be due to a faint galaxy (eg A1254+27, V = 14.7) superposed on the SW side 25″ from the center, and a mag. 15 star in the NE side. A close pair of galaxies, eg 4842AB (V = 13.9, 15.0) lies 2′.6 E.

eg 4841A dimen. 1′.7 × 1′.7 V = 12.7: sfc. br. 13.8
eg 4841B dimen. 1′.3 × 1′.3 V = 13.3: sfc. br. 13.9
eg 4841A + B dimen. 1′.9 × 1′.7 V = 11.5 sfc. br. 12.6
Only 32″ apart, these two Coma cluster galaxies are nearly in contact with 30 cm. The southwestern component, eg 4841A, is the larger and brighter of the two. At 225× it is about 40″ diameter, slightly elongated E-W, with a strong, even concentration to a conspicuous 8″ core. The northeastern component is similarly concentrated in the outer regions, but the core is substellar and a little fainter.

eg 4848 dimen. 1′.8 × 0′.6 V = 13.6 sfc. br. 13.5
This Coma cluster galaxy is distinctly elongated in 30 cm, reaching to 48″ × 30″ in pa 160°. The lenticular halo has a moderate, even concentration.

eg 4858 dimen. 0′.4 × 0′.3 V = 15.0 sfc. br. 12.4
eg 4860 dimen. 1′.0 × 0′.9 V = 13.5 sfc. br. 13.2
The brighter of this pair of Coma cluster galaxies is 45″ across in 30 cm. The light is evenly concentrated to an indistinct core and faint nucleus. The faint companion, 34″ SW, is visible only with difficulty using averted vision.

eg I3955 dimen. V = 14.3 sfc. br.
This galaxy forms the western vertex of a triangle with a mag. 15 star 45″ ENE and the galaxy pair eg 4864/67, *q.v.*, 2′.0 SE. It is very small in 30 cm, only 10″ diameter, with a very faint stellar nucleus.

eg 4864 dimen. 0′.6 × 0′.4 V = 13.6 sfc. br. 12.1
eg 4867 dimen. 1′.1 × 1′.1 V = 14.2 sfc. br. 14.4
Both of these Coma cluster galaxies are visible in 30 cm at 225×, forming a close pair only 35″ apart. eg 4864 is the larger and brighter of the two, appearing about 30″ across, somewhat elongated toward −67, i.e., in pa 120°. It is well concentrated to a small substellar core. The tiny eg 4867 is a sharply concentrated substellar spot clearly separated from the SE edge of −64.

eg 4865 dimen. 1′.4 × 0′.8 V = 13.3 sfc. br. 13.4
eg M+05-31-63 dimen. 0′.4 × 0′.4 V = 14.4 sfc. br. 12.5
Lying only 2′.8 NW of a mag. 7 star, eg 4865, a member of the Coma cluster, is nevertheless easy to see in 30 cm. The concentrated halo is 30″ long, elongated ESE-WNW. The faint companion 1′.4 WSW is 20″ across, a diffuse unconcentrated glow.

eg 4869 dimen. 1′.1 × 1′.1 V = 13.5 sfc. br. 13.7
This Coma cluster galaxy is located 2′ SW of a mag. 12.5 star, which lies an equal distance SW of eg 4874, *q.v.* It appears larger and brighter than eg 4872, *cf.*, in 30 cm, and has a mag. 13.5 star involved 17″ NW of its center. The halo is about 30″ across, just reaching this star, and is not strongly concentrated.

eg 4871 dimen. 0′.5 × 0′.4 V = 14.1 sfc. br. 12.1
eg 4872 dimen. 1′.0 × 1′.0 V = 13.7 sfc. br. 13.7
eg 4873 dimen. 0′.8 × 0′.6 V = 14.2 sfc. br. 13.1
eg I3998 dimen. 0′.6 × 0′.4: V = 14.7 sfc. br. 13.1:
These four galaxies are the brightest of a virtual cloud of tiny objects surrounding eg 4874, *q.v.*, in the core of the Coma cluster. eg 4872 is the closest to −74, about 50″ away in pa 200°. In 30 cm it is a small star-like spot with a weak halo no more than 15″ diameter. eg 4871, 1′.3 W of −74, is a small concentrated spot similar to −72, but somewhat fainter. eg 4873, at 1′.6 in pa 340° from −74, seems larger and less concentrated than the others. The well-defined portion is about 15″ across, elongated ESE-WNW. eg I3998, 2′.3 in pa 75° from −74, is fairly faint, only 10″ diameter, and moderately concentrated.

eg 4874 dimen. 2′.7 × 2′.7 V = 11.9 sfc. br. 14.0
This is the second brightest member of the Coma cluster and is located 6′.5 S of a mag. 7 star. It is pretty faint for 15 cm, a poorly concentrated, roughly circular patch with a mag. 12.5 star 2′ SW. In 30 cm the mag. 7 star interferes somewhat with viewing. The halo is about 1′.5 across and roughly circular, but the many tiny associated galaxies nearby make for a seemingly irregular outline. The halo has a moderate, even concentration to a 20″ core; there is no distinct nucleus. Some of the nearby companion galaxies are described immediately above.

eg 4881 dimen. 1′.0 × 1′.0 V = 13.5 sfc. br. 13.5
This is the first of three brighter galaxies located north and east of the center of the Coma cluster, including eg

Chart VIII. Coma cluster = Abell 1656.

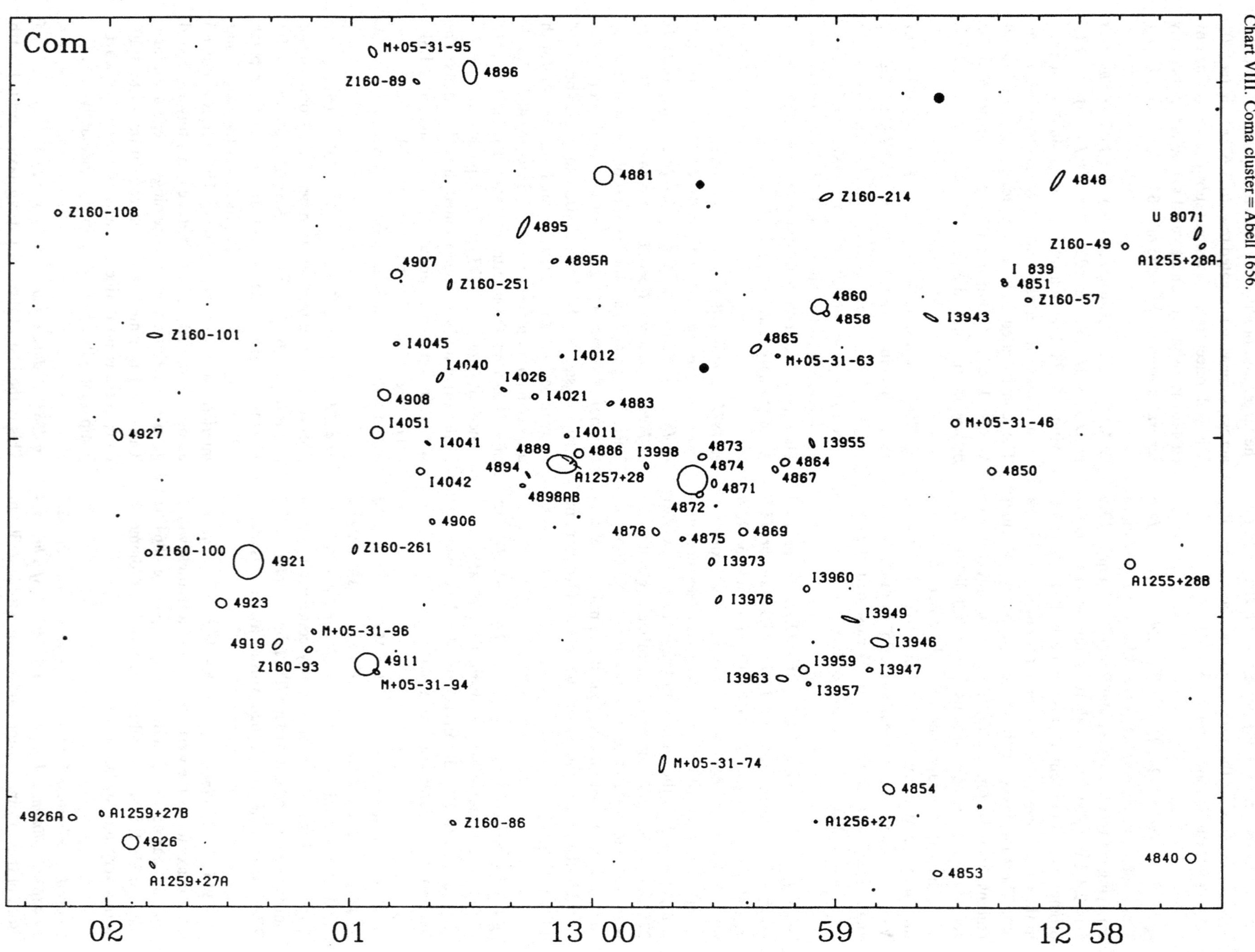

4895 and eg 4907, *q.v.* In 30 cm it is a 20″ circular glow, well concentrated to a sharp, conspicuous stellar nucleus.

eg 4883 dimen. 0′.4 × 0′.3: V = 14.3 sfc. br. 12.0:
Forming a triangle with eg 4874 and eg 4889, *q.v.*, this galaxy is typical of many in the Coma cluster. In 30 cm the diffuse, moderately low surface brightness patch is 15″ across.

eg 4886 dimen. 0′.8 × 0′.8 V = 13.9 sfc. br. 13.5
eg A1257+28 dimen. V = 14.9 sfc. br.
eg 4889 dimen. 3′.0 × 2′.1 V = 11.4 sfc. br. 13.4
eg 4889 is the brightest object in the Coma cluster of galaxies. It is distinctly brighter than eg 4874, *cf.* 9′.2 W, in 15 cm at 75×. The 1′.5 halo is evenly and well concentrated to a faint stellar nucleus. At times it seems elongated E-W, but a mag. 13.5 star 1′ SE may be influencing this. 30 cm shows a 1′ × 0′.6 core elongated E-W; here this seems due mostly to a very faint galaxy, eg A1257+28, superposed on the W side only 28″ from the center. The core is surrounded by a very faint more-or-less circular halo about 1′.4 diameter. About 65″ NW is eg 4886, which is visible in 30 cm as a faint spot. The faint star SE and this galaxy make −89 occasionally seem elongated NE-SW.

eg 4895 dimen. 2′.3 × 0′.9 V = 12.8 sfc. br. 13.4
This is the second of three galaxies in a line across the northeast side of the Coma cluster, including eg 4881 and eg 4907, *q.v.* It is conspicuous in the field with 30 cm. The 25″ × 10″ halo is elongated in pa 150°, of moderately high surface brightness, and well concentrated toward the center.

eg 4898AB dimen. V = 13.6 sfc. br.
In 30 cm this galaxy is a conspicuous object 2′.5 SW of eg 4889, *q.v.*, in the center of the Coma cluster. The high surface brightness 25″ × 15″ patch is elongated in pa 70°, and has little surrounding halo.

eg 4907 dimen. 1′.5 × 1′.3 V = 13.4 sfc. br. 13.9
This galaxy is the easternmost of three brighter objects north and east of the center of the Coma cluster, including eg 4881 and eg 4895, *q.v.* 30 cm shows it as a moderately faint object with a mag. 13 star 40″ SW of center. The 30″ halo is approximately circular and weakly concentrated to a faint but distinct stellar nucleus.

eg 4911 dimen. 1′.3 × 1′.2 V = 12.8 sfc. br. 13.1
eg M+05-31-94 dimen. 0′.6 × 0′.4 V = 15.3 sfc. br. 13.8
eg 4911 is just visible in 15 cm, forming the southeastern member of a string with two mag. 12.5 stars within 4′ to the northwest. In 30 cm it is circular, 1′.25 diameter, and broadly concentrated. A mag. 14.5 star is visible off the halo 45″ SE of center. The companion galaxy is on the SW edge of the halo 35″ from the center. It is very faint and unconcentrated, but well defined.

eg 4919 dimen. 1′.4 × 0′.8 V = 13.7 sfc. br. 13.7
eg M+05-31-96 dimen. V = 14.8 sfc. br.
eg 4919 is the brightest of the galaxies between eg 4911 and eg 4921, *q.v.*, in the Coma cluster. In 30 cm it appears 15″ across, and well concentrated to a stellar nucleus. The companion galaxy, 1′.9 away in pa 290°, is very faint substellar spot.

eg 4921 dimen. 2′.7 × 2′.4 V = 12.1 sfc. br. 14.0
One of the brighter Coma cluster members, this galaxy is just visible in 15 cm as an unconcentrated 2′ patch with a mag. 13 star about 3′ NE. With 30 cm the unconcentrated low surface brightness halo is elongated in pa 120°, reaching to 1′.2 × 0′.8. The core is brighter, however, and a stellar nucleus is clearly visible. eg 4923, *q.v.*, lies 2′.6 SE.

eg 4923 dimen. 1′.3 × 1′.2 V = 13.6 sfc. br. 13.8
In 30 cm this galaxy seems only a little fainter though quite a bit smaller than eg 4921, *cf.* 2′.6 NW. The diffuse 30″ halo is moderately concentrated to a bright substellar center.

eg 4961 dimen. 1′.7 × 1′.3 V = 13.5 sfc. br. 14.2
In 30 cm this is a very small, faint galaxy. The halo is 1′ × 0′.5, elongated E-W, and shows a slight, even concentration to a faint stellar nucleus. Photographs show eg 4957 (V = 12.9) 12′.5 SSW.

gc 5024 diam. 13′ V = 7.5 class V
Messier 53 appears about 3′ diameter in 6 cm. It is weakly concentrated to a broad core that has a mag. 11 stellaring in its center. A mag. 11.5 star is visible in the NE side, and a mag. 9,10 pair of 1′.4 separation in pa 67° lies 9′ SSE. At 200× 15 cm shows a diffuse grainy object that can be partially resolved on a steady night. 25 cm shows about 150 stars across the cluster, including many outliers. The core is about 1′.5 across, and the outliers extend to 4′ diameter. 30 cm shows it 6′ across with most of the outliers on the NW and SE. The cluster is well resolved overall, and in poor seeing the core looks roughly triangular.

gc 5053 diam. 11′ V = 9.9 class XI
This faint globular cluster is visible in 15 cm about 6′ WNW of a mag. 9.5 star. The unconcentrated, unresolved patch is about 5′ diameter. With 25 cm a few stars

are resolved at medium power, while 30 cm will resolve about 30 faint stars overlying a 5′ area. A slight concentration is evident to a 1.′8 core. The brightest cluster member, about mag. 14, lies SSE of the center.

eg 5116 dimen. 2.′2 × 0.′9 B = 13.2 sfc. br. 13.7
Though not visible in 15 cm, 30 cm will reveal this faint galaxy about 2′ ESE of a mag. 12.5 star. The unconcentrated halo extends to 1.′2 × 0.′6 in pa 40°. At high power the middle parts are unevenly bright, and a very faint star is visible 1.′5 NE.

eg 5172 dimen. 3.′3 × 1.′9 V = 11.9 sfc. br. 13.7
This galaxy is faintly visible in 15 cm, about 11′ NW of a mag. 7.5 star and 3.′5 SW of a mag. 12.5 star. The very weakly concentrated halo is about 2′ × 1′, elongated E-W. With 30 cm the halo is poorly and evenly concentrated to a distinct and broadly concentrated core about 1′ across containing a very faint stellar nucleus. The halo extends to 2.′25 × 1.′5, elongated in pa 80°. At 250 × the middle of the core seems elongated almost N-S. A mag. 14 star lies 43″ NNE of center. eg 5180, *q.v.*, lies 14′ S.

eg 5180 dimen. 1.′8 × 1.′2 mag.$_z$14.3 sfc. br.
This object is located 6.′2 SSW of the mag 7.5 star that lies SE of eg 5172, *cf.* In 30 cm the galaxy is very small, but of high surface brightness, and forms a pair with a mag. 14 star 18″ SE. The poorly concentrated halo is about 40″ × 20″, elongated N-S. Despite the poor overall concentration, several distinct sparkles are visible along the major axis. A very faint star, looking like an off-center stellar nucleus, is visible in the NW side of the elongated core.

Corona Australis

The southern crown is a small constellation below Sagittarius. gc 6541 is the most conspicuous object, while bright and dark nebulae can be found between γ and ε Coronae Australis. The center of the constellation culminates at midnight about 1 July.

gc 6496 ***diam. 6.′9*** ***V=8.5*** ***class XII***
6 cm will show this cluster as a faint and poorly concentrated spot. A threshold magnitude star is visible on the E side. 25 cm shows a diffuse and unresolved glow delimited by two mag. 11.5 stars. It is 2′ diameter with some faint field stars nearby. With 30 cm the ill-defined 2.′5 haze is dominated by superposed stars. The eastern mag. 11.5 star has three faint companions, and another faint star is visible on the S edge.

gc 6541 ***diam. 13′*** ***V=6.1*** ***class III***
Visible even in 3 cm aperture as a well-concentrated spot with a substellar center, 6 cm will show this cluster to have a diameter of about 2.′5. A prominent mag. 11 star is visible on the E side. The cluster extends to 5′ diameter and is partially resolved in 25 cm. The resolved stars thin quickly in passing from the sharply brighter core to the outliers, and seem more numerous on the W side: a dark patch lies against the east side. In 30 cm the core seems broader, about 1′ diameter, and outliers fill a 3′ area. Just off the S edge of the core is a small group of stars that makes the core appear irregular in outline. At 250× the cluster is pretty well resolved, with thinly scattered outliers extending to 5′ diameter.

gn 6726–27 ***dimen. 8′×8′*** ***star variable***
This nebula lies in a field of bright and dark patches interspersed with stars and is illuminated by the variable star TY Coronae Australis ($8.7 \leq$ mag. v ≤ 12.4). The nebula is smaller than nearby gc 6723 (*cf.* in Sagittarius) in 6 cm, but has a much higher surface brightness and appears brighter overall. The central star is conspicuous at about mag. 7. 15 cm reveals a small bright nebula lying 12′ NE of a bright pair (Brisbane 14: V=6.5,6.7; 13″; 281°). 25 cm shows it 3′–4′ diameter with indefinite borders. Medium powers are best for viewing. gn 6729, a variable reflection nebula, lies 5′ SW, associated with the stars R and T CrA.

pn I1297 ***dimen. 8″×6″*** ***nebula 10.7v*** ***star***
In 15 cm this planetary is nearly indistinguishable from a mag. 10.5 star. At high powers 25 cm shows a grey spot 8″–10″ diameter. A mag. 13 star lies 30″ NNW, and another fainter star is in contact with the nebula on the SE edge, making the object seem elongated.

Corona Borealis

Though it lies at high galactic latitude, there are few objects in this constellation likely to be visible with amateur telescopes. A few members of the Corona Borealis cluster of galaxies (centered at $15^h22^m + 27°40'$ for equinox 2000) are reportedly visible with very large instruments. The center of the constellation culminates at midnight about 19 May.

eg 5958 ***dimen. 1:2 × 1:1*** ***mag.$_z$13.2*** ***sfc. br.***
This object is visible at 75× in 15 cm about 11′ WSW of a mag. 9 star. The galaxy is very small and unconcentrated, but of high surface brightness. In 25 cm it is circular, about 35″ diameter, with a faintly granular texture. Magnitude 12.5 stars are visible 4′ E and 3:3 NW. 30 cm reveals an evenly concentrated galaxy with a 1′ × 0:7 oval halo elongated N-S. The more nearly circular core is 30″ across, brightening gradually to a faint stellar nucleus. A few mag. 14 stars are visible within 2′ radius.

eg 5961 ***dimen. 1:0 × 0:4*** ***mag.$_z$14.0*** ***sfc. br.***
Though it is a small, moderately faint object for 30 cm, this galaxy has a high surface brightness. The halo is about 45″ × 20″, elongated in pa 100°, and is little brighter across the center. eg U9920 (mag.$_z$15.1), a narrow spindle, lies 3:7 S, on the SE side of a mag. 13 star.

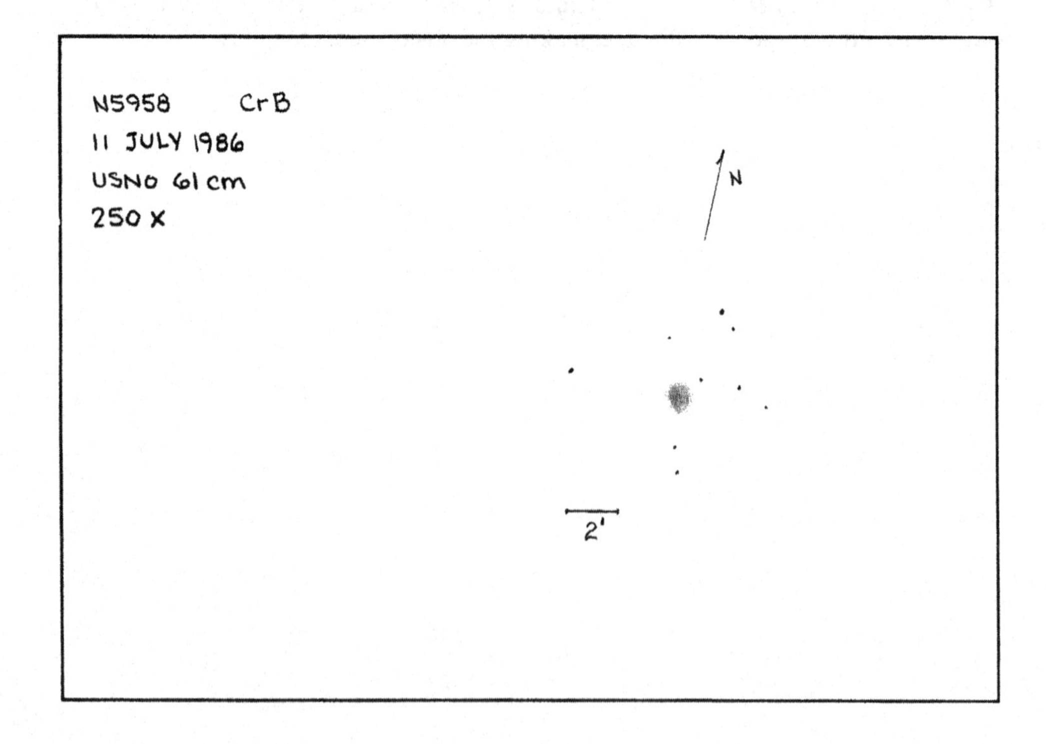

Corvus

This constellation holds a bright planetary nebula and several galaxies, including two interacting pairs. The center of the constellation culminates at midnight about 29 March.

eg 4024 dimen. 2'.3 × 1'.7 B = 12.6 sfc. br. 14.0
This is a small object of high surface brightness in 25 cm. The halo is about 50″ × 30″ in pa 65° and contains a stellar nucleus. 30 cm shows it 5′ ENE of a 4′ triangle of stars with another star in its center. It is small, 1'.5 × 0'.8, with a very small core and bright stellar nucleus. Stars are noted about 1'.75 SW and NW.

eg 4027 dimen. 3'.0 × 2'.3 V = 11.1 sfc. br. 13.1
This object is possibly interacting with eg 4038–39, *q.v.*, 41′ NE. Just visible to 15 cm, it appears as a circular spot about 2′ across. Some moderately faint stars are visible about 10′ E of the galaxy. It is elongated in pa 150° in 25 cm, about 2′ × 1′ with a granular circular core. Two mag. 11 stars about 3′ to the NE and NW form a triangle with it. 30 cm shows an irregular outline with a stellaring lying S of the middle. A mag. 14 star 45″ NE is associated with faint extensions to the pa 150° bar. Two other stars lie E, a third to the W.

eg 4033 dimen. 2'.5 × 1'.1 B = 12.4 sfc. br. 13.5
In 15 cm this object appears like a slightly fuzzy star. 25 cm shows a small concentrated spot 30″ diameter with a bright core and stellar nucleus. It grows to 2′ × 0'.75 in pa 40° with 30 cm. The thinly tapered halo surrounds a narrow core and a stellar nucleus. Three stars are visible about 3′ N.

eg 4038 dimen. 2'.6 × 1'.8 V = 10.7 sfc. br. 12.2
eg 4039 dimen. 3'.2 × 2'.2 mag. sfc. br.
eg 4038 + 4039 dimen. 7'.6 × 4'.9 B = 10.6 sfc. br. 14.4
This colliding system is possibly interacting with eg 4027, *q.v.* 15 cm shows only a large, bright, and round spot about 2'.5 across without much concentration to the center. A mag. 8 star is located 6′ NW. In 25 cm a semicircular outline is evident, a cashew-shape about 2'.5 across. The N lobe (−38) is brighter and has a stellaring on its S side. A bridge of light arcs around E and S to −39. In 30 cm, −38 is larger and a little brighter than its companion. It is elongated in pa 100° with a convex northern perimeter and trailing ends. Two distinct condensations are visible in the irregular core, and the halo trails S off the E end, reaching to −39. A mag. 14 star lies 1'.5 NW. eg 4039 is inclined toward −38 in pa 60°. About 3′ to the E is a faint star, and 6′ ESE lies an unequal pair with 12″ separation in pa 55°.

eg 4050 dimen. 3'.1 × 2'.2 B = 12.3 sfc. br. 14.2
This galaxy is not visible in 15 cm and is fairly faint for 25 cm, located 6′ NE of a mag. 8 star. It is about 1'.5 across with no central brightening. In 30 cm the large and ill-defined halo extends to 2'.5 × 1'.5 elongated E-W; the slightly concentrated oval core is 1'.2 across. The brighter part of the core is off-center to the N, and a faint star is visible at the NE edge of the galaxy.

eg 4094 dimen. 4'.2 × 1'.8 B = 12.5 sfc. br. 14.6
Forming a 2′ triangle with two mag. 11 stars, this is a faint galaxy for 30 cm. It is elongated in pa 65°, 2'.5 × 1'.5. There is little concentration with a few very faint stellarings in the middle. eg 4114 (mag.$_{VV}$13) lies 28′ NE.

pn 4361 dimen. 1'.9 × 1'.9 nebula 10.9v star V = 13.0
Viewed in 15 cm this planetary appears large and diffuse. The central star is visible at high power. In 25 cm the central star appears about mag. 13. It is surrounded by a bright area 40″ across; around this is a fainter oblong 60″ × 50″ patch that extends more to the S than N. The central star is easy to see in 30 cm. The grey nebula has a round outline and is fainter on the S.

eg 4462 dimen. 3'.7 × 1'.6 B = 12.6 sfc. br. 14.4
Viewed in 30 cm this object is elongated in pa 130°, 2'.5 × 1'.25, with several blobs along the major axis. The halo does not bear magnification well.

eg 4756 dimen. 2'.0 × 1'.7 B = 13.2 sfc. br. 14.3
This galaxy is visible only with difficulty in 15 cm. It is fairly faint in 25 cm, appearing as a circular spot about 1′ diameter with broad concentration but no sharp nucleus. In 30 cm the halo seems elongated NE-SW, but a mag. 14 star 2'.1 NE and a mag. 14.5 star 45″ away in pa 60° may be misleading. Overall it is about 2′ × 1'.5 with moderate concentration to circular inner parts and a faint nucleus.

eg 4763 dimen. 1′.7 × 1′.3 B = 13.3 sfc. br. 14.0
Not visible in 15 cm, this galaxy appears similar to eg 5017 (*cf.* in Virgo) with 25 cm. Viewed at high power, it is fairly faint, 30″ diameter with a nonstellar core. At low power in 30 cm it is slightly elongated E-W, about 1′ diameter. At high power a very faint stellar nucleus is usually visible, but otherwise the galaxy is hardly concentrated.

eg 4782 dimen. 1′.5 × 1′.5 V = 11.7 sfc. br. 12.7
eg 4783 dimen. 1′.7 × 1′.6 V = 11.8 sfc. br. 12.9
These interacting galaxies are 0′.7 apart and have a combined V magnitude of 11.2. They are difficult to view in 15 cm, appearing as a single amorphous glow. Separable at 100 × in 25 cm, both galaxies have uniform light, but −82 is brighter and larger. The pair is less than 1′ diameter overall. With 30 cm the galaxies lie in pa 10° and are enveloped in a common halo. The southern member (−82) is about 1′.5 across with even concentration but no obvious nucleus. At 250 × it seems elongated approximately N-S, and occasionally a faint stellar nucleus is visible within. The northern member (−83) is smaller and more sharply concentrated than its companion. Its halo extends to 1′ diameter and contains a nucleus more distinct than −82's at all powers. eg 4792 lies 8′ NE; eg 4794 (mag.$_{vv}$14) is 8′.7 ESE.

eg 4802 dimen. mag.$_{vv}$12 sfc. br.
With 15 cm this galaxy is visible as a faint haze surrounding the eastern star of a mag. 11 pair of 2′.2 separation in pa 115°. It is centered only 15″ NW of this star in 30 cm. The 1′ high surface brightness halo has a smooth texture. The galaxy is evenly concentrated to the center and without a distinct nucleus.

Crater

Lying immediately west of Corvus, this region is unmarked by bright stars, and deep-sky observing is limited to galaxies. The center of the constellation culminates at midnight about 13 March.

eg 3511 dimen. 5′.4 × 2′.2 B = 11.6 sfc. br. 14.1
In 15 cm this galaxy has a low surface brightness and is difficult to view. The halo is a dim 2′.5 × 1′ streak elongated nearly E-W. The broad uncondensed galaxy is 4′ × 1′ in 25 cm, elongated in pa 75°. A stellaring is visible N of center, and a mag. 12.5 star lies at the E tip. 30 cm shows it 5′ × 1′ with a broad core and no nucleus. Three stars (including the mag. 12.5 star) are visible in the nebula: in decreasing order of brightness they lie 1′.7 E, 2′.1 WSW, and 35″ NE of center.

eg 3513 dimen. 2′.8 × 2′.3 B = 12.0 sfc. br. 13.8
Located 11′ SE of eg 3511, *q.v.*, 15 cm shows this galaxy only as a faint circular haze without central brightening. In 25 cm the galaxy forms a triangle with two mag. 9.5 stars on the S and SW. The halo is 2′ × 1′ elongated SE-NW and contains a stellar nucleus. A faint star is visible on the E side 1′.9 from the center. It appears about 1′.5 diameter in 30 cm.

eg I2627 dimen. 2′.7 × 2′.6 V = 12.0 sfc. br. 13.9
Viewed in 30 cm this object is about 1′.5 × 1′ in pa 90° with a slight granular brightening in the center. Several stellarings appear in the middle regions, but none seem centered or prominent enough to be the nucleus.

eg 3571 dimen. 3′.3 × 1′.3 B = 12.8 sfc. br. 14.3
This galaxy is faintly but distinctly visible in 15 cm about 6′ S of a mag. 13.5 star. At 75 × it is elongated E-W. With 30 cm the halo extends to 2′.75 × 1′ in pa 90°. It is moderately well and evenly concentrated to elongated inner regions and a faint stellar nucleus. At 250 × the inner 30″ is knotty along the major axis.

eg 3636 dimen. 1′.1 × 1′.1 mag.$_{vv}$13 sfc. br.
eg 3637 dimen. 1′.7 × 1′.6 B = 12.8 sfc. br. 13.8
Lying 3′ NE of a mag 6.5 star, 15 cm shows eg 3637 as a small but easily visible spot. It is circular in outline and concentrated to a faint stellar nucleus. In 25 cm the mag. 6.5 star has a reddish cast, and the galaxy seems a little elongated NE-SW. 30 cm gives 1′.5 diameter; the circular halo is abruptly concentrated at a tiny knotted core about 20″ across, but no nucleus is discernable. Other than the core, the halo exhibits very little concentration and seems flattened on the SE side. Just 1′.7 NW of the bright star, 25 cm will show eg 3636 as an evenly bright circular spot about 45″ diameter. The bright star interferes with viewing in 30 cm. The small object is about 50″ diameter and has an even, moderate concentration to a stellar nucleus.

eg 3672 dimen. 4′.1 × 2′.1 B = 11.7 sfc. br. 13.9
In 15 cm this galaxy is a faint, unconcentrated haze. The low surface brightness halo is about 2′ × 1′, elongated N-S. It is much brighter in 25 cm, showing as a spindle elongated in pa 10° with a stellar nucleus. 30 cm shows a large oval with weak, even concentration to a faint nucleus. Overall the halo extends to 3′.25 × 1′.2. At 250 × the core is approximately 2′ × 0′.9 and appears asymmetric: the S side extends farther than the N.

eg A1131–09 dimen. mag. sfc. br.
eg M-01–30–03 dimen. 1′.5 × 1′.2 mag.$_{vv}$15 sfc. br.
eg M-01–30–04 dimen. 1′.3 × 0′.3 mag.$_{vv}$16 sfc. br.
30 cm shows eg M-01–30–03 and eg M-01–30–04 as a single object lying 16′ N of eg 3732, *q.v.* The galaxies are nestled among four faint stars, three of which lie in the NE quadrant, the other 2′ S. The halo is 1′.25 × 0′.6, elongated N-S, with moderate concentration to an elongated 25″ core and a distinct nucleus. eg A1131–09 is visible 7′ NNW. It is very faint, about 30″ diameter, and unconcentrated, except that a faint stellaring is occasionally visible in the center.

eg 3732 dimen. 1′.3 × 1′.2 B = 13.3 sfc. br. 13.6
In 30 cm this galaxy appears small, but it has a high surface brightness. The halo extends to only 45″ diameter, with even, moderate concentration to a small bright center. The center occasionally shows a faint stellar nucleus. A mag. 12 star lies 1′ SW. eg 3722, 3724, and 3730 (all mag.$_{vv}$15) lie in a close group 11′ NNE. eg M-01–30–03,04 *q.v.*, lie 16′ N.

eg 3865 ***dimen. 2′.3 × 1′.7*** ***B = 12.9*** ***sfc. br. 14.2***
eg 3866 ***dimen. 1′.4 × 1′.0*** ***mag.*** $_{VV}$***14*** ***sfc. br.***
The brighter of these two galaxies is just visible in 15 cm as an unconcentrated patch about 1′ diameter lying 6′ SSE of a mag. 9 star. In 30 cm the halo is about 1′.75 × 1′.25, elongated ENE-WSW. It is weakly concentrated to a nonstellar nucleus that is about 10″ across. At high power the middle is unevenly bright. Lying 7′ SE is a mag. 13 star, just E of which is eg 3866. In 30 cm it is about 45″ diameter with an even, moderate concentration to a faint stellar nucleus.

eg 3887 ***dimen. 3′.3 × 2′.7*** ***V = 11.0*** ***sfc. br. 13.2***
In 15 cm this galaxy is closely involved with two faint stars that make the elongation uncertain. The halo is about 3′ × 2′, elongated N-S. A mag. 13 star lies 1′.3 NE, and a very faint star is visible 2′ ENE. 25 cm shows an irregular oval outline without central brightening. The halo extends to 2′.25 × 1′ in pa 165°. With 30 cm the almost unconcentrated halo is 3′.5 × 2′.25. A darker area lies in the southern half of the core, and a faint stellar nucleus is occasionally visible in the northern half with averted vision.

eg 3892 ***dimen. 2′.9 × 2′.2*** ***B = 12.5*** ***sfc. br. 14.3***
Viewed in 15 cm, this galaxy is fairly faint. The halo is about 2′ diameter and has a weak, even concentration. A threshold magnitude star lies about 2′ NW. In 30 cm it is elongated in pa 100°, 2′ × 1′ in extent. The halo exhibits an even, moderate concentration to the center, but the nucleus is indistinct. At 250 × the nucleus is very small, but appears nonstellar. A mag. 14 star is visible 1′ SW.

eg 3955 ***dimen. 3′.2 × 1′.3*** ***V = 11.9*** ***sfc. br. 13.3***
This galaxy is just visible in 15 cm 5′ SW of a mag. 9 star. It is elongated approximately N-S, but needs more aperture for details. The object is small and quite elongated with 30 cm, about 1′.5 × 0′.5 in pa 165°. Overall the halo has only a slight broad concentration to the center, where a faint nonstellar nucleus can be discerned at 250 × . A threshold magnitude star lies 1′ NNW.

eg 3956 ***dimen. 3′.5 × 1′.2*** ***B = 12.5*** ***sfc. br. 14.0***
Viewed in 30 cm, this low surface brightness object is about 3′ × 1′, elongated in pa 65°. A very faint halo surrounds the unconcentrated oval core, which measures 1′ × 0′.7.

eg 3957 ***dimen. 3′.5 × 0′.8*** ***B = 12.7*** ***sfc. br. 13.7***
In 30 cm this galaxy is elongated in pa 170°, with a size of 2′.8 × 0′.6. The halo is moderately concentrated to a thin bar-shaped core and faint nucleus. A mag. 13.5 pair of stars with 15″ separation lies 6′.5 E.

eg 3962 ***dimen. 2′.9 × 2′.6*** ***V = 10.6*** ***sfc. br. 12.7***
This is a bright and well-concentrated galaxy in 30 cm, forming a triangle with two mag. 10.5 stars to the S. The halo is elongated in pa 10°, 1′.5 × 1′ in extent. The core is 30″ across and holds a bright substellar nucleus.

eg 3981 ***dimen. 3′.9 × 1′.5*** ***B = 12.4*** ***sfc. br. 14.2***
In 30 cm this galaxy has a tapered halo that is 3′.2 × 1′ in pa 20°. A mag. 12 star is visible 1′ E of center; a mag. 12.5 star lies about 2′.5 E.

Cygnus

The swan is an interesting constellation to the naked eye, binoculars, or telescope. Since it lies along the galactic plane, many open clusters are contained within its borders. The center of the constellation culminates at midnight on 31 July.

pn BD+30°3639 dimen. 13″×10″ nebula 11.4v star V=10.1
This planetary is bright enough to be visible in small apertures but needs high magnification to reveal its character. In 15 cm at 175× the planetary is just discernable as the southernmost of three "stars" of similar magnitude that are aligned NE-SW. Even with 30 cm the nebula is very small, with a stellar nucleus and nearly imperceptible halo.

oc 6811 diam. 20′ V=6.8 stars 249
This cluster is nicely resolved in 6 cm. At 25× about 20 stars are visible in a squarish group without any haze. In 15 cm it appears about 20′ diameter. Approximately 50 members of mag. 10 and fainter are resolved over a bit of haziness on a rich background. 25 cm shows 6 stars and some haze in a 20′ area, including several groups. In 30 cm the cluster appears elongated slightly E-W, with 75 stars visible in an 11′ area.

oc 6819 diam. 10′ V=7.3 stars 929
In 6 cm this cluster appears as a grainy, well-concentrated haze of moderately high surface brightness. At 25× only a few stars stand out. In 20 cm about 50 stars are visible in a 5′ area. The brighter stars form an "X" or "K." 25 cm shows about 75 stars, including a string of stars aligned SSE-NNW on the W side. A pair is visible on the S and an unresolved clump of stars is located on the E. 30 cm shows 80 stars without haze in an irregularly concentrated group about 5′ across. On the E and SE are two small concentrations; the southeastern one is larger and has 20 stars in it.

pn 6826 dimen. 27″×24″ nebula 8.8v star V=10.7
This planetary is distinguishable in 6 cm as a substellar spot at 25×. In 15 cm it is brighter than Messier 57, *cf.* pn 6720 in Lyra. The central star is clearly visible at 175× embedded in a nearly circular and almost uniform haze about 25″ diameter. At 250× in 25 cm it is circular with diffuse edges. The central star is a conspicuous pinpoint. In 30 cm it is elongated in pa 120°, about 25″×20″. The light is uneven on the W, with a dark patch about 8″ across visible near the mag. 11 central star. A mag. 12 star lies 1′.6 to the S.

oc 6834 diam. 5′ V=7.8 stars 128
In 25 cm this is a fairly small and surprisingly rich cluster. A line of evenly spaced stars, including a mag. 9.5 central star, goes through the cluster in pa 80°. Clumps of stars lie 2′.5 N and S of the central star, and a total of 40 stars is visible in an 8′ area. In 30 cm the brighter stars, about mag. 11, appear on the N side of the group.

gn 6857 dimen. 1′×1′ star V=13.3:
In 25 cm this small nebula is boxed in by four stars in an imperfect square: on the SW is a mag. 12 star; NW is one of mag. 12.5; NE is a mag. 13 star; SE is a mag. 14 star. The nebula is circular and fades toward the edges. The central illuminating star is faintly visible at about mag. 14. In comparison with pn 6894, *cf.*, 30 cm shows it to be smaller and of a different character. It is approximately 40″ across with a slight concentration.

oc 6866 diam. 7′ V=7.6 stars 129
Visible in 6 cm, this cluster is elongated SE-NW, with a few resolved stars embedded in a bright background haze. About ten stars are visible, including a conspicuous string of three or four running SE-NW across the center. 15 cm shows twice as many stars plus unresolved haze in a 5′ area, while 25 cm shows 40 stars in an 8′ area. Many bright stars are situated nearby, including a wide pair to the SSE (V=10.1,10.7; 50″; 140°) and an unequal pair on the N side of the center. 30 cm shows several small groups in a loosely concentrated cluster of 45 stars. The brightest concentration is on the W, elongated SE-NW, but most of the outliers extend in groups E and NE, giving an overall elongation in pa 60°. An arc of brighter stars runs SE-NW on the W side of the cluster. The overall size is 7′×4′.

oc 6871 diam. 25′ V=5.2 stars 66
In 6 cm this cluster is nearly indistinguishable from the Milky Way. Nine bright stars trail southward from

orangish 27 Cygni (V = 5.4), including two nice pairs (ADS 13374AF: V = 6.8,7.4; 36″; 28°, and V = 7.9,8.9; 21″; 235°). 25 cm shows a brilliant field at 50×. At 100× many close, colorful multiple stars are visible. 30 cm shows roughly 80 stars in a 25′ field, with dark areas visible to the E and W.

oc I1310 ***diam. 4.′0*** ***mag.*** ***stars 12***
In 15 cm this cluster is a fairly faint patch less than 2′ diameter, exhibiting some concentration to the center. One star, mag. 12.5, is visible near the center, but no others are resolved. Lying 9.′2 SE is a close, bright pair (ADS 13465: 9.2,9.4; 4.″1; 172°). Viewed in 30 cm, about 15 stars, including the conspicuous central star, are resolved in a 1.′5 area over some haze. On the NNE, 1.′5 from the center, is a mag. 12,13 pair with 9″ separation in pa 60°.

oc I1311 ***diam. 9′*** ***mag. 13.1p*** ***stars 60***
This faint cluster is located in a circlet of mag. 9 stars. In 25 cm at 200× it is an indefinite 3′ patch that appears slightly granular. A mag. 10 star is visible toward the W. 30 cm shows about ten stars overlying a diffuse patch about 2.′5 diameter with a slight brightening on the E side. The mag. 10 star is visible to the W, and another fainter one is on the E.

oc 6883 ***diam. 15′*** ***mag. 8.0p*** ***stars 30***
This object is located 1° E of 27 Cygni (involved in oc 6871, *q.v.*). To 6 cm it is an inconspicuous group of only a few stars. 25 cm shows a loose group of a dozen stars in a 4′ area. Three brighter stars form a triangle, the southern corner of which is a pair (ADS 13486AD: 10.4,10.5; 17″; 52°); a closer pair lies about 2′ NE (ADS 13490: 10.2,11.5; 8.″5; 359°). With 30 cm about ten stars are visible in an unconcentrated group about 5′ diameter.

gn 6888 ***dimen. 20′ × 10′*** ***star V = 7.5***
This nebula is fairly faint in 15 cm, involved with a triangle of unequally bright stars. Between the brightest star of the triangle and the faint point to the E is a diffuse hazy area with a convex northern perimeter. The brighter star is a wide, unequal pair (ADS 13515: V = 7.2,10.5; 14″; 60°). From this pair the brightest segment of the nebula leads away SW and is involved with some faint stars. With 30 cm the bright arc is aligned in pa 215°, and several faint stars are visible along its eastern edge. It has a fairly uniform width of about 1′, and reaches 7′ SE before fading. The portion curving NE from the pair grows wider and has a lower surface brightness. About halfway to the eastern star it fades, then brightens briefly, but fades again before reaching the eastern point of the triangle. This diffuse portion has very few faint stars attending compared with the brighter southwestern arc.

pn 6894 ***dimen. 44″ × 39″*** ***nebula 12.3v*** ***star 17.6p***
In 25 cm this planetary appears as a round, well-defined disk. At 250× it has a grey color and indefinite annularity. 30 cm shows a fuzzy but distinct annulus 45″ in diameter. The north side is a little brighter, and a faint star is suspected on the outer edge at pa 330° from the center.

oc I4996 ***diam. 6′*** ***V = 7.3*** ***stars 56***
The brightest stars in this cluster form a conspicuous triangle in 25 cm. The triangle is quite small, about 25″ long, with its long base facing N; the brightest star, mag. 8, forms the obtuse southern tip. Spreading N and E from this group are about 18 stars at the threshold of vision. Viewed with 30 cm, the central group has a fourth member; together they form a trapezium lying on the W edge of the cluster. The remainder, composed mostly of mag. 13 and fainter stars, is sprinkled in an uncondensed group to the E. A total of about 30 stars is visible in a 3.′5 area.

gn I1318A ***dimen. 45′ × 20′*** ***star***
Lying about 2° NW of γ Cygni, this faint nebulosity is visible in 6 cm, mixed with mag. 9 and 10 stars. It is elongated roughly N-S, with a close group of three stars on the N side.

oc 6910 ***diam. 8′*** ***V = 6.6*** ***stars 67***
This cluster is located ½° N of γ Cygni (V = 2.2). 15 cm shows a mag. 7.5 star at each end, with a dozen fainter ones running between them in an oblong shape. 25 cm shows 20 stars in a 5′ circle around the bright stars, including some faint isolated stars to the NE. Most of the members lie in a group elongated SE-NW that touches the bright star SE. 30 cm shows about 25 stars loosely scattered between the mag. 7.5 stars.

oc 6913 ***diam. 7′*** ***V = 6.6*** ***stars 81***
Messier 29 is a small, pretty cluster of bright stars in a blank field to 6 cm. It resembles Messier 18, *cf.* oc 6613 in Sagittarius. A dozen stars including six brighter ones are discernable over a little haze. 15 cm reveals at least 15 stars in an 8′ area, while 20 cm will show about 20 stars. The brighter stars give the cluster a box-shaped form. In 25 cm the box-shape is outstanding. About 25 stars are visible here in a 6′ area. The brightest part is 6′ diameter in 30 cm, and the outliers extend to 10′ diameter. A total of 21 stars is visible. The six brightest stars form two arcs concave outward on the NE and SW.

eg 6946 dimen. 11′ × 9.′8 V = 8.8 sfc. br. 13.8
In 6 cm this galaxy looks similar to oc 6939, *cf.* in Cepheus, but it is slightly smaller and totally unconcentrated. 15 cm shows the object well on the N side of a diamond-shaped asterism. In spite of a few stars superposed on the galaxy, it has a uniform, diffuse light. The halo is elongated E-W in 25 cm with a faint core. A wide pair of stars lies 8′ NNW with 50″ separation in pa 40°. For 30 cm, it is a very low surface brightness glow in pa 45°. The core is about 30″ diameter and unevenly bright, with knots visible particularly to the S. The overall size is 3′ × 2.′

gn 6960 dimen. 3.°5 × 2.°7 star
The Veil Nebula is a large, nearly circular network of nebulosity extending over almost 3° of the sky to the S of ε Cygni (V = 2.5). The NGC includes five entries for this supernova remnant: 6960, 6974, 6979, 6992, and 6995. The last two numbers apply to the brightest portion on the NE, which 6 cm will show easily in dark skies as a slightly textured arc about 1° long. The portion passing on the E side of 52 Cygni (6960) is not as conspicuous; a few very faint shreds of light are visible NE of a line connecting these two brighter portions (this is 6974). In 25 cm, 6960 runs 25′ N from the E side of 52 Cygni. This portion is brightest 15′–20′ away from the bright star, where it passes just E of a mag. 10 star. The nebula persists as a thin spike N of here. Extending S and curving E from 52 Cygni is a smooth band of nebulosity. At about 30′ distance it becomes broad and diffuse, fading into a rich field of mag. 11–13 stars. 52 Cygni is a double star (ADS 14259: V = 4.2,9.4; 6.″2; 67°) that is separable at low power in 25 cm.

oc 6996 diam. 7′ mag. 10.0p stars 40
gn 7000 dimen. 100′ × 60′
oc Collinder 428 diam. 14′ mag. 8.7p stars 20
oc 6996 lies on "Ohio" of the North America Nebula, gn 7000. The nebula is best seen with binoculars or the naked eye in dark skies. In 6 cm, oc 6996 is an elliptical hazy patch 15′ × 10′ elongated in pa 150°, with about ten faint stars cast over it. 15 cm shows a faint group of about 25 mag. 11 and fainter stars on a hazy background. It is easily picked out from the field at 50 × in 25 cm. About 60 stars can be counted in a slightly oval area 15′ diameter. 30 cm shows 40 stars of roughly equal brightness scattered evenly in a 15′ area. About 1° E, in the "Pacific Northwest" region of the nebula, is oc Collinder 428. In 25 cm it is a smaller group of about 25 stars just E of a mag. 8 star.

pn 7008 dimen. 98″ × 75″ nebula 10.7v star V = 13.2
A peculiar and richly detailed object in amateur telescopes, this nebula is located on the N side of a mag. 9.3,10.2 pair of 18″ separation in pa 186° (h1606). At 50 × in 25 cm it looks like a cluster of four "stars" associated with the pair. At 200 × and 250 × one of these is the central star surrounded by a dark area. On the NE is a brighter lobe, containing on its W side a bright spot that appears nearly stellar. This spot is directly north of the central star. The overall size is 1.′5 × 1′ in pa 45°. A mag. 14 star lies 25″ NE of center, near the edge but within the nebula. In 30 cm the lobes are aligned obliquely with the overall outline, not on the major or minor axes. The NE lobe is 45″ across; the SE lobe is fainter and more diffuse. The faintly bluish nebula is about 1.′5 × 1.′2 overall. A mag. 14.5 star lies 45″ W.

pn 7026 dimen. 27″ × 11″ nebula 10.9v star V = 14.2
In 15 cm this planetary is much smaller than pn 6826, *cf.*, appearing only about 8″ diameter. In the center is an exceedingly faint stellaring that occasionally seems double. The nebula lies only 25″ WSW of a mag. 10 star, and a threshold magnitude star lies 1′ NW. At 300 × in 25 cm it is round, about 15″ diameter, with a star-like center and diffuse edges. At 425 × in 30 cm the nebula is elongated N-S by 40 percent and appears divided into two bright lobes separated by a sharp dark line. The lobe nearest the mag. 10 star has a stellaring on its outer edge.

pn 7027 dimen. 18″ × 10″ nebula 8.5v star 15.6p
In 15 cm this nebula is faint but not difficult to see. With 25 cm it appears similar to pn 6826, *cf.*, but about two-thirds the size. It has an oval outline, elongated SE-NW, with diffuse edges and a star-like center. At 425 × in 30 cm it is bright with an elongated bar-like shape and a faint surrounding halo. On the NW tip is a bright spot.

oc 7039 diam. 16′ V = 7.6 stars 185
Lying between two mag. 7.5 stars, this cluster looks like a weak concentration of stars in a rich field with 25 cm. About 40 stars are visible, mostly mag. 11–12, spread in a 20′ × 7′ area, but half this number is contained in the central 10′ × 5′ region. oc I1369 (V = 8.8) lies 2° N.

oc 7044 diam. 3.′5 mag. 12.0p stars 60
This is a difficult object for 15 cm, which shows only a faint unresolved nebulosity 3′ across. 25 cm reveals a round diffuse patch with a few faint sparkles and a faint unequal pair on the ENE edge. About ten stars are resolved with 30 cm over a very low surface brightness haze.

pn 7048 dimen. 60″ × 60″ nebula 12.1v star 18:p
This is an annular planetary nebula in 25 cm, lying 3.′5 NE of a mag. 10 star. It is about 1′ diameter, but seems to trail N to a mag. 12 star. Threshold stars are visible on

the NE and W, the latter being involved with the nebula. The nebula is elongated N-S in 30 cm, and the W edge is brighter and more sharply defined. The stars NE and W are both superposed on the haze. A small Y-shaped asterism leads SE from a mag. 11 star off the SSE edge.

oc 7062 ***diam. 7′*** ***V=8.3*** ***stars 85***
This is a conspicuous cluster at low power with 15 cm. Elongated E-W, it is loosely concentrated with ten stars in a 5′ diameter. 25 cm shows about 30 stars within a 5′ circle. In addition, two small dark gaps are visible: one near the center about 1′.5 across, and another of the same size on the E end near a mag. 10 star. 30 cm shows about 35 stars between two mag. 10 stars on the ESE and WSW.

oc 7063 ***diam. 8′*** ***V=7.0*** ***stars 66***
Viewed in 25 cm, this cluster is loosely concentrated, containing 18 stars in a 10′ area. The cluster is poor, but the stars are fairly bright.

oc 7086 ***diam. 9′*** ***V=8.4*** ***stars 80***
In 6 cm this small cluster contains 10–15 stars on a hazy background with two brighter stars near the center. 15 cm shows a slightly concentrated cluster about 10′ diameter with 30 stars. 25 cm reveals 40 stars, including a dozen brighter members. In 30 cm about 50 stars are visible. The greatest concentration of stars is on the SE side, with outliers spreading to the N and W; the cluster is more sharply bounded on the E side. A dark patch in the Milky Way 25′ across is visible 15′ N of the cluster.

oc 7092 ***diam. 32′*** ***V=4.6*** ***stars 28***
Messier 39 is a nearby cluster of bright stars. 6 cm shows 25 or 30 stars in a large triangle pointing N. The brightest stars are on the SE and SW points. 15 cm shows a widespread group without central concentration. About 50 stars are visible, ranging from mag. 7 to 11, including many close pairs. 30 cm shows that the cluster is set against a rich background of mag. 12 and fainter stars.

oc 7127 ***diam. 2′.8*** ***mag.*** ***stars 12***
This small cluster is visible in 6 cm about 15′ NE of an orange mag. 7.5 star. The cluster contains a central star of mag. 9.5 and a few threshold stars to the NW. In 25 cm and 30 cm about ten stars are visible in a 2′ area round the conspicuous central star.

oc 7128 ***diam. 3′.1*** ***V=9.7*** ***stars 71***
In 6 cm this cluster looks similar to oc 7127, *cf.* It is a small, irresolved spot with a mag. 11.5 star on the SE edge. 25 cm shows 15 stars, ten of which are in a circlet including two pairs, a triplet, and three single stars. The cluster is 2′ diameter and roughly octagonal. In 30 cm the circlet is like a jeweled necklace hanging from the cluster. The brightest star in the circlet is reddish. The rest of the cluster is concentrated on the NW side of the circlet and contains about ten more stars.

oc-gn I5146 ***diam. 8′*** ***V=7.2*** ***stars 110***
dimen. 10′×10′
The Cocoon Nebula can be seen in 6 cm as a roughly circular 8′ diameter haze around two mag. 9.5 stars. An ill-defined dark streak leads away from the nebula to the

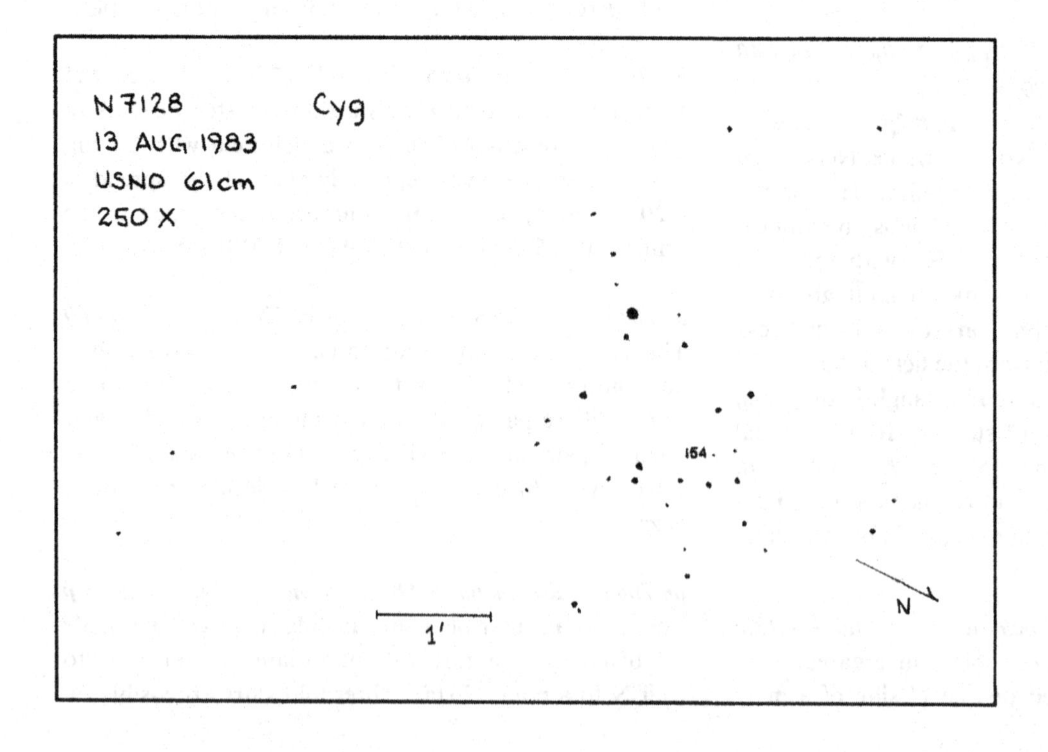

Chart IX. NGC 7128. A 30-minute exposure on 103a-O with the USNO 1 meter telescope taken by J. Christy on 19 July 1963. Magnitude sequence from: Hoag *et al.* 1961, Pubs. of the U.S. Naval Observatory, second series, volume 17, part 7. U. S. Naval Observatory photograph.

W. 25 cm shows the moderately faint nebula about 10′ across. Leading away from it, going W and a little N, is a 2° long "snake" of dark matter seemingly composed of a series of circular patches 10′–15′ across. The "snake" then curves abruptly S, ending as a dark splotchy area about 45′ across. 30 cm shows five fainter stars accompanying the two bright ones in the nebula. The "snake" is of variable width, and its N edge is sharply defined.

Delphinus

A fish-shaped pattern marks this region at the edge of the Milky Way. The constellation is small and contains only a few objects. The center culminates at midnight on 1 August.

pn 6891 dimen. 74″ × 62″ nebula 10.5v star V = 12.4
This planetary is easily seen in 15 cm, with a quite stellar appearance at low magnifications. At 125 × the nebula is a well-defined though featureless spot. There is some concentration evident to 25 cm, and the central star is distinguishable even though the nebula has a high surface brightness. A faint star lies on the E edge. This star is about mag. 13.5 in 30 cm, lying 15″ from the center. At 425 × the nebula is about 10″ across without any discernable elongation. It is broadly concentrated and has a smooth texture.

pn 6905 dimen. 42″ × 35″ nebula 11.1v star 14.2p
This planetary is distinguishable in 6 cm at 25 × . In 15 cm it lies on the W edge of a triangle of stars and appears circular with uniform light and diffuse edges. In 25 cm the nebula has a much lower surface brightness than pn 6891, *cf.* The unevenly bright 40″ disk grows generally brighter toward the center, where a faint sparkle can be seen at 200 × . Close N and S are the western members of the triangle, mag. 11 and 12. With 30 cm the central star is plain, and the nebula is slightly elongated N-S.

eg 6928 dimen. 2.′2 × 0.′8 V = 12.6 sfc. br. 13.1
eg 6930 dimen. 1.′4 × 0.′6 V = 13.1 sfc. br. 12.8
eg 6928 is visible in 15 cm as a small faint patch slightly elongated E-W with a mag. 13 star lying 1.′5 ENE. With 25 cm it appears 1′ × 0.′4 in pa 100° and is broadly concentrated to a "confused" center. 30 cm shows a faint star closely involved NE only 20″ from the center. The halo extends to 1.′2 × 0.′5, but an accurate pa is difficult to estimate because of the nearby star. The core is more circular, only slightly brighter, and has no central concentration of its own. At 125 × , eg 6930 is visible in 30 cm 3.′7 SSE, just N of a narrow triangle of mag. 12.5 stars. It has a very low surface brightness and weak concentration. The halo is 1′ × 0.′3, elongated in pa 15°, reaching almost to the northwestern sharp point of the triangle. Lying 6.′5 W of eg 6928 is a mag. 13.5 star with three close, faint companions (one SW, two SE) that looks a lot like a faint galaxy. eg 6927 (V = 14.5) lies 3′ from − 28 in pa 260°.

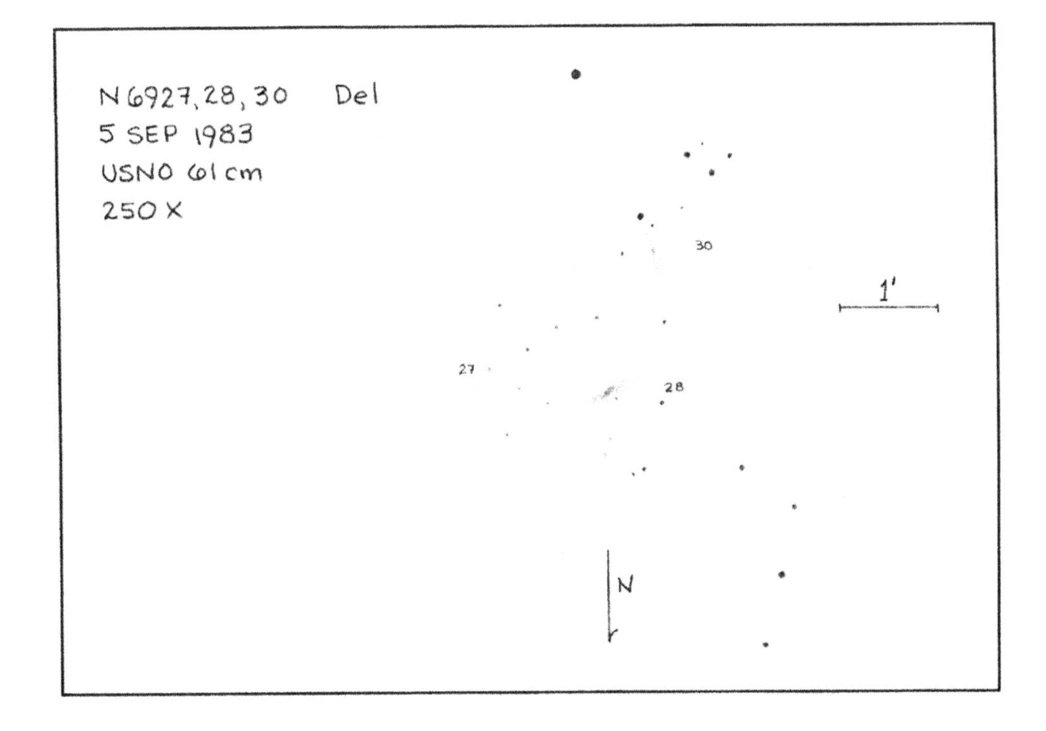

gc 6934 diam. 5′.9 V = 8.7 class VIII
This cluster is visible in 6 cm as a small, moderately faint spot lying 2′ E of a mag. 9.5 star. A typical globular cluster for 15 cm, it is elongated slightly N-S, 1′.5 diameter, and remains unresolved at high powers. At 250 × in 20 cm a grid-like pattern of resolved stars overlies the unresolved but granular central haze. In 25 cm at 300 × it is partially resolved with a conspicuous star near the center. Here the cluster appears about 3′ diameter and seems elongated NE-SW. At 425 × in 30 cm it remains only partially resolved. The core is 40″ across in a halo 1′.75 diameter. The most conspicuous member is on the NE side of the core.

eg 6944A dimen. 1′.2 × 0′.9 mag.$_z$15.0 sfc. br.
eg 6944 dimen. 1′.7 × 0′.8 V = 13.2 sfc. br. 13.4
With 25 cm, eg 6944 is difficult to view clearly. The halo is perhaps 30″ across with a brighter center and a very faint stellar nucleus. Lying 4′ N is a faint pair of stars that looks like a companion galaxy. eg 6944A lies 6′.5 SSW but is not visible in 25 cm.

eg 6954 dimen. 1′.1 × 0′.6 V = 13.2 sfc. br. 12.6
This galaxy is visible in 25 cm at medium power. At 200 × the circular glow is 25″–30″ across and has two threshold stars involved on the N and SE edges. The halo brightens slightly toward the center and a possible faint stellar nucleus, but the nearby stars are confusing.

eg 6956 dimen. 2′.1 × 2′.0 mag.$_z$13.5 sfc. br.
In 25 cm this galaxy is visible at 50 × on the W side of a mag. 10.5 star. At 200 × the nearby star interferes with viewing. The uniformly high surface brightness halo extends to about 45″ and contains a distinct mag. 14 stellar nucleus.

gc 7006 diam. 2′.8 V = 10.5 class I
At a distance of nearly 40 kiloparsecs, this globular cluster is faintly visible in 15 cm. The faint glow is about 1′ diameter with a relatively bright core. With 25 cm it is 1′.25 diameter and shows no resolution even in good seeing. The haze is broadly brighter across the center, exhibiting no sharp central condensation. A mag. 14 pair with 20″ separation in pa 85° is visible 1′.5 S. In 30 cm the cluster is 1′.5 diameter. A few faint stars are resolved over a granular background at 475 ×.

Draco

Winding through twelve hours of right ascension, this circumpolar constellation begins near Vega and twists north and west over the top of the Big Dipper. Several bright galaxies fall within its boundaries. The center of the constellation culminates at midnight about 10 May, but parts may be viewed throughout the year.

eg 3147 dimen. 4′.0 × 3′.5 V = 10.7 sfc. br. 13.4
Though only just visible in 6 cm, in 30 cm this galaxy is bright and well concentrated to a nearly stellar nucleus. The overall size is 2′.5 × 1′.75, elongated SE-NW, with an uneven circular or slightly elongated core 1′.25 across. At 225 × the nucleus is about 20″ across with a brighter center. Three mag. 13 stars enclose the galaxy.

eg 3329 dimen. 2′.1 × 1′.3 B = 13.0 sfc. br. 13.9
This galaxy is fairly easy to see in 30 cm at 150 ×. The halo is elongated in pa 140°, 1′.2 × 0′.6, with a broad, slight concentration. A mag. 13.5 star is visible 1′.4 NNW. The galaxy is the brightest of a group of twelve.

eg 3403 dimen. 3′.1 × 1′.3 B = 12.7 sfc. br. 14.1
30 cm shows a thin galaxy in pa 75°, 2′ × 0′.5 with very little concentration except for a few beads along the N side of the major axis. A faint star is visible about 3′.5 SE.

eg 3735 dimen. 4′.2 × 1′.0 B = 12.5 sfc. br. 13.9
In 25 cm this is a long, narrow galaxy, 3′.5 × 0′.75 in pa 130°. The box-shaped 30″ × 20″ core is displaced slightly NW of the middle of the spindle-shaped halo. A mag. 13.5 star is visible about 1′ NE, and a mag. 14 star lies 2′.5 W of center. With 30 cm the halo appears bar-like, about 3′.25 × 0′.5. Overall it has a slight but uneven concentration to the round core. About 1′ from the center on either side there seem to be slightly brighter patches in the halo.

eg 4121 dimen. 0′.6 × 0′.5 mag.$_z$14.6 sfc. br.
eg 4125 dimen. 5′.1 × 3′.2 V = 9.7 sfc. br. 12.8
eg 4125 is visible to 6 cm as a "companion" 2′.5 WNW of a mag. 10 star. The object is well concentrated to an almost stellar nucleus. In 30 cm it is elongated in pa 90°, 2′.2 × 1′, with an oval core and a bright nonstellar nucleus. eg 4121, located 3′.6 SSW, is visible in 30 cm as a moderately concentrated glow about 40″ across.

eg 4128 dimen. 2′.8 × 1′.0 B = 12.8 sfc. br. 13.7
Viewed in 25 cm this is a small, bright galaxy with a conspicuous stellar nucleus that dominates at 100 ×. The halo is 1′.5 × 0′.3 in pa 60° with a much brighter 40″ × 15″ core and a centrally embedded nucleus. In 30 cm the halo is about 1′ × 0′.5. The 20″ circular core contains a bright but nonstellar nucleus. A pretty kite-shaped asterism of four mag. 7.5–10 stars lies 15′ W.

eg 4236 dimen. 19′ × 6′.9 V = 9.6 sfc. br. 14.7
In 25 cm this large, low surface brightness object is a 20′ × 4′ patch of light elongated in pa 160°. The halo grows very broadly brighter across the center, but there is no distinctly bounded core. In the slightly brighter central region is a very faint star that is probably the nucleus; another lies 4′.5 SE along the E flank. The nebulosity reaches almost to a mag. 10 star that lies 13′ S. This galaxy is difficult to view in 30 cm. The faint, very low surface brightness haze extends to about 15′ × 4′. The halo has a slight, even concentration without a distinct nucleus. A faint stellaring in the halo is visible 4′ NNW of center, and a faint star lies about 3′ W.

eg 4256 dimen. 4′.6 × 1′.0 B = 12.7 sfc. br. 14.1
30 cm shows this galaxy as a thin streak in pa 40°, 4′ × 0′.5, with thinly tapering ends. The rectangular core is about 1′ long and seems to have two brighter areas on either side of it. No bright nucleus is visible.

eg 4291 dimen. 2′.2 × 1′.9 B = 12.2 sfc. br. 13.7
This galaxy forms the NW corner of a nearly perfect 2′ × 1′.5 rectangle of stars. In 25 cm it is small and bright, about 45″ across, with a nearly circular outline. The core is small but quite bright and contains a very faint occasionally stellar nucleus. A mag. 13 star lies 26″ from the center in pa 275° and looks about a magnitude fainter than the central region of the galaxy. A very faint star is visible at 25″ in pa 190°. In 30 cm the galaxy is very evenly concentrated, about 1′ diameter with an 8″ nucleus. eg 4319, *q.v.*, lies 7′.4 SE.

eg 4319 dimen. 3′.1 × 2′.5 mag.$_z$13.0 sfc. br.
eg Mk 205 dimen. V = 15.2 sfc. br.
Compared with eg 4291, *cf.* 7′.4 NW, eg 4319 presents a strong contrast between elliptical and spiral galaxies. In

25 cm the broadly concentrated oval halo is 1′.5 × 1′.25, elongated approximately N-S. The small, more circular core is about 15″ across. With 30 cm the diffuse, tapering halo extends to 2′.2 × 1′.8 around the rectangular core. A mag. 13.5 star lies 1′.5 NW, and lying 40″ from the center in pa 190° a very faint mag. 14 star is visible. This is Markarian 205, a quasar with a redshift of about 21,000 km/sec.

eg 4386 dimen. 3′.0 × 1′.7 B = 12.4 sfc. br. 14.0
This galaxy forms an equilateral triangle with two mag. 12 stars 3′ NE and NW. 25 cm shows the lenticular halo to 1′.5 × 0′.75 in pa 130°, with a more circular core 20″ across. The nucleus is conspicuous, but not quite stellar. Occasionally there appear to be spikes extending SE and particularly NW from the core into the halo. In 30 cm the galaxy has a similar brightness to eg 4291, *cf.* 15′ SW. The halo is about 1′.2 × 0′.75 and shows good concentration to an elongated core with a bright stellar nucleus.

eg 4572 dimen. 1′.7 × 0′.3 mag.$_z$14.9 sfc. br.
eg 4589 dimen. 3′.0 × 2′.7 B = 11.8 sfc. br. 14.1
eg 4589 is not too faint for 15 cm, located about 20′ SW of two similar unequal pairs of stars of about 30″ separation each. It is a moderately bright 2′ × 1′ oval in 25 cm, elongated in pa 110°. The core is conspicuous though small, and contains a substellar nucleus. In 30 cm the halo has a fat lenticular outline, like an eye. Overall it is 2′ × 1′.75 with a round 40″ core and a faint stellar nucleus. A faint star lies on the WNW edge 1′.0 from the center. Lying 7′.5 W is eg 4572, which is visible to 25 cm as a faint, thin lenticular object. It is perhaps 1′.5 long, elongated roughly SE-NW, and without much central brightening.

eg 4648 dimen. 2′.2 × 1′.7 mag.$_z$12.6 sfc. br.
In 25 cm this is a conspicuous galaxy lying among several mag. 10 and 11 stars. The halo has a fairly low surface brightness and is 1′.25 × 0′.75 in extent, elongated in pa 65°. The stellar nucleus rises suddenly in brightness from the halo.

eg 4750 dimen. 2′.3 × 2′.1 B = 12.3 sfc. br. 13.8
In 30 cm this object is elongated in pa 135°, 2′ × 1′.5, with a patchy 1′.2 × 1′ core. The halo is broadly concentrated toward the center, showing no nucleus at any magnification. A string of five stars with a bend in it is visible to the W.

eg 5678 dimen. 3′.2 × 1′.7 B = 12.1 sfc. br. 13.8
Located 2′.6 SE of a mag. 9 star, 30 cm shows a fairly bright object with a broad 3′ × 1′.5 halo elongated N-S. It brightens very gradually to the center with a nonstellar nucleus occasionally visible. A mag. 14 star lies 2′ NNE.

eg 5866 dimen. 5′.2 × 2′.3 V = 10.0 sfc. br. 12.6
This bright lenticular galaxy is easily seen in 6 cm. The high surface brightness halo is clearly elongated SE-NW, and a mag. 11.5 star is visible 1′.6 NNW. With 15 cm another very faint star is visible 1′.5 SW. The halo is 2′ × 0′.75, growing smoothly brighter to a small nonstellar nucleus. The major axis stands out as a brighter streak. In 25 cm the halo extends to 3′ × 0′.75 in pa 125°. The center is very bright, yet a stellar nucleus is just discernable over it. 30 cm shows the halo 2′.8 × 1′.2 in pa 125°. The core is asymmetric, with "wings" extending from its ends on opposite sides of the major axis. This gives the core an apparent pa of 115° or so. No distinct nucleus is visible here.

eg 5879 dimen. 4′.4 × 1′.7 V = 11.5 sfc. br. 13.5
Located 7′.5 SE of a mag. 7 star, this galaxy is moderately faint in 15 cm, appearing as an elongated streak with a brighter center and a very faint stellar nucleus. At 200 × in 25 cm the halo is 1′.5 × 0′.5 in pa 0°. The core has a relatively high surface brightness and is somewhat more elongated than the halo, which is spread faintly over a large area. A faint stellar nucleus is discernable in the center, and a mag. 14.5 star lies a little N of W, 40″ from the center. With 30 cm the extensive, irregular halo grows to 2′.5 × 0′.8. The core is quite elongated and contains a distinct but nonstellar nucleus.

eg 5905 dimen. 4′.2 × 3′.3 B = 12.3 sfc. br. 15.0
Despite its low surface brightness, 15 cm shows this galaxy lying 4′.2 NNW of a mag. 11.1,11.5 pair of 9″ separation in pa 17° (Wirtz 12). It appears as a faint, roughly circular glow about 1′ diameter with little central brightening. In 25 cm the halo is circular, about 1′.75 diameter with a mottled texture that grows only a little brighter across the middle. It is a faint but very unevenly bright glow with 30 cm. The halo extends to 2′.5 diameter with indefinite elongation, though the core seems to be more clearly elongated E-W. A mag. 13.5 star lies 1′.5 E. eg 5908, *q.v.*, lies 13′ SE.

eg 5907 dimen. 12′ × 1′.8 V = 10.2 sfc. br. 13.4
This galaxy is faintly visible to 15 cm as a 5′–6′ streak with vague hints of detail in the core. A faint star is visible 1′ W of center. In 25 cm it extends to as much as 9′ × 0′.75 in pa 150°. The middle parts around the elongated 2′ core don't bulge, though the W edge of the halo is more diffuse here. The N side of the halo extends 1′.5 past a mag. 14 star that lies off the NE flank; the S half ends just even with a mag. 14.5 star 2′ E of the S tip.

Stellarings are visible in the core, and another is discernable 3′ S of the center in the arms. In 30 cm the halo is 9′ × 0.′8, brightening evenly through the core: no nucleus stands out. Along the W flank a dark lane is visible for about 1.′5 on either side of the center.

eg 5908 dimen. 3.′2 × 1.′3 V = 11.9 sfc. br. 13.3
In 15 cm this galaxy is much brighter than its companion eg 5905, *cf.* 13 NW. The halo is circular, about 30″ diameter. 25 cm shows the lenticular halo to 2′ × 1′ in pa 150°. The major axis stands out as a conspicuous streak with stellarings in the center. Around the outer edge of the core the halo is brighter in a way that gives the impression that the center parts of the galaxy are illuminating the outer parts. In 30 cm this galaxy, along with eg 5905, lies in a rich field of mag. 12.5–13 stars. It is brighter but smaller than −05, about 2′ × 1′ in extent. The inner core is quite elongated inside the broad, slightly concentrated halo.

eg 5949 dimen. 2.′4 × 1.′2 B = 12.9 sfc. br. 13.9
Located 8′ SE of an unequal pair (ADS 9666: 9.6,11.1; 19″; 45°), 30 cm shows this galaxy fairly faintly, though it is not difficult to see. The mottled surface is elongated in pa 145°, about 1.′5 diameter. The halo has a pear-shaped outline, larger on the SE. A mag. 14 star is visible 2′ E.

eg 5981 dimen. 2.′8 × 0.′6 V = 13.0 sfc. br. 13.3
eg 5982 dimen. 2.′9 × 2.′2 V = 11.1 sfc. br. 13.2
eg 5985 dimen. 5.′5 × 3.′2 V = 11.0 sfc. br. 14.0
eg 5982 is the brightest of this group. In 15 cm the moderately faint circular 30″ halo grows gradually brighter across the center and shows a stellar nucleus at 125 ×. In 25 cm the halo is slightly elongated in pa 100°, about 45″ × 35″. The core is about half this size, but is not well defined against the halo. The western side of the core seems brighter. The halo seems circular in 30 cm, but the inner parts are elongated E-W. Located 7.′7 E of −82 is eg 5985, which is fairly faint in 15 cm, but not much harder than −82 because of its larger size. High power gives a 2.′5 × 1′ halo, elongated N-S. The core is inconspicuous, lying a little N of the geometric center of the halo. 25 cm shows the mottled halo to 3′ × 1.′2 in pa 20°. The inconspicuous core is quite eccentric and more elongated than the halo, about 20″ × 5″. Overall the object has a moderately low surface brightness, but fades suddenly to a relatively distinct edge. With 30 cm the mottled, poorly concentrated halo extends to about 4′ × 3.′5 with a 50″ core. eg 5981 lies 6.′3 W of −82 and is only just visible to 15 cm. With 25 cm the faint, thin object is about 1.′2 × 0.′3 in pa 135°. The halo is faintly mottled, but exhibits no general brightening toward center. Photographs show eg 5989 (mag.$_z$13.6) 32′ NE of eg 5982.

eg 6015 dimen. 5.′4 × 2.′3 V = 11.1 sfc. br. 13.6
This galaxy is faintly visible to 15 cm about 2.′5 E of a mag. 11 star. In 25 cm it is 3′ × 1.′25 in pa 30°, a fat oval broadly brighter to the center with a narrow central bar occasionally visible. It grows to 5.′5 × 1.′8 with 30 cm, with weak concentration to a broad core. Several stellarings are involved with the object, particularly on the SW. A mag. 13.5 star is visible within the halo 2′ S.

eg 6340 dimen. 3.′4 × 3.′0 V = 11.0 sfc. br. 13.4
15 cm shows a small, fairly faint object here 2′ SE of a mag. 11.5 star. The galaxy appears circular, about 1′ diameter, with a much brighter center. 25 cm shows a mag. 12.5 companion to the star NW. The nucleus is bright but nonstellar, and the overall view is similar to that in 15 cm. 30 cm shows it in pa 120°, 1.′75 × 1′ with a circular core 40″ across. The texture is smooth except for some stellarings around the stellar nucleus.

eg 6412 dimen. 2.′3 × 2.′1 V = 11.8 sfc. br. 13.4
25 cm shows this galaxy as a faint glow 2′ NW of a mag. 11 star with other fainter stars nearby. The halo is about 1′ diameter without central concentration. In 30 cm the object lies N of a string of three mag. 9–11 stars 7′ long in pa 30°. It appears 2′ across with a slight, smooth concentration. A threshold magnitude star is visible on the SSW edge 1′ from the center.

eg 6503 dimen. 6.′2 × 2.′3 V = 10.2 sfc. br. 12.9
6 cm will show this galaxy as a high surface brightness lenticular patch of light elongated in pa 120°. A star is visible nearby E. In 15 cm it has a brighter center with an overall size of 3′ × 0.′75. 25 cm reveals a bright lens that is much fainter on the E side: the core appears off-center to the W. The overall size is 4′ × 0.′75. At 150 × in 30 cm the core appears to have bright speckles on it. At high power the core seems elongated more directly E-W than the halo. The core is large and mottled and the N flank of the halo is more sharply defined. A mag. 13.5 star lies 2.′2 N.

pn 6543 dimen. 23″ × 17″ nebula 8.1v star V = 11.3:
This bright planetary nebula is visible in 6 cm as a starlike spot 3′ ESE of a mag. 8 star. The nebula is clearly discerned in 15 cm. No details can be seen though the outline is somewhat elongated. It appears elongated in pa 15° with 25 cm, and the outline is not symmetrically oval. The N end is rounded, while the S side is more sharply tapered but diffuse. On the N portion is a bright nonstellar spot. The outer edge is well-defined with a faint halo 20″ wide around it. At 225 × in 30 cm it is elongated in pa 20°, 20″ × 15″, with a very smooth texture. The northern two-thirds is brighter with a few very

small bright spots on the periphery. At 450× the southern end is diffuse while the northern side has a broken bright edge on the E and W sides.

eg 6643 ***dimen. 3.′9 × 2.′1*** ***V = 11.1*** ***sfc. br. 13.2***
A faint object for 15 cm, this galaxy forms a small triangle with two mag. 12 stars to the W. The halo is elongated NE-SW and is about 1′ diameter. It is broadly concentrated and without a sharp center in 25 cm. The overall size is 1.′5 × 1′ with a bright spot occasionally visible SW of center. 30 cm shows it to 3′ × 1.′75 in pa 40°. The halo has a mottled grainy texture and grows only a little brighter to the center, yet contains a faint stellar nucleus.

Equuleus

Equuleus is a small, inconspicuous region containing only two objects easily accessible to amateur instruments. The center culminates at midnight on 9 August.

eg 7015 ***dimen.*** $2'.0 \times 1'.8$ ***mag.*** $_z13.2$ ***sfc. br.***
This faint galaxy is not visible in 15 cm. In 25 cm it is a fairly faint cigar-shaped nebula with only a little concentration. Elongated approximately N-S, the halo is about $30'' \times 20''$. Lying 2.'5 S is a faint pair of stars with about 20″ separation that appears nebulous at low magnifications. Between the pair and the galaxy lies a very faint star. A good view is afforded by 30 cm, showing a smoothly textured oval elongated E-W. With uneven edges, the halo is $1' \times 0'.8$ with moderate concentration to an elongated core containing a faint sparkle in or just S of its center. The galaxy forms a triangle with the pair to the S and a single star to the W. A mag. 14.5 star is just visible to the NE, thus forming a triangle around the galaxy.

eg 7046 ***dimen.*** $2'.0 \times 1'.5$ ***mag.*** $_z14.2$ ***sfc. br.***
A faint object for 25 cm, this galaxy is visible at 100×. The halo is elongated N-S, $40'' \times 20''$, exhibiting almost no central brightening. A wide double star is visible 3′ SE. Closer to the S and SSE two fainter stars form an arc with the pair, curving up to the S end of the galaxy. Although it is difficult at low magnifications in 30 cm, higher powers show it 50″ diameter with uneven brightness.

Eridanus

Beginning near the foot of Orion, this large constellation winds south to nearly −60° declination, terminating at the bright star Achernar. The area contains many galaxies, including some at the edge of the Fornax I cluster. The center of the constellation culminates at midnight about 10 November.

eg 1084 dimen. 2.′9 × 1.′5 V = 10.6 sfc. br. 12.1
15 cm shows this galaxy as a fairly bright, round, and uniformly illumined patch about 1′ diameter. In 25 cm the high surface brightness halo is elongated NE-SW, about 1.′75 × 1′, and seems divided into two parts: a large circular area with a smaller, slightly fainter oval appendage on the NE side. A dark lane is occasionally visible between them. 30 cm shows it 2.′5 × 1′ in pa 40°. The broad, oval core is about 1.′3 × 0.′8 and fades more abruptly on the NW side.

eg 1140 dimen. 1.′4 × 0.′9 V = 12.5 sfc. br. 12.5
In 30 cm this tiny object is only 45″ diameter. The relatively large core is moderately brighter than the halo, about 20″ across, and a stellar nucleus is occasionally visible.

eg 1172 dimen. 2.′2 × 2.′0 V = 12.0 sfc. br. 13.6
Though not visible in 15 cm, this galaxy is not difficult to view in 25 cm. The unconcentrated 30″ diameter glow lies 2′ WSW of a mag. 9.5 star. With 30 cm it is circular, 45″ diameter, with little concentration and low surface brightness.

eg 1179 dimen. 4.′6 × 3.′9 B = 12.2 sfc. br. 15.2
While this galaxy is not visible in 15 cm, 25 cm shows a tiny star-like spot with a faint halo 30″ diameter. With 30 cm it is diffuse and unconcentrated, with very low surface brightness. Small mottlings are visible around the middle of the faint irregular halo that extends to 3′ diameter. A mag. 13.5 star just touches the halo at pa 120°.

eg 1187 dimen. 5.′0 × 4.′1 B = 10.9 sfc. br. 14.1
A large, diffuse object, this galaxy is faintly visible in 15 cm about 4.′5 SE of a mag. 8 star. It appears about 3′ across and a little brighter to the center, though some bright lumps are visible. The halo is 4′ × 3′ in 25 cm, elongated in pa 135°. The center is not brighter, but some patches are visible on the S side. With 30 cm the core seems elongated in pa 120°, tilted within the lumpy halo. A faint stellar nucleus is visible in the center.

eg 1199 dimen. 2.′2 × 1.′9 V = 11.5 sfc. br. 13.1
Lying about 3′ SW of a mag. 10 star, this galaxy is faintly visible in 15 cm. In 25 cm it is circular, 50″ diameter, with a diffuse border. It is somewhat brighter to the center with a nonstellar nucleus. With 30 cm it is just less than 1′ diameter. The broadly concentrated core does not have a sharp nucleus. eg 1209, *q.v.*, lies 35′ E; four other faint members of the eg 1209 group are 3′–5′ W and SW.

eg 1209 dimen. 2.′6 × 1.′5 V = 11.4 sfc. br. 12.8
A very faint object for 15 cm, this galaxy appears less than 1′ diameter with a moderately faint star-like center. 25 cm shows it 1′ × 0.′8 in pa 80°. The galaxy looks like an egg with a double stellar yolk: a double nucleus is pretty clearly seen, with the E component the fainter. The core looks very thin; the halo is rounded and more abrupt on the W. With 30 cm it is 1′ × 0.′5 with a 30″ diameter circular core and a nearly stellar nucleus. This galaxy is the brightest of a group including eg 1199, *q.v.*, 35′ W.

eg 1232 dimen. 7.′8 × 6.′9 V = 9.9 sfc. br. 14.1
15 cm shows this object SE of a mag. 8.5 star. It appears round and diffuse, about 3′ diameter. In 25 cm the diffuse, lumpy halo grows to 3.′5 diameter with a 30″ indistinct core. It is large and moderately concentrated in 30 cm, about 5′ diameter with a faint stellar nucleus. The core is quite unevenly bright with several dark areas. A mag. 13.5 star is visible off the NE edge. eg 1232A, a Magellanic-type companion, lies at 4.′2 in pa 110°.

eg 1241 dimen. 3.′0 × 1.′9 B = 12.7 sfc. br. 14.4
eg 1242 dimen. 1.′3 × 0.′7 mag.$_{vv}$14 sfc. br.
eg 1241 can be seen only with difficulty in 15 cm. It is moderately faint in 25 cm, located 3′ S of a mag. 9 star. The halo is circular, 1′ diameter, with irregular edges. In 30 cm it is 1.′5 diameter, possibly elongated SE-NW, and evenly concentrated. A very faint pair of stars (13,13; 12″; 185°) is visible 3′ ESE. eg 1242 is just visible in 30 cm 1.′5 NE. It is very small, extending to only 20″ diameter, but is well concentrated.

eg 1297 dimen. 2′.3 × 2′.0 B = 12.6 sfc. br. 14.2
A very small object, this galaxy is not observable in 15 cm. It is only 20″ diameter with 25 cm and quite faint, but with a brighter center. A mag. 13 star is visible 1′ N in pa 20°. In 30 cm the halo is 1′ diameter with a brighter 30″ core. The galaxy is moderately concentrated but has low surface brightness.

eg 1300 dimen. 6′.5 × 4′.3 V = 10.4 sfc. br. 13.8
Although this galaxy is often seen in photographs as a beautiful barred spiral, it is not as spectacular a sight in amateur instruments, and 15 cm will show it only marginally. 25 cm shows a faint bar in pa 100°, 4′ × 1′.5. It has a grainy texture and is little brighter across the center. In 30 cm the core is 50″ across with a stellaring at or near the center, and the halo extends out to a maximum of 2′ diameter.

eg 1309 dimen. 2′.3 × 2′.2 V = 11.6 sfc. br. 13.2
Located 4′ NE of a mag. 7.5 star, this object is fairly bright in 25 cm. A very faint stellar nucleus is visible inside a smooth 1′.5 diameter halo. The unevenly bright halo appears nearly circular in 30 cm, 1′.5 diameter, elongated NE-SW. It shows a moderate concentration to a small, slightly brighter nonstellar nucleus.

eg 1325 dimen. 4′.6 × 1′.8 V = 11.6 sfc. br. 13.6
Though not visible in 15 cm, 25 cm will show this galaxy as a diffuse unconcentrated glow about 1′ across lying just WSW of a mag. 11.5 star. A mag. 13 star is visible 1′.5 ESE, forming a triangle with the galaxy and the brighter star. 30 cm reveals a star superposed very close to the center in pa 100°.

eg 1331 dimen. 1′.1 × 1′.0 mag.$_{vv}$14 sfc. br.
eg 1332 dimen. 4′.6 × 1′.8 V = 10.3 sfc. br. 12.4
eg 1332 is a bright object for 15 cm. The halo is elongated E-W, 2′ × 0′.5, with a stellar nucleus in a very bright core. 25 cm shows it about the same size and elongated in pa 110°. A stellar nucleus is visible inside a very small, bright core. The halo is faint by comparison. The halo extends to 2′.5 × 0′.8 in 30 cm. The core is bright and round, 30″ diameter with a nearly stellar nucleus. eg 1331, though not visible in 25 cm, appears as a star-like spot in 30 cm, 2′.8 ESE of −32.

eg 1337 dimen. 6′.8 × 2′.0 V = 11.6 sfc. br. 14.3
This object is faintly observable in 25 cm, appearing elongated SE-NW, 3′ × 0′.5. The core is broadly concentrated and seems mottled with bright and dark points. 30 cm shows the halo to 4′ × 0′.6 in pa 150°. The brighter part is 1′.8 × 0′.6, the ends of the core seemingly broken into pieces. A mag. 13 star lies 2′.6 ESE.

eg 1353 dimen. 3′.4 × 1′.5 V = 11.4 sfc. br. 13.0
In 15 cm this galaxy appears as a faint patch elongated SE-NW. The halo is 1′.5 × 0′.75 in pa 135° in 25 cm, with a thin, spike-like southeastern perimeter and a comparatively oval northwestern boundary. 30 cm shows the halo 2′ × 0′.9, containing a circular 50″ core and a slightly brighter 20″ central region. A mag. 11.5 star lies 3′ SE.

eg 1355 dimen. 1′.8 × 0′.5 V = 13.1 sfc. br. 12.9
eg 1358 dimen. 2′.8 × 2′.1 V = 12.1 sfc. br. 13.9
Neither of this pair of galaxies is visible in 15 cm. 25 cm shows eg 1358 between two mag. 11.5 stars, the one E being a pair (14″; 25°). The faint halo is about 30″ diameter and exhibits no central concentration other than a faint stellar nucleus. 30 cm shows −58 in a string of mag. 11.5 stars leading SW from a mag. 8.5 star plotted on the AE. The narrow core seems elongated N-S with a bright inner part and a stellar nucleus. The halo is irregular, spreading mostly to the E side of the core. It is 1′.5 × 1′.2 overall, the core 50″ across. eg 1355 lies 6′.8 NNW. In 30 cm it is very small, but has moderate surface brightness. The halo is 30″ × 12″ in pa 80°, with a bright substellar nucleus on the E side of center.

eg 1357 dimen. 2′.5 × 1′.8 V = 11.7 sfc. br. 13.1
This galaxy is a fairly bright object for 25 cm, forming a flat triangle with two mag. 8 stars. It is 2′ × 0′.75 in pa 75°, an oval with a stellar nucleus. With 30 cm it is 1′.5 × 1′.3 with good concentration to a faint stellar nucleus that occasionally seems multiple. Two faint stars E and S make a box around the galaxy with the two brighter stars.

eg I1953 dimen. 2′.8 × 2′.3 B = 12.3 sfc. br. 14.2
Lying very close to τ^5 Eridani (V = 4.3), this object is not visible in 15 cm. 25 cm shows an unconcentrated blob 2′ diameter with some stellarings on it. It is a very diffuse 1′.1 diameter glow in 30 cm, brightening slightly to a 30″ core.

eg 1359 dimen. 1′.9 × 1′.7 V = 12.2 sfc. br. 13.3
Located 4′.5 SSW of a mag. 10.5 star, this galaxy is not visible in 15 cm. It is inconspicuous to 25 cm, which shows it about 1′ diameter and a little brighter at the center. It appears stellar in 30 cm, with a 30″ haze surrounding a mag. 12.5 stellar nucleus.

eg 1376 dimen. 2′.0 × 1′.9 B = 12.8 sfc. br. 14.1
This is a moderately faint object in 15 cm, appearing about 1′ diameter. 25 cm shows a round dot with no central concentration or features. The irregularly bounded and unevenly bright halo extends to 1′.25 diameter in 30 cm. It shows a slight, broad concentration, but no distinct core or nucleus.

eg 1386 dimen. 3.′5 × 1.′5 B = 12.3 sfc. br. 13.9
eg 1369 dimen. 1.′4 × 1.′3 mag.$_{VV}$13 sfc. br.
A member of the Fornax I cluster, eg 1386 is a bright galaxy in 25 cm. The halo extends to 2′ × 0.′6 in pa 25°. Broadly concentrated toward the center, the core is about 20″ × 12″ with a faintly mottled texture and a stellar nucleus. A faint star is visible 1.′8 NW. With 30 cm the halo is about the same size and has a low surface brightness. Lying 15′ S, 4.′5 NW of a mag. 7.5 star, is eg 1369, which is visible in 25 cm as a small, inconspicuous spot.

eg 1389 dimen. 2.′1 × 1.′4 V = 11.5 sfc. br. 12.7
On the SW corner of a quadrilateral of mag. 10–12 stars, 25 cm shows this galaxy easily. The halo is about 45″ diameter, slightly elongated NE-SW, with a broadly brighter, ill-defined core and a stellar nucleus. 30 cm shows the galaxy 3′ S of the brightest member of the quadrilateral. The halo is 45″ diameter here also with moderate concentration, but the stellar nucleus is not visible. The galaxy is a member of the Fornax I cluster.

eg 1395 dimen. 3.′2 × 2.′5 B = 11.1 sfc. br. 13.4
This is a bright object for 25 cm. The halo is 1.′25 × 0.′75 in pa 100°, with a bright and well-defined circular core and a stellar nucleus. A mag. 14 star is visible on the W edge 30″ from the center. With 30 cm the halo extends to 2.′2 × 1.′8, elongated E-W, but the faint star embedded to the W may be influencing this. Overall it is well concentrated to a bright nucleus that is nearly stellar at low magnifications. A mag. 14 star is visible 1′ N. eg 1401 (mag.$_{VV}$13.5) lies 21′ NE.

eg 1400 dimen. 1.′9 × 1.′7 V = 11.1 sfc. br. 12.2
15 cm shows this moderately faint galaxy less than 1′ diameter with a stellar nucleus; it appears fainter than eg 1407, *cf.* 12′ NE. In 25 cm it is 45″ across with a conspicuous stellar nucleus. The halo grows to 1.′2 diameter in 30 cm and is slightly concentrated to a 30″ core.

eg 1404 dimen. 2.′5 × 2.′3 V = 10.1 sfc. br. 12.1
This galaxy in the Fornax I cluster is bright in 25 cm. The high surface brightness object is 45″ diameter, brightening gradually and evenly to a stellar nucleus. A mag. 12.5 star lies 40″ SE. With 30 cm this galaxy has a higher surface brightness than eg 1399, *q.v.* 13′ NNW in Fornax. The halo extends to 1′ diameter with a fairly bright core, but no nucleus is discernable here.

eg 1407 dimen. 2.′5 × 2.′5 V = 9.8 sfc. br. 11.7
In 15 cm this galaxy appears larger and brighter than eg 1400, *cf.* 12′ SW. It is about 1′ diameter with a bright stellar nucleus. 25 cm shows a small core and a stellar nucleus in a 1.′5 diameter halo. The halo is about 1′ diameter in 30 cm, with a well-concentrated core 35″ across and a very small nucleus. This galaxy is the brightest of a group including eg 1400.

eg 1415 dimen. 3.′6 × 2.′1 B = 12.4 sfc. br. 14.4
This is a moderately faint object for 25 cm. The halo is 1′ × 0.′3 in pa 140° with pointed tips. The core is oval and has a dotted streak along its major axis. With 30 cm darkenings are visible in the core around the nucleus. The overall size is 1.′5 × 1′, and the E side seems to fade more abruptly.

eg 1417 dimen. 2.′8 × 1.′9 V = 12.0 sfc. br. 13.7
eg 1418 dimen. 1.′5 × 1.′2 mag.$_{VV}$14.5 sfc. br.
25 cm shows eg 1417 elongated N-S, 1.′2 × 0.′75, with a sparkle or two in the center, and a mag. 11 star nearby 1.′3 SE. It is small and faint in 30 cm. The halo is 1.′25 × 1′ with a moderately brighter core 10″ across but no nucleus. eg 1418, 5′ SE, is visible in 30 cm as a round, unconcentrated spot 30″ diameter 1.′3 NNW of a mag. 13 star. eg I 344 (mag.$_{VV}$15) lies 7.′1 NW of eg 1417.

eg 1421 dimen. 3.′6 × 1.′0 V = 11.2 sfc. br. 12.7
In 15 cm this galaxy is faintly visible, appearing about 2.′5 long. In 25 cm the halo is 3′ × 0.′5, elongated N-S, with a 1′ × 0.′25 core. The surface is knotty with stellarings, and three or four dark lanes seem to cut across the major axis. A mag. 12 star is visible W of the N tip. 30 cm shows a nice bar-like halo extending to 3′ × 0.′6 in pa 0°. The core is generally brighter on the N end and has three or four blotches along its major axis. A mag. 12.5 star is visible E of the S tip, 2.′6 from the center.

eg 1426 dimen. 2.′1 × 1.′5 V = 11.4 sfc. br. 12.7
A curious object for 25 cm, this galaxy shows little definite detail. The starry core is double: the central stellar nucleus has a nearly stellar spot on its N edge. The halo is elongated roughly SE-NW, about 1′ diameter. It appears small in 30 cm, about 55″ across with a 15″ core. The nucleus is more prominent at low power.

eg 1437 dimen. 2.′9 × 2.′1 B = 12.6 sfc. br. 14.4
A member of the Fornax I cluster, this object is difficult to see in 30 cm, having a low surface brightness. The faint, uncondensed patch extends irregularly to about 2′ × 1.′5 in pa 150° and has an unevenly bright texture.

eg 1439 dimen. 2.′3 × 2.′2 B = 12.5 sfc. br. 14.3
A small, fairly faint object for 25 cm, this galaxy appears circular, 45″ diameter, with a sharp bright nucleus. In 30 cm it is 1′ × 0.′5, elongated roughly N-S. The halo exhibits very little concentration except for the bright stellar nu-

cleus. At 225× the nucleus loses its stellar appearance, but a faint star is visible just N of it.

eg 1440 dimen. 2.'3 × 1.'9 B = 12.7 sfc. br. 14.2
This small object appears moderately faint in 15 cm and contains a stellar nucleus. The faint round halo is about 40″ diameter in 25 cm, and the nucleus still appears stellar. In 30 cm at low power the nucleus is bright but nonstellar. At 225× the halo is elongated in pa 30°, 1.'1 × 0.'7, and shows little concentration except for the small 10″ nucleus.

eg 1452 dimen. 1.'7 × 1.'5 B = 13.1 sfc. br. 14.0
This object appears similar to eg 1440, *cf.* 22′ N, in 15 cm, though it is somewhat larger. The halo is about 50″ across. It is 1′ diameter and roughly circular in 25 cm, with a stellar nucleus. The halo is slightly elongated SE-NW, but the brighter parts are in pa 45°. 30 cm shows it 1.'5 × 1′ in pa 45°. The nucleus is about 10″ across but appears almost stellar at 150×.

eg 1453 dimen. 2.'1 × 1.'6 V = 11.6 sfc. br. 12.9
This is a small but bright object in 25 cm. The halo is less than 1′ diameter with a well-defined core 30″ across and a stellar nucleus. With 30 cm the object is 50″ diameter and well concentrated to an 8″ core and nearly stellar nucleus. The light fades evenly to the edge. This galaxy is the brightest of a group.

eg 1461 dimen. 3.'3 × 1.'1 V = 11.7 sfc. br. 13.1
A pretty bright object for 25 cm, this galaxy is elongated in pa 160°, 1.'5 × 0.'5, with a distinct elongated core. A mag. 10 star lies 3.'3 NW. 30 cm gives it 1.'2 × 0.'5 with very pointed ends. The halo is well concentrated to a bright core and stellar nucleus that seems off-center to the N.

eg I2006 dimen. 2.'3 × 2.'1 V = 11.4 sfc. br. 13.0
Viewed in 30 cm, this galaxy appears as a small, roundish patch with splotchy concentration. The halo is about 1.'25 diameter, but at 225× it seems slightly elongated N-S. The nucleus appears nearly stellar at 150× within an irregularly bright core. A pair of stars with contrasting colors is visible 5′ SW (9,10; 20″; 150°).

eg 1507 dimen. 3.'4 × 1.'0 V = 12.2 sfc. br. 13.4
This object is barely observable in 15 cm, appearing as a faint 2.'5 × 1′ patch elongated N-S. It is moderately bright in 25 cm, extending to 3′ × 0.'5. The central 1.'5 × 0.'5 area is mottled. A bright star lies 3.'5 SE and a mag. 12 star is closer WNW. 30 cm shows it 3′ × 0.'5 in pa 10° with a more sharply defined E flank. There is hardly any concentration except for five or six beads strung out for 2′ length along the major axis.

eg 1518 dimen. 3.'0 × 1.'4 V = 11.8 sfc. br. 13.2
Lying just NE of a mag. 9 star, 25 cm shows this galaxy as an elongated patch about 2′ × 0.'75 in pa 40°. In 30 cm the bar-like halo extends to 3.'2 × 1′ and seems to fade more abruptly on the NW. Overall it is poorly concentrated to a faint nucleus. The inner region seems elongated about 10°–15° closer to N-S than the halo. eg 1521, *q.v.*, lies 22′ ENE.

eg 1521 dimen. 2.'9 × 1.'9 V = 11.4 sfc. br. 13.3
This is a fairly faint galaxy for 25 cm, located 4.'3 N of a mag. 8 star. It appears circular, about 1.'5 diameter, with weak concentration to the core. A mag. 12 star lies 1.'5 W. In 30 cm the nucleus is nearly stellar. The brightest part of the nebula is about 35″ × 25″, elongated NE-SW, and the halo extends out to 1′ diameter.

eg 1531 dimen. 1.'3 × 0.'8 V = 12.1 sfc. br. 12.2
eg 1532 dimen. 5.'6 × 1.'8 B = 11.5 sfc. br. 13.9
This pair of galaxies lies 1.'7 apart. In 25 cm eg 1531 is elongated SE-NW, 1′ × 0.'3, with no central concentration. It is moderately concentrated in 30 cm to a faint occasionally stellar nucleus. The overall size is 1.'2 × 1′ in pa 135°. eg 1532 is elongated NE-SW, 2.'5 × 0.'5, in 15 cm with a slightly brighter center. It grows to 3′ × 0.'5 in 25 cm with a grainy box-shaped core 1.'25 × 0.'5. 30 cm shows it 6′ long in pa 35° with tapering ends and a bulging middle. The core is about 1.'25 long and seems tilted within the halo at pa 45°. The brightness is uneven within 2′ of the center: a slightly condensed patch 35″ across is visible 1.'5 SW of the center; a prominent dark lane passes across the NW side of the core. The two galaxies make a beautiful contrasting pair.

pn 1535 dimen. 48″ × 42″ nebula 9.6v star V = 12.6:
This is one of the best planetaries for amateur viewing. It is visible in 6 cm as a mag. 9 star 30′ E of an equilateral triangle of mag. 8.5 stars plotted on the AE. The nebula is conspicuous in 15 cm, with a bright inner circle visible inside an incomplete narrower outer shell. The central star is suspected. With 25 cm the central star can be seen clearly. The bright well-defined core is 20″ across with a fainter halo out to 35″ diameter. The color is creamy blue. 30 cm shows it elongated in pa 10°, 25″ × 18″ overall. At 225× the bright inner zone surrounding the central star is 12″ × 8″. The fuzzy halo fades abruptly at its outer edge. At 450× the inner core has a bright outer rim, and its surface is riddled with holes. Three mag. 13.5 to 14.2 stars lie 2.'5 W.

eg 1537 dimen. 2.'4 × 1.'6 V = 10.7 sfc. br. 12.2
A moderately faint object for 15 cm, this galaxy appears round, less than 1′ diameter, with a stellar nucleus. In

25 cm the halo is about 1′.25 across, averted vision showing it elongated in pa 100° with thin spikes extending from the core. The core is very small and contains a stellar nucleus. Some faint stellarings can be seen occasionally over the object. In 30 cm the bright nucleus is less than 5″ across—substellar—at 225 ×. The inner core is quite elongated in pa 100°, as is the halo. The nucleus and inner core lie closer to the N side, making that side appear more abrupt. The overall size is 1′.25 × 0′.6.

eg 1600 ***dimen. 2′.5 × 1′.8*** ***V = 11.1*** ***sfc. br. 12.7***
eg 1601 ***dimen. 1′.0 × 0′.7*** ***V = 13.8*** ***sfc. br. 13.4***
eg 1600 lies on the S side of a thin isosceles triangle of mag. 7.5 stars 30′ long. It appears about 1′.5 across in 15 cm. In 25 cm the smooth oval is elongated in pa 10° with a somewhat concentrated core that contains a stellaring at its SSW end. In 30 cm the object forms an equilateral triangle with two stars of a four-star staggered string running SE-NW. It is 1′ × 0′.6 overall, with a 25″ circular core and a stellar nucleus. Lying 1′.6 NNE is a very faint companion, eg 1601, which is a threshold magnitude stellar spot in 25 cm. eg 1600 is the brightest of a group, including eg 1601 and eg 1603 (mag.$_{VV}$15.5), which lies 2′.5 ESE of eg 1600.

eg 1618 ***dimen. 2′.8 × 1′.2*** ***V = 12.6*** ***sfc. br. 13.7***
In 30 cm this galaxy is clearly visible lying 13′ N of ν Eridani. The smoothly textured glow is about 2′ × 1′, elongated in pa 30°. The halo is evenly concentrated to more circular inner regions and a faint nucleus. A symmetric 1′.8 × 0′.8 diamond of mag. 13–14 stars lies 1′.5 E, and a mag. 14.5 star lies 1′.2 N of center.

eg 1622 ***dimen. 3′.9 × 1′.1*** ***V = 12.3*** ***sfc. br. 13.7***
Lying 11′ NNE of ν Eridani, 30 cm shows this galaxy fairly easily despite the proximity of the bright star. The faint halo extends to about 3′ × 1′ in pa 35°, showing a sharp but uneven central concentration. The more circular core is about 1′.5 × 1′ with many stellarings within. The nucleus stands out clearly, but is not quite stellar and seems displaced a little NE from the center of the core.

eg 1625 ***dimen. 2′.7 × 0′.8*** ***V = 12.4*** ***sfc. br. 13.0***
This is the brightest of three similar galaxies near the bright star ν Eridani (V = 3.9), which interferes with observing, particularly in smaller apertures. In 25 cm it is faintly visible at 200 × as a slight brightening on the sky with a mag. 14.5 star on the NW. This star is clearly visible in 30 cm, which shows the galaxy as a broad, moderately concentrated 2′ × 0′.5 spindle elongated in pa 130°. The mag. 14.5 star lies 40″ NW of center, just N of the major axis. The core is about 1′.2 long and unevenly bright, comprised of two or three brighter parts. There is no sharp nucleus. At 250 × the SW side of the halo fades more abruptly.

eg 1637 ***dimen. 3′.3 × 2′.9*** ***V = 11.0*** ***sfc. br. 13.3***
This is a difficult object for 15 cm, which shows a 2′ patch with a slight brightening in the center. The galaxy is broadly concentrated in 25 cm, about 2′.5 across. The brighter part takes up about 75 percent of it, leaving a thin, faint rim on the outside. A mag. 13 star lies 2′ NE of the center. In 30 cm the halo is unevenly bright, with the most conspicuous lumps SE and NE of center. The overall size is 3′ × 2′ in pa 45° with a 25″ core and a stellar nucleus.

eg 1638 ***dimen. 2′.5 × 1′.9*** ***V = 12.1*** ***sfc. br. 13.6***
This is a faint object for 25 cm, appearing 45″ diameter and only slightly brighter toward a knotted center. 30 cm shows a poorly concentrated and smooth halo 1′.25 × 0′.6 in pa 50°. There is no distinct core, but a stellar nucleus is prominent.

eg 1640 ***dimen. 2′.8 × 2′.0*** ***V = 11.7*** ***sfc. br. 13.5***
Forming an equilateral triangle with two mag. 11.5 stars at 2′ distance, this is a moderately faint galaxy in 25 cm. The oval halo brightens to a circular core. 30 cm shows it 2′ × 1′ in pa 50°. There is not much concentration, but the surface brightness is moderately high. The core bulges somewhat and contains a faint nucleus.

eg 1659 ***dimen. 1′.7 × 1′.3*** ***V = 12.5*** ***sfc. br. 13.2***
A faint galaxy for 25 cm, this object is best viewed at medium power. The halo is elongated in pa 45°, 1′ × 0′.3, with some faint granularity in the center. Many stars are visible in a 40′ field around the nebula. High power gives the better view in 30 cm. The object is 1′.2 × 0′.8 in pa 45° with a stellar nucleus the only brightening except for a tiny region immediately around it. The halo is mottled, has fairly definite edges, and moderate surface brightness.

eg 1667 ***dimen. 1′.5 × 1′.3*** ***V = 12.1*** ***sfc. br. 12.6***
Faintly visible in 15 cm, this galaxy appears as a uniform 1′ glow, slightly elongated N-S. 25 cm shows an oval 1′.5 × 1′ in pa 170°, and some stellarings are visible in the core. A mag. 14.5 star lies 1′.7 SW. In 30 cm the halo is 1′ × 0′.8, with weak, uneven concentration to a faint nucleus. Photographs show eg 1666 (mag.$_{VV}$13.5) 15′ S.

eg 1700 ***dimen. 2′.9 × 2′.0*** ***V = 11.0*** ***sfc. br. 12.9***
With a bright star 7′ NW, this galaxy is an easy object for 15 cm. It is about 1′ diameter, brightening little toward the center. In 25 cm the core is nearly stellar in a

45″ × 30″ halo. 30 cm shows a high surface brightness halo and a small brighter core about 15″ across and a stellar nucleus. The halo extends to 1′ × 0.′75 in pa 80°, fading little from the center to the fairly definite edge. eg 1699 ($mag._{VV}$15) lies 6.′5 N.

eg 1726 dimen. 1.′4 × 1.′1 V = 11.7 sfc. br. 12.5
At high power 15 cm shows this galaxy as a 40″ spot without detail. A mag. 12 star lies 50″ S of center. 25 cm shows it 1.′2 × 0.′5, elongated N-S, with a stellar nucleus inside a smooth halo. In 30 cm the galaxy is less than 1′ diameter, but the star S interferes with determining the elongation. Two mag. 9 stars lie about 5′ SE and WSW. eg 1720 ($mag._{VV}$13) lies 8.′2 SW.

eg 1741AB dimen. 1.′7 × 0.′9 V = 13.4 sfc. br. 13.7
eg I 399 dimen. 0.′6 × 0.′5 V = 14.5 sfc. br. 12.8
This complex interacting system is too faint for 15 cm. In 30 cm eg 1741AB is a small, fairly faint glow about 40″ × 20″ in size, elongated roughly E-W. A substellar point stands out near the center, and a faint stellaring is discernable at the W edge; this latter point is the A component, the rest is the B component. A mag. 11.5 star is 1′ SE; 1.′4 farther SE is eg I 399. In 30 cm it is a small, weakly and evenly concentrated spot about 25″ diameter.

Fornax

Though low in the sky for many northern observers, Fornax contains a bright cluster of galaxies. The center of the constellation culminates at midnight about 2 November.

eg 922 dimen. 1′.9 × 1′.8 V = 12.2 sfc. br. 13.4
This object is only faintly visible in 15 cm. In 25 cm it lies 2′ SSE of a mag. 12 star. The galaxy is circular, 35″ diameter, with a granular, somewhat brighter core and occasionally stellar nucleus. The nucleus is quite conspicuous in 30 cm. The halo is 1′.2 across with a few knots particularly on the NE and SE. Photographs show eg E479-G04 (mag.$_{VV}$12) 34′ NE.

eg 986 dimen. 3′.7 × 2′.8 V = 11.0 sfc. br. 13.4
Viewed in 30 cm this galaxy is located 8′ N of a mag. 9 star halfway to a thin, S-pointing triangle of mag. 10.5–11 stars. The narrow N base of this triangle is a 12″ pair in pa 275°. The well-concentrated halo of the galaxy is elongated in pa 60°, 1′.8 × 0′.6, tapering at each end. The 10″ core shows a stellar nucleus with averted vision.

eg E356-G04 dimen. 20′ × 14′ V = 8.4 sfc. br. 14.5
gc E356-SC01 diam. 70″ V = 13.5 class VIII
gc 1049 diam. 50″ V = 12.6 class V
gc E356-SC05 diam. 35″ V = 13.6 class IV
gc E356-SC08 diam. 51″ V = 13.4 class III
The Fornax dwarf galaxy is a member of the Local Group, and, though it has a relatively bright integrated magnitude, it is not discernable even in 15 cm with a 70′ low-power field. The brighter globular clusters in the system, gc 1049 and gc E356-SC08, are, however, visible as threshold stellar points at 75 ×. With 30 cm, gc 1049 is a moderately faint and concentrated spot about 20″ across with a conspicuous stellar nucleus. Very similar in appearance is gc E356-SC08, which may be slightly brighter. gc E356-SC05 is smaller and somewhat fainter, extending to only 15″ diameter around its fainter but still conspicuous stellar nucleus. gc E356-SC01, though it is the largest of all the visible globulars, has a very low surface brightness and is difficult to see even at 250 ×. Overall it is 25″–30″ diameter, showing poor central concentration and no stellar nucleus.

eg 1079 dimen. 3′.1 × 1′.9 V = 11.4 sfc. br. 13.2
15 cm shows this galaxy 12′ S of a mag. 9 star as a faint patch about 1′ diameter with a brighter grainy center. A mag. 10.5 star is located 6′ SSW. In 25 cm it is 1′.25 × 0′.5 in pa 110°. The E tip is longer, narrower, and a bit fainter than the W end. The core is more circular than the halo and contains an occasionally stellar nucleus. With 30 cm the nucleus is bright and substellar. The overall diameter is about 1′.

eg 1097 dimen. 9′.3 × 6′.6 V = 9.3 sfc. br. 13.6
Lying 15′ N of a triangle of uneven magnitude stars about 4′ across, this is a pretty easy object for 6 cm. It appears elongated SE-NW, about 4′ × 1′.5 with a bright nonstellar center. 15 cm shows it 3′ × 1′ in pa 150°. The faint stellar nucleus is surrounded by a small oval core and a very thin sharply tapered halo. It grows to 4′ × 0′.75 with 25 cm. The core is 35″ across, nearly circular, but sometimes seems flattened on the N side. Lying 1′.5 NW of center, on the SW flank of the halo, is a faint star. With 30 cm it is 4′ × 1′.8. The halo is narrow by the core, but on the N and particularly the S end it feathers out to the W. The core is 50″ × 30″, elongated in pa 0°; the stellar nucleus is eccentric to the NNE side of the core. At high power the halo has a curdled texture. eg 1097A (V = 12.9) lies 3′.5 N.

eg 1201 dimen. 4′.4 × 2′.8 V = 10.6 sfc. br. 13.2
This is a moderately faint object for 15 cm, appearing less than 1′ diameter with a very small, nearly stellar center surrounded by a faint halo. It is moderately bright in 25 cm. The 10″ core and stellar nucleus are together very prominent, though the nucleus is not conspicuous over the core. The faint halo extends to 1′.25 diameter, possibly elongated N-S. 30 cm shows the halo to 2′.25 × 1′ in pa 10°. At 150 × the core is nearly stellar and seemingly multiple. Higher power shows a bright 5″ nucleus.

eg 1255 dimen. 4′.1 × 2′.8 V = 11.1 sfc. br. 13.6
Lying 2′ NE of a mag. 12 star, 25 cm shows this galaxy as a 2′ area of low surface brightness nebulosity with a coarse grainy texture. The brightest part is about 1′ across. It grows to 3′.2 × 2′.5 in 30 cm with irregular edges and a few lumps in the center.

eg 1288 ***dimen. 2.′3 × 2.′2*** ***V = 12.1*** ***sfc. br. 13.7***
In 30 cm this faint galaxy appears as a small, unconcentrated spot. The halo is about 1′ across, possibly elongated N-S, with a faint nuclear condensation. A mag. 13 star lies 2′ SSE.

eg 1292 ***dimen. 3.′2 × 1.′7*** ***B = 12.6*** ***sfc. br. 14.2***
Located 2.′5 SW of a wide mag. 11 pair (24″; 330°), 25 cm shows this galaxy as a 1′ × 0.′5 glow with a faint stellar center occasionally visible. In 30 cm the object is elongated in pa 10° with a faint stellaring in the center. The overall size is 1.′25 × 0.′75, the brighter part being 1′ long.

eg 1302 ***dimen. 4.′4 × 4.′2*** ***B = 11.4*** ***sfc. br. 14.4***
This is a moderately faint object for 15 cm, located about 2′ SE of a mag. 11.5 star. The halo appears circular, 1.′25 diameter, brightening suddenly at the center to a stellar nucleus. In 25 cm the faint, ill-defined halo extends to 1.′5 diameter. The stellar nucleus seems slightly off-center to the SE. At low power 30 cm shows the halo to 2.′5 diameter, and the inner regions show some N-S elongation. At 225× the nucleus is substellar, but the halo shrinks to 1.′5 × 0.′8 in pa 160°.

eg 1316 ***dimen. 7.′1 × 5.′5*** ***V = 8.8*** ***sfc. br. 12.7***
This probable member of the Fornax I cluster is visible in 6 cm. With 30 cm the halo is 2.′5 × 1.′5 in pa 50°, the core 40″ across containing a bright nonstellar nucleus. A mag. 14.5 star is visible 1′ E. eg 1317, *q.v.*, lies 6.′3 N.

eg 1317 ***dimen. 3.′2 × 2.′8*** ***V = 11.0*** ***sfc. br. 13.2***
A member of the Fornax I cluster, this galaxy is fairly bright in 30 cm. The halo is about 1′ across with a 25″ core and nearly stellar nucleus. eg 1316, *q.v.*, lies 6.′3 S.

eg 1326 ***dimen. 4.′0 × 3.′0*** ***V = 10.5*** ***sfc. br. 13.1***
Found 4.′3 SSE of a mag. 11 star, 30 cm shows this member of the Fornax I cluster as a well-concentrated spot 1.′2 diameter with a bright stellar nucleus. A group of brighter stars lies 10′ W.

eg 1339 ***dimen. 2.′3 × 1.′9*** ***B = 12.5*** ***sfc. br. 14.1***
30 cm shows this galaxy 5.′7 SE of a mag. 11,13 pair with 25″ separation in pa 35°. It is a small object, about 50″ diameter, with a stellar nucleus fading suddenly to the halo, which has indistinct edges.

eg 1344 ***dimen. 3.′9 × 2.′3*** ***V = 10.3*** ***sfc. br. 12.7***
This galaxy is just visible in 6 cm as a concentrated spot of light forming the SW point of an obtuse triangle. The halo is 2′ × 1′ in pa 165° with 30 cm, with strong, even concentration to a stellar nucleus.

eg 1350 ***dimen. 4.′3 × 2.′4*** ***V = 10.5*** ***sfc. br. 12.9***
A member of the Fornax I cluster, this galaxy is 3.′2 × 1.′5 in pa 20° with 30 cm. The small, bright core is mottled toward its center and shows a stellar nucleus off-centered to the W. Close on the W of center a dark patch is visible, which looks like a dark lane. Two stars lie 2.′8 SE of the object.

eg 1351 ***dimen. 1.′8 × 1.′2*** ***V = 11.7*** ***sfc. br. 12.7***
In 25 cm this member of the Fornax I cluster is conspicuous as a fairly bright but undetailed spot about 50″ diameter. With 30 cm the halo is about 1′ × 0.′5, elongated in pa 140°. The concentrated central regions are small and contain a conspicuous stellar nucleus. The innermost region near the nucleus seems to have some small bright spots.

pn 1360 ***dimen. 9′ × 5′*** ***nebula 9.4:v*** ***star V = 11.4***
Even 6 cm will show this planetary, one of the brightest in the sky, as a well-defined oval haze elongated NE-SW. The nebula is uniformly bright, and the central star is clear at 25×. In 15 cm the nebula is a large, smoothly textured glow elongated in pa 30°. At low powers the central star is conspicuous. With 30 cm it extends to about 7′ × 4′, delimited on the N by a mag. 13 star just off the edge, and on the NW edge by a mag. 14 star. Generally the nebulosity brightens somewhat toward the bright central star, but a slightly dimmer patch lies in the southern half of the otherwise seemingly transparent object.

eg 1365 ***dimen. 9.′8 × 5.′5*** ***V = 9.5*** ***sfc. br. 13.7***
This galaxy, though located in the direction of the Fornax I cluster, may be a foreground object. It is fairly conspicuous but faint to 25 cm at low power. The halo is a large circular haze about 3′ diameter around an evenly bright, almost circular core 20″–30″ across. A very faint band of haze extends N from the W side reaching to a mag. 13.5 star that lies 1.′3 NW of center; another faint band leads S off the E side. In 30 cm the halo has extremely low surface brightness. It is about 3′ diameter overall, with indefinite elongation in pa 60°. The core, however, has a moderate surface brightness. It shows an elongation in pa 0° with knobs on the N and S sides.

eg 1366 ***dimen. 2.′7 × 1.′2*** ***B = 12.8*** ***sfc. br. 14.0***
Located 6.′6 S of a mag. 6 star [van den Bos 52: 6.8,7.1; 0.″3; 320° (1985)], this galaxy has a very bright stellar nucleus in 30 cm. A small bright region surrounds the nucleus, and the halo extends to 1′ × 0.′5, elongated N-S.

eg 1371 ***dimen. 5.′4 × 5.′0*** ***B = 11.5*** ***sfc. br. 14.7***
Lying 4.′2 SW of a mag. 8 star, this is a pretty faint object for 15 cm, which shows a small, bright center. In 25 cm

the halo is 1′.3 × 0′.3 in pa 100° with thinly tapered tips. The core is rectangular but not easily distinguishable and contains a substellar nucleus. 30 cm shows it 3′.2 × 1′.5 overall, elongated in pa 140°. The core is quite elongated in pa 100°, 40″ × 15″, within which the nucleus resides off-center to the N. There seem to be dark areas N and S of the bar. Low power shows the halo best, higher power the details in the center.

eg 1373 ***dimen. 0′.9 × 0′.7*** ***mag.$_{VV}$15*** ***sfc. br.***
eg 1374 ***dimen. 1′.8 × 1′.8*** ***V = 11.2*** ***sfc. br. 12.6***
eg 1373 and eg 1374 are members of the Fornax I cluster. The brighter object is visible in 25 cm as an evenly concentrated 45″ spot with a bright stellar nucleus. With 30 cm the halo is about the same size, but the nucleus does not stand out in the small core. Lying 4′.8 NW is eg 1373, which is barely visible in 25 cm as a very small, low surface brightness smudge. eg 1375, *cf.*, lies 2′.4 S of −74.

eg 1375 ***dimen. 1′.9 × 0′.6*** ***V = 12.2*** ***sfc. br. 12.3***
In 25 cm this member of the Fornax I cluster appears about the same size as, but about a magnitude fainter than its close companion 2′.4 N, eg 1374, *cf.* The halo is nearly circular, brightening evenly and moderately to the center. In 30 cm it is 50″ × 15″, elongated E-W.

eg I 335 ***dimen. 1′.7 × 0′.4*** ***mag.$_{VV}$13*** ***sfc. br.***
This probable member of the Fornax I cluster is a tiny, thin spindle in 25 cm. Best viewed at medium and high powers, it is about 30″ × 6″, elongated E-W.

eg 1379 ***dimen. 2′.0 × 1′.9*** ***V = 11.1*** ***sfc. br. 12.5***
This Fornax I cluster member appears more diffuse than eg 1387, *cf.*, in 25 cm. The halo is about 45″ diameter, brightening gradually across the center. The surrounding field has many stars of mag. 13 and fainter. With 30 cm the halo is about the same size and shows little concentration for an elliptical galaxy. The broad core fills most of the object, and a faint stellar nucleus is occasionally visible within.

eg 1380 ***dimen. 4′.9 × 1′.9*** ***V = 10.2*** ***sfc. br. 12.5***
25 cm shows this Fornax I cluster member as a bright elongated object. The halo is 1′.75 × 0′.75 in pa 5°. Extending from the core along the major axis of the halo is a brighter streak. The core itself is more circular, about 30″ diameter, and shows strong concentration but no distinct nucleus. A mag. 14 star is visible 50″ SW of center. In 30 cm the halo extends to 2′.5 × 1′ and is fairly well concentrated to a nonstellar nucleus.

eg 1380A ***dimen. 2′.8 × 0′.7*** ***mag.$_{VV}$12.5*** ***sfc. br.***
This member of the Fornax I cluster is visible in 25 cm at medium power. The very small object is best viewed at high power, where the low surface brightness halo extends to 30″ × 10″, elongated N-S, growing only a little brighter across the center.

eg 1381 ***dimen. 2′.9 × 0′.8*** ***V = 11.6*** ***sfc. br. 12.5***
In 25 cm this member of the Fornax I cluster is a small, difficult object. The halo is 35″ × 20″ in pa 140°. The thin, starry core contains a stellar nucleus and is surrounded by very little halo. With 30 cm the halo is 1′.5 × 0′.5 with fairly strong concentration to a substellar nucleus. About 1′.8 ESE is a mag. 14 star.

eg 1382 ***dimen. 0′.8 × 0′.8*** ***V = 12.9*** ***sfc. br. 10.0***
This member of the Fornax I cluster is visible in 25 cm as a weakly concentrated circular patch about 30″ diameter with low surface brightness.

eg 1385 ***dimen. 3′.0 × 2′.0*** ***V = 11.2*** ***sfc. br. 12.9***
A fairly faint object for 15 cm, this galaxy appears 1′ diameter with little concentration to the center. In 25 cm it is not much brighter than eg 1371, *cf.* The surface is diffuse, slightly brighter to the center, and there seem to be dark radial lines extending outward, particularly on the SE side. 30 cm shows it 2′.8 × 1′.5 in pa 20°, the faint halo showing up best at low power. The broad, circular core is 1′.3 across with a stellar nucleus clearly off-center to the W.

eg 1387 ***dimen. 2′.4 × 2′.2*** ***B = 11.9*** ***sfc. br. 13.6***
With 25 cm this Fornax I cluster member lies in an empty high-power field. The overall size is 45″ × 35″ in pa 110°. It grows broadly brighter toward an elongated 6″ × 2″ nucleus, but there is no distinct core. It appears overall very similar to eg 1389, *cf.* in Eridanus, with 30 cm. The middle parts are fairly bright, and the overall size is the same as that in 25 cm.

eg 1398 ***dimen. 6′.6 × 5′.2*** ***V = 9.7*** ***sfc. br. 13.4***
This galaxy is just discernable in 6 cm as a patch 7′ E and a little N of a mag. 9.5 star. A moderately bright object for 15 cm, it appears about 1′.5 diameter with a stellar nucleus just visible over the bright center. The halo is about the same size in 25 cm, with a 20″ core and a nearly stellar nucleus. The SE edge seems sharply defined. 30 cm shows strong, even concentration to a bright inner core and stellar nucleus. The object is elongated N-S, with an overall size of 1′.5 × 1′.

eg 1399 ***dimen. 3′.2 × 3′.1*** ***V = 9.8*** ***sfc. br. 12.3***
25 cm shows this member of the Fornax I cluster in the same high-power field as eg 1404, *q.v.* 13′ SSE in Eridanus. The two galaxies have a similar magnitude, though eg 1399 is less concentrated and of lower surface

Chart X. Fornax I cluster.

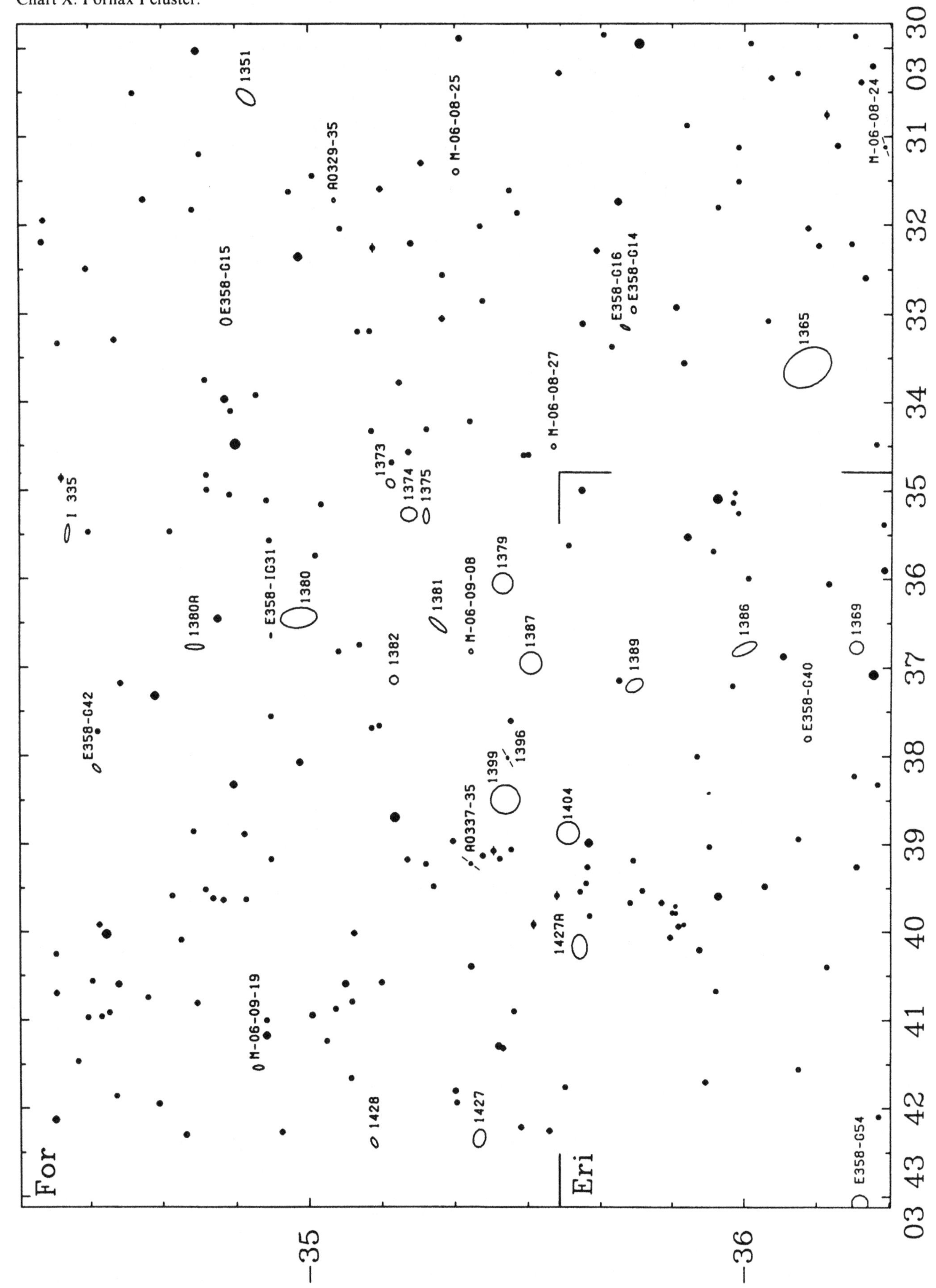

brightness. Its halo is 1.'25 diameter with a broadly brighter, mottled core and a faint stellar nucleus. A faint star is closely involved, lying only 20″ from the center in pa 15°. With 30 cm the halo is about the same size, but shows fairly good concentration to the unevenly bright 30″ core.

eg 1406 dimen. 3.'9 × 1.'0 B = 12.6 sfc. br. 13.9
With 30 cm this galaxy is a thin 3.'2 × 0.'6 spindle in pa 15°. The halo tapers slightly toward the tips and shows slight concentration to a roundish core and faint stellar nucleus. At times the core seems slightly curdled along its major axis.

eg 1425 dimen. 5.'4 × 2.'7 B = 11.7 sfc. br. 14.3
This galaxy is to the S side of a string of three stars aligned SE-NW, including a 30″ mag. 13,14 pair at the SE end. In 30 cm the halo is elongated SE-NW, 3.'2 × 1.'5, with a moderately concentrated center and faint ends. The nucleus is stellar, and two stellarings are visible in the core on either side of the center along the major axis.

eg 1427 dimen. 2.'8 × 2.'0 V = 11.0 sfc. br. 13.0
30 cm shows this object to have a moderately low surface brightness halo, brighter core, and distinct stellar nucleus. A mag. 13.5 star is visible 1.'6 W.

Gemini

This bright constellation holds many interesting objects. Messier 35 (oc 2168) and pn 2392 rate as showpieces, though some of the fainter clusters are also interesting. The center of the constellation culminates at midnight about 6 January.

oc 2129 diam. 7′ V=6.7 stars 73
This cluster is just visible in 6 cm around two bright stars aligned N-S, mag. 7.5 and 8.0. 25 cm shows a nice, bright cluster with 35 stars in a 5′ area. 30 cm gives a similar view, showing 21 stars weakly concentrated in a 4′ area.

oc I2157 diam. 7′ V=8.4 stars 56
This cluster is discernable in 6 cm as a faint round haze with three stars superposed. With 25 cm it is an indistinct cluster of about 15 mag. 10.5 and fainter stars in a 4′–5′ area. In 30 cm three mag. 11 stars stand out within the 4′.5 diameter cluster. At higher magnifications 15 stars can be counted over some clumpy haze.

oc 2158 diam. 5′ V=8.6 stars 973
With 6 cm this distant cluster (4.9 kpc) can be seen just SW of a group of ten bright stars on the outskirts of Messier 35, *q.v.* oc 2168. It is an unresolved circular glow with a mag. 10.5 star on the SE edge. In 15 cm it appears as a milky spot with 10 or 20 stars loosely scattered over it. 20 cm shows about 20–30 stars at the threshold of vision in a 4′ area. It is richer toward the center, and short arms extend S and W. In 25 cm it looks like a globular cluster, showing about 50 stars, mag. 13 and fainter, in a 4′ diameter. Very faint extensions to about 6′ diameter are visible with 30 cm when the telescope is jiggled. On the W side of the core are about eight brighter stars in a string aligned NE-SW. The broadly concentrated core is elongated E-W, its brighter part off-center to the E.

oc 2168 diam. 28′ V=5.1 stars 434
Messier 35 is a naked-eye cluster well resolved by 6 cm, with most of the bright stars strewn to the E of center. At least 40 stars can be seen in the doughnut-shaped cluster. On the NNE edge is a bright, conspicuous pair (ADS 4744: V=7.3,9.1; 31″; 188°). A grand sight in a 15 cm RFT, the outliers fill a 1°.5 field. 25 cm shows it 80′ across with the central part 20′ diameter. The hole in the center contains only a few faint stars. The pair NNE are contrastingly colored yellow and blue. Outliers extend E to the mag. 6 star 5 Geminorum. The central hole is not so obvious in 30 cm as the vacant area extends E-W most of the way through the cluster. Possibly 200 stars are visible overflowing a 23′ field. Marvelous!

pn J 900 dimen. 12″ × 10″ nebula 11.8v star V=16.3:
This planetary appears nearly stellar at 200× in 25 cm. At 250× and 300× it appears round, 10″ across with very high surface brightness. A mag. 13 star lies 12″ SSW, but no central star is evident. In 30 cm the nebula is 10″ × 8″, elongated in pa 45°. There is some central concentration, but no central star.

oc 2266 diam. 7′ mag. 9.5p stars 50
This is a moderately bright and compact cluster in 25 cm, with 50 stars lying within a 4′–5′ area. A distinguishing arc of seven stars begins on the SW with a mag. 8.5 star, crosses through the southern side of the cluster curving E, and fades to a mag. 12 star. In 30 cm the cluster exhibits weak, uneven concentration; otherwise the view is similar.

oc 2304 diam. 5′ mag. 10.0p stars 30
25 cm shows this cluster as a faint cloud of 20 stars in a 4′ area. The brightest star is mag. 11.5, and a chain of stars lies on the NW side. 30 cm shows 20 stars of mag. 13 and fainter with some unresolved haze; the overall size is 5′.5 × 4′, elongated NE-SW. The chain of stars to the NW is aligned NE-SW, forming a definite edge there: the outliers spread mostly to the E and NE.

eg 2339 dimen. 2′.8 × 2′.1 V=11.6 sfc. br. 13.4
Fairly faint in 25 cm, this galaxy is 1′.5 diameter with a brighter core within which a stellar nucleus is sometimes seen. In 30 cm it is 1′.5 diameter, of moderately low and uneven surface brightness, with a slightly brighter 20″ core. A mag. 14 star lies just within the nebula on the NE edge, and another is 2′ W of center.

eg 2341 dimen. 1′.1 × 1′.0 mag.$_z$13.7 sfc. br.
eg 2342 dimen. 1′.6 × 1′.5 mag.$_z$12.6 sfc. br.
The brighter of this close pair of galaxies, eg 2342, is only just visible in 15 cm. With 30 cm it is a fat 1′ × 0′.7 oval

elongated in pa 45°. There is only a slight, broad brightening across the center: no core or nucleus stands out. eg 2341 is visible in 30 cm 2'.5 SW. It has a similar surface brightness to its larger companion, though it is only about 20" across, with a circular outline. It has a slight, even concentration toward a small core. A mag. 13.5 star is visible 35" N.

oc 2355 ***diam. 9'*** ***mag. 9.7p*** ***stars 40***
This cluster is just visible to 6 cm with a mag. 8 star 7' NE and a mag. 10 star 3' E. It is elongated slightly N-S in 25 cm with 35 stars in an 8'–10' diameter. At least 30 stars are visible in a 7' area with 30 cm. It is elongated in pa 30°, mostly due to a string of three stars trailing to the SSW. Two tight groups lie within the boundary, and a brighter mag. 10 star marks the center.

pn 2371–72 ***dimen. 2'.2 × 0'.9*** ***nebula 11.3v*** ***star V = 14.8***
Though this nebula is discernable only as a faint hazy area with a star-like spot involved in 15 cm, 25 cm clearly shows two patches connected by faint haze. Aligned NE-SW, they exhibit differing character: the southwestern

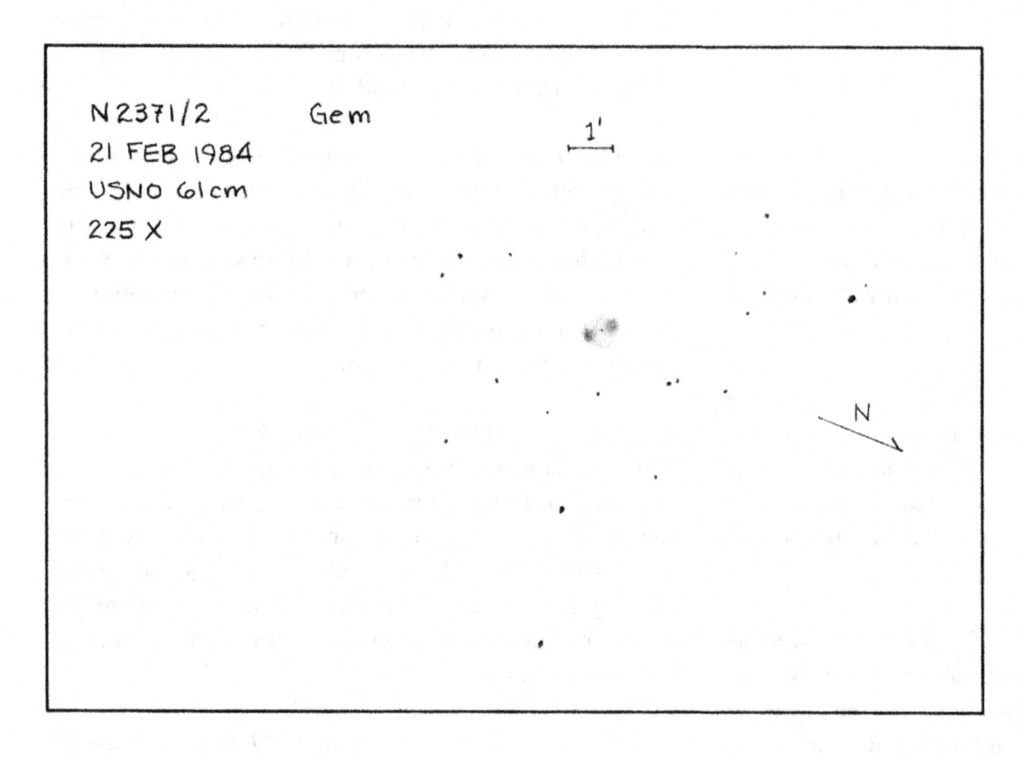

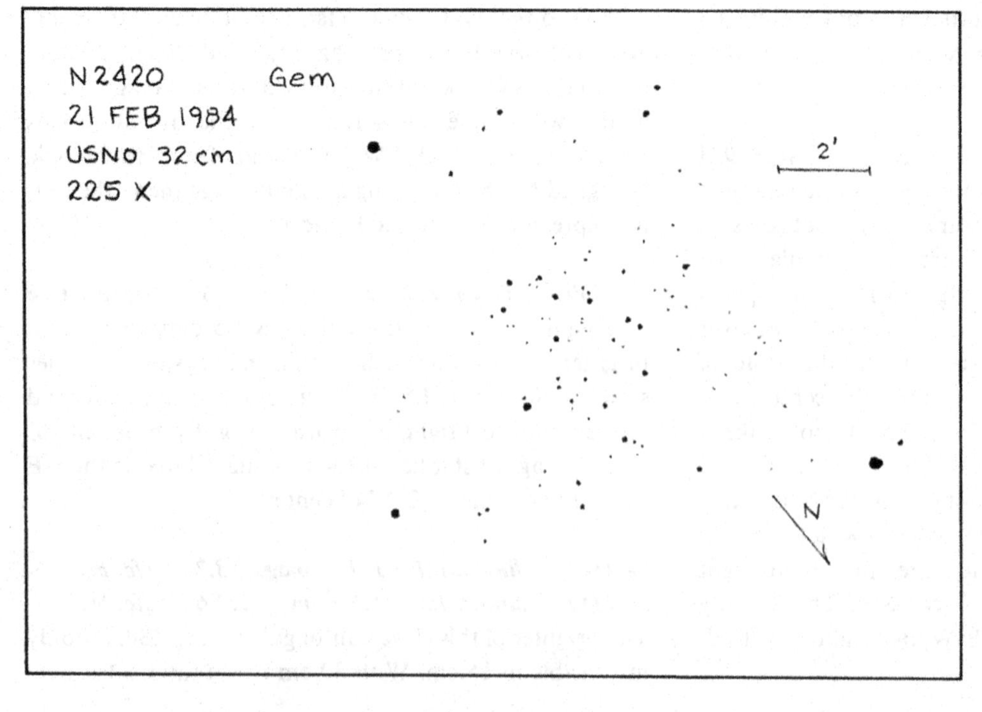

one (2371) is brighter and more concentrated with a stellar center. In 30 cm the faint haze fills a 50″ diameter roughly circular area surrounding the two lobes. Both of the parts are approximately the same size, but −71 is again much brighter with a stellar nucleus. At 225× the central star is visible, about mag. 14.5, sparkling faintly between the two bright lobes.

pn 2392 dimen. 47″×43″ nebula 9.2v star V=10.5
This planetary is easily discernable in 6 cm 1′.6 S of a mag. 8.5 star. It appears a bit fainter but clearly larger than this star. In 15 cm the central star is very prominent within the bluish nebula, and high magnifications seem to reveal a dark patch lying just W of the center. At 300× with 25 cm the round nebula seems granular and patterned with tightly intertwined coils. It is a bright 45″ diameter bluish glow with 30 cm. Surrounding the central star is a bright area 15″ across, and a darker ring is suspected about halfway from the center to the edge; it is particularly noticeable on the E.

oc 2395 diam. 12′ V=8.0 stars 53
While only just visible to 6 cm, 25 cm shows this cluster as a moderately bright, poorly concentrated group about 20′ across with 40 irregularly distributed stars of mag. 9 and fainter. Leading off the S end is a "tail" containing two mag. 10 stars. It is a loose and poor cluster to 30 cm with a triangular form elongated in pa 110°. Thirty-five stars are visible in a 15′ area.

oc 2420 diam. 10′ V=8.3 stars 304
6 cm will show this cluster as a broad 10′ diameter haze lying between two stars that are aligned N-S. 25 cm shows a compact 6′×4′ cluster, elongated roughly SE-NW. At 200× about 20 stars are resolved over a granular haze. 30 cm shows it 8′×6′ with about 35 stars, including seven brighter ones, the brightest of which is on the W, mag. 11.

Grus

This southern constellation, appearing like a backwards "λ" at culmination, contains a number of bright groups of galaxies, some of which are visible to northern observers. The center of the constellation culminates at midnight on 28 August.

eg 7410 ***dimen. 5′.5 × 2′.0*** ***V = 10.4*** ***sfc. br. 12.9***
This galaxy is fairly bright in 15 cm, appearing as a broadly concentrated 2′ × 1′ patch in pa 45°. 25 cm shows the mottled halo extending to 3′.5 × 0′.6. The core is rectangular, 40″ × 20″, with a grainy texture and a star or knot on its N edge. A star is visible 1′.7 NE of center.

eg 7418 ***dimen. 3′.3 × 2′.8*** ***V = 11.4*** ***sfc. br. 13.6***
This object can be seen very faintly in 15 cm. In 25 cm, due to its low surface brightness, it is best viewed at medium powers. The halo is elongated roughly E-W, 2′.5 × 1′. A few broad bright and dark areas are visible over the cigar-shaped object, but otherwise the surface is fairly smooth and without central concentration.

eg 7421 ***dimen. 2′.2 × 2′.1*** ***V = 12.0*** ***sfc. br. 13.5***
Located 20′ S of eg 7418, *q.v.*, this galaxy is not visible to 15 cm. 25 cm reveals a circular glow 1′.5 diameter with a slight central condensation. The 20′ diameter field around the object is fairly empty of stars.

eg I1459 ***dimen. 3′.4 × 2′.5*** ***V = 10.0*** ***sfc. br. 12.3***
This galaxy is visible in 6 cm as a well-concentrated substellar spot 3′ E of a mag. 11 star. 15 cm shows the object as a small, round, and fairly bright spot between the star W and a fainter one 3′ E. The halo is 1′ diameter and has a nearly stellar center. It is a bright object for 25 cm with high surface brightness, like a concentrated globular cluster. The circular 1′ glow has a bright center and a stellar nucleus. In 30 cm it is elongated approximately NE-SW, 1′.2 × 0′.8, with a round 30″ core. eg I5264 ($mag._{VV}$12) lies 6′.7 SW; eg I5269B ($mag._{VV}$13.5) lies 15′ NW.

eg 7552 ***dimen. 3′.5 × 2′.5*** ***V = 10.7*** ***sfc. br. 12.9***
Viewed in 30 cm this galaxy brightens evenly toward the center except for the stellar nucleus, which is embedded in a small bright spot perhaps 10″ across. The circular halo is 2′ diameter, but at 225 × the core is elongated in pa 100°, 1′.3 × 0′.9. A mag. 13 star is visible 1′.7 S.

eg 7582 ***dimen. 4′.6 × 2′.2*** ***V = 10.6*** ***sfc. br. 12.9***
eg 7590 ***dimen. 2′.7 × 1′.1*** ***V = 11.5*** ***sfc. br. 12.6***
eg 7599 ***dimen. 4′.4 × 1′.5*** ***V = 11.4*** ***sfc. br. 13.3***
In 30 cm eg 7582 is an elongated object, 3′ × 1′ in pa 160°. The elongated core is broken up into two or three parts, each about 30″ across, though this effect is not as obvious at high power. eg 7590 is smaller than −82, but similar in appearance. It is elongated in pa 40°, 2′ × 0′.6. The core is oval with several stellarings in it. A bright star is embedded in the SE flank of the NE end. eg 7599 is a broad, diffuse object, elongated in pa 60°, 3′ × 1′.5, with a few stellarings on the inner portions of the halo. A mag. 12.5 star is visible just outside of the halo on the ESE. eg 7632 ($mag._{VV}$13) lies 33′ ESE.

Hercules

This region contains many widely separated objects including the famous globular cluster Messier 13 (gc 6205). The center of the constellation culminates at midnight about 12 June.

eg 6052AB dimen. 1'.0 × 0'.7 V = 13.0 sfc. br. 12.5
While this small, faint colliding pair of galaxies is not visible with 15 cm, 25 cm shows them as a single object about 1′ across and somewhat brighter toward a faint stellar nucleus. Not difficult in 30 cm, it lies between two stars 2'.5 SSE and NNW. The object is moderately concentrated and small, about 50″ diameter.

pn 6058 dimen. 25″ × 20″ nebula 13.0v star V = 13.8
Found within a rectangle of mag. 10 stars, this nebula appears only as a faint star in 15 cm. The central star is clearly visible with 25 cm at 150 ×. The round, diffuse glow fades gradually to the edge with an overall diameter of about 20″. With 30 cm it looks much like a galaxy. The central star is easily seen with the grey nebula fading evenly to an indefinite outer edge at about 40″ diameter.

pn I4593 dimen. 12″ × 10″ nebula 10.7v star V = 11.3
This planetary is bright enough to be seen in 6 cm as a star. Starting from a mag. 8.5 double star (ADS 9959: V = 8.5,9.7; 7″.3; 147°) plotted on the AE, it is the first of two stars to the NW. It is very bright with 25 cm, about 10″ diameter with a sharply concentrated core that does not sparkle to indicate the central star. In 30 cm the central star is visible, the 10″–15″ diameter nebula fading suddenly away from it to "hairy," indefinite edges. The spectroscope shows one image for the nebula and a faint continuum with two emission lines for the central star.

eg 6106 dimen. 2'.6 × 1'.5 V = 12.2 sfc. br. 13.5
Not visible with 15 cm, this galaxy is about 1′ diameter and moderately faint in 25 cm. The core is broadly brighter, and a stellar nucleus is occasionally visible. With 30 cm it grows to 1'.2 × 0'.8 in pa 150° with very slight concentration to a faint stellar nucleus.

eg 6146 dimen 1'.6 × 1'.2 mag.$_z$13.8 sfc. br.
In 15 cm this galaxy is visible forming a pair with a star of similar magnitude 1'.1 E. It is very small, almost stellar at 100 ×. With 25 cm the halo is elongated roughly E-W, about 40″ × 25″, showing an even concentration to a circular core and a faint stellar nucleus. Photographs show eg 6145 (mag.$_z$15.1) 3'.5 NW.

eg 6158 dimen. mag.$_z$15.5 sfc. br.
Lying 2′ NW of the northern of two bright stars aligned N-S, this galaxy is faint in 25 cm. The halo extends to only 20″ diameter and contains a faint stellar nucleus. A mag. 13.5 star lies E, forming a "Y" with the two brighter stars and the galaxy.

eg 6160 dimen. 2'.1 × 1'.7 mag.$_z$14.8 sfc. br.
This galaxy is closely involved with two faint stars to its NE. In 25 cm these stars are not distinct at low power, but seem to extend the halo of the galaxy to the NE. At high power the galaxy itself is about 30″ diameter, and the closer of the two stars is distinguishable 20″ NE of center.

eg 6166 dimen. 2'.4 × 1'.8 V = 12.0 sfc. br. 13.5
In 25 cm this galaxy is a moderately and evenly concentrated glow without a distinct core or nucleus. The nearly circular halo extends to about 45″ diameter, slightly elongated NE-SW.

eg 6173 dimen. 2'.2 × 1'.7 mag.$_z$14.0 sfc. br.
This faint galaxy is visible in 25 cm at 50 × as an almost stellar spot. At high power the halo is 45″ diameter and circular, or slightly elongated SE-NW in the inner regions. eg 6175AB (mag.$_z$15.0) lies 11′ SSE.

eg 6181 dimen. 2'.6 × 1'.3 V = 11.9 sfc. br. 13.0
Forming a triangle with a moderately bright star 2'.4 W and a faint one 1'.4 SSW, this galaxy is faintly visible in 15 cm. Rectangular in form with 25 cm, it is 1'.5 × 1′ in pa 160°. A bright core with stellarings is visible, but no stellar nucleus is evident. With 30 cm the halo is 1'.75 × 0'.9, elongated in pa 0°, with the 50″ × 30″ core in pa 160°. The center is fairly concentrated with a nonstellar nucleus occasionally visible.

eg 6194 dimen. mag.$_z$14.6 sfc. br.
This galaxy looks similar to but fainter than eg 6196, *cf.* 17′ SE, in 30 cm. The halo is about 35″ across, but shows little concentration to the center except for the conspicuous stellar nucleus. eg U10473, *q.v.*, lies 13′ NNE.

eg U10473 ***dimen 1′.7 × 0′.5*** ***mag.*** $_z$***15.0*** ***sfc. br.***
In 30 cm this galaxy forms the southern oblique point of a triangle with mag. 12.5–13 stars. It is a faint, diffuse patch about 40″ diameter, elongated approximately N-S. The halo shows a weak, even concentration, and no nucleus is visible. eg 6194, *q.v.*, lies 13′ SSW.

eg 6196 ***dimen. 1′.7 × 1′.2*** ***V = 12.8*** ***sfc. br. 13.5***
eg 6197 ***dimen. 0′.8 × 0′.3*** ***mag.*** $_z$***15.4*** ***sfc. br.***
eg I4614 ***dimen. 1′.0 × 0′.9*** ***mag.*** $_z$***15.3*** ***sfc. br.***
In 25 cm, eg 6196 is visible at 100 ×. It is quite small, about 30″ diameter, growing much brighter to a substellar nucleus. With 30 cm the halo is about the same size, but elongated SE-NW. It exhibits little concentration here except for a conspicuous stellar nucleus. eg 6197 lies 4′.8 S and a little E. 30 cm shows it 1′ NW of a mag. 13 star. It is very small and diffuse, about 20″ diameter with a faint stellar nucleus. Lying 3′ NNW of eg 6196 is eg I4614. In 30 cm it is a very faint unconcentrated patch about 20″ across. This galaxy lies two-thirds of the way from −96 to a mag. 13.5 star that is the eastern point of an irregular 3′ × 2′ diamond-shaped asterism.

gc 6205 ***diam. 17′*** ***V = 5.7*** ***class V***
Visible to the naked eye on good nights, Messier 13 is one of the brightest objects of its class. 6 cm shows a round, bright ball without detail between two mag. 7 stars. In 15 cm outliers can be seen with medium power with partial resolution attainable in the core at high powers. 25 cm shows outliers to 15′ diameter, though the brightest part is about 8′ across. Long, curving strings of stars stand out at 250 ×. From the S side curving W is a curl of stars. Protruding into the SE side is a conspicuous dark wedge; another cuts less conspicuously into the NE side. Overall the outliers give the cluster an elongated appearance in pa 165°. With 30 cm the outliers are scattered more densely to the E and W sides of the 5′ core, whose E side is flattened. On the E side is a small 30″ diameter dark patch. Starry arms extend N and S, giving the cluster a bird-like appearance.

eg 6207 ***dimen. 3′.0 × 1′.4*** ***V = 11.6*** ***sfc. br. 13.1***
This galaxy lies about 40′ NNE of Messier 13 (*q.v.* gc 6205). In 15 cm it is an elongated and weakly concentrated patch with a small, starry center. With 25 cm the halo is 1′.75 × 0′.75, elongated in pa 25° with sharply tapered ends. A mag. 13.5 star is superposed just N of the brightest part, perhaps contributing to the irregular appearance of the middle portions of the nebula.

pn 6210 ***dimen. 48″ × 8″*** ***nebula 8.8v*** ***star V = 12.9:***
Located at the apex of a triangle of stars, this planetary nebula is quite bright in 15 cm. A uniformly illumined bluish disk is revealed at 200 ×. The nebula is recognizable at 50 × in 25 cm. High magnifications reveal a bluish 25″ × 15″ oval in pa 90°, with a sharp, starry point in the center.

gc 6229 ***diam. 4′.5*** ***V = 9.4*** ***class VII***
Lying E of two mag. 8 stars, this globular is small and faint in 15 cm. High magnifications with 25 cm show a 50″ diameter glow with a weakly granular texture. It is a small, circular 2′ diameter glow with broad concentra-

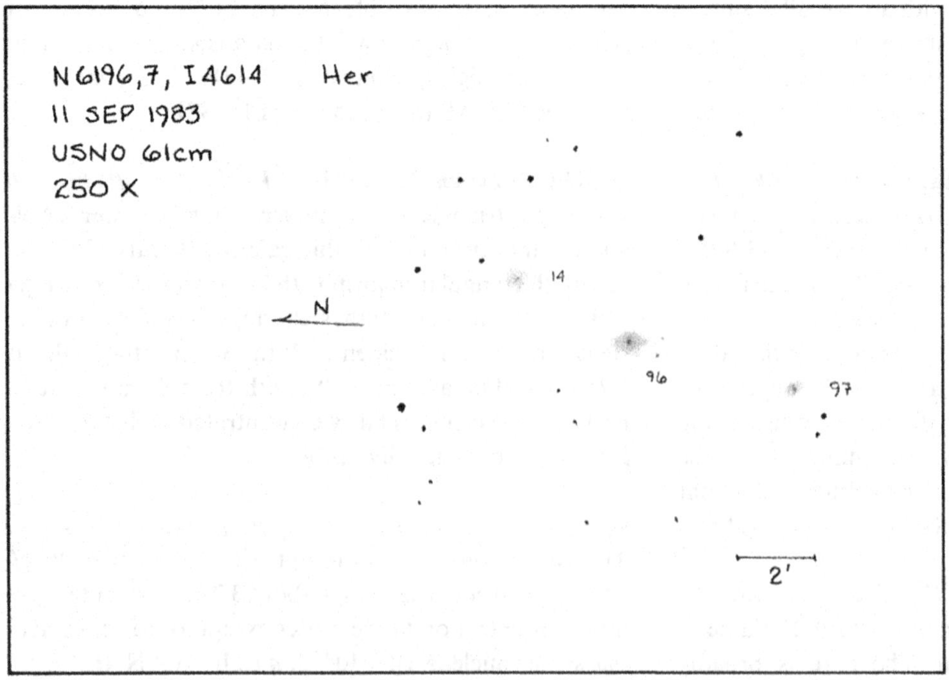

N6196,7, I4614 Her
11 SEP 1983
USNO 61cm
250 X
N
14
96
97
2′

tion in 30 cm. It exhibits no sparkling to indicate resolution, but is granular throughout.

eg 6239 dimen. 2'.8 × 1'.3 V = 12.3 sfc. br. 13.6
This faint galaxy is found within a 1°.5 circlet of mag. 6 stars. Though not visible in 15 cm, 25 cm shows an irregularly outlined elongated 1'.5 × 1' blur in pa 115°. There is a slight broad brightening to a circular, indistinct core, but no sharp nucleus is visible. The halo is 1'.5 × 0'.8 in 30 cm. It has a slight concentration to a broad core within which a faint substellar nucleus is occasionally visible.

eg 6255 dimen. 3'.5 × 1'.5 mag.$_z$13.8 sfc. br.
This galaxy is faintly visible in 25 cm at 50 ×. At 100 × the halo is clearly elongated in pa 90°, 2'.5 × 1', with a cigar-shaped outline. It shows weak concentration toward the center with faint mottlings and stellarings over its surface. A mag. 13.5 star lies nearby S.

gc 6341 diam. 11' V = 6.4 class IV
Messier 92 is easily found in 6 cm by moving about 6° N from e (69) Herculis (V = 4.7). It is more strongly concentrated than Messier 13 (*cf.* gc 6205), but appears a little smaller and fainter overall. Very bright in 15 cm, it is evenly graduated in intensity from dim outliers to an intense core. At 200 × the outer regions are well resolved, but the core shows much unresolved haze. The outliers show a distinct NNE-SSW elongation in 25 cm, spreading to an overall size of about 9' × 6'. The SE side is flattened more than the NW. The core is circular, about 2'.5 diameter, and loses some resolution in the central 1'.5 diameter. A mag. 10 star is visible 6' E, and a mag. 12 star is 3' SW. 30 cm resolves the entire cluster well to a diameter of 8'. The 4' core has an abrupt and pointed northern perimeter.

eg 6482 dimen. 2'.3 × 2'.0 V = 11.3 sfc. br. 12.9
A mag. 13 star is closely involved with this galaxy, making observation difficult in 15 cm. With 25 cm the galaxy is small and pretty faint, the mag. 13 star lying just W of center. It seems slightly elongated E-W with an overall size of about 40" × 30". In 30 cm the galaxy forms the southern tip of a 2'.5 triangle with mag. 12 stars. The halo shows little concentration and is less than 1' diameter.

eg 6574 dimen. 1'.4 × 1'.1 V = 12.0 sfc. br. 12.4
This faint and isolated galaxy is barely visible in 15 cm. With 25 cm the halo is approximately 1' across and is indefinitely elongated N-S. A stellar nucleus is occasionally visible within the broadly brighter core. The elongation is clearly visible at high powers with 30 cm: the overall size is 1'.5 × 1' in pa 20°. It is moderately and broadly concentrated to a round, bulging core, from which the halo tapers to pointed tips. A mag. 14.5 star lies within the halo about 30" S of center.

pn Humason 2-1 nebula 11.5v star V = 13.6
Indistinguishable from a star in 4" seeing with 30 cm, this small planetary has its character revealed easily by the spectroscope: with a single image, it is the northern and brighter of two stars found in the area.

Horologium

A dull group low in the south for northern observers, this constellation contains a few galaxies, of which one was easily accessible in our survey. The center of the constellation culminates at midnight about 10 November.

eg 1512 ***dimen. 4.′0 × 3.′2*** ***V = 10.6*** ***sfc. br. 13.3***
Reaching a maximum altitude of only 11° for the authors, observation of this galaxy requires a clear southern horizon. Viewed in 30 cm it is about 2.′1 × 1.′5 in pa 55°. The 30″ diameter core seems inclined to the halo, perhaps in pa 90°. The halo shows very little concentration, but a bright substellar nucleus is visible in the center. eg 1510 (V = 13.0) is located 5′ SW.

Hydra

This is the largest of the constellations, winding through seven hours of Right Ascension. There is a wide selection of galaxies throughout. The center culminates at midnight about 16 March, but portions of the constellation culminate from January until May.

oc 2548 diam. 54′ V = 5.8 stars 37
Messier 48 is a nice cluster for small telescopes. 6 cm will reveal perhaps 60 stars, the brighter ones aligned N-S through the center, including a few pairs. 15 cm shows about 50 stars in a 40′ area. A tight circlet is visible toward the center. In 25 cm about 100 stars are visible including several bright pairs near the center. Stars straggle to the NW, giving a total size of 60′ × 40′. 30 cm will show 100 stars in a 40′ diameter.

eg 2583 dimen. 0′.7 × 0′.7 V = 13.5 sfc. br. 12.7
eg M-01-22-06 dimen. mag.$_{vv}$16 sfc. br.
eg 2583 is the southwestern member of a three-galaxy string including eg 2584 and eg 2585, *q.v.* In 30 cm it is the most concentrated member of the group, though it is quite small, only about 30″ across. At times the halo seems elongated slightly E-W, but this impression may be caused by a conspicuous star superposed on the galaxy just W of center. Two mag. 13 stars aligned N-S about 1′.5 W form a thin triangle with the galaxy. Not quite 4′ N of these stars is a 1′ triangle of mag. 12–13.5 stars. Immediately N of the northernmost and faintest star in this triangle is eg M-01–22–06. 30 cm shows it as a very faint circular spot about 20″ diameter. eg 2584, *q.v.*, lies 2′.7 NE.

eg 2584 dimen. 1′.3 × 0′.7 mag.$_{vv}$14.5 sfc. br.
This galaxy is the middle member of a string of three galaxies: eg 2583 lies 2′.7 SW, and eg 2585 is 4′.2 NE, *q.v.* With 30 cm it is the faintest member in the string, difficult to see even at 250 ×. At best it is a small, low surface brightness patch with weak, even concentration to the center.

eg 2585 dimen. 2′.2 × 0′.8 mag.$_{vv}$14.5 sfc. br.
In 30 cm this galaxy is the largest member of a string of three galaxies including eg 2584, 4′.2 SW, and eg 2583, 6′.9 SW, *q.v.* The halo is 40″–45″ diameter, slightly elongated in pa 165°. It shows a moderate, broad concentration to the center, but no distinct core or nucleus. At 250 × the middle regions occasionally seem elongated ESE-WNW.

pn 2610 dimen. 50″ × 47″ nebula 12.8v star V = 15.8
15 cm will just reveal this planetary 3′.5 SW of a mag. 6.5 star, and just touching the western vertex of an oblique triangle of mag. 12 stars. In 25 cm the nebula is an even grey patch about 40″ diameter without annularity. In 30 cm it is about 45″ diameter with a slight broad concentration to the center: there is no indication of annularity. The nearby western star of the oblique triangle lies on the edge of the nebula, which seems to have a slightly brighter part elongated away from this star: the impression may be due only to the star.

eg 2612 dimen. 3′.2 × 0′.7 mag.$_{vv}$13.5 sfc. br.
This galaxy is visible in 15 cm at 75 × as a small spot of moderate concentration between two mag. 12 stars that are 1′.6 apart aligned N-S. The nebula is only 30″ away from the star N. 30 cm shows a fairly small and very elongated oval in pa 120°. The halo is pretty thin, almost a spindle, and moderately concentrated. The core is a fairly distinct, tiny oval perhaps 5″ × 2″.5; a very faint stellar nucleus is occasionally visible within.

eg 2642 dimen. 2′.3 × 2′.2 B = 12.5 sfc. br. 14.2
This object is faintly visible in 15 cm and lies just N of a 4′ triangle of mag. 8–8.5 stars. It is an inconspicuous circular object in 25 cm. The halo is about 1′.5 diameter and shows moderate concentration to a faint stellar nucleus. A faint trio of stars 10′ E may be mistaken for a nebula at low power; a mag. 13.5 star lies 1′.5 N of center. With 30 cm it appears about 1′.5 across with an unevenly bright core that is elongated SE-NW.

eg 2713 dimen. 3′.9 × 1′.7 V = 11.7 sfc. br. 13.6
In 25 cm this galaxy is moderately bright with a smooth outline. The halo is elongated E-W, but the core is circular. With 30 cm it extends to 2′ × 1′ in pa 115°. The core is 40″ across and contains a faint substellar nucleus. eg 2716 (V = 12.4) lies 11′ N.

eg 2763 dimen. 2′.3 × 2′.1 V = 12.1 sfc. br. 13.6
This galaxy has a pill-shaped outline in 25 cm, elongated roughly E-W. The unconcentrated low surface brightness

halo is about 1.'25 × 0.'5. Several stars can be seen around the object. It is pretty faint and weakly concentrated in 30 cm. The brighter parts are not quite centered in the 1.'5 diameter halo; lying instead more toward the SE side, they seem to be elongated roughly N-S. Four stars appear about 3' W; a mag. 11.5 star lies just off the N edge.

eg 2781 dimen. 3.'9 × 1.'9 V=11.5 sfc. br. 13.6
This object is faintly visible in 15 cm as a featureless 1' spot. It is elongated E-W in 25 cm, 1.'5 × 1.' The core is elongated and contains a couple of bright splotches. A faint 45" pair of stars lies 4' SE; another star is 2' N. 30 cm shows a stellar nucleus and a knot or star on the SW side of the galaxy.

eg 2811 dimen. 2.'7 × 1.'0 V=11.3 sfc. br. 12.2
This galaxy is discernable in 15 cm and appears less than 1' diameter. 25 cm reveals a faint circular object 1.'2 diameter with a small, slightly brighter core. A mag. 12 star is visible 1.'6 W. 30 cm shows it 1.'5 × 1', elongated NNE-SSW, with strong concentration to a circular core. A mag. 14.5 star is embedded in the halo about 35" NE of the nucleus.

eg 2848 dimen. 2.'7 × 1.'8 V=12.1 sfc. br. 13.6
Visible only as a faint, indefinite patch in 25 cm, 30 cm shows this galaxy in pa 30°, 1.'8 × 1' and barely concentrated toward the center. The inner regions are more circular and have some brighter spots. A threshold magnitude star is visible on the NE tip; a mag. 12.5 star lies 2.'7 NE. Photographs show eg 2851 (mag.$_{VV}$15) 5.'2 ENE.

eg 2855 dimen. 2.'7 × 2.'4 V=11.6 sfc. br. 13.4
Located 4' S of a mag. 9 star, this galaxy can be seen at 50× with 15 cm. It appears circular with no central concentration. 25 cm reveals a 1.'5 × 1' glow elongated SE-NW. The core is small and not outstanding. 30 cm shows a circular object about 2' diameter with a 1' core.

eg 2889 dimen. 2.'0 × 1.'8 V=11.8 sfc. br. 13.1
In 25 cm this object appears fainter than eg 2855, *cf.* The halo is circular, about 1' diameter, with a small central brightening. Two bright stars are visible S and SW. In 30 cm the halo reaches halfway to the mag. 11.5 star 1.'5 S. The circular object has its brightest part E of center. eg 2884 (mag.$_{VV}$13) lies 13' WNW, eg 2881AB (mag.$_{VV}$14) lies 28' SW, and eg I2482 (mag.$_{VV}$13) is 28' SSW.

eg 2902 dimen. 1.'3 × 1.'2 B=13.3 sfc. br. 13.6
25 cm shows this small object as a 30" diameter glow with a faint stellar nucleus. The overall brightness, however, seems greater than that of eg 2924, *cf.* It appears 40" diameter in 30 cm, with strong, even concentration to an occasionally stellar nucleus. A threshold magnitude star lies just off the NNW edge.

eg 2907 dimen. 1.'0 × 1.'3 B=13.0 sfc. br. 13.9
Moderately bright for 25 cm, this galaxy appears elongated in pa 110°, 1.'5 × 0.'6, with thinly tapered tips and a slightly brighter core. On the NW is a mag. 13.5 star. 30 shows it 1.'4 × 0.'8 with tapering, pointed ends. The elongated core has no distinct nucleus, but a few beads are visible parallel to the major axis but displaced a little S of center. A mag. 13.5,14.5 pair of stars is visible 2.'5 ESE; other stars can be seen to the SW.

eg 2924 dimen. 1.'6 × 1.'5 B=13.1 sfc. br. 14.0
Lying within a 15' triangle of mag. 8 stars, this galaxy is small and faint in 25 cm, appearing about 30" diameter with a small central brightening. A mag. 14 star is visible 1' from the center in pa 130°. The halo seems elongated in pa 40° with 30 cm. The oval core contains some stellarings, and a threshold magnitude star is visible on its SW side 20" from the center, perhaps influencing the observed elongation. eg I 546 (mag.$_{VV}$14.5) lies 5' W.

eg 2935 dimen. 3.'5 × 3.'0 B=12.0 sfc. br. 14.4
Forming a triangle with two stars 4' S and W, 15 cm shows this galaxy as a round spot about 1' across. In 25 cm the core is elongated SE-NW with a stellar nucleus, while the halo seems elongated E-W, extending to 1.'5 × 1.' In 30 cm the halo grows to 3' × 1.'8, elongated roughly E-W. The core seems elongated N-S, about 30" long, but this impression is probably due to a mag. 14.5 star that photographs show about 25" NNE of the nucleus. eg 2921 (mag.$_{VV}$13) lies 33' WNW.

eg 2962 dimen. 3.'3 × 2.'4 V=11.7 sfc. br. 13.8
This galaxy lies 5.'5 N of a short line of three stars with the brightest star in the middle. It is too faint for 15 cm, but 25 cm shows it 1.'75 × 0.'75, elongated N-S. The circular and somewhat brighter core is about 30" across. It appears 50" diameter in 30 cm, concentrated to a bright substellar nucleus.

eg 2983 dimen. 2.'6 × 1.'8 V=11.7 sfc. br. 13.2
Though not visible in 15 cm, this galaxy is moderately faint but conspicuous in 25 cm, located 7' SSE of a mag. 6.5 star. It is elongated NE-SW, 1' × 0.'75. The core is a little brighter, about 20" across, with an inconspicuous stellar nucleus. In 30 cm the halo is broadly concentrated, 1.'5 × 1', with a faint star 1' S of the center.

eg 2986 dimen. 2.'5 × 2.'3 V=10.9 sfc. br. 12.8
This object is visible in 15 cm as a dull uniform patch about 2' diameter lying 10' SE of a mag. 8.5 star. It is

pretty bright in 25 cm, appearing about 1′ across with a broad 40″ core and a stellar nucleus. A mag. 12.5 star lies 1′.2 SE. Viewed in 30 cm the bright core is 30″ across, the halo extending to 1′ diameter. In addition to the star SE are two to the W at 45″ and 2′ from the center.

eg 2992 ***dimen. 4′.1 × 1′.4*** ***V = 11.9*** ***sfc. br. 13.6***
eg 2993 ***dimen. 1′.6 × 1′.3*** ***V = 12.6*** ***sfc. br. 13.2***
Neither of this interacting pair is visible in 15 cm. In 25 cm eg 2992 is the brighter of the two, a circular spot about 1′ diameter with a stellar nucleus. eg 2993 lies 2′.9 SE of −92 and appears similar but fainter. In 30 cm −92 is oval and broadly concentrated to a faint nucleus. The halo extends very faintly to 2′ × 1′ in pa 20°. eg 2993 is smaller but more concentrated. Its halo is elongated in pa 10°, 1′ × 0′.5. At 150 × the nucleus is 10″ across and appears nearly stellar; at 225 × it seems multiple. A mag. 13 star is visible 1′.9 SSE.

eg 3052 ***dimen. 2′.1 × 1′.5*** ***B = 12.8*** ***sfc. br. 14.0***
Lying 23′ N of a red mag. 5 star, this object is pretty faint for 25 cm. The halo is 1′.5 × 0′.75, elongated E-W; the core is 30″ × 20″ and looks like a lobed star. A mag. 11.5 star is visible 3′.5 W; a threshold magnitude star is visible 20″ WSW of center. A pretty field of stars lies 20′ NE. 30 cm shows the halo to 1′.5 × 1′ in pa 100° with a brighter core but no nucleus. On the N side of the star W is a fuzzy spot that photographs show to be a double star. eg 3045 (mag.$_{VV}$14) lies 16′ W.

eg 3091 ***dimen. 2′.2 × 1′.7*** ***B = 12.5*** ***sfc. br. 13.9***
eg M-03-26-06 ***dimen. 0′.4 × 0′.4*** ***mag.$_{VV}$15*** ***sfc. br.***
In 25 cm, eg 3091 is elongated SE-NW, 2′ × 0′.75. The core is concentrated and circular, about 30″ across. The halo looks like a pair of skinny triangles attached to the rounded central bulge. A trio of stars is visible 4′ NW. 30 cm shows the galaxy W of a wide pair of stars. The halo extends to 2′ × 1′ in pa 145° with a bright core and stellar nucleus. Two faint stars are visible 2′ and 3′.5 SE. In 30 cm eg M-03-26-06 is just visible as a faint star-like spot 1′.3 NW of eg 3091. eg 3096 (mag.$_{VV}$15) lies 4′.8 ESE.

eg 3124 ***dimen. 3′.2 × 2′.7*** ***B = 12.4*** ***sfc. br. 14.5***
Lying 5′.5 N of a bright pair (ADS 7641: 8.8,10.0; 9″.5; 146°), 25 cm shows this galaxy as an amorphous glow about 1′.5 diameter. A group of faint stars 11′ W appears nebulous at 100 × . 30 cm shows a circular spot 2′ across.

eg 3200 ***dimen. 4′.8 × 1′.8*** ***B = 12.3*** ***sfc. br. 14.5***
Located 2′ E of a mag. 10.5 star, this is a fairly faint object for 25 cm. It has low surface brightness and no central concentration. The halo is 2′.5 × 1′, elongated N-S. A very faint star is visible 55″ N of center, within the halo. 30 cm shows it 4′ × 1′ in pa 165°; the core is 1′ across with a stellar nucleus occasionally visible. A faint pair of stars is visible 3′.6 N; another star is barely visible on the W side of the nucleus.

pn 3242 ***dimen. 45″ × 36″*** ***nebula 7.8v*** ***star V = 11.9***
This bright planetary appears pale blue in 15 cm at 50 × . At 200 × the surface is uniform, but a bright spot can be

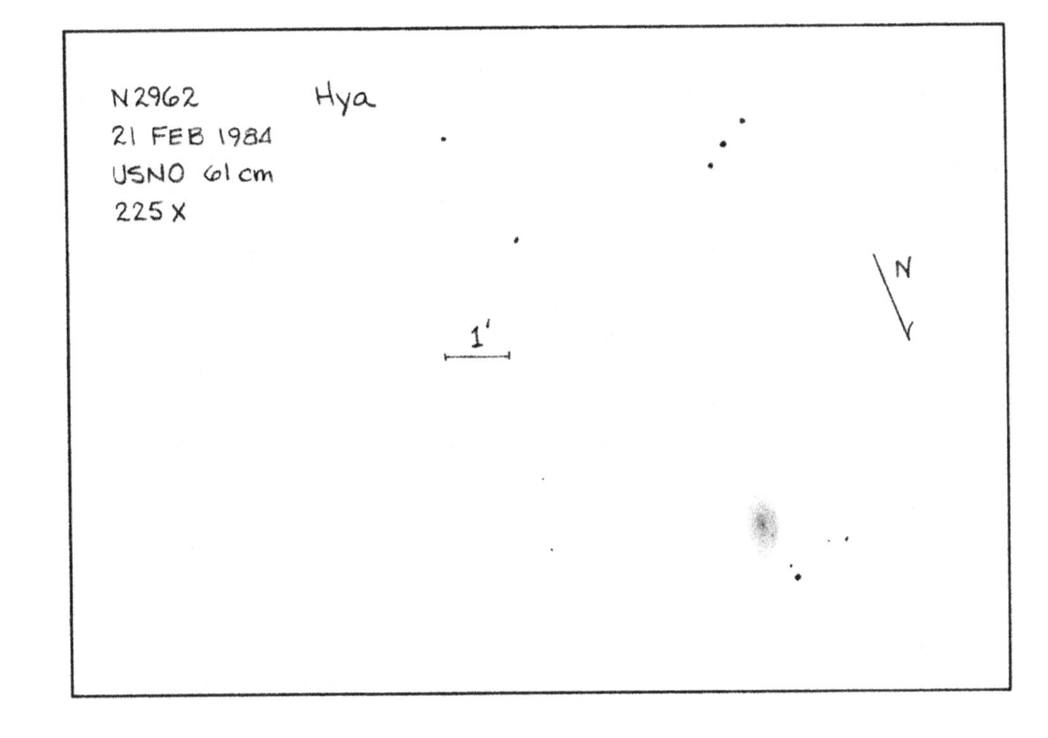

seen occasionally on the NW side. It is elongated E-W in 25 cm, an oval with diffuse edges. The bright spot NW is seen clearly; otherwise, the object is featureless.

eg 3285 dimen. 2′.5 × 1′.7 B = 13.0 sfc. br. 14.4
In 30 cm this galaxy is visible as a poorly concentrated patch with a faint stellar nucleus lying 7′ SW of a mag. 7.5 star. At 250 × the halo extends to 2′ × 1′ in pa 110°.

eg 3308 dimen. 2′.0 × 1′.4 V = 12.2 sfc. br. 13.3
This member of the Hydra I cluster appears similar in brightness to eg 3312, *cf.*, in 30 cm. The halo is about 2′ × 1′ in pa 35°, showing moderate concentration to a small and knotty 30″ core. Occasionally a knot will stand out near the center, but any nucleus is inconspicuous.

eg 3309 dimen. 1′.9 × 1′.7 V = 11.9 sfc. br. 13.2
In 30 cm this member of the Hydra I cluster forms the western point of a triangle with two mag. 12 stars that lie 3′.5 from each other. The halo is about 1′.5 across with weak concentration to a broad core. At the edge of the halo only 30″ from the center in pa 120°, a mag. 13 star is visible. eg 3311, *q.v.*, lies only 1′.7 ESE.

eg 3311 dimen. 2′.5 × 2′.2 V = 11.6 sfc. br. 13.3
30 cm shows this member of the Hydra I cluster between the two stars that form the eastern base of the triangle including eg 3309, *cf.* 1′.7 WNW. This galaxy is larger than −09, but of lower surface brightness. The halo is approximately circular, about 2′ diameter, with barely any central concentration.

eg 3312 dimen. 3′.6 × 1′.5 B = 12.7 sfc. br. 14.5
Lying about 5′ SE of eg 3309 and eg 3311, 30 cm shows this member of the Hydra I cluster to have the lowest surface brightness of the group. The poorly concentrated halo is 2′.5 × 1′.5 in pa 175°. At low power the broadly concentrated core contains a stellar nucleus. At high power the nucleus is still distinct but loses its stellar appearance.

eg 3390 dimen. 4′.0 × 0′.7 V = 12.2 sfc. br. 13.1
This galaxy is easy to find in 30 cm, lying 10′ N of a mag. 6 star. The galaxy forms a triangle with this star and a mag. 7.5 star a little closer to the SW. It is elongated N-S, 2′.25 × 1′, with poor concentration to an unevenly bright core about 1′ × 0′.5. No nucleus is discernable even at high power. A mag. 14 star lies 1′ N of center on the W edge of the northern arm; another slightly fainter star lies 2′ S.

eg 3585 dimen. 2′.9 × 1′.6 V = 10.0 sfc. br. 11.7
Easy to see in 15 cm, this oval object is about 2′ diameter and uniformly bright except for a stellar nucleus. The halo is elongated roughly E-W in 25 cm, the outlying haze extending to 2′.25 × 1′.5. The core is 1′ × 0′.5 and has a stellar nucleus. A threshold magnitude star is visible 2′.5 ENE. The light is evenly concentrated in 30 cm, rising considerably in intensity to an elongated nonstellar nucleus. The halo is 2′.25 × 1′.25 in pa 115°. Additional mag. 13.5 stars can be seen on the NNW and WNW, both within 5′ distance.

eg 3673 dimen. 3′.5 × 2′.5 B = 12.4 sfc. br. 14.6
Viewed in 30 cm this object lies 1′.6 W of an unequally bright pair (10.5,12;39″;240°). The galaxy is elongated in pa 70°, 1′.75 × 0′.8, with tapering ends. The inner parts are more elongated, but little brighter, and no nucleus is visible.

gc 4590 diam. 12′ V = 7.7 class X
Messier 68 is visible in 6 cm as a weakly concentrated circular glow about 5′ diameter. In 15 cm a few sparkling outliers are resolved. It is about 10′ across in 25 cm. It appears granular at 50 ×, but higher power resolves it well across the center with some underlying haze. 30 cm shows a cottony center 3′ across. The central stars are bright, the brightest being S of center. The overall diameter is 10′.

eg 5085 dimen. 3′.4 × 3′.0 B = 11.9 sfc. br. 14.3
15 cm will just show this galaxy about 4′.5 N of a mag. 8.5 star. In 30 cm it is an unconcentrated low surface brightness patch about 2′ diameter with an indefinite N-S elongation. A pretty, equal pair of mag. 12.5 stars lies 3′.3 NNE with 45″ separation in pa 155°.

eg 5236 dimen. 11′ × 10′ B = 8.2 sfc. br. 13.2
Messier 83 is a mottled, poorly concentrated glow in 6 cm. The core appears about 30″ across in a 5′ × 2′ inner halo elongated in pa 70°. The halo extends to a mag. 11 star on the SW with an overall diameter of 7′. It is bright in 15 cm, which shows a bright core within a large faint halo. The NE edge is sharply defined in 25 cm. The circular core is 1′ across. South of the core is a broad dark band curving E to W. 30 cm reveals a brightening on the NW side of the center, another farther to the NE.

eg 5328 dimen. 1′.7 × 1′.4 V = 11.8 sfc. br. 12.7
In 30 cm this galaxy is visible as the northern member of a chain with two mag. 13.5 stars. The halo is about 1′ diameter with a fairly strong, even concentration to a distinct but nonstellar center. About 16′ S and a little E is a close 12″ pair (the penultimate star in another string headed S) that looks persistently hazy, as if the western member were a galaxy. Photographs show only a faint

Chart XI. Hydra cluster = Abell 1060.

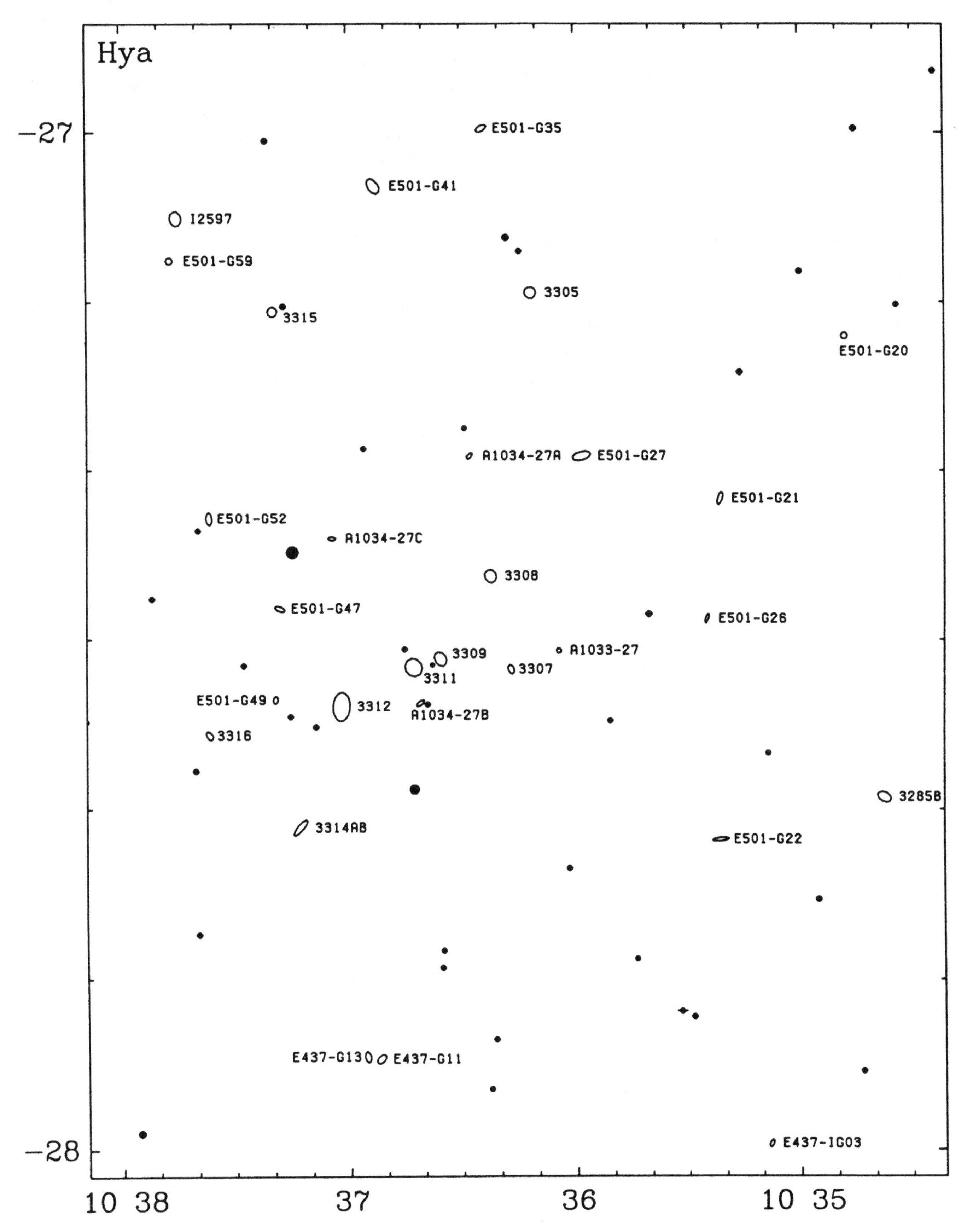

third star in the pair. eg 5330 (mag.$_{vv}$15) lies 1.'7 NE of −28.

eg I4351 dimen. 5.'6 × 1.'2 B = 12.3 sfc. br. 14.2
This galaxy is 7′ E and a little south of a mag. 9 star plotted on the AE. The galaxy is very elongated in 30 cm, 3.'5 × 0.'5 in pa 15°. It shows concentration to a spotty core, but no nucleus stands out at any magnification. A mag. 14 star lies 45″ from the center in pa 105°, and a mag. 13,13.5 pair of 1′ separation in pa 300° lies 3′ SW. At higher power a threshold magnitude star is visible just less than 1′ WSW.

eg 5556 dimen. 3.'1 × 2.'7 B = 11.9 sfc. br. 14.0
In 30 cm this galaxy has a very low surface brightness and is hard to see. It lies 5′ W of an elongated symmetrical 3′ × 1′ trapezoid of mag. 12.5–13 stars. The halo is about 2.'5 diameter, but any elongation is indefinite, as the many associated stars are confusing: a mag. 14.5 star is in the E edge 50″ from center, two more threshold stars are in the N edge, and a mag. 14 star lies NW with a mag. 14.5 star 45″ N of it. At 250 × a 20″ diameter spot is visible about 50″ W of the star in the E edge: this is probably the core of the galaxy.

gc 5694 diam. 3.'6 V = 9.2 class VII
In 25 cm this cluster appears similar to gc 5634, *cf.* in Libra, though the core is more strongly concentrated. The cluster is 2′ diameter and shows no resolution. 30 cm shows a strong broad concentration toward the center. The total diameter is 1.'8 with a fairly distinct 50″ core. The core and inner halo appear granular. Two mag. 10.5 stars are visible to the SW.

Lacerta

Lacerta, the lizard, is an oft neglected constellation lacking in bright offset stars. Star fields, however, are rich and contain some fine open clusters. The center of the constellation culminates at midnight about 28 August.

oc 7209 diam. 25′ V=7.7 stars 98
This is an attractive but moderately faint cluster in 15 cm. About 40 stars forming many right-angled patterns lie in a 20′ area. The cluster contains about 75 stars in a 15′ area in 25 cm, with more scattered outliers to 25′ diameter. The members are mostly fainter than mag. 10 and irregularly scattered, showing no concentration to the center. A mag. 8.5 star lies to the SW. About 100 stars are visible in a 25′ area with 30 cm. On the E side are three triplets with the same orientation.

oc I1434 diam. 8′ mag. 9.0p stars 40
This is a good cluster for 15 cm at medium power, though it is too faint to show up well at high power. In 25 cm at least 100 stars, mostly fainter than mag. 12.5, are visible over a little haze in a 20′ area. Three brighter stars form a conspicuous triangle on the S edge. At 50 ×, long curving rays stand out, particularly to the NE and WNW. With 30 cm about 60 stars can be counted in a 12′ area. The conspicuous string leading NE from the N side contains about 20 stars.

eg 7240 dimen. 0.′8 × 0.′8 V=13.8 sfc. br. 13.1
eg 7242 dimen. 2.′6 × 2.′0 V=12.0 sfc. br. 13.7
In 25 cm eg 7242 is visible as a diffuse hazy spot, elongated NE-SW and embedded in a rich field with many mag. 13 stars. Lying 3.′8 SW is eg 7240, which is visible in 25 cm as a faint patch that is much smaller than −42. Photographs show eg I1441 ($mag._z$15.3) lying 1.′3 NNW of −40, or 3.′5 W of −42.

oc 7243 diam. 21′ V=6.4 stars 38
15 cm shows about 40 stars in this cluster, including many bright ones, in a 20′ area. Four sections are grouped in a semicircle opening on the N side. In 25 cm the brighter stars form a parabola opening N, including a nice pair on the E side (ADS 15785: V=9.3,9.6; 9.″4; 11°). About 25 stars are in or involved with this arc; spreading mostly E and N are another 40 stars. The overall size including the outliers is about 30.′ With 30 cm it is about 25′ × 18′ in pa 60°. The stars are irregularly grouped, but two concentrations NE and SW stand out,

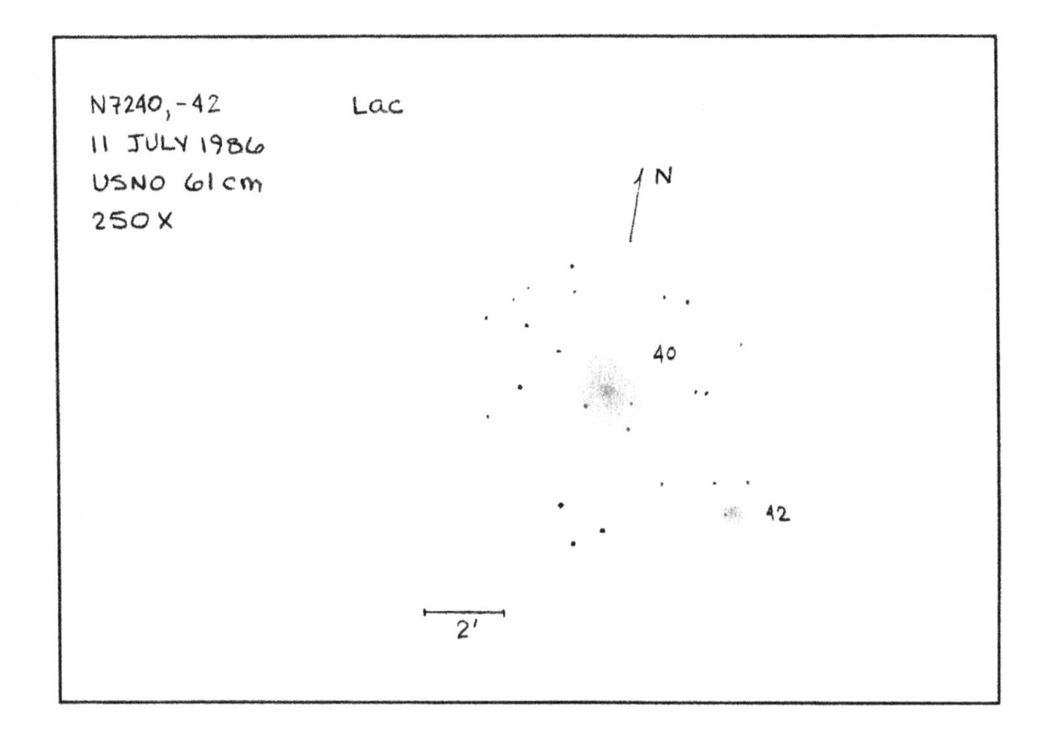

the one SW being larger. About 80 stars can be counted overall.

oc 7245 diam. 7′ V=9.2 stars 169
This is an interesting cluster of faint stars in 25 cm. The brightest members are fainter than mag. 13, but the cluster stands magnification well. It is 6′ diameter with about 40 stars resolved. A dark bar about 1′ wide and 5′ long aligned in pa 30° crosses the E side: a few stars lie on its far side. 30 cm shows 20 stars in a 2′ area bounded on the W and SSE by mag. 11 stars. The compact hazy center is about 45″ across. oc King 9 lies 10′ NE.

oc I1442 diam. 3′.5 V=9.1 stars 104
This cluster is an unconcentrated group of about 35 stars in a 10′ area with 25 cm, spreading mostly N and W of a mag. 10.5 star. The members range widely in brightness. To the N are loosely scattered mag. 11–12 stars that bridge across to oc 7245, *q.v.* 22′ NNW.

eg 7248 dimen. 2′.1 × 1′.1 mag.$_z$13.6 sfc. br.
This galaxy is not visible in 15 cm and appears fairly faint even in 25 cm at 200×. The halo extends to only 40″ × 20″ in pa 130°, with some concentration to a small oval core. A faint pair is visible 2′.6 SW (12,13; 7″; 320°); another slightly wider pair (13,13.5; 11″; 77°) lies 2′.2 E.

eg 7250 dimen. 1′.4 × 0′.7 mag.$_z$13.1 sfc. br.
Difficult to view in 25 cm, this galaxy lies 1′ NNW of a mag. 11 star. The halo has a lenticular outline, elongated in pa 165°. The middle two-thirds shows some indistinct mottling. A very faint star is visible at the N tip.

pn I5217 dimen. 7″ × 6″ nebula 11.3v star V=15.6
Care is needed in finding this tiny planetary without the use of a spectroscope. Located 1′.9 N of a mag. 11.5 star, at 200× in 25 cm it is elongated N-S and has a stellar nucleus. The S tip has a sharp taper like a football. At 425× in 30 cm it has a brighter core and a blue-grey color. A bright spot is visible W of the center.

oc 7296 diam. 4′.0 mag. 9.7p stars 20
6 cm will show this cluster as a small, moderately high surface brightness patch with marginal resolution at 25×. A single brighter star stands out on the W side. In 25 cm the group is about 4′ diameter including about 20 scattered mag. 12–13 stars. With 30 cm 35 stars can be counted, filling a roughly club-shaped area.

Leo

The borders of this group extend a good deal south of the familiar star pattern. Dozens of bright and faint galaxies occur here. The center of the constellation culminates at midnight about 2 March.

eg 2872 ***dimen. 2.′1 × 1.′9*** ***mag.$_Z$13.0*** ***sfc. br.***
eg 2873 ***dimen. 0.′8 × 0.′2*** ***mag.*** ***sfc. br.***
eg 2874 ***dimen. 2.′5 × 0.′9*** ***mag.$_Z$13.5*** ***sfc. br.***
The brightest member of this possibly interacting group, eg 2872, is faint but not difficult to see in 15 cm, lying 5.′5 SE of a mag. 10.5 star. With 30 cm it is clearly the brightest and most strongly concentrated member. The halo is about 45″ diameter and sharply concentrated to a small starry core. It is slightly elongated SE-NW, pointing roughly toward eg 2874. In 15 cm −74 is only just visible as a tiny stellar spot 1.′4 ESE of the brighter galaxy. 30 cm will show its low surface brightness halo extending to 50″ × 30″, elongated in pa 45°. The small, moderately brighter circular core is similar in size to but fainter than that in −72. eg 2873 is just visible in 30 cm using averted vision. It is a tiny spot about 10″ diameter with a substellar nucleus lying 1.′8 N of −74.

eg 2903 ***dimen. 13′ × 6.′6*** ***V = 8.9*** ***sfc. br. 13.5***
Easily visible in 6 cm, this bright galaxy is elongated in pa 15°, about 8′ × 4′. The core appears very elongated and some stellarings are associated with the object. In 15 cm the middle parts of the nebula appear unevenly bright. The halo grows to 10′ × 4′ in 25 cm. A bright stellar nucleus is visible in the center, and a star is visible off the S side. The core is 30″ × 15″ in 30 cm, but does not show a distinct nucleus. The E flank of the nebula is less extensive, fading more abruptly from the brighter central regions. A bright patch is visible 1.′5 N of center; a fainter symmetrical counterpart is visible S of center. The unevenly bright halo extends to 10′ × 4′.

eg 2911 ***dimen. 4.′3 × 3.′2*** ***V = 11.6*** ***sfc. br. 14.3***
This galaxy has a very low surface brightness and is not visible in 25 cm. It is quite faint for 30 cm, located 15′ S of a long triangle of stars pointing S. About 45″ diameter overall, it is weakly concentrated and without a stellar nucleus. The galaxy is in a group including eg 2914 (V = 13.1) 4.′8 SE, eg 2919 (mag.$_Z$13.6) 19′ ENE, and other fainter galaxies.

eg 2916 ***dimen. 2.′6 × 1.′9*** ***V = 12.0*** ***sfc. br. 13.5***
In 15 cm this galaxy is a moderately faint circular spot about 2′ N of two mag. 11.5 stars. With 30 cm it is moderately bright, extending to 55″ × 40″ in pa 25°. The halo shows weak, even concentration to a faint but distinct stellar nucleus. At 250 × the halo has a slightly mottled texture.

eg 2964 ***dimen. 3.′0 × 1.′7*** ***V = 11.3*** ***sfc. br. 12.9***
eg 2968 ***dimen. 2.′2 × 1.′5*** ***V = 11.8*** ***sfc. br. 13.0***
eg 2964 is the brightest of a group, its brightest companion being eg 2968, 5.′8 SE. eg 2964 is a difficult object for 15 cm, appearing about 1.′5 diameter with no apparent elongation or central brightening. 25 cm shows a 1.′5 × 1′ oval elongated nearly E-W. There is little central brightening except for a faint stellar nucleus. 30 cm reveals a broad oval 1.′5 × 1.′2 in pa 95°. The core is slightly concentrated with small mottlings and stellarings. eg 2968 is not visible in 15 cm. With 25 cm it is about 1′ diameter with a nebulous extension on the NW side. Bright spots are visible on the core as well as a very faint stellar nucleus. In 30 cm it appears smaller and fainter but better concentrated than −64. The halo is elongated NE-SW, 1′ × 0.′6, with a 10″ core. eg 2970 (mag.$_Z$14.7) lies 5.′1 NE of −68.

eg 3016 ***dimen. 1.′3 × 1.′1*** ***mag.$_Z$13.7*** ***sfc. br.***
In 30 cm this galaxy is the second brightest member of the eg 3020 group. Its moderately low surface brightness halo is about 30″ diameter, slightly elongated ENE-WSW, with weak, even concentration to the center. At 250 × there occasionally seems to be a string of a few faint stellarings along the major axis.

eg 3019 ***dimen. 1.′0 × 0.′7*** ***mag.$_Z$15.0*** ***sfc. br.***
eg 3020 ***dimen. 3.′2 × 1.′8*** ***mag.$_Z$13.2*** ***sfc. br.***
eg 3020 is the brightest of a group including eg 3016, eg 3019, and eg 3024, *q.v.* 15 cm will faintly show eg 3020 lying 8′ E of a mag. 8 star. In 30 cm it has a broad, diffuse halo reaching to 1′ × 0.′5, elongated E-W. The central region has a granular texture, but shows no distinct nucleus at 250 × . A faint companion, eg 3019, is just visible in 30 cm 4′ S, forming a pair with a mag. 13.5 star 45″ NE.

eg 3024 ***dimen. 2.′2 × 0.′6*** ***mag.$_z$13.7*** ***sfc. br.***
Lying 5.′8 SE of the brightest member of the group, eg 3020, *q.v.*, 30 cm shows this galaxy as a tiny 1′ × 0.′2 spindle elongated in pa 125°. The halo doesn't reach to a mag. 13.5 star 1.′5 SE. At the middle the core bulges to about 15″ width and is distinctly brighter than the narrow, spike-like extensions of the halo.

eg 3032 ***dimen. 2.′5 × 2.′1*** ***V = 11.9*** ***sfc. br. 13.6***
Located between mag. 8 and 10 stars, this galaxy is not visible in 15 cm. In 25 cm it appears roughly circular, about 1′ diameter with a conspicuous granular core. A beady stellar nucleus is prominent in 30 cm, and the faint halo is about 1′ diameter.

eg 3041 ***dimen. 3.′7 × 2.′5*** ***V = 11.5*** ***sfc. br. 14.5***
This object is not visible in 15 cm. The halo is 2′ × 1′ in pa 100° in 25 cm, but it shows no central brightening. A mag. 12.5 star is visible 1.′5 SW. The mottled low surface brightness halo grows to 2.′8 × 1.′75 in 30 cm, weakly concentrated to a very elongated inner region. The core is 2′ × 0.′75 and contains faint stellarings but no nucleus. A mag. 14 star is visible at the edge of the halo 1.′25 WNW; another mag. 14 star lies farther W, forming a triangle with the mag. 12.5 and 14 stars.

eg 3067 ***dimen. 2.′5 × 1.′0*** ***V = 11.7*** ***sfc. br. 12.5***
This galaxy is just visible in 15 cm with a star-like center. It is circular in 25 cm, about 1.′5 diameter with a faintly mottled core. The halo is elongated in pa 110°, 2′ × 0.′6, in 30 cm, with rounded ends. The uneven core is 1′ long; there is no distinct nucleus.

eg 3098 ***dimen. 2.′6 × 0.′8*** ***V = 12.0*** ***sfc. br. 12.5***
Though not visible in 15 cm, 25 cm will show this narrow galaxy faintly. It is elongated in pa 90°, 1′ × 0.′25, a small sharp spindle in a notably blank field. The core is bright and has sharply pointed tips. 30 cm shows the halo to 1.′75 × 0.′8 with a small elongated core and stellar nucleus. A few bright beads stand out along the major axis. A mag. 13 star is visible 2.′5 E.

eg 3153 ***dimen. 2.′3 × 1.′1*** ***mag.$_z$13.6*** ***sfc. br.***
While not visible in 15 cm, 30 cm will show this faint object as an unconcentrated low surface brightness glow. The halo is 1.′25 × 0.′75, elongated in pa 165°.

eg 3162 ***dimen. 3.′1 × 2.′7*** ***V = 11.6*** ***sfc. br. 13.7***
An interesting object for 15 cm, this galaxy appears large and faint with low surface brightness. A mag. 11 star is visible 3.′7 to the NE; another brighter star lies 3.′5 W. It is 1.′3 diameter in 25 cm, and the broad core contains a faint stellar nucleus. In 30 cm the irregular halo is 1.′6 diameter with a slight, broad concentration. Large mottlings appear across the surface and a very faint nucleus is occasionally visible. A mag. 14 star is visible 1′ SE.

eg 3177 ***dimen. 1.′7 × 1.′3*** ***V = 12.3*** ***sfc. br. 13.0***
15 cm will show this object as a faint round spot. It is uninteresting in 25 cm, an amorphous glow without concentration about 45″ across. 30 cm reveals a 1′ diameter halo with a moderate concentration to a distinct core and nonstellar nucleus.

eg 3185 ***dimen. 2.′3 × 1.′6*** ***V = 12.2*** ***sfc. br. 13.4***
This galaxy, lying 11′ SW of eg 3190, *q.v.*, is just visible in 15 cm as a faint but strongly concentrated circular spot with a stellar nucleus. In 25 cm it is about 1′ diameter, while with 30 cm it grows to 1.′2 × 1′, elongated in pa 125°. It shows little overall concentration here except for a small inconspicuous core around the stellar nucleus. A mag. 14 star lies near the W edge, 40″ from the center; another slightly brighter star lies 1.′3 SW.

eg 3187 ***dimen. 3.′3 × 1.′5*** ***V = 13.1*** ***sfc. br. 14.7***
This galaxy lies 4.′9 NW of eg 3190, *q.v.*, and is not visible in 15 cm. With 30 cm it is faint, diffuse, and unconcentrated toward the center. The overall size is 1′ × 0.′5, elongated in pa 120°, pointing directly toward eg 3190.

eg 3190 ***dimen. 4.′6 × 1.′8*** ***V = 11.0*** ***sfc. br. 13.1***
In 15 cm this is the brightest galaxy of the group including eg 3185, eg 3187, and eg 3193, *q.v.* With 25 cm it is elongated SE-NW, about 1.′5 × 0.′5, with a brighter elongated core. 30 cm shows the halo extending to 2.′5 × 0.′8, elongated in pa 120°. The tiny oval core is much brighter than the halo and contains a stellar nucleus. Both the core and the nucleus are crowded against the SW flank of the halo, which is sharply defined, fading more abruptly than the opposite flank. A mag. 13.5 star lies 1.′2 W of center.

eg 3193 ***dimen. 2.′8 × 2.′6*** ***V = 10.9*** ***sfc. br. 13.1***
This galaxy is the second brightest in the eg 3190 group in 15 cm, though it is only a little fainter than eg 3190, *cf.* In 25 cm it is a small, round, concentrated spot less than 1′ diameter lying 1.′4 S of a mag. 8.5 star. With 30 cm it grows to 1.′2 diameter, nearly circular in outline, but slightly elongated in pa 10°. It shows a classic zoned brightness profile, with a distinct core and bright stellar nucleus.

eg 3221 ***dimen. 3.′3 × 0.′9*** ***mag.$_z$14.3*** ***sfc. br.***
This galaxy is visible in 15 cm as a faint, hardly concentrated streak elongated N-S. 30 cm shows a moderately

faint 3′ × 0′.4 spindle elongated in pa 170°. There is a slight, broad, somewhat irregular concentration and no central bulge.

eg 3222 dimen. 1′.3 × 1′.1 V = 12.8 sfc. br. 13.1
This galaxy lies 12′ W of eg 3226–27, *q.v.* In 25 cm it appears as an oval granular blob 3′ N of a mag. 12 star. In 30 cm the halo is elongated NE-SW, 40″ × 30″, and is well concentrated to a faint nearly stellar nucleus. eg 3213 (mag.$_Z$14.3) lies 23′ SW.

eg 3226 dimen. 2′.8 × 2′.5 V = 11.4 sfc. br. 13.5
eg 3227 dimen. 5′.6 × 4′.0 V = 10.8 sfc. br. 14.0
eg 3227, a Seyfert galaxy, forms an interacting pair with eg 3226 2′.2 NW. Both objects are easily detectable in 15 cm, though their forms are indefinite. eg 3226 is circular in 25 cm, fading gradually from a stellar nucleus to the edge. The S edge contacts −27, which appears a bit brighter than −26. eg 3227 is 2′ × 1′ with two star-like nuclei. In 30 cm the pair provide a nice contrast between elliptical and spiral galaxies. The elliptical (−26) is smooth and circular, 1′.2 diameter, with a substellar nucleus. The spiral (−27) is elongated in pa 155°, 2′.5 × 1′, rising to an indistinct 45″ × 15″ core and a prominent stellar nucleus. The pair of galaxies is aligned in pa 165°.

eg 3274 dimen. 2′.2 × 1′.1 V = 12.8 sfc. br. 13.6
This galaxy is not visible to 15 cm and is fairly faint in 25 cm. Its halo is elongated in pa 90°, extending to 1′.25 × 0′.5. The center is only a little brighter and has a faint stellar nucleus just off-center to the SW. Two mag. 13 stars are visible 1′.8 NNW and SW. It is faint and unconcentrated in 30 cm, with a 1′.1 × 0′.75 halo. A mag. 11.5,13.5 pair with 13″ separation in pa 210° is visible 3′.7 ESE.

eg 3287 dimen. 2′.2 × 1′.1 B = 12.9 sfc. br. 13.7
25 cm will show this galaxy only faintly. It appears circular, about 40″ across, with a faint stellar nucleus occasionally visible. 30 cm shows an elongated object 2′.1 × 0′.75 in pa 20°. The light is unconcentrated, but a very faint stellaring is visible near the center. A bright pair of stars with contrasting colors is visible 6′.5 to the WSW (ADS 7836: 7.6,9.0; 11″; 259°). eg 3301, *q.v.*, lies 34′ ENE.

eg 3300 dimen. 2′.1 × 1′.1 B = 13.1 sfc. br. 13.9
In 25 cm this galaxy is faint, but easy to find because of nearby bright stars. It appears as a small, patchy spot less than 1′ diameter with a brighter core. It is fairly easy to see in 30 cm, lying in a 9′ box of five mag. 8–12 stars. The halo extends to 2′ × 1′, elongated N-S, with moderate concentration and a few stellarings in the core.

eg 3301 dimen. 3′.6 × 1′.2 V = 11.4 sfc. br. 12.8
This galaxy is located on the S side of a 3′ triangle of mag. 10 stars. It is faintly visible in 15 cm, a diffuse glow without central brightening. 25 cm shows it 1′ × 0′.5 in pa 55° with a conspicuous stellar nucleus. The halo extends more to the NE (i.e., the nucleus appears off-center to the SW). It is a bright, well-concentrated object in 30 cm. The halo is elongated in pa 55°, 3′ × 0′.75, while the core is in pa 65° or so due to a bright patch on the NW flank

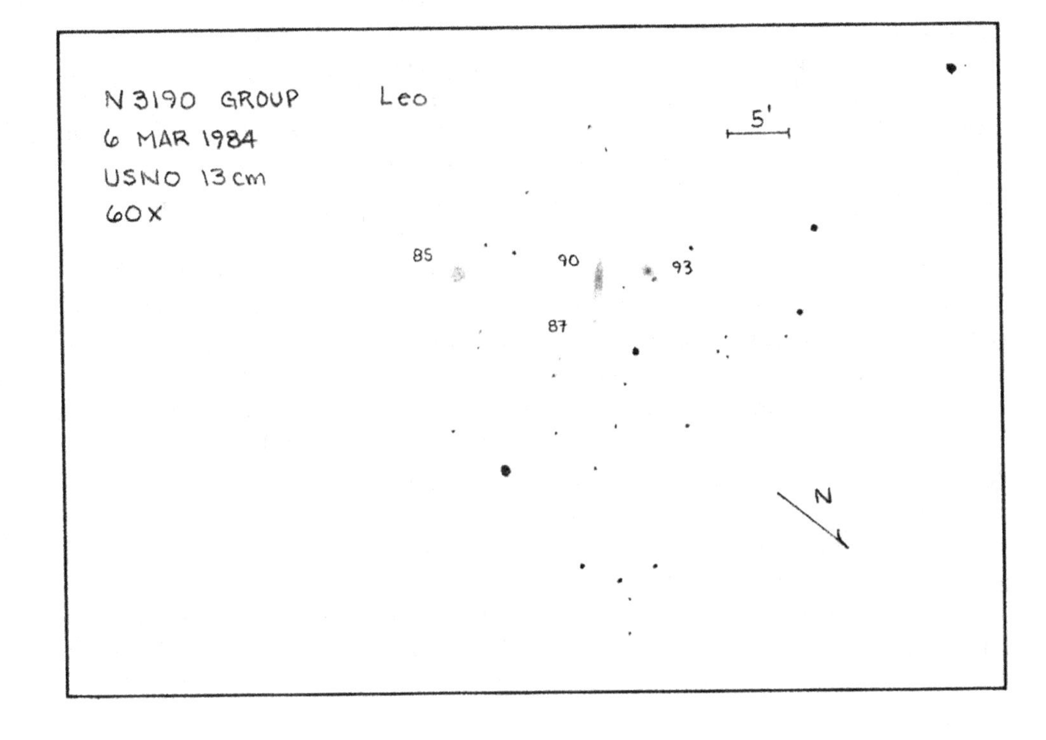

of the SW end. The SE flank fades more abruptly than the NW flank. eg 3287, *q.v.* lies 34′ WSW.

eg 3338 ***dimen. 5′.5 × 3′.7*** ***V = 10.8*** ***sfc. br. 13.9***
eg U5832 ***dimen. 1′.2 × 1′.1*** ***mag.$_z$13.8*** ***sfc. br.***
Lying 2′.7 E of a mag. 8 star, eg 3338 is difficult to see in 15 cm. The diffuse glow is elongated E-W in 25 cm, extending to about 3′.5 × 1′. The center is little brighter, and a few stellarings are visible along the major axis. A mag. 11.5 star is visible 5′ N. In 30 cm the halo reaches to 5′.5 × 2′, almost to the bright star on the W. It shows only a little broad brightening toward the center. A very faint nucleus is more conspicuous at low power; otherwise the galaxy is very smooth across the core. Four mag. 13.5 stars are visible 2′–3′ distant to the N, NE, SW, and NW. Lying 22′ SE is eg U5832, only just visible in 30 cm as a diffuse 1′ diameter patch without concentration. eg 3338 is a member of the Leo (M96) group.

eg 3346 ***dimen. 2′.8 × 2′.5*** ***B = 12.2*** ***sfc. br. 14.2***
This object is just visible in 15 cm as a 2′ patch with two mag. 11 stars nearby on the W. It is moderately bright in 25 cm and seems elongated in the same position angle as the two stars 3′ W. Except for an occasionally visible stellar nucleus, the 2′ × 1′ halo is without features. In 30 cm it is a large diffuse object with moderately low surface brightness and very poor concentration. The halo is about 3′ × 2′.5 in pa 100°, showing no core or nucleus even at 250 ×, though the middle two-thirds is unevenly bright.

eg 3351 ***dimen. 7′.4 × 5′.1*** ***V = 9.7*** ***sfc. br. 13.5***
Messier 95 is visible in 6 cm as an unevenly bright and well-concentrated glow about 3′ diameter. It appears in the same low-power field with Messier 96 (eg 3368) and Messier 105 (eg 3379), *q.v.* It is bright in 15 cm, the diffuse halo rising to a bright core and sharp nucleus. It appears slightly fainter than M96 in 25 cm. The bright halo is elongated in pa 120°, 2′.5 × 2′, with a small, more circular core. In 30 cm the inner 2′ × 1′ is elongated in pa 130° and contains a small 15″ nucleus but no stellar center. Immediately N and S of this bar is darker than farther out, this showing best at 250 ×. The unevenly bright halo extends to 5′ × 3′.5, elongated NNE-SSW. This galaxy is a member of the Leo (M96) group.

eg 3367 ***dimen. 2′.3 × 2′.1*** ***V = 11.5*** ***sfc. br. 13.0***
15 cm shows this object 10′ N of a pair of mag. 9 stars of about 45″ separation in pa 265°. The faint halo is elongated E-W, and two stars are visible in the SW quadrant. The halo is circular in 25 cm, about 1′.25 diameter. A very faint stellar nucleus is visible in a bar-shaped core running E-W through the halo. In 30 cm it is slightly elongated in pa 80°, extending to 2′.5 × 2′ with a very slight broad concentration to the center. A very faint stellar nucleus is occasionally visible, more steadily at 250 × than at 125 ×. The nearby mag. 12.5 stars lie 2′.1 WSW and 2′.6 SSW.

eg 3368 ***dimen. 7′.1 × 5′.1*** ***V = 9.2*** ***sfc. br. 12.9***
Messier 96 is the brightest of the Leo group, which includes Messier 95 (eg 3351), Messier 105 (eg 3379), and many fainter objects. In 6 cm it appears about the same brightness as M105. The core is much brighter than the halo in 15 cm. The halo is elongated SE-NW in 25 cm, extending to 4′ × 2′.5. The core is 2′ × 0′.5 and contains a stellar nucleus. In 30 cm it is moderately well concentrated with a sharply brighter nonstellar nucleus, though the nucleus is smaller than the one in M95, *cf.* The halo is about 5′ × 3′, elongated in pa 140°, but the central regions are quite circular. At 250 × the halo extends to about 5′.5 along the major axis, and the SW side fades more abruptly than the NE. The nucleus is substellar at this magnification, but a bright region about 10″ across surrounds it.

eg 3370 ***dimen. 3′.1 × 1′.9*** ***B = 12.3*** ***sfc. br. 14.0***
Viewed in 25 cm this galaxy is pretty faint and exhibits no central concentration. No details are visible in the circular 1′ diameter halo.

eg 3377 ***dimen. 4′.4 × 2′.7*** ***V = 10.2*** ***sfc. br. 12.9***
This bright object is visible in 15 cm as a circular spot with a concentrated core. 25 cm shows a circular central bulge about 1′ across within a bright 2′ × 1′ halo in pa 45°. With 30 cm the halo extends to 4′.5 × 2′.5, with strong sharp concentrations to elongated inner regions and a bright nucleus. Two mag. 14 stars are visible lying 2′ ESE and 2′.5 SE. Photographs show a low surface brightness companion. eg 3377A (mag.$_z$15.0), 7′.0 NW.

eg 3379 ***dimen. 4′.5 × 4′.0*** ***V = 9.3*** ***sfc. br. 12.4***
Messier 105 is easily visible in 6 cm in the same low-power field with Messier 95 (eg 3351) and Messier 96 (eg 3368), *q.v.* 15 cm shows a stellar nucleus in the center of a bright circular spot. In 25 cm the brightness increases evenly up to the substellar nucleus: the core and halo are not clearly delineated. The overall size is 1′.5 × 1′.2, elongated E-W. With 30 cm it grows to 4′ × 3′, elongated in pa 80°. The smoothly textured halo is moderately concentrated to the core, which shows strong, sharp concentration to a bright 10″ nucleus.

eg 3384 ***dimen. 5′.9 × 2′.6*** ***V = 10.0*** ***sfc. br. 12.9***
In 15 cm this object appears similar to Messier 105 (eg 3379), *cf.*, though it is slightly smaller. The nucleus is

stellar. 25 cm shows it 3′ × 1′, elongated NE-SW, and clearly fainter than M105. The core is circular, about 40″ across, with a stellar nucleus. A mag. 13 star lies 2.′3 SE of center. In 30 cm the halo is very elongated, extending to 5.′5 × 1.′75 in pa 50°. It is abruptly concentrated to circular central regions and a more nearly stellar nucleus than in M105.

eg 3389 dimen. 2.′7 × 1.′5 V = 11.8 sfc. br. 13.2
This galaxy is discernable in 15 cm as a faint uniform glow a little smaller than its nearby companions eg 3387 and M105 (eg 3379), *cf.* A faint stellar center is visible in 25 cm; the halo extends to 2′ × 1′, elongated ESE-WNW. 30 cm shows the unconcentrated halo to 3′ × 1.′5 in pa 110°. No nucleus is discernable even at high power, though the middle parts show some mottling. A mag. 13.5 star is visible 2.′7 NNE.

eg 3412 dimen. 3.′6 × 2.′0 V = 10.6 sfc. br. 12.6
Viewed in 15 cm, this galaxy is easily visible at 50 ×, showing a prominent stellar nucleus. In 25 cm it is moderately bright, about 1.′5 diameter, elongated a little SE-NW. The core looks like a bright planetary nebula 20″ across. At times a dark arc around the north side of center is suspected. The halo extends to 3.′5 × 1.′75 in 30 cm, elongated in pa 150°. It is abruptly concentrated to a circular 50″ core and a bright nonstellar nucleus. At 250 × the E flank of the galaxy is darker, and a dark lane is suspected along this side. Lying 1.′3 N of center is a mag. 14 star on the E edge of the halo.

eg 3433 dimen. 3.′5 × 3.′2 B = 12.3 sfc. br. 14.8
In 25 cm this galaxy appears as a very faint, indefinite sparkly patch about 1.′5 diameter. 30 cm shows a low surface brightness patch about 2′ diameter that exhibits no central concentration other than a faint stellar nucleus. High power shows the nucleus more distinctly, but the halo shrinks to about 1.′5 diameter. Two similar wide unequal pairs of stars lie 5′ NE and NNW.

eg 3437 dimen. 2.′6 × 0.′9 B = 12.8 sfc. br. 13.6
In 25 cm this galaxy is moderately faint, elongated in pa 120°, 1.′25 × 0.′5. The NW tip is much more sharply pointed. The oval core has coarse stellarings along the major axis. The halo is 2.′5 × 0.′75 in 30 cm, with a broad, moderate concentration and no nucleus. The NW half of the halo extends farther than the SE half, reaching toward a mag. 13.5 star 2.′3 NW. The bulging core lies to the NE of the center and several faint stellarings dot its major axis; one stands out above the others. Another mag. 13.5 star is visible 3′ SW.

eg 3443 dimen. 2.′6 × 1.′3 mag.$_z$14.3 sfc. br.
This is a very faint, very low surface brightness galaxy in 30 cm, difficult to see even at 250 ×. It is about 45″ diameter at best, and shows no central brightening. A mag. 12.5 star lies 1.′5 E.

eg 3447 dimen. 3.′8 × 2.′3 mag.$_z$14.3 sfc. br.
eg 3447A dimen. 1.′7 × 1.′0 mag. sfc. br.
While barely visible in 15 cm, this galaxy remains difficult even in 30 cm. Lying 3′ NE of a mag. 9 star, it forms a 1′–1.′25 unconcentrated glow that is slightly enlongated roughtly N-S. eg 3447A is just visible in 30 cm as a faint stellaring 1.′6 E. The magnitude listed is for both objects combined.

eg 3454 dimen. 2.′2 × 0.′5 mag.$_z$14.1 sfc. br.
eg 3455 dimen. 2.′8 × 1.′8 B = 12.8 sfc. br. 14.4
With 15 cm, eg 3455 is a faint, low surface brightness spot lying 2′ S of a mag. 10.5 star. At 75 × it has no discernable central brightening. In 25 cm the halo is roughly circular, about 1′ diameter, and occasionally shows a faint stellar nucleus. 30 cm reveals a 1′ × 0.′8 halo slightly elongated ENE-WSW with a broad central concentration and no conspicuous nucleus. eg 3454 is visible in 30 cm 1.′7 NNW of the mag. 10.5 star. It is a 1.′3 × 0.′2 spindle elongated in pa 115°, showing only a little brightening along its major axis. Overall it appears about one magnitude fainter than eg 3455. Photographs show eg U6035 (mag.$_z$15.2) 16′ SE.

eg 3457 dimen. 1.′3 × 1.′3 mag.$_z$13.0 sfc. br.
This is a moderately faint object in 15 cm, though it has a fairly high surface brightness. At 75 × it is a small, concentrated spot with a bright center. With 30 cm the halo has a circular outline about 40″ diameter and a terraced brightness profile, fading from a conspicuous mag. 13.5 stellar nucleus to a distinctly bounded core and a fainter halo. A faint pair (14,14; 11″; 45°) 4.′5 E appears nebulous at 250 ×.

eg 3485 dimen. 2.′5 × 2.′2 B = 12.6 sfc. br. 14.3
This galaxy is moderately faint in 25 cm, lying 1.′7 E of a mag. 11 star. The halo is about 1.′25 diameter, circular, and little concentrated toward the center.

eg 3489 dimen. 3.′7 × 2.′1 V = 10.3 sfc. br. 12.5
Viewed in 25 cm this member of the Leo I cloud has a distinct core and nearly stellar nucleus. The halo is 1.′5 × 0.′75 in pa 75° with tapered tips. A mag. 12.5 star is visible just S of the W tip.

eg 3495 dimen. 4.′6 × 1.′3 B = 12.4 sfc. br. 14.2
Located about 10′ E of d (58) Leonis (V = 4.8), this is a moderately faint object for 25 cm. The halo appears elon-

gated in pa 20°, 1'.25 × 0'.5; the core is circular but not well defined against the halo. It is 3'.5 × 1' in 30 cm, the S end somewhat wider than the N. It is very weakly concentrated, and a few lumps are visible along the major axis, but there is no distinct core or nucleus.

eg 3501 ***dimen. 3'.7 × 0'.6*** ***mag.$_Z$13.8*** ***sfc. br.***
This very thin spindle is not visible in 15 cm. At 225× in 30 cm the galaxy is 4' × 0'.4 in pa 30°. There is no central bulge, only a patchy irregular brightening along the major axis. The most conspicuous stellaring is about halfway from the center to the SW tip.

eg 3506 ***dimen. 1'.3 × 1'.3*** ***B = 13.4*** ***sfc. br. 13.9***
This small object can be seen without much trouble in 25 cm. The halo is 30" diameter, showing some concentration to the center. With 30 cm it grows to about 1'.5 diameter with slight, broad concentration to the center, where a faint stellar nucleus is occasionally visible.

eg 3507 ***dimen. 3'.5 × 3'.0*** ***mag.$_Z$11.4*** ***sfc. br.***
In 15 cm this galaxy is only just distinguishable as a slight haze around a mag. 11 star. With 30 cm it is a moderately bright galaxy with the bright star superposed 30" NE of a distinct, concentrated 15" core. A grainy-textured halo reaches to 1' × 0'.8, clearly elongated E-W despite the ordinarily confusing influence of a superposed star.

eg 3521 ***dimen. 9'.5 × 5'.0*** ***V = 8.9*** ***sfc. br. 12.9***
6 cm shows this galaxy easily 11' NW of a mag. 8 star. It appears elongated in pa 160°, 4' × 2', and moderately concentrated to a bright nucleus. 25 cm shows it 5' × 2', the diffuse halo brightening to a 30" × 20" core and stellar nucleus. It is a beautiful object for 30 cm, extending to 6' × 2'.5. The ends are ragged, the core and halo mottled. The oval core is roughly centered, but the brighter parts become progressively more eccentric to the W edge, where a dark lane 20" wide passes. The W flank is generally dimmer, though it extends as far from the center as the E side.

eg 3547 ***dimen. 2'.2 × 1'.1*** ***V = 12.8*** ***sfc. br. 13.6***
This object is fairly faint for 25 cm, which shows a circular 1' glow with a slight concentration to the core. In 30 cm the halo extends to 1'.5 × 1' in pa 5°. It shows a broad, weak concentration to a 1' × 0'.5 core, but no further brightening except for an occasionally visible stellaring in the center and another 15" S of center.

eg 3593 ***dimen. 5'.8 × 2'.5*** ***V = 11.0*** ***sfc. br. 13.7***
This galaxy is a member of the eg 3627 (M66) group. It is moderately faint in 25 cm, 3' × 2', elongated E-W. The halo is oval while the 30" core is circular. 30 cm shows it to be 3'.5 × 1', evenly concentrated without distinct zones. A thin dark lane passes just N of the center.

eg 3596 ***dimen. 4'.2 × 4'.1*** ***B = 11.6*** ***sfc. br. 14.6***
In 25 cm this galaxy has a circular 2'.25 diameter halo with low surface brightness but fairly distinct edges. It brightens only a little toward the center. The central 1' area shows several faint stellarings, the most distinct one lying W of center. Three mag. 13 stars lie 3'.5–4'.5 N, E, and SW.

eg 3598 ***dimen. 2'.1 × 1'.4*** ***mag.$_Z$13.5*** ***sfc. br.***
This galaxy forms a close pair with a faint star. In 25 cm it looks like a hazy double star at low power. At 200× the galaxy is distinguishable from the mag. 14 star that lies 40" N. It is very small, about 20" diameter, with a brighter center of a similar magnitude to the star. eg 3592 (mag.$_Z$14.8) lies 11' due W.

eg 3599 ***dimen. 2'.8 × 2'.8*** ***V = 11.9*** ***sfc. br. 14.0***
With 25 cm this galaxy is moderately faint. The halo is about 45" × 20", elongated roughly NE-SW. Overall it shows a moderate, even concentration but a sharp, clearly stellar nucleus of mag. 13.5 stands out in the center.

eg 3605 ***dimen. 1'.7 × 1'.0*** ***B = 13.1*** ***sfc. br. 13.6***
eg 3607 ***dimen. 3'.7 × 3'.2*** ***V = 10.0*** ***sfc. br. 12.6***
eg 3607 is the brightest of a group including eg 3599, eg 3605, and eg 3608, *q.v.* In 15 cm, −07 appears as a circular spot 1' diameter. It is very bright in 25 cm, elongated NE-SW, about 1'.5 × 1', with a stellar nucleus. eg 3605, 2'.7 SW, is visible in 25 cm. Photographs show a mag. 11 star superposed on the center of eg 3607. eg 3608 lies 5'.8 NE, *q.v.*; eg U6296 (mag.$_Z$14.3) lies 15' S.

eg 3608 ***dimen. 3'.0 × 2'.5*** ***V = 11.0*** ***sfc. br. 13.2***
In 15 cm this galaxy appears similar to but fainter than eg 3607, *cf.* 5'.8 SW. Two mag. 12 stars are visible 1'.4 NW and 1'.9 NE. 25 cm shows a bright circular core within a 1' diameter halo.

eg 3611 ***dimen. 2'.4 × 2'.0*** ***V = 12.2*** ***sfc. br. 13.8***
This object is moderately faint for 25 cm. The halo is elongated roughly N-S, 1'.25 × 0'.75, and seems to extend more to the S of center. The core is very conspicuous and sharp, but nonstellar. 30 cm shows it 1'.3 × 0'.8 in pa 15°, poorly concentrated to a faint nonstellar nucleus, which is on the N side of the core. A pair of stars of 55" separation in pa 245° is located 7' S.

eg 3623 ***dimen. 10' × 3'.3*** ***V = 9.3*** ***sfc. br. 12.9***
Messier 65 is bright in 6 cm, a very elongated and concentrated nebula. Low powers show a stellar nucleus. It

appears fainter than Messier 66 (eg 3627), *cf.* 20′ E, in 15 cm, and extends to 5′ × 1′. 25 cm shows it up to 8′ × 2′ with a 3′ × 2′ core. The N side of the halo seems quite knotted; the core is granular. A mag. 12 star is visible 2′ SW; a fainter star lies about the same distance NE of center. In 30 cm the halo is about 7′ × 2′ in pa 170°. It tapers somewhat from the middle outward, but the ends are rounded. Passing E of the core is a dark lane, making the E side of the nebula generally fainter. Overall it shows a strong, even concentration toward mottled inner regions and a distinct nonstellar nucleus. The mag. 12 star to the SW lies 1′.5 W of the S end; near this star a very faint stellaring is visible in the halo. Lying 30″ E of center in the dark lane is a threshold magnitude star.

eg 3626 dimen. 3′.1 × 2′.2 V = 10.9 sfc. br. 12.8
A stellar nucleus is visible in this object in 15 cm. 25 cm shows a sharp central condensation in a circular 1′ halo that fades abruptly at its outer edge.

eg 3627 dimen. 8′.7 × 4′.4 V = 9.0 sfc. br. 12.8
Messier 66 is a large, lenticular object in 6 cm, located at the SE end of a crooked string of three mag. 9 stars. The surface is irregularly concentrated and without a prominent nucleus. 15 cm shows a large bright object with a bright concentrated core. The halo is elongated roughly N-S with 25 cm, extending to 5′ × 2′. The core is a 2′ × 0′.25 streak with an inconspicuous 30″ nucleus. A faint star is visible 3′ SW. The halo is about 5′ × 2′ in pa 0° with 30 cm, and the NE side is particularly mottled. The core, which brightens evenly to a nonstellar nucleus, is chopped-off on its NE and SW sides by dark blotches, giving it a position angle of 160°.

eg 3628 dimen. 15′ × 3′.6 V = 9.4 sfc. br. 13.5
This long galaxy is bright in 25 cm. The halo extends to 10′ × 1′, elongated E-W; it extends more W than E from the center. The core is rectangular, 3′ × 0′.75, with a circular, more concentrated spot inscribed in it. A star is visible just S of the E end. In 30 cm it is at least 10′ × 1′.5 in pa 105°, up to 12′ × 2′ at 225 ×. The elongated core is crowded against the N flank of the halo and contains a faint nucleus.

eg 3629 dimen. 2′.2 × 1′.7 B = 12.8 sfc. br. 14.1
Viewed in 25 cm this galaxy is circular, a bit larger than 1′ diameter, and well concentrated to a faint stellar nucleus.

eg 3630 dimen. 2′.3 × 0′.9 B = 12.9 sfc. br. 13.5
This object is fairly bright but very small in 25 cm. The halo is sharply tapered, 50″ × 15″ in pa 20°, with a conspicuous stellar center. In 30 cm it grows to 1′.5 × 0′.5; the faint halo is elongated in pa 40°, the core in pa 20°. A nearly stellar nucleus is visible in the N side of the 55″ × 25″ core. The galaxy is a member of a group, the brightest of which is eg 3640, *q.v.*

eg 3640 dimen. 4′.1 × 3′.4 V = 10.3 sfc. br. 13.2
eg 3641 dimen. 1′.1 × 1′.1 mag.$_z$14.4 sfc. br.
eg 3640 is easily visible in 25 cm. The tenuous halo is elongated in pa 80°, 1′.5 × 1′, with a bright 20″ core and a faint stellar nucleus. In 30 cm the halo is approximately circular, about 2′.25 diameter. It shows fairly strong concentration to a distinct 15″ core that has a faint stellaring N of its center. At 225 × there seem to be small dark spots around the core, particularly on the N side. A mag. 14 star lies 2′.3 NNW. eg 3641 is faintly visible in 25 cm 2′.5 S and a little E of −40. In 30 cm it is about 45″ diameter with good concentration to a faint stellar nucleus.

eg 3646 dimen. 3′.9 × 2′.6 V = 11.2 sfc. br. 13.6
This is a faint low surface brightness object for 25 cm, best viewed at medium power. The overall size is about 2′. The core is barely distinguishable from the featureless halo. eg 3649 (mag.$_z$14.7) lies 7′.8 E.

eg 3655 dimen. 1′.6 × 1′.1 V = 11.6 sfc. br. 12.1
This galaxy can be seen faintly in 15 cm. The halo is elongated NE-SW in 25 cm, with a faint stellaring visible within the SW edge. Overall the galaxy appears about the same brightness as a star 2′.4 E.

eg 3659 dimen. 2′.1 × 1′.2 B = 12.8 sfc. br. 13.6
On the threshold of visibility in 25 cm, this galaxy appears circular, about 1′ diameter, with irregular edges.

eg 3664 dimen. 2′.0 × 1′.9 B = 12.7 sfc. br. 14.1
This object is barely noticeable in 25 cm. 30 cm reveals a faint unconcentrated patch 2′ diameter. At 225 × a couple of brighter patches can be discerned aligned in pa 30°. The galaxy is situated along the hypotenuse of a 10′ right triangle of mag. 11.5 stars.

eg 3666 dimen. 4′.2 × 1′.4 B = 12.4 sfc. br. 14.1
In 30 cm this galaxy is elongated in pa 100°, 2′.8 × 1′.2, a bar-shaped object with a bright beaded streak running along the bar near the center. The bar seems more sharply defined on the S flank, perhaps giving rise to the streaky appearance. A mag. 14 star is visible 1′.5 N of the E side.

eg 3681 dimen. 2′.5 × 2′.4 V = 11.7 sfc. br. 13.5
This object seems elongated E-W in 15 cm, about 1′ diameter. It is small and moderately faint in 25 cm. The

halo is 1′ diameter, concentrated to a round core. eg 3684, *q.v.*, lies 14′ NE.

eg 3684 dimen. 3′.2 × 2′.3 V=11.7 sfc. br. 13.7
This galaxy is very faint in 15 cm, appearing as a circular spot with a substellar nucleus. With 25 cm it is not as bright nor as concentrated as −81, *cf.* 14′ SW; however, it does seem larger. eg 3686, *q.v.*, lies 14′ NE; eg 3691, *q.v.*, lies 16′ SE.

eg 3686 dimen. 3′.3 × 2′.6 V=11.4 sfc. br. 13.6
15 cm shows this object faintly 2′.5 S of a mag. 10.5 star. It is large and fairly bright in 25 cm. There is little concentration, but a few stellarings are visible on the nebula. The halo extends to 3′ × 2′, elongated NNE-SSW. eg 3684, *q.v.*, lies 14′ SW.

eg 3689 dimen. 1′.6 × 1′.2 V=12.3 sfc. br. 12.8
This galaxy is moderately faint but well defined in 25 cm. At 200 × the halo extends to 1′ × 0′.3, elongated in pa 90°. The center is brighter along the major axis.

eg 3691 dimen. 1′.3 × 1′.0 B=13.5 sfc. br. 13.6
Very faintly visible in 25 cm, this galaxy appears as a 1′ diameter glow that grows gradually brighter toward the center.

eg 3705 dimen. 5′.0 × 2′.3 B=11.8 sfc. br. 14.3
This bright object is visible at low power in 25 cm. The irregularly oval halo is 2′ × 0′.75, elongated in pa 130°, with a conspicuous nucleus. 30 cm reveals a bright stellar nucleus on the W edge of a 5″ inner core. The outer 1′.3 × 1′ core is conspicuously mottled and fades suddenly at its outer edge to the halo. The halo extends to a overall size of 3′ × 1′.5. eg 3692 (mag.$_z$12.9) lies 27′ WNW.

eg 3720 dimen. 1′.1 × 1′.0 V=13.3 sfc. br. 13.3
Lying 5′.5 NNE of a mag. 9.5 star, this galaxy is just visible in 25 cm. It appears only 30″ across with a stellar nucleus in a diffuse halo. With 30 cm the halo is about 1′.2 diameter, showing moderate concentration to a 20″ core and very faint nucleus. eg 3719 (mag.$_z$13.8) lies 2′.3 NW.

eg 3745 dimen. 0′.4 × 0′.2 mag.$_{vv}$15.5 sfc. br.
eg 3746 dimen. 1′.3 × 0′.7 mag.$_z$15.3 sfc. br.
eg 3748 dimen. 0′.8 × 0′.3 mag.$_z$15.5 sfc. br.
eg 3750 dimen. 0′.9 × 0′.7 mag.$_z$15.2 sfc. br.
eg 3751 dimen. 0′.4 × 0′.3 mag.$_{vv}$15.5 sfc. br.
eg 3753 dimen. 2′.0 × 0′.6 } mag.$_z$14.6 sfc. br.
eg 3754 dimen. 0′.4 × 0′.4 } sfc. br.
At least four members of Copeland's Septet are visible in 25 cm. Three members lie about 1′.5 S of a mag. 12 star that lies 9′ NNW of a mag. 8.5 star plotted on the AE. Together they form an elongated haze with two brighter spots aligned NE-SW: the southwestern spot is eg 3750, the northeastern one being eg 3753 and −54, which are not visually separated here. Lying 2′.6 due S from this group is a very faint isolated spot, eg 3751. The other three members of the Septet lie from 1′.9 to 2′.6 NW of the mag. 12 star.

eg 3773 dimen. 1′.6 × 1′.4 B=13.1 sfc. br. 13.8
This faint object is not visible in 25 cm, and it appears quite small in 30 cm. At high power a prominent stellar nucleus is visible in the otherwise unconcentrated 40″ halo.

eg 3801 dimen. 3′.2 × 1′.9 V=12.1 sfc. br. 13.9
eg 3802 dimen. 1′.4 × 0′.4 V=13.6 sfc. br. 12.7
In 25 cm, eg 3801 is much larger and brighter than its companion. Its halo is about 45″ diameter with a low surface brightness and indefinite elongation in pa 120°. eg 3802 is a threshold magnitude spot about 20″ across lying 2′.3 N of −01. A mag. 14.5 star lies just over 1′ E. Photographs show eg 3806 (mag.$_z$14.6) 8′.1 NE of −01, 4′.7 N of a mag. 9 star in the field, and eg 3790 (mag.$_z$14.5) 7′.1 W.

eg 3808 dimen. 1′.8 × 1′.0 mag.$_z$14.1 sfc. br.
eg 3808A dimen. 0′.7 × 0′.4 mag.$_{vv}$15 sfc. br.
eg 3808 is a pretty faint galaxy for 25 cm, found 3′.6 NNE of a mag. 9 star. The low surface brightness halo extends to about 45″ diameter, exhibiting no visible concentration toward the center. eg 3808A lies 1′.9 N of −08, but it is not discernable in 25 cm.

eg 3810 dimen. 4′.3 × 3′.1 V=10.7 sfc. br. 13.4
This galaxy is just visible in 6 cm at low power. It is bright for 25 cm, 1′.75 × 1′.25 elongated in pa 30°. The brighter core contains a stellar nucleus at the threshold of visibility. In 30 cm the halo grows to 2′.5 × 1′.8, with broad moderate concentration to an irregularly shaped core. A faint star or stellaring is visible in the NE side of the halo. A triangle of mag. 11–12 stars lies 9′ S.

eg 3821 dimen. 1′.6 × 1′.4 mag.$_z$13.8 sfc. br.
Located about 3′ E of a mag. 11 star, this galaxy is visible as a very faint hazy spot in 15 cm. At first sight in 30 cm it looks like a mag. 13 double star of 16″ separation in pa 35°: the galaxy is the northeastern component, and its very faint halo just envelopes the star, making it about 30″ across. The galaxy is fairly well concentrated to a small granular core.

eg U6697 dimen. 1′.7 × 0′.4 mag.$_z$14.3 sfc. br.
A distorted object located near the center of Abell 1367, this galaxy is visible in 30 cm as a 1′.2 × 0′.15 spindle

Chart XII. NGC 3745 group—Copeland's Septet.

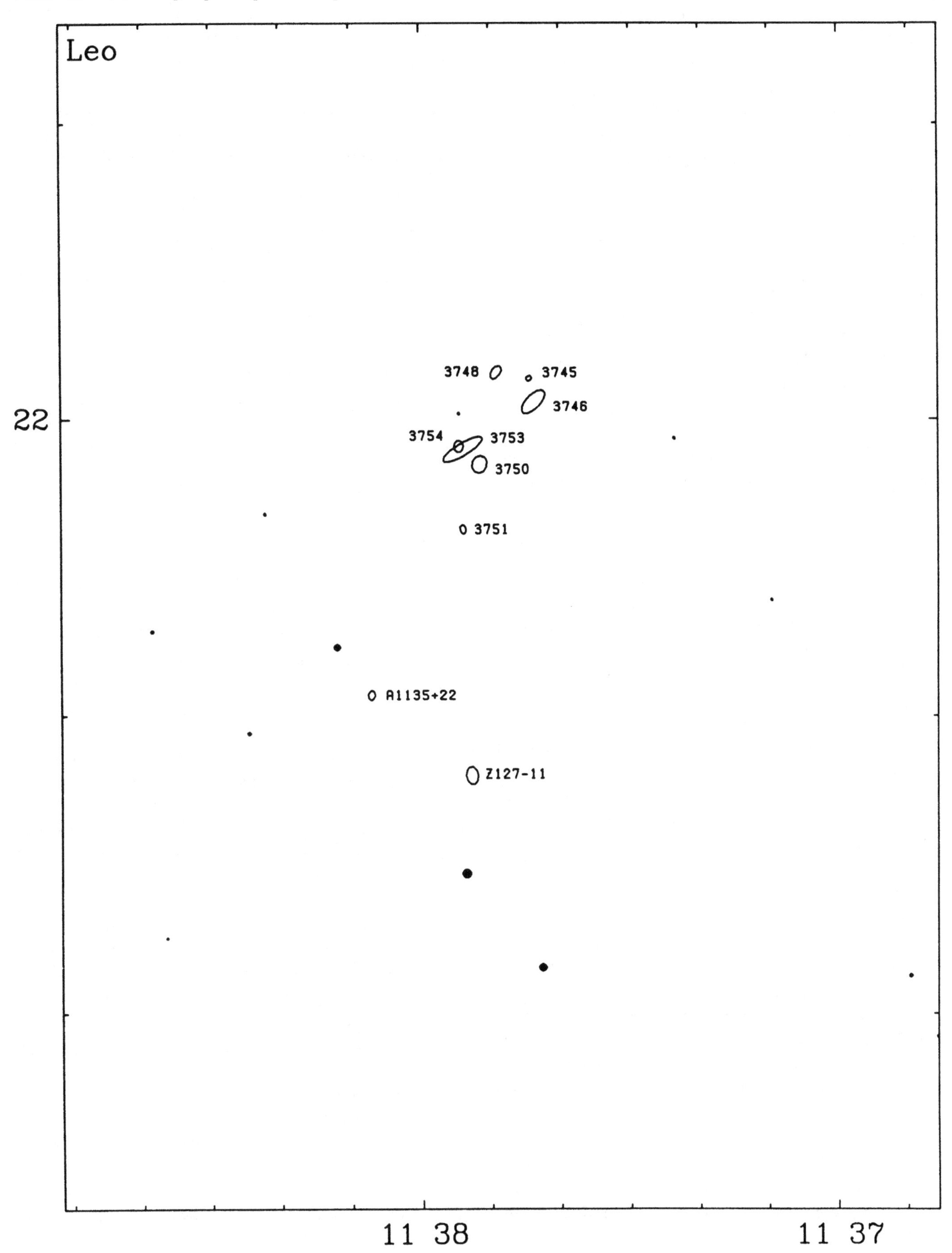

elongated in pa 135°. The brightness along the major axis is approximately uniform except where it fades at the very tips.

eg 3837 ***dimen. 1.′0 × 0.′9*** ***mag.$_Z$14.2*** ***sfc. br.***
Lying near the center of Abell 1367, this galaxy appears only a little fainter in 30 cm than the bright central galaxy, eg 3842, *q.v.* 3.′6 NNE. The 30″ halo is not quite as well concentrated as that of −42.

eg 3840 ***dimen. 1.′2 × 0.′8*** ***mag.$_Z$14.7*** ***sfc. br.***
eg 3844 ***dimen. 1.′7 × 0.′3*** ***mag.$_Z$14.9*** ***sfc. br.***
eg Z97–86 ***dimen.*** ***mag.$_Z$15.7*** ***sfc. br.***
eg 3844 lies in a crowded region of Abell 1367 and is the second of three similar galaxies in a line N of eg 3842, including eg 3841, *q.v.*, and eg 3840. In 30 cm the well-concentrated 30″ glow is slightly elongated NE-SW. eg 3840, at the north end of the line of three, is a bit larger and better concentrated than −44. The very faint galaxy to the W, Z97–86, forms a nearly equilateral triangle with −40 and −44. It is visible with difficulty in 30 cm.

eg 3841 ***dimen. 0.′9 × 0.′4*** ***mag.$_Z$15.0*** ***sfc. br.***
eg 3842 ***dimen. 1.′2 × 1.′0*** ***mag.$_Z$13.3*** ***sfc. br.***
eg Z97–90 ***dimen. 0.′2 × 0.′2*** ***mag.$_Z$15.3*** ***sfc. br.***
eg 3842 is the brightest galaxy in the cluster Abell 1367. With 15 cm it appears as a conspicuous but diffuse spot 2.′7 SW of a mag. 11 star. In 25 cm the moderately faint circular halo is 40″ across, and moderately concentrated, but no distinct core or nucleus stands out. At high power in 30 cm the galaxy lies in the midst of a field crowded with fainter objects. Here the 50″ halo shows a small indistinct core and stellar nucleus. eg Z97–90 lies 1.′1 in pa 280° from −42. 30 cm shows it as a concentrated substellar spot with a mag. 15 star 35″ NW. eg 3841, 1.′3 N of −42, is a conspicuous well-concentrated glow in 30 cm. eg U6697, *q.v.*, is 3.′5 WNW of −42, while eg 3845 (mag.$_Z$15.1) lies at 2.′9 in pa 15°.

eg M+04–28–43 ***dimen. 0.′7 × 0.′4*** ***mag.$_Z$15.1*** ***sfc. br.***
eg M+04–28–44 ***dimen. 0.′4 × 0.′3*** ***mag.*** ***sfc. br.***
This is a difficult pair of galaxies for 30 cm. At low power the two patches are visible in contact about 30″ apart in pa 200°. At 225× the southern one (−43) is substellar, while the companion is visible only intermittently with averted vision. The magnitude listed is for both objects combined.

eg 3851 ***dimen. 0.′3 × 0.′3*** ***mag.$_Z$15.2*** ***sfc. br.***
Typical of the Abell 1367 cluster, in 30 cm this is a small concentrated galaxy about 20″ across lying 2.′2 E of the mag. 11 star that is 2.′7 NE of eg 3842, *cf.*

eg U6719 ***dimen. 1.′3 × 0.′9*** ***mag.$_Z$14.6*** ***sfc. br.***
This galaxy appears as large as eg 3842, *cf.*, in 30 cm. The diffuse halo is not well concentrated and is elongated NE-SW.

eg 3857 ***dimen. 1.′4 × 0.′7*** ***mag.$_Z$15.1*** ***sfc. br.***
Lying in the south side of Abell 1367, in 30 cm this galaxy appears as a 20″ spot concentrated to a stellar nucleus.

eg 3860 ***dimen. 1.′3 × 0.′7*** ***mag.$_Z$14.5*** ***sfc. br.***
One of the larger objects in Abell 1367, this galaxy is easily visible in 15 cm. In 30 cm the broadly concentrated halo is elongated slightly NE-SW.

eg 3861 ***dimen. 2.′4 × 1.′5*** ***mag.$_Z$14.0*** ***sfc. br.***
eg M+03–30–94 ***dimen. 0.′6 × 0.′2*** ***mag.*** ***sfc. br.***
eg 3861 is visible as a diffuse patch in 15 cm. It is nearly 1′ across in 30 cm, exhibiting a relatively large halo with a small, sharp center. A very faint superposed companion, 50″ in pa 115° from the center, seems only to be a faint extension to that side of the larger galaxy's halo. The magnitude listed is for both objects combined.

eg 3862 ***dimen. 1.′6 × 1.′6*** ***V = 12.6*** ***sfc. br. 13.7***
eg I2955 ***dimen. 0.′4 × 0.′3*** ***mag.$_Z$15.2*** ***sfc. br.***
eg 3862 is an easy object for 15 cm at 75× and is located in the same Abell 1367 field with eg 3860, eg 3861, and eg 3873/75, *q.v.* In 25 cm, −62 seems a little easier to view than eg 3842, *cf.*, but it is of lower surface brightness. The core is moderately concentrated and contains a faint stellaring in the center. 30 cm shows both −62 and its companion 55″ away in pa 345°. The brighter galaxy is 45″ diameter with a moderately strong but azonal brightness profile; a stellar nucleus is occasionally visible. eg I2955 is fainter, but not difficult to see, clearly separated from −62. It is only 20″ across and rather more diffuse.

eg 3872 ***dimen. 2.′2 × 1.′5*** ***V = 11.7*** ***sfc. br. 13.1***
This galaxy is small and has a high surface brightness in 25 cm. The halo is less than 1′ diameter and contains a bright nucleus that looks like a mag. 12 star. In 30 cm the halo is elongated in pa 20°, showing sharp concentration to a small circular core and stellar nucleus.

eg 3873 ***dimen. 1.′1 × 1.′0*** ***mag.$_Z$14.2*** ***sfc. br.***
eg 3875 ***dimen. 1.′3 × 0.′3*** ***mag.$_Z$14.8*** ***sfc. br.***
The brighter of these two Abell 1367 members, eg 3873, is easily visible in 15 cm as a concentrated spot. In 25 cm it appears as a mag. 13 patch with a fainter star-like spot 55″ E, which is eg 3875. Both are visible at low power. 30 cm gives a brighter version of this: −73 is clearly nebu-

Chart XIII. Abell 1367.

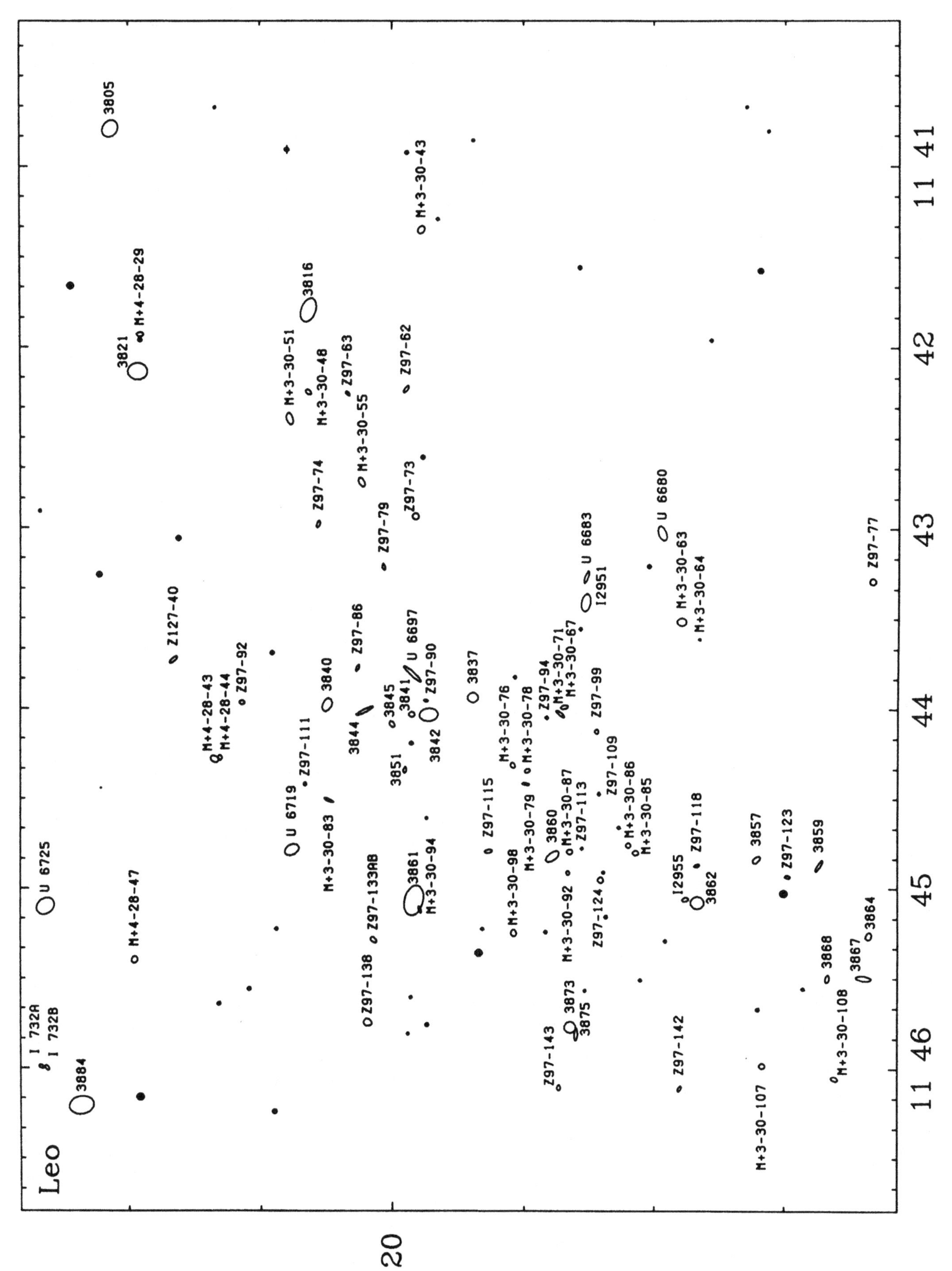

Chart XIV. NGC 4005 group.

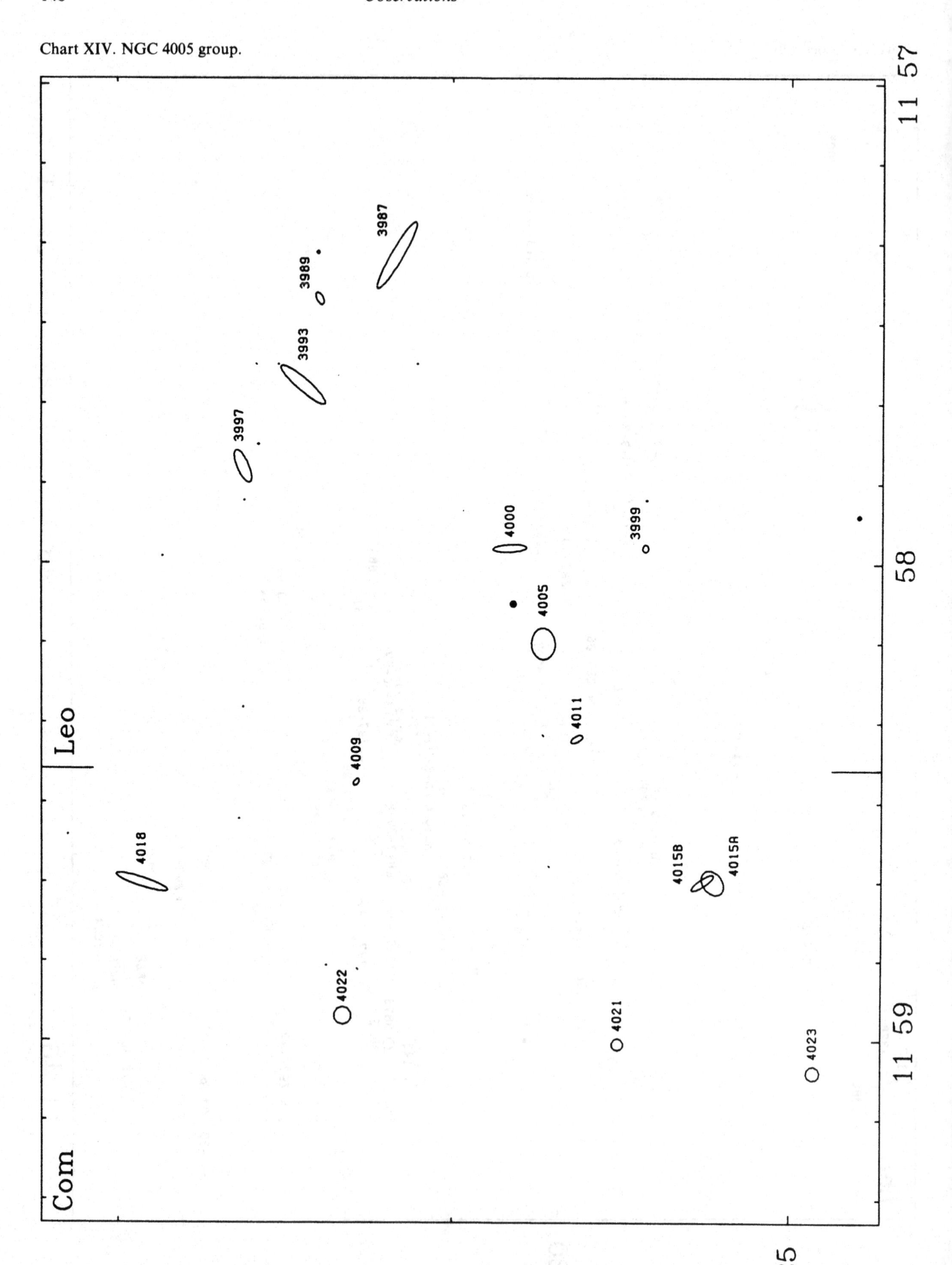

lous and is a bit elongated SE-NW, while −75 appears similar to a field star nearby to the SE, but becomes a small, well-concentrated nebula with attention at 225×.

eg 3900 ***dimen. 3!5×1!9*** ***V=11.4*** ***sfc. br. 13.3***
eg U6791 ***dimen. 1!9×0!4*** ***mag.$_Z$15.2*** ***sfc. br.***
15 cm needs medium power to show the brighter of these two galaxies easily on the W side of a 6′ triangle of mag. 10–12 stars. It is fairly bright in 25 cm, showing a narrowly oval halo to 1!75×0!75, elongated N-S. It grows broadly brighter to a faint stellar nucleus, but the core is not well defined against the halo. The halo is 2!5×1!5 in pa 0° with 30 cm. The nearly circular core and inner halo are mottled along the major axis, and a stellar nucleus is visible in the center. eg U6791, 17′ S, is just discernable in 25 cm as a tiny, unconcentrated patch.

eg 3902 ***dimen. 1!8×1!5*** ***mag.$_Z$14.0*** ***sfc. br.***
Though not visible in 15 cm, 25 cm will show this galaxy as a diffuse circular patch about 1′ diameter. Overall it is more weakly concentrated than nearby eg 3912, *cf.*, showing no distinct core or nucleus.

eg 3911 ***dimen. 1!3×1!0*** ***mag.$_Z$15.4*** ***sfc. br.***
eg 3920 ***dimen. 1!4×1!3*** ***mag.$_Z$14.1*** ***sfc. br.***
In 25 cm, eg 3920 is a faint circular patch about 30″ diameter with strong, even concentration and a threshold stellar nucleus. eg 3911 is barely visible as an unconcentrated spot 9!9 W, only 45″ E of a mag. 12.5 star.

eg 3912 ***dimen. 1!7×1!0*** ***B=13.3*** ***sfc. br. 13.8***
This galaxy is only marginally visible in 15 cm and is fairly faint in 25 cm. Lying in a blank high-power field, the moderately low surface brightness nebula is about 45″ diameter, showing a very slight N-S elongation. It grows broadly brighter across the center, where a stellar nucleus is occasionally visible. In 30 cm the halo extends to 1′×0!5, elongated in pa 5°, with rounded ends. A stellar nucleus stands out in or just S of center at both low and high power. eg 3944 (mag.$_Z$14.3) lies 44′ ESE.

eg U6806 ***dimen. 2!2×0!8*** ***mag.$_Z$14.4*** ***sfc. br.***
Forming a triangle with mag. 12 stars about 2′ S and SE, this galaxy is discernable in 25 cm at medium power. The diffuse circular halo shows a slight concentration, but is otherwise undetailed.

eg 3987 ***dimen. 2!5×0!5*** ***mag.$_Z$14.4*** ***sfc. br.***
This galaxy is the largest and brightest of three in a line including eg 3993 and eg 3997, *q.v.* In 30 cm it is a 1′×0!2 sharply pointed spindle elongated in pa 60° and shows no central bulge. The broadly brighter center contains some spotty stellarings.

eg 3993 ***dimen. 1!9×0!6*** ***mag.$_Z$14.8*** ***sfc. br.***
Set amongst some faint stars, this galaxy is the faintest of three in a line including eg 3987 and eg 3997, *q.v.* The halo is about 50″ across in 30 cm, slightly elongated SE-NW. A mag. 14 star is visible 55″ S. eg 3989 (mag.$_Z$15.7) lies 2!5 WSW.

eg 3997 ***dimen. 1!8×1!0*** ***mag.$_Z$14.3*** ***sfc. br.***
This object is the easternmost of a line of three galaxies including eg 3987 and eg 3993, *q.v.* In 30 cm it is the smallest of the three, but in the middle in brightness. The 40″ halo has a weak, even concentration toward the center. A mag. 12 star lies less than 1′ E; another is a similar distance SW.

eg 4000 ***dimen. 1!4×0!3*** ***mag.$_Z$15.2*** ***sfc. br.***
eg 4005 ***dimen. 1!2×0!7*** ***mag.$_Z$14.1*** ***sfc. br.***
eg 4005 is the brightest of a group of galaxies straddling the Leo/Coma Berenices border. It lies only 1!6 SE of a mag. 7.5 star, but is not difficult to view in 30 cm at high power. The moderately concentrated halo is 20″ diameter. A mag. 13.5 star is visible off the halo 45″ NNW. eg 4000 lies 3!2 WNW, 1!6 past the bright star. 30 cm shows a diffuse object that is slightly larger, but much fainter than −05.

eg 4008 ***dimen. 2!5×1!5*** ***V=12.0*** ***sfc. br. 13.4***
With 25 cm this moderately faint object is elongated nearly N-S, reaching to 1′×0!6. The bright circular core has a sharp nucleus. 30 cm shows it in pa 165°, 1!3×0!75, with a lenticular outline and sharp concentration to a conspicuous stellar nucleus. eg 4004 (mag.$_Z$14.0) lies 20′ SSW; eg 4017 (mag.$_Z$13.5) lies 45′ SSE; eg 4016 (mag.$_Z$14.6) lies 40′ SSE.

Leo Minor

This small group is situated just north of its larger namesake. Though its stars are faint, a score of galaxies mark it for the observer. The center of the constellation culminates at midnight about 23 February.

eg 2859 ***dimen. 4′.8 × 4′.2*** ***V = 10.7*** ***sfc. br. 13.9***
A fairly bright object for 25 cm, this galaxy is elongated nearly N-S with a sharply defined 30″ core and strongly concentrated nonstellar nucleus. The overall size is 2′ × 1′. 30 cm shows a faint halo extending to 2′ × 1′.5 in pa 165°, with strong concentration to a 30″ core and a substellar nucleus.

eg 2942 ***dimen. 2′.2 × 1′.8*** ***B = 12.8*** ***sfc. br. 14.2***
This galaxy is visible faintly in 30 cm, lying in the northern side of a long triangle of mag. 11–12 stars pointing WNW. The galaxy is about 50″ across with indefinite edges. A faint star is visible 1′.3 W of center.

eg 2955 ***dimen. 1′.8 × 1′.0*** ***V = 12.7*** ***sfc. br. 13.1***
This galaxy is just visible with 15 cm. 25 cm shows it 1′ × 0′.5 in pa 160°. A mag. 12 star lies 2′.2 S, and a threshold magnitude star is visible 35″ W. In 30 cm the galaxy is an unconcentrated and featureless spot about 50″ diameter.

eg 3003 ***dimen. 5′.9 × 1′.7*** ***V = 11.7*** ***sfc. br. 14.0***
Visible at 100 × in 25 cm, this faint streak extends to 3′ × 0′.6 in an E-W direction. The center is slightly brighter, and some brightenings are faintly visible along the major axis. It is 4′ × 1′ in pa 80° in 30 cm, with a bulging middle and pointed ends. A thin line of brightenings is clearly visible extending along the entire major axis.

eg 3021 ***dimen. 1′.7 × 1′.0*** ***B = 13.1*** ***sfc. br. 13.5***
In 25 cm this galaxy is situated about 1′ NW of a mag. 10.5 star. The moderately bright object is 1′ × 0′.6, elongated SE-NW, with well-defined edges. A faint stellar nucleus is occasionally visible at the center; otherwise it is without detail. The halo is 1′.2 × 0′.8 in 30 cm, elongated in pa 105°. The core has a bright bar running through it along the major axis. A star is visible on the edge of the galaxy, 20″ NE of center.

eg 3158 ***dimen. 2′.3 × 2′.1*** ***V = 11.8*** ***sfc. br. 13.6***
eg 3163 ***dimen. 1′.4 × 1′.4*** ***V = 13.1*** ***sfc. br. 13.7***
eg 3158 is a very small, moderately faint object for 25 cm, appearing no more than 30″ diameter with a faint stellar nucleus. 30 cm shows the galaxy contained within a 4′ triangle of mag. 13 stars. It is elongated in pa 145°, 1′.2 × 0′.75, the core showing the elongation best. While the nucleus is stellar at 150 ×, it is not so at 225 ×. eg 3163 lies 7′.5 SE. In 30 cm it is a small, well-concentrated spot about 45″ diameter with a bright middle. eg 3158 is the brightest of a group of galaxies.

eg 3245 ***dimen. 3′.2 × 1′.9*** ***V = 10.8*** ***sfc. br. 12.6***
This is a bright galaxy for 25 cm. The lenticular halo is 1′.5 × 0′.75 in pa 175°, with a bright core and stellar nucleus. The galaxy is 2′ × 1′ with 30 cm. The distinct circular core is about 40″ across and holds a substellar nucleus that occasionally seems multiple. With averted vision the tips of the halo are sharply tapered.

eg 3254 ***dimen. 5′.1 × 1′.9*** ***V = 11.5*** ***sfc. br. 13.8***
Lying about 6′ W of a mag. 10 pair of stars, 25 cm will show this galaxy as a diffuse 2′ × 0′.75 glow with a stellar nucleus. The halo extends to 2′.25 × 0′.8 in pa 45° with 30 cm, showing little concentration except for a nonstellar nucleus. In the area immediately surrounding the nucleus are a few slightly brighter spots. A mag. 14 star lies 2′.8 W.

eg 3277 ***dimen. 2′.0 × 1′.9*** ***V = 11.7*** ***sfc. br. 13.0***
This is a small, moderately faint galaxy in 25 cm, but it shows an exceptionally strong concentration toward the center. Overall the halo is about 1′ diameter. In 30 cm the halo extends to 1′.25 diameter around a 45″ core and a stellar nucleus. At high power the nucleus occasionally seems multiple.

eg 3294 ***dimen. 3′.3 × 1′.8*** ***V = 11.7*** ***sfc. br. 13.5***
Visible at 50 × in 25 cm, this galaxy appears 2′.5 × 1′, elongated SE-NW, with little concentration toward the center. A mag. 11 star is visible about 5′ S. In 30 cm the broad mottled core seems inclined to the halo, with a position angle of 110°; the halo is 3′ × 1′ in pa 120°. A faint star is visible 2′.5 W.

eg 3344 ***dimen. 6.′9 × 6.′5*** ***V = 10.0*** ***sfc. br. 14.0***
This bright object is visible in 6 cm. Two stars are involved on the E side: the brighter one (mag. 10.5) is farther from the center and lies near the edge of the 5′ diameter halo. There is a small sudden central brightening, but otherwise the nebula is weakly concentrated. In 25 cm the halo seems much smaller, reaching only to the closer, fainter star of the two E. On the SSE is a mag. 14 star with a brightness comparable to that of the stellar nucleus. 30 cm shows the pair E to be entirely within the halo. Overall the halo extends to about 5′ diameter, showing a distinctly spotty texture. At the center is a small 20″ core with a distinct but nonstellar nucleus. The star to the SE lies about 30″ from the center, forming a triangle with the nucleus and the fainter member of the pair in the E side of the halo. At times the inner parts of the nebula seem elongated E-W, but the involved stars make this questionable.

eg 3395 ***dimen. 1.′9 × 1.′2*** ***V = 12.1*** ***sfc. br. 12.9***
eg 3396 ***dimen. 2.′8 × 1.′2*** ***V = 12.2*** ***sfc. br. 13.4***
This interacting pair of galaxies is moderately bright in 25 cm, appearing nearly in contact, with a separation of

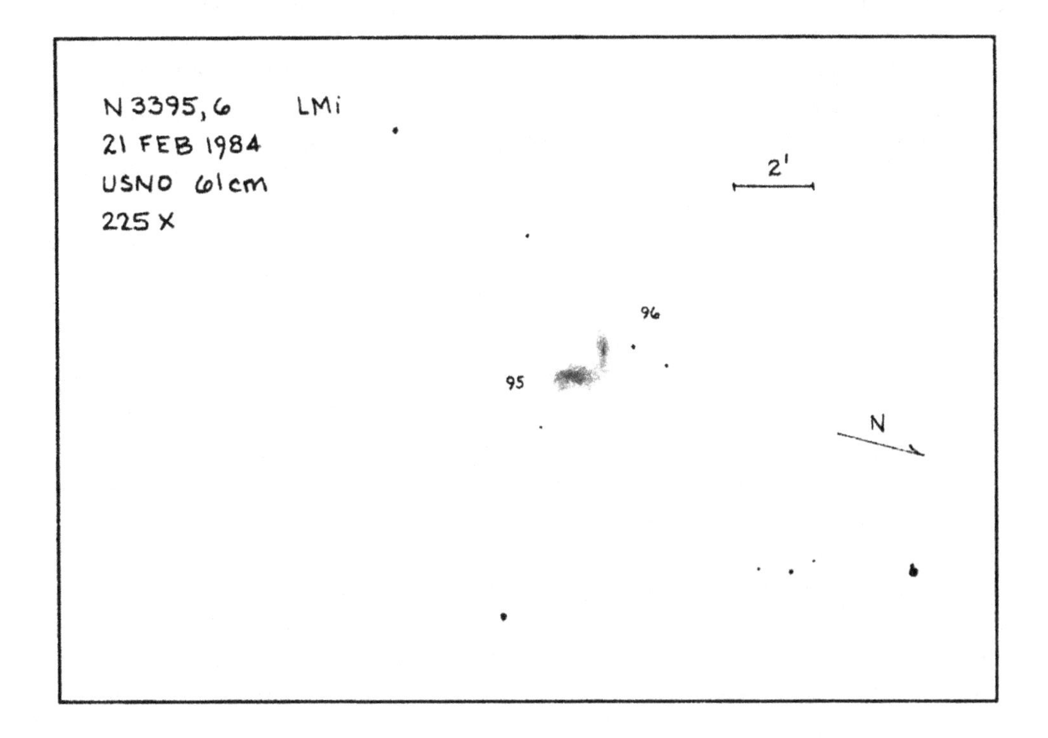

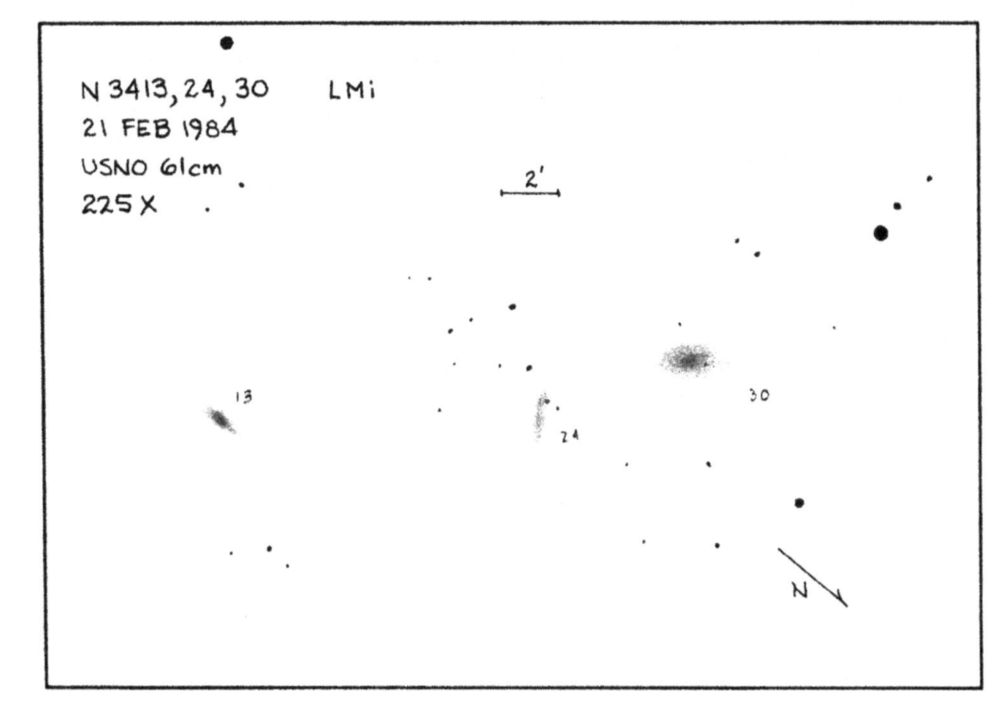

1'.2 in pa 60°. eg 3395 is somewhat larger, elongated in pa 30°, 1' × 0'.75; −96 has a prominent stellar nucleus in a circular halo less than 1' diameter. At first glance in 30 cm the galaxies appear quite similar. However, eg 3395 shows a stronger, even concentration to a nonstellar nucleus and a more extensive halo than −96, with an overall size of 1' × 0'.7. eg 3396 is 50" × 35", elongated approximately E-W, and has little concentration except for the bright stellar nucleus.

eg 3414 ***dimen. 3'.6 × 2'.7*** ***V = 10.8*** ***sfc. br. 13.1***
Plainly visible at 50× in 25 cm, this object has a small very prominent core with a barely distinguishable stellar nucleus. The light fades rapidly outward to a 1'.25 circular halo. In 30 cm the halo extends to 2'.25 × 1'.5 in pa 20°. There is a strong, even concentration to a bright 1' core and a stellar nucleus that occasionally seems multiple. eg 3418 ($mag._Z$14.5) lies 8'.2 N.

eg 3430 ***dimen. 3'.9 × 2'.3*** ***V = 11.5*** ***sfc. br. 13.8***
In 25 cm this galaxy has a fat oval form, extending to 2'.5 × 1'.5 in pa 35°. At 200× the galaxy seems to have a thin, slightly brighter streak along the major axis. A threshold magnitude star is visible on the NE tip, 1'.3 from the center. In 30 cm the nebula has fairly low surface brightness. The halo is about 2' × 1'.2, showing some concentration to the center, but no nucleus. Two mag. 12.5 stars lie 3'.7 ESE. Photographs show eg 3413 ($mag._Z$13.1) 15' SW, and eg 3424 ($mag._Z$13.2) 6'.2 SW.

eg 3432 ***dimen. 6'.2 × 1'.5*** ***V = 11.3*** ***sfc. br. 13.5***
This is a fairly bright spindle-shaped object in 25 cm. The halo extends to 4'.5 × 0'.75 in pa 40°, growing slightly brighter to a thin 2' × 0'.1 core. Three stars are involved with the object: on the SW tip is a pair, including the brightest and faintest of the three; the third star lies close to the SE flank. 30 cm shows a blotchy 3' × 0'.75 bar, the NW flank of which seems more sharply defined. The star E is 30" off the bar; directly N of this star is the brightest patch in the halo, another is visible W of the star.

eg 3486 ***dimen. 6'.9 × 5'.4*** ***V = 10.3*** ***sfc. br. 14.1***
This galaxy is pretty bright with 25 cm. The halo is 3' × 1', elongated in pa 80°, with a patchy 40" × 30" core; the halo is diffuse by comparison. 30 cm shows a broadly oval 3'.5 × 2' halo with a 2' × 1'.5 core. The core is distinctly mottled, and several bright spots surround the stellar nucleus.

eg 3504 ***dimen. 2'.7 × 2'.2*** ***V = 11.1*** ***sfc. br. 12.9***
This galaxy is 1'.5 × 1' in 25 cm, elongated in pa 150°. The halo has a high surface brightness and shows a bright, nearly stellar nucleus in the center. 30 cm shows the halo to 1'.75 × 1'.25 with little concentration except for the bright nearly stellar nucleus. With averted vision a few spots are visible surrounding the nucleus. Two mag. 14 stars lie 2'.6 SSW and 1'.8 NNW. eg 3512, *q.v.*, lies 12' ENE.

eg 3510 ***dimen. 3'.8 × 0'.9*** ***V = 12.9*** ***sfc. br. 14.1***
This object can barely be detected in 25 cm; it is faint and difficult in 30 cm. The halo is elongated in pa 165°, 2' × 0'.4, with a granular texture and a very slight central brightening.

eg 3512 ***dimen. 1'.7 × 1'.5*** ***V = 12.4*** ***sfc. br. 13.3***
In 25 cm this galaxy appears much fainter than eg 3504, *cf.* 12' WSW. The circular 1' halo grows little brighter toward the center. It is elongated in pa 100° with 30 cm, 1'.5 × 1'.2 in extent. The halo is poorly concentrated, but a few stellarings are visible along the S side of the major axis. Several stars are visible within 4' radius. Photographs show eg 3515 ($mag._Z$14.8) 14' NE.

Lepus

Located at Orion's feet, this region holds some fine galaxies and a globular cluster. The center of the constellation culminates at midnight about 14 December.

eg 1744 dimen. 6′.8 × 4′.1 V = 11.2 sfc. br. 14.6
This is a fairly faint, unconcentrated object in 25 cm, with many stars associated. The halo is poorly defined, extending to about 4′ × 1′, elongated nearly N-S. 30 cm shows a generally unconcentrated but mottled patch that is up to 5′ × 2′ in pa 175°, though the steadily visible part is only 1′ × 0′.8. A faint star lies 1′.3 N and another only 20″ SW of center.

eg 1784 dimen. 4′.2 × 2′.8 V = 11.8 sfc. br. 14.3
Situated in the W side of an irregular quadrilateral of mag. 12 stars, 25 cm shows this object as a faint 2′.5 × 1′ haze in pa 90°. The core looks like a spotted bar inside the oval halo. A very faint stellar nucleus is visible with 30 cm inside an unconcentrated 3′ × 2′ halo. A mag. 14 star lies 50″ SE of the center.

eg 1832 dimen. 2′.8 × 1′.9 V = 11.4 sfc. br. 13.0
Located 1′.1 W of a mag. 10.5 star this moderately faint object has a well-defined 1′.5 × 1′.2 halo in 25 cm, with a central bar in pa 165°. A faint stellar nucleus is visible in an inconspicuous 20″ core. 30 cm shows the halo to 2′.8 × 1′.4 in pa 20°. Overall the core looks circular, about 1′.2 diameter, but the brightest part is elongated in pa 165° and contains a faint stellar nucleus.

gc 1904 diam. 8′.7 V = 7.8 class V
Messier 79 is a bright cluster in 6 cm, lying between two mag. 9 stars about 10′ N and S. It appears about 3′ diameter with an almost stellar center. A mag. 12 star is visible on the N edge. The cluster can be partially resolved at 100 × in 15 cm. It is well resolved in 25 cm, about 4′.5 diameter, the brighter core 1′ across. 30 cm shows sparse but relatively bright outliers spreading to 4′ × 3′.5 in pa 15°. The bright, broadly concentrated core is about 1′.25 across.

pn I 418 dimen. 14″ × 11″ nebula 9.3v star V = 10.3
This planetary is clearly visible in 6 cm, appearing as an undistinguished mag. 9 star: longer focal lengths are required to show its nebular character. In 15 cm the central star becomes visible, while 25 cm shows it clearly at 200 ×. The surrounding nebula has a high surface brightness, making a poor contrast for the central star. The overall size is about 10″. With 30 cm the central star is prominent, especially at high power. Here the nebula shows a slight elongation N-S.

eg I2132 dimen. 1′.8 × 0′.8 mag. $_{vv}$14.5 sfc. br.
eg 1954 dimen. 4′.1 × 2′.3 mag. $_{vv}$13 sfc. br.
eg 1957 dimen. mag. sfc. br.
eg A0530–14 dimen. mag. sfc. br.
eg 1954 is the brightest of this small group of galaxies. It is faintly visible in 15 cm, which shows a small patch elongated roughly E-W with two mag. 12.5 stars involved on the NW side. In 30 cm the halo is 55″ × 35″ in pa 100°, not quite reaching the closest mag. 12.5 star 45″ NNW. The light is evenly brighter toward the center, where there is a distinct stellar nucleus. eg I2132, 9′.3 NW, is marginally visible in 15 cm. 30 cm shows a small 25″ × 15″ oval elongated in pa 170°. Like eg 1954, the halo is evenly concentrated, but to an indistinct stellar nucleus. A mag. 13.5 star lies 1′.2 SE. eg 1957, 4′.5 SSE of − 54 and 1′.2 NNE of a mag. 11.5 star, is quite small in 30 cm. The circular 20″ halo shows a strong, even concentration to a faint stellar nucleus. The anonymous galaxy, 4′.8 ESE of − 54, is occasionally visible in 30 cm as an unconcentrated spot just N of a faint, unequal pair of stars.

eg 1964 dimen. 6′.2 × 2′.5 V = 10.7 sfc. br. 13.5
This galaxy lies SE of the southern tip of a thin triangle of mag. 10–11 stars pointing S. 25 cm shows a 1′.5 × 0′.6 lenticular halo in pa 30°, with a prominent nucleus like a mag. 12.7 star. On the W, 40″ from the center is a mag. 13.5 star; 55″ NNE of center, on the NW edge of the halo, is a mag. 14 star. In 30 cm the halo is 3′.2 × 1′ with a bright substellar nucleus in a very bright 10″ × 6″ core. An additional star is visible on the SW side of the halo, 1′.25 from the center.

oc 2017 diam. mag. stars
In 15 cm this cluster is an attractive group of four stars with a few faint ones in the background. The members are lightly colored yellow, blue, and orange. Two close pairs are included in the group: ADS 4254AB

(V=6.4,7.8; 0″.6; 155°) and ADS 4254CD (8.5,9.2; 1″.5; 357°).

eg 2139 ***dimen. 2′.2 × 1′.9*** ***V=11.7*** ***sfc. br. 13.1***
Visible at 50 × in 25 cm, this galaxy appears 1′.6 diameter with a slightly brighter center and sharply defined edges. A mag. 13 star is visible 1′.2 N. 30 cm shows it evenly concentrated to a faint nucleus. The halo is about 1′.8 × 1′.4, elongated roughly N-S. A threshold magnitude star is visible 35″ S of center.

eg 2179 ***dimen. 1′.5 × 1′.0*** ***V=12.5*** ***sfc. br. 12.8***
In 25 cm this galaxy is quite faint and lies between two mag. 14 stars that are aligned in pa 160°. It seems elongated roughly parallel to the stars, about 1′ × 0′.25 in extent. The overall size in 30 cm is difficult to estimate due to the stars, but seems about 1′.2, reaching almost to the nearby stars.

eg 2196 ***dimen. 2′.8 × 2′.2*** ***V=11.2*** ***sfc. br. 13.0***
At 200 × in 25 cm this galaxy is a circular patch about 1′.25 diameter lying in a field crowded with faint stars. The small knotted core is comparatively bright within the smooth faint halo. 30 cm shows the halo to 1′.5 × 1′, elongated NE-SW. Overall the galaxy is moderately well concentrated to an irregular core. A mag. 14 star is visible 1′.2 W of the center.

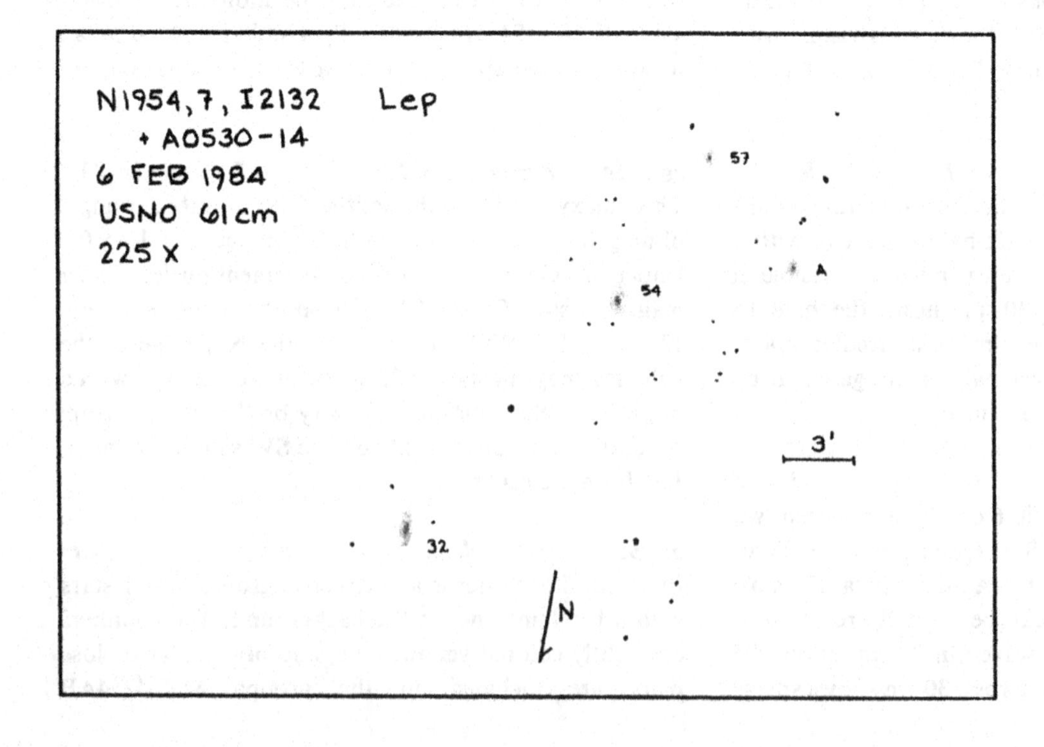

Libra

This constellation contains a large globular cluster and many galaxies. The center culminates at midnight about 9 May.

eg 5595 ***dimen. 2'.0 × 1'.2*** ***B = 12.7*** ***sfc. br. 13.5***
eg 5597 ***dimen. 2'.0 × 1'.8*** ***V = 12.1*** ***sfc. br. 13.3***
In 25 cm eg 5595 seems brighter than its companion 4'.2 SE. The halo is about 1'.25 × 1', elongated NE-SW, with a brighter core and stellar nucleus. eg 5597 is also 1'.25 × 1', but elongated in pa 90°, with a small bright core. Both objects have fairly low surface brightness. With 30 cm −95 extends to 2' × 1'.25 in pa 50°, a lenticular object with weak concentration to circular inner regions. Its companion has much lower surface brightness. Its faintly mottled halo is about 1'.8 diameter, slightly elongated E-W, showing a faint nearly stellar nucleus, but otherwise little concentration.

eg 5605 ***dimen. 1'.8 × 1'.5*** ***B = 13.2*** ***sfc. br. 14.0***
In 25 cm this is a moderately faint galaxy with low surface brightness. The halo is 1'.25 diameter, rising evenly in brightness to a faintly mottled, slightly brighter core. With 30 cm the broad, barely concentrated halo is irregularly shaped, growing to 1'.75 × 1'.5, elongated E-W. No nucleus is visible, but a spur extends through the halo from the center to the SE. A mag. 12.5 star lies 3' W; a mag. 13.5 star is visible 4' SSW.

eg 5716 ***dimen. 1'.9 × 1'.5*** ***mag.$_{vv}$13*** ***sfc. br.***
Lying 23' SSW of eg 5728, *q.v.*, this galaxy is invisible to 15 cm and with 25 cm it is visible only at higher powers. The fairly faint unconcentrated halo is 1'.25 diameter with an irregular outline. Two stars of mag. 11 and 12 are visible on the NE flank, about 40″ from each other.

eg 5728 ***dimen. 2'.8 × 1'.6*** ***V = 11.3*** ***sfc. br. 12.8***
Though barely visible to 15 cm, this galaxy is moderately bright for 25 cm. The low surface brightness, sharply tipped halo is 2'.5 × 0'.75 in pa 30° with a broadly brighter 1' × 0'.5 core. The center has two star-like points on it: photographs show one to be the nucleus, the other a mag. 14 star about 12″ NE of the center along the major axis. Another mag. 14 star is visible 1' SSW of the nucleus. In 30 cm the galaxy also appears about 2'.5 × 0'.75, reaching just to the faint star SSW. At 225 × the beady nucleus appears stellar at times. eg 5716, *q.v.*, lies 23' SW.

eg 5756 ***dimen. 2'.0 × 1'.0*** ***B = 13.2*** ***sfc. br. 13.9***
This galaxy is visible at 100 × in 25 cm, showing an elongated 2' × 0'.75 halo with tapered ends much like eg 5728, *cf.* The core is broadly brighter, exhibiting no sharp nucleus. In 30 cm the oval halo is about 2' × 1', elongated in pa 40°, with a weak broad concentration. There is still no prominent nucleus, but several bright knots are visible in the core along the major axis; these may look stellar at low power. eg M-02–38–15, *q.v.*, lies 35' N; photographs show eg M-02–38–13 (mag.$_{vv}$15), an object of high surface brightness, 18' S.

eg 5757 ***dimen. 2'.1 × 1'.9*** ***B = 12.6*** ***sfc. br. 13.9***
This galaxy is visible in 25 cm 1'.5 S and a little W of a mag. 12.5 star. The halo appears nearly circular, 1'.25 diameter, rising smoothly to a slightly brighter 20″ core and a very faint stellar nucleus. In 30 cm it appears about 1'.4 diameter with moderate concentration to a core that seems elongated E-W.

eg M-02–38–15 ***dimen. 2'.4 × 2'.4*** ***mag.$_{vv}$13*** ***sfc. br.***
Located 35' N of eg 5756, *q.v.*, this faint galaxy is not visible in 15 cm. At 200 × with 25 cm it is pretty faint, about 45″ diameter, with a slight SE-NW elongation. The galaxy has low surface brightness and no central concentration.

eg 5768 ***dimen. 2'.0 × 1'.6*** ***B = 12.9*** ***sfc. br. 14.0***
Though not visible in 15 cm, 25 cm shows this galaxy centered only 30″ N of a mag. 12 star. It appears faint and unconcentrated, about 1' diameter. In 30 cm it is about 1'.5 × 1', elongated E-W. The surface is distinctly granular and weakly concentrated toward the center.

eg I1077 ***dimen. 1'.4 × 1'.2*** ***mag.$_{vv}$13.5*** ***sfc. br.***
This galaxy is a distant companion to eg 5791, *q.v.* 20' ESE. It is not visible with 15 cm, and with 25 cm it is moderately faint and seems large but indefinite. Overall the halo is about 1'.25 × 0'.75, elongated NE-SW, with little central brightening.

eg 5791 ***dimen. 2'.4 × 1'.4*** ***B = 12.6*** ***sfc. br. 14.0***
This galaxy lies NW of a 10' triangle of mag. 8 and 9 stars. 25 cm reveals a very faint elongated halo, 1'.5 × 0'.3

in pa 170°. The nearly circular core is 15″ across and contains a bright stellar nucleus. A mag. 11 star 3′ SE forms an oblique triangle with the galaxy and the nearest star of the bright triangle. The small object is not at all faint for 30 cm. The halo is roughly rectangular, 1′ × 0′.5, showing moderate concentration to a prominent stellar nucleus. eg I1077, *q.v.*, lies 20′ WNW; eg I1081, *q.v.*, lies 2′.6 NE.

eg 5792 dimen. 7′.2 × 2′.1 B = 11.7 sfc. br. 14.5
This galaxy is visible in 15 cm as a very faint, diffuse blob E of a mag. 9.5 star. 25 cm shows the star 1′ from the center, in contact with the WNW edge of the halo. The nebula appears slightly elongated E-W, 1′.25 diameter, with a broad concentration and no sharp nucleus. The halo grows to 2′.5 × 1′ in pa 90° with 30 cm, though the nearby star interferes with viewing. The broad, irregular core has cusps on the E and S sides, and seems displaced from the center of the halo toward the star, but this is probably only only a contrast effect.

eg I1081 dimen. 0′.9 × 0′.4 mag.$_{vv}$14.5 sfc. br.
Lying 2′.6 NE of eg 5791, *q.v.*, this galaxy is not visible with 15 cm, though it is clearly seen with 25 cm at 200 ×. The tiny object shows a 45″ × 10″ spindle-shaped form elongated nearly parallel to eg 5791. The oval core is small and inconspicuous.

eg 5793 dimen. 1′.8 × 0′.6 V = 13.2 sfc. br. 13.2
This galaxy lies 4′.3 S of eg 5796, *q.v.*, and is not visible with 15 cm. 25 cm shows a moderately faint spot about 20″ across with little central concentration. A faint star lies 1′.2 SE, the nearest of four stars from NE through S.

eg 5796 dimen. 1′.9 × 1′.7 B = 12.7 sfc. br. 13.9
This galaxy is visible at 50 × in 25 cm. The circular halo is 45″ diameter, rising evenly to a bright core and stellar nucleus. In 30 cm this small object is well concentrated to a prominent 10″ nucleus, with an overall diameter of 50″. eg 5793, *q.v.*, lies 4′.3 S.

eg 5812 dimen. 2′.4 × 2′.2 V = 11.2 sfc. br. 13.0
In 15 cm this galaxy is a small concentrated spot with a stellar nucleus. In 25 cm it is circular and pretty sharply defined, like a 1′ planetary with a central star. The nucleus is nonstellar in 30 cm, appearing about 5″ across. The halo extends to 50″ × 45″, elongated in pa 70°. A little cluster of stars is visible about 6′ to the E. eg I1084 (mag.$_{VV}$14.5) lies 4′.7 ESE.

eg I1091 dimen. 1′.3 × 0′.9 mag.$_{vv}$14.5 sfc. br.
Forming a triangle with a mag. 10 star 2′.5 WNW and a mag. 12 star 2′ NW, this galaxy is not visible with 15 cm. 25 cm reveals a roughly circular low surface brightness halo 45″ across, growing little brighter toward the center. No core or nucleus is discernable. The surface brightness is much lower than eg 5858, *cf.* 9′.6 SE. A brighter galaxy, eg 5861, *q.v.*, lies 19′ SE.

eg 5858 dimen. 1′.4 × 0′.6 mag.$_{vv}$14 sfc. br.
Though invisible to 15 cm, this galaxy is not hard for 25 cm. The object appears very small, about 30″ diameter, and grows much brighter to a 15″ core and a mag. 13.5 stellar nucleus. 30 cm shows the halo to 45″ diameter, slightly elongated SE-NW, with strong concentration to a stellar center. eg 5861, *q.v.*, lies 9′.2 SE; eg I1091, *q.v.*, lies 9′.6 NW.

eg 5861 dimen. 3′.0 × 1′.8 B = 12.3 sfc. br. 14.0
This galaxy is barely shown by 15 cm as a faint and unconcentrated 1′ × 0′.5 patch elongated in pa 150°. 25 cm shows a little central concentration in the tapered halo, but otherwise the view is similar. The halo extends to 2′ × 1′.25 in 30 cm, with an oval outline and a mottled texture. A few faint beads of light are visible near the center, but none stands out as the nucleus. A mag. 10 star lies 2′.5 SSW. eg 5858, *q.v.*, lies 9′.2 NW; eg I1091, *q.v.*, lies 19′ NW. Between −58 and −61, the two larger apertures show a mag. 11,11.5 pair, 15″ apart, aligned nearly parallel to −61.

eg 5878 dimen. 3′.5 × 1′.7 V = 11.5 sfc. br. 13.2
25 cm shows this moderately bright galaxy with a 1′.75 × 0′.5 halo elongated in pa 0°. The brighter oval core contains a stellar nucleus. A mag. 14.5 star is visible 45″ from the center in pa 170° on the very edge of the S flank. Viewed in 30 cm, the galaxy is 2′ × 0′.75 with slight concentration to a circular 10″ core and stellar nucleus. The halo and the outer parts of the core are unevenly bright.

eg M-02-39-07 dimen. 3′.2 × 2′.7 B = 12.4 sfc. br. 14.6
This object is a very low surface brightness patch in 25 cm, yet visible at 100 ×. The large unconcentrated blob is about 2′.5 diameter with two stars nearby to the SE. In 30 cm the galaxy appears as an exceedingly faint 2′ patch without concentration. The nearby stars are 1′.2 SE and 2′.2 SSE.

eg 5885 dimen. 3′.5 × 3′.2 V = 11.7 sfc. br. 14.2
At 100 × in 15 cm this galaxy appears as a diffuse spot 1′ diameter, lying 1′.5 SW of a mag. 8.5 star. 25 cm reveals a circular patch 1′.5 diameter with a faint stellar nucleus. 30 cm shows a broadly oval form elongated in pa 50°, 1′.8 × 1′.5. The halo is unevenly bright, with a slightly brighter band running N-S on the W side, but it exhibits no overall concentration toward the center.

gc 5897 diam. 13′ V=8.6 class XI
In 25 cm this broad, diffuse globular cluster has low surface brightness like Messier 55 (*cf.* gc 6809 in Sagittarius). It can be partially resolved, showing a slight E-W elongation, with an overall size of 10′ × 8′. 30 cm shows the cluster centered about 5′ SSE of a triplet of mag. 11.5–13.5 stars. It appears 8′ diameter with outliers spreading more to the E and W. The resolved stars are smoothly distributed over the cluster, though the brightest lie in the E side of the core.

eg 5898 dimen. 1′.7 × 1′.7 V=11.5 sfc. br. 12.6
eg 5903 dimen. 2′.0 × 1′.7 V=11.5 sfc. br. 12.8
eg 5898 is moderately bright in 25 cm, showing a 1′ halo with a small, highly concentrated core that is elongated roughly E-W. The galaxy is more broadly concentrated than eg 5903 in 30 cm, with an overall size of 1′.5. eg 5903, lying 5′.6 E and a bit N, appears a little brighter than −98 in 25 cm, mostly due to its stellar nucleus. The halo extends to 1′.25 × 0′.75 in pa 170° around a bar-shaped core. In 30 cm it appears similar to −98 except for the sharp central condensation. A mag. 12 star is visible 1′.5 NW.

eg I4538 dimen. 2′.7 × 2′.0 mag.$_{VV}$13 sfc. br.
Lying about 15′ W of pn Merrill 2–1, *q.v.*, this galaxy cannot be seen with 15 cm. With 25 cm it appears as a faint patch of very low surface brightness. At 200 × the roughly circular outline is 1′.5–2′ across, showing no concentration other than a few faint stellarings. 30 cm shows a faint, slightly mottled glow 1′.5 diameter. A mag. 13 star is visible 2′.3 S of center.

eg 5915 dimen. 1′.6 × 1′.2 B=12.9 sfc. br. 13.5
eg 5916 dimen. 2′.9 × 1′.0 mag.$_{VV}$14 sfc. br.
In a rich field of stars, these objects are the brightest of a triple interacting system; the third member is designated eg 5916A (mag.$_{VV}$15) and lies 4′.6 W of eg 5915. With 25 cm eg 5915 is 45″ × 30″, elongated in pa 120°. The core is nearly circular, 20″ across, and contains a conspicuous mag. 13.5 stellar nucleus. A mag. 12 star lies 2′.1 NE. In 30 cm the halo extends to 1′ × 0′.75, with a slightly brighter knotted core. A mag. 14 star is visible 30″ SSW of center, on the southern edge of the halo. eg 5916, 4′.8 S, is not visible to 15 cm, and with 25 cm it is fairly faint. The poorly concentrated halo is elongated in pa 30°, showing faint mottling over its surface. In 30 cm it is 50″ × 25″ with some concentration to a small faint nucleus.

pn Merrill 2-1 dimen. 6″ × 6″ nebula 11.6v star V=16.7:
This nebula is barely distinguishable from a mag. 12 star in 15 cm, lying 50″ E of a mag. 9 star. With 25 cm the planetary is discernable from the star by its lower surface brightness. It is fairly bright and small, with a star-like center standing out at 200 ×. In 30 cm it has a greyer color than the nearby star, but it otherwise is still nearly stellar in appearance. eg I4538, *q.v.*, lies 15′ W.

Lupus

A bright group of stars delineates this constellation, which includes three globular clusters, planetaries, and some galaxies. The center of the constellation culminates at midnight about 10 May.

eg 5530 dimen. 4.′1 × 2.′2 B = 12.0 sfc. br. 14.2
This object is visible to 15 cm as a diffuse 1′ patch with a slight central concentration and a faint star near the center. The halo is slightly elongated in pa 130°. 25 cm shows a very diffuse glow with a faint stellar nucleus, but otherwise no central brightening. The mag. 12 star lies about 10″ NE of the nucleus.

pn I4406 dimen. 1.′7 × 0.′6 nebula 10.3v star 17:p
In 15 cm this planetary appears about 30″ across, elongated in pa 80°. The eastern and southern edges seem more definite. The core is elongated N-S and shows an occasional stellaring in the center. 25 cm shows the central star and a mag. 13 star 35″ from the center on the W edge. The nebula is unevenly bright, with faint extensions tapering to the E and W.

gc 5824 diam. 6.′2 V = 7.8 class I
Easily visible in 6 cm, this globular looks similar to Messier 80 (*cf.* gc 6093 in Scorpius), with strong concentration to a bright, almost stellar center. In 30 cm the 1′ core is well concentrated, but without a sharp nucleus. The surrounding halo extends to 2′ diameter. No resolution is indicated even at high powers.

pn 5873 dimen. 3″ × 3″ nebula 11.2v star
Visible in 15 cm, this tiny planetary is hardly distinguishable from a star even in 30 cm. It forms the northern and brightest point of a 2′ equilateral triangle of mag. 11, 11.5, and 12 stars. A fainter star SW makes the triangle into an uneven diamond.

gc 5986 diam. 9.′8 V = 7.5 class VII
This cluster can be partially resolved with 15 cm. The freckled blaze has an ill-defined halo 3.′5 diameter with a large, broadly brighter core. A mag. 12 star is visible on the E edge. The cluster is broadly concentrated and still only partially resolved in 25 cm. Two mag. 13 stars stand out in the NE side. In 30 cm it is fairly well concentrated to the center, with outliers spreading to about 4.′5 diameter.

pn 6026 dimen. 54″ × 36″ nebula 12.9:v star
In 15 cm this planetary is just discernable as a faint hazy "star" 7.′3 NW of a mag. 8 star: it is the only object of its magnitude in the area. With 30 cm it appears about 55″ × 35″, elongated E-W, with diffuse edges and no central concentration. The central star, about mag. 13, stands out sharply.

Lynx

This area contains some interesting galaxies and the distant globular cluster gc 2419. The center of the constellation culminates at midnight about 20 January.

gc 2419 ***diam. 4'.1*** ***V = 10.3*** ***class VII***
This distant globular cluster is visible in 25 cm just E of two mag. 7.5 stars. The hazy unresolved cluster is boxed in by four mag. 13.5–14 stars, the brightest and furthest of them being 2'.5 NW. At 100× and 200× the cluster appears about 2' across and shows some blotches in the central region. It is broadly concentrated in 30 cm, about 2' diameter with a granular texture and some large clumps of light within 25" of the center. The outer edge is somewhat irregular.

pn PK 164+31°1 ***dimen. 6'.8 × 6'.0*** ***nebula 12.1:v star V = 17.3:***
Nothing of this faint planetary is visible in 25 cm. In 30 cm an exceedingly low surface brightness patch about 1'.5 diameter is visible. A mag. 12.5 star lies about 5' WSW. Photographs show eg 2474 and eg 2475 (combined mag.$_z$13.9), a close pair of elliptical galaxies only 45" apart, 35' S.

eg 2500 ***dimen. 2'.9 × 2'.7*** ***V = 11.6*** ***sfc. br. 13.7***
This galaxy is visible in 25 cm as an amorphous 1'.5 glow among some faint stars. The halo is of fairly low surface brightness and is slightly elongated in pa 90°. In 30 cm the object is set in the S side of a loose group of about ten stars, between three stars to the N and two S. It appears about 1'.75 across with a slight broad concentration toward the center. A faint star lies just off the edge in pa 115°; another is visible farther off to the NW.

eg 2537 ***dimen. 1'.7 × 1'.5*** ***V = 11.7*** ***sfc. br. 12.5***
Lying 2' NW of a mag. 11 star, this galaxy appears about 1'.5 diameter in 25 cm, with the brighter parts occasionally seeming elongated SE-NW. In 30 cm the smooth, high surface brightness halo is unconcentrated and has a sharply defined boundary. The object is slightly elongated in pa 120°, 1'.1 × 1', looking much like a planetary nebula. At 225× the NW edge contains a faint star or stellaring and the edge of the halo seems a little brighter there. eg I2233 (V = 13.0), a thin spindle, lies 19' SE.

eg 2541 ***dimen. 6'.6 × 3'.5*** ***V = 11.8*** ***sfc. br. 15.0***
This object is not visible in 25 cm and is only faintly visible in 30 cm. The low surface brightness halo extends to 2'.5 × 1'.5 in pa 170°, showing only a slight concentration to an irregularly round core and a faint nucleus.

eg 2549 ***dimen. 4'.2 × 1'.5*** ***V = 11.1*** ***sfc. br. 13.0***
In 25 cm this galaxy has a lenticular outline, 1'.5 × 0'.5, elongated N-S. The core is small and well concentrated. The nucleus is stellar in 30 cm and is set within a bright 20" × 8" core. The halo has thin, tapered ends and extends to 2' × 0'.5 in pa 175°.

eg 2552 ***dimen. 3'.2 × 2'.2*** ***V = 12.2*** ***sfc. br. 14.1***
Though not visible in 25 cm, 30 cm shows this faint galaxy as a 2' patch at 150×. At high power the halo is harder to see, but the object shows some brighter spots on it. A mag. 13 star lies 3' NE; threshold stars are visible 1'.3 NNE and 1'.7 ENE.

eg 2683 ***dimen. 9'.3 × 2'.5*** ***V = 9.7*** ***sfc. br. 12.9***
Situated 10' N of a trapezium of mag. 9–11 stars, this galaxy is easily visible to 6 cm. It is elongated NE-SW, 6' × 2', with a mag. 12 star 2'.7 E. 25 cm shows the brighter part 5' × 1'.5 with a bright spot NE about halfway between the center and a mag. 13 star that lies 2' NE, on the NW edge of the halo. Some dark spots appear on the periphery of the core, particularly to the SW, and the halo is somewhat fainter there. 30 cm shows the halo to 9' × 1'.5 in pa 45°. The light is broadly and moderately concentrated, showing no sharp nucleus. The core has extensions on the N and S sides, making it elongated in approximately pa 15°. The smooth core and inner halo stand out well at 225×, which also reveals a dark lane about 1'.5 long on the NW flank directly N of center.

eg 2712 ***dimen. 2'.9 × 1'.7*** ***V = 12.0*** ***sfc. br. 13.6***
Viewed in 25 cm, this galaxy appears elongated in pa 10°, 1'.5 × 0'.6, with a very faint stellar nucleus. The stellar nucleus is more prominent at high power in 30 cm and seems a little N of center. The halo extends to 2'.2 × 1', showing moderate concentration to an irregularly shaped core.

Chart XV. NGC 2419. A 60-minute exposure on 103a-D plus GG14 filter with the USNO 1.55 meter telescope taken by H. Guetter on 29 January 1971. Magnitude sequence from: Racine and Harris 1975, Ap. J. 196: 413. U.S. Naval Observatory photograph.

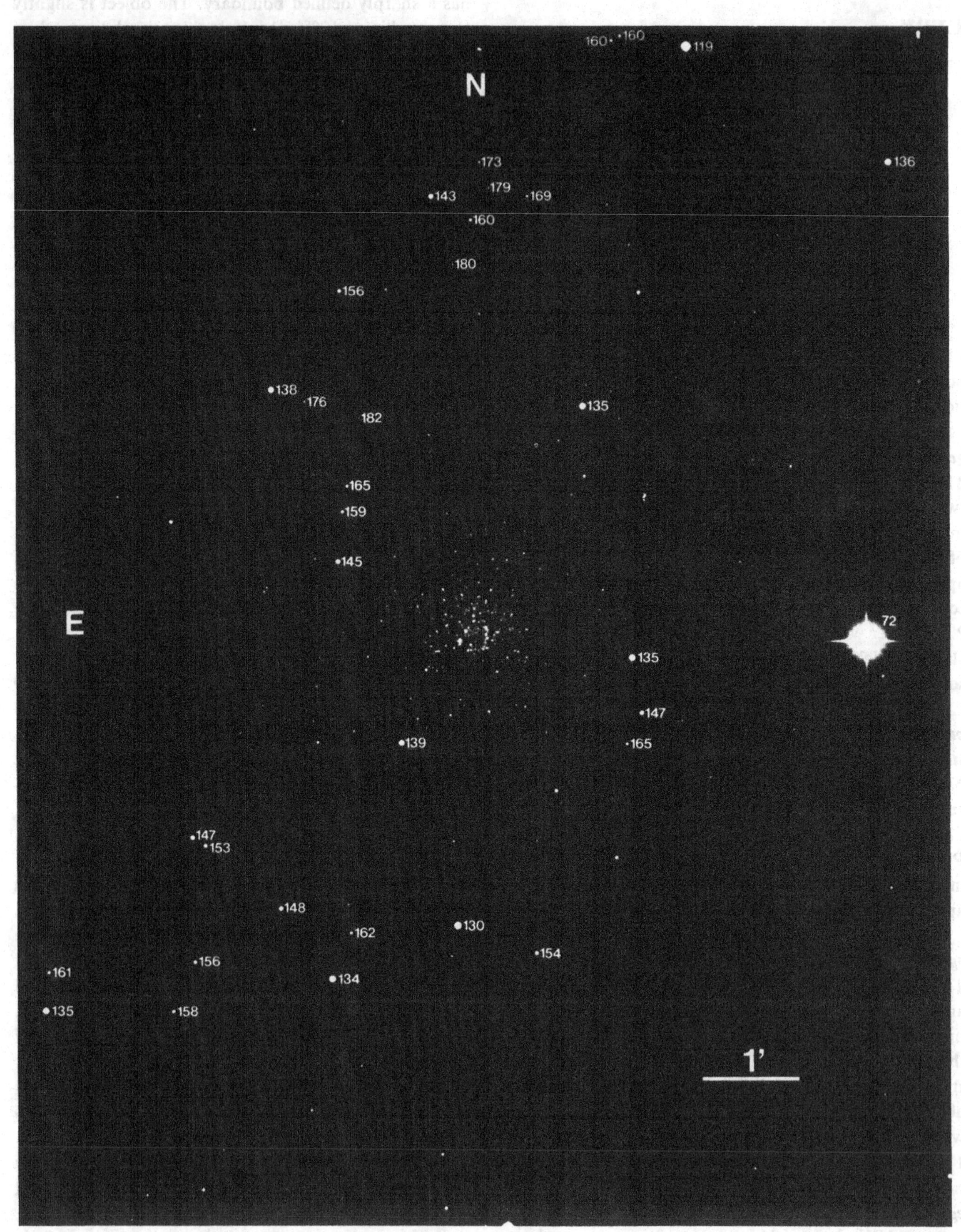

eg 2770 ***dimen. 3.'7 × 1.'3*** ***mag.$_z$12.1*** ***sfc. br.***
15 cm will show this galaxy faintly 2′ S of a mag. 10 star. It grows to 2.'5 × 0.'7 in 30 cm, a broadly concentrated and faintly mottled spindle elongated in pa 45°. At high power a mag. 14 stellaring sparkles faintly in the SE side of the halo. Lying 3′ NW, clearly beyond the maximum extent of the halo, is a mag. 13 star; a brighter star is off the E side.

eg 2776 ***dimen. 2.'9 × 2.'7*** ***V = 11.6*** ***sfc. br. 13.7***
This galaxy is faint and ill-defined in 25 cm, appearing round, about 1.'75 diameter, with an inconspicuous 30″ core. The galaxy looks like a broad, poorly concentrated globular cluster in 30 cm. The halo is about 1.'5 diameter with a 1′ core and a faint nonstellar nucleus. The mottled core is elongated in pa 130°.

eg 2782 ***dimen. 3.'8 × 2.'9*** ***V = 11.5*** ***sfc. br. 13.9***
A fairly bright object in 25 cm, this galaxy appears about 1′ diameter with a stellar nucleus. On the S at about 3′ are two mag. 13 stars. In 30 cm it is brighter and has a more striking nucleus than eg 2844, *cf.* At 150 × the nucleus seems off-center to the NW. At 225 × the halo is elongated in pa 165°, 1.'5 × 1′, lenticular in form with brightenings along the major axis. The core is about 25″ across and contains a substellar nucleus. A mag. 13 pair (11″; 95°) is visible 4.'3 NW.

eg 2793 ***dimen. 1.'3 × 1.'1*** ***B = 13.0*** ***sfc. br. 13.2***
In 25 cm this galaxy is a faint patch within a 15′ × 8′ diamond of mag. 10 stars, the northern apex of which is a pair (h2491: 10.2,10.3; 15″; 200°). It appears round and uniformly bright in 30 cm and is only about 40″ diameter. A faint star is visible 1.'7 SW.

eg 2798 ***dimen. 2.'8 × 1.'1*** ***V = 12.3*** ***sfc. br. 13.4***
eg 2799 ***dimen. 1.'9 × 0.'6*** ***mag.$_z$14.4*** ***sfc. br.***
eg U4904 ***dimen. 1.'0 × 0.'7*** ***mag.$_z$15.0*** ***sfc. br.***
eg 2798 is visible at 50 × in 25 cm, appearing like an unresolved double star. At 200 × it is nebulous, less than 30″ across with an inconspicuous stellar nucleus. On the E side is eg 2799, visible as a star-like appendage to the halo. In 30 cm, −98 is a thin 1.'5 × 0.'8 streak elongated in pa 160°. The small sharp core is quite elongated and contains a stellar nucleus. At 1.'5 in pa 105° is eg 2799, a low surface brightness, little-concentrated glow with a faint stellar nucleus. Overall it is 45″ × 30″, elongated SE-NW, and seems connected to −98 by a faintly luminous bridge. Lying 5.'4 S, just NW of two mag. 12.5 stars, is a very small and concentrated galaxy, eg U4904. It is about 20″ diameter and without a distinct nucleus.

eg 2825 ***dimen. 1.'1 × 0.'5*** ***mag.$_z$15.3*** ***sfc. br.***
eg 2832 ***dimen. 3.'3 × 2.'2*** ***V = 11.5*** ***sfc. br. 13.6***
eg 2832 is the brightest galaxy in Abell 779. In 25 cm it is a moderately faint round spot about 1′ diameter with an indistinct central brightening. eg 2825 is steadily visible at 200 × about 5′ W of −32. It is elongated E-W. In 30 cm, −32 forms a triangle with two pairs of stars 2.'5 SSE and 3.'2 ESE: the former pair is fairly close (h2493: 10.1,11.7; 10″; 157°). The galaxy is weakly concentrated to a faint stellar nucleus. The halo seems slightly elongated in pa 45°, 50″ × 40″. eg 2825 is visible SE of a 2.'7 triangle of unequally bright stars. It is about 20″ long

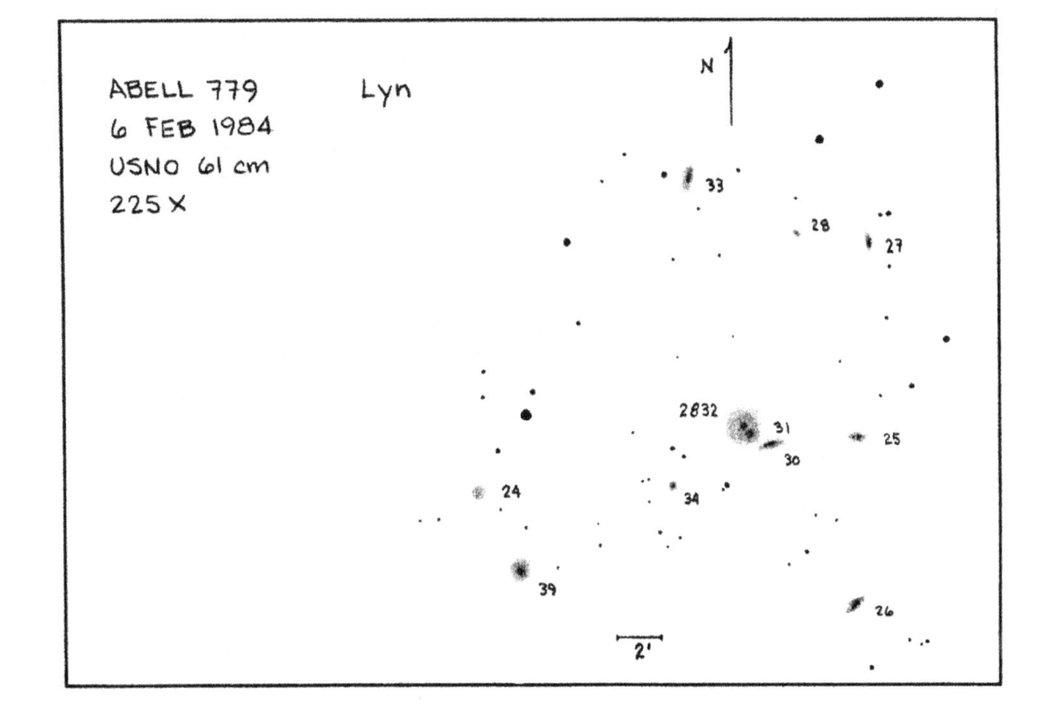

Chart XVI. Abell 779.

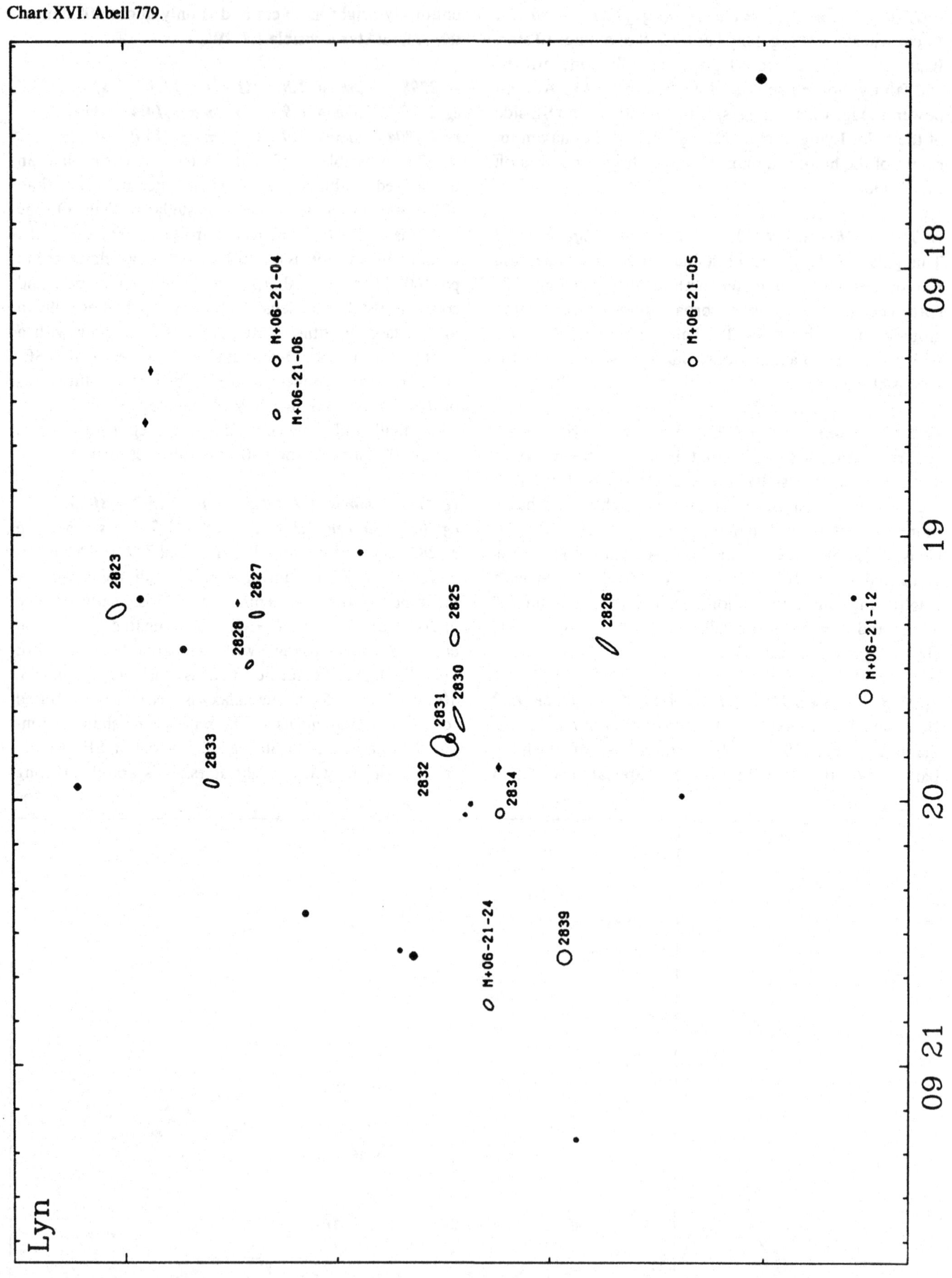

with some concentration to a small core. Photographs show that the apparent elongation of −32 is probably due to eg 2831 (V = 13.6), which is attached to the SW side of −32.

eg 2844 dimen. 1.′9 × 1.′0 V = 12.9 sfc. br. 13.4
This faint object appears round in 25 cm, 45″ diameter, with a very faint stellar nucleus. A mag. 12.5 star lies 2.′2 NW. In 30 cm it is elongated N-S, extending to 1.′2 × 0.′8. The halo has a lumpy texture and exhibits strong concentration to a well-defined 10″ core within which a stellar nucleus is occasionally visible. Photographs show eg 2852 ($mag._z$14.0) and eg 2853 ($mag._z$14.6), a pair at 2.′2, lying 17′ E.

Lyra

The lyre is a small configuration led by the brilliant star Vega. Star fields are rich, particularly on the east side. The Ring Nebula, Messier 57, the most familiar object of its class, is a fascinating deep-sky object. The center of the constellation culminates at midnight about 4 July.

eg 6702 dimen. 2′.1 × 1′.6 V = 12.2 sfc. br. 13.6
This faint galaxy is not visible in 15 cm and is fairly faint even in 25 cm. In this aperture it is a little smaller than eg 6703, *cf.* 12′ SE, but much fainter. The halo is about 35″ diameter, with a smooth texture and even concentration toward the center, though no distinct nucleus is visible. With 30 cm the halo extends to about 50″ × 30″ in pa 65°, showing poor overall concentration except in the inner regions. The core is about 15″ diameter and seems more strongly elongated than the halo. An arc of four or five mag. 13.5–14 stars starts off the NE side, then curves across the N side.

eg 6703 dimen. 2′.6 × 2′.5 V = 11.4 sfc. br. 13.4
In 15 cm this galaxy is visible in a starry field as a fairly faint and evenly concentrated spot nearly 1′ diameter. Two mag. 12.5 stars to the S and one to the NW form a curved string that includes the galaxy. With 25 cm it is a moderately high surface brightness object extending to 45″ × 35″, elongated roughly E-W. The brightness rises evenly toward the center, where a substellar nucleus is visible. In 30 cm it appears circular, showing an even, moderate concentration to a fairly prominent nucleus that usually seems substellar, but occasionally appears both stellar and nonstellar. Overall the halo extends to about 1′.25 diameter.

eg 6710 dimen. 2′.0 × 1′.2 V = 12.8 sfc. br. 13.6
This galaxy is a difficult object for 25 cm. At 125 × it is a faint 1′ × 0′.7 patch elongated NE-SW, lying in a rich star field.

pn 6720 dimen. 86″ × 62″ nebula 8.8v star V = 15.0
The grey torus of Messier 57 is visible with 6 cm even at low power. With 15 cm, the ring appears uniformly bright and slightly oval in shape. A mag. 12 star lies off the E edge, about 1′ from the center. At 250 ×, 25 cm shows that the center is not completely dark, and the acute ends of the ring appear dimmer than the rest. Faint stars are visible 75″ N and NW; another lies about 2′ SE. At 425 × in 30 cm the ring is elongated in pa 65°, and the obtuse sides have a slightly brighter outer edge. A faint

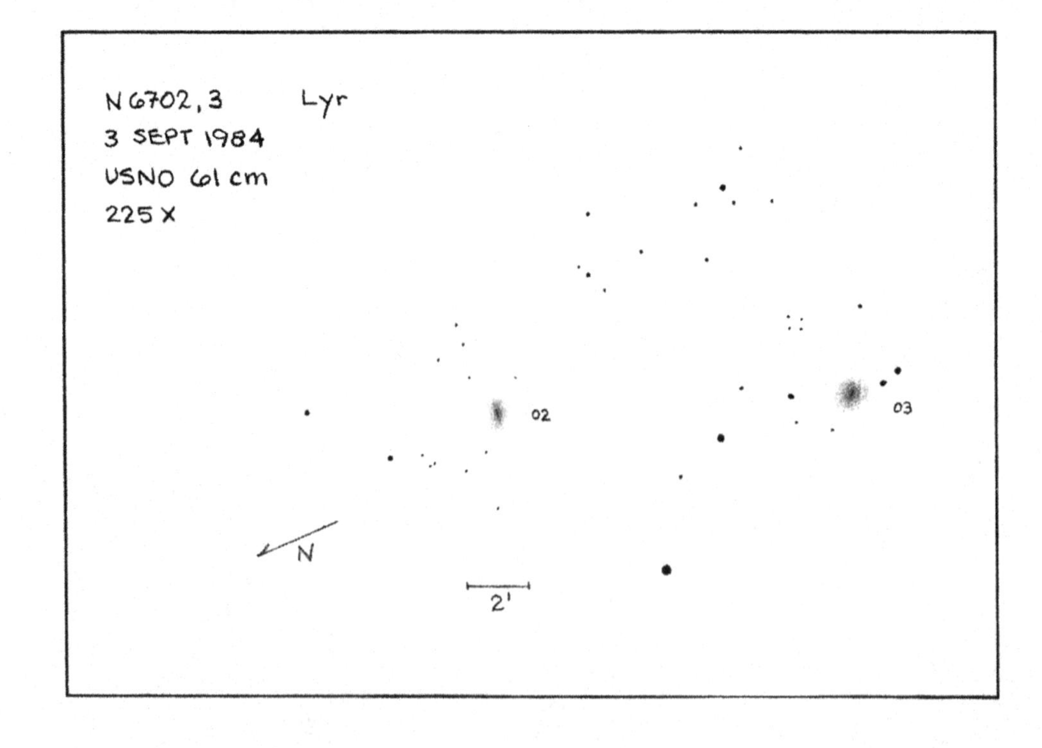

star is embedded in the SW acute end of the ring, but the central star is not discernable. eg I1296 (mag.$_z$15.4), a low surface brightness spiral, lies 4′.8 WNW.

gc 6779 diam. 7′.1 V=8.3 class X

Messier 56 is visible in 6 cm in a rich star field with a mag. 10 star off the W edge. The cluster appears about 2′ diameter, showing a weak, broad concentration to the center. In 15 cm it is granular at 50×, and high powers will partially resolve the cluster in good seeing. A mag. 11.5 star is visible 2′.7 NE, at about the same distance as the brighter star W. Bright and sparkling at 50× in 25 cm, it is well resolved with haze at 200×. Many bright stars overlie the field, but the cluster extends to only 3′ diameter. 30 cm shows it a little more than 5′ diameter, with an irregularly round 1′.75 core.

oc 6791 diam. 16′ V=8.2 stars 380

6 cm will show this cluster as a moderately faint and slightly concentrated glow with an overall brightness similar to that of oc 188 (*cf.* in Cepheus), though it is a little smaller. Three mag. 11 stars stand out in a line on the S side. In 25 cm the cluster stars just begin to be resolved. The hazy, partially resolved area is 10′ diameter, overlain with about two dozen mag. 11–13 field stars.

Microscopium

Though it lies far to the south, this constellation contains several galaxies accessible to northern observers. The center culminates at midnight about 5 August.

eg 6923 *dimen. 2.'5 × 1.'4* *V = 12.1* *sfc. br. 13.2*
Difficult at 150× in 30 cm, this object is located only 1′ SE of a mag. 13 star. At 225× it is elongated in pa 80°, 1.'5 × 0.'6. The halo is weakly concentrated to an elongated core.

eg 6925 *dimen. 4.'1 × 1.'6* *V = 11.3* *sfc. br. 13.2*
This galaxy is located between the last two stars on the SW end of a 45′ string of mag. 11 stars aligned NE-SW. In 30 cm the object is 2.'8 × 0.'8, elongated in pa 10°. The mottled halo is moderately concentrated to an uneven core and a faint substellar nucleus.

eg 6958 *dimen. 2.'4 × 2.'2* *V = 11.5* *sfc. br. 13.2*
This galaxy forms the eastern corner of a 3′ equilateral triangle with two mag. 10 stars. 30 cm shows only a faint 40″ diameter haze that is slightly concentrated to a broad core. No nucleus is visible.

eg I5105 *dimen. 2.'5 × 1.'5* *V = 11.7* *sfc. br. 13.2*
While not visible to 15 cm, 25 cm shows this galaxy as a 1.'5 × 1′ glow elongated in pa 45°. The core is more nearly circular and contains a faint stellar nucleus. Photographs show eg I5105A (B = 13.5) 20′ NE, and eg I5105B ($mag._{VV}$14.5) 26′ SE.

Monoceros

Lying along the winter Milky Way, this constellation is rich in open clusters, and many crowded fields of stars can be found while sweeping this area at low power. The center of the constellation culminates at midnight about 6 January.

gn 2170 ***dimen. 2′ × 2′*** ***star 10.2v***
gn 2183 ***dimen. 2′ × 2′*** ***star 13.5v***
gn 2185 ***dimen. 2′ × 2′*** ***star 12.5v***
Nothing but a few mag. 11–12 stars are visible with 15 cm in the area of gn 2183–85, while 25 cm reveals some moderately faint nebulosity strewn E-W among six or seven stars. In 15 cm and 25 cm gn 2170 is a brighter nebulosity forming a roughly circular glow around a mag. 9.5 star. Lying 13′ NE is another, very slightly brighter star surrounded by a similar nebula that merges with −70, forming an elongated patch in pa 30°.

oc 2215 ***diam. 10′*** ***V = 8.4*** ***stars 40***
This cluster is visible in 6 cm 35′ NE of 7 Monocerotis (V = 5.3). About seven stars are visible with some haziness around them. In 15 cm eight stars attract attention in an 8′ area with ten fainter members visible. It is moderately bright but poorly populated in 25 cm. A brighter central star has two dozen more around it in a 10′ area, including a trapezoidal figure on the W side. Some unresolved haze is visible at 100 ×. 30 cm shows 30 stars in a 10′ area with two brighter members on the E and W. No background haze is noticeable.

oc 2232 ***diam. 45′*** ***V = 3.9*** ***stars 43***
This large group is visible in 15 cm as a slight concentration of mag. 7.5 and fainter stars straggling S from the bright star 10 Monocerotis (V = 5.0). A total of perhaps a dozen stars seem part of the group, not including others scattered to the N and W.

oc 2236 ***diam. 9′*** ***V = 8.5*** ***stars 243***
15 cm shows this cluster as a hazy area surrounding a mag. 11 central star. The cluster members are at the threshold in 25 cm. In 30 cm an impressively faint assemblage is grouped in an acute triangle pointing SE, with the brightest and densest portion at the tip. The mag. 11 star is about 1′ from the tip. The cluster is about 5′ long with 25–30 stars resolved into knotty groups. The fainter stars have fairly uniform magnitudes of about mag. 13.5 and fainter.

gn I2167 ***dimen. 2′ × 1′*** ***star***
oc-gn I2169 ***dimen. 22′ × 17′***
The reflection nebula oc-gn I2169 is visible in a 15 cm RFT as a 10′ diameter haze associated with about five faint stars. On the N side is an unequal pair (8.5,10; 20″; 40°). gn I2167, 23′ N, is faintly visible in 25 cm around a mag. 11 star.

gn 2237 ***dimen. 90′ × 90′***
oc 2244 ***diam. 27′*** ***V = 4.8*** ***stars 46***
The Rosette Nebula, which has several NGC numbers, is a faint circular patch surrounding a sparse naked-eye open cluster, and easily visible in small binoculars from a dark site. 6 cm will show at least 15 stars in the cluster, including two bright stars off the SE end and several pairs. The nebula is faintly visible off-centered from the cluster and extends to about 1° diameter at 25 ×. 15 cm shows about 30 stars. The six brightest stars form two rows in the middle; the two central members are wide pairs (50″; 295° and 64″; 235°). Higher powers show many fainter stars grouped around these bright stars. The brightest star is 12 Monocerotis (V = 5.8), a yellow star on the SE side. Between the middle pairs of bright stars about 15 mag. 11–12 stars are visible in 25 cm, while a total of 40 can be counted in the whole cluster. To the W and N is a sharply bounded kidney-shaped dark area about 20′ across. In 30 cm the central condensation of fainter stars is about 5′ across, and a total of 50 stars is visible loosely scattered in a 25′ area. The nebulosity is not distinct at higher powers in any aperture.

gn 2245 ***dimen. 2′ × 2′*** ***star 10.8p***
gn 2247 ***dimen. 2′ × 2′*** ***star 10.8p***
In 6 cm, gn 2245 is visible as a faint patch of nebulosity associated with a mag. 11 star that lies 2′ SW of a mag. 8 star. 25 cm shows the mag. 11 star on the NE edge of a comet-shaped nebula. In 30 cm it appears similar to Messier 78 in Orion (*cf.* gn 2068). The irregularly round 1.′25 nebula fades smoothly from the star on the NE. About 12′ NNE, gn 2247 is visible in 25 cm and 30 cm as some low surface brightness nebulosity around and S of a mag. 8.5 star.

Chart XVII. NGC 2251. A 30-minute exposure on 103a-O with the USNO 1 meter telescope taken by A. Hoag on 12 October 1958. Magnitude sequence from: Hoag *et al.* 1961, Pubs. of the U.S. Naval Observatory, second series, volume 17, part 7. U. S. Naval Observatory photograph.

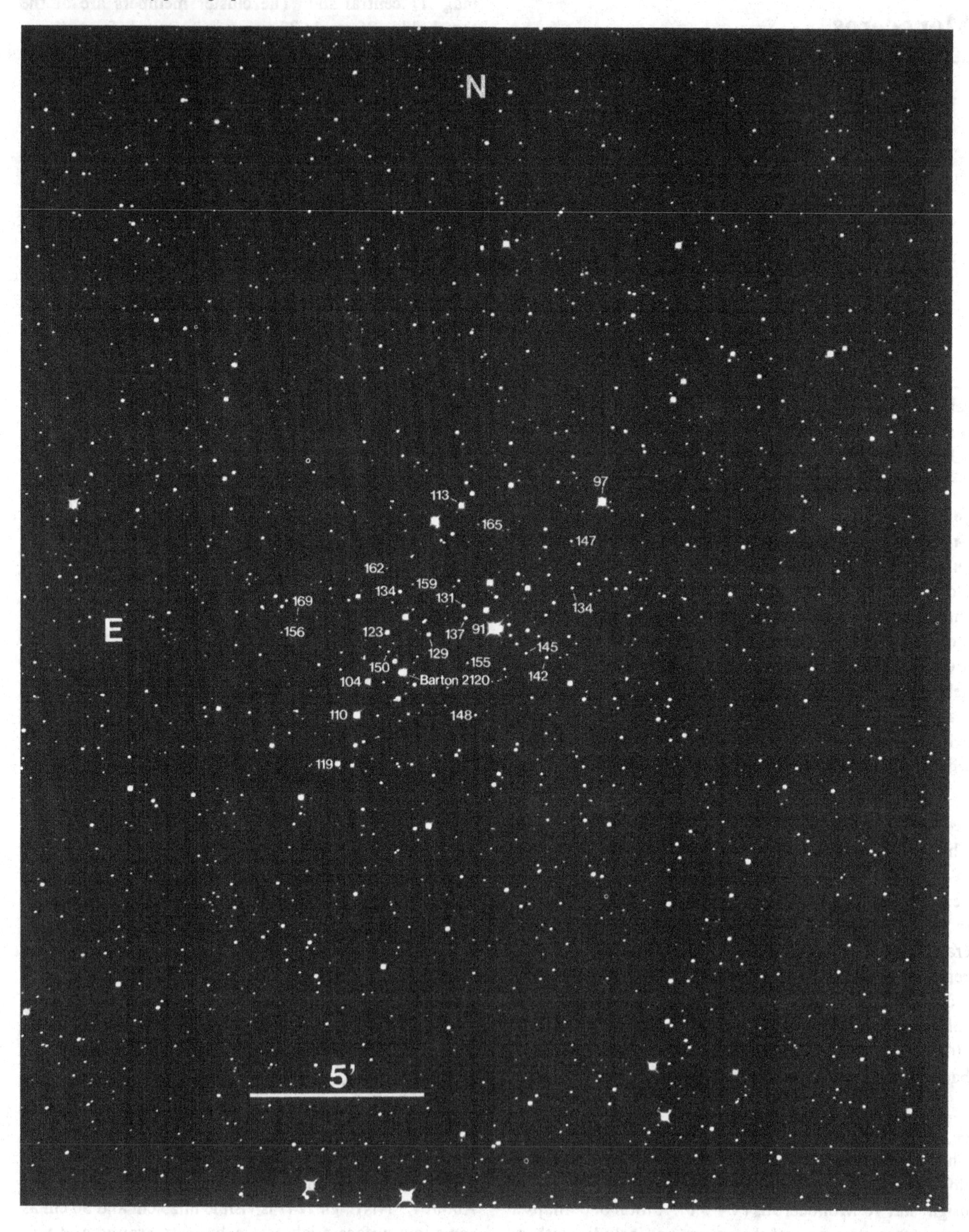

oc 2251 ***diam. 7′*** ***V=7.3*** ***stars 92***

An inconspicuous cluster is visible here in 6 cm, which shows 15 stars in a loose group elongated SE-NW. Few more stars are visible with 15 cm. The string of stars is 6′ long in 25 cm, and a bright star on the W has a close companion (10″; 270°). 30 cm shows 25 stars in a 10′ × 6′ cluster composed of three groups about 2′ diameter each. The southeastern group has a close pair of mag. 10.5 and 12 with 4″.8 separation in pa 81° (Barton 2120).

oc 2252 ***diam. 20′*** ***mag.7.7p*** ***stars 30***

6 cm shows 10–15 stars in this λ-shaped cluster with the brightest star on the NW end. 15 cm shows 18 stars in the λ or wishbone-shaped group. In 25 cm an arc of stars is visible leading NE from oc 2244, *q.v.*, to the N side of this cluster. The cluster members lead SE then S from here, with 22 stars visible, mostly about mag. 11.5. 30 cm shows around 20 stars with a close pair (10,10; 10″; 350°) near the middle. The E leg of the λ consists of only a few stars and is not conspicuous in this aperture.

oc 2254 ***diam. 4′.0*** ***V=9.1*** ***stars 93***

15 cm will show this cluster as a small fuzzy spot at low power. It appears less than 1′.5 diameter in 25 cm, with four or five stars visible at 100 ×. The overall outline is irregular with arms extending E and S. 30 cm shows a very small cluster of ten stars, the brightest star, on the N, having a companion 15″ W. The rest of the stars are grouped in a 40″ long N-S arc that is concave to the E.

oc 2259 ***diam. 4′.5*** ***mag.10.8p*** ***stars 25***

This is a small, moderately faint cluster in 25 cm, lying 14′ E of a mag. 6.5 star, 6′ past a mag. 8.5 star. On the N side is a pair (11.5,12; 7″; 85°), about 2′ from the center. The cluster itself is about 4′ diameter and granular to partially resolved at 200 ×. In 30 cm about fifteen mag. 13 and fainter stars are resolvable over a rich and irregularly concentrated background haze.

gn 2261 ***dimen. variable*** ***star variable***

Hubble's Variable Nebula is easily visible in 6 cm as a very small patch about 2′ SW of a mag. 11 star. At the time of viewing in 25 cm, R Monocerotis, the illuminating star, was barely visible at the southern tip of the nebula. The object is shaped like a comet, the tail fanning to the N. In 30 cm, R Mon appeared about mag. 12 with the high surface brightness fan broadly diverging to the N and curving to the W. The nebula is quite pointed at the S end at the star and is about 1′.5 long, the brighter part being about 1′ in length. As one considers it from S to N, the nebula is initially bright (around R Mon), but gets slightly brighter about 1′ from the tip before fading rapidly to the N end. Visible details in the nebula will change somewhat depending on the time of observation.

oc-gn 2264 ***diam. 20′*** ***V=3.9*** ***stars 222***
dimen. 10′ × 7′

This cluster is led by 15 Monocerotis (V=4.6) and has some irregularly shaped nebulosity associated with it, known as the Cone Nebula. 6 cm shows a long, thin triangular group of stars pointing SSE. About 30 stars are visible in a 25′ area, including nine stars in the immediate vicinity of 15 Mon. With 15 cm about 35 stars are countable, and some faint nebulosity can be seen. 25 cm shows many bright stars strewn uninterestingly in a 60′ × 40′ area. In 30 cm the cluster is hardly distinguishable, but a little nebulosity is discernable around 15 Mon.

oc 2269 ***diam. 3′.0*** ***V=10.0*** ***stars 12***

In 15 cm this cluster appears as a moderately faint, partially resolved patch on the E side of a fine 1° field. At 75 × a few stars are resolved, including a single mag. 11.5 star near the center. The cluster shows 25 stars in a 5′ area to 30 cm. Most of the members are concentrated into a 5′ × 1′ area elongated SE-NW. The central star is only slightly brighter than the rest, which are mag. 12 and fainter.

oc Collinder 115 ***diam. 10′*** ***mag.9.2p*** ***stars 50***

With 15 cm this cluster looks more like a hole through obscuring clouds than a *bona fide* cluster. About 15 stars are resolved over a weakly hazy 20′ × 10′ area elongated NE-SW. The brightest star, mag. 11, stands out on the N side. The cluster fills a 20′ field in 30 cm with about 40 stars, including some wide pairs. With this aperture the cluster does not stand out well from the field.

gn 2282 ***dimen. 3′ × 3′*** ***star 9.7v***

This isolated nebula is visible in 25 cm as a small patch of light spreading mostly NW of a mag. 10 star. With 30 cm it is an unevenly bright glow about 2′ across. oc Collinder 115 lies 29′ N, *q.v.*

oc 2286 ***diam. 15′*** ***V=8.5*** ***stars 80***

6 cm will show 15 stars with three pairs and a little haziness in this cluster. With 25 cm the fainter stars are concentrated on the NW side of a large group of loosely scattered bright stars about 40′ across. About 25 stars mag. 12 and fainter are resolved in a 6′ diameter with some underlying haze. 30 cm shows 35–40 stars weakly concentrated into a 10′ core with outliers to 25′. A few wide pairs are notable among the brighter stars S and SE.

oc 2301 ***diam. 15′*** ***V=6.0*** ***stars 75***

In 6 cm this is a striking N-S string of stars about 20′ long with a clustering in its middle. About ten stars, including four bright ones just NE of center, are resolved among many threshold sparkling points. Some stragglers trail to the E. 15 cm shows about 50 stars to mag. 11, including an arc of four mag. 8.5–9 stars leading S from the S side. 25 cm shows 50 stars in an 8′ area near the center of the string. With 30 cm most of the 70 stars visible are within 5′ diameter, the stragglers extending E to about 10′. Of two bright stars near the center, the southern one is reddish.

oc 2302 ***diam. 5′*** ***V=8.9*** ***stars 16***
This cluster does not stand out well from the field in 15 cm, showing only 8 stars including three brighter ones in a line on the NW side. With 30 cm 24 stars are visible in a 4′ × 3′ area, elongated NE-SW. The three brighter members on the NW side have a fourth faint companion immediately N of the westernmost star. On the S side of center, which is itself empty of bright stars, is an unequal pair (12,13.5; 10″; 315°). The NE side of the oval cluster is sparsely populated, consisting mostly of an isolated group of about a half-dozen stars.

oc Biurakan 10 ***diam. 4′.0*** ***V=10.4:*** ***stars 20***
15 cm will show a single mag. 11 star in this cluster surrounded by a small, slightly hazy spot at 60 ×. With 30 cm the mag. 11 star lies in one arm of a V-shaped asterism of nine mag. 13–14 stars that opens to the E. A few others of the same brightness lie about in a 2′ area. At 250 × the V remains slightly hazy, and the star immediately NW of the brightest star is a pair.

oc 2309 ***diam. 3′.0*** ***mag.10.5p*** ***stars 40***
This is a rather inconspicuous cluster in 15 cm, lying 4′ SSW of a mag. 8.5 star. About a dozen stars are resolvable in an area a few arcminutes across. The cluster is a moderately faint and compact group in 25 cm, showing 25 stars of mag. 11 and fainter in a 4′ area. A faint pair (12.5,12.5; 8″; 97°) lies 1′.5 SW of the bright star to the NNE of center. 30 cm shows 35–40 stars in 2′.5 diameter with a compact 1′ core. The brightest star to the W has faint companions at 10″ in pa 100° and at about 4″ in pa 35°. Two more pairs are visible on the N and S sides of the core of the cluster (12,12; 6″; 0° and 11,12; 10″; 60°).

oc Biurakan 9 ***diam. 3′.0*** ***mag.*** ***stars 30***
Immediately S of a mag. 12 star, 15 cm shows this cluster only as a small, fairly faint, and unresolved spot. With 30 cm, 18 stars fainter than mag. 13 are visible in a 3′ area, showing some concentration to the center.

oc 2311 ***diam. 7′*** ***mag.9.6p*** ***stars 50***
6 cm shows this cluster as a hazy patch 10′ W of a trio of mag. 8 stars aligned almost E-W. A few faint stars are visible over the haze with a brighter star on the N edge. In 15 cm about 15 stars are visible scattered in a 15′ area. 25 cm shows 20 stars of mag. 11 and fainter in a 4′ × 3′ area elongated roughly N-S. About 25 stars are shown by 30 cm in a 7′ × 5′ area. A group of mag. 11 stars trails E, reaching S of the trio of bright stars, making the overall elongation E-W, but these do not seem to be cluster members. The field to the W is dark.

oc 2323 ***diam. 16′*** ***V=5.9*** ***stars 152***
Messier 50 is a fine sight in 6 cm. At least 25 stars can be resolved in a 10′ diameter circular cluster including several brighter stars on the NE and a single one on the S. Many threshold magnitude stars can be seen in the haziness at 50 ×. While it is only partially resolved at 25 × in 15 cm, 100 × reveals at least 75 stars of mag. 8 and fainter within 20′ diameter. 25 cm shows 150 stars in 25′ diameter with a dark patch near the center. A conspicuous pair (V=9.4,10.8; 20″; 60°) lies 1′.1 NW of the bright star on the S edge. A faint group of seven stars is visible on the SW, about 8′ from the center. 30 cm reveals 90 stars in 10′ diameter with outliers extending to 30′ diameter. The middle of the cluster is relatively empty except for one small group. The mag. 8 star on the S is red. Lying about 12′ SE of center is the fine pair ADS 5740 (9.8,9.9; 5″.0; 38°).

oc 2324 ***diam. 8′*** ***V=8.4*** ***stars 133***
While this cluster is not visible in 6 cm, this aperture shows a scattered array of a dozen mag. 8–9 stars in an area 25′ S of a wide pair (OΣΣ82: V=6.6,7.4; 90″; 318°). The cluster is located nearby to the SW. 25 cm shows the cluster faintly with averted vision, a granular glow about 5′ across with many bright and faint field stars. 30 cm resolves at least 50 stars of mag. 12.5 and fainter in a 7′ area. The cluster appears rich but is unconcentrated toward the center. Two brighter mag. 10 stars mark the cluster: one 4′.5 NE, another in the NW edge of the cluster.

oc Haffner 3 ***diam. 5′*** ***mag.*** ***stars 20***
Though barely discernable in 15 cm, 30 cm shows this as a faint cluster in a rich field. A total of perhaps 25 stars is visible in a 4′ area, all fainter than mag. 12.5. A few brighter field stars lie about on the periphery.

oc 2335 ***diam. 12′*** ***V=7.2*** ***stars 57***
This cluster is situated about 8′ WSW of a mag. 7 star. 6 cm reveals a rhombus of two brighter and two fainter stars in the center of a 5′ diameter haze. The asterism is conspicuous in 25 cm as well, along with 25 other stars fainter than mag. 11 in a 10′ area. 30 cm shows 40 stars in the 10′ area with some central concentration. A pretty, almost equal pair lies on the N side (V=11.5,11.7; 25″; 255°).

oc Collinder 465 ***diam.9′*** ***mag.10.1p*** ***stars 30***
oc Collinder 466 ***diam.4.′0*** ***mag.11.1p*** ***stars 25***
oc 2343 ***diam. 13′*** ***V=6.7*** ***stars 55***

Of the brighter cluster, 6 cm reveals about ten stars in a 5′ area with some haziness. Three of the brighter stars form a large triangle. 25 cm shows a 7′ group that stands out well at low power; 22 stars are visible, including an unequal pair on the E (ADS 5817: V=8.5,10.7; 11″; 302°). The cluster appears squarish in 30 cm. The members are unconcentrated toward the center, with half a dozen brighter stars amongst a total of 25. The brighter of the pair on the E appears orange. The Collinder clusters lie about 20′ W of oc 2343, N and S of each other. In 25 cm, Collinder 465 is a loose collection of mag. 12 stars scattered about in a 5′ area showing no central concentration. A small clump on the S is Collinder 466.

pn 2346 ***dimen. 60″ × 50″*** ***nebula 11.8v*** ***star variable***

This planetary is visible without difficulty in 15 cm, though it is small and fairly faint. Two mag. 13 stars are visible 50″ E and 1.′1 W. With 30 cm the high surface brightness nebula has a squared-off outline, extending to 50″ × 40″ in pa 80°. The central star (V651 Monocerotis) can be difficult at minimum light (it ranges between V=11.3 and 13.5 or fainter over a period of about 16 days), but is usually visible at high power.

oc 2353 ***diam. 20′*** ***V=7.1*** ***stars 106***

6 cm gives a good view of this cluster, which is located between two orange mag. 6 stars. About 30 stars are visible in a 25′ area, including a bright star (V=6.0) on the S edge and a pair 2′ NE of it (Σ1052: 9.1,9.3; 20″; 21°). Several pairs and groups are shown by 25 cm, but the cluster is not concentrated overall. About 35 stars are resolved in a 10′ area. With 30 cm 50 stars are visible in 8′ diameter nearby the mag. 6 star on the S. Many more mag. 10.5 stars appear N of this group, extending to 25′ diameter, elongated somewhat E-W.

oc 2368 ***diam. 5′*** ***mag.11.8p*** ***stars 15***

In 25 cm this is a faint cluster of about 20 mag. 12.5 and fainter stars in a 3′–4′ area. A dark band passes through the group from SSE to NNW, dividing the cluster unequally into two triangular sections. The western section is more equilateral and contains the majority of the stars, including the brightest member at its W tip. The eastern portion forms a narrow isosceles triangle pointing to the SSE.

eg 2377 ***dimen. 1.′8 × 1.′6*** ***mag.*** ***sfc. br.***

Viewed in 30 cm, this galaxy is a fairly faint, unconcentrated haze involved with two stars: a mag. 13 star embedded 30″ SSW of center, and a mag. 14.5 star on the NE edge, about 25″ from center. The low surface brightness halo extends to about 1′ diameter, a bit larger than the separation of the two stars. At 250× a very faint stellaring is visible immediately WSW of the fainter star.

oc Melotte 72 ***diam. 9′*** ***mag.10.1p*** ***stars 40***

This cluster is just visible to 6 cm as a faint spot 6′ SE of a red mag. 7.5 star. In 25 cm the cluster members fill a triangular form about 4′ across. 30 cm shows a rich cluster of faint stars in an isosceles triangle pointing N. A total of 35 stars is visible, including two slightly brighter stars marking both corners of the southern base. The stars are most densely packed at the western side of the triangle's base in an area about 2′ across.

oc 2506 ***diam. 10′*** ***V=7.6*** ***stars 807***

Though exhibiting no resolution in 6 cm, this cluster is nonetheless easily visible with this aperture. Two mag. 11 stars stand out on the W side, and a few other faint stars can be seen in a 5′ area. 15 cm will show about a dozen stars over a granular haze. Two wide pairs of mag. 11–11.5 stars lie in the W and E sides, separated by a slightly darker area. In 25 cm the cluster is irregularly round with a blank spot on the S side. About half a dozen mag. 11–12 stars are superposed on a mass of stars of mag. 12.5 and fainter. In 30 cm many faint stars bridge the gap between the northern members of the two wide pairs, forming a squared U-shape opening to the SSW. The densest portion of the cluster is about 8′ across and shows 100 stars, most very faint. Scattered outliers spread to about 14′, adding another 30–40 stars to the total.

Ophiuchus

This large constellation separates Serpens into two parts. Though few galaxies penetrate the Milky Way here, the area is dense with globular clusters, seven of which are Messier objects. The center of the constellation culminates at midnight about 12 June.

gc 6171 ***diam. 10′*** ***V=8.1*** ***class X***
Messier 107 is visible to 6 cm as a diffuse and poorly concentrated patch 4′ diameter with a slightly brighter central pip. In 15 cm at high power the cluster has a granular texture, but shows a few faint stars against a milky background. In 25 cm the outer regions are elongated E-W, about 4′ × 3′, around a circular 2.′5 core. The cluster is not condensed, and at 200 × a modicum of stars is resolved against a granular background. Magnitude 11 and 12 stars frame it on the E, S, and W, the latter being the brightest and farthest. Outliers spread to 5′ diameter in 30 cm, with the core about 2′ across. The brightest cluster stars lie in the NW side of the core, which gives the core an abrupt edge there. About 50 stars can be resolved over a granular haze.

gc 6218 ***diam. 15′*** ***V=6.8*** ***class IX***
Messier 12 is easily visible in binoculars and small telescopes. In 6 cm it lies among several mag. 10.5–11 field stars: one is visible in the halo SSE, another farther N. The overall appearance is similar to Messier 10 (*cf.* gc 6254) about 3° SE, though M12 is less concentrated and has a larger halo. The texture seems uneven, but the foreground stars involved may be causing this rather than incipient resolution. At 200 × 15 cm resolves perhaps 75 stars over a background haze. With 25 cm the cluster is well resolved, showing about 20 brighter stars and hundreds of fainter ones. The outliers spread unevenly in arms and chains, leaving dark spaces in between. In 30 cm the cluster is smaller than M10, and has a smaller central condensation. Outliers extend to about 11′ diameter around a 3′ core and a brighter nuclear region that is 1′ across. A mag. 12 star is conspicuous in the southeastern side of the core, lying about 30″ N of the mag. 10.5 star in the halo SSE.

gc 6235 ***diam. 5.′0*** ***V=10.0*** ***class X***
This cluster is just visible to 6 cm using averted vision. In 15 cm it is situated in a triangle of mag. 12 stars. The cluster is weakly concentrated, 2′ diameter, with an elongated core. There is no resolution, but a few faint stars can be seen on the core; a mag. 13 star lies on the E edge. It is not resolved with 25 cm: at 200 × it is granular at best. The field E and S of the cluster is vacant. In 30 cm at 225 × the core is elongated roughly N-S and has a distinct granular texture.

gc 6254 ***diam. 15′*** ***V=6.6*** ***class VII***
Messier 10 is a bright companion to Messier 12 (*q.v.* gc 6218), which lies about 3° NW. In 6 cm it is a bright and moderately concentrated glow with faint stars at is SE and NW edges about 4.′5 from the center. At 25 × the core has a granular texture, and the halo seems flattened on the NE side, making it appear elongated SE-NW. 15 cm will resolve the cluster fairly well, showing scores of stars over a hazy background. In 25 cm it is well resolved, with an even grid of mag. 12 stars overlying many fainter ones in a residual haze. The well-defined core seems elongated NE-SW, about 5′ × 4.′ Outliers spread to 10′ diameter. 30 cm reveals outliers to almost 15′ diameter and a bright central ball about 4.′5 across. The brighter stars in the outliers form two arms loosely spiraling N and S, unwinding counterclockwise on the sky. Several other bright stars stand out against the inner regions.

gc 6266 ***diam. 14′*** ***V=6.7*** ***class IV***
Messier 62 has a similar brightness to Messier 19 (*cf.* gc 6273) in 6 cm, though it is not elongated. In 15 cm the core is very bright in comparison to the halo, which extends to about 7′ diameter. The core is not centered in the halo, lying instead somewhat SE of center. At 150 × the cluster is granular throughout. At 200 × in 25 cm stars just begin to shine out. The well-concentrated core fades suddenly into the 4′ halo, which is flattened on the SE side. In 30 cm the outliers are well resolved, while the core is partially resolved at 225 ×. The SE side shows very few outliers, giving the cluster an overall NE-SW elongation. The stars here are fainter than those in M19; the brightest one lies on the NW side.

pn I4634 ***dimen. 10″ × 8″*** ***nebula 10.9v*** ***star 17.4p***
In 25 cm this planetary is clearly visible, but hardly distinguishable from a mag. 11.5 star. It forms the western

apex of a 2′.3 × 0′.8 triangle with two mag. 12 stars. At 250 × the white nebula is elongated roughly N-S. With 30 cm the nebula is also white and small, but occasionally shows a sparkle in the center.

gc 6273 diam. 14′ V=6.7 class VIII
Messier 19 is found easily in 6 cm 7°.5 directly E of Antares. It is clearly elongated N-S and has a large, irregularly bright core. 15 cm gives partial resolution at 150 ×. Overall the cluster spreads to 6′ × 4′.5 in pa 20°. The core is about 4′ across and is concentrated off-center to the N. The cluster has a nearly rectangular appearance in 25 cm. At 200 × the surface is dense with threshold magnitude stars on a milky background haze. In 30 cm the cluster is well resolved and extends to 7′.5 × 5′ around a 4′ × 3′ core. A conspicuous mag. 12 star lies just beyond the core to the NE; another slightly fainter star is visible on the NW edge of the core. The outer core seems rippled on the SE by a concentric condensation of stars and a dark patch. The density of stars drops off rapidly outside the core.

gc 6284 diam. 5′.6 V=8.9 class IX
This is a fairly faint cluster for 6 cm, lying 10′ ENE of a wide mag. 8.5 pair of stars. 15 cm shows a bright star-like center in an extensive surrounding haze. The circular cluster is 1′.5 diameter with 25 cm. At 200 × the stellar nucleus is visible within a granular core. The edges of the haze are ragged where a few clumps of stars are resolved; a mag. 12 star lies on the E side 2′ from the center. In 30 cm the cluster seems elongated in pa 120°, 1′.8 × 1′.5. The halo is evenly concentrated to the bright nucleus, which only occasionally appears stellar.

gc 6287 diam. 5′.1 V=9.3 class VII
This cluster is just visible in 6 cm, appearing about the same size but fainter than gc 6284, *cf.* In 15 cm the cluster lies at the S edge of a dark, starless area. Overall it is 2′ diameter and has a broad central concentration. At 200 × in 25 cm it is granular in the slightly brighter central regions, but no stars are resolved. 30 cm also shows no resolution, but at 225 × a faint sparkle is occasionally visible in the center.

gc 6293 diam. 7′.9 V=8.2 class IV
Viewed in 6 cm this cluster is brighter than gc 6316, *cf.*, and shows a strong, even concentration to the center. It appears about 2′ diameter in 15 cm, with a brighter 30″ core. At 175 × the cluster is slightly granular, and occasionally shows a faint stellaring in the bright center. 25 cm shows granulation across the cluster at 200 ×. The halo spreads to 2′.5 diameter, reaching to a faint pair on the NE side. In 30 cm the outliers show partial resolution, spreading particularly on the W and NW sides. The 40″ core has a triangular outline, with one apex pointing S. The overall diameter is 2′.5.

gc 6304 diam. 6′.8 V=8.4 class VI
This cluster is visible in 6 cm as a small, poorly concentrated spot. Faint stars lie 6′–8′ NE, S, and SW. With 15 cm the cluster appears brighter than gc 6316, *cf.* At 125 × it is 2′ diameter and shows some concentration to a broad core; no stars are resolved. 25 cm reveals some granulation at the edges. The oblong core is elongated in pa 100° and its W side is brighter and flattened. The overall diameter is 1′.6. With 30 cm the entire cluster is partially resolved. Faint stars spread to 1′.8 diameter, showing only a weak irregular concentration to a broad core.

pn 6309 dimen. 52″ × 52″ nebula 11.5v star V=13.7
25 cm will show this planetary closely involved with a mag. 11.5 star. The greenish 20″ × 10″ nebula is elongated roughly N-S, and the central star is just visible at 250 ×. The bright star to the N lies only 25″ from the center of the nebula. In 30 cm the planetary is broadly oval and shows much faint haze feathering off the edges along the obtuse sides. The brighter part of the nebula is cigar-shaped, elongated in pa 165°. This brighter portion seems divided into two distinct lobes: the northern one is both larger and brighter. At 425 × the central star is clearly visible.

gc 6316 diam. 4′.9 V=8.8 class III
This globular is faint but not difficult in 6 cm. It grows broadly and only moderately brighter across the center and has a mag. 11.5 star on its SE edge. In 15 cm the cluster is about 1′.8 diameter and the star SE lies 1′.2 from the center. The cluster remains unresolved in 25 cm, but in 30 cm a few faint sparkles are visible on the broad core. Two mag. 13 stars are visible 1′ and 1′.8 SW of the center.

gc 6325 diam. 4′.3 V=10.6 class IV
15 cm shows this globular as faint and diffuse spot about 1′.8 diameter with a slight central concentration. At 200 × in 25 cm it is about 1′.5 diameter, while it is only 1′.3 diameter in 30 cm. Here the low surface brightness patch is weakly and broadly concentrated to a faint stellaring that is occasionally visible in the center.

gc 6333 diam. 9′.3 V=7.6 class VIII
Messier 9 is visible in 6 cm as a concentrated ball with a slightly elongated halo. In 15 cm it is 2′.8 diameter and granular to partially resolved, while it is nicely resolved in 25 cm. The halo is elongated in pa 110°, extending to 4′.5 × 3′.5 around a barred core elongated N-S. The cluster

boundary is more sharply defined on the S. In 30 cm the outliers spread to about 4′ diameter. Here the 2′ core has a triangular outline with one apex pointing N. A brighter star stands out on the SE edge of the core; another lies 2′ ESE of the center.

gc 6342 diam. 3′.0 V=9.8 class IV
In 6 cm this cluster is small and faint but not difficult to see at 25×. 15 cm and 25 cm reveal a circular and weakly concentrated patch about 1′ diameter and a mag. 12 star 1′.2 S. In 30 cm the cluster extends to 1′.5×1′.3, elongated NE-SW. The light is fairly well concentrated toward the center but there is no sharp nucleus.

gc 6355 diam. 5′.0 V=9.7 class
A fairly faint object for 15 cm, this cluster appears about 1′.5 diameter, moderately concentrated and without resolution. 25 cm shows it up to 2′ diameter with a distinctly elongated core. The field surrounding the cluster has very few stars, especially to the S. With 30 cm the core is 1′.25×1′, elongated N-S, while the faint circular halo spreads to 1′.5 diameter. A few faint stars sparkle across a granular background glow.

gc 6356 diam. 7′.2 V=8.2 class II
6 cm will reveal this cluster as a small but fairly bright and well-concentrated glow. In 15 cm it is about 2′ diameter, and the central regions seem elongated roughly N-S. In 25 cm the core is about 45″ across while the overall size is about the same as in 15 cm. 30 cm will resolve a few outliers and show granulation across the center. The outliers extend to 2′.8×2′.3, elongated in pa 70°, but the core is elongated in pa 165°.

gc 6366 diam. 8′.3 V=8.9 class XI
This cluster is a moderately faint and weakly concentrated spot in 6 cm. Two mag. 9 stars lie on the W side; the nearest is at the edge of the halo. In 15 cm a fainter pair is visible on the SSW edge (11,11; 50″; 245°). Here the faint haze is about 2′.5 diameter and shows a little concentration toward the center, but no stars. With 25 cm at 200× a few faint stars are visible over a granular haze about 4′ diameter. In 30 cm a thin overlay of very faint stars spreads evenly through a 10′ area, reaching from a mag. 13.5 star 5′ E almost to the more distant mag. 9 star W. In the central area is a very faint, slightly concentrated haze about 4′ across.

oc Trumpler 26 diam. 7′ V=9.5 stars 40
In 6 cm this cluster is a faint blob with a half-dozen resolved stars lying 25′ NE of 45 Ophiuchi (V=4.3). About 25–30 stars can be resolved in a 5′ area with 15 cm, including many at the threshold of vision. A conspicuous pair lies near the center (11,11; 5″; 45°). 25 cm shows about 40 stars rather irregularly clumped together in a 15′×8′ area elongated NE-SW. The pair near the center has a fainter member at 16″ in pa 30°. 30 cm reveals at least 20 stars scattered in a 15′ area, including a tight group of four on the W side.

pn 6369 dimen. 58″×34″ nebula 11.4v star 16:p
Clearly annular in 15 cm, this nebula is 20″ diameter and has a bright spot on the N edge. In 25 cm the central hole is about 10″ across. The northern perimeter of the torus is lined with a brighter arc of nebulosity that contains a bright, star-like spot at 200×. The overall diameter is 25″ in 30 cm. The bright spot in the northern edge is conspicuous at 250×.

eg 6384 dimen. 6′.0×4′.3 V=10.6 sfc. br. 14.0
This galaxy is barely visible with 15 cm and is a faint diffuse patch in a starry field to 25 cm. The uniformly bright core is 1′.5×1′, elongated NE-SW, and is surrounded by a faint, ill-defined halo. 30 cm reveals a moderately but sharply concentrated halo extending to 3′.5×3′ in pa 30°. The nucleus is about 20″ across and has several faint stellarings about it in the inner core. The core and halo are both unevenly bright, with small light and dark spots spread over their surfaces. Two mag. 13 stars are visible 1′.7 NE and 1′.3 SE.

gc 6401 diam. 5′.6 V=9.5 class
15 cm reveals this cluster as a small, weakly concentrated patch about 1′.5 diameter. A conspicuous mag. 12.5 star lies embedded near the SE edge of the core. With 25 cm the cluster is 1′×0′.75, elongated in pa 100°. 30 cm gives no resolution, though the cluster lies in a starry field. The moderately concentrated circular core is 50″ across and the halo extends to 2′ diameter.

gc 6402 diam. 12′ V=7.6 class VIII
Messier 14 is a very broadly and poorly concentrated cluster in 6 cm, seemingly all "core" with a thin surrounding rind as the halo. A very faint sparkle is just distinguishable in the center, but the cluster shows no resolution. It is a large cluster in 15 cm, and high power will begin to show some granularity. In 25 cm the cluster is slightly elongated in pa 45°, extending to 4′×3′.5. At 200× a few faint stars are resolved. At 225× in 30 cm it remains poorly resolved, especially for so bright an object. Faint stars are spread evenly over a 9′ area, showing no concentration over the cottony 4′.5 core.

gc 6426 diam. 3′.2 V=11.1 class IX
This faint globular is not visible in 15 cm, and it is a very faint 2′.5 patch in 25 cm, which shows a small central

brightening. At 200× the surface is unevenly bright. With 30 cm the cluster is a faint and weakly concentrated spot of low surface brightness still without granularity.

oc I4665 diam. 70′ V=4.2 stars 57
This nearby cluster (430pc) is best viewed in wide-field instruments. 6 cm will show 18 stars in a 40′ roughly circular area. Six of the brightest stars form a Y-shaped asterism on the NW side. In a 15 cm RFT about 30 fairly bright stars are visible, uniformly scattered in a low-power field.

gc 6517 diam. 4′.3 V=10.3 class IV
This cluster is a small and fairly faint object in 6 cm, forming a triangle with a mag. 10 star 5′.6 S and a mag. 11.5 star 3′.7 W. In 15 cm it is a difficult, tiny object with diffuse edges, showing no resolution or central brightening. A faint but sharp central condensation is visible in 25 cm. At 225× in 30 cm the halo is about 1′.25 diameter, slightly elongated NE-SW. The small, slightly brighter central region seems more sharply concentrated at low power.

pn 6572 dimen. 16″×13″ nebula 8.1v star V=13.6:
This planetary is distinguishable in 6 cm as a mag. 8.5 star, but the nebular nature is revealed only by the use of a direct-vision spectroscope: a mag. 9.5 star lying 3′.7 E affects a good contrast with the nebula here. It is a very bright object in 15 cm, an oval spot without noticeable color. In poor seeing, the nebula can be easily mistaken for a star. The nebula appears greenish in 25 cm. It is elongated approximately N-S, with diffuse edges and no central star or internal brightening. 30 cm shows it to be elongated in pa 15°, about 15″ diameter with little central brightening. The nebula is too bright for the central star to be visible directly, but the spectroscope will reveal its presence: the spectrum shows four or five beads from the nebula strung on the faint continuum of the central star.

oc 6633 diam. 20′ V=4.6 stars 159
This naked-eye cluster is a fine sight in 6 cm, which shows a 20′×10′ group elongated NE-SW; the brightest stars are on the SW end. On the W side is an isolated clump of half a dozen stars that seems nebulous at low power. About 50 stars are visible with outliers spreading to 1° diameter. The cluster is 50′×20′ in 25 cm, the center lying 20′ NNW of a mag. 5.5 star. In 30 cm the cluster is too widespread to be appreciated. At least 80 stars can be counted in the area, the members showing no concentration to the center.

Orion

One of the brightest groups in the sky, this constellation presents an array of interesting deep-sky objects. Many bright and dark nebulae in association with the Orion Nebula are visible in addition to several open clusters. The center of the constellation culminates at midnight about 15 December.

oc 1662 ***diam. 12′*** ***V=6.4*** ***stars 59***

This is a conspicuous but loose cluster of 25 stars in a 15′ area to 15 cm. The brightest star is in a small 1′.3 × 0′.5 group (=h684) with four others, two of which form an unequal pair (h684CD: V=9.6,11.3; 11″; 307°). 30 cm reveals a faint sixth member in the central group lying 16″ in pa 210° from the brightest star.

eg 1729 ***dimen. 1′.6 × 1′.5*** ***mag.***$_{vv}$***13*** ***sfc. br.***

15 cm will show this galaxy as a small, threshold glow bordered on the E side by a mag. 10 star and on the N by a mag. 11 star. With 30 cm both stars lie about 1′ from center, clearly beyond the border of the halo, which is a little less than 1′ across. The halo grows broadly brighter to the center and a faint substellar nucleus. Very faint stellarings are occasionally visible in the halo E and NNE of center.

eg 1740 ***dimen.*** ***mag.***$_{vv}$***15*** ***sfc. br.***

Viewed in 30 cm this galaxy appears smaller and a bit fainter than eg 1729, *cf.* It lies on the N side of a 15′ field of mag. 10–13 stars, 25″ NE of a mag. 11 star. The halo is about 40″ diameter, slightly elongated NE-SW, and strongly concentrated to a tiny, bright core. If the major axis were extended to the SW, it would pass N of the nearby star.

pn J 320 ***dimen. 22″ × 22″*** ***nebula 12.0v*** ***star V=14.4***

In 25 cm this planetary is easily visible, though nearly indistinguishable from a star. The nebula forms the eastern end of a curved 3′ string of three stars. At 250× the star-like spot has slightly diffuse edges and is elongated roughly E-W. At 425× in 30 cm the nebula is a small, irregularly bounded greyish spot that brightens unevenly to the center, where the central star sparkles. Overall, though the nebula has an irregular outline, it is elongated in pa 120°.

gn 1788 ***dimen. 8′ × 5′*** ***star V=10.4***

This bright nebula is a pretty sight in 6 cm, which shows a mag. 10 star connected to a bright spot about 2′ SE by a band of nebulosity. Faint feathery extensions spread E and NE. In 15 cm three stars are visible involved in the nebulosity: the bright spot at the SE contains a mag. 11.5 star and another faint star lies E of the bright star. The nebula spreads to 4′ × 2′ in 25 cm. The southwestern edge is relatively sharply defined and lies much closer to the two bright stars. A faint mag. 12.5–13 star is visible about 50″ WNW of the mag. 11.5 star. In 30 cm the southeastern portion of the nebula is very bright and well concentrated, such that the embedded star is hardly distinguishable. The brightest part is shaped like a hot dog stretching between the two bright stars. The brightest star has a faint companion about 15″ N.

eg 1875 ***dimen.*** ***mag.***$_{z}$***15.2*** ***sfc. br.***

In 30 cm this is a fairly faint galaxy, visible 1′ E of a mag. 12.5 star. The halo is circular, about 30″ diameter, and shows moderate concentration to a small core.

oc-gn 1976 ***diam. 45′*** ***V=3.7*** ***stars 50***
dimen. 90′ × 60′

Messier 42, the Orion Nebula, is one of the most beautiful objects in the sky and is visible to the naked eye as a pink fuzzy patch in Orion's Sword. The four brightest stars (θ^1 Orionis: V=6.7,8.0,5.1,6.7 from E to W) form a tight group called the Trapezium, which is easily resolvable in 6 cm; the separation between the closest stars is about 9″. The nebula is sharply bounded on the NE side, but fades smoothly to irregular edges from the SE around to the W. Near the center and the trapezium the nebula is mottled, but the irregularity of the edges is more conspicuous. A mag. 10 star is visible 1′.75 NE of the trapezium on the W edge of the "Fish's Mouth," a dark intrusion from the NE side of the nebula ending close to the trapezium stars. Five stars can be seen in the trapezium with 15 cm, the faint member (mag. 11.5) lying at 4″ in pa 351° from the westernmost star. The Trapezium stars are embedded in the N side of a lump of bright nebulosity, and a long bright arm curves E and S from this area. At 200× much detail can be seen in the bright greenish area around the Trapezium stars and an E-W line of three stars to the E (θ^2 Ori: V=5.1,6.4 and a companion on the E, V361 Ori, V~8.2). In 25 cm the brightest star in the trapezium has a greenish hue, while the star to the NW is a deep blue. The bright streamer that curves E and S is sharply defined on its southern edge and shows a purplish tinge on steady nights. The brightest part of the nebula is a well-defined rectangular area around the trapezium and also shows a purplish cast on good nights. This area is riddled with streaks and dark patches in 30 cm. The streamer extending E and S is about 2′ wide and at least 20′ long, passing over the westernmost of two mag. 7 stars to the SE. The Fish's Mouth merges into a distinct dark bar about 2′ wide that extends out of the nebula for at least 10′ in pa 110° and has a little bright "island" in it. A conspicuous dark patch stands out about 2′ S of the trapezium: it is elongated SE-NW and is wider on the NW end. Four stars are visible against the nebula W of the Fish's Mouth; two fainter ones lie in a dark band that flares as it leads out of the nebula to the NW. 25 cm and 30 cm reveal a sixth trapezium member (mag. 11.0) about 4″ in pa 122° from the brightest star.

oc-gn 1977 ***diam. 20′*** ***V=4.2*** ***stars 35***
dimen. 40′ × 25′

Lying between Messier 42 on the S and oc 1981, *q.v.*, immediately N, this nebulous cluster is a large unevenly bright patch in 6 cm, associated with a loose group of about a dozen stars. The brightest nebulosity is gathered around the three brightest stars on the S. In 15 cm this brighter portion feathers S and particularly W to a few fainter stars in these directions. Overall this section forms a distinct if uneven half-circle concave to the S. North of this area are two pairs of stars [V=7.3 (KX Orionis), 11.1; 33″; 12° and V=9.8,10.1 (LZ Ori); 14″; 219°] associated with fainter nebulosity. Lying just E of these two pairs, and N of the westernmost bright star of the southern group is a very faint, unconcentrated patch of nebulosity not evidently associated with any stars. Farther E is an isolated pair (V=10.0,12.7) of 30″ separation in pa 25°.

oc-gn 1980 ***diam. 20′*** ***V=2.5*** ***stars 37***
dimen. 14′ × 14′

In 15 cm this nebulous cluster is a loose but fairly distinct group of 21 stars. The nebulosity is gathered about the brightest star (ι Orionis, V=2.8), which has two small, sharp companions (at 11″ in pa 140° and 50″ in pa 103°) and a nearly equal pair to the SW (ADS 4182: V=4.8,5.7; 36″; 223°). Another conspicuous pair lies W of the last pair (Σ745AB: V=8.4,9.3; 29″; 347°). Overall the group is elongated in pa 30°, formed into a blunt spearpoint tipped on the NNE. Most of the cluster members are bright, so larger apertures reveal only a few additional stars: with 25 cm about 25 stars are counted; 30 cm will show perhaps five more. The nebulous regions near the bright stars exhibit little or no brightening toward the embedded star, unlike most such nebulae.

oc 1981 ***diam. 28′*** ***V=4.6*** ***stars 40***

This cluster is a loose grouping of 18 stars in 6 cm. Five brighter ones stand out, including three somewhat isolated on the E side. With 15 cm about 45 stars can be counted, not including those obviously associated with nebulosity on the S (which belong to oc-gn 1977, *q.v.*).

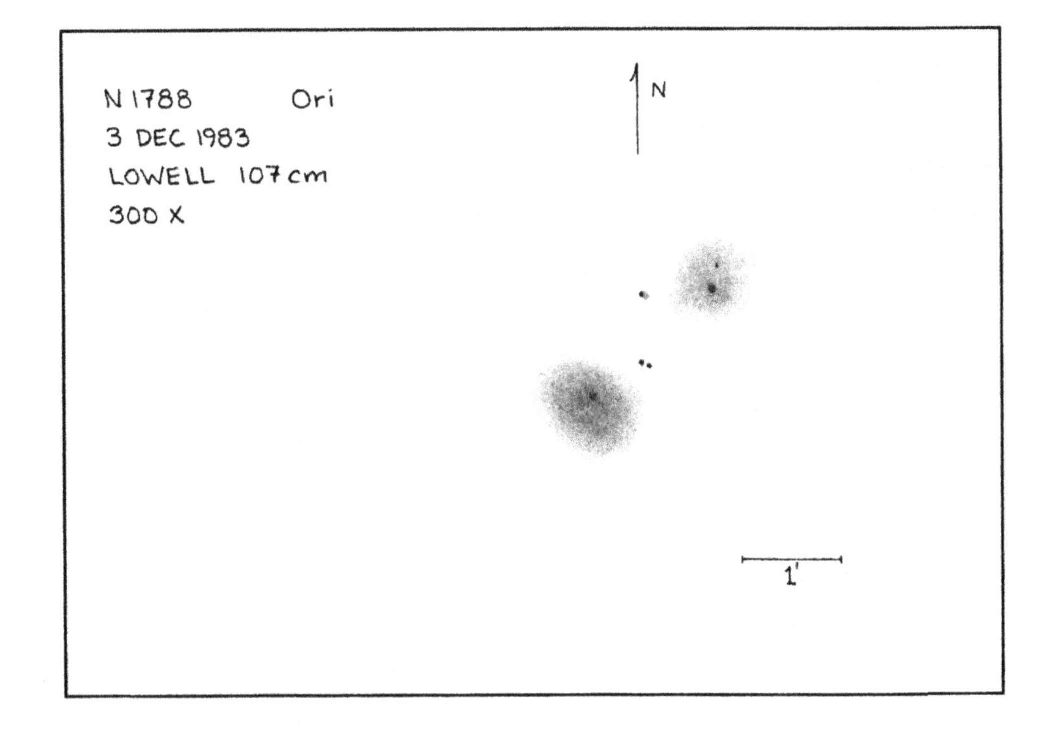

Many faint members spread to the NW. The northernmost bright star on the E side is a close pair (ADS 4192: V = 6.6,8.4; 4″.2; 60°).

gn 1982 ***dimen. 20′ × 15′*** ***star V = 6.8***
Messier 43 is the bright roundish nebula lying just N of the dark lane bounding the N and NE sides of Messier 42. In 6 cm the nebula is a smoothly textured glow with an eccentrically embedded mag. 7 star (NU Orionis). A mag. 11.5 star (NQ Orionis) lies in the nebula 1′.5 from the bright star at pa 300°, and another lies on the NE edge about 3′.5 from the mag. 7 star. The smooth texture persists in 15 cm, but the NE quadrant seems unevenly darkened. A right-angled dark patch cut out of the NE side is visible in 25 cm, and the nebula is not as bright on the SW side as in the eastern half. The surface of the nebula is mottled in 30 cm, and a faintly glowing arc curves across the N side of the dark patch. The nebulosity is brighter and smoother in the direction of M42, which it almost touches except for a thin dark strip between the two objects.

gn 1999 ***dimen. 2′ × 2′*** ***star V = 10.4***
This small but interesting nebula lies about a degree S of the Orion Nebula complex. In 15 cm it is a small distinctly fuzzy spot with an embedded mag. 10 star. With 25 cm the star actually lies a little E of center, and a small, slightly darker patch lies just W of it, giving the nebula an annular form. In 30 cm the nebula has a circular overall outline. The mag. 10 star is closely bordered on the W by the 10″ diameter, slightly irregular hole. Nebulosity is clearly visible circling around the far side, though it is brightest around and ENE of the star.

pn 2022 ***dimen. 28″ × 27″*** ***nebula 11.9v*** ***star V = 14.9***
Though this object is faintly visible at 50 × in 25 cm, high power shows a grey annular disk about 25″ across that is not completely dark inside. A very faint threshold stellaring is visible in the S edge. In 30 cm the nebula is 25″ × 20″, elongated in pa 15°. Slightly brighter spots stand out at the acute ends of the ring, the southern one being more apparent. The grey nebula is clearly annular, but shows no central star at 450 × .

gn 2023 ***dimen. 10′ × 8′*** ***star V = 7.8***
This is an easily visible nebula in 15 cm, a smoothly textured glow without detail fading evenly away from a mag. 8 star. The nebulosity spreads farther toward the E; a mag. 12 star about 2′.5 distance from the central star in pa 120° about marks the maximum extent of the nebula. In 30 cm the nebula has a fairly high surface brightness and is not uniformly bright: dark shadows lie on the northeastern and western sides. To the NE, beyond the shadow, the nebula brightens just a little again, forming a faint, isolated patch. The western perimeter is fairly sharply defined. Two indistinct dark fingers extend for a short way into the nebula from the shadow on this side, one just W of the bright star, the other farther N.

gn 2024 ***dimen. 30′ × 30′*** ***star V = 1.8***
This large, detailed nebula lies ENE of ζ Orionis (V = 1.8) and is difficult to observe because of its proximity to the bright star. Two distinct parts can be seen in 15 cm,

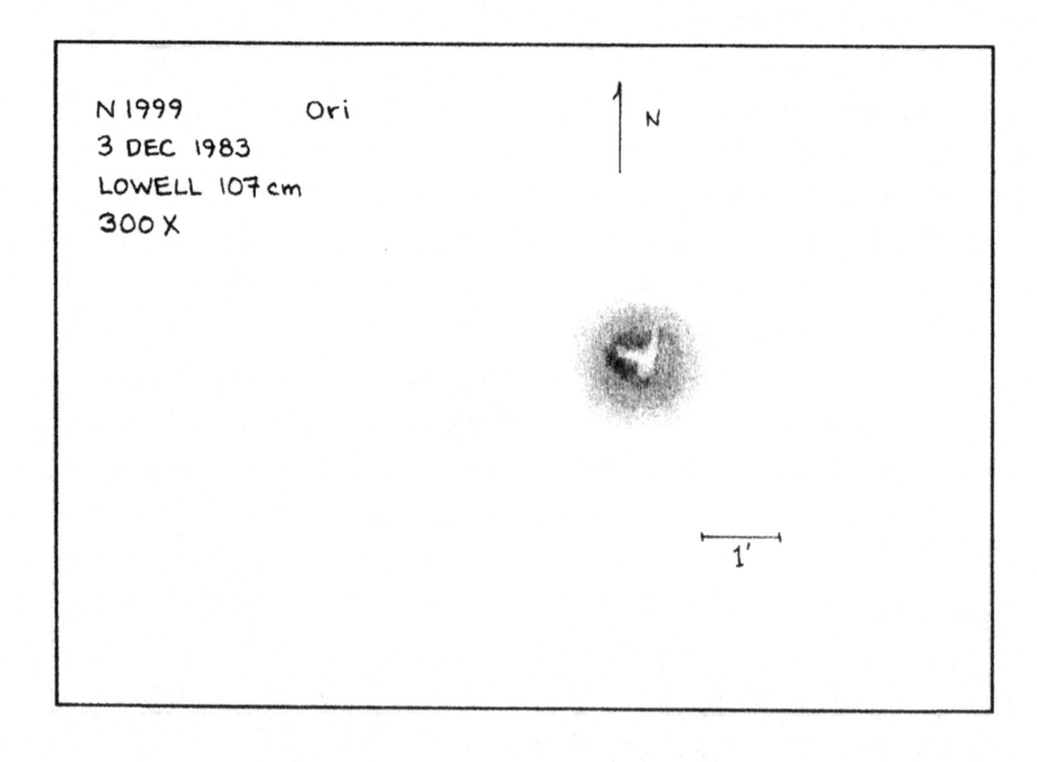

divided by an irregular dark lane about 3′ wide. The SW section, which is closer to ζ Ori, contains a star at each end; the NE section has a star on the N side. In 25 cm the dividing lane is 12′ × 3′, ending on the SE with a sharp curve to the NE. The bright nebula lining the E side forms a narrow-nosed horsehead shape. To the NW the dark lane opens widely. 30 cm shows three parts to the nebula, the largest being on the E. The southwestern section is teardrop-shaped, the point toward the N, and of lower surface brightness than the other sections. The prominent dark lane stretches in pa 150° and separates this section from the two eastern ones. The curving dark streak leading NE from the S end divides the eastern section into two unequal parts and becomes narrower as it penetrates. The northern section has several lighter and darker streaks crossing it in pa 60°. The southern section is moderately concentrated, has the highest surface brightness of all, and is elongated in pa 60°.

gn 2068 ***dimen. 8′ × 6′*** ***star V = 10.2***
gn 2071 ***dimen. 7′ × 5′*** ***star V = 10.1***

Messier 78 (gn 2068) and its companion nebula are easy to see in small telescopes. In 6 cm M78 is a grey, roughly circular blob with two embedded stars (ADS 4374AC: V = 10.2,10.6; 51″; 202°). The nebula is brighter along the N side, where the edge is more sharply defined. In 15 cm the object is an oval 6′ × 4′ haze. A third star, mag. 13.5, is visible on the SSE edge of the nebula. The outline is distinctly flattened on the sharp NW edge in 25 cm and spreads in a diffuse fan shape to the SE. 30 cm shows the haze spreading to 7′, elongated somewhat N-S. The N through NW edge is bounded by a bright semi-circular rim. About 15′ NNE is gn 2071, visible in 6 cm with averted vision as a faint haze around a mag. 10 star. With 30 cm the nebula is a smoothly fading light lying mostly S of the illuminating star, which has a faint companion at 15″ in pa 200°.

oc 2112 ***diam. 18′*** ***V = 8.4*** ***stars 100***

6 cm will show this cluster as a faint hazy patch about 5′ across with a brighter star off the NW side. Three stars are visible within at 50 ×. The cluster lies in a barren field with 25 cm. The core is 7′ across with outliers spreading to 12′, including a total of 35 stars and much haziness. On the N side of the cluster is a flat triangle of brighter stars. With 30 cm the cluster stars are slightly concentrated to the center, with a total of 35 members in a 12′–15′ area.

oc 2141 ***diam. 10′*** ***V = 9.4*** ***stars 365***

This cluster contains only very faint stars and is not resolvable in 15 cm, though a faint glow is discernable overlain by many foreground stars. In 25 cm it remains unresolved, while 30 cm will show two small clusterings of very faint stars lying within a nearly complete 10′ circlet of mag. 11–12 stars. About 20 stars of mag. 14 and fainter are resolved, including a brighter star in the S side and a group spreading N of the circlet. A group of six stars 1′ diameter lies 8′ NW of the cluster.

gn Cederblad 62 ***dimen. 2′ × 1′*** ***star V = 12.7***

Appearing as a distinctly nebulous star in 15 cm at low power, this interesting object lies 3′ W of a mag. 8 star. An easy mag. 10 pair (20″; 100°) is resolved 7.′7 SW. In 30 cm the nebula reaches a size of 2′ × 0.′75, elongated N-S. Although it extends symmetrically from the mag. 12 central star, the N side is significantly brighter.

oc 2169 ***diam. 6′*** ***V = 5.9*** ***stars 25***

6 cm will show this large multiple star system with a total of 13 stars: four lie in a string aligned N-S on the W side and the rest form two triangles on the E and S. In 15 cm the brightest star is a close pair (ADS 4728AB: V = 7.4,8.0; 2.″5; 109°), and 16 stars can be counted in a 5′ area. 20 cm shows 18 stars in two separate groups: a line of five on the W with an isolated star E of them, and a dozen others in the two triangles, including the bright bluish pair. A total of 19 stars is visible with 30 cm in the two groups. The two northern members of the western string are distinctly red, as is the eastern member of the compact group involving the bright pair.

oc-gn 2175 ***diam. 22′*** ***V = 6.8*** ***stars 75***
dimen. 40′ × 30′

Faintly visible in 6 cm, this nebula appears about 20′ diameter and surrounds a mag. 7.5 star. Several faint stars in a string lead N, beginning just E of the central star and ending near the edge of the nebulosity. 25 cm shows a fair amount of undetailed nebulosity about 20′ diameter around the bright star. Within 2° of the nebula are a number of groups: a large group lies about 30′ W that has a small knot of stars in it, and 20′ ENE is a concentration of 15 stars in a 3′ area. Three mag. 9.5 stars forming a line 25′ N have a few fainter ones amongst them at high power. No nebulosity is evident in the cluster with 30 cm. The small group of stars to the ENE includes three stars in a close arc concave to the S; the westernmost of these is an unequal pair (10″; 165°).

oc 2186 ***diam. 5′*** ***V = 8.7*** ***stars 21***

This cluster appears as a hazy region about 4′ diameter in 6 cm. A mag. 10 star appears in the N side, with similarly bright stars to the SE and WSW. In addition to the mag. 10 star, three or four stars at the limit of visibility can be seen within the cluster. 15 cm shows six stars in a zigzag shape elongated E-W. The fourth star from the E has several companions. 25 cm reveals about a dozen faint stars in the central part of the 4′ cluster. With 30 cm 25 stars can be counted in a rectangular area elongated E-W. The members are most densely packed on the NE, including the brightest star and an unequal pair (h2301: V = 10.9,11.8; 9″.6; 352°).

oc 2194 ***diam. 8′*** ***V = 8.5*** ***stars 194***
oc A0612 + 12 ***diam. 2′*** ***mag.*** ***stars***

In 6 cm oc 2194 is visible as a broadly concentrated patch of haze without resolution. A single mag. 11 star sparkles faintly in the SE edge, and several faint field stars lie around the periphery. 15 cm will show a sprinkling of perhaps a half dozen stars over the background haze, while 25 cm shows at least a dozen on a dense granular background. With 30 cm 50 stars are visible in the central 5′ × 3′.5 region, where most of the stars form a broken, angular "U" opening to the N. Many outliers scatter around this to 8′ diameter. Lying 15′ E is an uncatalogued but conspicuous group, A0612 + 12. In 25 cm it appears as an irresolved hazy patch about 2′ diameter. 30 cm resolves it into 20 mag. 13 and fainter stars in a 2′.5 × 1′.25 area elongated N-S. Two brighter stars stand out, one near the center, another off the S end.

Pegasus

Delineated by the Great Square and a few stars to the west, this region contains many bright and interesting galaxies. The center culminates at midnight about 1 September.

eg 1 dimen. 1′.9 × 1′.5 V = 12.9 sfc. br. 13.9
In 30 cm this galaxy is not difficult to see lying 2′ S of the E tip of a triangle of mag. 10.5 stars. It is fairly faint and of low surface brightness, but shows moderate concentration to the center. Overall it is about 45″ diameter, slightly elongated in pa 95°. Photographs show eg 2 (V = 14.0) 1′.8 S.

eg 16 dimen. 2′.1 × 1′.2 V = 12.0 sfc. br. 12.8
A moderately faint object in 25 cm, this galaxy appears as a circular spot less than 1′ diameter with stellar nucleus. In 30 cm the moderately concentrated halo extends to 1′ × 0′.5 in pa 20°. Two mag. 12 stars are visible about 5′ S and SE.

eg 23 dimen. 2′.3 × 1′.6 V = 12.0 sfc. br. 13.3
This galaxy is easily visible in 25 cm, though details are difficult to distinguish because of a mag. 12.5 star lying only 20″ from the nucleus in pa 130°. The faint halo is less than 1′ across and seems elongated roughly N-S. The nucleus is bright and nearly stellar, about a magnitude brighter than the attending star. In 30 cm the halo is about 45″ × 30″, elongated nearly toward the star SE. Overall the galaxy shows a fairly strong concentration to a conspicuous, still substellar nucleus.

gc 7078 diam. 12′ V = 6.0 class IV
Messier 15 is a very bright cluster in 6 cm, appearing slightly brighter than Messier 2 (*cf.* gc 7089, Aquarius), though it is clearly more concentrated. Overall it is about 4′ diameter, brightening suddenly at the center to a nonstellar nucleus. 15 cm shows the halo to about 5′ diameter around an intense core. At 200 × the cluster has a granular texture, while it is partially resolved at 250 × in 20 cm. Higher powers with 25 cm reveal a bright central pip 30″ across surrounded by a large well-resolved mass of stars about 7′ diameter. With averted vision stars extend in appreciable numbers to 15′ diameter. 30 cm resolves it well to the center, leaving only a little haze. The bright center is less than 1′ diameter, surrounded by a 5′ core and a very liberal sprinkling of mag. 12.5 stars out to 13′ diameter; the border of the cluster nearly reaches a mag. 7.5 star to the NNE. The distribution of the stars across the cluster is quite irregular.

eg 7137 dimen. 1′.5 × 1′.4 V = 12.4 sfc. br. 13.0
This galaxy is just visible in 25 cm at 100 ×. At 200 × it is a broad circular spot about 1′.5 diameter occasionally showing a stellar center. With 30 cm it appears about 1′ diameter and has a smooth texture. Averted vision at 225 × shows occasional lumps over the surface and a faint star on the NW edge about 25″ from center.

eg 7177 dimen. 3′.3 × 2′.3 V = 11.2 sfc. br. 13.2
In 25 cm this galaxy appears as a smooth oval haze elongated E-W, about 1′.5 × 0′.75. A faint, nearly stellar nucleus stands out at the center. In 30 cm the nucleus appears stellar at 150 ×, and about the same magnitude as a star 1′.3 SSW. At 225 × the nucleus is nonstellar, and the core seems elongated obliquely to the halo in pa 65°. The halo extends to 1′.9 × 0′.8 with a smooth texture and a more sharply defined northern edge.

eg 7217 dimen. 3′.7 × 3′.2 V = 10.2 sfc. br. 12.7
This is a bright galaxy in 25 cm, appearing as an extended circular glow about 2′ diameter. The large, moderately faint halo contains a small, nearly stellar nucleus. A mag. 10.5 star lies 3′.2 SE. 30 cm shows the smooth halo to 2′.2 diameter with even concentration to a small bright nucleus less than 10″ across.

eg 7315 dimen. 1′.9 × 1′.9 mag.$_z$13.8 sfc. br.
In 30 cm this galaxy is a circular 1′ spot. The light shows a sharp moderate concentration to the center, but there is no sparkling nucleus. The center is however very small, about 10″ diameter, and seems knotted. Faint stars are visible at about 50″ in pa 35° and 40″ in pa 270°; the western one is fainter.

eg 7317 dimen. 1′.0 × 0′.8 V = 13.6 sfc. br. 13.3
eg 7318A dimen. 1′.0 × 1′.0 V = 13.3 sfc. br. 13.2
eg 7318B dimen. 1′.9 × 1′.3 V = 13.1 sfc. br. 13.8
eg 7318A + B dimen. 1′.7 × 1′.3 V = 12.6 sfc. br. 13.3
eg 7319 dimen. 1′.7 × 1′.3 V = 13.1 sfc. br. 13.9
eg 7320 dimen. 2′.2 × 1′.2 V = 12.7 sfc. br. 13.5
Known as Stephan's Quintet, this group of five galaxies, at least four of which are interacting, lie packed within a

Chart XVIII. NGC 7331 group and Stephan's Quintet.

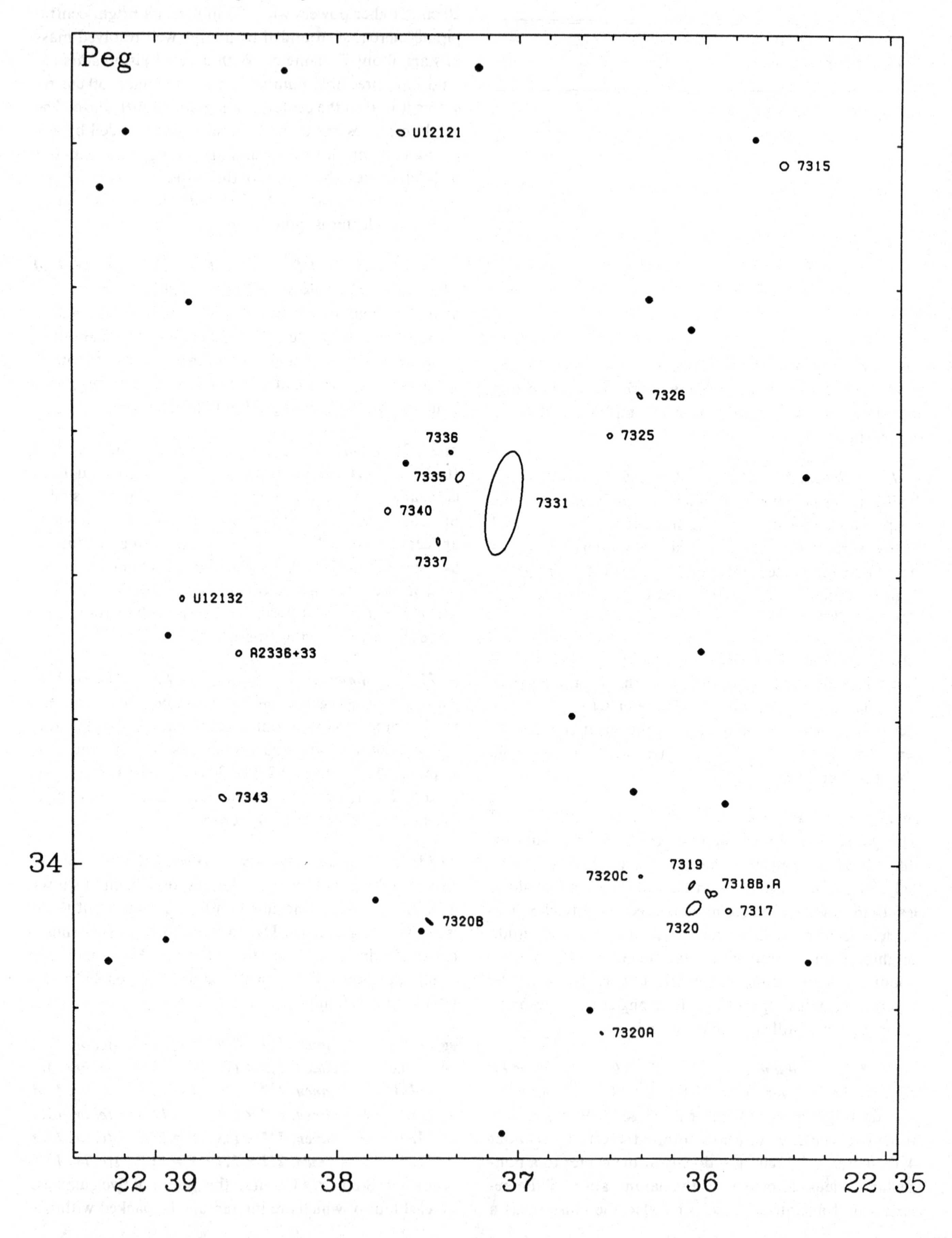

circle little more than 3.'5 across. With care 25 cm will show four of the members, but it takes 30 cm to show the fifth even faintly. In 25 cm eg 7317 looks like a hazy double star, with the real star (V = 13.4) lying on the NW edge of the faint galaxy, 15″ from its center. With 30 cm its overall size is only 20″, and shows only weak central concentration. The closely interacting pair, eg 7318AB, appears in 25 cm as a single 1′ × 0.'6 object elongated E-W and containing a single faint stellar nucleus. 30 cm shows two distinct concentrations of light merged by their halos, separated in pa 80°. The eastern member, eg 7318B, is larger, brighter, and more broadly concentrated, occasionally showing a faint stellar nucleus. The brighter parts of the halo are elongated NNE-SSW. eg 7318A is smaller, round, and more sharply concentrated. Together the galaxies are about 1′ × 0.'7. In 25 cm eg 7320 is elongated in pa 135°, about 50″ × 35″, and seems to have a faint stellar nucleus. With 30 cm this is the largest of the group, with the halo spreading to 1.'25 × 0.'75. It is weakly and broadly concentrated toward the center and shows no distinct nucleus: a faint star (V = 14.6) visible about 15″ SE of center is what gives the impression of a nucleus in 25 cm. The last member of the quintet, eg 7319, is the faintest of the group to 30 cm. It lies 1.'5 SSE of a mag. 13 star, the nearest of two to the NE of the quintet. The galaxy is an unconcentrated 40″ × 20″ patch elongated in pa 145°. A very faint star is visible about 45″ NE.

eg 7331 ***dimen. 11′ × 4.'0*** ***V = 9.5*** ***sfc. br. 13.5***
eg 7335 ***dimen. 1.'7 × 0.'8*** ***mag.$_Z$14.7*** ***sfc. br.***
eg 7337 ***dimen. 1.'3 × 1.'0*** ***mag.$_Z$15.7*** ***sfc. br.***
eg 7340 ***dimen. 1.'1 × 0.'8*** ***mag.$_Z$14.9*** ***sfc. br.***

eg 7331 is a large spiral galaxy that has several faint companions. In 6 cm it is bright and easy to see at low power as a moderately concentrated lenticular glow elongated approximately N-S and containing a stellar nucleus. With 15 cm the halo is about 4′ × 1′ and has a slightly grainy texture. In 25 cm the halo has a pearly lustre and extends to 8′ × 2′ in pa 170°. The oval core is 1.'5 × 1.' In 30 cm it grows to about 9′ × 2.' The halo is well concentrated to a 1′ core and stellar nucleus. Some mottling is visible in the inner halo, and the W side of the core has a dark lane crossing it. Two mag. 13.5 stars lie W of the N tip of the halo. The brightest companion galaxy is eg 7335, which lies 3.'5 NE. With 25 cm it appears 40″ × 25″, elongated in nearly the same position angle as − 31. eg 7337 lies 5.'2 SE of − 31. With 25 cm it looks like a double star (similar to eg 7317 in Stephan's Quintet, *cf.*), with the real star involved on the SE side only 9″ from the center of the galaxy. Two mag. 10 stars about 8′ E of − 31 point S to eg 7340, which is just discernable in 25 cm as a faint, nearly stellar spot.

eg 7332 ***dimen. 4.'2 × 1.'3*** ***V = 11.0*** ***sfc. br. 12.6***

This is a difficult object for 15 cm, viewed best at moderate powers. It is about 1′ × 0.'5 and evenly bright, though a stellar center is visible at times. In 25 cm it extends to 1.'5 × 0.'5, elongated roughly N-S. The distinct core is broadly condensed, about 20″ diameter, and contains a conspicuous stellar nucleus. The halo is diffuse and much fainter. The galaxy is a thinly tapered and well-concentrated spindle in 30 cm. The halo reaches to 1.'9 × 0.'4 in pa 155° around an elongated core and a bright, nearly stellar nucleus. eg 7339, *q.v.*, lies 5′ E.

eg 7339 ***dimen. 3.'0 × 0.'9*** ***V = 12.1*** ***sfc. br. 13.0***

In 25 cm this galaxy is about 1.'5 × 0.'5, elongated in pa 90°. The halo is quite diffuse, exhibiting little central brightening, but some granular detail appears near the center. In 30 cm it is similar in shape to eg 7332, *cf.*, though the tips are not so sharply tapered. The halo is 1.'5 × 0.'5 with a weak broad concentration and no distinct nucleus.

eg 7448 ***dimen. 2.'7 × 1.'3*** ***V = 11.7*** ***sfc. br. 12.8***

While only barely visible in 6 cm, 25 cm will show this galaxy as a uniform 1.'5 × 1′ glow of moderately low surface brightness elongated SE-NW. In 30 cm the broad, weakly concentrated patch extends to 2′ × 1′ in pa 150°. A faint pair is visible 2.'2 NW (13.5,13.5; 16″; 280°). eg 7454 (mag.$_Z$13.6) lies 29′ NE, the brightest of a group of three.

eg 7457 ***dimen. 4.'4 × 2.'5*** ***V = 10.6*** ***sfc. br. 13.2***

This galaxy is found just W of a 2.'5 triangle of stars, also 1.'5 N of a 45″ pair of mag. 12,12.5 stars. It is moderately bright in 25 cm, 1′ × 0.'3 in pa 130°, with a brighter core and nearly stellar nucleus. In 30 cm the halo has a broadly oval outline, extending to 1.'3 × 0.'8 and showing moderate concentration to the center.

eg 7469 ***dimen. 1.'8 × 1.'3*** ***V = 11.9*** ***sfc. br. 12.7***

The nucleus of this Seyfert galaxy varies in brightness from B = 14.7 to 15.5. With 15 cm the galaxy is only barely detectable, forming a flat triangle with mag. 8.5 stars 6′ NE and SE. Despite the published magnitudes, in 25 cm the nucleus looks like a mag. 12 star. Around it the halo fades smoothly. In 30 cm the object is only about 45″ diameter, fading rapidly from the center to a low surface brightness halo. eg I5283 (V = 13.8) lies 1.'3 NE.

eg 7479 ***dimen. 4.'1 × 3.'2*** ***V = 11.0*** ***sfc. br. 13.6***

In 6 cm this galaxy is faintly visible, lying just off the N end of a 1° string of five mag. 8–10 stars. The object itself is a small, concentrated spot 3′ N of a mag. 10 star, the last in the string. In 15 cm the poorly concentrated but mottled halo is about 3′ × 2′, elongated N-S. With 25 cm a mag. 13 star lies at the N edge. The galaxy still shows no overall central concentration, but the surface seems mottled with faint star-like spots. 30 cm shows an unevenly bright bar-like object whose S end curves toward the W. A faint star is visible 45″ SW of the small, slightly brighter core; faint knots are visible about 40″ SSE and SW of this star.

eg 7615 **dimen. *1′.2 × 0′.7*** ***mag.$_z$15.3*** ***sfc. br.***
In 30 cm this member of the Pegasus I cluster is a very faint patch about 40″ diameter, showing no concentration toward the center. A mag. 14 star is visible 1′ E.

eg 7619 ***dimen. 2′.9 × 2′.6*** ***V=11.1*** ***sfc. br. 13.3***
This is the brightest member of the Pegasus I cluster. Viewed in 25 cm it is a 1′ diameter haze with a stellar nucleus. In 30 cm the halo extends to 1′.2 × 0′.8, elongated in pa 40°, around a circular 30″ core. The light is evenly concentrated except for a sharp stellar nucleus.

eg 7623 ***dimen. 1′.9 × 1′.3*** ***V=12.4*** ***sfc. br. 13.3***
This is the third brightest member of the Pegasus I cluster to 30 cm. It lies 6′ N of a mag. 10 star that itself is 6′ NW of eg 7626, *q.v*. The galaxy is a small 30″ × 20″ moderately concentrated glow elongated N-S and contains a stellar nucleus at 250 ×. A mag. 14 star is visible at 1′.1 in pa 295°.

eg 7625 ***dimen. 1′.8 × 1′.7*** ***V=12.1*** ***sfc. br. 13.1***
Located in a rich field 7′ WSW of a mag. 6.5 star, 25 cm shows this galaxy as a circular high surface brightness spot with a small core and an inconspicuous stellar nucleus. Averted vision gives the impression of narrow radial lines extending from the core in different directions. In 30 cm the halo extends to 1′.5 diameter around the 20″ core.

eg 7626 ***dimen. 2′.5 × 2′.0*** ***V=11.2*** ***sfc. br. 13.0***
In 20 cm this member of the Pegasus I cluster is about the same size and brightness as eg 7619, *cf.* 6′.9 E, though its halo and core are more diffuse. The core is smaller than that of −19 in 25 cm, and overall the galaxy shows less central concentration. With 30 cm the galaxy forms a nearly equilateral triangle with −19 and a mag. 10 star 6′ NW. It has an appearance similar to its companion galaxy: the circular halo is about 1′ diameter, showing even concentration to a faint stellar nucleus that is less conspicuous than that in −19. A mag. 14 star lies 55″ W.

eg 7631 ***dimen. 1′.9 × 1′.0*** ***mag.$_z$13.8*** ***sfc. br.***
In 30 cm this member of the Pegasus I cluster is a 1′.1 × 0′.6 oval elongated in pa 75° with weak concentration to elongated inner regions, where no nucleus is discernable. A pretty triangle of mag. 11.5–12 stars 1′ on a side lies 5′ NW.

eg 7634 ***dimen. 1′.5 × 1′.1*** ***mag.$_z$13.7*** ***sfc. br.***
This faint galaxy lying NE of the core of the Pegasus I cluster is visible with 30 cm. The faint halo is circular, 40″ diameter, around a small 10″ core and an inconspicuous stellar nucleus, which seems to lie in the N side of the core. A mag. 13.5 star is closely involved, lying 20″ from center in pa 165°. Two mag. 13 and 14 stars 1′.3 SW and

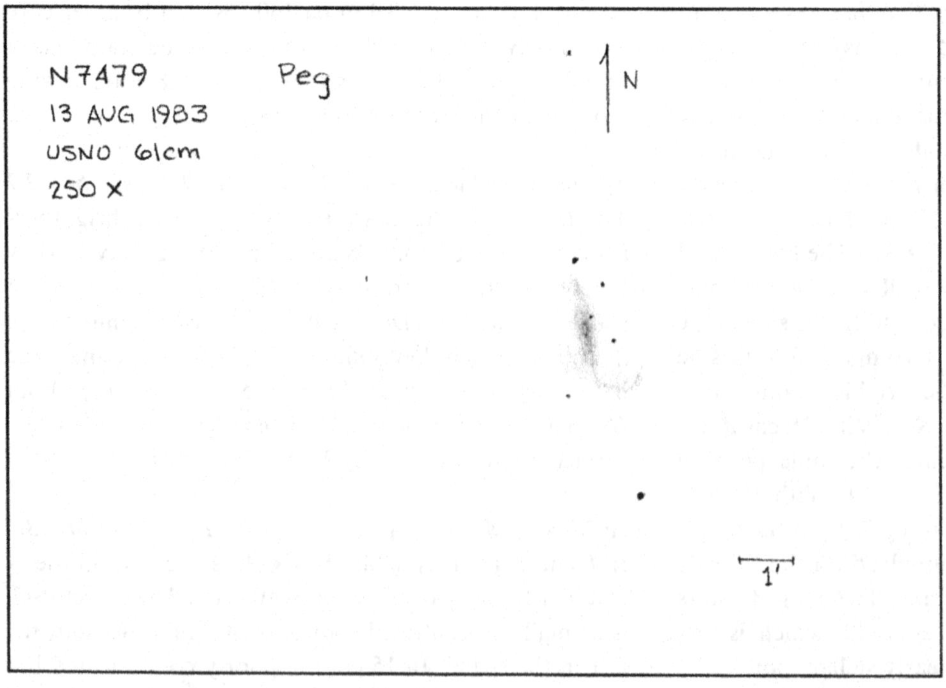

N7479 Peg
13 AUG 1983
USNO 61cm
250 X
N
1′

Chart XIX. Pegasus I cluster.

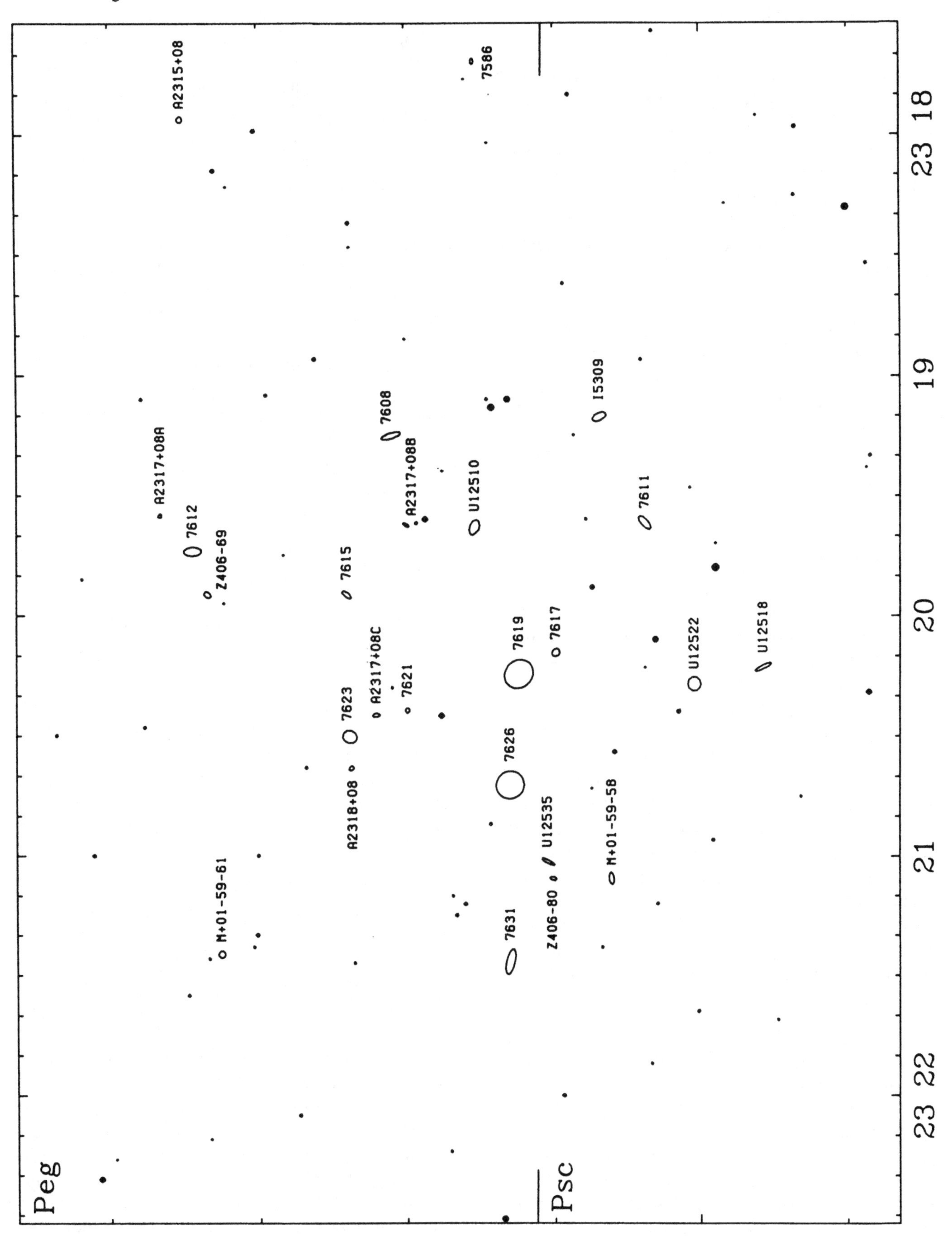

1′.0 W form an elongated quadrilateral with the galaxy and its close companion star.

eg 7648 dimen. 1′.9 × 1′.3 V = 12.9 sfc. br. 13.8
This is a moderately faint galaxy in 30 cm. The halo extends to about 1′ × 0′.4 in pa 90° and shows a moderate concentration to an indistinct core and a faint stellar nucleus. A mag. 14 star of comparable brightness to the nucleus lies 30″ E, and another slightly brighter star is visible 1′.1 SE. A threshold star 45″ W of center, with the star nearby E, makes the position angle of the elongation a little uncertain.

eg 7673 dimen. 1′.7 × 1′.6 V = 12.7 sfc. br. 13.6
Lying 3′ W of a wide mag. 8.0,8.5 pair, 25 cm shows this galaxy with a smooth, sharply tipped 1′ × 0′.3 halo elongated in pa 170°. The core is small but fairly bright and contains a faint stellar nucleus. In 30 cm the halo extends to only 40″ × 20″, but shows strong concentration to a circular 15″ core and a bright nucleus. eg 7677, *q.v.*, lies 6′ SE.

eg 7674 dimen. 1′.2 × 1′.1 V = 13.3 sfc. br. 13.5
eg M+01–59–81 dimen. 0′.3 × 0′.2 mag.$_{VV}$16 sfc. br.
eg 7675 dimen. 0′.7 × 0′.4 V = 13.9 sfc. br. 12.5
eg 7674 is just visible in 15 cm as a faint unconcentrated patch forming the N end of a 9′ arc with two bright stars. In 30 cm the two brighter galaxies are visible, forming an unequal pair 2′.2 apart in pa 105°. eg 7674 has a roughly 1′ diameter halo with weak central concentration to a circular 15″ core and a faint stellar nucleus. In the NE side of the halo 15″ from center is a very faint star. Beyond this is an even fainter stellaring that photographs show to be a small attached lenticular galaxy (eg M+01–59–81); together these two objects give the galaxy an elongated appearance. Another mag. 14.5 star is visible 1′.4 NW. eg 7675 is a very small companion in 30 cm. It is about 15″ diameter and unconcentrated except for a conspicuous substellar nucleus. A mag. 14.5 star to the SW of −75 forms a nearly equilateral triangle with the two galaxies.

eg 7677 dimen. 1′.9 × 1′.2 mag.$_{Z}$13.9 sfc. br.
Located 6′ SE of eg 7673, *q.v.*, this galaxy is barely visible in 25 cm as a circular 20″ spot with a small, brighter core. In 30 cm the galaxy lies 55″ SE of an equal pair of mag. 12 stars with 40″ separation in pa 320°. The halo extends to only 25″ diameter, showing a very slight concentration to a faint stellaring close to the center. The closer star of the pair NW has a faint companion at 25″ in pa 105°.

eg 7678 dimen. 2′.3 × 1′.8 V = 12.2 sfc. br. 13.6
Found within a 2′.5 × 1′.2 triangle of mag. 11 stars pointing to the S, this galaxy has fairly low surface brightness in 25 cm. The halo is about 1′.5 × 1′, elongated in pa 40°, with a small and only slightly brighter core. On the W edge is a threshold magnitude star, and the triangle star to the S nearly touches the object. No overall concentration is evident in 30 cm, but the middle parts have a granular texture.

eg 7741 dimen. 4′.0 × 2′.8 V = 11.4 sfc. br. 13.8
This is an exceedingly faint galaxy for 25 cm lying just SE of a bright, unequal pair (9,12; 24″; 155°). The halo seems elongated roughly E-W, extending faintly to about 3′ × 2′. Within is an indistinct 20″ core only slightly brighter than the halo. The galaxy has a very low surface brightness in 30 cm. The halo is about 2′ diameter overall, showing very slight concentration to a bar elongated in pa 100°. Three other stars are visible nearby: ESE 3′ is a mag. 10.5 star, with a mag. 14 star between it and the galaxy; ENE 2′.3 is a mag. 14.5 star.

eg 7742 dimen. 2′.0 × 2′.0 V = 11.5 sfc. br. 12.9
Located 1′.2 W of a mag. 11 star, this galaxy is a roundish patch about 1′ diameter in 15 cm with some central brightening. In 30 cm the circular halo shows moderate concentration to a stellar nucleus.

eg 7743 dimen. 3′.1 × 2′.6 V = 11.2 sfc. br. 13.3
A fairly bright object for 25 cm, this galaxy appears round, about 1′ diameter, with a well-condensed core. A mag. 12 star lies 1′ SSE. 30 cm shows a conspicuous stellar nucleus like a mag. 13.5 star in a bright 30″ core. The halo extends to an overall diameter of 1′.5.

eg 7769 dimen. 1′.8 × 1′.8 V = 12.1 sfc. br. 13.2
eg 7770 dimen. 1′.0 × 0′.9 mag.$_{Z}$14.5 sfc. br.
eg 7771 dimen. 2′.7 × 1′.3 V = 12.3 sfc. br. 13.5
In 25 cm eg 7769 is a bright galaxy reaching to about 1′ × 0′.75, elongated roughly E-W. The core is small and concentrated with a very faint stellar nucleus. A mag. 13 star is visible 2′ S. The halo appears circular in 30 cm, about 1′ diameter, with a brighter but unconcentrated core. Another fainter star is visible here lying 50″ SE of center. eg 7771, 5′.4 ESE of −69, appears nearly as bright as −69 in 25 cm, with a 1′.2 × 0′.7 halo elongated in pa 75°. The broadly concentrated core has a brighter central bar. A mag. 12 star lies 2′ NE, and 1′.1 SW is a faint companion, eg 7770. In 25 cm it is a small, very faint spot with a stellar nucleus. With 30 cm it seems elongated NE-SW, but otherwise appears about the same.

eg 7814 dimen. 6′.3 × 2′.6 V = 10.5 sfc. br. 13.3
This is a bright, easy object for 15 cm, lying 8′ NE of a wide mag. 9.5 pair (2′.9; 20°). The halo is elongated SE-NW, 2′.75 to 3′ long with a granular 1′ core. In 25 cm the core is broadly condensed but a star or stellar nucleus appears off-center to the N. With 30 cm the galaxy is elongated in pa 135°, about 3′ long with a circular 1′ core. The halo has very low surface brightness, and the core is well concentrated but without a conspicuous nucleus. A pair of mag. 13.5 stars (45″; 210°) is resolved 4′ S.

Perseus

Partially immersed in the Milky Way, the constellation contains an array of clusters, galaxies, and nebulae. The leading star, α Persei, is surrounded by a moving association of stars that is a fine sight in binoculars. The center culminates at midnight about 7 November.

pn 650 dimen. 2′.7 × 1′.8 nebula 10.1v star V = 15.9:
Messier 76 is faintly visible in 6 cm as an unconcentrated patch elongated NE-SW. A small grouping of faint stars stands out about half a degree NE. 15 cm will show three stars nearby, enclosing the nebula in a right triangle pointing toward the E. The southwestern lobe of the dumbbell-shaped nebula is brighter and more well defined than the northeastern lobe. With 25 cm the lobes seem to be two fairly isolated irregular patches with fainter nebulosity strewn between. The overall size is about 1′.3 × 1′ in 30 cm, elongated in pa 40°. The unevenly bright lobes are about 50″ apart, 30″ diameter each, and set in a faint surrounding halo. This halo fills a more nearly circular area, but is more easily visible on the SE. A mag. 13.5 star is visible within the halo on the S edge of the southwestern lobe.

oc 744 diam. 5′ V = 7.9 stars 99
In 6 cm this cluster is a moderately faint group of about ten stars lying S of a mag. 8 star. The cluster is small and moderately dense in 25 cm, centered 7′ SSW of the bright star. About 20 stars are visible, with the brighter ones on the N side. 30 cm shows about 25 stars loosely strewn over a 7′ area. A bright pair (Stein 1741: V = 11.4,11.4; 14″; 235°) is on the SE side, separated from the densest part of the cluster by a 2′ void.

oc 869 diam. 18′ V = 3.5 stars 350
oc 884 diam. 18′ V = 3.6 stars 300
These two clusters form an incredibly rich and beautiful 2°.5 field. oc 869 has two bright stars near its center and seems more sharply concentrated than its companion in 6 cm. The brightest central star (h Persei, V = 6.6) has a bowl-shaped arc of five stars on its E side among a total of 40 stars in the cluster. About 65 stars can be seen with 15 cm; h Per and its bright companion NE seem to be in the foreground with a stratum of faint stars behind. The bowl of five stars near h Per is prominent in 25 cm: opposite h Per is a larger bowl of stars opening in the same direction. 30 cm shows about 100 stars filling a 23′ low-power field and many mag. 9–10 stars scattered far to the W. The arc close by E of h Per outlines a nearly perfect parabola. oc 884 is larger but less concentrated than −69, and about 50 stars can be seen in it with 6 cm. A bright knot lies in the SW side, including two close trios of stars. A circular pattern is apparent with 15 cm: from a bright orange star (RS Persei) on the SW, a chain of stars leads N and E, finally doubling back to this star. The cluster stars are packed in the W and SW sides around the two triple stars. It contains about as many stars as −69 with 20 cm, but overall the cluster is not as bright. The cluster spreads beyond 25′ diameter in 30 cm; at least 160 stars can be counted.

oc 957 diam. 10′ V = 7.6 stars 119
This is a loose cluster scattered between and N of two bright stars, the eastern of which is a pair (ADS 1937: V = 8.0,9.9; 23″; 19°). 25 cm shows about 20 stars in a 10′ × 6′ area elongated E-W. With 30 cm a sprinkling of mag. 11–12 stars and a few fainter ones fills a 12′ × 5′ area, including a total of about 35 stars.

oc Trumpler 2 diam. 17′ V = 5.9 stars 109
Viewed in 25 cm this cluster is elongated E-W, about 10′ × 5′, containing about 20 stars of mag. 8 and fainter.

eg 1003 dimen. 5′.4 × 2′.1 V = 11.5 sfc. br. 14.0
Lying 1′.8 NE of a mag. 10.5 star, this galaxy is easily visible in 15 cm. The halo is elongated E-W, and exhibits moderate concentration to a small substellar nucleus. On the NE side of the core, at the edge of the halo, is a mag. 13 star. With 30 cm the halo extends to 3′ × 0′.9 in pa 95°, showing moderate even concentration toward the center: there is no distinctly bounded core or nucleus. The star NE lies 45″ from center, and another threshold magnitude star or stellaring is visible embedded 20″ NW of center. At 250 × the southern flank of the halo seems to fade more abruptly, placing the concentrated central regions off-center toward that side.

eg 1023 dimen. 8′.7 × 3′.3 V = 9.2 sfc. br. 12.7
This galaxy is visible in 6 cm as a small but high surface brightness football-shaped spot elongated approximately E-W. Two unequally bright stars lie nearby SW. 15 cm

shows the halo extending to 3′ × 1′ around a fairly distinct core and a stellar nucleus. In 25 cm the brighter part of the halo is about 3′ × 1′, but averted vision will show it extending faintly beyond this. In the center the stellar nucleus stands out within a 20″ core. The halo reaches 6′ × 1.′5 in pa 85° with 30 cm, showing fairly strong and even concentration to a substellar nucleus. Photographs show this galaxy to be the brightest of a group including eg I 239 (V = 11.2), a low surface brightness face-on spiral 46′ W, eg 1003 nearly 2° N, *q.v.*, and eg 1058, *q.v.*

oc 1039 ***diam. 25′*** ***V = 5.2*** ***stars 50***
Messier 34 is a bright, naked-eye cluster, a splendid object in 6 cm, showing many geometric patterns among its stars. About 40 stars are visible, half of them in a well-defined core. 15 cm will show about 50 stars of mag. 7–11 in a 45′ area, including a close pair in the S side (ADS 2052: V = 8.4,9.1; 1.″4; 55°). It is a pleasing cluster in 25 cm, with 75 stars appearing in groups and lines. With 30 cm at least 50 stars can be counted, spilling out of a 23′ low-power field.

eg 1058 ***dimen. 3.′0 × 2.′9*** ***V = 11.5*** ***sfc. br. 13.7***
Though it is not difficult to see in 15 cm, this galaxy appears only as an undetailed 1′ diameter glow lying 2.′3 NNE of a mag. 11.5 star. With 25 cm the halo grows to about 1.′5 diameter. There is little central concentration and no distinct nucleus. 30 cm shows the very low surface brightness halo spreading to about 1.′75 diameter, but otherwise the view is similar to that in 25 cm.

eg 1160 ***dimen. 1.′6 × 0.′8*** ***mag.$_z$13.0*** ***sfc. br.***
This galaxy is not difficult to see with averted vision in 15 cm as a small, unconcentrated spot lying 3.′5 N of eg 1161, *q.v.* In 30 cm the galaxy is a little smaller than its nearby companion, extending to 40″ × 30″ in pa 50°. It shows only slight central concentration. A mag. 14 star lies 1.′1 SSW, and a small triangle of mag. 13–14 stars is visible 1.′8 N.

eg 1161 ***dimen. 3.′1 × 2.′2*** ***mag.$_z$12.6*** ***sfc. br.***
In 15 cm this galaxy appears as a substellar spot lying immediately NE of a wide pair (h2167: 9.4,9.9; 37″; 34°). 30 cm shows the halo to 1′ × 0.′6 elongated in pa 20° with moderate, even concentration to a more circular core and a faint substellar nucleus. The nucleus lies almost due east of the northern member of the pair, and at a similar distance to the pair's separation.

eg 1169 ***dimen. 4.′4 × 3.′0*** ***V = 11.7*** ***sfc. br. 14.3***
In 15 cm this galaxy is a small 1′ diameter spot with a brighter, beady center. 25 cm shows the galaxy lying 1.′4 W of a mag. 12.5 star. The core is 45″ across with a stellar nucleus and fades rapidly to an overall diameter of 1.′5. In 30 cm it is sharply concentrated to a small core. At 225 × a faint star (V = 13.7) lies toward the SW about 10″ from the center, appearing like an off-centered nucleus. This gives the central regions of the galaxy an elongated appearance, but the halo is circular, about 1.′8 diameter.

eg 1175 ***dimen. 2.′5 × 0.′8*** ***V = 12.8*** ***sfc. br. 13.4***
This is a faint galaxy for 25 cm, which shows a tiny 45″ × 10″ spindle with a stellar nucleus and a faint outer halo. In 30 cm it is small enough and of the proper form to be mistaken for a faint double star. The sparkly core is very elongated, appearing almost multiple at low power. At 225 × the fuzzy-ended halo extends to only 50″ × 15″ in pa 155°. eg I 284, *q.v.*, lies 18′ E; eg 1184, *q.v.*, is 30′ NNE.

eg 1186 ***dimen. 3.′3 × 1.′4*** ***mag.$_z$12.5*** ***sfc. br.***
Though not visible in 15 cm, 30 cm will show this galaxy as a weakly concentrated low surface brightness glow extending to 1.′5 × 0.′8 in pa 130°. The core is small and broadly concentrated, exhibiting no distinct nucleus. A bright mag. 12.5 star is superposed just SW of center, and one of several nearby fainter stars lies against the SE flank 55″ from center.

oc 1193 ***diam. 3.′0*** ***mag. 12.6p*** ***stars 40***
In 15 cm this cluster is a moderately faint hazy spot lying about 4′ ESE of a mag. 7.5 star. At 75 × the haze extends from a mag. 11.5 star on the WNW edge to a mag. 13 star on the ESE. With 30 cm the cluster is about 2′ diameter. Lying 30″ W of the mag. 13 star is a pair (14,14; 8″; 290°); this pair lies S of the geometric center of the hazy region. Including the pair, a dozen stars are resolvable, all faint. Overall the cluster has a circular outline if the brighter stars WNW and ESE are disregarded.

eg I 284 ***dimen. 4.′0 × 2.′2*** ***V = 11.8*** ***sfc. br. 14.0***
eg VZw319 ***dimen. 0.′2 × 0.′2*** ***mag.$_z$17.5*** ***sfc. br.***
In 30 cm eg I 284 is visible 2.′5 SE of a nearly equal mag. 11 pair (17″; 15°). The broadly concentrated core is about 1′ across with fainter haze spreading to 2′ length in pa 20°. A mag. 13 star lies against the halo 1.′2 NE of center. At 225 × the core is unevenly bright, and a small brighter spot is visible off the SW side of the core embedded in the halo—this is the compact elliptical companion VZw319.

oc 1220 ***diam. 2.′0*** ***mag. 11.8p*** ***stars 15***
This is a very small cluster in 25 cm, showing perhaps a dozen stars without haze in a 1.′5 × 1′ area elongated N-S. Six or eight of the brighter stars form an arrowhead

pointing N. At high power 30 cm shows 13 stars of mag. 13 and fainter in a 1′.5 area. A string of three stars stands out in the N side, and two brighter stars lie on the SW and S, the latter somewhat isolated from the rest of the group.

eg 1224 **dimen. 1′.7 × 1′.5** **mag.$_z$15.5** **sfc. br.**
This galaxy in the Perseus I cluster is small but not faint in 30 cm. The circular halo is about 45″ diameter, showing moderate, even concentration to a faint nucleus. A close, equal pair of stars is visible a little over 1′ ENE (14,14; 13″; 90°).

oc King 5 **diam. 6′** **mag.** **stars 40**
This is a loosely spread and hazy group in 25 cm, which will resolve about 20 stars of mag. 12.5 and fainter in a 5′ area. Two mag. 10 stars lie nearby, one at the NE edge, the other off about 5′ N.

oc 1245 **diam. 10′** **V = 8.4** **stars 156**
Lying just N of a mag. 8 star, this cluster is visible in 6 cm as a faint hazy patch, with about half a dozen faint stars resolved in the N side. 25 cm shows a moderately faint but rich cluster, containing 50 stars in a 10′ area. 30 cm will show about 60 stars, mostly mag. 12.5 and fainter, in a 10′ area. Most of the stars are concentrated into a clumpy ring, leaving only three or four stars in the center.

eg 1250 **dimen. 2′.7 × 1′.1** **mag.$_z$14.2** **sfc. br.**
This member of the Perseus I cluster is a weakly and broadly concentrated glow in 30 cm. The halo extends to 1′.2 × 0′.8 in pa 150°. The core is circular, about 30″ diameter, and occasionally shows a faint stellar nucleus. A faint star is visible 1′ N.

eg I 310 **dimen. 1′.6 × 1′.6** **V = 12.7** **sfc. br. 13.6**
In 30 cm this member of the Perseus I cluster is a moderately faint spot about 50″ diameter. It shows weak, uneven concentration to a 10″ core, but has no distinct nucleus. The galaxy forms the northern tip of a nearly equilateral triangle with two stars, each at about 1′; the eastern one is quite faint.

eg 1259 **dimen. 1′.1 × 1′.1** **V = 14.0** **sfc. br. 14.3**
eg 1260 **dimen. 1′.7 × 0′.8** **V = 13.2** **sfc. br. 13.3**
Both of these members of the Perseus I cluster are visible in 30 cm. At 250 × eg 1260 extends to about 25″ × 20″, elongated E-W. Lying 1′.5 S is a tight triangular asterism of mag. 13.5 stars, while about 3′.5 NNE is a larger triangle of mag. 12.5 stars 1′.5 on a side. The very faint companion is just discernable 2′.3 SW. The tiny patch is only 10″ diameter and has a mag. 14.5 star involved at its W edge.

eg Z540–087 **dimen.** **mag.$_z$15.3** **sfc. br.**
In 30 cm this Perseus I cluster member is a tiny circular spot that forms the western apex of an uneven quadrilateral with three faint stars, the closest lying 50″ NNE.

eg 1267 **dimen. 1′.4 × 1′.1** **V = 13.0** **sfc. br. 13.4**
eg 1268 **dimen. 1′.2 × 0′.9** **V = 14.2** **sfc. br. 14.1**
Together with eg 1270, *q.v.*, these two members of the Perseus I cluster are involved in an incomplete circlet of nine mag. 13.5–14 stars 2′.5 diameter. 30 cm shows eg 1268 as an unconcentrated, low surface brightness object in the NW side of the circlet. The halo extends to 45″ × 30″, elongated SE-NW. A faint stellaring sparkles a little S of center, which may be the nucleus. Lying 1′.4 S is eg 1267. In 30 cm it is a tiny diffuse spot only 10″ diameter centered just 5″ W of one of the circlet stars.

eg 1270 **dimen. 1′.1 × 0′.9** **V = 12.9** **sfc. br. 12.9**
This Perseus I cluster member is visible in 30 cm about 1′ S of three close members of the circlet of stars described under eg 1267–8, *q.v.* It has a small, bright center about 5″ across within a faint 30″ diameter halo.

eg 1271 **dimen.** **V = 13.9** **sfc. br.**
In 30 cm this member of the Perseus I cluster is a very small spot about 20″ diameter. The halo seems slightly elongated toward a threshold magnitude star lying 35″ from center in pa 290°. The galaxy is slightly concentrated to a starry center that is distinctly brighter than this star.

eg 1272 **dimen. 2′.5 × 2′.3** **V = 12.5** **sfc. br. 14.4**
This is the second brightest member of the Perseus I cluster to 30 cm. At 250 × it appears as large as eg 1275, *cf.*, though of lower surface brightness. The halo is 1′ × 0′.75, elongated E-W, showing even concentration to the center and no distinct nucleus. A very faint star lies toward the E edge of the halo about 20″ from center.

eg 1273 **dimen. 1′.4 × 1′.4** **V = 12.9** **sfc. br. 13.6**
30 cm shows this member of the Perseus I cluster as an evenly, weakly concentrated spot about 35″ diameter with a faint substellar nucleus.

eg M+07–07–61 **dimen.** **V = 13.7** **sfc. br.**
In 30 cm this galaxy, a member of the Perseus I cluster, is a very small spot only 15″ diameter. It forms the northern apex of a symmetrical diamond-shaped asterism 40″ on a side with three mag. 14 stars.

eg 1274 **dimen. 0′.6 × 0′.4** **V = 14.0** **sfc. br. 12.4**
This member of the Perseus I cluster is a very small object in 30 cm. The halo is about 15″ × 10″ in pa 35°.

eg 1275 dimen. 2′.6 × 1′.9 V = 11.6 sfc. br. 13.2
This Seyfert galaxy is the brightest member of the Perseus I cluster. In 15 cm the halo appears about 30″ across and contains a stellar nucleus. With 25 cm the galaxy looks like a fuzzy star: a bright stellar nucleus surrounded by a 40″ halo. In 30 cm the halo extends to 1′ diameter, showing moderate concentration to an irregularly bright core and stellar nucleus. At 250 × the nucleus seems to lie a little W of the of the geometric center of the halo. A mag. 13 star is visible 50″ NE of center.

eg 1277 dimen. 0′.8 × 0′.3 V = 13.5 sfc. br. 11.8
eg 1278 dimen. 1′.7 × 1′.4 V = 12.6 sfc. br. 13.5
30 cm will show both members of this pair of galaxies in the Perseus I cluster. The larger object, eg 1278, has a 40″ diameter halo slightly elongated E-W, with weak broad concentration. A faint stellar nucleus is occasionally visible in the center. eg 1277 lies about 50″ distant in pa 320°. It is a very small 15″ diameter spot with a starry center.

eg 1281 dimen. 0′.7 × 0′.5 V = 13.5 sfc. br. 12.2
In 30 cm this member of the Perseus I cluster is visible 1′ ENE of a mag. 11 star. It is very small, about 20″ diameter, and unconcentrated except for a few faint sparkles near the center; at times it appears elongated E-W because of these.

eg 1282 dimen. 1′.7 × 1′.3 V = 12.9 sfc. br. 13.8
eg 1283 dimen. 1′.3 × 0′.8 V = 13.6 sfc. br. 13.6
This pair of galaxies in the Perseus I cluster lie 2′ apart in pa 15°. In 30 cm eg 1282, the southern and brighter of the two, extends to 35″ × 20″, elongated roughly NE-SW. It shows a slight central concentration to a circular 10″ core and a faint stellar nucleus. A mag. 13.5 star is visible 40″ W. eg 1283 forms the southern point of a 50″ triangle with two mag. 14 stars. The galaxy is very small, about 10″ diameter, and has a starry center.

eg I 313 dimen. 1′.6 × 1′.1 V = 13.5 sfc. br. 13.9
With 30 cm this galaxy, a member of the Perseus I cluster, appears about 20″ diameter. Occasionally it seems to be elongated toward a close faint pair lying 30″ SE (14,14; 7″; 50°). eg I 316, *q.v.*, lies 4′.7 NE.

eg I 316AB dimen. 1′.8 × 1′.0 V = 14.8 sfc. br. 15.1
Lying 4′.7 NE of eg I 313, *q.v.*, 30 cm will show this Perseus I cluster member as a 25″ diameter spot of very low surface brightness. The halo brightens hardly at all toward the center, but shows a faint nonstellar nucleus. Photographs show this to be a close pair of galaxies separated by 0′.2; the magnitude listed is for the combined light.

eg 1293 dimen. 1′.1 × 1′.1 V = 13.3 sfc. br. 13.6
eg 1294 dimen. 1′.8 × 1′.6 V = 13.1 sfc. br. 14.1
Both of these Perseus I cluster members are visible to 30 cm. eg 1293 has a fairly well-concentrated 15″ halo that seems elongated toward its companion 2′.1 SSE. The inner parts appear circular and contain a stellar nucleus. eg 1294 is of a similar brightness, but is a little larger, extending to about 20″ × 12″ in pa 0°. The nearly circular core is small, about 5″ across, and occasionally shows a stellar nucleus. Overall it appears almost as well concentrated as eg 1293. A very faint star is visible 25″ SSE, just off the E edge of the halo.

gn 1333 dimen. 9′ × 7′ star V = 10.5
In 15 cm this reflection nebula is not difficult to see as an elongated haze associated with a mag. 10.5 star. A mag. 12 star 10′ NE also has some faint, barely discernable nebulosity associated with it. With 25 cm the larger nebula is about 8′ × 6′, elongated in pa 45°. At 100 × the surface shows some indistinct mottling, and the brightest part is centered about 2′.5 SW of the mag. 10.5 star. The nebulous patch surrounding the mag. 12 star NE has a nearly circular outline and a slightly lower surface brightness.

oc 1342 diam. 17′ V = 6.7 stars 99
6 cm will show at least 15 stars in this cluster, including two bright stars on the N, and a third at the S edge. About 60 stars of mag. 8.5 and fainter are visible in 25 cm. The main group is about 12′ across, with stragglers spreading to 35′ diameter. In 30 cm about 45 stars appear in a box shape. The brightest members run E-W across a diagonal of the box, then trail in a curving arc toward the E. The cluster overall looks like a stingray swimming to the W. A nice pair (Ali 516: V = 10.4,11.2; 14″; 234°) on the W side of the box forms one eye of the creature.

pn I 351 dimen. 8″ × 6″ nebula 12.0v star 15:p
This planetary nebula is visible in 15 cm, though it is indistinguishable from a star at 75 ×. It lies about 3′.5 NW of a mag. 10,12,13 triplet. With 30 cm the nebula is definitely nonstellar, appearing about 8″ diameter with a circular outline and a fairly high surface brightness. At 250 × it seems unevenly bright at times, with star-like spots on the periphery. A mag. 14.5 star lies 20″ away in pa 295°; another threshold magnitude star is visible at 35″ in pa 350°.

oc 1444 diam. 4′.0 V = 6.6 stars 57
This small cluster surrounds the bright pair ADS 2783AB (V = 6.9,9.1; 8″.6; 253°). 25 cm shows about eight other stars in a 6′ area down to mag. 13.5. At least a

Chart XX. Perseus cluster = Abell 426.

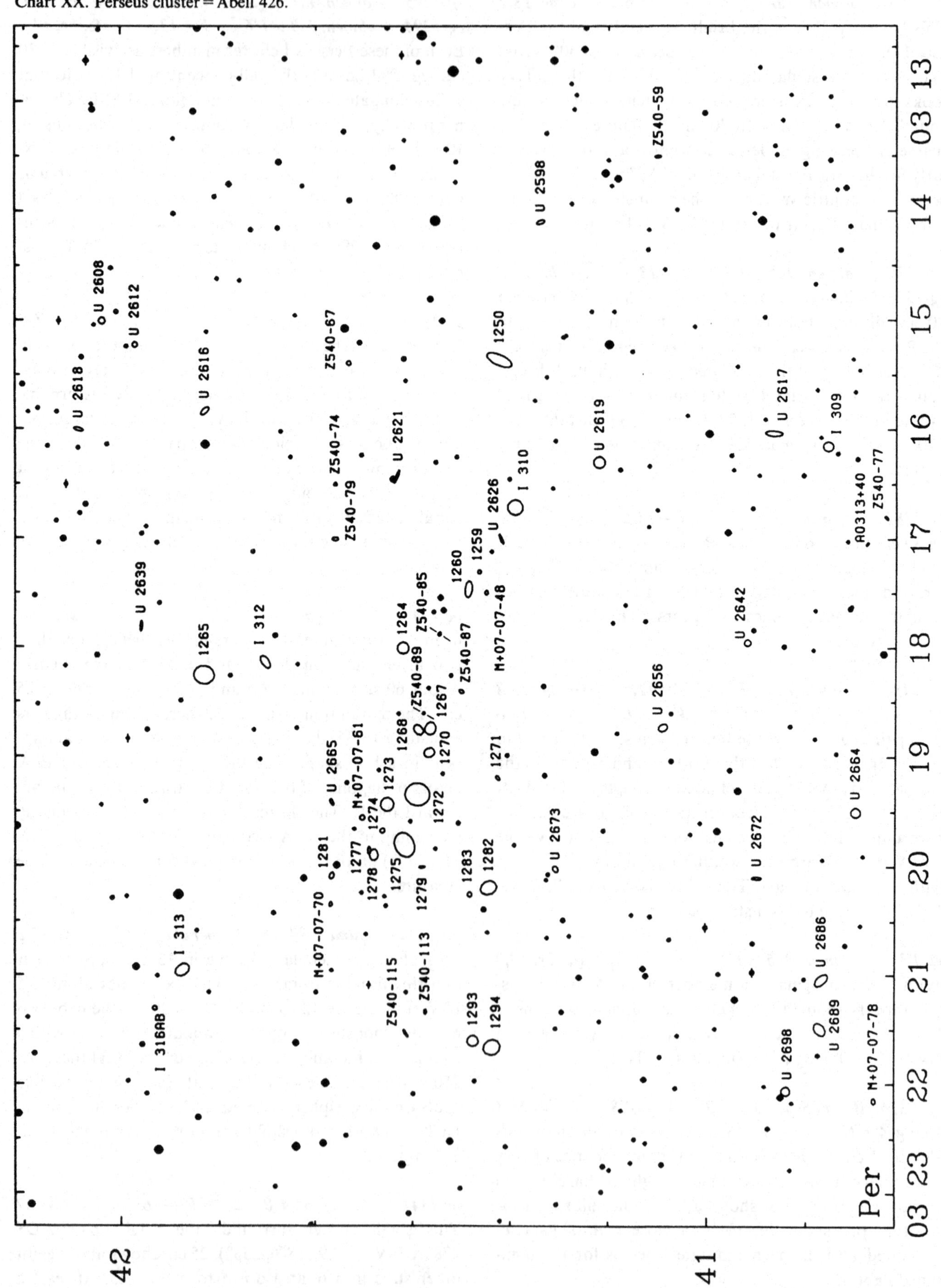

dozen can be seen in 30 cm, including a string of four stars aligned NE-SW to the NW of the bright star, which has a third companion of mag. 12 at 12″ in pa 39°. The NE member of this string is a close pair (ADS 2783DE: 10.3,10.7; 2″.6; 231°).

pn I2003 dimen. 7″ × 6″ nebula 11.5v star V = 16.5:
In 15 cm this tiny planetary appears as a 12th magnitude star lying in the middle of a group of about ten mag. 11–13 stars. 30 cm shows a small nebula that is similar to nearby I 351, *cf.*, though it is a little bigger, perhaps 10″ diameter. At 250 × the nebula shows some central concentration to an occasionally star-like point. A mag. 14 star is visible 16″ SW.

gn 1491 dimen. 9′ × 6′ star V = 11.2
With 15 cm this emission nebula forms a roughly triangular haze lying against the W side of a mag. 11 star. A fainter star, about mag. 12, is 3′ NNW, just off the northern apex. 25 cm will show several other faint stars involved, two or three of them embedded in nebulosity. Overall the nebula has a moderate surface brightness and is easily visible at low power.

oc 1496 diam. 3′.0 mag. 9.6p stars 10
This small cluster is easily discernable from the field with 15 cm at low power, though its stars are not bright. About ten members are resolved within 5′ diameter. 30 cm shows about 20 stars in a 5′ × 3′ area elongated E-W. The brightest star, mag. 10, is on the north side of a concentrated clump to the W side of the cluster. At least three pairs are visible, the closest being 2′ S of the bright star (11.5,13; 8″; 290°).

gn 1499 dimen. 160′ × 40′ star V = 4.0
The California Nebula is visible in 6 cm as a faint, very low surface brightness glow lying N of ξ Persei (V = 4.0). Overlying a rich 2° field, at 25 × it forms an elongated, slightly curved patch convex on its northeastern perimeter.

oc 1513 diam. 12′ V = 8.4 stars 43
This cluster is barely discernable in 6 cm, which will resolve about half a dozen faint stars over a hazy region just S of a mag. 10 star. 25 cm shows about 30 moderately faint stars in a group elongated SE-NW about 10′ across. In 30 cm about 30 stars are visible in an irregular group. Six of the brightest members form an empty 2′.5 circlet in the SE side. The mag. 10 star on the N has a mag. 11.5 companion 45″ NNW.

oc 1528 diam. 18′ V = 6.4 stars 165
In 6 cm this cluster appears club-shaped, with over 30 stars resolved in a 25′ area. The handle of the club points SE, and the head is formed by an elongated, irregular circlet that includes the brightest member of the cluster. About 75 stars are resolved by 15 cm with little central concentration. 25 cm shows outliers to 40′ diameter, and many geometric shapes and pairs can be discerned. In 30 cm 75 stars are visible in a 20′ field, showing some concentration toward the center. The brighter stars lie on the W side.

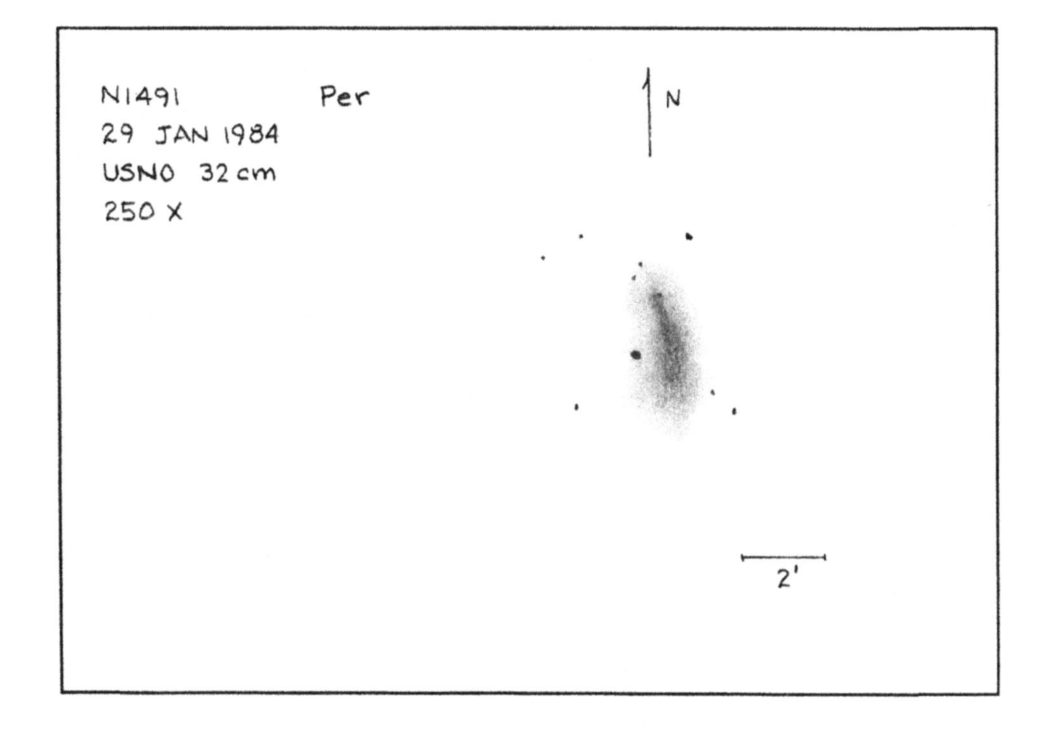

oc 1545 ***diam. 12′*** ***V=6.2*** ***stars 65***
Near the center of this cluster 6 cm shows a pretty 2′.5 triangle pointing SW, formed by blue, orange, and yellow stars (moving clockwise from the SW apex). A few faint stars lie to the E. 25 cm shows 45 stars loosely scattered through a 25′ area. In the N side is a relatively bright unequal pair (ADS 3136: 7.9,9.4; 18″; 346°), the southern, brighter member of which is orange. In 30 cm about 35 stars are visible in an 18′ area.

gn 1579 ***dimen. 12′×8′*** ***star 17v***
Located about 12′ E of a mag. 8 star, 25 cm shows this reflection nebula easily at low power. A diffuse nebulosity spreads irregularly through a 5′×3′ area elongated roughly N-S. Near the center is the brightest portion, looking much like a 1′.5 diameter mag. 12 galaxy. A total of eight stars is involved, including a mag. 12 star on the N side and a wide, faint pair on the S. The central part is concentrated toward its center, but no stars sparkle within at 200×. About 13′ NE is a mag. 12 star with a little nebulosity around it.

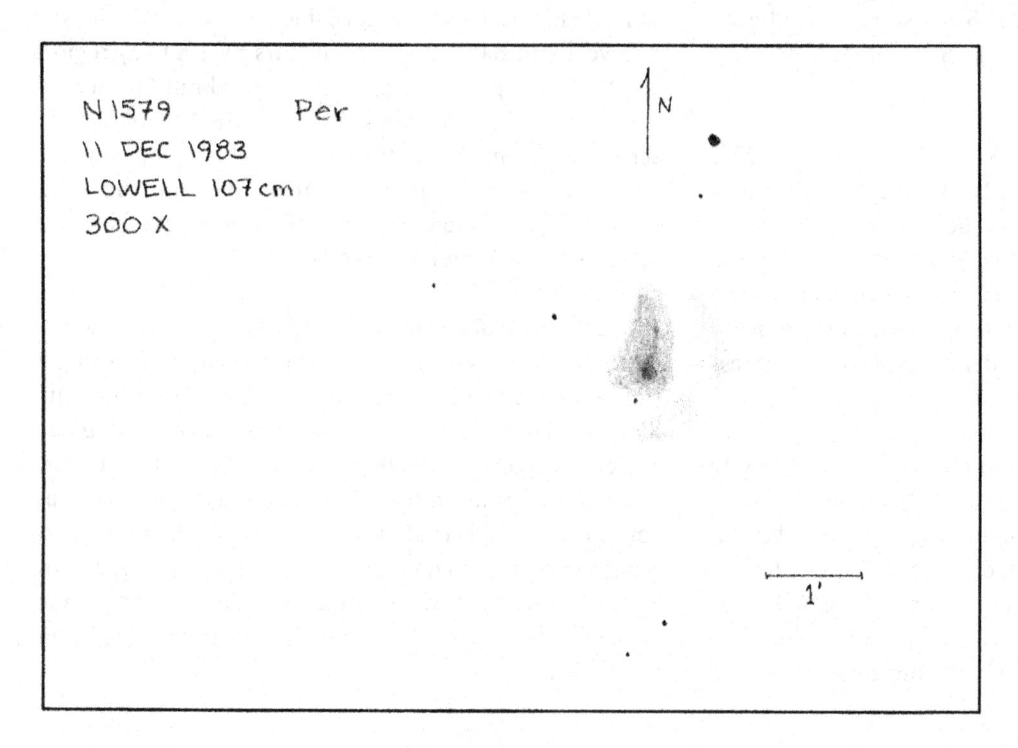

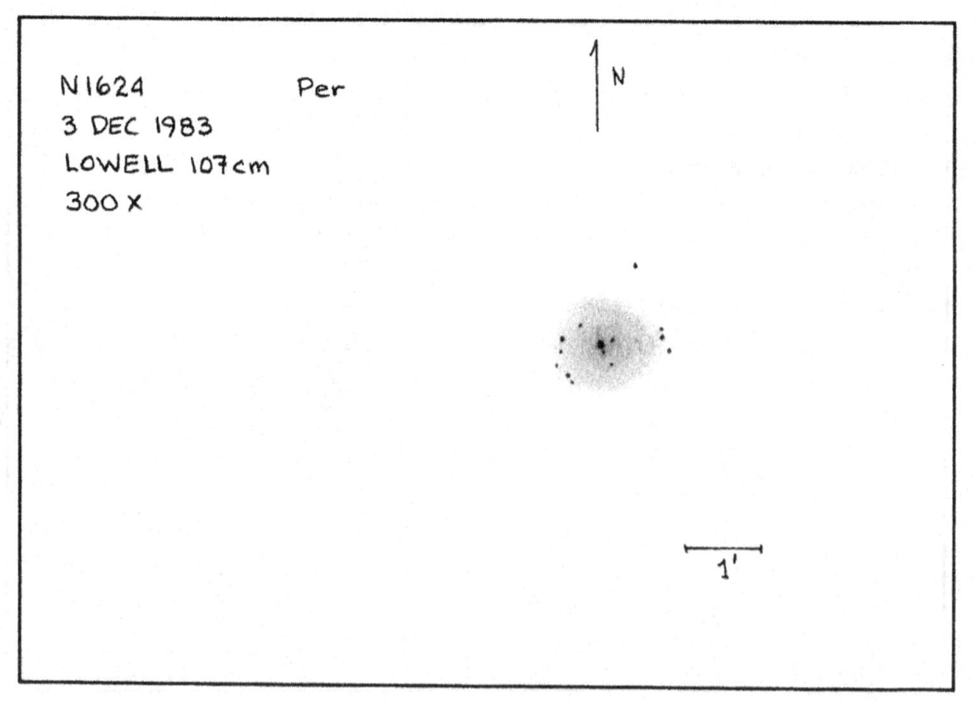

oc-gn 1624 ***diam. 4!0*** ***V = 10.4*** ***stars 14***
dimen. 5′ × 5′

15 cm reveals this nebula faintly as a diffuse glow lying mostly N of a mag. 12 star that has two fainter companions E and SE. With 25 cm the nebula remains fairly faint. The cluster is poor, containing only a dozen stars in an L-shaped asterism with the brightest star at the bend.

Phoenix

Occupying the region immediately south of Sculptor, this constellation holds a few galaxies easily accessible to northern observers. The center of the constellation culminates at midnight about 5 October.

eg 625 dimen. 3'.0 × 1'.3 B = 12.2 sfc. br. 13.6
This galaxy appears about 1'.5 across in 30 cm. The halo has an indefinite shape and shows no central concentration. A mag. 13.5 star lies 1'.4 N.

eg I5325 dimen. 2'.5 × 2'.5 B = 12.3 sfc. br. 14.2
30 cm shows this galaxy 1'.2 NE of the eastern of two mag. 11 stars that lie 5'.5 apart. The halo extends to about 1'.5 diameter, at times seemingly elongated toward the nearby star.

eg 7744 dimen. 2'.3 × 1'.8 B = 12.5 sfc. br. 13.9
In 30 cm this galaxy has a 1'.1 × 0'.6 halo elongated NE-SW. At 225 × it is moderately concentrated to a mottled core and a bright stellar nucleus.

eg 7764 dimen. 1'.5 × 1'.0 V = 12.3 sfc. br. 12.6
This is a faint and unconcentrated object for 30 cm. The halo reaches to about 1′ diameter and is slightly elongated N-S. At 225 × a few faint stellarings are visible N of center.

Pisces

This long constellation winds around the Square of Pegasus and contains many galaxies. The center culminates at midnight about 28 September.

eg 95 dimen. 1′.9 × 1′.5 V = 12.6 sfc. br. 13.5
Located 2′ SW of a mag. 11 star, this galaxy is a fairly faint object for 25 cm. It has a small inconspicuous core with a granular texture and a very faint stellar nucleus. The halo extends to 1′.25 × 0′.6 in pa 70°. 30 cm shows a circular glow about 50″ diameter with weak concentration to the center. The nucleus appears to have some stellarings associated with it.

eg 125 dimen. 2′.0 × 1′.8 V = 12.3 sfc. br. 13.5
Located 6′.4 WSW of eg 128, *q.v.*, this galaxy is visible in 30 cm 1′ N of a wide mag. 12 pair of stars with 20″ separation in pa 345°. The halo is about 50″ diameter and weakly concentrated to a nucleus that appears nearly stellar at low powers.

eg 128 dimen. 3′.4 × 1′.0 V = 11.6 sfc. br. 12.8
This is the brightest of a group of galaxies including eg 125, *q.v.*, 6′.4 WSW. In 25 cm it has a lenticular outline and a knotted core. With averted vision brighter radial arms seem to extend from the center. 30 cm shows the halo to 2′ × 0′.75 in pa 0°. It has moderate concentration to an unevenly bright core and nearly stellar nucleus. Photographs show eg 126 (mag.$_Z$15.5) 3′.6 SW, eg 127 (V = 14.0) 50″ NW, and eg 130 (V = 14.3) 1′ NE.

eg 467 dimen. 2′.4 × 2′.3 V = 11.9 sfc. br. 13.7
Located 11′ SW of eg 479, *q.v.*, and 3′ W of a mag. 8 star, this galaxy is moderately faint in 25 cm. The halo is circular, about 1′ across, and brightens to a core that shows a few stellarings at 200 ×. A mag. 13.5 star is visible just over 2′ SW; another slightly fainter star lies closer to the W.

eg 470 dimen. 3′.0 × 2′.0 V = 11.9 sfc. br. 13.7
This galaxy appears as a round, diffuse blob in 25 cm, about 1′.5 diameter. The core is brighter and occasionally shows a faint stellaring in its center. Two faint stars separated by about 40″ lie 2′.2 S, and another star is visible 1′.7 N. In 30 cm the smoothly textured halo is elongated in pa 150°, extending to about 2′ × 1′ and showing a moderate central concentration. A third faint star is visible 2′ SE. eg 474, *q.v.*, lies 5′.5 E.

eg 473 dimen. 2′.2 × 1′.4 B = 13.0 sfc. br. 14.2
This is a moderately faint galaxy in 25 cm. The halo is 1′ × 0′.5, elongated in pa 150°. With 30 cm the halo is well concentrated clear to a stellar nucleus.

eg 474 dimen. 7′.9 × 7′.2 V = 11.1 sfc. br. 15.3
In 25 cm this galaxy appears slightly brighter but a little smaller than eg 470, *cf.* 5′.5 W. It is diffuse and has a very small nucleus. With 30 cm it has a bright nonstellar nucleus, and a low surface brightness halo extending to 1′.2 diameter.

eg 488 dimen. 5′.2 × 4′.1 V = 10.3 sfc. br. 13.4
In 6 cm this galaxy is visible 9′ W of a mag. 8.5 star and just N of the E end of a string of three mag. 11.5 stars. 25 cm shows a bright circular haze about 1′.25 across rising to a bright, concentrated center. The nearest mag. 11.5 star is on the SSE side of the galaxy 1′.7 from the center. In 30 cm the halo extends to 3′ × 1′.8, elongated N-S, with a 1′ core and a stellar nucleus.

eg 514 dimen. 3′.5 × 2′.9 V = 11.9 sfc. br. 14.2
Located 3′ W of a mag. 9.5 star, 25 cm shows this galaxy as an indefinite, low surface brightness glow about 1′.5 diameter. A 20″ core stands out weakly, occasionally exhibiting a stellar nucleus. The halo is unevenly bright in 30 cm, and slightly concentrated to a very faint stellar nucleus. A mag. 13.5 star is visible 3′.5 SW.

eg 516 dimen. 1′.6 × 0′.6 mag.$_Z$14.3 sfc. br.
This faint companion to eg 524, *q.v.*, is visible in 30 cm about 10′ W of the bright galaxy. The low surface brightness halo is about 50″ diameter and shows weak concentration toward the center.

eg 520 dimen. 4′.8 × 2′.1 V = 11.2 sfc. br. 13.6
This interacting pair of galaxies is an irregular 1′.2 × 0′.5 glow in 25 cm, elongated in pa 135°. Its overall outline is squarish, composed of two brighter parts separated by a slight darkening; the NW lobe is brighter than the SE and has several bright spots within it. 30 cm shows the overall size as 3′ × 0′.8. The dark blotch in the core divides about two-thirds of the light into the NW portion. Sev-

eral faint stars are visible nearby, the two brightest lying 3.′5–4′ SW.

eg 524 ***dimen. 3.′2 × 3.′2*** ***V = 10.6*** ***sfc. br. 13.0***
This galaxy is just visible in 6 cm as a small concentrated spot 2.′5 N of a mag. 11 star. In 15 cm a second star is visible slightly closer to the N. The galaxy is about 1′ across and has a small intense core that fades rapidly toward the halo. In 30 cm the halo extends to about 1.′75 diameter, showing strong concentration to a nonstellar nucleus set in a large core. Occasionally the halo seems elongated slightly NNE-SSW. A third star, mag. 14, lies 1.′3 ESE. eg 532, *q.v.*, is located 18′ SSE, and eg 516, *q.v.*, lies 10′ W.

eg 532 ***dimen. 2.′8 × 1.′0*** ***mag.$_z$13.5*** ***sfc. br.***
Viewed in 30 cm, this faint galaxy lies 18′ SSE of eg 524, *q.v.* It has a very low surface brightness halo extending to 1′ × 0.′7 in pa 30°.

eg 628 ***dimen. 10′ × 9.′6*** ***V = 9.2*** ***sfc. br. 14.0***
Messier 74 is visible in 6 cm as a circular weakly condensed patch about 5′ diameter. 25 cm shows the halo to about 8′ diameter, slightly elongated E-W. The 1′ diameter core seems off-center and has a granular texture. Nine or ten stars lie about the periphery, mostly to the W. With 30 cm the irregular halo is unevenly bright with many stellarings and knots appearing almost unconnected by the faint background haze. The core is very broadly condensed, about 2.′5 across, and has at least two faint stars superposed on it. A particularly bright knot is visible in the center of a string of three stars 3.′5 E of center. Another fainter knot is visible between this string and the nucleus.

eg 660 ***dimen. 9.′1 × 4.′1*** ***V = 10.7*** ***sfc. br. 14.5***
In 15 cm this galaxy is a large, low surface brightness patch with weak concentration. The halo is about 8′ × 4′, elongated roughly NE-SW. The middle parts are unevenly bright, showing two or three faint stellarings at 75 ×, though none stands out prominently like a nucleus. A faint star is visible 1.′7 ESE of center. With 30 cm the diffuse halo extends to only 6′ × 2′ in pa 30°. The 2.′5 × 1′ core region grows broadly brighter toward the center and is distinctly uneven: a knotty streak runs unsymmetrically through it (not really bisecting it) with a prominent 5″ knot in it NE of center. Many other stellarings are visible in the core. The mag. 13.5 star to the ESE is an unequal pair with 16″ separation in pa 25°. Photographs show U1195 (mag.$_z$13.9) 22′ NNW.

eg 676 ***dimen. 4.′3 × 1.′5*** ***mag.$_z$10.5*** ***sfc. br.***
This galaxy is hardly distinguishable in 15 cm as a slight, smoothly textured haze extending N and S from a mag. 9.5 star. In 30 cm the bright star lies nearly perfectly centered on the galaxy, which is elongated in pa 175°. The 3.′5 × 1′ halo has a lenticular form, with tapering tips N and S. The surface brightness is highest around the N through E sides of the bright star. A very faint star is visible 25″ E of center; another is 1.′6 S, lying W of the maximum southern extent of the halo. Photographs show eg 693 (mag.$_z$13.5) 27′ NE, and eg 706 (mag.$_z$13.2) 45′ NE.

eg 718 ***dimen. 2.′8 × 2.′5*** ***V = 11.7*** ***sfc. br. 13.7***
This is a fairly faint object for 25 cm. The halo is only 1′ diameter, but contains a conspicuous stellar nucleus that seems off-center to the N. With 30 cm the halo extends to 1.′5 × 0.′8 in pa 45°. The relatively bright core is circular, about 40″ across, and contains a bright stellar nucleus that seems almost multiple at 150 ×. At 250 × the nucleus lies distinctly N of center.

eg 741 ***dimen. 3.′2 × 3.′2*** ***V = 11.3*** ***sfc. br. 13.8***
eg 742 ***dimen. 0.′2 × 0.′2*** ***mag.$_z$14.8*** ***sfc. br.***
eg 741 is a small, round object in 25 cm, with a stellar nucleus like a mag. 13 star and a 45″ diameter halo. 30 cm shows a well-concentrated nearly circular glow within which a stellar nucleus is occasionally visible. The halo is about 1′ diameter, slightly elongated E-W, with a 20″ core. A mag. 11 star is visible 2.′5 NW. eg 742 lies 50″ in pa 100° and is visible in 30 cm as a mag. 14 stellar spot. eg 741 is the brightest of a group.

eg 7537 ***dimen. 2.′3 × 0.′7*** ***V = 13.2*** ***sfc. br. 13.5***
Located 3.′1 SW of eg 7541, *q.v.*, in 25 cm this galaxy is elongated in pa 80°, 1.′2 × 0.′25. It is faint, but has an edge-on appearance with a slightly brighter central bulge. 30 cm shows it only 45″ × 30″ with a 10″ core and an occasionally stellar nucleus.

eg 7541 ***dimen. 3.′5 × 1.′4*** ***V = 11.7*** ***sfc. br. 13.3***
A moderately bright object for 25 cm, this galaxy shows a 3′ × 0.′75 halo elongated in pa 100°. The brighter elongated core has an occasional stellar flash in it. A mag. 12 star lies beyond the E tip, 2.′1 from center. With 30 cm the halo is about 2.′5 × 1.′ The core is unevenly bright, but without a distinct nucleus. The S flank of the bar seems more sharply defined.

eg I5309 ***dimen. 1.′6 × 0.′7*** ***mag.$_z$15.0*** ***sfc. br.***
This member of the Pegasus I cluster is faintly visible in 30 cm about 7′ WNW of eg 7611, *q.v.*, and 7′ S of three mag. 8–10 stars. The halo is 40″ × 20″ in pa 15°, with almost no central brightening. A mag. 13.5 star is involved on the S side about 20″ from center.

eg 7611 ***dimen. 1′.5 × 0′.7*** ***V = 12.6*** ***sfc. br. 12.5***
In 30 cm this member of the Pegasus I cluster has the most conspicuous nucleus of the group. The galaxy lies 5′.5 NW of a mag. 7 star. The halo extends to 55″ × 20″ in pa 135°, but shows poor concentration except for the prominent stellar nucleus. The core, though it doesn't stand out well, is clearly more circular in outline than the halo. Two faint stars are nearby, one 1′.6 NW, the other a little closer NNW. eg I5309, *q.v.*, lies 7′ WNW.

eg 7617 ***dimen. 1′.0 × 0′.8*** ***V = 13.8*** ***sfc. br. 13.4***
Located 2′.7 in pa 210° from eg 7619, *q.v.* in Pegasus, 30 cm will show this galaxy as a 20″ × 10″ glow elongated in pa 25°. The galaxy is quite faint overall, but has a relatively bright stellar nucleus.

eg 7679 ***dimen. 1′.9 × 1′.3*** ***V = 12.7*** ***sfc. br. 13.5***
eg 7682 ***dimen. 1′.2 × 1′.0*** ***V = 13.4*** ***sfc. br. 13.4***
eg 7679 is visible in 30 cm about 5′.5 SE of a mag. 9 star (ADS 16777: 10.0,10.1; 1″.3; 48°). At 250× the halo is about 45″ diameter, with very faint wisps extending to perhaps 1′ × 0′.75, elongated roughly E-W. A faint stellar nucleus is occasionally visible at the center, and at times a very faint star or stellaring sparkles just SW of it. Two mag. 11.5 stars lie about 2′ WNW and NNE; the latter is a little fainter and closer. eg 7682, 4′.5 ENE, is just visible in 30 cm as a faint circular glow without detail.

eg 7716 ***dimen. 2′.3 × 1′.9*** ***V = 12.2*** ***sfc. br. 13.7***
Located 2′ N of a mag. 9 star, this galaxy is about 2′ across in 25 cm, with a small slightly brighter core. 30 cm shows the halo to only 1′ diameter, with very weak concentration to an unevenly bright core and a nearly stellar nucleus.

eg 7778 ***dimen. 1′.4 × 1′.3*** ***mag.$_z$13.8*** ***sfc. br.***
eg 7779 ***dimen. 1′.6 × 1′.3*** ***mag.$_z$13.6*** ***sfc. br.***
30 cm shows this pair of galaxies lying 9′.5 SW of eg 7782, *q.v.* The two galaxies are 1′.8 apart, aligned nearly E-W. The eastern member, eg 7779, is brighter than its companion in 30 cm, and about 1′ diameter. eg 7778 is about 50″ diameter and has a brighter substellar nucleus. A mag. 12.5 star is visible 1′.7 NNW, and a mag. 14 star lies 1′.7 S. Photographs show eg 7781 (mag.$_z$15.2) 4′.9 from −79 in pa 100°.

eg 7782 ***dimen. 2′.4 × 1′.4*** ***B = 13.1*** ***sfc. br. 14.3***
This galaxy is faint in 25 cm, appearing as an elongated glow with rounded ends aligned N-S. With 30 cm the low surface brightness halo extends to 2′ × 0′.5 in pa 175°. It brightens only slightly to the center, where a faint stellaring shows as the nucleus. A mag. 14 star is visible 3′ SSW; eg 7778 and 7779, *q.v.*, lie about 9′.5 SW.

eg 7785 ***dimen. 2′.3 × 1′.4*** ***V = 11.6*** ***sfc. br. 12.8***
25 cm will show this galaxy easily 4′.3 E and a little S of a mag. 8 star, and about 3′ NW of two mag. 9 stars. The halo extends to 1′.2 × 0′.5 in pa 145°. Relative to the halo, the core is small and concentrated. A mag. 14 star is visible 1′ N. With 30 cm the nucleus stands out fairly prominently, but is clearly nonstellar.

Pisces Austrinus

Led by the bright star Fomalhaut, this constellation contains many galaxies, most of which are very faint. The center of the constellation culminates at midnight about 26 August.

eg 7135 ***dimen. 2.′8 × 2.′0*** ***V = 11.7*** ***sfc. br. 13.4***
This galaxy is located just SE of a 4′ triangle of mag. 10–11 stars that points S. It is not easily visible in 15 cm; 25 cm shows a 2′ × 0.′75 halo elongated E-W with a starry nucleus. In 30 cm the nucleus is a faint sparkle within a 25″ core and a diffuse 1.′3 halo. At high power the inner portions are elongated E-W. eg I5135 (V = 12.3) lies 18′ WSW; eg I5131 (V = 12.4) lies 29′ W.

eg 7172 ***dimen. 2.′2 × 1.′3*** ***V = 11.9*** ***sfc. br. 12.9***
With 25 cm this galaxy, the northernmost of a string of four galaxies including eg 7173, 7174, and 7176, *q.v.*, forms a diffuse 2′ × 1′ glow elongated in pa 90° without central brightening. In 30 cm the diffuse halo is about 1.′3 × 1′, showing a little concentration and occasionally a faint stellar nucleus. A mag. 11 star lies 2.′7 SE; 1.′5 SW is a mag. 13 star. Photographs show eg 7163 (V = 13.4) 30′ W.

eg 7173 ***dimen. 1.′3 × 1.′1*** ***V = 12.1*** ***sfc. br. 12.4***
This galaxy, lying 1.′3 NW of the interacting pair eg 7174/76, *q.v.*, is a conspicuous round spot in 25 cm. The halo is about 1′ diameter and well concentrated, though without a sharp nucleus. The nucleus is inconspicuous in 30 cm, the brightness fading evenly from the center into the circular halo.

eg 7174 ***dimen. 1.′3 × 0.′7*** ***V = 12.6*** ***sfc. br. 12.3***
eg 7176 ***dimen. 1.′3 × 1.′3*** ***V = 11.9*** ***sfc. br. 12.4***
15 cm shows this interacting pair of galaxies as a single circular spot with a stellar nucleus. With 25 cm the two galaxies are still not separated, showing only an irregular 1.′2 halo and a bright stellar nucleus. The pair is distinguishable in 30 cm: at first inspection they appear as a single object with elongated inner regions and an off-center stellar nucleus. At 250 × the core separates into two condensations aligned NE-SW. The northeastern component, eg 7176, is sharply concentrated to a bright stellar nucleus. The southwestern member, eg 7174, is broader, without a nucleus, but with a smooth 20″ core that is elongated E-W. The area between the two galaxies is darker, but not so dark as the sky. Overall the enveloping halo extends to 1.′25 × 0.′6, elongated NE-SW. This pair of galaxies is the southernmost of a string including eg 7173, *q.v.* 1.′3 NW, and eg 7172, *q.v.* 7.′3 N.

eg 7314 ***dimen. 4.′6 × 2.′3*** ***V = 10.9*** ***sfc. br. 13.3***
Though this is the brightest galaxy in the constellation, it is a faint object for 15 cm, showing only a diffuse, unconcentrated 2′ × 1′ glow elongated N-S. In 25 cm the halo grows to 3′ × 1.′2 in pa 0° with broad concentration across the mottled center. A mag. 12 star is visible 2′ to the W. 30 cm shows an irregular oval with slightly pointed ends. The core is mottled and seems tilted somewhat with respect to the axis of the halo: the halo is in pa 0°, the core in pa 10°. No nuclear condensation is visible. A threshold magnitude star lies on the E side 1′ from the center.

eg 7361 ***dimen. 3.′5 × 1.′0*** ***V = 12.5*** ***sfc. br. 13.8***
Faintly visible in 25 cm 5.′5 E of a mag. 8 star, this galaxy appears as a diffuse streak elongated N-S with a grainy texture. Two faint stars are visible about 2.′8 S and WSW. 30 cm shows a thin 2.′5 × 1′ spindle elongated in pa 10°. The halo shows weak, broad concentration to more strongly elongated inner regions. A threshold magnitude star is visible 3′ E.

eg I5271 ***dimen. 2.′3 × 0.′8*** ***mag.$_H$12.6*** ***sfc. br.***
A faint object for 25 cm, this galaxy appears roughly circular, about 1.′5 diameter, with little concentration toward the center. A mag. 12 star is visible 2.′3 N. The halo is elongated in pa 130° in 30 cm, extending to about 2′ × 1.′ The broadly concentrated 1′ core is nearly circular and occasionally shows a faint stellaring at, or just E of, the center.

Puppis

This rich constellation lies in the southern winter sky for northern observers. Puppis holds examples of every type of object, though open clusters dominate and galaxies are largely obscured by the Milky Way. The center of the constellation culminates at midnight about 9 January.

gc 2298 ***diam. 6′.8*** ***V=9.2*** ***class VI***

In 6 cm this cluster is a small, moderately faint glow with moderate, broad concentration. 15 cm shows a few faint stars on the periphery, but otherwise the cluster is only granular. A mag. 12.5 star lies near the NNE edge. With 25 cm it lies in a rich Milky Way field and within a triangle of faint stars. The cluster appears about 2′ diameter and shows only mild concentration toward the center. At 200× it has a granular texture, and many stars are resolved at the threshold of visibility around the irregular border. The star on the NNE is a pair in 30 cm (V=12.5,14.1; 8″.5; 115°). The halo extends to 3′.5 diameter, reaching two-thirds the way to a mag. 11 star 2′.8 SSE. The globular is partially resolved at 225× into about 30 stars; at the very center is a faint stellaring.

oc 2401 ***diam. 2′.0*** ***mag. 12.6p*** ***stars 20***

This tiny cluster looks like a typical mag. 13 galaxy in 15 cm at 75×: the nebulous spot is about 2′.5 across and lies N and a little E of two mag. 10 stars. Four or five mag. 13 stars are resolved in 25 cm in a 1′.5 area over a faint irregularly shaped background haze. A triangular form is evident in 30 cm, created by a mag. 13 star on the W and a mag. 13 pair (15″; 0°) off to the N side of the well-defined cluster. About a dozen other stars are resolved at 225× in a 1′.5×1′ area elongated E-W.

oc Ruprecht 24 ***diam. 8′*** ***mag.*** ***stars 15***

In 25 cm this cluster is a conspicuous grouping with fairly bright stars at 50×. In a 12′ area are about 30 mag. 11.5 and fainter stars, showing some concentration toward the center. Three mag. 10 stars form a 5′.5×2′.5 triangle over the cluster pointing to the W.

oc 2414 ***diam. 4′.0*** ***V=7.9*** ***stars 27***

Grouped around a mag. 8 star, 15 cm shows about eight stars of mag. 12 and fainter in this small cluster. Most of the stars form a U-shaped asterism opening NNW on the NW side of the lucida. Among a total of 15 stars in a 4′ area, 25 cm resolves a faint pair 30″ S of the lucida (V=12.4,12.5; 6″; 135°). In 30 cm about 15 stars are visible in the irregular 2′ cluster. The U-shaped asterism has nine or ten stars, including the three brightest other than the mag. 8 star.

oc 2421 ***diam. 8′*** ***V=8.3*** ***stars 28***

This cluster is faintly visible in 6 cm, lying 10′ SE of a wide mag. 8.5 pair of stars. The irresolved patch is about 10′ across and shows perhaps a half dozen stars at 25×. About 35 stars are resolved by 15 cm in a 10′ area, while 25 cm will show 45 stars, including some pairs. With 30 cm at least 40 stars are visible in a triangular form about 6′ on a side, with a blunt northern tip; the SE and SW apices are represented by fairly isolated mag. 10.5 stars. A wide, nearly equal pair lies near the center (V=10.9,11.5; 18″; 355°), resolvable without difficulty in 15 cm; 30 cm shows two mag. 13 companions to these.

oc 2422 ***diam. 25′*** ***V=4.4*** ***stars 117***

Messier 47 is a bright naked-eye cluster and a pretty sight in 6 cm, which shows about 30 stars irregularly grouped around half a dozen members. The brightest star (ADS 6208AC: V=5.7,9.7; 20″; 41°) is in the W side, and an arm of fainter stars trails N to oc 2423, *q.v.* In a 30′ area 15 cm shows 50 stars, including three pairs and a few lightly colored stars. With 25 cm about 80 members are visible. Near the center of the cluster, on the S side of a group of bright stars, is a nice pair (ADS 6216: V=7.0,7.3; 7″.4; 305°).

oc 2423 ***diam. 12′*** ***V=6.7*** ***stars 86***

In 6 cm this is a fairly rich cluster with about two dozen stars in a 15′ area. A mag. 9 star dominates N of center (ADS 6223: 9.6,10.2; 0″.5; 159°). It is ill-defined from the Milky Way background in 25 cm, showing 30 stars in a 20′ area. About 60 stars are visible in 30 cm, including several pairs and a few faintly red stars. The main part of the cluster is 13′ across, but an extension to 20′ NE of center includes another 15 stars.

oc Ruprecht 26 ***diam. 4′.0*** ***mag.*** ***stars 20***

This is a moderately faint cluster in 25 cm. About 25 members are visible in a slightly hazy 8′ area, showing some weak central concentration. A mag. 9 star is on the

E side; N of this star and the cluster is a loose group of mag. 10 stars apparently not related to the cluster.

oc Melotte 71 ***diam. 8′*** ***V=7.1*** ***stars 216***
This cluster is a fairly high surface brightness patch in 6 cm, showing only a handful of stars in the E side at 50×. A south-pointing triangle of mag. 10 stars stands out on the SW edge. 25 cm shows about 50 stars of mag. 11 and fainter in a circular, moderately concentrated group. The NE apex of the triangle to the SW is a close pair (10.5,10.5; 9″; 110°). 30 cm reveals about 60 stars in an 8′ area including a small knot in the center. Two brighter stars stand out SE of center.

oc 2425 ***diam. 5′*** ***mag.*** ***stars 30***
This is a small, fairly faint cluster in 25 cm, with 20 stars spread over a 5′×2′ area elongated roughly E-W. A crossbar of four or five stars caps the E end, aligned NE-SW.

oc 2432 ***diam. 7′*** ***mag. 10.2p*** ***stars 50***
In 15 cm this cluster is visible as a conspicuous string of about a dozen mag. 12–13 stars elongated N-S. 25 cm shows most of the brighter stars in a slightly curved 5′×1′ group, opening toward the W. About 30 stars are visible in and around the string, including some nice pairs.

oc 2437 ***diam. 20′*** ***V=6.1*** ***stars 186***
pn 2438 ***dimen. 73″×68″*** ***nebula 11.0v*** ***star V=17.8***
Messier 46, a faint naked-eye object, is a rich patch of faint stars in 6 cm. The stars are uniformly distributed, showing no distinct central concentration. The brightest star (V=8.7) shines out on the W side. With 15 cm about 75 stars are visible in a 25′ area over a bright background haze. Outliers spread more to the E and W in 30 cm, giving an overall size of 30′×25′. The stars are remarkably uniform in distribution except for a 2′ diameter void near the center that has only a few faint stars in it. The embedded planetary, pn 2438, is easily visible in 6 cm at 50× and 100× in the N side of the cluster and just NW of a mag. 11 star. In 25 cm a mag. 13 star is visible near the center of the nebula, though this is not the central star. The nebula itself is about 1′ diameter. A 2.′5 long string of mag. 12 and fainter stars runs roughly E-W about 1.′5 S of the nebula. With 30 cm a second very faint star lies about 15″ SW of the brighter star within the nebula. At times the nebula seems slightly elongated SE-NW; the acute ends are a bit fainter than the NE and SW sides. The annularity of the nebula is subtle, showing little contrast at all magnifications.

oc 2439 ***diam. 9′*** ***V=6.9*** ***stars 181***
This bright cluster is a conspicuous partially resolved haze in 6 cm on the SW side of a bright reddish star (R Puppis, V=6.7). It is well resolved in 15 cm, showing 30 stars in an 8′ area; R Pup lies N and a little E of the center. A couple of pairs stand out, though with 30 cm they are too wide to be interesting. Here a total of 60 members is visible, evenly scattered within 8′ diameter.

pn 2440 ***dimen. 74″×42″*** ***nebula 9.4v*** ***star V=14.3:***
This planetary is visible in 6 cm as a star-like object 7′ NW of a mag. 9 triplet and 3′ W of a star of similar

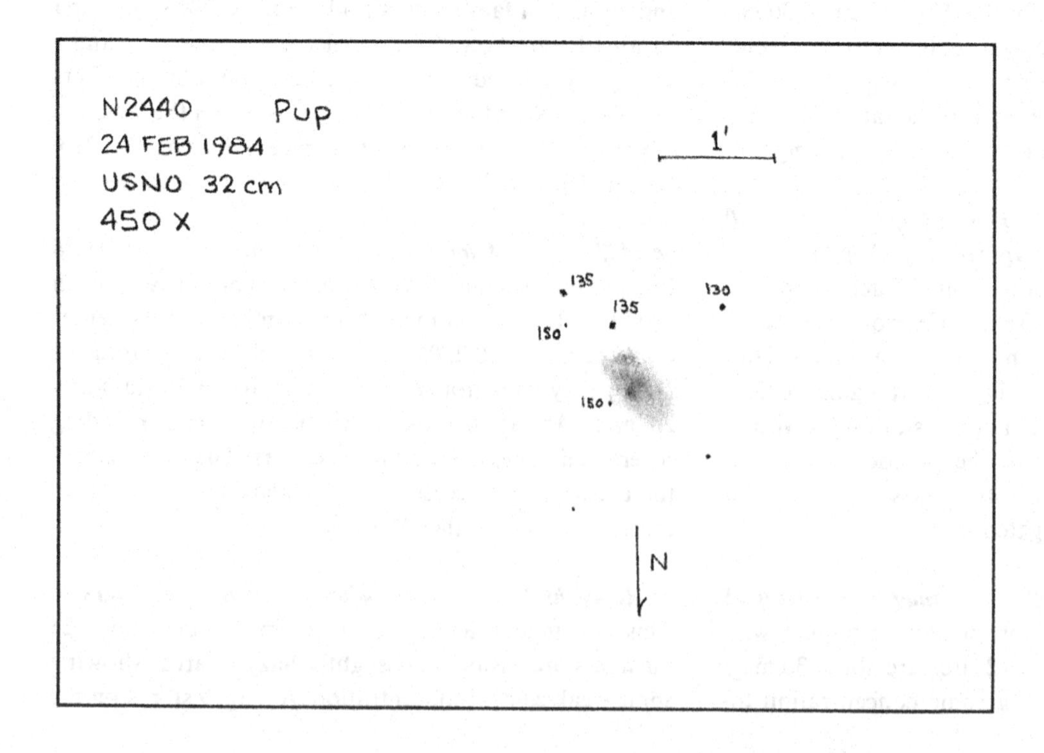

brightness to the nebula. In 25 cm the nebula has a 30″ × 20″ lenticular form elongated in pa 40°. The brighter central regions are more circular, 15″–20″ across and uniformly bright. The central star is barely discernable at 200 ×. With 30 cm the nebulosity extends to 45″ × 20″. At 450 × the central part has bright spots on opposite ends of the minor axis (pa 130°). The nebula is encircled by very faint stars: two on the S, one each to the SE, NE, and N, all within 1′. The triplet SE actually has four resolvable components: ADS 6309AB: 9.4,9.9; 1″.8; 291°, ADS 6309AC: 9.4,10.3; 25″; 2°, and ADS 6309AE: 9.4,11.1; 22″; 334°.

oc 2447 diam. 10′ V = 6.2: stars 80
Messier 93 is a sparkling sight in 6 cm, which shows about two dozen stars, the two brightest standing out on the SW. 15 cm reveals about 30 stars in a 10′ area. A group in the center includes three lines of stars arranged like trident spikes pointing NW. With 25 cm the cluster members show some concentration to the center, where two or three groups with brighter stars lie. The cluster fills a 23′ field in 30 cm with 110 stars, 50 of them in the central 4′, and the outliers spreading mostly W and S. The trident asterism is not so obvious here, but eight of the brighter stars form a "V" pointing SW.

oc 2451 diam. 50′ V = 2.8 stars 153
Surrounding c Puppis (V = 3.6), this cluster appears as a loose, bright aggregation in 6 cm. About two dozen stars can be seen, including the very red c Pup (B-V = 1.7). At least 30 mag. 6–10 stars are visible in a 45′ area with 15 cm, including many blue and yellow stars among the brighter members. 30 cm shows 50 stars in a 40′ area with a sprinkling of mag. 11–12 stars filling the spaces between the bright stars.

pn 2452 dimen. 31″ × 24″ nebula 12.2v star 19:p
This nebula is visible as a concentrated mag. 10.5 spot about 20″ diameter in 15 cm. It is clearly annular in 30 cm at 225 ×. Brighter elongated lobes aligned approximately E-W appear on the N and S sides; the E and W sides are dimmer. The northern lobe has a stellaring in it, but no central star is visible within the annulus. A mag. 13 star lies about 50″ SW, and a mag. 14 star is 20″ N. oc 2453, *q.v.*, lies 8′ N.

oc 2453 diam. 4′.8 V = 8.3 stars 76
In 6 cm this cluster appears as a 3′ diameter concentrated glow with a few sparkles in the center. It is small and fairly compact in 25 cm, typical of many of the clusters in this part of the sky. About 15 stars, mostly mag. 12 and fainter, are resolved in a 1′.5 area. 30 cm shows a central kernel of about ten stars packed into a 1′ × 0′.75 area elongated N-S. A total of about 20 stars is visible within 2′.5 diameter, including some brighter isolated outliers. A mag. 9.5 star lies on the NW side; a mag. 10 star lies farther S.

oc 2455 diam. 15′ mag. 10.2p stars 50
This is a small and faint cluster in 6 cm that lies on the W side of a mag. 8.5 star. Overall it is a little smaller than oc 2421, *cf.*, but much fainter. It appears 7′ diameter in 25 cm with about 20 stars of mag. 12 and fainter. An elongated group of 15 mag. 11–12 stars lies 10′ NE. 30 cm

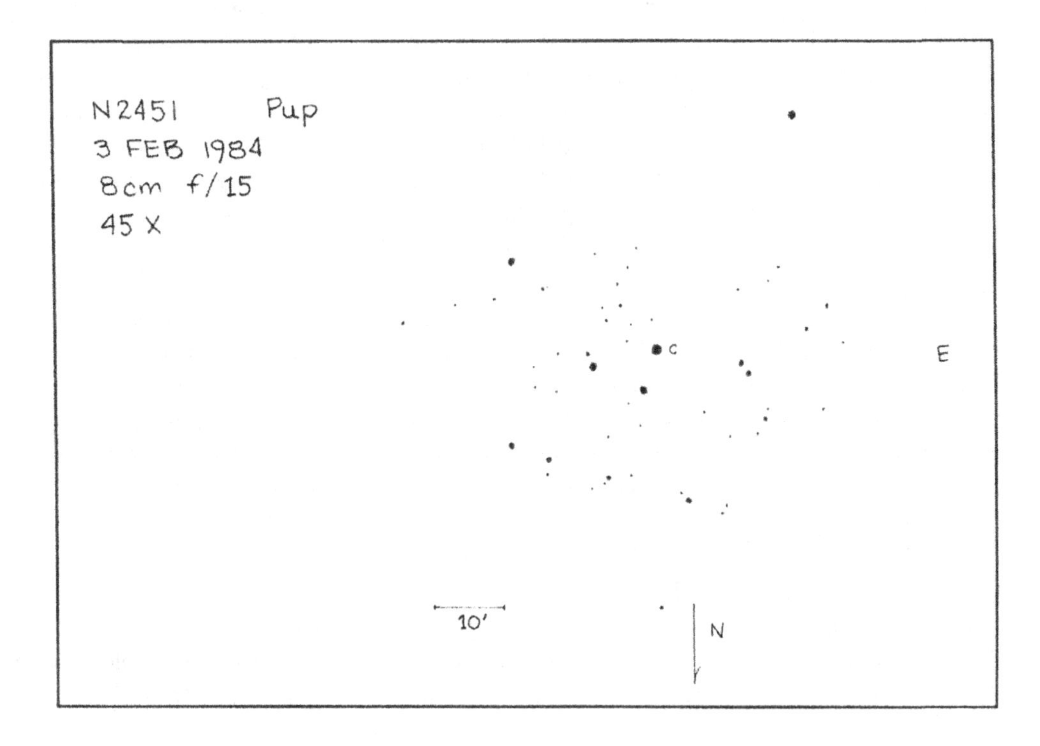

resolves more than 25 stars in the unconcentrated cluster. The string NE shows about 20 stars here in an 8′ × 1′ area elongated E-W.

oc Haffner 16 ***diam. 5′*** ***V = 10.0*** ***stars 15***

This cluster is visible at 100 × in 25 cm, but is better viewed at high power. About ten stars are visible over some haze in a 3′ diameter area. A few brighter stars give it a triangular appearance, although the bulk of the cluster is roughly circular.

oc-gn 2467 ***diam. 15′*** ***mag. 7.1p*** ***stars 50***
dimen. 8′ × 7′ ***star V = 7.8***
oc Haffner 18ab ***diam. 2.′0*** ***V = 9.3*** ***stars 25***
oc Haffner 19 ***diam. 3.′2*** ***V = 9.4*** ***stars 39***

In 15 cm at 75 × this region is dominated by a circular nebula with a mag. 8 star N of its center. The nebula is about 5′ diameter, bounded on the SE by a mag. 11 star and on the NNW by a curved string of three mag. 11.5 stars. A slightly dimmer bar intrudes into the nebula from the SE, ending just S of the bright star. There is a slight increase in the density of stars around the nebula due to oc Haffner 18ab/19, but these are not conspicuous as clusters in this aperture. The nebula is not quite as big in 30 cm at high power as in 15 cm because the portion N of the illuminating star is all but invisible. The dark bar extends nearly all the way through the nebula. The southern periphery has a broadly brighter rim. About a dozen stars are visible within the boundary of the object. oc Haffner 18ab is an elongated group 4′ NE of the bright star in the nebula. With 30 cm fifteen stars are in the string, which is aligned SE-NW, and they get generally brighter toward the NW. In the line near a mag. 12 star at the SE end are three mag. 14.5 stars. oc Haffner 19 is a tiny clump of faint stars 8.′5 NE of the nebula, at the N side of the starry region. The cluster contains a dominant central pair (Barton 2926: V = 10.8,12.6; 4.″4; 106°) around which are ten faint stars in a 1′ clump. A 3′ diameter region centered on the clump is faintly nebulous due either to other very faint stars or genuine nebulosity.

oc 2477 ***diam. 43′*** ***V = 5.8*** ***stars 1911***

Lying N of b Puppis (V = 4.5), 6 cm shows this cluster as an unresolved haze with slight concentration to a core elongated in pa 40°. A faint unequal pair is visible in the N side (V = 9.7,10.7; 35″; 165°). The cluster looks like a loose globular cluster in 15 cm, showing at least 100 stars of mag. 10.5 and fainter at 75 × . A string of brighter stars leads out of the cluster NE. In 25 cm the cluster is distinctly oval, reaching to about 30′ × 20′, elongated NE-SW, and contains many star chains and dark gaps. It is a very rich assemblage in 30 cm, which will show too many stars to count. The brighter part of the cluster is 15′ across, containing perhaps 250 stars, with stragglers spreading to 25′ diameter. Knots and larger lumps of stars together with maze-like dark passages are visible in the S half of the core, and several slightly brighter stars overlie the center and NE side of the cluster, giving the central regions their apparent NE-SW elongation.

oc 2479 ***diam. 11′*** ***mag. 9.6p*** ***stars 45***

This cluster is barely visible in 6 cm as a 10′ diameter occasionally granular patch 12′ ENE of a mag. 9 pair (8.9,9.1; 20″; 230°). The cluster members are mostly fainter than mag. 12 in 25 cm, which resolves about 40 of them in an 8′ area showing some concentration toward the center. A dark patch cuts into the center from the E side, giving the cluster an arced form opening toward the ESE. 30 cm reveals 50 stars in a loose group about 10′ diameter. The northern arm of the arc is composed of six brighter stars, and leads away E from the NE side of the main clustering.

oc 2482 ***diam. 10′*** ***V = 7.3*** ***stars 39***

This is a fairly rich but faint cluster in 6 cm, showing about 15 mag. 10–11.5 stars over a faint, broadly concentrated haze. A brighter star (V = 8.0) stands out on the W side, and a thin triangle of mag. 8 stars lies just NE. 15 cm will show about 30 stars in 20′ diameter, while 25 cm shows 50 stars within a 12′–15′ area, including many wide pairs. With 30 cm the cluster has a roughly triangular outline 20′ across. About 70 stars are visible with the densest part on the E side.

oc Trumpler 9 ***diam. 6′*** ***V = 8.7:*** ***stars 19***

In 6 cm this cluster forms a conspicuous haze just SW of a mag. 8.5 star, and bracketed on the W side by a N-S string of mag. 7.5–9.5 stars. Two or three faint stars sparkle within at 25 × . 25 cm will show about 15 stars in an irregular group, while to 30 cm 16 stars are visible in an 8′ area elongated SE-NW.

oc 2489 ***diam. 6′*** ***V = 7.9*** ***stars 112***

Located 13′ N of a red mag. 6 star, 6 cm shows this cluster as a broadly concentrated granular glow about 4′ across. About 30 stars can be resolved by 25 cm. The cluster is circular, about 5′ diameter, and is somewhat denser in the center. Two zones of different concentration stand out in 30 cm: a distinct core 4′ across is surrounded by outliers spreading to 8′ diameter for a total of 50 stars.

oc 2509 ***diam. 12′*** ***mag. 9.3p*** ***stars 70***

This cluster is visible in 6 cm as a faint patch 6′ NW of a mag. 8 star. In a 6′ area about ten stars are visible, including several on the SE side and a prominent mag. 10 star on the S side of a small unresolved clump on the N. 25 cm shows a compact irregular cluster with about 50 stars within 6′ diameter. 30 cm shows 45 stars in the 6′ area. Most of the 6′ diameter is due to a largely empty ring of about eight mag. 10–12 stars. There is a small 1.′5 diameter knot on the NNW side of this ring that contains about 15 stars and some unresolved haze. A bright unequal pair stands out to the NE (10.5,12; 25″; 110°), and the easternmost member of the ring is a closer, fainter double (12.5,13; 13″; 195°).

eg 2517 ***dimen. 0.′8 × 0.′6*** ***V = 12.3*** ***sfc. br. 11.3***
Though this galaxy is visible in 15 cm, it is hardly discernable from a star at 75 × . It lies between a mag. 11.5 star 1′ NW and a wide mag. 12 pair 1.′1 S (12,12.5; 25″; 95°). In 30 cm it is a small and faint but strongly concentrated spot about 35″ diameter. The center is very bright and usually appears nonstellar, though a stellar nucleus would be difficult to distinguish in the bright inner core.

eg 2525 ***dimen. 2.′9 × 2.′0*** ***V = 11.6*** ***sfc. br. 13.4***
This galaxy is a faint diaphanous glow in 15 cm between two mag. 8 stars. In 25 cm it is a low surface brightness glow about 3′ diameter without central brightening, occasionally showing an indistinct NE-SW elongation. The faint halo extends to 3′ × 1.′5 in 30 cm, elongated in pa 70°, with weak broad concentration to an unevenly bright 1.′25 core. A mag. 13 star is at the ESE edge of the halo, and two similarly bright stars about 20″ apart are on the WSW side aligned pointing toward the center of the galaxy; the nearest lies just outside the halo.

oc 2527 ***diam. 16′*** ***V = 6.5*** ***stars 45***
In 6 cm this object is an unconcentrated cluster of about 20 stars without haze. 25 cm shows a loose group of stars scattered in a 25′ diameter. Many relatively bright stars lie about in a 60′ area around this, but the cluster proper is probably only the central condensation. 30 cm reveals 50 stars in the central 12′ area including a dozen brighter members. Many more seem associated out to a total diameter of 25.′ A tight group elongated NNE-SSW, containing a dozen mag. 11–12 stars, is visible on the E side.

oc 2533 ***diam. 10′*** ***V = 7.6*** ***stars 124***
In 25 cm this cluster contains about 11 stars within 3′ diameter surrounding a mag. 9 star. In 30 cm the mag. 9 star is near the SW apex of the roughly triangular cluster; the second brightest star, mag. 11, is SW of this, forming the apex. With only 15 other stars visible, the cluster does not stand out well against the Milky Way.

oc 2539 ***diam. 15′*** ***V = 6.5*** ***stars 59***
With 6 cm this cluster appears as an irregularly concentrated group of faint stars NW of 19 Puppis (V = 4.7). In 20′ diameter 25 stars are visible in two or three large condensations. The brightest star is a pair (V = 10.0,11.1; 23″; 200°) on the N side. The cluster appears similar to Messier 46, *q.v.* oc 2437, in 15 cm, showing 100 stars spreading to almost 40′ diameter. Within a 30′ area 25 cm shows 120 stars. At least 130 stars are visible in 30 cm within 25′ diameter, including several other pairs and tight groups.

oc 2546 ***diam. 70′*** ***V = 6.3*** ***stars 40***
This is a fairly rich cluster for 6 cm, splayed out in a 50′ × 25′ area elongated SSE-NNW. About 50 stars are visible, with several brighter ones on the NW end and a bright star on the SE end. Two wide pairs are conspicuous NW of center. The cluster is overwhelmingly large in 30 cm, with perhaps 80 stars countable in a 45′ diameter. The two pairs in 6 cm are part of the densest group, including 20 stars in an 8′ × 2′ area elongated in pa 30°.

eg I2311 ***dimen. 1.′3 × 1.′3*** ***mag.$_{vv}$14*** ***sfc. br.***
Lying in a rich field, this galaxy is distinctly brighter in 15 cm than eg 2566, *q.v.* 8′ S, though it is quite small, appearing only as a stellar point at 75 × . 30 cm shows a high surface brightness spot 30″ across with strong even concentration to a mag. 13 stellar nucleus embedded in a bright core.

eg 2566 ***dimen. 3.′1 × 2.′0*** ***mag.$_{vv}$13*** ***sfc. br.***
This galaxy is fainter than eg I2311, *q.v.*, in 15 cm. It appears as a small diffuse spot amongst a rich field of very faint stars. With 30 cm it is the larger of the two galaxies. The diffuse halo is 1′ × 0.′8, elongated approximately E-W. There is no central concentration except for a tiny 5″ core that seems like a superposed star; it is brighter than a mag. 13.5 field star 22″ from the center in pa 100°.

oc 2567 ***diam. 11′*** ***V = 7.4*** ***stars 117***
This cluster is faint in 6 cm, a conspicuously elongated group of faint stars 5′ NE of a mag. 9 star. The brightest star in the cluster stands out on the SW side; averted vision will show many faint members NE of this star. The group extends to 10′ × 5′ in 25 cm, encompassing 35 stars. Running nearly N-S through the center are six stars in a line with a seventh star missing from its center. On the W is a V-shaped asterism pointing S, the tip of which is a pair (V = 11.1,11.2; 13″; 70°). 30 cm shows 50 stars in a 12′ × 8′ area. Most of the brighter stars are

concentrated in the "V" and N-S string; elongation in the cluster is expressed mostly by these stars, rather than by the fainter outliers.

oc 2571 ***diam. 7′*** ***V=7.0*** ***stars 49***
6 cm shows this cluster as a small hazy area with a half-dozen faint stars around a wide mag. 9 pair. About 22 stars are resolved by 25 cm, loosely spread in a 15′ × 10′ area. With 30 cm 25 stars are visible in a 13′ × 5′ area elongated SE-NW. The mag. 9 stars in the center are 1′ apart, and the southeastern star has a mag. 12 companion 16″ SE.

oc 2580 ***diam. 8′*** ***mag. 9.7p*** ***stars 50***
This cluster is a small faint spot in 6 cm all but swamped by a very rich and bright background. About 17 stars are visible, mag. 10 and fainter, in a 7′ cluster. The brightest stars form an "X" across the group.

oc 2587 ***diam. 10′*** ***mag. 9.2p*** ***stars 40***
6 cm will just show this cluster as a faint hazy patch with a few faint stars on the W side of a mag. 9 star. 25 cm shows a pretty group of about 30 stars in 5′ diameter plus a few stragglers trailing N. With 30 cm the stars trailing N give the cluster a sort of teardrop shape.

Pyxis

Located at the edge of the winter Milky Way, galactic and extragalactic objects intermingle in this constellation. Of particular interest is the open cluster oc 2818, which contains a planetary nebula. The center of the constellation culminates at midnight about 4 February.

eg 2613 dimen. 7.′2 × 2.′1 V = 10.4 sfc. br. 13.3
This galaxy is a moderately faint spindle in 15 cm, elongated in pa 115°. The brighter center is quite small, and three mag. 12–13 stars are visible N, S, and W. With 25 cm the galaxy lies in a starry field and reaches an overall size of 5′ × 2.′ The oval 1.′5 × 1′ core shows a few mottlings at 125 ×. The long halo extends more to the E than W in 30 cm, with a total size of 4.′5 × 0.′8. The 30″ × 15″ core contains a very faint stellar nucleus. A mag. 14 star is visible 27″ NE of the center near the edge of the halo, and the star near the W tip is a pair (12,13; 13″; 295°).

oc 2627 diam. 9′ mag. 8.4p stars 60
15 cm shows this cluster as a bright haze with about a dozen faint stars contained in a 15′ × 10′ area elongated E-W. A wide pair of mag. 11 stars is resolved on the WNW side (26″; 20°). The cluster shows about 35 stars within a 10′ × 5′ area in 25 cm. There is some central concentration and a little faint haze amongst the stars. It is still hazy in 30 cm, where 60 mostly faint stars can be seen in the central 6′ × 3′ area. A mag. 9.5 star on the ENE side has a faint companion 16″ away in pa 40°.

oc 2658 diam. 10′ mag. 9.2p stars 80
This is a moderately faint and irresolved cluster in 15 cm. Among several stars appearing in the cluster is a mag. 10 star on the SW edge. 30 cm shows an irregular, broad concentration of stars and haze. At 225 × about 20 stars are visible in a 4′ area.

oc 2818 diam. 8′ V = 8.2 stars 298
pn 2818A dimen. 35″ × 35″ nebula 11.9v star V = 18.5
In 15 cm the open cluster is a poorly resolved object without much haze that does not stand out well from the field at 75 ×. About a dozen stars are visible within 10′ diameter. The brightest star (V = 11.6), which is near the center on the concentrated part of the cluster, is roughly the same brightness as the planetary nebula, which appears as an unconcentrated blob on the NW side of the cluster core. In 25 cm, 50 stars are resolved in 7′ diameter. The brighter stars are arranged in the shape of a five-pointed star; the planetary lies in the W point. The nebula is about 40″ across and has a faint star or stellaring embedded in its S side. 30 cm shows 75 stars in an 8′ area, moderately concentrated across the 4′ core. The planetary is roughly circular, about 50″ across. A dimmer band running in pa 100° nearly splits it, the southern lobe, containing the star or stellaring, being slightly brighter.

eg 2888 dimen. 0.′8 × 0.′7 V = 12.5 sfc. br. 11.8
Small and moderately faint in 15 cm, this galaxy is elongated approximately SE-NW and shows some concentration toward the center. It is an oval capsule-shaped object in 30 cm, extending to 45″ × 30″ in pa 165°. The halo has a moderate surface brightness and is evenly concentrated to a stellar nucleus. Photographs show a mag. 15 star 12″ from the center in pa 165°, which probably influences the observed elongation in larger apertures.

Sagitta

Situated between Vulpecula and Aquila, this small constellation lies in the heart of the Milky Way. The center of the constellation culminates at midnight about 17 July.

oc Harvard 20 diam. 9′ V=7.7 stars 28
This cluster is hardly distinguishable in 6 cm as a small hazy spot with a few faint sparkles on the E side of two mag. 9 stars. In 15 cm about a dozen stars are visible over haze within a 6′ × 3′ area elongated E-W. 25 cm will show about 30 stars, most of mag. 12 and 13. The cluster has a flat triangular appearance in 30 cm with the long base on the S.

gc 6838 diam. 7.′2 V=8.0 class
Messier 71 is visible in 6 cm as a broadly and weakly concentrated glow about 15′ E of a Y-shaped asterism. At low power nearby field stars give the cluster an irregularly round appearance. A mag. 11 star is visible 2′ S of center. At 200 × 15 cm will give partial resolution, with starry arms extending from the center. Many stars are visible in 20 cm, and dark patches in the cluster give the impression of "eyes." In 25 cm about 50 stars are resolved at 200 ×, including straggling arms on the N, E, and SW. The W side is more sharply defined, lending to an out-of-round shape. Well resolved in 30 cm, it is about 4′ diameter with some brighter stars on the NE periphery that seem to be in the foreground. The most prominent star, about mag. 12, is on the NE side of the center.

pn 6886 dimen. 9″ × 9″ nebula 11.4v star V=15.7:
This tiny planetary lies at the NW apex of a thin triangle of stars pointing ENE. At 250 × in 25 cm it appears starlike, with a magnitude similar to the triangle star 50″ S. At 225 × in 30 cm it has a greenish hue, in contrast to the other stars, and remains stellar in appearance until 450 ×. The nebula is less than 10″ diameter, slightly elongated SE-NW, and has a brighter but nonstellar center.

Sagittarius

This spectacular constellation contains a large and varied selection of deep-sky objects. Every type of object is represented, globular clusters and planetary nebulae being especially numerous. The center of the constellation culminates at midnight about 8 July.

gc 6440 ***diam. 5.′4*** ***V = 9.1*** ***class V***
15 cm shows this cluster in a 12′ string with four mag. 11–12 stars aligned SSE-NNW. It is about 2′ diameter and fairly well concentrated to the center. With 25 cm it is 1.′75 across, the diffuse halo rising to a moderately brighter core. The nearest star of the 12′ string is 1.′7 NNW, and a fainter star lies at the same distance SW. In 30 cm the cluster is about the same size as in 25 cm. The brightness increases broadly across the core, but a faint nucleus is visible at 225×. At this high power the cluster remains unresolved, though the core and extensive halo have a slightly granular texture.

pn 6445 ***dimen. 3.′1 × 0.′9*** ***nebula 11.2v*** ***star V = 18.9***
This fine planetary is easily visible in 15 cm, lying 5′ W of an unequal pair (h2810: 7.7,10.5; 41″; 189°). At 125× it is a fairly bright glow elongated in pa 160° with well-defined edges. The acute ends of the nebula are brighter, and a mag. 12 star is visible on the NW about 30″ from center. With 25 cm the nebula appears about as bright as gc 6440, *cf.* 22′ S, though less than half the size. At 250× the nebula is 45″ × 30″ and the N lobe is brighter than the S. At 225× in 30 cm the planetary looks like a miniature Messier 27, *cf.* pn 6853 in Vulpecula, with bright lobes on each side and dark space between them. The bright area in the N lobe, which appears stellar at low powers, is multiple at 225×. At this power a faint irregular outer haze elongated in pa 45° surrounds the object.

oc 6469 ***diam. 8′*** ***mag. 8.2p*** ***stars 50***
In 6 cm this cluster appears as an irresolved haze about 10′ diameter with a mag. 9.5 star on the SE edge. Six or seven stars encircling the cluster can be resolved at 50×. In 15 cm about 40 stars are visible in a 15′ × 10′ area contained in a horseshoe of six brighter stars opening N. One of the members of this asterism is the mag. 9.5 star on the SE, which is an unequal pair (h4990: 9.5,11.0; 23″; 300°). 25 cm will show 75 stars in a 20′ × 10′ area elongated approximately N-S, including two clumps on the N side. 30 cm shows a loose clustering of 50 stars, mostly mag. 12 and fainter, in a 15′ region that barely stands out from the background.

oc 6494 ***diam. 30′*** ***V = 5.5*** ***stars 131***
Messier 23 is a bright and sparkling cluster in 6 cm, showing about 50 stars in a 50′ diameter lying SE of a mag. 6.5 star. The brightest member (V = 8.2) is on the NE edge. 15 cm shows 75 stars in a 20′ area. In 25 cm outliers spread to 40′ diameter around the main body of the cluster, which is about 25′ across. About 120 stars are visible, including many pairs, loops, and curves lines of stars with remarkably uniform brightness. With 30 cm the bright stars are evenly sprinkled through a 20′ central area, while many faint outliers spread N to 20′ radius.

oc Ruprecht 136 ***diam. 3.′0*** ***mag.*** ***stars 40***
This cluster is visible at low power in 25 cm. At 200× only eight cluster stars are visible over a conspicuously hazy background, not including two brighter field stars on the E and SE edges. Overall the cluster is about 4′ × 3′, elongated SE-NW. oc 6506, *q.v.*, a more widespread group, is centered about 7′ E.

oc 6506 ***diam. 6′*** ***mag.*** ***stars 20***
In 25 cm this cluster contains 30 stars of mag. 11–14 in a 7′–8′ area. Several radially aligned strings stand out, but the stars show no general concentration toward the center. oc Ruprecht 136, *q.v.*, lies 7′ W.

oc-gn 6514 ***diam. 13′*** ***V = 6.3*** ***stars 67***
dimen. 20′ × 20′ ***stars V = 7.6,7.4***
Messier 20, the Trifid Nebula, is discernable in apertures as small as 3 cm at the SW end of a loose association of stars trailing from Messier 21, oc 6531, *q.v.* In 6 cm the nebula appears as a hazy area surrounding two mag. 7.5 stars about 8′ N and S of each other. The northern portion is fainter. At 100× the southern star is a fairly close pair (ADS 10991AC: V = 7.6,8.7; 11″; 212°), and the surrounding nebula shows some detail. Three indistinct dark lanes lead from near the center outward to the W, NE, and S, the northeastern one showing the most contrast. 25 cm shows the lanes clearly: none is really straight, and the one leading NE passes close to the W side of the central pair. Each of the members of the central pair has a fainter companion in this aperture: the

northern pair is ADS 10991AB (V=7.6,10.4; 6″.0; 22°), the southern is ADS 10991CD (V=8.7,10.5; 2″.3; 281°). In 30 cm the nebulosity is of surprisingly low surface brightness, but close scrutiny brings out the detail. The three parts of the nebula divided by dark lanes are unequal, with the southeastern portion being the most generous, the southwestern the least. The dark lane leading W is shorter and broader than the others. Overall the nebulosity is about 15′ diameter and shows only a slight concentration to the central multiple star. About ten faint stars are visible in the N and E sides of the nebula, hardly standing out as a cluster.

oc 6520 ***diam. 5′*** ***mag. 7.6p*** ***stars 60***
This cluster is superposed against the Sagittarius star cloud and has a bright luminous background as well as a sharp dark cloud (Barnard 86) nearby to the W. The cluster is easy to see but very small in 6 cm. At low powers it appears as a streak elongated in pa 135°, but high power shows two brighter stars in pa 135° with the fainter cluster members surrounding them and spreading to the NE. These stars are at the limit of visibility: only about ten are resolved at 100×. 15 cm shows it about 5′ diameter with 20 stars resolved, including three or four brighter ones. The dark cloud on the W is hard to see. 20 cm resolves the cluster well, showing about 30 stars including many pairs in a 5′×3′ area set in a very rich surrounding field. 25 cm gives color to some of the pairs, and shows the dark matter well. The dark patch is trapezoidal, about 5′×3′, elongated roughly NE-SW. An orange mag. 7 star lies on its N side. A thin dark line about 12′ long and 2′ wide leads away from the patch, passing along the southern edge of the cluster. 30 cm shows 50 stars in the cluster, about half of them packed into a 1′.25 core surrounding a bright reddish star. The rest are spread loosely over 5′ diameter, mostly N of this group.

gc 6522 ***diam. 5′.6*** ***V=8.4*** ***class VI***
Easily visible in 6 cm, this cluster appears as a small, broadly concentrated spot with two faint stars 1′.2 E and SW. In 15 cm the cluster has a much brighter center surrounded by a circular halo 1′ diameter, while 25 cm shows it 1′.25 diameter. The core is small and very bright, and has a prominent mag. 12 star on its NE edge. A dark lane aligned N-S cuts through the cluster on the W side. Occasionally the core seems to have a bar-like brightening aligned approximately in pa 120°. Overall the cluster has a clumpy texture. Viewed in 30 cm the halo extends to 1′.9 diameter. At 225× the edges appear granular, but no resolution is indicated. This cluster appears brighter and larger than its nearby companion, gc 6528, *cf.* about 16′ E.

gn 6523 ***dimen. 90′×40′*** ***stars V=6.0,10.5:***
oc 6530 ***diam. 15′*** ***V=4.6*** ***stars 113***
Messier 8, the Lagoon Nebula, and oc 6530 are visible to the naked eye as a small bright patch in the Milky Way. 6 cm shows a dark bar slashing obliquely between the cluster and the bright, condensed portion of the nebula. The nebulosity extends through the cluster, and a few faint stars appear in the bright part of the nebula. In 15 cm the dark lane is a clearly defined, slightly curved bar about 40″ wide aligned NE-SW. The bright part of the nebula extends SW from two bright stars of mag. 6 and 7 that lie 3′ apart. In 25 cm a fainter band of nebulosity runs E-W across the north side of the area, passing just S of a mag. 7.5 star, and a faint general illumination extends S and E over the cluster. 30 cm shows a mag. 10.5 star on the W edge of the condensed core of the nebula. The dark lane is fairly sharply defined, and some dark streaks run parallel to it in the adjacent nebulosity, particularly on the NW side. The lane has at least five stars in it, including a wide pair to the NE (V=9.8,10.4; 23″; 270°). The texture of the nebula is fairly smooth, though some mottlings are visible on the fainter SE portion. 6 cm shows about 20 stars in oc 6530 enveloped in the outer haze of the nebula. 15 cm shows 30 bright stars, 20 cm 50 stars in a 30′ area. The cluster is generally unconcentrated and many bright stars are bluish.

gc 6528 ***diam. 3′.7*** ***V=9.5*** ***class V***
This globular cluster is visible in 6 cm as a fainter and smaller version of gc 6522, *cf.* 16′ W. In 25 cm it is about the same size as −22, with a diffuse halo rising much to the center. The surrounding field is dramatically darker here than around −22, the edge of the dark area lying roughly halfway between the two clusters. With 30 cm it is only 1′.2 diameter and not very well concentrated, showing little outerly haze. A mag. 13 star is visible on the SW edge 40″ from the center.

oc 6531 ***diam. 15′*** ***V=5.9*** ***stars 63***
Messier 21 is visible in 6 cm as a small sparkling companion to the Trifid Nebula, *cf.* gn 6514. Several brighter stars are in a streak on the SE side, ending on the SW with a pair (South 698: V=7.3,8.8; 30″; 316°). About 25 stars are visible in 15 cm including a third star forming a triangle with the central pair. The cluster is a little concentrated, about 15′ diameter, but this is difficult to estimate since the cluster merges smoothly with the background. With 25 cm the triangle in the center has three more stars involved. A total of about 45 stars is visible; many of the brighter ones are bluish. 30 cm shows about 30 stars in a 10′ area. The brighter members are on the E side of the group, and faint outliers trail to the SW.

oc 6540 diam. 1′.5 mag. 14.6p stars 10
In 25 cm this cluster is just visible as a tiny, faint patch involved with a partially resolved 1′ string of stars that bends S at its W end. With 30 cm the cluster seems quite elongated at low power due to this string. At 225× the string is resolved into seven or eight faint stars in an arc concave to the N, running through the center of the cluster and out the E and W ends. The cluster is a 30″ spot centered on the arc and is without resolution.

gc 6544 diam. 8′.9 V=8.1 class
This cluster is visible in 6 cm as a small hazy spot concentrated to a faint star-like center. A mag. 8 star lies 6′.5 SSE, and rich fields of faint stars S and E give a false impression of resolution of outliers. With 15 cm some faint resolved stars on the ESE edge give the cluster an irregular outline. A conspicuous mag. 12 star is visible just W of center. In 25 cm the cluster is 1′.5×1′.25, elongated SE-NW, showing moderate concentration and a granular texture at 200×. In 30 cm it is a pretty sight at low power, sparkling throughout, with bright stars near the center looking like a nucleus. At 225× two or three brighter stars near the center form an E-W dash; the brightest one is at the W end. Outliers are unevenly distributed to almost 2′ diameter. A mag. 11.5 star lies 1′.5 SW that has a faint companion about 10″ W.

oc 6546 diam. 15′ V=8.0 stars 53
In 6 cm this cluster is an irresolved circular spot on the E side of the diffusely luminous area N of the Lagoon Nebula, *q.v.* gn 6523. It is a fairly conspicuous cluster in 25 cm, which shows about 30 stars of mag. 11 and fainter in a circular hazy region 10′ diameter. Three brighter mag. 9 stars form a 4′ triangle on the E side; two of them are orange.

gc 6553 diam. 8′.1 V=8.1 class XI
This cluster is a tiny, diffuse dot in 6 cm; it appears about 2′ diameter and has irregular edges. With 25 cm it extends to about 2′.5×2′, elongated ESE-WNW. A mag. 11.5 star lies in the NW side 50″ from center. In 30 cm the cluster appears similar to gc 6569, *cf.*, but has a more extensive, wispy halo. At low power some faint sparkling to the SW side looks like outliers, but 225× shows only one mag. 13 star there and another in the E side. Averted vision shows an irregularly bright granular halo 2′.1 diameter with poor concentration.

gc 6558 diam. 3′.7 V=9.8 class
25 cm shows this cluster as a 1′ spot that is somewhat fainter than gc 6569, *cf.* At 200× the irregularly bounded halo has a slightly granular texture. In 30 cm it is 1′×0′.8, elongated in pa 30°. At 150× it appears much like a galaxy, but at 225× it is granular and evenly concentrated to a faint nucleus.

pn 6563 dimen. 54″×41″ nebula 11.0v star V=17.0
With 25 cm this planetary appears as a round, uniformly illumined spot about 40″ diameter in a field rich with faint stars. In 30 cm the nebula is about 50″×40″, elongated NE-SW, and is slightly brighter on the NW side. At 225× the surface is somewhat unevenly bright, occasionally appearing annular. A faint star lies just off the S edge.

pn 6565 dimen. 10″×8″ nebula 11.6v star V=19.2
In 30 cm this planetary appears as a very small, uniformly bright disk at 150×. At higher powers it is slightly elongated N-S, showing a darker center at moments of good seeing. A faint star or stellaring is visible on the N side.

pn 6567 dimen. 11″×7″ nebula 11.0v star 15.0p
This object lies at the boundary between a very rich field to the E and a dark cloud to the W. In 30 cm it is very small and seems elongated E-W due to a mag. 12.5 star on the E edge. It brightens evenly to a bright, almost stellar center. The nebula is approximately round, but interference from the star hinders certain determination of the elongation.

oc 6568 diam. 12′ mag. 8.6p stars 50
25 cm shows this cluster as an irregular group of 40 stars in a 20′ area, including a Y-shaped asterism of four stars on the W. The brighter stars are on the southern perimeter; the center is less populated. In 30 cm it is large and unconcentrated with 35–40 stars in a 15′×10′ area elongated N-S. Several pairs stand out at low power.

gc 6569 diam. 5′.8 V=8.7 class VIII
Located 8′.5 N of a mag. 7.5 star, this cluster is visible in 6 cm as a round disk without concentration, like a silver dollar on the sky. 25 cm shows the halo to 1′.5 diameter with a brighter, broadly concentrated core. It has a slightly granular texture at 200×, and several moderately bright stars are nearby on the S. In 30 cm the slight halo grows to 1′.8 diameter around the large, broadly concentrated core. At 250× it remains unresolved, showing a very smooth, cottony texture without granulation.

oc Markarian 38 diam. 2′.0 V=6.9 stars 14
Lying within the Messier 24 star cloud (*cf.* oc 6603), 15 cm will show this cluster as a small group of seven or eight stars around a mag. 7 star. 25 cm shows a dozen stars in a 1′.75×1′ area elongated N-S, including a close

pair 20″ N of the lucida (ADS 11193BR: V=11.3,12.5; 5″.0; 324°). The surrounding field is extremely rich with stars.

oc 6583 ***diam. 5′*** ***mag. 10.0p*** ***stars 35***
In 25 cm this cluster is a moderately bright granular spot about 3′ across. At high power only a few stars are resolved. 30 cm shows 15–20 faint stars in 2′.25 diameter packed so closely they are difficult to count. A string of five or six stars runs N-S through and out the cluster to the N. A 2′ arc of three stars aligned SE-NW is 3′ S of the cluster.

gn 6589 ***dimen. 5′ × 3′*** ***star 9.5p***
gn 6590 ***dimen. 3′ × 2′*** ***star 10.0p***
Viewed in 30 cm, gn 6590 is visible around a mag. 10 pair (h2827: 10.0,10.1; 20″; 254°); 6′ NNW is gn 6589, around a mag. 9.5 star with a faint companion about 1′ ESE. Each of the nebulae are about 1′ diameter and best viewed at low power. The field to the SE has very few stars.

oc 6596 ***diam. 10′*** ***mag.*** ***stars 30***
This cluster is inconspicuous in 25 cm, lying in the E edge of a fine low-power field of many mag. 8–10 stars scattered among small dark patches. The cluster proper is an oval 6′ × 4′ circlet of brighter stars with a vacant center, elongated N-S.

oc 6603 ***diam. 4′.5*** ***mag. 11.1p*** ***stars 100***
Messier 24 is a rich extensive cloud of stars visible to the naked eye. In 6 cm the large star cloud dominates the field with a scattered array of mag. 7 and fainter stars. oc 6603 is a rather inconspicuous concentrated glow embedded in the northeast end of the star cloud with mag. 6.5 stars about 15′ W and S. With 15 cm the cluster is quite small, an oblong, irresolved fuzzy spot showing only a few sparkles near the center. 25 cm resolves the cluster well, showing a concentrated cloud that stands out well from the rich background. About 75 stars of mag. 12 and fainter are countable in 10′ diameter. Intruding into the NW side is a dark, V-shaped shadow outlined by strings of stars. In 30 cm the string of stars along the NE edge of the shadow stands out as a tightly condensed bar running SE-NW through the middle of the cluster. A total of 80 stars is visible here in an 8′ area, with some haze in the central bar. A bright, very red star lies about 4′ SSW.

oc 6613 ***diam. 8′*** ***V=6.9*** ***stars 40***
Messier 18 is a small, poor cluster of bright stars visible with 6 cm in the same low-power field as Messier 17, *cf.* oc-gn 6618. About ten stars are in the group, the three brightest forming an arc across the N and NW sides. With 15 cm the cluster members outline a squarish pattern about 4′ across. The pattern is less conspicuous in 25 cm, where about 20 stars are visible within 10′ diameter. 30 cm shows about ten brighter stars, with perhaps ten more in the area that could be counted as cluster members. The cluster has very few faint stars, and does not show any overall concentration.

oc-gn 6618 ***diam. 25′*** ***V=6.0*** ***stars 660***
dimen. 40′ × 30′
Messier 17 is the brightest galactic nebula in the sky for northern observers except Messier 42, the Orion Nebula. 6 cm shows a bright bar elongated in pa 110° that fades to a faint extensive haze on the S. The W end of the bar has thin extensions to the S with two mag. 10 stars on its W edge. The bar is not uniformly bright; a particularly bright lump lies two-thirds of the way from the E end. The field W is dark and starless, while the field N is rich in stars. In 15 cm a "2" shape stands out, with an elongated base and a mottled texture in the hook. The cluster, which lies mostly to the north of the nebula, contains 50 stars in a 25′ area. 25 cm shows bright and dark patches in the bar and distinct mottlings in the hook. In 30 cm the bar is about 13′ long, and faint wisps of nebulosity spread east of it.

gc 6624 ***diam. 5′.9*** ***V=8.0*** ***class VI***
This is a moderately bright cluster in 25 cm. It is 2′ diameter, and the small core appears granular at 200 ×. A mag. 11.5,13 pair with 10″ separation in pa 270° lies 1′.7 WSW of center. 30 cm shows it to 2′.2 diameter with hints of granulation in the core. The halo shows a fair degree of concentration to an occasionally visible nonstellar nucleus.

gc 6626 ***diam. 11′*** ***V=6.8*** ***class IV***
Messier 28 appears as a small, sharply concentrated spot with a conspicuous stellar nucleus in 6 cm. In 15 cm it is 2′ across and seems clumpy around the edges, though no resolution is indicated. With 20 cm it is well resolved at the edges at 250 ×, with star chains extending to about 4′ diameter. The core is 2′ across in 25 cm, dropping off sharply in brightness to a rich scattering of stars in the 4′ halo. With 30 cm the overall size is about the same, but the core shows a distinct elongation in pa 30°. At 250 × the core has an overlay of faint sparkling stars on a very hazy background.

pn 6629 ***dimen. 16″ × 14″*** ***nebula 11.3v*** ***star V=12.8***
This planetary looks like a mag. 10 star in 25 cm at 100 ×. At 250 × it is about 15″ diameter, fading abruptly at the edge, and occasionally shows a faint sparkle in the

center. The distinctly bounded nebula has a fairly high surface brightness in 30 cm, but does not brighten toward the center. The central star stands out well at higher powers.

gc 6637 diam. 7.′1 V=7.6 class V
Messier 69 is visible in 6 cm 4.′5 SE of a mag. 8 star. It looks similar to Messier 70, *cf.* gc 6681, though it is somewhat brighter and more broadly concentrated. At 100× it is unresolved in 15 cm, but 200× shows outliers clearly. In 25 cm it is circular, 2.′5 diameter, with a broadly concentrated core but a sharp, nearly stellar nucleus. It is partially resolved at 200×. In 30 cm it is granular at 150× and about 3′ diameter. The broadly concentrated core is elongated N-S and is surrounded by only a little outerly haze. A conspicuous mag. 13 star lies on the E edge of the core.

gc 6638 diam. 5.′0 V=9.1 class VI
Faintly visible in 6 cm, this cluster appears as a well-concentrated spot 3.′5 N of a mag. 10 star. In 15 cm it is faint and round, about 1′ diameter, with a brighter center. In 25 cm a smooth, diffuse halo extends to 1.′5 diameter, and the small, sharply concentrated core seems off-center to the S. With 30 cm it is 2′ diameter and very evenly concentrated to an occasional sparkle in the center. Some granulation is visible around the edges of the core, but there is no real resolution.

oc I4725 diam. 30′ V=4.6 stars 600
Messier 25 is a naked-eye object and a fine cluster for 6 cm, which shows at least 35 stars in a 30′ area. The members are loosely scattered over the field except for a tight 1′ knot of about eight stars near the center. The brightest star in the cluster, the Cepheid U Sagittarii (6.4 < V < 7.0), stands out just E of the knot. In 15 cm about 50 stars are visible. Another slightly fainter star NW of the knot forms, together with the knot and U Sgr, an elongated 3′×1′ group. A group of fainter stars S of this forms a second string running more nearly E-W than the northern group. 25 cm shows about 75 stars in a 45′ area, including many pairs and colored stars. With 30 cm, 80–90 stars are visible spilling out of a 23′ low-power field. U Sgr has a distinct reddish hue.

gc 6642 diam. 4.′5 V=9.4 class
6 cm will just show this small cluster among a few faint stars. In 15 cm it is about 1′ diameter; higher powers reveal no details in the cluster, only more faint stars associated nearby. It is 1.′5 diameter in 25 cm, still showing no real resolution, though the granular texture and many nearby field stars give the impression of partial resolution. 30 cm will show a few faint stars resolved over a granular haze about 1.′2 diameter. Overall the cluster shows a strong, sharp central concentration.

oc 6645 diam. 15′ mag. 8.5p stars 40
With 15 cm this is a rich cluster of mostly faint stars. At higher powers at least 30 stars are visible irregularly scattered through 15′ diameter, a couple standing out above the others. 25 cm will show about 70 stars within 20′ diameter. Most of the brighter members are in the N side. The outliers fill a 23′ low-power field in 30 cm with a total of perhaps 60 or 70 stars. The denser core is about 10′ across and dominated on the N side by a 2.′5 circlet of a dozen stars with an empty center. About 3′ S of the circlet, at the S edge of the core, is a close group of four stars.

gc 6652 diam. 3.′5 V=8.8 class VI
This cluster is discernable as a small nearly stellar spot in apertures as small as 6 cm, lying just over 7′ SE of a mag. 7 star. In 25 cm it is about 45″ diameter with an irregular shape and a slightly granular texture. On the W edge is a mag. 13 star. The faint, extensive halo extends to 1.′8 diameter in 30 cm, though the cluster remains unresolved at 250×. Overall it exhibits relatively weak concentration to a 30″ core, which occasionally shows a faint stellaring in its center.

gc 6656 diam. 24′ V=5.1 class VII
Messier 22 is a naked-eye object that can be partially resolved even in 6 cm. At 75× it is a large cottony ball, slightly elongated NE-SW, showing many faint sparkling stars, particularly in the outer regions. Conspicuous field stars lie about 10′ NNE and SE. In 15 cm many dozens of stars are resolved clear across the hazy core. At 200× faint outliers spread irregularly to 20′ diameter. With 25 cm the core still seems elongated NE-SW, but the outliers define a roughly circular area. On the SE edge of the 4′ core is a nice pair (10.5,11.5; 11″; 0°), and a small clump of several stars is visible on the NE edge. 30 cm shows outliers spreading to at least 23′ diameter, though they are more numerous toward the S and E sides. The cluster is well resolved in the outliers and across the broadly concentrated core. The clump of brighter stars on the NE edge of the core has about five members, all packed into a 30″ area.

gc 6681 diam. 7.′8 V=8.0 class V
Messier 70 is a well-concentrated glow in 6 cm, similar in appearance to Messier 69, *cf.* gc 6637, though somewhat fainter. Two stars of mag. 10 and 11 lie 5′–6′ N and NNE. In 15 cm it appears a little smaller than M69 and shows a few faint stars at 200×. 25 cm also shows a few sparkles, but most of them are probably foreground stars: the clus-

ter itself has only a granular texture and extends to about 1.′75 diameter. Three of the brighter resolved stars lead away N from the NE edge; many of the rest are in the southern half. With 30 cm many faint stars are resolved, spread evenly over 3′ diameter, showing no concentration over the 1.′25 core. Occasionally the cluster seems elongated slightly N-S.

gc 6715 diam. 9.′1 V=7.6 class III
Messier 54 appears as a small but strongly concentrated cluster in 6 cm, which shows a star-like center. 15 cm shows a very intense 15″ core within a circular 1′ halo. On the SE is a mag. 12 star just within the halo. With 25 cm it is about 2′ diameter. At 250× the cluster has a granular texture and shows a few scattered stars in the halo, but these show no central concentration and are probably only field stars. In 30 cm the cluster looks like Messier 80, *cf.* gc 6093 in Scorpius, with a very bright, smooth, cottony core. Otherwise it is about the same size as in 25 cm, and shows no resolution.

oc 6716 diam. 10′ V=6.9 stars 38
oc Collinder 394 diam. 54′ V=5.6 stars 51
6 cm shows oc 6716 as a loosely scattered group of about ten stars in two groups that barely stand out from the Milky Way. 25 cm shows a rectangular grouping on the SW and a circlet bounding the cluster on the NE, together containing about 30 stars in a 10′×8′ area elongated NE-SW. 30 cm reveals 25 stars in the two groups: the arc NE forms an L-shaped asterism with the majority of the stars; in the southwestern group are two brighter stars with about ten fainter companions. oc Collinder 394 forms a rich low-power field in 25 cm about 40′ SW.

gc 6717 diam. 3.′9 V=9.2 class VIII
This globular cluster is visible in 25 cm about 2′ S of ν^2 Sagittarii (V=5.0), whose glare interferes with viewing. Overall the cluster is 45″ across, rising sharply in brightness toward the center and an occasionally stellar nucleus. A close pair of mag. 13 stars is off the NE edge (8″; 115°); a single mag. 13 star is on the edge at pa 290°, 20″ from the center.

gc 6723 diam. 11′ V=7.2 class VII
In 6 cm this object is a large, broadly and weakly concentrated cluster similar to Messier 4, *cf.* gc 6121 in Scorpius. A conspicuous mag. 10.5 star is on the NE side. With 15 cm it is fairly well resolved at higher powers, while 25 cm resolves it well even with lower magnifications. Overall it is about 4′ diameter, slightly elongated in pa 110°. In 30 cm it is a fine sight, well resolved across the center, with a little haze in the core. Outliers spread to 5.′5 diameter around a 1.′9 core that has no obvious nucleus, though a brighter star embedded just N of center looks like one at low power. The brighter part of the cluster seems displaced N of center, though the brighter star near the center makes this impression uncertain.

gc 6809 diam. 19′ V=6.4 class XI
Messier 55 is a large, irregularly bright haze in 6 cm, showing some granularity on the broad core at 50×. 15 cm resolves the cluster well at 100×. Outliers are visible to about 10′ diameter, but a large bite is missing from the SE side of the halo, where only a few stars are visible. 25 cm shows hundreds of stars over a very uniform non-granular haze. With 30 cm it is a pretty cluster, well resolved even at low power. The stars range widely in brightness; the brighter ones form a uniform grid across the cluster that seems superposed on hundreds of tiny, faint stars in the central parts. The dark shadow on the SE side has an evenly curved semicircular outline and borders against the edge of the broadly concentrated 4′ core. At 225× outliers reach to about 14′ diameter.

pn 6818 dimen. 22″×15″ nebula 9.3v star 14:p
This object is a fairly bright bluish spot in 15 cm. At 100× the edges are sharply defined, and an indistinct dark streak seems to run through the center. At 250× 25 cm shows a smoothly textured circular nebula with a slightly dimmer center. A mag. 12.5 star is visible 40″ NW; another lies 1′ ENE. In 30 cm the nebula is nice at all powers. Low power shows a small, roundish dot with a bluish-grey hue and a dark spot in the center. At 225× the dark spot is more distinct, but is definitely not as dark as the sky, showing less contrast with the ring than that in Messier 57, *cf.* pn 6720 in Lyra. The nebula is elongated roughly N-S here. At 450× the elongation is most obvious, the nebula extending to 20″×15″ in pa 10°, with an 8″ hole. The acute ends of the annulus have faint extensions, also like M57. The E side of the ring seems to have a brighter spot embedded in it. A third faint star is visible 1.′3 SW.

eg 6822 dimen. 10′×9.′5 B=9.3 sfc. br. 14.2
This member of the Local Group of galaxies is a weak but definite glow in 6 cm, where it appears elongated N-S and shows a very slight central concentration. In 25 cm motion of the field helps in showing the low surface brightness galaxy, but it is difficult and ill-defined at best. Many field stars are superposed on an indefinite haze of faint stellarings near the threshold of vision. The overall size is approximately 15′×5.′ With 30 cm it is even more difficult, since the galaxy is of such low surface brightness. Two brighter patches stand out aligned roughly NE-SW, each about 1.′25 diameter. The field is rich with faint stars, particularly to the N.

eg 6835 ***dimen.*** $2'.7 \times 0'.7$ ***V = 12.5*** ***sfc. br. 13.1***
In 25 cm this galaxy is a lenticular object elongated in pa 80°, 1′ × 0′.6 in size. The center is only slightly brighter, and a mag. 13 star lies 1′ E. In 30 cm the halo extends to about 1′.2 × 0′.9, showing moderate, broad concentration and no distinct nucleus. eg 6836 (V = 12.9) lies 7′.3 SSE.

gc 6864 ***diam.*** $6'.0$ ***V = 8.5*** ***class I***
Messier 75 appears similar to Messier 54, *cf.* gc 6715, in 6 cm, though it is smaller, fainter, and more sharply concentrated toward the center. 15 cm shows it about 2′ diameter with a small, intense core. A faint star is visible about 2′ N, but no resolution is evident. In 25 cm a few outliers are resolved that spread thinly to 2′.5 diameter, as far as a mag. 12.5 star on the SE side. 30 cm shows it 2′.2 across with a 30″ core. With averted vision the outer haze seems elongated slightly NE-SW. At high power the cluster is granular throughout, and the core is slightly brighter on the N side. A faint star is visible 25″ NE of the center.

Scorpius

The scorpion is a prominent constellation resembling its namesake perhaps better than any other group. Over two dozen open clusters appear in the region of the tail, the sparkling naked-eye cluster Messier 7 being the brightest of these. The center of the constellation culminates at midnight about 4 June.

pn 6072 dimen. 70″ × 70″ nebula 11.8:v star 17.5p
This planetary is only moderately faint with 15 cm, but no details are discernable. Two faint stars are visible 2.′5–3.′5 W. In 25 cm the nebula is conspicuous at 100 × as a 50″ circular spot with a slight central darkening. In 30 cm it is about 60″ diameter and has a ragged border.

gc 6093 diam. 8.′9 V=7.3 class II
Messier 80 is easily found in 6 cm by moving east from β Scorpii and lies 4.′5 SW of a mag. 8.5 star. The cluster is 2.′5 diameter and strongly concentrated, though the nucleus is not sharp at 50 × . It is about 4′ diameter in 15 cm and will show some granulation around the edge at high power. At 200 × 25 cm will partially resolve the cluster, showing many faint stars spreading over a well-defined 4′ area around a brighter 45″ core. In 30 cm a well-resolved mass of stars extends to about 6′ diameter. Here the core appears about 1.′5 across and has a band of stars aligned E-W off its S side.

gc 6121 diam. 26′ V=5.8 class IX
A faint spot to the naked eye, Messier 4 appears as a broad and weakly concentrated glow in 6 cm. At 75 × the irregular core sparkles with a few stars, the brightest ones lying on the S side. With 15 cm a conspicuous string of brighter stars is visible running N-S through the center. 20 cm gives partial resolution at 50 × , while 250 × reveals a well-resolved mass about 10′ diameter with only a bit of haze remaining in the center. Faint outliers spread to at least 15′ diameter in 30 cm. On the E side of the core about 1′ from center is a small knot of stars; 1.′7 S of this is a wide, bright pair (V = 10.8,10.9; 17″; 220°). A curved string of mag. 12–13 stars is visible in the WNW side of the cluster about 3.′5 from center.

oc 6124 diam. 40′ V=5.8: stars 100
In this cluster about 75 stars can be counted in a 15′ area with 15 cm. The densest portion, containing about 25 stars, is about 6′ across. Outliers spread especially E and W to an overall diameter of perhaps 30.′ In 25 cm at least 90 stars are visible in a 35′ area. An equal pair of mag. 10 stars with 20″ separation in pa 80° is isolated about 9′ SE of center. With 30 cm the central 6′ area contains few more members than with 15 cm, showing only 28 stars or so: there are not many faint stars.

gc 6139 diam. 5.′5 V=8.9 class II
Though this cluster is only just visible to 6 cm, it is pretty bright for 25 cm. The circular haze is 2′ diameter and moderately concentrated toward the center. At 200 × it has a slightly granular texture at moments of good seeing. In 30 cm the brightness rises evenly from the 2′ halo to a 50″ core and an inconspicuous nucleus. No real resolution is indicated even at high powers.

gc 6144 diam. 9.′3 V=9.0 class XI
This cluster is fairly faint in 15 cm, a circular, poorly concentrated glow about 40′ NW of Antares. At 150 × it has a granular texture, but a single mag. 12 star sparkles on the W edge. With 25 cm the cluster is partially resolved at 150 × , showing a few mag. 13 stars sprinkled over a hazy background. Overall it is 3′ diameter, reaching just beyond the bright star on the W. It remains only partially resolved in 30 cm, but here outliers spread to about 4′ diameter. A band of stars aligned nearly N-S stands out in the core.

pn 6153 dimen. 28″ × 21″ nebula 10.9v star V=16.1
This nebula is a bright object in 25 cm, forming a diamond with three mag. 10 stars to the N, NE, and NW. The nebula is a circular, well-defined spot with a slightly brighter center, though no central star is evident.

oc 6178 diam. 5′ V=7.2 stars 19
Viewed in 6 cm this cluster shows five or six faint stars grouped around the northern of two mag. 8.5 stars. 15 cm reveals nine stars and 25 cm a dozen within a 5′ area. An unequal pair is off the E side, slightly separated from the cluster (V=9.9,12.2; 12″; 80°).

oc 6192 diam. 14′ mag. 8.5p stars 60
In 6 cm this cluster is a pretty object, partially resolved and showing a dozen stars in an irregular 4′ area. With

15 cm about 20 stars are visible, while 25 cm will show 35 stars of mag. 9 and fainter. Most of the members are packed into a 5′ area, but stragglers spread to an overall diameter of about 10′. In 30 cm about 40 stars are visible in the central 6′ area. These show no overall concentration toward the center, but are instead grouped into two or three clumps.

oc 6222 ***diam. 4′.0*** ***mag. 10.2p*** ***stars 35***
6 cm will show this cluster as a faint, slightly concentrated patch about 4′ across. At times it seems elongated N-S, but this is probably due to a faint star at the N edge. With 15 cm about ten stars are visible over a very hazy background. In 25 cm about 25 stars are resolved in a 5′ area, still leaving much background haze.

oc 6231 ***diam. 26′*** ***V=2.6*** ***stars 93***
This pretty cluster is a tight knot of many bright stars in 6 cm, located at the S edge of a large scattered association of stars that includes oc Trumpler 24, *q.v.* At least 20 stars are countable without any discernable haze, including nine brighter members and a wide pair of bright stars off the NW side (V=5.5,7.7; 65″; 13°). With 15 cm about 100 stars are visible in a 10′ area. Many blue stars and small geometric patterns stand out at low power. In 25 cm at least 120 stars are visible, spreading to about 15′ diameter. Most of the outliers are in a dense region of fainter stars on the S side of the central 10′ group.

oc 6242 ***diam. 9′*** ***V=6.4*** ***stars 23***
This cluster is visible in 6 cm, which will show ten stars within a distinctly hazy cloud. The brightest stars form a group in the middle that is elongated N-S; the brightest member (V=7.3) is SE of the group. About 35 stars are visible in a 10′ area with 15 cm. 25 cm reveals 50 stars, including 15 brighter members and many faint ones: few are of intermediate brightness. 30 cm shows 40 stars in a 7′ area. The three brightest stars form a triangle pointing SE.

oc Trumpler 24 ***diam. 60′*** ***mag. 8.6p*** ***stars 30***
This huge association has some nebulosity on its N end (I4628) and is best viewed in wide-field instruments. In 6 cm the object is characterized by an even sprinkling of mag. 8–10 stars in an irregularly bounded area about 50′ across. The NE edge is fairly sharply defined, and several fainter stars trail S to oc 6231, *q.v.* 15 cm shows two smaller gatherings on the N and W sides, and well over 100 stars are visible in a 90′ field.

oc 6249 ***diam. 6′*** ***V=8.2*** ***stars 15***
oc 6259 ***diam. 15′*** ***V=8.0*** ***stars 162***
oc 6259 is a large diffuse patch in 6 cm, showing no resolution even at high power, though many foreground stars are superposed on it. The smoothly textured hazy area is about 12′ diameter and shows some brightening across the middle. 15 cm begins to show some member stars, but they are all faint. Perhaps 30 stars can be counted at high power over the granular haze. With 25 cm the cluster is reminiscent of Messier 11, *cf.* oc 6705 in Scutum. About 100 stars of mag. 12 and fainter are uniformly distributed through a 15′ × 12′ area elongated E-W. oc 6249 is 34′ WSW, and visible in 25 cm as a small group of about ten fairly bright stars.

oc 6268 ***diam. 6′*** ***mag. 9.5p*** ***stars 24***
In 6 cm about ten stars are resolvable in this cluster over much haziness. Most of the visible stars are grouped into a bar running NE-SW on the N side; immediately SE of this is a dark area, then a small knot of stars. 15 cm shows these ten stars and about as many more fainter ones. In 25 cm it is a loosely scattered group of about 25 stars within 10′ diameter. With 30 cm 35 stars are visible irregularly condensed into an 8′ × 5′ area elongated N-S. Fifteen stars, including most of the bright members, are in the 5′ × 1′ bar; others straggle to the N.

oc 6281 ***diam. 8′*** ***V=5.4*** ***stars 70***
With 20 stars arranged over haze like Christmas tree lights, this is a pretty cluster for 6 cm. A wide pair including the brightest star stands out on the NE side, and a number of field stars trail off to the S. In 25 cm the cluster is partly enclosed by eleven stars in two lines on the N and W aligned in pa 70° and 140°, respectively. The wide pair forms part of the northern line; the brighter component is itself a closer pair (V=7.9,10.7; 7″; 330°). About 20 other stars are visible, mostly mag. 11 and fainter. With 30 cm the cluster has a rough teardrop shape, the double star at the pointed tip of the drop. About 30 stars are visible in a 7′ × 6′ area with no hint of haziness and plenty of space between the stars.

gn 6302 ***dimen. 83″ × 24″*** ***star***
This is a curious, high surface brightness nebula in 25 cm, extending to 1′.5 × 0′.3 in pa 80°. Offset to the E side is a bright circular core that has a conspicuous stellar nucleus in its center. The fainter nebulosity spreads farther toward the W and has a slightly brighter spot in it at that end. In 30 cm fainter extensions reach to 2′.25 × 0′.8. The sharply concentrated portion seems elongated in pa 30°, and with averted vision the fainter nebulosity along the flanks pinches in here. Moving away from the stellar center to the E and W sides the nebula brightens a little, particularly on the W, before fading rapidly to the much fainter halo.

oc 6318 diam. 4'.0 mag. 11.8p stars 20
With 15 cm this cluster appears about 5′ diameter and is granular to partially resolved at 100×. About 35 stars are shown by 25 cm, mostly in a 5′ area, but outliers extend to 10′ diameter. The group is elongated N-S, and a line of three closely spaced stars stands out at the S end. In 30 cm a string of seven stars aligned N-S runs across the E side, connecting at its S end with the shorter E-W string. In a 4′×3′ area are about ten other stars, most of them quite faint.

oc 6322 diam. 9′ V=6.0 stars 38
This cluster is an inconspicuous group in 6 cm, showing about six stars and a little haze within a triangle of mag. 7.5 stars 6'.5 on a side. With 15 cm about 15 stars are visible including two pairs of similar position angle but different separation: V=9.0,9.5; 22″; 25° and V=11.0,11.0; 10″; 20°. 25 cm will show perhaps 20 stars. Each of the bright triangle members has a faint coterie of stars attending. The members are a scattered group in 30 cm, spilling out of the triangle on the SE side. Thirty stars are visible within the triangle; a total of 40 includes the stragglers.

pn 6337 dimen. 49″×45″ nebula 12.3:v star 17:v
This is a bright nebula in 25 cm, a diffuse, annular glow 40″ across at 200×. Many stars are associated. At pa 30° on the inner edge of the annulus is a mag. 12 star; WSW 45″ is another. A third star lies 1′ S, not so intimately involved. With 30 cm the nebula is about 50″ across, with a 20″ hole that is not completely dark within. Superposed on the SW side of the annulus is a threshold star. The mag. 12 star to the WSW has a faint companion at 12″ in pa 130°.

oc Trumpler 25 diam. 4'.0 mag. 11.7p stars 40
This cluster is a faint granular glow in 15 cm, located N of two widely spaced mag. 9.5 stars. With 25 cm it is about 7′ diameter, a bright irresolved haze at 50×, but well resolved at 200× into about 40 stars. A dark lane intrudes from the S, and another crosses the N side aligned E-W. In 30 cm the cluster is an unconcentrated scatter of faint stars that stands out poorly from the Milky Way. At 125× about 35 stars are visible in a 12′–13′ area.

gn 6357 dimen. 12′×11′
oc Pişmiş 24 diam. 5′ V=9.6 stars 15
15 cm shows a moderately faint nebula here at the north end of a string of five mag. 6–7 stars. An inconspicuous cluster of about five resolved stars is embedded. In 30 cm the 5′ cluster contains eighteen stars. The brightest, mag. 10.5, is at the N end; a mag. 11.5 pair (4″; 135°) is just SE of it. The cluster is bounded sharply on the N side by the sharp-edged 5′×2′ nebula, which is elongated E-W and fades slowly farther north.

oc Collinder 333 diam. 8′ mag. 9.8p stars 30
This is a small, conspicuous group in 15 cm, including a total of eight or nine stars in 5′ diameter. 25 cm will show about 15 stars between mag. 9 and 12. A mag. 9.5 star lies off the S side, and two pairs are resolved on the NE: 11.5,11.5; 11″; 95° and 11,11.5; 16″; 215°.

oc Harvard 16 diam. 15′ mag. stars 70
Located just N of the stinger of Scorpius, this group is visible in 6 cm, but hardly distinguishable as a cluster in this aperture. At 75× about ten stars are resolved, including three or four brighter ones. 15 cm will show 30 members within a 10′×5′ area elongated ENE-WSW. With 25 cm 40 stars are visible in a 15′×10′ field, while 30 cm shows 70 stars, about half of them concentrated into a 20′×5′ band elongated ENE-WSW. There are about a half dozen brighter members and scads of mag. 12 stars.

oc 6383 diam. 4'.0 V=5.5 stars 27
This cluster is just distinguishable to the naked eye as a tiny patch 1°.25 W and a little S of Messier 6, oc 6405 *q.v.* In 15 cm most of the cluster members spread W of the bright lucida (V=5.6), totaling 20 stars in a 20′ area. With 25 cm the brighter stars to the W do not seem to be members. In 5′ diameter around the bright star are 19 others, including two close companions to the lucida (h4962: V=5.6,10.9,10.2; 5″.4,13″; 102°, 83°). With 30 cm the cluster seems even more restricted, encompassing only ten stars within 3′ diameter. Still, nevertheless, most of them lie W of the bright central star.

gc 6388 diam. 8'.7 V=6.7 class III
This globular cluster is small but not difficult to see in 6 cm. At 75× it is a circular spot roughly 1'.5 diameter with a small core and extensive halo. In 15 cm it is 1'.25 across and strongly concentrated to a bright center. A mag. 10.5 star stands out at the N edge. The cluster remains unresolved in 25 cm, though the halo has a finely granular texture and the core is unevenly bright. Overall it is 2'.25 diameter and has a sharply brighter nucleus. With 30 cm it is evenly concentrated from a 3′ halo clear to the center. At 250× it is still unresolved, showing only a granular texture. The mag. 10.5 star 1'.3 N has a very faint close companion, and a faint star is visible near the E edge of the halo.

oc Trumpler 27 diam. 7′ V=6.7 stars 82
This cluster is not conspicuous in the field with 15 cm: a small concentration of stars is gathered around and SE

of two mag. 9 stars. 30 cm shows a total of 40 stars irregularly grouped within 10′ diameter. Dark gaps intervene among the stragglers spreading S and E from the two bright stars. The northeastern mag. 9 star has four faint stars very close to it and forms a close, unequal pair (V = 8.8,13.0; 8″.4; 20°) with one of them.

oc Trumpler 28 diam. 13′ V = 7.7 stars 85

This cluster lies between oc 6383 and oc 6405, *q.v.* Viewed in 25 cm it is an unconcentrated group about 10′ diameter including 30 stars, but nonetheless stands out well from the field.

oc 6396 diam. 3′.0 V = 8.5 stars 22

Viewed in 25 cm this cluster is a compact 3′ group with 15 stars of mag. 10 and fainter. Four stars form an unevenly spaced 1′ string through the middle pointing NW to a bright pair (h4966AC: V = 9.8,10.8; 12″; 273°; each of these is also a close pair: the primary is Innes 1007AB: 10.0,10.3; 1″.0; 117°; the secondary is Innes 1007CD: 11.7,12.0; 2″.5; 245°). An isolated mag. 9 star is about 3′ W of center.

oc 6400 diam. 12′ mag. 8.8p stars 60

6 cm shows this cluster as a faint haze with one or two threshold sparkles about 10′ N of two mag. 9 stars. With 15 cm about 25 stars are visible. From a central, roughly circular clump, straggling arms extend NE, S, and W. It is a fairly bright group in 25 cm, showing 40 stars in an 8′ area. A blank space separates several stars on the W side. 30 cm reveals about 30 stars in an uncondensed, strung-out group 8′ × 4′ in extent elongated N-S.

oc 6404 diam. 6′ mag. 10.6p stars 50

Viewed in 25 cm this cluster is a very faint unresolved glow about 5′ diameter visible best at lower powers. A mag. 10 star is superposed on the W edge. It remains unresolved in 30 cm and seems only about 3′ diameter. A second, fainter star marks the E edge. oc Trumpler 27, *q.v.* lies 45′ SW.

oc 6405 diam. 33′ V = 4.2 stars 331

Messier 6 is easily visible with the naked eye and is resolvable even in binoculars. In 6 cm about 40 stars are visible in a glittering group about 20′ across elongated NE-SW. Half a dozen brighter members stand out, the brightest a conspicuously red star on the E side (BM Scorpii). Fainter stars fill out a clearly defined 30′ area in 15 cm, giving a total of 50 stars. Four of the brightest stars (including BM Sco) outline a 10′ × 6′ parallelogram on the NE side, the long dimension oriented E-W. In 25 cm the main group is about 30′ × 15′, including 100 stars at 100 ×, though outliers extend to 45′ diameter. A V-shaped asterism of mag. 10–11.5 stars lies between the two stars that mark the western vertices of the parallelogram.

oc Trumpler 29 diam. 12′ mag. 7.5p stars 30

This cluster is a hazy irresolved patch in 6 cm, lying just S of an arc of three bright stars. About half a dozen faint stars are visible at 75 ×, including a wide pair in the W side (56″; 270°). In 15 cm the western member of this pair is resolved into a close triplet of three mag. 11 stars aligned E-W, each about 13″ apart. Between this multiple (and its more distant fourth component) and the rest of the cluster to the E is an area devoid of stars. 25 cm will show about 30 stars counting stragglers to the S. The cluster is about 7′ diameter, and the blank area E of the multiple is conspicuous at 200 ×. In 30 cm around 25 or 30 stars seem to be members, most of them grouped in a 4′ area.

oc 6416 diam. 30′ V = 5.7 stars 304

Though visible in 6 cm, this cluster is hardly distinguishable from the field in this aperture. About 20 stars are visible, including several brighter ones and a reddish star on the SSW side. In 25 cm it is a large widespread group of about 60 stars in a 35′ × 15′ area elongated roughly N-S. Most of the brighter stars are in the S side, but the fainter members are concentrated in the N. 30 cm reveals about 50 stars loosely scattered in a 25′ × 13′ area. Two brighter stars are located on each side of the N end of the cluster; the two on the E side are red.

oc 6425 diam. 15′ V = 7.2 stars 73

25 cm shows about 35 stars from mag. 11 to 13 in this cluster. It is about 10′–12′ diameter and has a hole near the center that contains only two stars. In 30 cm the cluster has a roughly triangular outline and contains 23 stars. The members are uniformly distributed except for a knot of five stars at the center that includes a 30″ triangular asterism of mag. 10–11.5 stars.

gc 6441 diam. 7′.8 V = 7.2 class III

Located just over 4′ E of G Scorpii (V = 3.2), 15 cm shows this cluster as a bright central blaze surrounded by a halo to 1′.25 diameter. A mag. 10 star is just off the SW edge 1′.5 from the center; a mag. 12.5 star is the same distance NNW. In 25 cm it grows to nearly 3′ diameter; the star SW is on the exact edge of the visible halo. The bright core is 30″ across. The core/halo and halo/sky boundaries are well defined. With 30 cm the cluster is a little more than 2′ diameter with a broadly concentrated 55″ core. At high power it is unresolved, but the outer edges have a granular texture.

oc 6444 diam. 12′ mag. stars 40
Located 50′ W of Messier 7, oc 6475 *q.v.*, 15 cm shows this cluster as an unconcentrated group of about 30 stars in a 20′ area. With 25 cm 40 stars are visible, most of them mag. 11.5–12. The stars show no general concentration toward the center, but form several small clumps of two to six members each.

oc 6451 diam. 8′ mag. 8.2p stars 80
6 cm shows this cluster as a faint granular patch with an irregular outline. 25 cm reveals a 4′–5′ cluster of 30 stars in a nebulous haze that does not stand out well from the rich background. With 30 cm about 35 stars are visible in an 8′ area. It has an irregular form: most of the stars are concentrated into a narrow N-S bar with many others in a group to the E.

gc 6453 diam. 3.′5 V=9.8 class IV
This globular lies behind the outliers on the NW side of Messier 7, oc 6475 *q.v.* It is 1.′25 diameter and unresolved with 25 cm at 200×. In 30 cm it is about 1.′8 diameter, poorly concentrated, and slightly elongated in pa 120°. A mag. 13 star sparkles at the edge of the core 20″ from the center; another is about 40″ SE. A few faint stars are sprinkled over the cluster, but they may only be in the foreground.

oc 6475 diam. 80′ V=3.3 stars 54
Messier 7 is a conspicuous, sparkling patch in the Milky Way to the naked eye; even the smallest telescopes afford good views of this cluster. 6 cm will reveal about 35 stars in an irregular and slightly condensed group. A distorted K-shaped asterism stands out. In 15 cm the cluster fills a 90′ field with many blue stars and a single orange star on the SW side of the central group. It fills a 70′ field in 25 cm with at least 100 stars. In the center is a box about 10′ across with 30 stars involved; inside it is the K-shaped asterism.

oc Trumpler 30 diam. 8′ mag. 8.8p stars 20
In 15 cm this cluster is a triangular group of 50 stars over an evenly distributed haze about 10′ diameter. The N apex is marked by a mag. 8 star. It is a rich cluster in 25 cm, showing about 120 stars uniformly spread over a 20′ diameter. The brighter stars are clumped into the triangle, but many fainter stars are scattered to the S.

Sculptor

This inconspicuous region contains some of the finest galaxies in the sky, belonging to a group nearby to the Milky Way Galaxy. eg 55 is the largest and brightest, but eg 253 is much better placed for northern observers. The center of the constellation culminates at midnight about 27 September.

eg 24 dimen. 5′.5 × 1′.6 V = 11.5 sfc. br. 13.7
While this galaxy is only faintly visible with 15 cm, 25 cm shows it to nearly 5′ × 1′ in pa 45°, a lenticular form with low surface brightness. At 200 × the core is off-center toward the NE, leaving a more extensive halo to the SW. A mag. 12.5 star is visible just E of the NE tip. In 30 cm it is about 3′ × 0′.7 with thin tips and a bulging middle. At 225 × the extensions of the halo to the NE and SW are unevenly but distinctly brighter along the major axis.

eg 55 dimen. 32′ × 6′.5 B = 7.9 sfc. br. 13.5
This galaxy is the brightest of the Sculptor group and presents a fine view in 6 cm at 25 ×, although at higher powers the view rapidly deteriorates. The large halo is elongated in pa 110° and is distinctly brighter on the W side. This galaxy is one of the most interesting and detailed in the sky with 25 cm. The overall size is 30′ × 4′ with a much-detailed 5′ × 3′ core off-center to the W. A knot is visible a little SE of the core as well as two smaller patches well E from the center, past a star on the S flank. 30 cm shows it 18′ × 4′, a long rod-shape with a more sharply defined northern flank. The patch off the E end of the core (designated I1537) has a stellar center surrounded by a small hazy area.

eg 131 dimen. 2′.1 × 0′.6 mag.$_{vv}$13 sfc. br.
In 30 cm this galaxy is a fairly faint, roughly circular splotch about 1′ diameter, though a faint star 1′.8 NE occasionally makes it look elongated. There is some concentration to a moderately brighter 15″ core. eg 134, *q.v.*, lies 9′.4 E.

eg 134 dimen. 8′.1 × 2′.6 V = 10.1 sfc. br. 13.2
This galaxy is a fine sight with 25 cm. The halo extends to 6′ × 1′.2 in pa 45°. The core appears roundish but confusing, about 1′ across, with a mag. 13 star on its NNW about 35″ from the center. It is a beautiful object in 30 cm, appearing like a miniature eg 4565, *cf.* in Coma Berenices. The thin 6′.5 × 1′ bar is concentrated to a faint nucleus. At high power a dark lane is clearly visible crossing the broadly oval core on its NW side, with a faint bit of core on the far side near the star NNW. eg 131, *q.v.*, lies 9′.4 W.

eg 148 dimen. 2′.4 × 1′.2 V = 12.1 sfc. br. 13.1
Viewed in 25 cm, this galaxy lies about 5′ W of a 3′ triangle of mag. 12 stars. It appears very small, 30″ diameter, with high surface brightness and a stellar center. With 30 cm the halo extends to about 1′ × 0′.6 elongated E-W, and is well concentrated to a nucleus that looks stellar at low power.

eg 150 dimen. 4′.2 × 2′.3 V = 11.1 sfc. br. 13.4
In 30 cm this galaxy is a broadly and weakly concentrated glow with a fairly high surface brightness. At high power the halo extends to 2′.8 × 1′.8 in pa 110°. The core is a large, unevenly bright region that occasionally seems elongated in pa 150°. The nucleus is inconspicuous at all magnifications.

eg 253 dimen. 25′ × 7′.4 V = 7.1 sfc. br. 12.6
This galaxy is easily visible in binoculars and a fine sight in any aperture. With 6 cm it is a high surface brightness, broadly concentrated spindle elongated NE-SW lying just N of two mag. 9 stars. In 15 cm the halo can be traced out to 25′ × 5′ in pa 50°. At least three stars are visible superposed on the nebula itself. Toward the closer mag. 9 star S is a brighter area in the halo like a smaller secondary core. 25 cm shows the halo extending to at least 30′ length. In 30 cm the galaxy is a huge, mottled glow showing hardly any overall concentration except for an indistinct 3′ × 1′ core. At 225 × a dark band lies against the NW side of the core. Other than the two bright stars off the S flank, about a dozen faint stars are intimately involved with the galaxy.

eg 254 dimen. 2′.1 × 1′.1 V = 11.8 sfc. br. 12.6
Located 5′ SW of a mag. 7 star, this is a moderately bright object in 25 cm. The halo is elongated in pa 135°, extending to 1′.5 × 0′.3 around a tiny circular core and a stellar nucleus. 30 cm shows a small, moderately faint

glow about 50″ diameter with a bright center and a conspicuous stellar nucleus.

gc 288 diam. 14′ V=8.1 class X
This cluster, located 1°.8 SE of eg 253, *q.v.*, is easy to see in 6 cm as a broad, poorly concentrated glow without resolution. It seems elongated SE-NW in 25 cm, about 10′ × 8′, and is well resolved at 200 × . In 30 cm the cluster is similar to but not as rich as Messier 4, *cf.* gc 6121 in Scorpius. At high power the 3′.5 core is resolved into many very faint stars, outliers extending irregularly around them to about 7′ diameter.

eg 289 dimen. 3′.7 × 2′.7 B=11.8 sfc. br. 14.2
This is quite a bright object for 25 cm, which shows a moderately low surface brightness 2′ × 1′.5 halo elongated in pa 155° with a brighter, sparkly center. In 30 cm it is about the same size, but shows a broad concentration from a mottled halo to a faint nonstellar nucleus. A mag. 13.5 star is just off the N end, 1′.2 from center.

eg 300 dimen. 20′ × 15′ B=8.7 sfc. br. 14.7
A member of the Sculptor group, this galaxy has very low surface brightness for 6 cm, but is discernable just NW of a mag. 10 star plotted on the AA. At low power it is a diffuse glow about 10′ diameter, elongated ESE-WNW, but with no central brightening. In 25 cm it is about 15′ × 10′. Several stars are involved with the nebula: a mag. 11.5 star is about 1′.2 E of center; another, brighter star lies 2′.5 SW. 30 cm shows the galaxy with difficulty at 150 × in the NW side of the 5′ triangle formed by the three bright stars visible in 25 cm. The indefinite glow is about 6′ diameter and without central concentration.

eg 439 dimen. 1′.8 × 1′.1 B=12.7 sfc. br. 13.3
eg 441 dimen. 1′.1 × 0′.8 mag.$_{vv}$14 sfc. br
eg 439 is a moderately faint object for 25 cm, appearing as a 1′ circular spot with a starry core. In 30 cm it is about the same size but brightens evenly to a very bright 20″ core and stellar nucleus. Two mag. 14 stars are visible 2′.3 SW and SE. eg 441 lies 2′.6 S, and with 25 cm it is an unconcentrated, diffuse glow about 30″ diameter lying 1′.3 SW of the star that is SE of eg 439.

eg 491 dimen. 1′.5 × 1′.3 B=13.2 sfc. br. 13.8
Viewed in 30 cm this galaxy is small but not difficult to see just 50″ ENE of the E tip of a 4′ equilateral triangle of mag. 11–12 stars. At high power it is a broadly concentrated oval elongated roughly E-W with a bright nonstellar nucleus. Photographs show eg E352-G41 (mag.$_{vv}$12.5) 27′ W.

eg 613 dimen. 5′.8 × 4′.6 V=10.0 sfc. br. 13.4
Located 2′.4 SW of a mag. 8.5 star, this galaxy is quite bright to 25 cm. The halo is elongated in pa 120°, 4′ × 0′.75, with a tiny flattened core of high surface brightness. A knot is visible about 45″ SE of center on the major axis. 30 cm shows a broadly elliptical 4′ × 2′ halo. The unevenly bright core is elongated in pa 60° and contains a brighter 25″ nucleus. A faint arm originates near the S end of the core, curving E and then N before fading.

eg 7507 dimen. 2′.6 × 2′.6 V=10.4 sfc. br. 12.5
In 15 cm this galaxy is visible 6′ NW of a mag. 10 star, and 3′.2 SE of a mag. 11.5 star. It is a small, concentrated spot about 30″ diameter with a stellar nucleus. Overall the galaxy has a brightness similar to the star NW. At 50 × in 25 cm, the galaxy is about 45″ across with a 20″ core and a stellar nucleus. At 225 × in 30 cm the halo seems slightly elongated NE-SW, though the brighter regions are quite circular. The small, bright core is about 10″ across; a stellar nucleus barely stands out within it. eg 7513 (V=11.8) lies 18′ NE, 6′ N of an unequal pair of stars.

eg I5332 dimen. 6′.6 × 5′.1 V=10.6 sfc. br. 14.2
This galaxy is too faint for 15 cm and remains difficult even in 25 cm and 30 cm. In 25 cm it is a faint, barely concentrated patch about 1′.5 diameter. With 30 cm the view is a little better, though here a smaller region near the center about 20″ across stands out steadily, yet shows no indication of a nucleus.

eg 7713 dimen. 4′.3 × 2′.0 B=11.6 sfc. br. 13.8
In 15 cm this galaxy is detectable as a diffuse patch just E of two mag. 11 stars. With 25 cm the faint halo extends to 3′ × 1′.5, elongated N-S, growing broadly brighter toward the center and a faint stellaring that lies a little W of center. Off the N tip is a mag. 12.5 star. 30 cm shows a large, low surface brightness glow with little or no central brightening and an overall size of 2′.9 × 1′.8. Photographs show eg 7713A (mag.$_{vv}$13) 17′ NE.

eg 7755 dimen. 3′.7 × 3′.0 B=12.0 sfc. br. 14.5
This galaxy is visible in 15 cm as a faint spot 9′.5 ESE of a mag. 9 star. With 25 cm a smoothly textured, low surface brightness halo spreads to 1′.25 × 0′.75, elongated NE-SW, around a small, grainy core. In 30 cm the halo grows to 2′.5 × 0′.9. The core has a circular outline about 50″ diameter and contains a conspicuous nucleus that is stellar at low power. At times there seem to be dark spots in the inner core near the nucleus, but these are probably merely contrast effects caused by the bright nucleus.

eg 7793 ***dimen. 9!1 × 6!6*** ***V = 9.1*** ***sfc. br. 13.4***
In 15 cm this galaxy is large and fairly bright overall, but of low surface brightness. The smoothly oval halo extends to about 6′ × 4′, elongated in pa 100°. It shows little general concentration, though it has a small, slightly brighter spot near the center. A mag. 12.5 star is off the N edge 2!8 from center. With 25 cm the halo grows to 8′ × 4′, dominated on its S side by a large, slightly brighter patch. 30 cm shows the halo to about 6′ × 3!5, a large, broad oval with irregular edges. The halo is spotted with discrete patches while the core has a more finely mottled texture. The nucleus, a distinct nonstellar spot at 150 × , is not so apparent at 225 × . A close triplet of stars lies 8!5 N.

Scutum

T. W. Webb said, ". . . the very ground of the Milky Way seems here resolvable." The Scutum star cloud is a prominent feature in the summer Milky Way for northern observers. The center of the constellation culminates at midnight on 1 July.

oc 6631 ***diam. 7′*** ***mag. 11.7p*** ***stars 30***
This cluster is an elongated, hazy group of about 17 stars in 15 cm, the brighter members grouped toward the NNW and SSE ends. A mag. 11,11.5 pair of 12″ separation in pa 230° is isolated off the SSE edge. 25 cm will show 35 stars in a 6′ × 3′ area elongated in pa 150°. The NW and SE ends are each marked by mag. 10.5 stars; the rest of the members are mostly fainter than mag. 12.

oc Ruprecht 141 ***diam. 6′*** ***mag.*** ***stars 20***
In 25 cm this cluster stands out as a condensation in the more-or-less uniform background of the Milky Way. This is the most conspicuous group of three in the area (including oc Ruprecht 142 and oc Ruprecht 143, *q.v.*) since it contains a relatively bright star near the center. At 200 ×, 25 stars are visible in a 7′ area, most of them north of the central mag. 8.5 star. The members are mag. 11 and fainter, and include a few close pairs.

oc Ruprecht 142 ***diam. 5′*** ***mag.*** ***stars 15***
This cluster lies between oc Ruprecht 141 and oc Ruprecht 143, *q.v.*, and is the least conspicuous of the group in 25 cm, difficult to distinguish from the rich background of the Milky Way. At high power two mag. 10 stars stand out amongst a dozen or more of mag. 12 and fainter scattered irregularly through a 4′–5′ area.

oc Ruprecht 143 ***diam. 6′*** ***mag.*** ***stars 30***
This cluster is an arc-shaped group opening to the W in 25 cm, lying 23′ NE of oc Ruprecht 141 and 9 ′ NE of oc Ruprecht 142, *q.v.* At the midpoint of the arc is the brightest star, mag. 9. At 200 ×, 15 other mag. 12.5–14 stars are visible in a slightly hazy 4′ × 3′ area elongated N-S. About 9′ NNW is a small group of six or eight stars of mag. 10 and fainter.

oc Ruprecht 144 ***diam. 2.′0*** ***mag.*** ***stars 10***
On the southern border of an extensive dark cloud, this cluster is a group of 15 stars in 15 cm, uncondensed except for a small knot on the SW side. With 25 cm the brightest members are in this knot, a 1.′5 × 0.′5 group elongated SE-NW. The rest of the members are loosely spread to the N and E, forming an irregular circlet about 4′ across. A total of 20 stars is visible, one of mag. 10.5, the rest mag. 11.5 and fainter.

oc 6649 ***diam. 7′*** ***V = 8.9*** ***stars 477***
This is a faint cluster in 15 cm, lying in a blank, nearly starless field. The brightest member is a close, unequal pair on the S edge (ADS 11441: V = 9.7,11.4; 4.″1; 90°); the rest of the members are too faint to be seen easily at low power. At 125 ×, 18 others are visible in a 4′ area. With 25 cm the brighter members lie around the periphery; inside is a single mag. 12 star near the center and about 20 others of mag. 12.5 and fainter. In 30 cm about 20 stars are visible loosely scattered in a 5′ area.

oc 6664 ***diam. 12′*** ***V = 7.8*** ***stars 60***
In 6 cm this cluster is a partially resolved cloud 20′ E of α Scuti (V = 3.9). 15 cm will show 35 stars in a 25′ area, the brighter members lying mostly in the N side. Medium powers show at least 60 stars in 25 cm, all fainter than mag. 10. Overall it exhibits no concentration toward the center and is slightly elongated N-S. An irresolved patch on the N side is resolved by 30 cm into a small group of seven or eight mag. 14 stars. On the E side a string of about ten stars trails S, giving the cluster its overall N-S elongation; without this appendage, the rest of the cluster is elongated E-W.

oc Trumpler 34 ***diam. 5′*** ***V = 8.6*** ***stars 87***
Viewed in 25 cm this is a small, moderately concentrated cluster 4′–5′ diameter, including a total of about 30 stars. A few mag. 11.5 stars lie near the center; the rest of the members are much fainter.

oc 6683 ***diam. 3.′0*** ***V = 9.4*** ***stars 27***
This is a pretty cluster in an interesting field with 25 cm. About 20 stars are visible in a 4′ × 2.′5 group elongated NE-SW. Just over 10′ W is the edge of the Great Rift: the border is sharply defined and makes a striking contrast at low and medium powers, where the exceedingly rich and nearly blank fields split the field of view. Two elongated groups of field stars lie about 40′ and 50′ N, each

Chart XXI. NGC 6649. A 30-minute exposure on 103a-O with the USNO 1 meter telescope (diaphragmed to 0.5 meters) taken by A. Hoag on 30 April 1959. Magnitude sequence from: Madore and van den Bergh 1975, Ap. J. 197: 55.

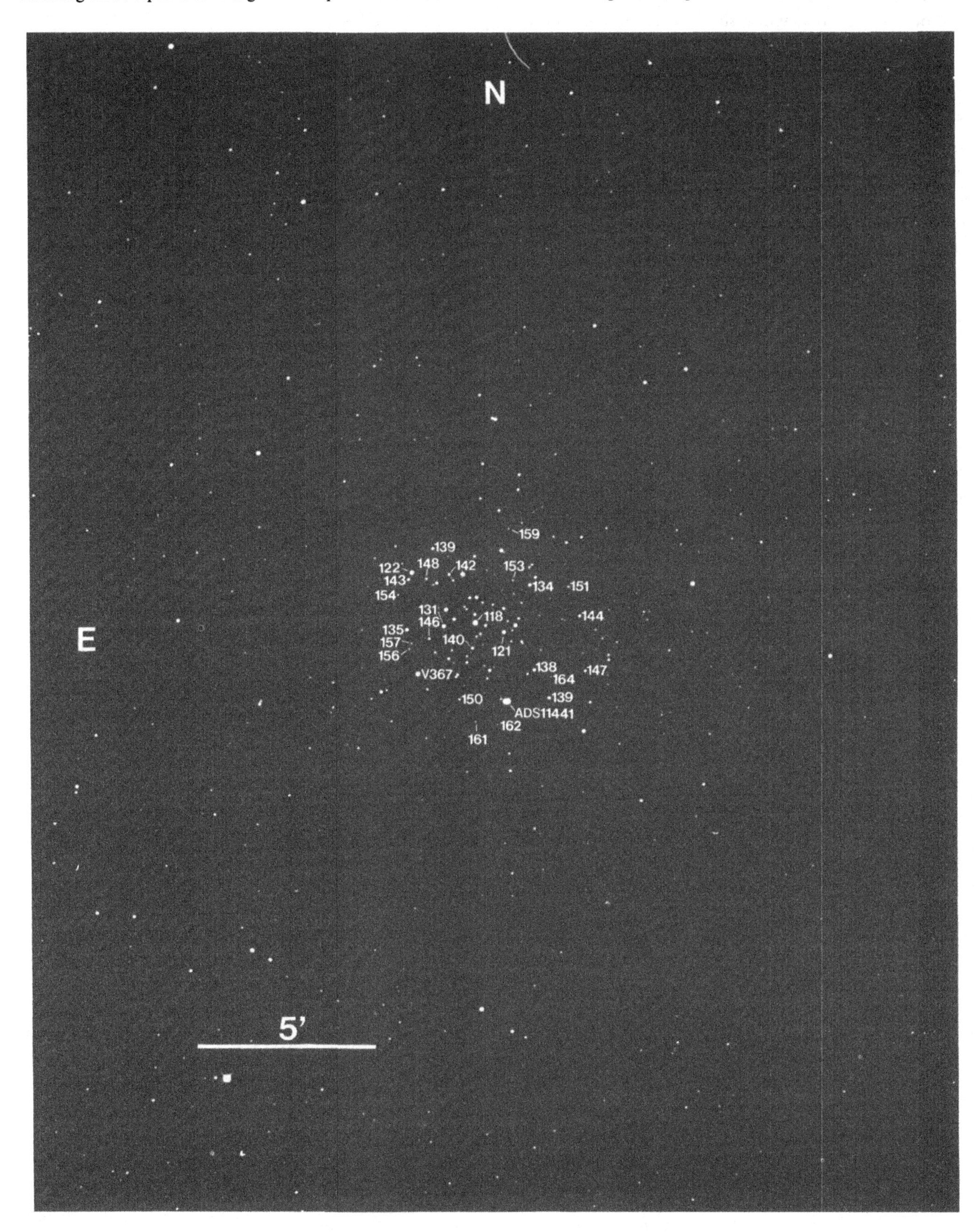

hazy and well concentrated at low power, and revealing large numbers of faint stars at 200 ×. A large, quite circular group of slightly brighter stars 20′ across stands out from the background about 25′ S.

oc Trumpler 35 ***diam. 6′*** ***V = 9.2*** ***stars 65***
15 cm will just show this cluster as a slightly hazy patch associated with five or six faint stars. In 25 cm it is a conspicuous, hazy group at 50 ×, which medium powers resolve into two dozen mag. 10 and fainter stars within a circle 5′–6′ diameter. The members are moderately concentrated toward the center, and the cluster stands out well from the field.

oc 6694 ***diam. 8′*** ***V = 8.0*** ***stars 120***
Messier 26 is a small, faintly sparkling patch in 6 cm. A conspicuous mag. 9 star shines from the SW edge; perhaps a half dozen others are visible scattered over the 4′ haze. In 15 cm about 20 stars are resolved in a 7′ area, while 25 cm shows 35 stars in a 5′ area. The group has some central concentration and a gaping hole in the N side. The cluster is about 9′ across with 30 cm. An arc of mag. 12.5–13 stars is on the NW, and a dipper-shaped asterism lies near the center.

oc 6704 ***diam. 6′*** ***V = 9.2*** ***stars 71***
In 6 cm this cluster is a faint, unresolved haze in a dark, nearly starless field. A single mag. 11.5 star lies off the SSE side. 15 cm will resolve about seven stars, six of them forming a prominent "V" pointing N. With 25 cm, 15 stars seem members in a 6′ area. The eastern arm of the "V" is formed from three of the brightest members. In 30 cm few more stars are visible. The bright star about 2′ SSE is red.

oc 6705 ***diam. 25′*** ***V = 5.8*** ***stars 682***
Messier 11 is one of the best open clusters in the sky for northern observers and is a fine sight in any instrument, visible faintly even with the unaided eye. Higher powers with 6 cm will partially resolve it, 100 × showing about two dozen stars over a thick haze around a prominent mag. 8 star near the center. A wide mag. 9 pair with 43″ separation lies just off the SE edge. In 15 cm the cluster is resolved into a beautiful mass of perhaps 150 stars from mag. 11 to the threshold. Many of the stars are grouped into smaller knots and clumps with dark lanes winding amongst them: two more prominent ones extend for some distance through the cluster, starting near the center and leading out to the N and W. Where these lanes reach the edges of the cluster, the outline seems indented, giving the group an irregular form. With 30 cm a nearly uncountable horde of stars is visible. The central, bright portion of the cluster is about 8′ across; this portion is very rich but shows little central concentration. Beyond this, outliers spread out to an overall diameter of 15′, elongated a little N-S, thinning irregularly and indefinitely into the rich background of the Scutum Star Cloud.

gc 6712 ***diam. 7.′2*** ***V = 8.2*** ***class IX***
This cluster is a small but distinct spot in 6 cm. At low power in 15 cm it stands out well from the sparkling, starry foreground, but will show only a few stars cast over it at 200 ×. With 25 cm at higher powers it is granu-

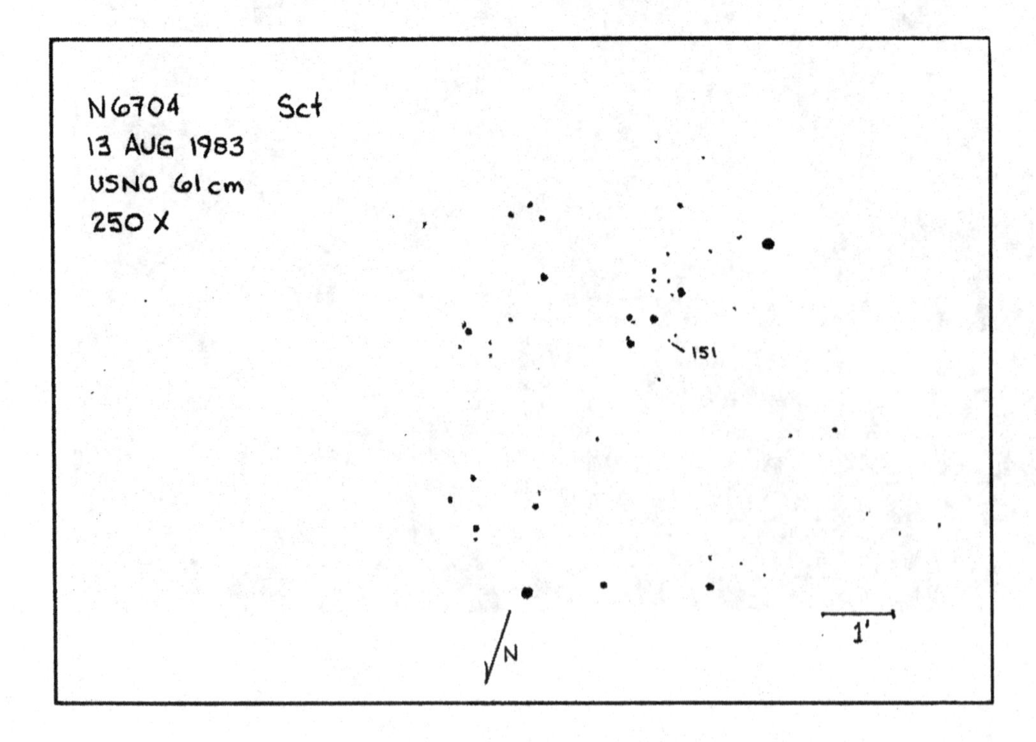

lar to partially resolved and has an overall size of 2!5. 30 cm resolves many stars and shows faint outliers to 5!5 diameter around a 2!5 core. Very few outliers are to the S, flattening the halo on that side.

pn I1295 dimen. 1!7 × 1!4 nebula 12.7v star 15:p
In 25 cm this is a large and diffuse planetary, seeming to lie in front of the rich star field: two superposed mag. 13.5 stars near the E and W edges seem to be behind the nebula. The ghostly blob is a nearly circular, almost uniformly bright glow. At 200× a faint stellaring is visible on the NW edge. With 30 cm the nebula is about 2′ diameter, slightly elongated E-W, and has a fairly distinct 40″ hole in the center.

Serpens

The Serpent is a long string of stars in the grip of Ophiuchus. The head holds some galaxies, while the tail is in the Milky Way. Serpens Caput culminates at midnight about 18 May, then Serpens Cauda does so about 23 June.

gc 5904 ***diam. 17′*** ***V = 5.7*** ***class V***
Messier 5 is a bright cluster easily visible in 6 cm 22′ NNW of 6 Serpentis (V = 5.4). At low power it is a bright, well-concentrated glow with granular edges. In 15 cm it is about 10′ diameter and very strongly condensed at the center. At 200 ×, many stars are resolved, leaving little haziness. 25 cm shows granularity at 50 × and resolves it well at 200 ×. The resolved stars seem to form a many-armed pattern over the cluster. A string of four or five mag. 12.5 stars runs across the N side of the core. It fills a 13′ field in 30 cm, giving the impression of a rich open cluster superposed on a bright galaxy. The core is granular, about one-fourth the total diameter. More outliers spread to the S, W, and N than to the E, giving the halo an asymmetric form. A mag. 10 star is on the SW, 2.′8 from center.

eg 5921 ***dimen. 4.′9 × 4.′2*** ***V = 10.8*** ***sfc. br. 14.0***
Located 2.′9 NW of a mag. 9 star, this galaxy is faintly visible in 15 cm as a concentrated, roughly circular glow closely involved with a mag. 12 star on its SW edge. In 25 cm the star at the SW is the northernmost of a group of four mag. 12 stars. The galaxy extends to 1′ × 0.′75 in pa 30° and is well concentrated to a bright but nonstellar center. In 30 cm the four stars to the S form a pretty arc leading to the galaxy, which forms the fifth "star" in the chain. The halo has a lenticular form and contains a small core within which a faint stellar nucleus is occasionally discernable.

eg 5936 ***dimen. 1.′5 × 1.′4*** ***V = 12.4*** ***sfc. br. 13.1***
25 cm shows this galaxy as an unconcentrated but somewhat mottled circular spot about 1′ diameter. In 30 cm the halo extends to 1.′2 diameter around slightly brighter 45″ core. Two or three brightenings are visible near the center, particularly on the E side.

eg 5957 ***dimen. 3.′0 × 2.′9*** ***mag.$_z$13.3*** ***sfc. br.***
Though barely visible in 15 cm, 25 cm will show this galaxy at 100 × as a moderately faint, low surface brightness glow lying 2.′5 S of a mag. 10 star. At high power the halo is slightly elongated E-W, though the small, slightly brighter core is nearly circular. No sharp nucleus is discernable.

eg 5962 ***dimen. 2.′8 × 2.′1*** ***V = 11.4*** ***sfc. br. 13.1***
In 25 cm this galaxy has an irregular, 1.′5 × 1′ halo elongated in pa 120° and a conspicuous stellar nucleus like a mag. 12 star. With 30 cm the high surface brightness and slightly mottled halo extends to 2′ × 1′ around a large, broad core. Here the nucleus is not so conspicuous, though it remains stellar. The brightness fades more abruptly to the NE, placing the nucleus slightly off-center to the SW.

eg 5970 ***dimen. 3.′0 × 2.′1*** ***V = 11.4*** ***sfc. br. 13.3***
In 25 cm this galaxy is fairly bright. The squarish halo is 1.′5 × 0.′75, elongated in pa 90°. The core is even more strongly elongated, a 50″ × 10″ spindle-shaped brightening with several brighter splotches giving it a nearly grainy texture. Centered in the core is a nearly circular region about 10″ across that occasionally shows a faint stellar nucleus. 30 cm shows a 1.′2 × 0.′6 halo and a broadly concentrated core about 45″ long. eg I1131 (mag.$_z$14.8) lies 8.′2 SE.

eg 5984 ***dimen. 3.′0 × 0.′9*** ***B = 12.9*** ***sfc. br. 13.9***
This is a faint but interesting galaxy in 25 cm. The halo is about 2.′5 × 0.′5, elongated SE-NW. The core is only a slightly brighter region around the stellar nucleus. With 30 cm the halo grows to 2′ × 0.′8 in pa 140° around a broadly mottled core about 1′ long. The nucleus is a faint star-like spot that lies a little N of the geometric center of the core. An elongated triangular asterism of mag. 12.5 stars lies 2.′2 N.

eg 6070 ***dimen. 3.′6 × 2.′1*** ***V = 11.7*** ***sfc. br. 13.7***
Viewed in 30 cm this galaxy is a faint, low surface brightness 2′ × 1′ glow elongated in pa 60°. There is a slight, broad central brightening, but no sharp nucleus.

eg 6118 ***dimen. 4.′7 × 2.′3*** ***B = 11.9*** ***sfc. br. 14.3***
In 25 cm this galaxy is a faint, uncondensed patch 3.′5 NNW of a mag. 11.5 star. The irregular halo extends to

Chart XXII. NGC 5904 = Messier 5. A 10-minute exposure on IIa-O plus GG13 filter with the USNO 1.55 meter telescope taken by C. Luginbuhl on 2 June 1982. Magnitude sequence from: Arp 1955, A. J. 60: 317 and Arp 1962, Ap. J. 135: 311.

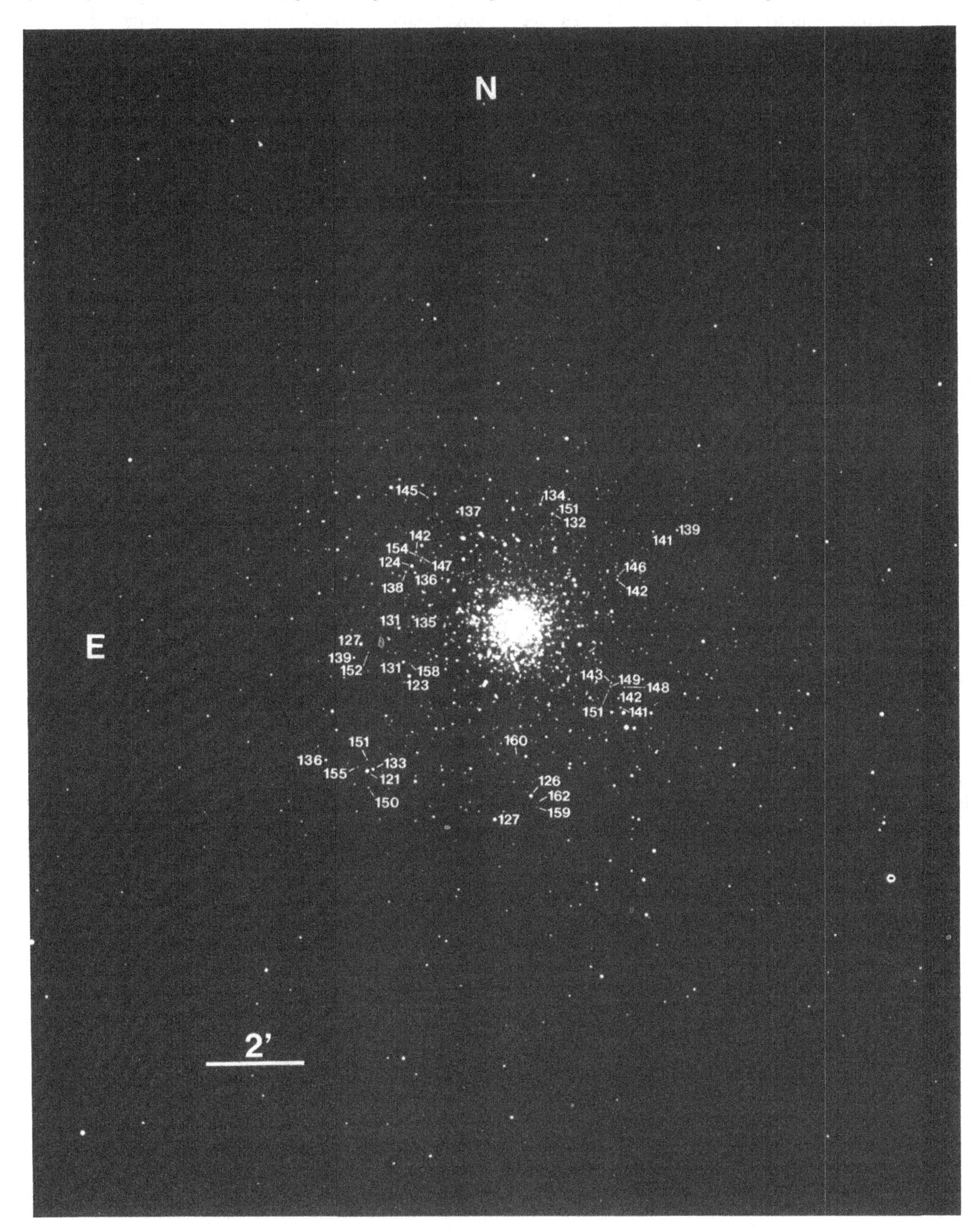

about 4′ × 2′, elongated NE-SW. A very faint star is just off the SE flank, 1′.5 E of the center. 30 cm shows the low surface brightness halo to 3′ × 1′.8 in pa 60°. The surface is slightly mottled, but there is no general brightening toward the center.

gc 6535 ***diam. 3′.6*** ***V = 10.5*** ***class XI***
This faint, unconcentrated globular cluster lies in a rich field, and 25 cm will show it as an irregularly bounded, granular spot perhaps 2′.5 diameter. In 30 cm three mag. 13–13.5 stars arc across the W side, giving the impression of partial resolution even though so few stars are visible. At 225 × several other very faint stars sparkle from the central regions; the three brighter stars seem to lie in the foreground. Overall the cluster is about 1′.5 diameter.

gc 6539 ***diam. 6′.9*** ***V = 9.8*** ***class X***
In 6 cm this cluster is a small, broadly concentrated spot that is a little easier to see than gc 6517, *cf.* in Ophiuchus, because it is larger. 25 cm shows it as a very diffuse unconcentrated glow about 3′.5 diameter. A mag. 12.5 star is visible on the NW edge, 1′.7 from center. In 30 cm the cluster remains unresolved, though the central portions are unevenly bright at 225 ×. A second, fainter star lies 1′.9 W.

oc Trumpler 32 ***diam. 6′*** ***mag. 12.2p*** ***stars 50***
This is a fairly faint cluster in 25 cm, showing about two dozen stars in a roughly circular 5′–6′ area. The southern third is split off from the main body by a 1′–2′ wide dark band aligned SE-NW. This smaller portion contains eight or nine members; the rest has about 20. 30 cm shows about 15 stars over a faint patchy haze 3′–4′ across. The brighter members lie on the N side near the edge.

oc 6604 ***diam. 4′.0*** ***V = 6.5*** ***stars 105***
Viewed in 25 cm, this cluster is a small group of fairly bright stars embedded in the S side of a large association of about thirty mag. 9–11 stars in a 30′ area elongated N-S. In 30 cm the cluster itself is an arc-shaped asterism of five stars, convex to the S. Three or four others in the 235immediate vicinity seem members, bringing the total population to less than a dozen. The brightest member (V = 7.5, slightly variable) is the second star from the E end of the arc.

oc 6611 ***diam. 21′*** ***V = 6.0*** ***stars 543***
gn I4703 ***dimen. 120′ × 25′***
Messier 16 is a nebulous cluster visible to the naked eye in which 6 cm shows about 15 stars, including a bright pair on the W side (V = 8.2,8.8; 27″; 350°). The nebulosity extends through and beyond the cluster to the S, while the N side is abruptly cut off by a triangular dark intrusion. With 25 cm the nebula is a faint haze over the cluster that fans out toward the E to an overall diameter of about 30′. The immersed cluster is 15′ diameter and contains about 30 stars. 30 cm will show 50 stars in a 20′ area, the densest, most concentrated portion being on the NW side. A large loop of stars to the S and SE encircles a fairly starless area wherein the nebula shows best.

oc I4756 ***diam. 40′*** ***V = 4.6*** ***stars 466***
6 cm shows this naked-eye cluster as a widespread group enclosed within a large trapezoid of mag. 5–7 stars. About 75 stars are visible, mag. 7.5 and fainter, with many pairs and clumps in the central portions. It is a magnificent sight in a 15 cm RFT, showing many scores of stars even at low powers. In 25 cm it fills a 70′ low-power field comfortably, showing 80 stars brighter than mag. 11. The cluster is generally unconcentrated toward the center, though some small groups and pairs stand out.

Sextans

This constellation is a group of faint stars at the feet of Leo, and some bright galaxies are included within its boundaries. The center of the constellation culminates at midnight about 24 February.

eg 2967 dimen. 3′.0 × 2′.9 V = 11.6 sfc. br. 13.8
25 cm shows this galaxy as a faint glow about 2′ diameter with moderate central brightening. The core is broadly concentrated and without a nucleus in 30 cm, and there is little outer haze. The overall size is about 1′ diameter.

eg 2974 dimen. 3′.4 × 2′.1 V = 10.8 sfc. br. 12.9
This galaxy is visible at low power in 15 cm as a concentrated spot closely involved with a mag. 10 star. At higher powers the halo seems elongated directly toward the nearby star, which lies only 40″ SW of the center. In 25 cm a mag. 13 stellar nucleus sparkles inside a moderately bright 1′.2 × 0′.5 halo elongated NE-SW. The faint extension of the halo ends just S of the star in 30 cm, giving an overall size of 1′.5 × 0′.7. At 225× the halo shows even concentration to the center, where a distinct but nonstellar nucleus stands out.

eg 2990 dimen. 1′.3 × 0′.8 V = 12.8 sfc. br. 12.6
In 25 cm this galaxy is a fairly faint circular spot 1′.5 SW of a mag. 13 star in an otherwise barren field. Overall it is perhaps 1′ diameter and has a broad, uniformly bright core. The halo is only 40″ across in 30 cm and shows weak, even concentration to the center.

eg 3044 dimen. 4′.8 × 0′.9 V = 12.0 sfc. br. 13.5
A moderately faint object in 25 cm, this galaxy is a thin, unconcentrated spindle extending to 4′ × 0′.75 in pa 115°. A mag. 14 star lies about 1′.5 N of the W tip. In 30 cm the halo is about 3′ × 0′.5 and shows some central concentration to an elongated, bar-like core that is about 2′ long. At 225× this brighter portion is unevenly bright, showing several spots and beads along its major axis.

eg 3055 dimen. 2′.2 × 1′.4 V = 12.1 sfc. br. 13.1
Located 5′.8 SSE of a mag. 10.5 star, 25 cm shows this galaxy as a broadly concentrated circular spot about 1′ diameter. A mag. 13.5 star is visible on the WSW edge. In 30 cm the halo grows to 1′.2 × 1′ in pa 65° with a slight, even concentration toward the center.

eg 3115 dimen. 8′.3 × 3′.2 V = 9.1 sfc. br. 12.6
This galaxy is easy to see even in 6 cm, which shows a 3′ × 1′ haze with a faint stellar nucleus in a thin, elon-

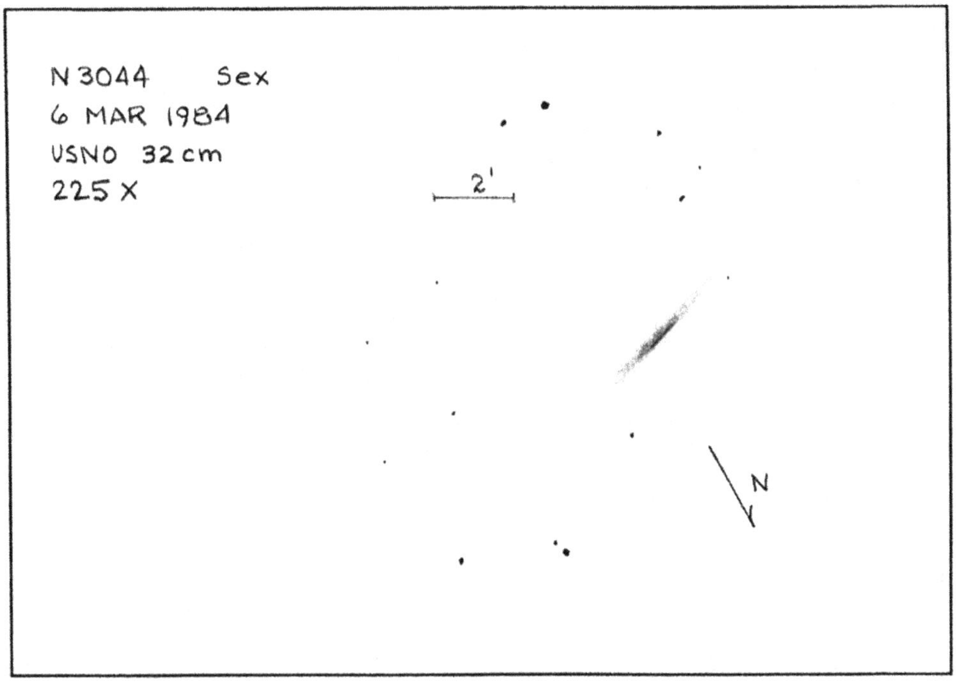

gated core. Two mag. 9.5 stars about 7′ E and SE form a triangle with the galaxy. In 15 cm the distinctly bounded halo is 4′ long and contains a brighter core that occasionally seems to have a rectangular outline. 25 cm shows the halo to 3′ × 1′, elongated in pa 45°. The thickened central bulge is more distinctly rectangular here and contains a bright 45″ core with a stellar nucleus. A mag. 11.5 star lies 3.′3 S. The galaxy is a beautiful sight in 30 cm, the smoothly textured glow growing to 5.′5 × 1′ with faint extended tips. The inner parts are strongly elongated and well concentrated to a bright inner core and stellar nucleus. A mag. 14 star lies 1.′3 S of center.

eg 3156 ***dimen. 2.′1 × 1.′2*** ***B = 13.0*** ***sfc. br. 13.9***
This galaxy lies 2′ NW of the western member of a 6′ triangle of bright stars. In 25 cm it is an evenly and moderately concentrated 40″ × 20″ spot elongated in pa 45°. A faint stellar nucleus sparkles unsteadily in the center. 30 cm shows a circular 1′ glow with moderate concentration to a mag. 14 stellar nucleus.

eg 3165 ***dimen. 1.′6 × 0.′9*** ***mag.$_z$14.5*** ***sfc. br.***
eg 3166 ***dimen. 5.′2 × 2.′7*** ***V = 10.6*** ***sfc. br. 13.3***
eg 3166 is the brightest of a group of galaxies including eg 3165 and eg 3169, *q.v.*, the latter of which is an interacting companion 7.′7 ENE. In 25 cm eg 3166 is a bright object indefinitely elongated E-W. The halo extends to about 1′ × 0.′8, growing evenly brighter to an indistinct core and a bright but inconspicuous stellar nucleus. With 30 cm the halo is 2′ × 1.′5, elongated in pa 90°. Here the 20″ core seems more distinct and contains a substellar nucleus. Two mag. 12.5 stars lie 2.′9 and 4.′9 WSW and SW, respectively, lying pretty much N and S of each other. On the W side of the half-way point between these two stars, forming a flat triangle with them, is eg 3165. In 25 cm it is a faint, difficult object but clearly visible with averted vision at medium powers. It is a diffuse patch elongated N-S with a broad central brightening.

eg 3169 ***dimen. 4.′8 × 3.′2*** ***V = 10.5*** ***sfc. br. 13.2***
In 25 cm this galaxy, an interacting companion with eg 3166, *cf.*, is about the same size but a little fainter than its companion 7.′7 WSW. The 1′ × 0.′8 halo is elongated roughly NE-SW around an oval, 20″ × 15″ core and a very faint stellar nucleus. The NE portion seems more sharply defined, though a mag. 11 star just off the E edge of the halo may be influencing this. With 30 cm the halo still does not reach quite to the star 1.′5 E. Overall the galaxy is similar to eg 3166, though it is a little smaller and the nucleus not as nearly stellar.

eg 3423 ***dimen. 3.′9 × 3.′5*** ***V = 11.2*** ***sfc. br. 13.9***
In 25 cm this galaxy is a large, moderately bright object. The rounded halo extends to 3.′5 × 1.′5 in pa 30°, not quite reaching a mag. 11.5 star to the NE; a mag. 10 star with a fainter companion lies about 5′ E. At 100 × the core is a bar-shaped brightening with embedded stellarings. In 30 cm the large, low surface brightness halo grows to 4′ × 2.′5, showing only very slight brightening toward the center. Two faint stars or stellarings lie on the SW edge of the halo.

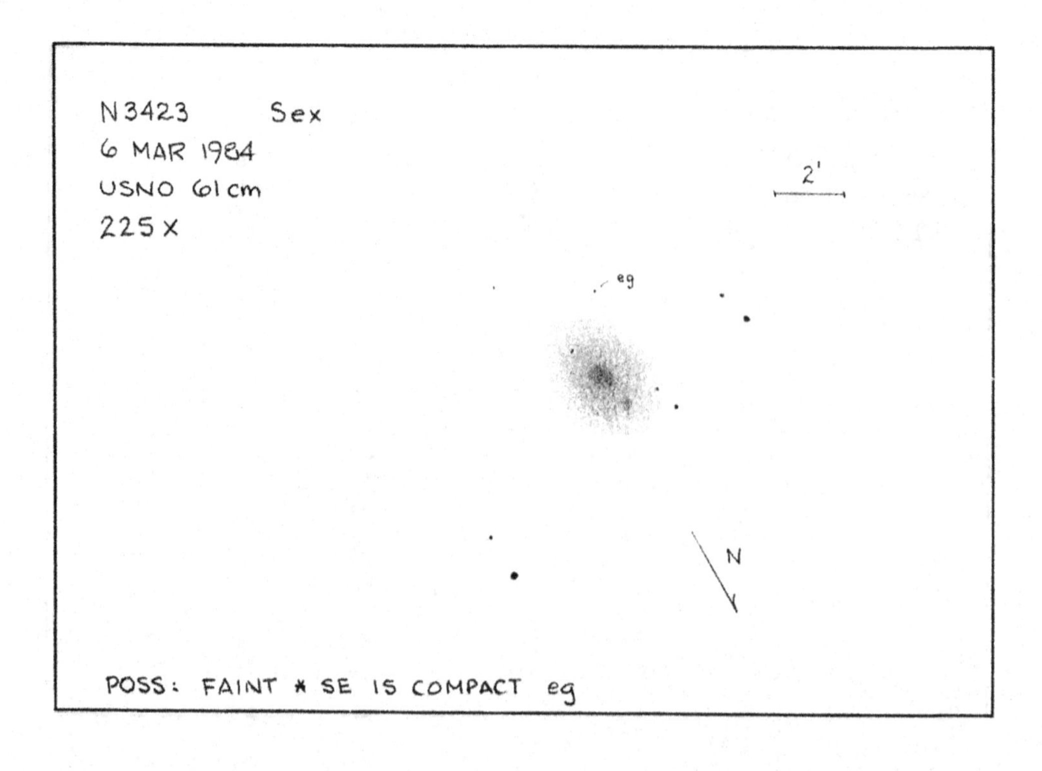

Taurus

This large, widespread constellation is marked by the naked-eye Hyades and Pleiades clusters, two of the closest clusters to the Sun. Besides the Crab Nebula, Messier 1, most of the rest of its deep-sky objects are mediocre. The center of the constellation culminates at midnight about 1 December.

oc Pleiades ***diam. 120′*** ***V=1.2*** ***stars 277***
gn 1435 ***dimen. 30′×30′*** ***star V=4.2***
Messier 45 is a bright, pretty cluster in 6 cm, showing at least 60 stars scattered through a 2°.5 low-power field. The brightest members form the familiar dipper-shaped asterism about a degree across that is discernable to the naked eye. A small triangle of mag. 6.5–8.5 stars lies just NW of Alcyone, and fairly isolated near the middle of the dipper is a mag. 8 pair (41″; 125°), the northwestern member of which is reddish in contrast to the very blue stars all around. Beginning SE of the midpoint between Merope and Alcyone is a bent string of six mag. 9–10 stars that leads S and then SE, each star fainter than the last. 15 cm is the largest aperture in which the cluster presents a nearly entire view. Here about 175 stars are visible in a 90′ area at 75×. Merope, the bright Pleiad at the S corner of the dipper, is the only star associated with a clearly visible nebulosity (gn 1435): the other stars show nothing that is definitely distinguishable from light scattered within the telescope optics. The faint haze fans S from the bright star, bounded on the E by three mag. 8–11 stars in an arc, the brightest at the S end. At the southwestern extremity is a mag. 12 star, the first in a curved group of three.

Many double stars are contained within the cluster: it is these that larger apertures show best. In 15 cm the first star in the bent string of six beginning E of Merope is a close unequal pair (ADS 2767: V=7.3,10.4; 6″.1; 265°). More difficult to resolve is a pair at the N side of the cluster, ADS 2766 (8.5,11.0; 6″.8; 330°). With 30 cm a close triple is separable about 15′ SE of Atlas, the brighter star at the end of the dipper's handle: two form an easy pair (ADS 2795AC: V=6.7,9.0; 10″; 236°), the brighter of which is also a more difficult pair (ADS 2795AB: V=6.7,9.8; 3″.2; 238°). The star about 6′ N of this triplet is also an unequal pair (9,12.5; 9″; 190°), as is the northwestern member of the wide pair near the center [ADS 2755CD: 8.1,12.1; 18″; 8°(1915)].

pn 1514 ***dimen. 2′.3×2′.0*** ***nebula 10.9v*** ***star V=9.4***
While only a trace of this nebula is discernable in 6 cm, its central star is easily found as a grey-hued mag. 10 star, the middle star in a three-star string with a mag. 8.5 star 8′ N and a mag. 10 star 8′ S. The nebula is more easily visible in 15 cm using averted vision, which shows a faint, undetailed glow about 1′ diameter. With 25 cm it is 1′.5 diameter, a diffuse circular haze fading smoothly from the central star. The nebula grows to 2′×1′.3 in 30 cm, elongated in pa 135°. Even at low powers the nebula has an unevenly bright surface; at times a slightly brighter bar seems to run along the major axis, ending near the NW edge with a brighter spot about 20″ across.

oc Hyades ***diam. 35°*** ***V=0.5*** ***stars 380***
This cluster forms the conspicuous V-shaped asterism representing the head of Taurus, the Bull. The very bright red star Aldebaran (V=0.75–0.95) is not a member of the cluster, lying in the foreground. The group is best viewed with the unaided eye and wide-field binoculars. From a dark site, the naked eye shows about two dozen stars in an area about 5° across. Outliers are difficult to distinguish from field stars in the sky nearby. $\theta^{1,2}$ Tauri (V=3.4,3.8; 5′.6; 349°) and $\sigma^{1,2}$ Tauri (V=4.7,5.1; 7′.2; 194°) are, respectively, fairly easy and relatively difficult naked-eye pairs.

eg 1587 ***dimen. 2′.0×1′.9*** ***V=11.6*** ***sfc. br.13.1***
eg 1588 ***dimen. 1′.8×1′.0*** ***V=13.0*** ***sfc. br.13.6***
Both of these galaxies are visible in 15 cm at 75×. eg 1587 is a small, concentrated spot with a conspicuous stellar nucleus, a little fainter overall than eg 1589, *cf.* 12′ N. The faint companion, eg 1588, is a very faint substellar spot on the ENE side, nearly touching the halo of the brighter galaxy. With 30 cm eg 1587 seems brighter than eg 1589, since it is smaller and more strongly concentrated. Overall it is 1′.1 diameter and nearly circular, though it occasionally shows slight elongation in its core toward the nearby companion, i.e., in pa 80° or so. The core itself is about 15″ across but irregular: the elongation seems expressed mostly by an indefinite extension on the side away from eg 1588 to the WSW. eg 1588 has a very low surface brightness halo extending to 45″ diameter; the core is almost as large as that in eg 1587, but

about two magnitudes fainter. A faint stellar nucleus is occasionally visible within at 225×.

eg 1589 ***dimen. 3′.1 × 1′.2*** ***mag.$_z$13.8*** ***sfc. br.***
This galaxy is a small, moderately faint glow in 15 cm, forming a pair with a mag. 13 star 1′ NE. The halo is slightly elongated in pa 160° and contains a faint mag. 13.5 stellar nucleus. With 30 cm the extensive halo grows to 1′.5 × 0′.6 around a fairly sharply concentrated 15″ × 10″ core and a stellar nucleus. A mag. 14 star lies 1′.2 SSW, and eg 1587 and eg 1588, *q.v.*, lie 12′ S.

eg 1590 ***dimen. 1′.3 × 1′.2*** ***mag.$_z$14.6*** ***sfc. br.***
Though easily located in a 1° string of mag. 8.5–9 stars running N-S, this galaxy is only barely visible with 15 cm. In 30 cm it is a small, 40″ spot with very slight elongation in pa 100°. The halo shows moderate even concentration to a poorly defined core and a faint substellar nucleus.

eg 1615 ***dimen. 1′.6 × 0′.9*** ***mag.$_z$15.0*** ***sfc. br.***
In 30 cm this galaxy is a small, moderately faint object. The halo is about 50″ × 30″, elongated in pa 115°. Within is a well-defined, more circular core that exhibits moderate concentration to a conspicuous stellar nucleus.

eg 1642 ***dimen. 2′.0 × 1′.8*** ***mag.$_z$13.6*** ***sfc. br.***
This galaxy is a threshold spot in 15 cm, lying among several faint stars 3′.5 E of a mag. 8.5 star. In 30 cm it remains fairly faint, a 50″ diameter spot with moderate, even central concentration and a small inconspicuous core. About a half dozen mag. 13 and fainter stars are within an area 4′ diameter centered on the galaxy.

oc 1647 ***diam. 40′*** ***V = 6.4*** ***stars 117***
In 6 cm this is a large, widespread cluster of about 40 stars in a 45′ area. A bright pair of mag. 9 stars lies near the center (32″; 110°), with a fainter companion just S of it. In larger instruments the stars appear in pairs or groups of three. In 15 cm at least 45 stars are visible within 40′ diameter, the groups separated by a minute or two of arc. Most of the stars are within a sharply bounded circle 30′ across in 25 cm, with outliers spreading to 60′. In the smaller area are 60 stars distributed without concentration. In 30 cm the cluster is too widespread to view in its entirety: at low powers it is a loose collection of widely separated small groups.

oc 1746 ***diam. 40′*** ***mag. 6.1p*** ***stars 20***
This cluster is a loose, moderately concentrated group in 6 cm, showing about 40 stars in a 45′ area. In general, the members are fainter than those in oc 1647, *cf.*, though six or seven brighter stars stand out, most of them in the W side. With 15 cm the cluster presents a nice field of about 75 stars within a 30′ circle, though the majority of the stars lie within a 30′ × 20′ oval elongated in pa 60°. In a group of brighter stars on the SW side is a wide pair, easily resolved at 75× (9.1,9.2; 19″; 189°).

oc 1807 ***diam. 12′*** ***V = 7.0*** ***stars 37***
6 cm will show this cluster as a loose, irregular group of about a dozen stars arranged into a "T" lying 25′ WSW of oc 1817, *q.v.* A fairly bright, close pair marks the spot

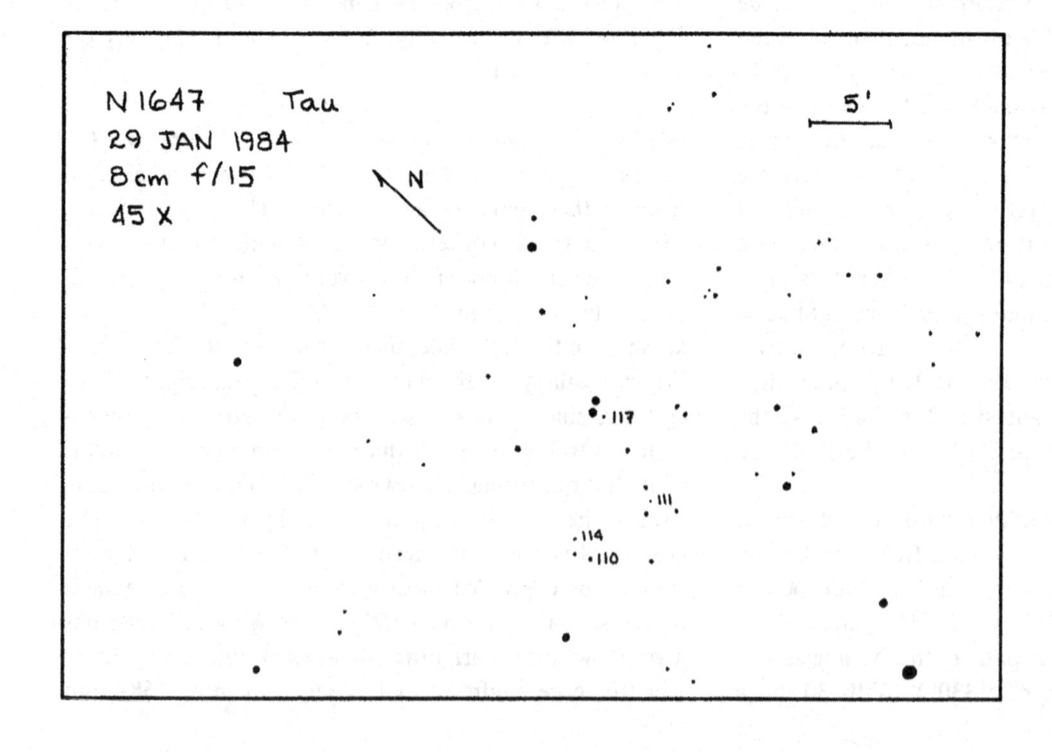

where the stem and top of the T join (ADS 3745: V = 9.6,11.0; 9″.9; 267°). 15 cm shows about 17 stars in the group, while with 25 cm a total of 21 is visible. Two stars near the central pair form a small triangle with it. With 30 cm 20 or 30 stars seem members of the cluster, most of them concentrated in the N-S arc that forms the top of the T.

oc 1817 ***diam. 16′*** ***V = 7.7*** ***stars 283***

This cluster is a faint group in 6 cm, which shows only five stars in an 8′ area elongated N-S. The brightest star (V = 8.6) lies to the NW. In 15 cm several of the brighter stars are grouped into a string running N-S with a hazy cloud of 20 faint stars to the E. The cloud is better resolved in 25 cm, which reveals about 75 stars fainter than mag. 11 in a 15′ diameter. The brighter stars on the W side form a conspicuous semicircular boundary in 30 cm, separating the rich field of the cluster to the E from a much darker, nearly starless field to the W.

gn 1952 ***dimen. 6′ × 4′*** ***star 15.9p***

Messier 1 is a distinct grey patch in 6 cm, broadly concentrated to the center and elongated SE-NW. With 15 cm the nebula is a moderately concentrated lenticular form extending to 5′ × 3′ in pa 135°. With scrutiny the brighter portions give it an "S" shape: the NE edge of the NW half and the SW edge of the SE half are brighter. In 30 cm the "S" shape is quite distinct, and the whole nebula is very unevenly bright. The SE half is generally fainter and more diffusely edged than the NW, the arm of the "S" being fainter and thinner here also. A large dark patch intrudes into this half from the NE, extending up against the SE side of the core and the back side of the arm. The core is elongated in pa 160°, forming the middle part of the "S." This is the brightest part of the nebula, and distinctly uneven in texture. The N edge fades more abruptly than any other place around the periphery. A mag. 13.5 star is on the NE edge of the core; another of the same brightness is on the W edge. Two mag. 13.5–14 stars are embedded in the N arm, one of which is at the tip; two others are embedded fairly symmetrically in the S arm. Several other faint stars are involved.

Triangulum

This is a small constellation, but it contains several bright galaxies, including Messier 33 of the Local Group. This galaxy is a good indicator of clear, dark skies if it is visible to the naked eye. The center of the constellation culminates at midnight about 24 October.

eg 598 ***dimen. 62′ × 39′*** ***V = 5.7*** ***sfc. br. 14.1***
Messier 33 is visible in 6 cm as a bright, weakly concentrated haze, extending to about 55′ × 35′ in pa 30° at low powers. With higher magnification the central regions dominate the view; here the indefinite core shows extensions that make it seem elongated nearly N-S. With 25 cm the halo can be traced to 60′ × 30′, sparkling all over with faint foreground stars. At low powers the halo is distinctly uneven, showing broad, splotchy arms curving out from the core, particularly prominent on the N and S flanks. A mag. 13 star 55″ NE of the mag. 13 stellar nucleus lies near the edge of the indistinctly bounded core. In 30 cm the halo spreads entirely across a low-power field, leaving no dark sky against which to determine its maximum extent. At 225 × the core is a generally smooth, moderately concentrated region with some fine detail. The field S of this has six or eight mag. 13 stars sprinkled over it. The whole surface of the galaxy is covered with faint stellarings and splotches.

Many faint clusters, associations, and nebulae are visible within the galaxy to larger apertures. The brightest, a giant HII region 12′ NE of the nucleus known as NGC 604, is visible even to 6 cm as a small concentrated spot in the halo 1′.1 NW of a mag. 10.5 star. In 25 cm it is a 30″ × 20″ spot elongated in pa 120° containing two stars: between these the nebulosity is fainter. 30 cm shows it to about 1′ diameter, a smooth, concentrated spot like an elliptical galaxy.

About 50″ ESE of the mag. 10.5 star near NGC 604 is A85 (see photo), visible in 25 cm as a faint stellaring of about mag. 13.8. Way out in the NNW part of the halo, A137 is faintly visible as an elongated patch with what looks like a faint star at its southern tip: this is actually a whole cluster. Just north of this association is a close mag. 13 pair with 11″ separation in pa 140°.

Moving westward along the spiral arm from NGC 604, past a mag 12.7 star, A75 is a brighter smudge with a mag. 13.5 star at its SE edge. A67 is a stellaring surrounded by a trace of haze; A34 is a faint star-like spot 1′.6 NNW of a mag. 14.5 star. Northwest of the nucleus is A62, a nebulous spot about 15″ across with a brightness similar to a mag. 13 star 1′.8 NE. Farther SW is A59, a brighter spot lying almost directly W of the nucleus. Isolated between spiral arms, this feature is the most conspicuous region of the galaxy other than NGC 604. At 200 × it appears elongated SE-NW due to a star-like spot on its SE edge; again, this is actually an entire cluster of stars. Moving farther W we find A27, which is discernable as a stellaring surrounded by a little low surface brightness haze. Near the western edge of the halo, A131 is hardly discernable.

The several bright associations immediately S of the core form an especially mottled area, though the individual groups are not clearly distinguishable from one another. A14, though bright, is barely discernable from the background, yet at 200 × it contains a few faint stellarings.

About 11′ SE of the nucleus is a mag. 12 star, N of which are two associations. A101 is the most conspicuous of the two, a large, low surface brightness feature with a roughly circular outline. A100 is clearly elongated in pa 45°, and its SE flank is sharply defined.

Far to the S, near a mag. 8 star 16′ S of the nucleus, is A116. This association is visible as a faint spot 5′.5 W of the star; 6′.5 N of this, A127 is a faint patch elongated E-W with a mag. 13 star embedded in its center. Finally, A128 is faintly visible just NW as an extended patch 15″ across.

eg 661 ***dimen. 2′.2 × 1′.8*** ***mag.$_z$13.0*** ***sfc. br.***
In 15 cm this galaxy is visible as a faint substellar spot. 30 cm shows a circular, high surface brightness glow about 40″ diameter with strong even concentration to a conspicuous stellar nucleus. Two mag. 13.5 stars are nearby, one about 1′ NE, the other the same distance SW.

eg 669 ***dimen. 3′.4 × 0′.8*** ***mag.$_z$12.9*** ***sfc. br.***
This galaxy is a faint spindle in 15 cm, located a few arcminutes S of a crooked line of three mag. 11 stars. A mag. 13 star is faintly visible almost directly S of the galaxy, about 1′.5 from center. With 30 cm a high surface brightness halo extends to 1′.8 × 0′.3 in pa 35°, showing weak, broad concentration across the center. The bright-

Chart XXIII. NGC 598 = Messier 33. A 55-minute exposure on 103a-D plus GG13 filter with the USNO 1.55 meter telescope taken by J. Priser on 15 August 1969. Object names and magnitude sequence from: Humphreys and Sandage 1980, Ap. J. Suppl. 44: 319 and Sandage and Johnson 1974, Ap. J. 191: 63.

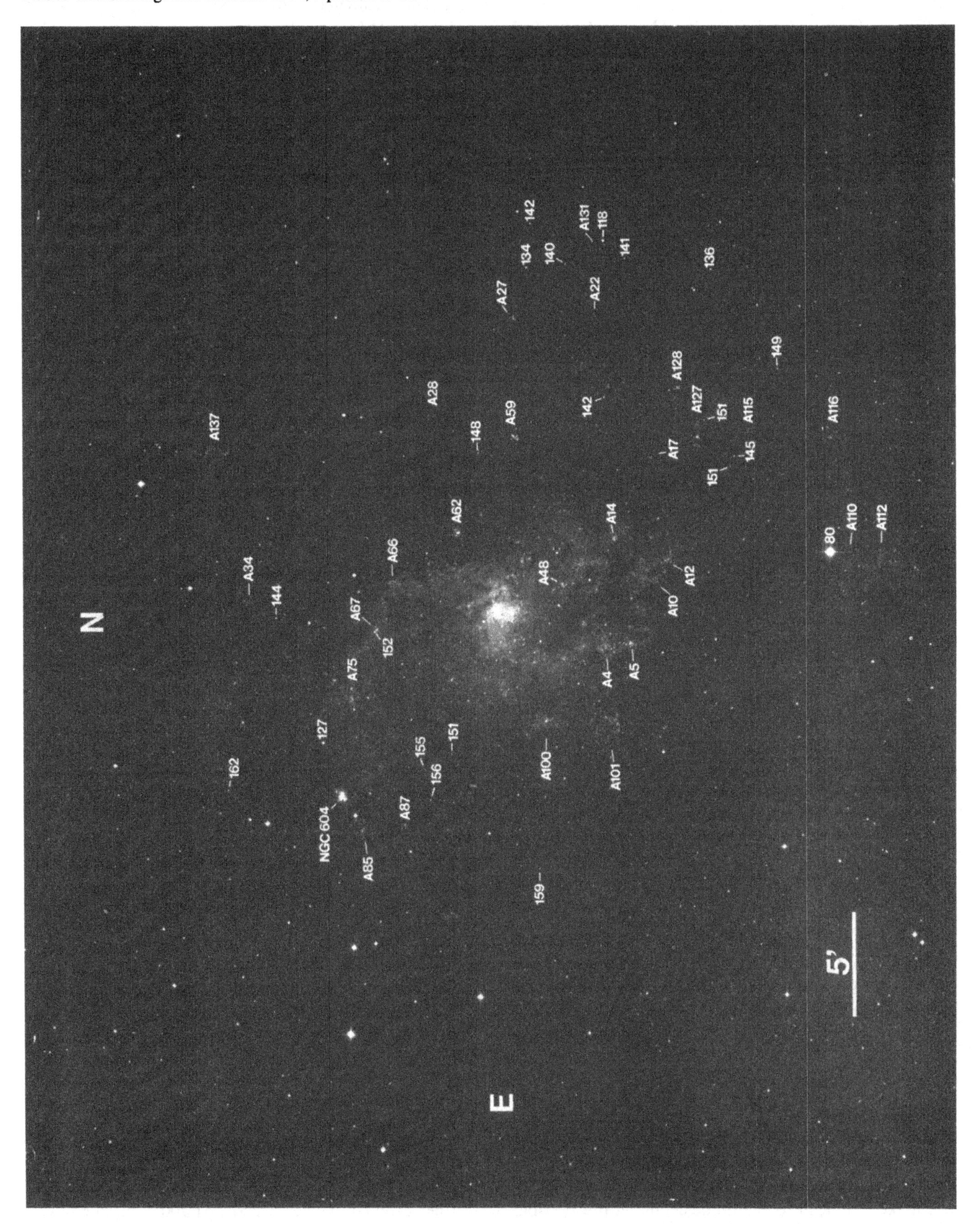

ness profile along the major axis is quite flat, though it fades sharply from the center along the minor axis.

eg 670 dimen. 2′.5 × 1′.1 V=12.1 sfc. br. 13.1
Viewed in 15 cm this galaxy is a small, concentrated spot forming a 3′ isosceles triangle with two mag. 11 stars. In 25 cm the lenticular, high surface brightness object extends to only 40″ × 15″ in pa 170°. The small elongated core contains a faint stellar nucleus that seems to lie off-center toward the S.

eg I1727 dimen. 6′.2 × 2′.9 V=11.6 sfc. br. 14.6
While not visible in 15 cm, 25 cm will show this galaxy as a very low surface brightness, unconcentrated glow extending to about 3′ × 1′.5 in pa 135°. At 200 × a few very faint stellarings are sprinkled across the surface. eg 672, *q.v.*, lies 8′ NE.

eg 672 dimen. 6′.6 × 2′.7 V=10.8 sfc. br. 13.7
Faintly visible even in 6 cm, this galaxy is easy to see in 15 cm as an elongated, smoothly textured glow with slight central concentration. A mag. 13.5 star is about 2′.25 NW, another lies farther E. In 25 cm the low surface brightness halo extends to 4′ × 1′.5 in pa 75°. The western half of the halo is distinctly broader, while the eastern end tapers nearly to a point. Averted vision shows a slight concentration to an elongated, bar-like core that contains several faint stellarings, though none of them stands out like a nucleus.

oc Collinder 21 diam. 6′ mag. 8.2p stars 14
This cluster is a small, unconcentrated group in 6 cm associated with a mag. 8 star. At 25 × a half-dozen faint stars are visible N of the bright member. With 25 cm fourteen stars are visible forming a circlet 6′ across with a bite missing on the NE side. On the NW side is an equal pair (Σ172: 10.5,10.7; 18″; 194°). eg I1731, *q.v.*, lies 4′ N of the northernmost star in the cluster.

eg I1731 dimen. 1′.7 × 1′.1 mag.$_z$14.2 sfc. br.
Located just 4′ N of the northernmost relatively bright member of oc Collinder 21, *q.v.*, this galaxy is too faint for 15 cm. In 30 cm it is still fairly faint, a slightly concentrated 1′.5 patch with a 20″ core but no distinct nucleus. A mag. 13.5 star lies on the E edge 40″ from the center.

eg 684 dimen. 3′.9 × 0′.9 mag.$_z$13.2 sfc. br.
This is a very faint galaxy in 15 cm, a tiny, stellar spot 1′.7 W of a mag. 11.5 star. With 30 cm it is about 40″ diameter, but the nearby mag. 11.5 star and a faint star 40″ N of it make it difficult to determine details. A pretty triple star lies 5′.4 SW (12,13,12; 11″,32″; 243°,246°).

eg 735 dimen. 2′.0 × 1′.0 mag.$_z$13.9 sfc. br.
Though not visible in 15 cm, 30 cm will show this galaxy as a faint, lenticular spot 1′.4 NE of a mag. 10 star. At 225 × it is 55″ × 20″, elongated in pa 140°. A mag. 13.5 star is embedded in the NW tip 20″ from center.

eg 736 dimen. 2′.0 × 1′.9 V=12.2 sfc. br. 13.5
This galaxy is a tiny, barely distinguishable spot in 15 cm, located about 15′ SW of eg 750/751, *q.v.* With 25 cm it looks very similar to eg 750, having nearly the same brightness and size. A mag. 13 star is closely involved, lying off the NNE edge of the halo 30″ from the center.

eg 750 dimen. 1′.6 × 1′.3 V=12.2 sfc. br. 13.0
eg 751 dimen. 1′.3 × 1′.3 V=12.5 sfc. br. 13.1
eg 750+751 dimen. 2′.9 × 2′.2 V=11.5 sfc. br. 13.3
Forming an interacting pair separated by 25″ in pa 165°, these galaxies are an elongated glow in 15 cm, not clearly separable at 100 ×. With 25 cm two nuclei are distinctly visible: the one on the N, eg 750, is a little brighter and substellar; the southern one, eg 751, is stellar, about mag. 13.5. The surrounding halo extends to 1′.5 × 1′; together the two objects form a filled-in figure eight, the northern half a little larger than the southern. In 30 cm both galaxies are well concentrated, though only eg 750 shows sharp concentration to a stellar nucleus; eg 751 is more broadly concentrated. Here the combined size of the halos is 1′.2 × 0′.7. A bright, close, and nearly equal pair lies 4′.6 SE [ADS 1562: 8.9,9.2; 1″.5; 229°(1978)].

eg 761 dimen. 1′.6 × 0′.6 mag.$_z$14.5 sfc. br.
This galaxy is too faint for 15 cm and is only just visible in 25 cm about 11′ NNE of eg 750/751, *q.v.* At 200 × it is a small spot elongated SE-NW with a stellar nucleus. A 1′ triangle of faint stars is immediately NE; the closest star, which is its S apex, lies 1′ E.

eg 777 dimen. 3′.0 × 2′.5 B=12.4 sfc. br. 14.5
eg 778 dimen. 1′.4 × 0′.6 mag.$_z$14.2 sfc. br.
In 15 cm eg 777 is a faint object with a stellar nucleus forming the N apex of a 6′ triangle with two mag. 8.5–9 stars SE and SW. With 25 cm the diffuse, low surface brightness halo is about 1′ across, slightly elongated SE-NW. The small, slightly brighter core contains a stellar nucleus. eg 778, 7′.1 S, is not difficult to see in 25 cm as a 20″ roughly circular spot with a bright, nearly stellar nucleus.

eg 783 dimen. 1′.8 × 1′.6 mag.$_z$12.8 sfc. br.
With 15 cm this galaxy is faintly visible at 125 × as a small spot with a mag. 13 star near its WNW edge, and another brighter star farther SE. At 200 × in 25 cm it is roughly circular, less than 1′ diameter, showing only the

slightest central concentration except for a faint stellar nucleus. The mag. 13 star nearby WNW is 25″ from the nucleus, just off the edge of the halo; the brighter star SE is 1′.2 from center. In 30 cm the halo is 55″ diameter, extending a little beyond the star WNW. Here the galaxy shows a moderate central concentration in addition to a faint stellar nucleus.

eg 784 ***dimen. 6′.2 × 1′.7*** ***V = 11.8*** ***sfc. br. 14.1***
This galaxy is a fairly faint spindle in 15 cm, elongated N-S and lying in the hypotenuse of a 9′ × 5′ right triangle of mag. 10–12 stars. With 30 cm the broadly concentrated halo grows to at least 6′ × 0′.8 in pa 0°, reaching a little beyond two mag. 13 stars on its SW flank and also a similar distance N past a mag. 13 star touching the E flank. A slightly brighter core one-fourth the total length is fairly distinct, though it contains no nucleus. At 225 × the entire surface is sprinkled with faint stellarings. Two other stars are visible nearby; one about 1′.5 ESE of center, another a similar distance NNW, just off the W flank.

eg 785 ***dimen. 1′.8 × 1′.1*** ***mag.$_z$13.9*** ***sfc. br.***
This galaxy, a faint companion 7′.7 SE of eg 783, *q.v.*, is not visible in 15 cm. With 30 cm it is a moderately faint spot about 40″ diameter, brightening evenly to the center where a stellar nucleus sparkles.

eg I1784 ***dimen. 1′.6 × 1′.0*** ***V = 13.2*** ***sfc. br. 13.5***
Though not visible in 15 cm, this galaxy is discernable in 25 cm at 100 × . At 125 × it is most distinct, forming a lopsided glow elongated in pa 90°: the W end seems fatter, more rounded than the E, and contains a few stellarings.

eg 890 ***dimen. 2′.9 × 1′.9*** ***V = 11.5*** ***sfc. br. 13.0***
At 125 × in 15 cm this galaxy is easy to see as a small, concentrated spot elongated in pa 60° with a bright stellar nucleus. With 25 cm averted vision will show a very faint halo extending to 2′ × 1′, though less attention shows only 1′.25 × 0′.75. The core is a small, more circular region around the stellar nucleus. In 30 cm the 1′.1 × 0′.6 halo brightens suddenly at the center to a stellar nucleus surrounded by a bright inner core. Three or four faint stars lie nearby to the W.

eg 925 ***dimen. 9′.8 × 6′.0*** ***V = 10.0*** ***sfc. br. 14.3***
This galaxy is faintly visible in 6 cm, which shows a small, round core and a halo seemingly elongated N-S, though larger apertures show that this impression is caused by some faint associated stars. The brightest of these, a mag. 10 star about 3′.5 S of center, is discernable here. In 15 cm three fainter stars, all about mag. 12.5, are visible that are responsible for the 6 cm elongation: two lie embedded in the halo just NNW of the core, another is near the S edge, about 1′.3 from the center. At 125 × the halo is a very low surface brightness haze surrounding a clearly elongated bar-like core aligned in pa 100°. With 25 cm the halo also shows elongation, though not parallel to the core; it extends to about 6′ × 4′ in pa 120°, reaching not quite to the mag. 10 star on the S. There are some vague light and dark areas in the halo, but most of the details are in the distinct 2′ × 0′.5 core. At high power it shows several stellarings over a conspicuously mottled

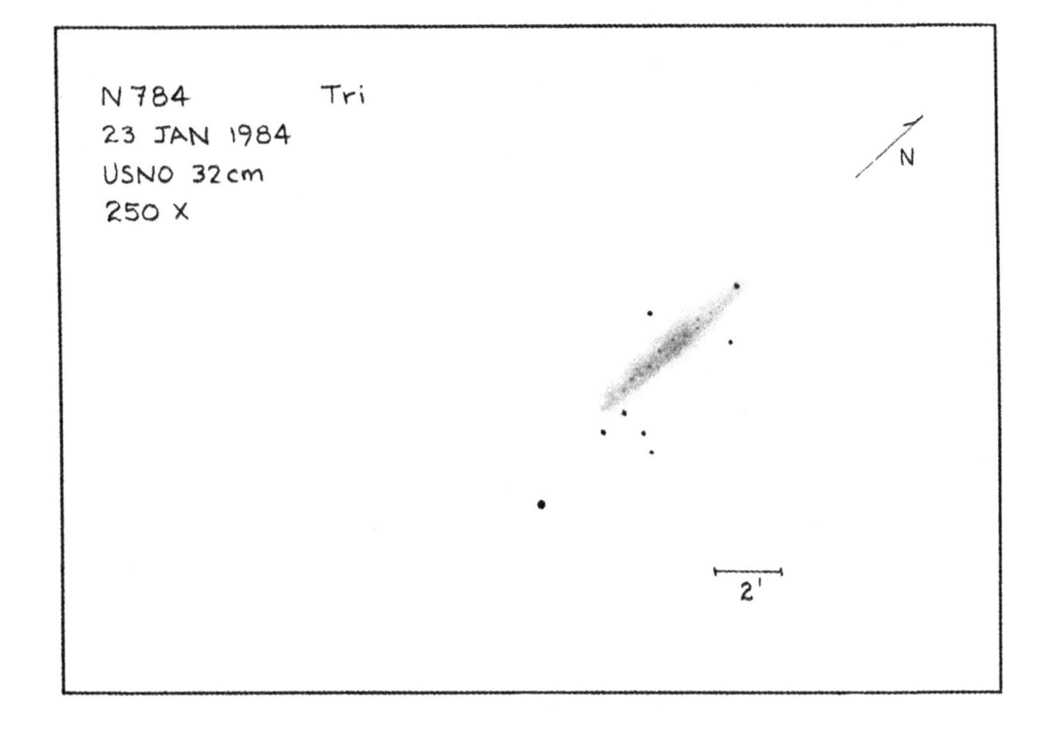

surface. A wide pair of mag. 12.5 stars lies 5′ W (41″; 215°); forming a triangle with the pair 1.′9 closer to the galaxy is a faint star-like spot, which photographs show to be a bright association of young stars and gas in an outer spiral arm of the galaxy. With care 30 cm shows the faint, diaphanous halo to about 6′ × 2′, sparkling with an overlay of foreground stars; at high powers it is nearly invisible, leaving only the elongated core and foreground stars. Here the core extends to 2′ × 1′, showing a broad, slight brightening to the center. The relatively bright stars nearby interfere, but with scrutiny the W end of the core curves a little N while the E end curves to the S.

eg 949 dimen. 2.′8 × 1.′7 V=11.9 sfc. br. 13.4
In 15 cm this galaxy is a 1′ × 0.′75 streak, elongated SE-NW and with a much brighter, nearly stellar center. With 25 cm the lenticular halo extends to 1.′5 × 1′ in pa 145°. The brighter, less elongated central regions show a grainy texture and a faint, occasionally stellar nucleus at 200 ×. A mag. 13 star is visible on the SW flank about 40″ from the nucleus. In 30 cm the thinly tapered halo grows to 2.′5 × 0.′75. The core is broadly concentrated and unevenly bright, occasionally showing a nucleus off-centered to the SW side of the bar. With averted vision there is a thin, slightly brighter streak along its major axis and the SW flank of the halo fades more abruptly.

eg 959 dimen. 2.′2 × 1.′4 V=12.4 sfc. br. 13.5
This galaxy is visible in 15 cm as a faint, weakly concentrated patch in a starry field. With 25 cm it is a sharply defined oval of moderately high surface brightness, extending to 1.′25 × 1′ in pa 65°. The central two-thirds is faintly speckled with stellarings at 200 ×.

Ursa Major

This constellation extends far beyond the familiar figure of the Big Dipper. A large number of galaxies is contained within its borders, including some of the brightest specimens in the sky. The center of the constellation culminates at midnight about 12 March.

eg 2639 ***dimen. 2:0 × 1:3*** ***V = 11.8*** ***sfc. br. 12.6***
15 cm shows this galaxy as a faint circular glow less than 1′ diameter with a stellar nucleus. Two mag. 10–11 stars 4:6 E and 5:2 NE form a nearly equilateral triangle with the galaxy. With 25 cm the sharply tapered lenticular halo extends to 1:25 × 0:75 in pa 140°. The bright, circular core contains an inconspicuous stellar nucleus. In 30 cm the halo grows to 1:7 × 1′, showing strong concentration to a bright core that seems multiple at low power. At 225 × the core has an unusual appearance: several stellarings crowded unsymmetrically around it within 10″ radius make the inner core seem quite elongated, but in a different position angle than the halo, i.e., in about pa 130°. A mag. 13 star lies 2:3 SE.

eg 2654 ***dimen. 4:3 × 0:9*** ***V = 11.8*** ***sfc. br. 13.2***
In 15 cm this galaxy is visible as a small, faint spot 4:5 S of a mag. 11 star. With 25 cm it is a strongly elongated streak extending to 2′ × 0:3 in pa 65° around a very thin 30″ × 5″ core. In 30 cm the halo grows to 2:5 × 0:8. The core is very irregular, about 40″ across, and contains a bright nucleus, which is stellar at 150 ×. Several stellarings lie in the brighter inner regions, particularly to the SW.

eg 2681 ***dimen. 3:8 × 3:5*** ***V = 10.3*** ***sfc. br. 12.9***
This object is a circular 2′ glow in 25 cm with a very small concentrated core. On the W, 2:1 from center, is a faint pair of stars (12,12; 31″; 320°); fainter stars appear 1:6 to the E and about 3′ SE. 30 cm shows the halo to about 1:75 diameter. The nucleus is a small, concentrated spot about 5″ across with a stellar center. The core, which seems slightly elongated E-W at 150 ×, shows two or three brighter spots within at 225 ×: the most obvious one is 20″ from the center in pa 100°; another fainter one is directly opposite the nucleus from the first at about the same distance.

eg 2685 ***dimen. 5:2 × 3:0*** ***V = 11.1*** ***sfc. br. 14.0***
Located 2:4 S of a mag. 11 star, this galaxy is faintly visible in 15 cm as a round, concentrated spot less than 1′ diameter. The halo extends to 1:25 × 0:3 in 25 cm, elongated in pa 40°, with smoothly tapered tips and a bulging oval core. The inner parts show a few stellarings at 200 ×. At 225 × in 30 cm the core seems elongated in pa 40° due to irregular wings including bright spots extending from the center; the overall outline is more nearly circular, about 2′ diameter. A nonstellar nucleus stands out just E of the geometric center.

eg 2693 ***dimen. 2:2 × 1:7*** ***V = 11.7*** ***sfc. br. 13.1***
This galaxy appears as a low surface brightness circular spot about 1′ diameter in 25 cm. The halo is almost uniformly bright, but the 20″ core shows mild concentration toward the center. In 30 cm it also shows poor overall concentration, but the suddenly brighter, nearly stellar nucleus stands out in the center. The halo is only 40″ across here, slightly elongated roughly SE-NW. A mag. 14 star is visible 1′ W.

eg 2701 ***dimen. 2:1 × 1:4*** ***V = 12.4*** ***sfc. br. 13.4***
In 15 cm this galaxy is faintly visible as a small patch with a faint star on its W edge. The halo appears 50″ diameter with 25 cm, extending beyond the mag. 13 star, which is about 20″ WNW. A stellar nucleus is visible at the threshold of brightness. In 30 cm the halo grows to 1:5 × 1:2 in pa 25°. The mottled core is very slightly concentrated but without a distinct nucleus.

eg 2742 ***dimen. 3:1 × 1:7*** ***V = 11.7*** ***sfc. br. 13.4***
In 15 cm this galaxy is a circular 1′ patch that is only a little brighter toward the center. 25 cm reveals a 1:75 × 1:25 glow with slight mottling in the core. A mag. 9 star lies 4:6 NW; a thin triangle of stars pointing S is about 3′ SW. The galaxy grows to 3′ × 2′ in 30 cm, elongated in pa 85°. At 225 × some stellarings are visible in the core, though overall it has a smooth texture.

eg 2768 ***dimen. 6:3 × 2:8*** ***V = 10.0*** ***sfc. br. 13.1***
Lying in a field of many bright stars, this object is a bright, broadly concentrated glow in 15 cm. The halo extends to 2′ × 1′, elongated E-W, and contains a conspicuous stellar nucleus. The halo is 2:5 × 1:5 with 25 cm, elongated in pa 95°. The sharply concentrated circular

core has a stellar nucleus; at times there seems to be another, fainter concentration embedded in the NE side. With averted vision, a narrow bar is visible along the major axis, extending more to the W than E. The halo grows to 3′ × 1′.6 in 30 cm around a well-concentrated circular core and bright nucleus. The core seems more sharply defined on the S side.

eg 2787 dimen. 3′.4 × 2′.3 V = 10.8 sfc. br. 12.9
Located about 14′ NW of a wide, equal pair (10,10; 38″; 100°), this galaxy is visible in 6 cm as a low surface brightness glow 1′–2′ diameter. With 25 cm it is a bright circular patch less than 1′ diameter that fades quickly from center to edge. The halo is about 2′ diameter in 30 cm, and evenly concentrated from a slightly irregular halo to a smoothly textured core. A faint star is visible in the halo 45″ from the center in pa 155°.

eg 2805 dimen. 6′.3 × 5′.0 V = 11.3 sfc. br. 14.9
This galaxy is difficult to see in 30 cm, though it is the largest of the group including eg 2814 and eg 2820, *cf.* The very low surface brightness, poorly concentrated halo has a circular outline about 1′.8 diameter delimited on the NE edge by a mag. 13.5 star. A mag. 9 star 6′.5 NNW is a wide, unequal pair (9,11; 14″; 240°).

eg 2814 dimen. 1′.4 × 0′.4 V = 13.8 sfc. br. 13.1
In 30 cm this galaxy is smaller but overall seems a little brighter than nearby eg 2820, *cf.*; a mag. 11 star 1′ SSW prevents a fine comparison. The halo extends to 1′.2 × 0′.6 in pa 0°, showing moderate, even concentration to a faint stellar nucleus.

eg I2458 dimen. 0′.5 × 0′.3 mag.$_z$15.1 sfc. br.
eg 2820 dimen. 4′.3 × 0′.7 V = 12.8 sfc. br. 13.8
eg 2820 is a faint, sharply tipped spindle in 30 cm, reaching to 2′ × 0′.2 in pa 60°. Faint stellarings are occasionally visible along the major axis, but there is no central bulge or distinct nucleus. Just off the SW tip is eg I2458, a small faint spot about 10″ diameter with a faint stellar nucleus.

eg 2841 dimen. 8′.1 × 3′.8 V = 9.3 sfc. br. 12.9
This galaxy is easily visible in 6 cm about 22′ SE of 37 Lynxis [ADS 7303: V = 6.1,10.2; 5″.7; 133° (1973)]. The well-concentrated halo is elongated SE-NW, about 3′ long; it does not reach a mag. 10 star to the NW, though it extends almost directly toward it. In 15 cm the core is about 1′ across; the surrounding halo extends to 5′ × 1′.5, almost reaching the mag. 10 star on the NNW end 2′.7 from center. With 25 cm the core is about 1′ across, fading rapidly at its edges to a smooth oval 7′ × 4′ halo. A mag. 13 star is embedded in the halo on the E side of the major axis, and 1′.8 NNW of center. 30 cm shows the halo to 6′ × 1′.8 in pa 150°. At 150 × two brighter spikes seem to extend from the core into the halo along the major axis, reaching about 2′ beyond the nucleus on either side; at 225 × this is less distinct. The halo is generally dimmer on the E side, yet there is a thin dark streak passing just W of the inner core.

eg 2880 dimen. 2′.6 × 1′.6 V = 11.6 sfc. br. 13.0
In 15 cm this galaxy is a star-like spot 1′.9 WSW of a mag. 11 star. With 25 cm it is a circular object about 40″ diameter, showing strong central concentration to a stellar nucleus. The halo is about 50″ diameter in 30 cm and

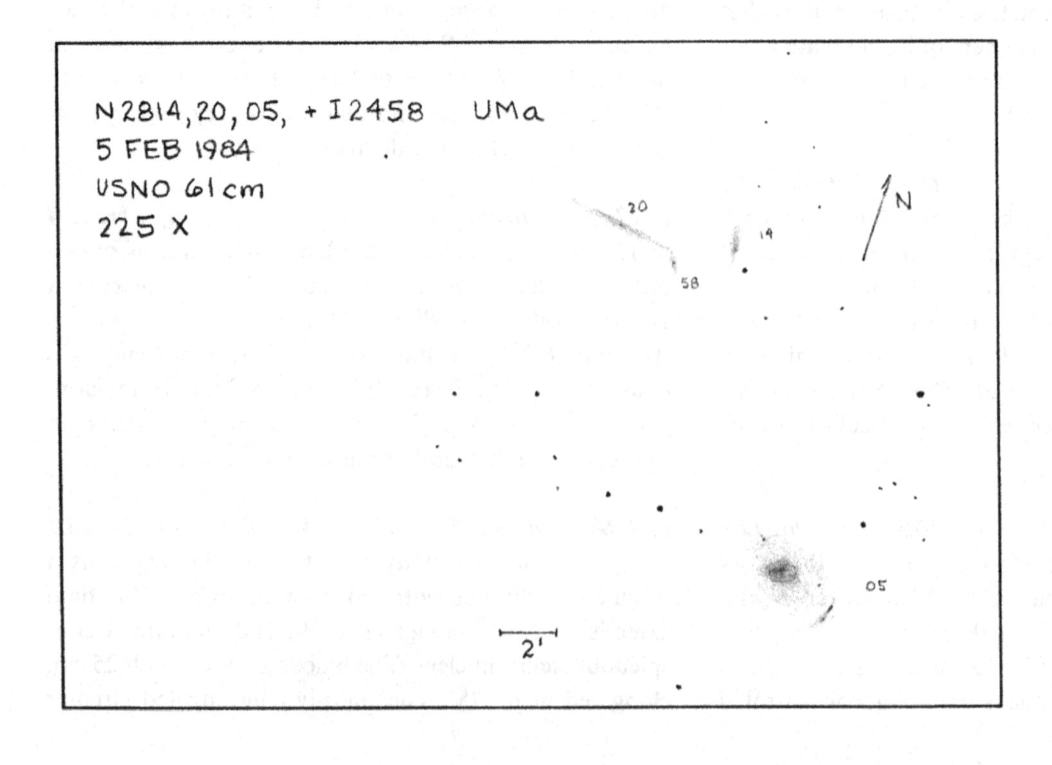

fairly well concentrated to a small core, though no stellar nucleus is visible here. Four faint stars lie in the NW quadrant within 2′.5. A faint companion, eg Z312–12 ($mag._Z$15.1) is 3′.2 N.

eg 2950 dimen. 3′.2 × 2′.1 V = 11.0 sfc. br. 12.9
This galaxy is a small, star-like spot in 15 cm surrounded by a very faint halo about 1′ diameter. 25 cm shows an exceptionally bright nucleus in a small bright core. The halo extends to 1′.25 × 0′.75 in pa 135°. With 30 cm the halo is only slightly elongated, about 1′ diameter, and well concentrated to a bright mag. 12 stellar nucleus embedded in a circular core.

eg 2976 dimen. 4′.9 × 2′.5 V = 10.2 sfc. br. 12.7
A member of the Messier 81 group, this galaxy is visible in 6 cm as a weakly concentrated glow 3′–4′ diameter. The galaxy has low surface brightness and is not bright in 25 cm. The halo is 2′.5 × 1′, elongated SE-NW, and seems to have a line in the core as though it were a barred spiral. 30 cm shows the halo to about 4′ × 2′.5 in pa 150° with a weak, very broad concentration and no nucleus. At 225 × the surface is distinctly uneven: three large, slightly brighter clumps dominate the central regions, showing the best contrast in the NW side. A mag. 13 star lies against the W side; a mag. 14 star is off the NW side 2′.5 from center.

eg 2985 dimen. 4′.3 × 3′.4 V = 10.5 sfc. br. 13.3
This galaxy is visible in 6 cm as a fuzzy, nearly stellar spot located about 10′ N of a mag. 10 star. In 15 cm it is still small, about 1′ across, and contains a stellar nucleus. A mag. 12.5 star is visible on the E edge 1′.0 from the nucleus. With 25 cm the halo extends just beyond this star to an overall diameter of almost 2′. The nucleus is stellar here too, but in 30 cm it loses its stellar character, though the center of the galaxy is very bright. At 225 × the circular halo is 2′ diameter and the core is slightly mottled. This galaxy is the brightest of a group including eg 3027 (B = 12.5) 25′ E and eg 3065–66, *q.v.*

eg 2998 dimen. 3′.0 × 1′.5 B = 12.6 sfc. br. 14.2
25 cm reveals this galaxy as a low surface brightness 1′.5 × 0′.5 haze elongated in pa 55°. The core is only slightly brighter and has a mottled texture at 200 ×. A mag. 12 star lies nearby N, 1′.8 from the center. With 30 cm the halo extends to only 1′.1 × 0′.75, showing a slight, even central concentration. Two additional stars are visible within 3′ to the SW. This galaxy is the brightest of the eg 2998–3010 group, the other members of which are very faint.

eg 3031 dimen. 26′ × 14′ V = 6.8 sfc. br. 13.1
Messier 81 forms a bright, interesting pair for amateur telescopes with its companion Messier 82 (*q.v.* eg 3034) 37′ N. Messier 81 is easily visible in 6 cm, a large, well-concentrated glow extending to as much as 15′ × 7′ in pa 155°. The nucleus is prominent but not quite stellar. Two bright stars aligned roughly N-S lie off the SSW side of the halo; two others, each about mag. 11, are superposed on the halo near the S edge of the core. With 25 cm the halo has an oval outline, but appears only about 11′ × 5′ in extent. A third mag. 11 star is near the edge of the halo on the WNW. The nearest star of the two off the SSW side is an equal pair (ADS 7565: 10.9,10.9; 9″.0; 271°); scrutiny at high power shows that the farther is also a close pair (ADS 7566: 9.5,9.5; 2″.1; 112°). The overall size is difficult to estimate in 30 cm since the bright inner parts fade to an extensive but low surface brightness halo. With care the nebula can be detected to an overall length approaching 20′, reaching at its SSE extremity to a mag. 14 star 9′ from the center. The central portions are well concentrated, brightening strongly to a 3′ core and a smooth, bright nuclear region 1′ across. At 225 × the core takes on a mottled appearance: a slightly brighter area about 45″ diameter stands out on the major axis 3′–3′.5 SSE of the nucleus.

eg 3034 dimen. 11′ × 4′.6 V = 8.4 sfc. br. 12.5
In 6 cm Messier 82 is definitely fainter than its companion, Messier 81, *cf.* eg 3031 37′ S, but much more interesting. At low powers it is a smoothly textured, fairly high surface brightness 8′ × 2′ streak elongated in pa 60°. With high power the middle part of the core has two brighter portions separated by a slight darkening. A mag. 10 star lies just S of the SW end; another lies farther SSW. With 15 cm the silvery, high surface brightness halo extends from a mag. 13.5 star S of the E tip past the mag. 10 star to the SSW, for an overall length of almost 8′. The dark band crossing the bar near the center is clearly visible at 75 ×; another, less distinct, is visible about one-third the distance from the center to the E tip. The brightest portion of the galaxy lies between these two dark bands. In 25 cm the halo extends about 2′ beyond the mag. 10 star, and about 1′ past the mag. 13.5 star, giving a length of over 9′. The entire central portion is distinctly mottled; only the tips of the halo seem smooth. 30 cm shows the halo to about 10′ × 2′. The irregular core is about 5′ long and has very little halo along its NNW and SSE flanks. The western "half" of the galaxy, that part lying W of the nearly central dark bar, is distinctly brighter overall than the eastern "half," though the brightest lump is in the eastern side. At high power this lump is elongated nearly E-W. Immediately W of the bar, near the S edge of the galaxy, is a circular, substellar

spot about 5″ across; another less conspicuous and larger brightening lies two-thirds the way from here to the W tip. The S edge of this brighter portion is sharply defined and bows inward. A very faint star is visible here just off the SSE flank.

eg 3043 dimen. 2′.0 × 0′.7 B = 13.5 sfc. br. 13.7
This galaxy is not visible with 15 cm and is quite faint even in 25 cm. At 200× it is a small, slightly concentrated 1′ × 0.5 glow elongated nearly E-W located 8′ S of a mag. 9.5 star. With 30 cm the halo extends to 1′.2 × 0′.6 in pa 80°. The core is a circular region that is slightly but abruptly brighter than the surrounding halo.

eg 3065 dimen. 2′.0 × 1′.9 V = 12.0 sfc. br. 13.3
eg 3066 dimen. 1′.2 × 1′.1 V = 12.9 sfc. br. 13.2
15 cm shows this interacting pair of galaxies as two small spots separated by 3′.0 about 15′ S of a mag. 8.5 star. The northern galaxy, eg 3065, forms a pair with a similarly bright star 1′.4 NW. In 25 cm both galaxies are about the same size, 40″ across, but eg 3065 is distinctly brighter and contains a stellar nucleus. With 30 cm they remain quite small, less than 1′ diameter. eg 3065 is well concentrated from a 40″ halo to a bright substellar nucleus. eg 3066 is by comparison only moderately concentrated from a 50″ halo to a large core that shows no distinct nucleus.

eg 3077 dimen. 4′.6 × 3′.6 V = 9.9 sfc. br. 12.7
Located just under 4′ SE of a mag. 8 star, this member of the Messier 81 group is easily visible at 50× in 6 cm as a circular glow about 1′ diameter. With 15 cm it is about 1′.5 diameter, while in 25 cm it is 2′ diameter and evenly concentrated to a broad core. In 30 cm a faint extensive halo extends to as much as 4′ diameter, spreading more to the E side of the 2′ core, giving an overall elongation in pa 55°. The core itself is fairly well concentrated and occasionally shows a stellar nucleus. Two mag. 13.5 stars are visible 3′–3′.5 S and SE.

eg 3079 dimen. 7′.6 × 1′.7 V = 10.6 sfc. br. 13.2
Located just N of a 6′ triangle of mag. 9–11 stars, 15 cm shows this galaxy easily as a thin, undetailed 5′ × 1′ streak elongated in pa 165°. In 25 cm the nebula extends a little beyond a mag. 13 star embedded in the W side of the N tip, for an overall length of about 7′. At 200× the surface is mottled but shows no distinct nucleus in the 2′ × 1′ core. With 30 cm the halo extends to 7′ × 2′ and exhibits moderate central concentration to an elongated, unevenly bright core. Three faint stars lie off the W side of the nebula; two of them are less than 1′ from the flank of the galaxy. At 225× the E side of the halo seems more sharply defined, though the nearby stars on the W may be causing this impression. Photographs show eg 3073 (mag.$_Z$13.8) 10′ WSW and eg M+09–17–09 (mag.$_Z$14.6) 6′.5 NW.

eg 3184 dimen. 6′.9 × 6′.8 V = 9.8 sfc. br. 13.8
This galaxy is visible in 6 cm as a faint, unconcentrated and smoothly textured glow about 4′ diameter with a mag. 11 star on its N edge. In 15 cm it is only a little larger, though the star N lies just within the halo. 25 cm reveals a diffuse object with starry patches around a small, barely distinguishable core. At 200× it occasionally seems horseshoe-shaped: the brightest part in the SW side is connected by a faint arc to a small, fainter appendage on the NE; a darker space lies between them on the NW side of the center. No distinct core or nucleus is visible in 30 cm, the galaxy showing almost no general brightening to the center. The halo is still about 4′ diameter. With averted vision the surface is unevenly bright; some brighter portions make the central regions seem elongated N-S.

eg 3198 dimen. 8′.3 × 3′.7 V = 10.4 sfc. br. 14.0
In 6 cm this galaxy is just discernable as a faint, indefinite patch located 3′.5 SSW of a mag. 11.5 star. In 15 cm the halo is 6′ × 1′.5, elongated in pa 40°. At high power the central regions exhibit a broadly mottled texture, but no nucleus stands out. The halo grows to 7′ × 2′ in 25 cm. At times the outline seems somewhat irregular, in particular the SE flank of the NE extension appears indented. With 30 cm the halo extends to about 8′ × 2′.8, showing slight concentration to a broad, very mottled core and an exceedingly faint nucleus. The brighter, most definite part of the object is about 5′ long. Two faint stars are to the S, 1′.9 and 2′.7 from the center.

eg 3259 dimen. 2′.3 × 1′.4 B = 12.9 sfc. br. 14.0
This galaxy, lying about 10′ W of a mag. 9 star, is faintly visible to 15 cm. With 25 cm it is a small, roundish spot about 1′ diameter with a tiny, sharply concentrated center. In 30 cm the irregular, low surface brightness halo grows to almost 2′ diameter, slightly elongated in pa 20°. The central regions are only a little brighter, and the nucleus is inconspicuous here. At high powers the galaxy stands out better, and the halo seems unevenly bright, particularly on the SW and NE sides. eg 3266 (mag.$_Z$13.5) lies 18′ S and a little E.

eg 3310 dimen. 3′.6 × 3′.0 V = 11.1 sfc. br. 13.4
Lying 10′ SSW of a reddish mag. 5.5 star, 6 cm will show this galaxy easily. At 50× it is a small, evenly bright spot about 1′ diameter that looks much like a planetary nebula. A faint star is visible 3′ N. In 25 cm it grows to about 2′ diameter and occasionally seems a bit elongated

E-W. The core is a fairly large and broadly concentrated region, but no nucleus stands out. With 30 cm the 1.′5 core forms most of the object; the halo is only a very faint irregular extension to the ESE and WNW sides to an overall length of about 2.′ Within the core is a somewhat brighter region about 40″ across that has an uneven texture but no distinct central concentration.

eg 3319 ***dimen. 6.′8 × 3.′9*** ***V = 11.3*** ***sfc. br. 14.8***
Though faintly visible in 15 cm, this low surface brightness galaxy needs larger apertures to be seen with any clarity. It is located about 20′ SSW of an arc of five mag. 9.5–11 stars. With 25 cm the diffuse halo is an irregular, knotted bar extending to 3.′5 × 0.′5 in pa 35°. Two or three slightly brighter patches aligned along the major axis are the only parts that stand out. 30 cm shows the very faint halo to 5′ × 2.′ The core is an elongated 2′ × 1′ region; immediately beside it along the major axis the halo seems fainter than farther away. Five mag. 11–13 stars lie within 10′ N; a single mag. 12 star is 5′ SW.

eg 3320 ***dimen. 2.′2 × 1.′2*** ***B = 12.9*** ***sfc. br. 13.9***
In 15 cm this galaxy is a spindle-shaped glow elongated in pa 20°, located 2′ S of a mag. 12 star. At high power a faint stellar nucleus sparkles in the center with a brightness similar to a mag. 13.5 star at the SSW tip. With 25 cm the halo is 1.′5 × 0.′3 around a 15″ × 5″ core, which here seems to be fainter than the star on the SSW. 30 cm shows an oval 1.′3 × 0.′6 halo that just reaches the star SSW. At 225 × it exhibits poor central concentration and no distinct nucleus.

eg 3348 ***dimen. 2.′2 × 2.′2*** ***V = 11.2*** ***sfc. br. 12.9***
At low powers in 30 cm this galaxy appears quite circular, a little over 1′ diameter, and with a small but nonstellar center. With high powers the halo seems elongated E-W, extending to about 1.′2 × 0.′7. At first glance the core looks peculiar: it appears elongated E-W and has a bright spot at each end. Closer scrutiny shows the eastern spot to be a mag. 13 star superposed right at the edge of the core 11″ from the center. This intimately involved star is probably the source of the elongated appearance of the core and halo. A mag. 10.5 star is 1.′6 WNW. eg 3364, *q.v.*, lies 26′ S.

eg 3353 ***dimen. 1.′5 × 1.′1*** ***V = 12.7*** ***sfc. br. 13.1***
In 25 cm this galaxy appears as a faint circular patch less than 1′ diameter. A mag. 13.5 star lies 1.′7 S. With 30 cm the halo is about 1′ diameter and contains a relatively bright but nonstellar nucleus in a large indistinct core. The star S is of comparable magnitude to the nucleus.

eg 3359 ***dimen. 6.′8 × 4.′3*** ***V = 10.5*** ***sfc. br. 14.0***
15 cm shows this galaxy as a diffuse undetailed glow extending to perhaps 2′ × 1.′5 elongated N-S. The nebula has a blockish or rectangular outline in 25 cm, but overall it is about the same size. The central regions are in contrast nearly circular but do not stand out well. At 200 × averted vision shows a slightly mottled texture in the brighter parts. The evenly concentrated halo grows to 6′ × 4′ in 30 cm, though it has a low surface brightness and is hard to trace in the outermost regions. Here the core is a more distinct but irregular region that shows best a low power, while high power will reveal faint mottling around its outer edges. At no power does the core contain a distinct nucleus. Two mag. 14 stars are visible about 3′ W and SE.

eg 3364 ***dimen. 1.′9 × 1.′8*** ***B = 13.5*** ***sfc. br. 14.7***
This galaxy is a faint, low surface brightness object in 30 cm, located 4′ S of a 25″ unequal pair of mag. 13 stars. The halo extends to 1′ × 0.′7 elongated E-W and shows a very weak central concentration. At 225 × the core seems composed of two or three knots. eg 3348, *q.v.*, is 26′ N.

eg 3415 ***dimen. 2.′4 × 1.′5*** ***B = 12.9*** ***sfc. br. 14.1***
In 30 cm this galaxy forms the northern vertex of a 1.′5 quadrilateral of mag. 12.5 stars; the nearest star lies just under 1′ S. The object is small and well concentrated to a prominent, nearly stellar nucleus. The halo extends to about 30″ × 20″ in pa 10°.

eg 3445 ***dimen. 1.′6 × 1.′5*** ***V = 12.4*** ***sfc. br. 13.2***
Though faintly visible to 15 cm, this galaxy is fairly hard to see even in 25 cm. At 200 × it is a circular 1.′25 glow that is broadly brighter toward the center. A mag. 9 star lies 2.′2 NE. With 30 cm the nebula is poorly concentrated and shows no nucleus at any power. The halo is elongated ESE-WNW, extending to 1.′5 × 1.′ At 250 × the N and NW edges are more sharply defined. eg 3440 (mag.$_z$14.0) lies 9.′9 NW.

eg 3448 ***dimen. 5.′4 × 1.′9*** ***V = 11.7*** ***sfc. br. 14.1***
With 15 cm this galaxy is a faint, slightly elongated glow with a low surface brightness and poor central concentration. 25 cm will show a lenticular halo extending to 2′ × 0.′3 in pa 65°. The middle parts are roughly circular, only a little brighter than the halo, and show a slightly mottled texture at high power. In 30 cm the faint halo grows to at least 3′ × 1.′ The core still is quite round, but seems to be centered a bit toward the SE side of the major axis, causing a more conspicuous bulge to that side. At 250 × the core has an irregular appearance but no distinct nucleus, and a few faint stellarings are visible

scattered through the core and halo. Three mag. 13.5–14 stars are at about 3′ ESE, WNW, and NNW.

eg 3458 ***dimen. 1′.7 × 1′.1*** ***B = 13.1*** ***sfc. br. 13.6***
Though this galaxy is hardly distinguishable from a star in 15 cm, 25 cm will show a small, faint halo surrounding the prominent stellar nucleus. The overall size is about 45″ × 15″, elongated nearly N-S. In 30 cm the nucleus appears substellar and is not quite centered within the halo, lying a bit NW of center. At 225 × the halo extends to somewhat over 50″ diameter, elongated roughly N-S, while the core is elongated more nearly SE-NW.

eg 3478 ***dimen. 2′.8 × 1′.4*** ***B = 13.0*** ***sfc. br.14.3***
Viewed in 30 cm this galaxy is faintly visible about 4′ N of two mag. 11 stars separated by 2′ in pa 50°. At 225 × the halo extends to 1′.75 × 1′ in pa 130°. No overall concentration is evident, but a very faint stellaring is visible near the center.

eg 3516 ***dimen. 2′.3 × 1′.8*** ***V = 11.7*** ***sfc. br. 13.2***
This galaxy is easy to see with 30 cm as a small, well-concentrated glow lying S of two mag. 10 stars aligned E-W 6′ apart. With higher magnification the halo extends to about 1′.2 × 0′.8 in pa 55°. Here the core has a circular outline about 35″ across and is much brighter than the halo, but shows no distinct nucleus. A mag. 13.5 star is 1′.2 SE. The nucleus of this Seyfert galaxy varies in brightness from B = 14.0 to 16.0.

eg 3549 ***dimen. 3′.2 × 1′.3*** ***B = 12.7*** ***sfc. br. 14.1***
Though this galaxy is quite faint in 25 cm, 100 × will show it fairly well. The halo has a lenticular form that extends to 2′ × 0′.5 in pa 40°. With 30 cm the halo seems more bar-shaped, tapering very little toward the ends. At 225 × it appears only 1′.8 × 0′.8 and seems unevenly concentrated along the major axis. A mag. 14 star lies 1′.7 E.

eg 3556 ***dimen. 8′.3 × 2′.5*** ***V = 10.0*** ***sfc. br. 13.2***
Messier 108 is faintly visible in 6 cm and is one of the most difficult objects in Messier's list for this aperture. With scrutiny at 75 × it is about 4′ × 1′.5, elongated nearly E-W. The central regions are quite elongated and considerably brighter than the halo. Several mag. 8–10 stars lead away NW from the W end. With 15 cm it grows to almost 8′ length. Medium powers show an uneven texture but no prominent central brightening. 25 cm reveals a faint star just W of a small central brightening. The 7′ × 2′ halo is very uneven; dark bars seem to cross it perpendicular to the major axis. In 30 cm the bar-shaped halo extends to 8′ × 1′.5 in pa 80°, ending just short of a mag. 12 star that is just S of the W tip. The surface is obviously uneven at all magnifications. A bright lump about 1′ across is 1′.5 W of the superposed mag. 13 star; a small condensation lies just 30″ E of the star. With averted vision at 225 × a very faint star is visible on the SE edge of the latter condensation, and another star or stellaring sparkles near the eastern end of the halo. A mag. 14 star lies 2′.5 N; Messier 97 (pn 3587), *q.v.*, is 48′ SE.

eg 3583 ***dimen. 2′.8 × 2′.0*** ***B = 12.2*** ***sfc. br. 13.9***
Viewed in 30 cm, this galaxy is a glow of moderate surface brightness that is slightly elongated SE-NW. At 225 × the halo extends to about 1′.3 diameter around a somewhat brighter, mottled core. No nucleus stands out at any magnification. Photographs show eg 3577 (mag.$_z$14.7) 5′.0 SW.

pn 3587 ***dimen. 3′.4 × 3′.3*** ***nebula 9.9v*** ***star V = 16.0***
Messier 97, the Owl Nebula, is faintly but steadily visible in 6 cm at 50 ×. A mag. 11 star lies beyond the edge on the NNE side about 2′.5 from the center of the uniform, circular nebula. In 15 cm it is easily visible at 25 ×, though no distinct detail is discernable even at 75 ×. With 25 cm the nebula is bright and has fuzzy edges. Two slightly darker patches are indistinctly visible aligned SE-NW on either side of the center, forming the eyes of the owl. At 150 × in 30 cm the eyes are more distinct but still show very poor contrast with the rest of the nebula. Each "socket" is about 50″ across. The axis of the nebula perpendicular to these eyes has a brighter patch at either end, giving the nebula a four-parted appearance. At 225 × the texture is uneven, particularly in the brighter lobes and in the slightly brighter area between the eyes. Here also the nebula seems slightly elongated in pa 150°. Five mag. 13.5–14 stars encircle the grey nebula on the N through E to SW sides; the northernmost is furthest at 4′.5, the southwesternmost is closest at 3′. Photographs show a faint galaxy, eg M + 09–19–14 (mag.$_{VV}$17), just S of the mag. 13.5 star 3′.7 SE of the nebula's center.

eg 3610 ***dimen. 3′.2 × 2′.5*** ***V = 10.8*** ***sfc. br. 13.0***
This galaxy is the brightest of a group of three including eg 3613 and eg 3619, *q.v.* eg 3610 is visible even at 25 × in 6 cm as a circular spot 1′.5 diameter with a bright stellar center. The stellar nucleus is conspicuous in 15 cm and 25 cm. In 30 cm the halo extends to 1′.75 × 1′ in pa 135°. The inner regions are also conspicuously elongated and grow sharply brighter right to the center; the bright stellar nucleus stands out at all magnifications.

eg 3613 ***dimen. 3′.6 × 2′.0*** ***B = 11.6*** ***sfc. br. 13.7***
Lying on the W side of a 12′ box of mag. 10–11 stars, this galaxy is faintly visible in 6 cm. It is easily observable in 15 cm, appearing as a well-defined 1′.5 streak with a

brighter core. With 25 cm the galaxy is brighter and more distinctly elongated than eg 3619, *cf.* 16′ SE. The halo is 2′.25 × 0′.75, elongated in pa 100°, and contains a circular 20″ core and a stellar nucleus. In 30 cm the halo is fairly well and evenly concentrated to the center, but here the nucleus is not quite stellar.

eg 3614 dimen. 4′.6 × 2′.9 B = 12.2 sfc. br. 14.9
This object is only faintly visible with 15 cm. In 25 cm a faint halo extends to 2′ × 1′.25 in pa 100°. The center is little brighter but a faint stellar nucleus is occasionally visible. The halo grows to 2′.5 × 1′.2 in 30 cm and shows a broadly mottled texture in the central regions. At 225 × the nucleus seems off-center toward the S. A faint star is visible 3′.5 E.

eg 3619 dimen. 3′.1 × 2′.6 B = 12.6 sfc. br. 14.7
eg 3625 dimen. 2′.2 × 0′.8 mag.$_Z$13.9 sfc. br.
eg 3619 is the faintest of three galaxies including eg 3610 and eg 3613, *q.v.* 15 cm shows a roundish, ill-defined spot 1′ diameter. At 200 ×, 25 cm reveals a lenticular object elongated roughly E-W with a somewhat brighter center. With 30 cm the halo extends to 1′.5 × 1′ in pa 110°. The central region appears only a little brighter but contains a stellar nucleus. A faint companion, eg 3625, is just visible in 30 cm 9′.5 E.

eg 3631 dimen. 4′.6 × 4′.1 V = 10.4 sfc. br. 13.5
In 25 cm this galaxy has a faint halo about 2′ diameter and a sharply defined 30″ core. At high powers the halo is very difficult to see. 30 cm will show the halo spreading faintly to at least 4′ diameter. The central regions are well concentrated through a fairly distinct 1′ core to a stellar nucleus. At 225 × the inner core and halo have a finely mottled texture. Two faint stars are visible WNW and NE, each at about 3′.5.

eg 3642 dimen. 5′.8 × 4′.9 V = 11.1 sfc. br. 14.6
While just visible in 15 cm, this galaxy is a large but still faint object for 25 cm. At 200 × a faint stellar nucleus sparkles within an indefinite halo that is elongated in pa 120°. In 30 cm the halo extends to 3′.5 × 2′.5. The overall concentration is quite weak, but several brighter lumps and stellarings are discernable in the irregular 2′ core: none of these is in the exact center. Several mag. 12.5 and fainter stars are within 5′ to the NE and NW.

eg 3665 dimen. 3′.2 × 2′.6 V = 10.8 sfc. br. 13.0
In 15 cm this galaxy is about 1′.5 × 0′.75, elongated in pa 30°. The bright core has a circular outline and contains a stellar nucleus. A wide pair lies 13′ S (11,11.5; 25″; 155°). With 25 cm the inner regions are bright enough so that the nucleus is barely distinguishable. A threshold magnitude star is off the N edge 1′.7 from center. 30 cm shows the halo to 1′.75 × 1′.25. The whole object has a smooth appearance and is well concentrated to the stellar nucleus. eg 3658 (mag.$_Z$13.3) lies 15′ SW.

eg 3675 dimen. 5′.9 × 3′.2 B = 10.9 sfc. br. 13.9
This galaxy is visible in 6 cm as a faint patch about 4′ × 2′.5 in extent elongated N-S. The small, concentrated center has a nearly stellar appearance at 100 ×. With 15 cm a star is visible off the S tip 2′ from the nucleus. The smooth, high surface brightness halo extends to 3′.5 × 1′ in pa 0° with 25 cm. A bright oval core stands out in the center; within this the substellar nucleus is still prominent. Several faint stars are nearby: 2′.5 SE is a mag. 13.5 star; three brighter stars form a 2′.5 × 1′ triangle centered 4′ W. In 30 cm the halo grows to 4′.5 × 2′. The core is a thin bright streak with a distinctly knotted appearance. A faint star or stellaring occasionally sparkles just N of its center, but otherwise there is no prominent nucleus here. At 225 × especially a dark lane is visible passing along the eastern flank of the core; the halo to this side is somewhat fainter than to the W.

eg 3683 dimen. 2′.0 × 0′.9 B = 13.4 sfc. br. 13.8
This is a faint object for 15 cm. At 150 × it appears as a faint, uniform streak about 1′.5 long, elongated roughly SE-NW. With 25 cm the halo is 1′.5 × 0′.5, elongated in pa 130°; the southeastern tip seems sharply tapered in comparison to the northwestern. At 200 × two stellarings are visible on the major axis near the center. 30 cm presents a good view of the galaxy at 225 ×, though it is easily visible at lower powers. The halo extends to 2′.2 × 0′.8. The core is a very thin brightening with bright spots strung along it, including the substellar nucleus. Thin spikes extend from the ends of the brighter parts of the core into the halo, which seems unevenly bright and irregularly bounded with averted vision. The bright spots in the core occasionally seem to lie along the northern edge, giving the impression of a dark lane hugging the core to this side. Photographs show eg 3674 (mag.$_Z$13.1) 14′ NW, and eg 3683A (mag.$_Z$12.6) 21′ NE.

eg 3687 dimen. 2′.0 × 2′.0 B = 12.8 sfc. br. 14.2
Located 6′.3 NE of a mag. 10 star, this galaxy is a faint unconcentrated glow in 15 cm, difficult to see at low powers. It is much easier with 25 cm. The halo extends to about 1′.5 diameter around a slightly brighter, circular core that contains a substellar nucleus.

eg 3690 dimen. 2′.4 × 1′.9 B = 12.0 sfc. br. 13.5
eg I 694 dimen. 1′.2 × 1′.0 mag. sfc. br.
This nearly merged interacting pair of galaxies is a large and fairly bright object for 25 cm. A bright irregular bar

extends along the major axis of an oval, 2′ × 1′.25 halo that is elongated in pa 80°. The E side of the halo seems poorly defined. In 30 cm the core is about 1′ × 0′.5 and has a very uneven texture: a faint stellar spot lies at its W end, and a 10″ knot marks the E tip. At low power an extremely faint halo is visible surrounding the object with an overall diameter of about 2′.5, elongated roughly N-S. A mag. 10 star lies 3′ S. Photographs show a small compact companion, eg M + 10–17–02 (mag.$_{VV}$16), 1′.1 NW.

eg 3718 dimen. 8′.7 × 4′.5 V = 10.5 sfc. br. 14.4
eg 3729 dimen. 3′.1 × 2′.1 V = 11.4 sfc. br. 13.3
These galaxies form an interacting pair 12′ apart. In 25 cm, eg 3718 is 4′ × 2′ in pa 30° and exhibits almost no central brightening. Averted vision will reveal a faint stellar nucleus and an outer halo elongated in pa 165°. 30 cm shows a large and poorly concentrated object 4′ across. The core is elongated NE-SW and is about 1′ wide. A mag. 11 pair of 35″ separation in pa 260° lies 2′.2 SSW. eg 3729 is ENE of – 18, and 25 cm shows a mag. 11 star on its S side 55″ from the center. The inconspicuous core is 2′ × 1′; the halo extends mostly NW to about 3′ length. In 30 cm the halo is about 1′.8 × 1′, elongated roughly N-S; the mag. 11 star lies on its southern edge. At times a broad spike seems to extend from the N side in pa 30°. Photographs show a peculiar chain of five galaxies, eg Z268–49 (combined mag.$_Z$14.7), 7′ S of eg 3718.

eg 3726 dimen. 6′.0 × 4′.5 V = 10.4 sfc. br. 13.9
With 15 cm this galaxy is about 5′ × 3′, elongated N-S. A mag. 11 star is visible on the N tip and another fainter one lies farther SW. 25 cm reveals a small, grainy 30″ × 10″ core within a 6′ × 2′ halo. The star on the SW seems fuzzy as though double. In 30 cm the halo is large, but very diffuse. At 225 × the surface is distinctly uneven; some of the brighter patches form an "S" shape across the center.

eg 3738 dimen. 2′.6 × 2′.0 V = 11.7 sfc. br. 13.3
In 25 cm this is a fairly bright object. A smoothly textured outer envelope extends to 1′.5 × 1′, elongated SE-NW. The core is more strongly elongated, about 45″ × 15″, and has a brighter spot on its NW side. With 30 cm the galaxy forms a 5′ × 4′ kite-shaped asterism with two mag. 10–11 stars E and a fainter one SE. The galaxy itself has a bar-shaped outline with broadly rounded ends and an overall size of about 1′.2 × 0′.6. The central region is slightly brighter, but no nucleus is discernable at the center. Threshold magnitude stars are visible N and E at 1′.5–2′; the latter forms the nail at the cross of the kite. eg 3756, *q.v.*, lies 16′ SE; eg 3733 (mag.$_Z$13.2) lies 21′ NNW.

eg 3756 dimen. 4′.4 × 2′.4 V = 11.5 sfc. br. 13.9
This is a fairly faint galaxy in 25 cm, though easily located 4′ S of a mag. 10 star. The outline is irregularly round, about 1′.5 diameter. At 200 × the surface appears to be broken into four or five slightly brighter patches grouped around a small nucleus. With 30 cm the halo grows to 4′ × 2′, elongated in pa 0°. The nebula has a low surface brightness and is poorly concentrated toward the center.

eg 3769 dimen. 3′.2 × 1′.1 B = 12.5 sfc. br. 13.8
In 25 cm this galaxy is faintly visible at low power as a 2′.5 × 0′.5 patch of moderately low surface brightness elongated in pa 150°. At 100 × a stellar nucleus stands out, while high power will show only a narrow, broadly concentrated core with a faintly mottled texture. A mag. 13.5 star lies 2′.2 NE. 30 cm shows a pointed, 2′ × 0′.7 lenticular form with an elongated though unconcentrated core. The surrounding field is mostly devoid of stars. Photographs show eg 3769A (mag.$_Z$14.7), an interacting companion 1′.2 SE.

eg 3780 dimen. 3′.1 × 2′.6 B = 12.3 sfc. br. 14.4
Only with difficulty will 15 cm show this galaxy as an ill-defined 2′ patch lying 24′ ENE of a bright close pair [ADS 8236: 7.9,8.4; 6″.0; 166° (1977)]. 25 cm reveals a smoothly textured squared-off oval extending NE-SW to 3′ × 2′. A mag. 13 star is visible 2′ ENE. With 30 cm at 150 × the broadly and very slightly concentrated halo extends to 3′.25 × 2′.5 in pa 40°. At 225 × the inner core is distinctly elongated E-W and has a small brightening off its eastern side. Photographs show eg 3804 (mag.$_Z$13.8) 13′ ESE, closely involved with a mag. 13 star.

eg 3782 dimen. 1′.7 × 1′.2 B = 13.1 sfc. br. 13.7
This galaxy is just visible in 15 cm as a tiny, unconcentrated spot closely involved with three mag. 11–12 stars. In 25 cm the galaxy seems elongated roughly N-S, extending to perhaps 1′ × 0′.4, though the nearby stars make the pa uncertain. A somewhat brighter elongated core stands out but contains no distinct nucleus. The nearest star is on the SSW edge, 30″ from the center; the second nearest is a little farther WNW. 30 cm shows a faint halo to about 1′ diameter, within which is embedded a slightly brighter, elongated core. A very faint star lies 1′.3 NNE.

eg 3813 dimen. 2′.3 × 1′.2 V = 11.7 sfc. br. 12.7
In 15 cm this galaxy is an oval, slightly concentrated object extending to 1′.5 × 0′.3, elongated E-W. The halo grows to about 2′.5 × 0′.5 in 25 cm, with a considerably brighter 1′ × 0′.4 core. Two mag. 14 stars bracket the

galaxy off the E and W ends; the eastern one is a little closer at 1.′4 from the center.

eg 3877 dimen. 5.′4 × 1.′5 B = 11.8 sfc. br. 13.9
Located 17′ S of χ Ursae Majoris (V = 3.7), this galaxy is visible in 15 cm as a faint but distinctly elongated glow. At 100 × the halo extends to about 4′ × 1′ in pa 35° and shows a weak, even concentration toward the center. With 25 cm the core, though not much brighter than the outer parts, begins to show a distinctly mottled texture. In 30 cm the galaxy appears well concentrated and fairly bright overall. Even at 150 × the middle parts are faintly granular. At 225 × the halo is 4.′5 × 1.′ The overall outline is bar-shaped: the ends of the halo do not taper, though the core bulges slightly. A faint nucleus barely stands out among several stellarings within the inner 1′ diameter.

eg 3888 dimen. 1.′8 × 1.′4 B = 13.1 sfc. br. 14.0
A fairly bright object in 25 cm, this galaxy appears 1′ × 0.′5 in pa 110°. The central regions are broadly concentrated even though the edges of the halo are sharply defined. In 30 cm the galaxy forms a triangle with mag. 13 stars about 2′ N and W. No sharp nucleus is visible within the unconcentrated oval core; the surrounding halo tapers rapidly to pointed ends, extending to 1.′75 × 0.′8. This galaxy is a member of the eg 3846–98 group; eg 3898, *q.v.*, the brightest member, lies 16′ NE.

eg 3893 dimen. 4.′4 × 2.′8 B = 11.1 sfc. br. 13.7
eg 3896 dimen. 1.′7 × 1.′3 mag.$_z$14.0 sfc. br.
eg 3906 dimen. 1.′9 × 1.′8 mag.$_z$14.1 sfc. br.
Only the brightest of this group (eg 3893) is visible in 15 cm, lying 3′ NE of a mag. 10 star. The moderately bright 2′ × 1′ oval is elongated in pa 165° and contains a faint stellar nucleus. With 25 cm a mag. 12.5 star is visible on the NW tip. At high power the knotty, circular core is conspicuous within the otherwise smooth halo. It is a nicely mottled object in 30 cm. The weakly concentrated, unevenly bright halo reaches to 2.′25 × 1.′5. The core is little brighter and has extensions into the halo in pa 0°, giving the impression of spiral arms winding counterclockwise on the sky. A faint stellaring is in the very center, and a few others are spread across the core. At 225 × a dark patch is occasionally visible on the NE side of the core. The mag. 12.5 star is 1′ NW, within the halo, but W of the major axis. eg 3896, 3.′9 SE, is visible in 30 cm on the S side of a mag. 13 star. eg 3906, 20′ SE of −93, appears about 40″ diameter in 30 cm, but shows little concentration or detail. A mag. 13.5 star lies 2.′1 S.

eg 3898 dimen. 4.′4 × 2.′6 V = 10.8 sfc. br. 13.3
This galaxy is the brightest of a group including eg 3888, *q.v.* 16′ SW. With 25 cm a conspicuous stellar nucleus is visible within a 40″ core and an oval 2′ × 1′ halo in pa 110°. The halo grows to 3′ × 1.′2 with 30 cm, but it is relatively faint compared to the inner regions. The brightness rises abruptly in distinct zones going from the halo to the circular 45″ core and bright nucleus. At high power a 60″ × 15″ dark lane is visible running E-W along the N edge of the core.

eg 3917 dimen. 4.′9 × 1.′4 B = 12.4 sfc. br. 14.3
eg 3931 dimen. 1.′4 × 1.′1 mag.$_z$14.6 sfc. br.
eg 3917 is a distinctly elongated galaxy in 25 cm. The 3′ × 0.′5 halo brightens gradually toward the center but is without a distinct nucleus. In 30 cm it is 3.′5 × 0.′75 in pa 80°, with a few faint stellarings on the slightly bulging middle. Two faint stars are visible about 1.′25 S and SE. eg 3931 lies 11′ NNE and is visible in 30 cm as a 25″ spot with a faint yet steadily visible stellar nucleus.

eg 3938 dimen. 5.′4 × 4.′9 V = 10.4 sfc. br. 13.8
In 15 cm this galaxy exhibits a large 3.′5 × 2.′5 halo elongated N-S. The small 30″ core is only slightly brighter than the halo. The core seems elongated in pa 120° with 25 cm, appearing 1.′75 × 0.′25 within the smooth, squarish 3.′5 × 3′ halo. It is of similar size in 30 cm, which reveals mottlings on the core and a faint stellar nucleus.

eg 3941 dimen. 3.′8 × 2.′5 B = 11.3 sfc. br. 13.6
This galaxy is easily visible in 15 cm as a 1.′5 × 0.′5 concentrated glow elongated in pa 175°. A stellar nucleus sparkles in the center. With 25 cm the halo extends to 2′ × 0.′75 in about pa 10° in contrast to the bright core, which still shows elongation in pa 175°. A mag. 13 star lies 1.′5 E.

eg 3945 dimen. 5.′5 × 3.′6 V = 10.6 sfc. br. 13.7
In 15 cm this galaxy is easily visible at low powers, showing a small, well-defined core with a bright center. The nucleus is quite stellar in 25 cm and of a similar brightness to a mag. 12 star 1.′3 SW. Surrounding the nucleus is a 30″ diameter core and a faint, nebulous halo that seems elongated roughly E-W. Two other mag. 12.5–13 stars are visible nearby: one lies off the edge of the halo to the S, the other is about 1.′5 NW. With 30 cm the galaxy is bright, round, and well concentrated to the center. At 225 × a stellar nucleus occasionally stands out in the broadly concentrated, 50″ core, which is elongated in pa 160°. The surrounding halo fades smoothly to an overall diameter of 2.′

eg 3949 dimen. 3.′0 × 1.′8 V = 11.0 sfc. br. 12.7
Moderately faint in 15 cm, this galaxy appears as a 1.′5 × 1′ oval patch elongated SE-NW. The halo is broadly concentrated, but contains a faint stellar nucleus. In 25 cm it

grows to 2′.5 × 1′.25 in pa 120°. No distinct nucleus is evident here, but a number of faint stellarings are visible in the core at 200 ×. 30 cm shows a weakly concentrated 2′ × 1′.1 oval with a stellar nucleus at its center. Brighter spikes extend along the major axis from the core into the otherwise smooth halo.

eg 3953 ***dimen. 6′.6 × 3′.6*** ***V = 10.1*** ***sfc. br. 13.3***
15 cm shows this galaxy easily, lying to the W of a 20′ N-S string of mag. 10–11 stars. The halo is about 5′ × 2′, elongated nearly N-S, and contains a broadly concentrated core. In 25 cm a stellar nucleus is visible, as well as a mag. 13 star on the W edge of the 1′.5 core, 50″ from the nucleus. The surrounding halo extends to 5′ × 3′ in pa 15°. 30 cm will show the halo to 4′.2 × 2′.1 around a 1′ oval core. At 225 × the core and inner halo exhibit a mottled texture, and the whole E side of the halo is generally dimmer than the W. At times there seems to be an indistinct dark lane pressing up against the E side of the core. A second, fainter star is visible against the halo, 45″ NE of the nucleus.

eg 3958 ***dimen. 1′.6 × 0′.8*** ***mag.$_z$13.1*** ***sfc. br.***
In 30 cm this galaxy appears as a small oval glow of fairly high surface brightness extending to 50″ × 30″ in pa 30°. It grows slightly and evenly brighter toward the center, where a stellar nucleus is visible. A faint star lies off the N side, 50″ from the nucleus. eg 3963, *q.v.*, is 8′.2 NNE.

eg 3963 ***dimen. 2′.8 × 2′.6*** ***B = 12.4*** ***sfc. br. 14.4***
With 25 cm this galaxy is a faint circular glow about 1′.5 diameter with little central concentration. A few faint stellarings are scattered over the core, but none stands out as a nucleus. In 30 cm the galaxy is very interesting. At 150 × the faint halo extends to 2′ × 1′ in pa 135°, but exhibits no real overall concentration toward the center. Even at low power, but especially at 225 ×, the middle two-thirds of the galaxy is strongly mottled. A threshold magnitude star lies on the SSW edge at pa 195°; on the opposite side of the galaxy from this is a bright knot. Several other bright and dark spots stand out with close attention. eg 3958, *q.v.*, lies 8′.2 SSW.

eg 3977 ***dimen. 1′.7 × 1′.6*** ***mag.$_z$14.7*** ***sfc. br.***
Viewed in 30 cm this galaxy is a very small and faint object lying 16′ NNW of eg 3982, *q.v.* The halo has virtually no central concentration, though a faint stellaring is occasionally visible at the center. A single mag. 13 star lies 2′.7 N; a wide pair of mag. 13 stars is 2′.5 ESE.

eg 3982 ***dimen. 2′.5 × 2′.2*** ***B = 11.9*** ***sfc. br. 13.6***
This galaxy is a fairly bright object in 25 cm, located about 3′.5 NNE of two mag. 12 stars. At 200 × the indistinct halo extends to 1′.5 × 1′, rising broadly toward the center and exhibiting no distinct nucleus. The halo grows to 1′.8 × 1′.5 in 30 cm, elongated in pa 5°. The broadly concentrated core is about 1′ across and still shows no distinct nucleus, though many faint slightly brighter spots are scattered over its surface. Photographs show eg 3972 (mag.$_z$12.9) 13′ NW.

eg 3985 ***dimen. 1′.2 × 0′.8*** ***B = 13.0*** ***sfc. br. 12.9***
This tiny object is not visible with 15 cm and is moderately faint with 25 cm. At 200 × it appears as a weakly concentrated spot 30″ across with an indistinct nucleus. In 30 cm the halo is elongated in pa 60° and has broad, moderate concentration to a 15″ core that seems multiple, the parts aligned along the major axis. A mag. 14 star lies 2′.4 SE.

eg U6930 ***dimen. 4′.3 × 3′.0*** ***mag.$_z$14.2*** ***sfc. br.***
Viewed in 30 cm, this galaxy is a faint patch about 1′.5 diameter. It shows moderate concentration to an inconspicuous 30″ core but has no distinct nucleus. A mag. 11 star lies 2′.3 NW; a mag. 13 star lies 1′.3 SW.

eg 3990 ***dimen. 1′.7 × 1′.0*** ***V = 12.6*** ***sfc. br. 13.0***
eg 3998 ***dimen. 3′.1 × 2′.5*** ***V = 10.6*** ***sfc. br. 12.7***
In 25 cm eg 3998 appears circular, about 1′ diameter, and has a stellar nucleus that stands out from the otherwise faint haze. 30 cm shows an oval 2′.1 × 1′.2 halo elongated in pa 130° rising little to an abruptly brighter 15″ core that is off-center to the SW side of the halo. The bright nearly stellar nucleus is off-center in the core to the NE, placing it nearer to the center of the halo. eg 3990 lies 3′ to the W. In 25 cm the galaxy is less than 30″ diameter and contains a stellar nucleus. With 30 cm the halo is sharply concentrated to a bright nucleus that seems multiple. A faint star lies halfway between the two galaxies, a little N of the line connecting them.

eg 3991 ***dimen. 1′.4 × 0′.5*** ***V = 13.2*** ***sfc. br. 12.5***
eg 3994 ***dimen. 1′.1 × 0′.7*** ***V = 12.7*** ***sfc. br. 12.3***
eg 3995 ***dimen. 2′.8 × 1′.1*** ***V = 12.6*** ***sfc. br. 13.7***
25 cm will show all three members of this multiple interacting system clearly. eg 3995 appears only slightly brighter than its companions: its halo is 1′.5 × 1′.25 in pa 35°, diffuse and little concentrated to a core that is off-center to the N. eg 3991 is 3′.7 NW of −95. It extends to 2′ × 0′.3, also in pa 35°, and has a small, sharp central brightening with a stellar nucleus. eg 3994 lies 1′.8 SW of −95. It appears elongated in pa 10°, 1′ × 0′.25, and also has a stellar nucleus.

eg 3992 ***dimen. 7′.6 × 4′.9*** ***V = 9.8*** ***sfc. br. 13.6***
Messier 109 is one of the two or three difficult objects in Messier's catalogue for 6 cm: even so, it is faintly visible

at 50×. The very low surface brightness, slightly concentrated halo has an indefinite extent, but is elongated toward a mag. 10 star that lies about 5′ SW. In 15 cm the galaxy appears slightly smaller and fainter than Messier 108, *cf.* eg 3556. The faint, diffuse halo extends to about 5′×3′ in pa 70°. With 25 cm three fainter stars are visible superposed against the galaxy and around its periphery: a mag. 12.5 star lies 50″ NNW of center; another is 3′.5 NE, near the maximum extent of the halo; a third, somewhat fainter, lies 3′ W. The halo is quite faint and hardly concentrated except for the small, 1′.5×1′ core, which is itself fairly well concentrated. The halo extends to nearly 8′×5′ in 30 cm, showing best at 150×. The core here seems about 30″ diameter and concentrated to multiple stellarings at its center. The inner parts have a mottled texture at low power, and the region immediately around the core is elongated in pa 30°.

eg 4013 dimen. 5′.2×1′.3 B=12.0 sfc. br. 13.9
15 cm reveals this galaxy as a very thin spindle about 2′ long, elongated in pa 65°. At the center is a bright stellaring that looks like the nucleus, but 25 cm shows this to be a mag. 12.5 star superposed on the galaxy less than 10″ E of the center. Here the halo is 2′.25×0′.25 and not strongly concentrated: no nucleus is distinguishable though the nearby star interferes. In 30 cm the halo grows to 3′×0′.5, a bar-shaped glow with a smooth texture.

eg I 749 dimen. 2′.5×2′.1 V=12.2 sfc. br. 13.9
eg I 750 dimen. 2′.9×1′.4 V=11.8 sfc. br. 13.1
Located about 3′ NE of a mag. 8.5 star, this pair of galaxies lies 3′.3 apart in pa 105°. With 15 cm eg I 750 is about 1′ long, elongated NE-SW, and has a bright nonstellar center. 25 cm shows the halo to 1′.25×0′.4 in pa 45°. The halo has tapered tips and the oval core has a sharp, seemingly multiple center. 30 cm reveals a well-concentrated 2′×0′.6 object with two or three bright spots along the major axis near the center. eg I 749 is only just visible with 15 cm due to its proximity to −50. With 25 cm it appears as a circular, diffuse blob about 1′.5 diameter without concentration. 30 cm shows it to about 1′.25 diameter with indefinite edges, a smooth texture, and a very slight concentration to the center.

eg 4026 dimen. 5′.1×1′.4 B=11.7 sfc. br. 13.7
Lying 7′ SSW of a mag. 9 star, this galaxy is just visible in 6 cm. 25 cm shows a bright, thin glow about 3′.5 long, elongated nearly N-S. The core is an oval brightening about 30″ across with a stellar nucleus. In 30 cm the bar-shaped halo is 2′.8×0′.8 in pa 175°, with very faint extensions to 3′.5 length. The brighter part of the halo is well and sharply concentrated to a bright but nonstellar nucleus. The E flank fades more sharply.

eg 4036 dimen. 4′.5×2′.0 V=10.5 sfc. br. 12.8
While this galaxy is faintly visible to 6 cm, it is an easy object for 15 cm. The uniformly bright 2′.5×1′ streak is elongated nearly E-W, and the E end appears a little irregular in shape. 25 cm reveals a very small nucleus in an oval core; the halo is 1′.7×0′.8 in pa 85°. 30 cm shows a smooth, moderately concentrated object with a prominent substellar nucleus that seems elongated. The long, tapering halo reaches to 3′×1′. eg 4041, *q.v.*, lies 15′ NNE.

eg 4041 dimen. 2′.8×2′.7 V=11.2 sfc. br. 13.2
In 15 cm this galaxy is a circular, uniformly bright patch about 1′.5 diameter. 25 cm will reveal some concentration in the broad core and a very faint stellar nucleus. At 200× the halo seems irregular, spreading more extensively toward the NE side for an overall size of about 1′ diameter. With 30 cm the entire object is irregularly bright and shows a slight but uneven brightening toward the center. At 150× the halo extends to about 2′ diameter, slightly elongated NE-SW. Many mag. 12 and fainter stars are nearby: two lie about 3′.5 SE; four fainter ones form a box 1′ on a side about 4′.5 WSW. eg 4036, *q.v.*, lies 15′ SSW.

eg 4047 dimen. 1′.5×1′.3 B=13.1 sfc. br. 13.7
Small and moderately bright in 25 cm, this galaxy appears about 1′.5 across, slightly elongated E-W. The weakly concentrated but relatively bright core dominates the object: the halo forms only a thin rim surrounding it. With 30 cm this is a small, faint object less than 1′ diameter. The halo grows moderately and broadly brighter toward the center. No distinct nucleus is visible, though the middle portions are scattered with faint stellarings.

eg 4051 dimen. 5′.0×4′.0 V=10.3 sfc. br. 13.7
15 cm shows this Seyfert galaxy as a 2′.5×1′ glow with a small core and bright stellar nucleus. A mag. 11 star lies 2′.2 W. The smooth haze grows to 3′×1′.5 in pa 135° with 25 cm. In 30 cm the galaxy appears larger than eg 4013, *cf.*, and with a similar texture, but it has a more prominent stellar nucleus. At 225× the mottled core has a bright center 10″ across and a stellar nucleus. The halo extends to 4′×2′.5 and the NE flank is dimmer.

eg 4062 dimen. 4′.3×2′.0 V=11.2 sfc. br. 13.4
In 15 cm this galaxy has a 2′×0′.75 halo elongated in pa 100°. The middle regions are a bit brighter and have a slightly granular texture at 200×. With 25 cm the halo

grows to 4′ × 1′. The galaxy has a long oval outline with rounded rather than tapered tips.

eg 4085 dimen. 2′.8 × 0′.9 V = 12.3 sfc. br. 13.2
Much fainter than its neighbor eg 4088 11′ N, *q.v.*, this galaxy appears 2′.5 × 0′.5 in 25 cm, elongated nearly E-W. The central concentration seems mainly expressed by several small bright splotches lying along the major axis; the brightest one is at the center. In 30 cm the spindle extends to 2′.25 × 0′.5 in pa 80°. There is a little concentration along the major axis, making the brighter part very thin. The ends of the halo taper slightly, and the E half is larger and better concentrated than the W.

eg 4088 dimen. 5′.8 × 2′.5 V = 10.5 sfc. br. 13.2
25 cm shows this galaxy as a fairly bright object with an unconcentrated, blocky core and a faint halo. At times the halo seems to extend to as much as 6′ × 2′.5 in pa 45°; the brighter, more distinct core is about 4′ × 2′ at 150 ×. With 30 cm the halo is about 5′ × 2′.5; the oval 4′ × 2′.5 middle part constricts suddenly to thin, pointed extensions that reach to the 5′ overall length. The middle half to two-thirds of the galaxy is mottled with 20″ light and dark patches. A barely distinguishable, elongated core about 15″ across sits near the SE edge of the otherwise unconcentrated halo.

eg 4096 dimen. 6′.5 × 2′.0 V = 10.6 sfc. br. 13.3
This galaxy is a slightly irregular, strongly elongated glow in 25 cm, reaching to about 5′ × 1′.5 in pa 20°. The brightest part is N of the geometric center. In 30 cm the galaxy has a smooth, regular texture. The halo grows to 6′ × 1′.1, and shows a moderate concentration to rounder inner parts. The core is about 1′.3 × 0′.9 and contains a faint, elongated nucleus. A wide unequal pair of faint stars lies 4′ NW.

eg 4100 dimen. 5′.2 × 1′.9 B = 11.6 sfc. br. 14.0
In 25 cm this galaxy has an elongated, tapering form. The halo is about 5′ × 1′.5, elongated in pa 165°, and contains a more circular core but no distinct nucleus. The halo is about 4′.5 × 1′.25 in 30 cm. The middle bulges out to the sides, and the halo is fairly extensive along either side of the bar, especially on the W side. The core, which is only slightly brighter than the halo, seems displaced toward the S end, making the northern extension appear longer than the southern. At 225 × several very faint stellarings are visible within the central 1′.75 length.

eg 4102 dimen. 3′.2 × 1′.9 B = 12.3 sfc. br. 14.1
This is a small, bright galaxy in 25 cm, with a prominent mag. 11.5 star close to the W edge, 50″ from the nucleus. The halo is about 1′.25 × 0′.75, elongated in pa 40°, not quite reaching to the star on the W. At the center is a small, strongly concentrated core about 20″ across that has a nonstellar nucleus. In 30 cm the galaxy grows to 1′.9 × 0′.8, the halo extending farther NE of center, away from the involved star. Here the core is an elliptical region about 20″ across that has its southern edge obscured by a dark patch on the inner halo. A stellar nucleus stands out in the center. Photographs show eg 4068 (mag.$_Z$13.3) 22′ WSW.

eg 4144 dimen. 5′.9 × 1′.5 B = 12.0 sfc. br. 14.3
Though faint in 25 cm, this galaxy is visible at 150 × as a distinctly elongated but poorly concentrated patch extending to about 3′ × 1′ in pa 105°. A few irregularly distributed brighter patches in the middle mark the core; there is no distinct nucleus. Three mag. 13 stars lie about 2′.5 SE; two of them form a 10″ pair aligned SE-NW. With 30 cm the object is slightly concentrated to a very thin patchy line running the length of the major axis; the brightest spots are in the western half. Overall the bar-shaped halo extends to 5′ × 0′.6. Many faint stars are within 5′ radius; another brighter pair of stars with a similar separation to the pair SE is 8′ N, aligned nearly perpendicularly to the first pair.

eg 4157 dimen. 6′.9 × 1′.7 B = 11.6 sfc. br. 14.1
In 25 cm this galaxy is an elongated object with tapering ends and a bulging middle. Overall it extends to about 4′ × 1′ in pa 65°. Along the major axis are a few inconspicuous patches and the SE edge seems more sharply defined. The halo extends to 4′.5 × 0′.6 in 30 cm, a very thin tapering spindle with an elongated core. At 225 × the core is about 1′.25 long and extends farther to the NE side of center. Several faint stellarings in the core and out toward the tips of the halo define the major axis. In contrast to the 25 cm view, the NW flank seems more sharply defined here.

eg 4284 dimen. 2′.9 × 1′.4 mag.$_Z$14.7 sfc. br.
eg 4290 dimen. 2′.5 × 1′.9 B = 12.6 sfc. br. 14.1
ds Messier 40
eg 4290 is a faint object in 25 cm. The circular core is 40″ across and is surrounded by a broken annular ring that extends to about 1′.5 × 1′.25, elongated NE-SW. It appears nearly circular in 30 cm, with well-defined edges and a very faint stellar nucleus. Messier 40 (the double star Winnecke 4) is a pair of mag. 9 stars about 12′ E. eg 4284, 4′.5 W, is barely detectable in 30 cm as a tiny patch of light less than 30″ diameter.

eg 4605 dimen. 5′.5 × 2′.3 B = 11.0 sfc. br. 13.6
This galaxy is just visible in 6 cm as a small patch slightly elongated ESE-WNW. It is a bright and interesting ob-

ject in 25 cm. The broad, mottled core takes up the central third of the 4′.5 × 1′ halo. Unlike most strongly elongated objects, the acute ends of the halo are well defined and do not extend with averted vision. With 30 cm the halo occasionally extends as far as 5′ × 1′.2 in pa 125°. The core is about 3′ × 1′ and shows no distinct nucleus at its center. At 150 × the bar-shaped object seems crooked, and the SE half is narrower than the NW half. A mag. 14 star is visible 1′.3 SSE.

eg 4814 ***dimen. 3′.2 × 2′.5*** ***B = 12.8*** ***sfc. br. 14.9***
In 25 cm this galaxy seems to have an egg-shaped outline, fatter on the SE side. The halo is elongated SE-NW, extending to 1′.25 × 1′. It is about 1′.5 diameter and roughly circular in 30 cm. The halo shows a broad, poor concentration to a faint stellaring at the center. About 11′ S is an interesting double star of contrasting colors (9,10; 20″; 272°).

eg 5204 ***dimen. 4′.8 × 3′.0*** ***V = 11.3*** ***sfc. br. 14.0***
A member of the Messier 101 (eg 5457) group, this galaxy is fairly faint in 25 cm. The well-defined 4′ × 2′ halo is elongated N-S and shows no central brightening. With 30 cm it also seems quite well defined, but extends only to 3′ × 2′ in pa 5°. At 150 × the low surface brightness halo shows a very slight concentration toward the center.

eg 5308 ***dimen. 3′.5 × 0′.8*** ***V = 11.3*** ***sfc. br. 12.2***
This galaxy appears moderately small in 25 cm, extending to 1′.5 × 0′.3 in pa 60°. The central regions brighten slightly and have a more nearly circular outline. At 200 × the NW flank of the halo seems more sharply defined, though this is not evident in 30 cm. With 30 cm at 150 × the halo has a 1′.5 × 0′.5 spindle-shaped outline. Here the galaxy shows fairly strong concentration to a circular 15″ core.

eg 5322 ***dimen. 5′.5 × 3′.9*** ***V = 10.0*** ***sfc. br. 13.3***
25 cm shows this galaxy as a very bright, concentrated glow with a nearly stellar nucleus and smoothly fading halo 1′.25 diameter. At 200 × there seems to be a star-like appendage on the SSE side of the core, which 30 cm confirms as a mag. 14 star 20″ from the nucleus. In this aperture the galaxy is a fairly bright and broadly concentrated object extending to 1′.5 × 1′ in pa 95°.

eg 5376 ***dimen. 2′.1 × 1′.4*** ***B = 13.0*** ***sfc. br. 14.0***
This is a fairly faint object for 25 cm, which will show an unconcentrated 1′.5 × 1′ oval elongated roughly E-W. It is still quite faint in 30 cm, but grows to 1′.8 × 1′.5 in pa 70°. At 225 × it has a skimpy halo surrounding a broad, poorly concentrated core that shows a faint stellar nucleus. eg 5379 (mag.$_{Z}$14.1) and eg 5389 (mag.$_{Z}$13.2) lie 15′ NE, 4′ apart.

eg 5422 ***dimen. 3′.9 × 0′.9*** ***B = 12.6*** ***sfc. br. 13.8***
25 cm will show this galaxy faintly at 100 × forming the NW apex of a 4′ × 2′.5 rectangle of mag. 10–11 stars. At 200 × it is a 1′.5 × 0′.3 glow elongated SSE-NNW with a stellar nucleus. In 30 cm it is easy to see at 150 ×. The halo extends to 2′ × 0′.5 in pa 150° around a distinct, circular 30″ core and a prominent nucleus. Photographs show eg 5443 (mag.$_{Z}$13.2) 41′ NNE.

eg 5430 ***dimen. 2′.4 × 1′.5*** ***B = 12.8*** ***sfc. br. 14.1***
In 25 cm this galaxy is a fairly faint object with an irregularly round outline about 55″ diameter and a broadly brighter center. At 150 × in 30 cm it is still fairly faint and seems elongated in pa 150°. Higher power, however, shows a faint star closely involved lying about 25″ from the center in pa 150°. The low surface brightness halo surrounds a 1′ diameter circular core, extending in pa 0° to an overall size of 2′ × 1′.

eg 5448 ***dimen. 4′.2 × 2′.0*** ***B = 12.2*** ***sfc. br. 14.4***
This galaxy is a 2′ × 0′.5 nearly uniformly bright streak in 15 cm, elongated roughly E-W. With 25 cm it is about the same length and only slightly wider. Here the halo brightens a little to a more circular core. 30 cm shows the halo to only 1′.5 × 0′.6 in pa 115°, exhibiting very little central concentration and no distinct nucleus.

eg 5457 ***dimen. 27′ × 26′*** ***V = 7.7*** ***sfc. br. 14.7***
Messier 101 is clearly visible in 6 cm as a large smoothly textured glow about 15′ diameter with slight broad concentration to a small core. In good conditions two or three small brightenings are discernable in the halo. No stars are visible superposed on the nebula in this aperture, but several faint stars are near the periphery, particularly across the S side. With 15 cm the galaxy shows an irregularly bright, slightly oval halo extending to about 20′ × 15′ in pa 20°. At the center is a small circular nucleus of about mag. 12.5; 1′.3 N of this is a mag. 12.5 star, and a fainter star lies toward the NW edge, 3′.5 from the nucleus. About 8′ SW, just off the SW edge, is a small, isolated patch about 15″ diameter with a stellar center. At low powers the overall impression is that the halo is more extensive toward the N and E sides, placing the brightest parts SW of the geometric center. 30 cm will show many details in the core and arms, particularly on the N, E, and SE sides. The core here appears about 2′.5 across, though it is irregularly shaped. Several faint stellarings are contained within it; one occasionally stands out as the nucleus. Forming a box 4′–5′ on a side with the core are three brighter patches or associations in the halo

to the NE, E, and SE. The southeastern one is the brightest, best concentrated, and most conspicuous of the three; the one to the NE is the least conspicuous. The eastern association is distinctly elongated NE-SW. The distant association to the SW is elongated roughly SE-NW and has a faint star about 30″ N of it; at 225× this association has an uneven, sparkly texture. A spiral arm is distinguishable leaving the SE side of the nucleus, curving NE then N, where it passes through the association on the NE. From here a uniform but low surface brightness band continues across the N. Another less distinct arm connects the associations on the SE and E sides. Though the NE through SE sides of the halo have more and brighter details than elsewhere, this side fades more abruptly to a relatively distinct edge.

eg 5473 ***dimen. 2′.6 × 1′.8*** ***V = 11.4*** ***sfc. br. 12.9***
In 25 cm this galaxy is fairly bright but appears little more than 30″ diameter. A large, well-defined core dominates the central three-fourths of the object, and a faint star lies about 20″ from the center on the NE edge. The halo grows to 1′.2 diameter with 30 cm, and shows fairly strong concentration to a substellar nucleus.

eg 5474 ***dimen. 4′.5 × 4′.2*** ***V = 10.9*** ***sfc. br. 13.9***
In 25 cm and 30 cm this member of the Messier 101 (eg 5457) group is a diffuse, poorly concentrated object about 3′ diameter without any distinct detail.

eg 5480 ***dimen. 1′.8 × 1′.3*** ***B = 12.9*** ***sfc. br. 13.6***
eg 5481 ***dimen. 1′.7 × 1′.4*** ***mag.$_Z$13.5*** ***sfc. br.***
eg 5480 is moderately faint in 15 cm, showing a roughly circular halo about 1′ across and a stellar nucleus. With 25 cm the halo extends to 1′.25 × 0′.75, elongated in pa 0°. Directly E 3′.1 is eg 5481, which has a brighter stellar nucleus than eg 5480. In 30 cm −80 is the larger of the pair, though it has weaker concentration and a similar overall brightness. eg 5481 is a circular patch about 1′ diameter with a stellar nucleus.

eg 5485 ***dimen. 2′.6 × 2′.1*** ***V = 11.5*** ***sfc. br. 13.2***
Lying 8′ N of a 1′ pair of mag. 12 stars, this galaxy appears similar to eg 5473, *cf.*, in 25 cm, though it is a bit larger. The halo is about 1′ diameter and contains a distinct core. In 30 cm the galaxy seems overall a bit fainter than and not as evenly concentrated as eg 5473. The distinctly bounded core is 45″ diameter and holds a faint, occasionally stellar nucleus at its center. At 225× the core has an uneven texture; a faint stellaring is occasionally visible just N of the nucleus. The halo extends to about 1′.2 diameter, slightly elongated N-S. eg 5484 (mag.$_Z$15.6) is 3′.7 NW; eg 5486 (mag.$_Z$14.0) lies 6′.3 NNE.

eg 5585 ***dimen. 5′.5 × 3′.7*** ***V = 10.9*** ***sfc. br. 14.0***
A member of the Messier 101 (eg 5457) group, this galaxy is faintly visible in 15 cm as a circular spot about 1′ diameter forming a triangle with stars to the NE and SE. With 25 cm it grows to 3′ × 1′ in pa 30°. The central regions are only slightly brighter. 30 cm shows an indistinct, low surface brightness glow about 2′.5 diameter without any central concentration or details. A mag. 13.5 star lies 2′ SSW.

eg 5631 ***dimen. 2′.2 × 2′.1*** ***B = 12.5*** ***sfc. br. 14.0***
This galaxy is a star-like spot with a faint halo in 15 cm. With 25 cm the halo has a fairly high surface brightness and is about 1′ diameter. It is quite a small but bright object in 30 cm, appearing no more than 50″ diameter. Here the nucleus is a bright nonstellar spot about 12″ diameter.

Ursa Minor

Though this constellation is circumpolar, the Guardians of the Pole, β Ursae Minoris (V = 2.1) and γ Ursae Minoris (V = 3.1), culminate at midnight in early May. Only one object in our survey falls within the boundaries of the constellation.

eg 6217 dimen. 3′.1 × 2′.7 V = 11.2 sfc. br. 13.4
This galaxy is a faint round spot in 15 cm about 1′ diameter. There is a starry point at the center, but no general brightening. It is about 1′.5 diameter in 25 cm, slightly elongated in pa 155°. Here the halo shows broad concentration to a slightly brighter core and stellar nucleus. 30 cm shows a bright stellaring near the center that appears double and a little SE of center at 225 ×. The halo extends to 2′.5 × 1′.2, slightly concentrated to the center and strongly mottled with brighter patches. Several faint stars are visible within 5′.

Vela

Representing the sails of the old constellation Argo Navis, this group lies in the Milky Way, but well to the south for northern observers. The center of the constellation culminates at midnight about 16 February.

oc 2547 **diam. 74′** **V = 4.7** **stars 112**
In 6 cm this cluster is a large unconcentrated group of about 20 stars led by a prominent mag. 6.5 star near the center. 25 cm shows only a few faint stars spread evenly among the brighter members, which are mostly mag. 8–11. Overall the group is about 30′ diameter.

gn 2626 **dimen. 5′ × 5′** **star V = 10.0**
In 15 cm this small nebula appears as a weak circular glow around the mag. 10 central star. With 30 cm the outline is roughly triangular with one apex pointing S. The illuminating star is at the N end near the base aligned E-W; the nebula is very slightly brighter near the star.

oc 2659 **diam. 15′** **V = 8.6:** **stars 16**
At 25× in 6 cm this cluster appears as a weak fuzz amongst some faint stars. It is moderately faint in 15 cm at 75×. About 15 stars are resolved in a group elongated NE-SW, with a brighter mag. 9.5 star off the SE side of the bar. In 30 cm it is 15′ × 7′, including 60 stars of mag. 10.5 and fainter, among which are a few pairs.

oc 2671 **diam. 5′** **mag. 11.6p** **stars 40**
This cluster is discernable in 6 cm as a faint, unresolved patch about 10′ diameter. The unconcentrated haze contains a dozen mag. 12 and fainter stars in 15 cm, including four brighter ones forming a triangle on the E side. With 25 cm at least 20 stars are visible in a 6′ area. In the center of the group is a faint unresolved group about 3′ across. About 30 stars can be seen within 10′ diameter with 30 cm.

pn 2792 **dimen. 13″ × 13″** **nebula 11.7v** **star**
Viewed in 25 cm, this object appears nearly stellar at 100×. Higher powers show a bright uniformly illumined spot characteristic of small planetaries. A matched pair of mag. 11 stars (36″; 100°) is conspicuous in the field about 4′ SE of the nebula.

oc 2849 **diam. 3′.0** **mag. 12.5p** **stars 40**
This very faint cluster is not visible in 15 cm, but it lies immediately E of a N-S line of three mag. 9 stars about 1′ long. In 30 cm the cluster is enclosed in a triangle formed by the line to the W and two other mag. 9 stars SE and NE. Ten very faint stars are resolved within 4′ diameter at high power.

pn 3132 **dimen. 84″ × 53″** **nebula 9.4v** **star V = 10.0**
Clearly visible in 15 cm at 75×, this conspicuous planetary is dominated by its mag. 10 central star. In 25 cm the nebula is elongated SE-NW, with the central star off center to the E side. With 30 cm the object has a slightly brighter rim, but the middle parts around the central star are not dark: the darkest area is in the NW side, where the otherwise ring-like structure is broken. This ring is brightest along the SW edge, and the outer rim is sharply defined there. At 450× a very faint star is visible at the very edge of the nebula in pa 210°.

gc 3201 **diam. 18′** **V = 6.7** **class X**
Even 6 cm shows this globular cluster as a bright unresolved haze of moderate, even concentration. It can be resolved well in 25 cm, where it appears about 8′ across and is very broadly brighter across the center. The core is about 5′ × 4′, elongated roughly NE-SW. In 30 cm about 50 stars are resolved over a grainy haze that is without much concentration. The resolved stars are almost uniformly distributed over the cluster.

Virgo

This is the second largest of the constellations, but by far it holds the largest number of bright galaxies of any in the sky. While most of the bright objects are grouped in the cluster near the Coma-Virgo border, many fine galaxies can be found in all corners of the constellation. The center of the constellation culminates at midnight about 12 April.

eg 3818 dimen. 2′.1 × 1′.4 V = 11.8 sfc. br. 13.0
In 15 cm this is a small, moderately faint galaxy with a relatively bright center; the overall diameter is less than 1′. Though the galaxy is faintly visible at low power in 25 cm, high power will show that the starry core is elongated in pa 110°. At best the halo reaches to only 45″ diameter. The peculiar galaxy eg M-01-30-22 (mag.$_{VV}$13.5) lies 21′ SSW.

eg 3976 dimen. 3′.9 × 1′.4 B = 12.2 sfc. br. 13.9
Visible at 50 × in 25 cm, this galaxy is about 1′ diameter, slightly elongated in pa 55°, and contains a stellar nucleus.

eg 4030 dimen. 4′.3 × 3′.2 B = 11.1 sfc. br. 13.8
Lying between two mag. 10 stars that are less than 4′ apart, this galaxy is an easy object for 15 cm. The halo is about 1′.25 long, elongated roughly NE-SW, and has a bright circular center. The mag. 1 5star SSW has a mag. 11.5 companion 50″ SE. In 25 cm the halo extends to 2′ × 1′.25 in pa 30°, the brightness rising smoothly but abruptly from the weak halo to a circular core and faint stellar nucleus. Photographs show eg U7000 (mag.$_{Z}$14.4) 17′ SE.

eg 4045 dimen. 2′.8 × 2′.0 V = 11.8 sfc. br. 13.6
In 25 cm this galaxy is visible as a diffuse and unconcentrated spot about 1′ diameter. Photographs show eg 4045A (mag.$_{Z}$15.2) 1′.5 S, and many other galaxies in a group about 30′ to the E, including eg 4073 (B = 12.7).

eg 4116 dimen. 3′.8 × 2′.4 V = 11.9 sfc. br. 14.1
This galaxy is visible in 15 cm as a circular spot 45″ diameter with a brighter center. It appears brighter than eg 4123, *cf.* 14′ NE, in 25 cm, though smaller. The 1′.5 halo has a threshold stellar nucleus at the center.

eg 4123 dimen. 4′.5 × 3′.5 V = 11.2 sfc. br. 14.1
Located 14′ NE of eg 4116, *q.v.*, this galaxy is visible in 15 cm as a circular spot about 1′ diameter. 25 cm reveals a large, moderately faint and diffuse object that is only slightly brighter toward the center. The overall size is 2′.5 × 1′ in pa 135°.

eg 4124 dimen. 4′.6 × 1′.7 B = 12.0 sfc. br. 14.1
A faint object for 15 cm, this galaxy appears more or less circular with a small, brighter core. The halo is 2′.5 × 0′.75 in extent with 25 cm, elongated in pa 115°. The small oval core contains a faint stellar nucleus. Several mag. 10–12 stars lie scattered in the medium-power field surrounding the object.

eg 4129 dimen. 2′.6 × 0′.8 V = 12.6 sfc. br. 13.2
Appearing nearly stellar at low power in 25 cm, higher power reveals this galaxy as a small, narrow object. The high surface brightness halo is 1′ × 0′.2, elongated nearly E-W.

eg 4164 dimen. 0′.3 × 0′.3 B = 15.1 sfc. br. 12.5
eg 4165 dimen. 1′.5 × 1′.1 V = 13.2 sfc. br. 13.6
eg 4168 dimen. 2′.8 × 2′.6 V = 11.3 sfc. br. 13.4
The brightest of this small group of galaxies is visible in 15 cm as a conspicuous haze with a bright core. The circular glow is about 1′ across in 25 cm, which also shows a very faint stellar nucleus. With 30 cm the outer halo is about 1′.5 diameter, but its maximum extent is very faint and thus indefinite; the outline seems slightly elongated SSE-NNW. The smooth inner regions exhibit a nearly azonal brightness profile reaching to the bright but nonstellar center. eg 4165, 2′.8 NNW, is marginally visible in 15 cm. In 30 cm it is a diffuse, very low surface brightness area 45″ across. The weakly and broadly concentrated halo is slightly elongated in about pa 160°, nearly toward − 68. eg 4164, 2′.9 W of − 68, is a tiny haze with a conspicuous stellar center in 30 cm.

eg 4179 dimen. 4′.2 × 1′.2 V = 10.9 sfc. br. 12.5
With 15 cm this galaxy is visible as a small 1′ × 0′.5 glow elongated SE-NW, with a stellar nucleus. It is fairly bright in 25 cm, reaching to 1′.75 × 0′.5 in pa 145°. Just

beyond the concentrated 40″ × 10″ core is a mag. 14 star, 33″ NW of center.

eg 4193 dimen. 2′.3 × 1′.2 V=12.4 sfc. br. 13.3
A difficult object for 15 cm, this galaxy appears as a concentrated substellar spot at the threshold of visibility. Moderately faint for 30 cm, the halo extends to about 2′ × 1′.2 in pa 90°. The very faint stellar nucleus seems to be slightly N of center, and the core is more sharply defined on the S, suggesting a dark lane there. A mag. 13.5 star is visible 1′.8 NW.

eg 4206 dimen. 5′.2 × 1′.2 V=12.1 sfc. br. 14.0
This spindle-shaped galaxy is pretty faint in 15 cm, but the N-S elongation is distinct. 30 cm shows the halo to more than 4′ length, but no more than 30″ wide. At high power the irregular spots along the major axis show a weak brightening trend across the center. Two mag. 12 stars 4′.1 apart lie just SE of the object. The galaxy is the first of three spindles in the field, including eg 4216, *q.v.* 11′ NE, and eg 4222, *q.v.* 23′ NE in Coma Berenices.

eg 4215 dimen. 1′.9 × 0′.8 B=13.1 sfc. br. 13.4
Visible as a faint diffuse glow in 15 cm, this galaxy lies 3′ SE of a mag. 12.5 star. It is not large in 25 cm, the high surface brightness halo only 35″ × 20″ in size, elongated nearly N-S. At the center it is suddenly brighter, the nucleus represented by a small pip. 30 cm shows thin, pointed tips to the halo reaching to 1′.75 length, though the width is no more than 30″. The elongated inner regions are moderately and evenly brighter, but contain an outstanding stellar nucleus, which occasionally appears multiple.

eg 4216 dimen. 8′.3 × 2′.2 V=9.9 sfc. br. 12.9
This is a bright object for 15 cm, a large 5′ × 1′ spindle elongated in pa 20°. The small sharply brighter center is surrounded by a smooth faint halo. 25 cm reveals a stellar nucleus within the 5′.5 × 0′.8 object. In 30 cm the faintest extension of the halo reaches halfway to a mag. 12 star 9′.8 NNE; overall it has a very smooth texture. The 45″ × 15″ core is capsule-shaped and of unusually high surface brightness, yet a bright stellar nucleus is still visible within it. A dark lane passes on the E side of the core, but the halo extends farther on this side than on the W. A mag. 13.5 star shines near the E edge 36″ from the core. eg Z69–113 (B=14.9), a tiny elliptical, lies 3′.9 N of center. eg 4216 is the brightest of three spindle-shaped galaxies including eg 4206, *q.v.* 11′ SW and eg 4222, *q.v.* 12′ NE in Coma Berenices.

eg 4224 dimen. 2′.4 × 1′.0 V=11.8 sfc. br. 12.7
While only faintly visible in 15 cm, this galaxy is conspicuous at 100 × in 25 cm. The oval 1′.25 × 0′.5 halo is elongated in pa 60°, and broadly brighter toward a threshold stellar nucleus. A mag. 13.5 star is visible 1′.6 N; a mag. 14.5 star lies at 1′.8 in pa 250°. The halo grows to 2′.5 × 1′ with 30 cm, with rounded rather than tapering ends. The galaxy has a moderate even concentration to elongated inner regions and a distinct stellar nucleus.

eg 4233 dimen. 2′.3 × 1′.1 V=12.0 sfc. br. 12.9
Viewed in 25 cm, this galaxy appears smaller and fainter than nearby eg 4224, *cf.* The halo is 25″ across, elongated very slightly N-S. There is little concentration except for a faint, sharp stellar nucleus. A mag. 12.5 star is visible 2′.8 NE. More of the halo is discernable in 30 cm, where it reaches 1′.25 × 0′.8 in pa 175°. The distinct 20″ core is itself without concentration except that a very faint stellar nucleus is occasionally visible at high power.

eg 4234 dimen. 1′.3 × 1′.2 V=12.9 sfc. br. 13.2
This fairly faint galaxy is visible at 100 × in 25 cm, but it appears diffuse, and is without concentration toward the center.

eg 4235 dimen. 4′.3 × 1′.1 V=11.6 sfc. br. 13.1
A moderately bright object for 25 cm, this galaxy appears as a large, diffuse glow at 50 × . The sharply tapered 2′ × 0′.75 halo is elongated in pa 50°. The core is somewhat brighter and contains a weak nonstellar nucleus. 30 cm shows the low surface brightness halo to 4′.25 × 1′. The inner brightness contours of the distinct 1′ × 0′.7 core become progressively more circular as they approach the center. A small nucleus is visible at 225 × , but it is not conspicuous. Three faint stars lie within 3′ to the NW, N, and ENE. eg 4235 is the brightest of a widely separated triple interacting system including eg 4246, *q.v.* 12′ E, and eg 4247, *q.v.* 13′ ENE.

eg 4241 dimen. 2′.5 × 1′.4 V=12.0 sfc. br. 13.2
Located 5′.4 N of a mag. 9 star, this galaxy is visible at low power in 25 cm. the circular 45″ halo has a well-defined border. The light is broadly concentrated with an uneven texture; no core or nucleus is discernable. In 30 cm a faint but much larger halo is visible, extending to 2′.2 × 0′.8, the tapered tips elongated in pa 130°. The core is more nearly circular, about 35″ across, and contains a sharp but weak concentration to a stellar nucleus. A mag. 11.5 star lies 2′.2 SSW.

eg 4246 dimen. 2′.5 × 1′.5 V=12.7 sfc. br. 14.0
This galaxy is a pretty faint object for 25 cm, which shows a very diffuse low surface brightness glow about 1′ diameter slightly elongated E-W. In 30 cm the halo is 1′.75 × 1′.25, elongated in pa 85°. The very broad concentration is barely noticeable; a faint nonstellar nucleus is

occasionally visible at high power. A mag. 14.5 star lies 2.′0 ENE.

eg 4247 dimen. 0.′7 × 0.′6 B = 14.8 sfc. br. 13.6
In 25 cm this galaxy is more strongly concentrated than eg 4246, *cf.*, but it is still a faint object. The small circular halo brightens to a conspicuous stellar nucleus. 30 cm shows it only 30″ across and evenly concentrated to a 15″ core. The core itself is almost unconcentrated, though it contains a faint nonstellar nucleus. A mag. 13 star is visible 2.′1 NNW.

eg 4255 dimen. 1.′5 × 0.′8 B = 13.6 sfc. br. 13.6
30 cm shows this galaxy without difficulty. At high power the 1.′2 × 0.′4 halo is elongated in pa 115°; thin, pointed extensions reach to 1.′5 length. The brightness rises evenly toward an oval 30″ × 15″ core, then abruptly to a distinct stellar nucleus. Overall the appearance is similar to eg 4215, *q.v.* A mag. 14 star is visible 2.′3 E.

eg 4259 dimen. 1.′1 × 0.′5 V = 13.6 sfc. br. 12.8
A faint member of the eg 4273 group, this galaxy is a small object less than 30″ across in 30 cm. The high surface brightness halo is well concentrated to a nonstellar nucleus. A mag. 14.5 star lies close by 25″ NE of center.

eg 4260 dimen. 2.′6 × 1.′4 V = 11.7 sfc. br. 13.0
In 15 cm this galaxy appears simply as a very faint spot. 25 cm reveals a 1′ × 0.′75 glow elongated in pa 60°. The fat lenticular halo contains a small, nearly circular core with a few stellarings though no distinct nucleus. 30 cm shows a 1.′2 × 0.′8 halo with a broadly concentrated core. At 225 × a faint stellar nucleus is occasionally visible at the center. A faint star is visible 1.′3 NE.

eg 4261 dimen. 3.′9 × 3.′2 V = 10.3 sfc. br. 13.1
This galaxy appears as a round spot of high surface brightness in 15 cm. The 45″ halo grows brighter to a stellar center. With 25 cm the appearance is similar, but the nucleus is substellar. 30 cm shows the halo a bit more than 1′ diameter. The core is relatively bright, broadly concentrated, and contains a nonstellar nucleus. eg 4264, *q.v.*, lies 3.′4 ENE; eg 4257 (B = 14.9) is 7.′1 SW; eg M + 01–31–53 (B = 14.4) is 5.′2 N.

eg 4264 dimen. 1.′1 × 0.′9 V = 12.9 sfc. br. 12.7
Located 3.′4 ENE of eg 4261, *q.v.*, this galaxy is barely visible with 15 cm. 25 cm shows only an unconcentrated 30″ spot. In 30 cm the halo grows to about 50″ diameter around a relatively large core that contains a stellar nucleus. A faint star is visible about 1′ NE.

eg I3153 dimen. B = 14.7 sfc. br.
This faint galaxy lies within a triangle formed by eg 4259, eg 4270, and eg 4273, *q.v.* It is barely visible in 30 cm as an indefinite 40″ patch of low surface brightness with slight concentration to a very faint nucleus.

eg 4266 dimen. 2.′1 × 0.′5 B = 14.5 sfc. br. 14.4
Lying only 40″ SE of a mag. 8.5 star (ADS 8510: 8.5,12.3; 1.″4; 274°), this galaxy is difficult to view in 30 cm. The faint unconcentrated patch is about 20″ across.

eg 4267 dimen. 3.′5 × 3.′2 V = 10.9 sfc. br. 13.4
This galaxy is a moderately faint object for 15 cm, appearing as a circular spot at 100 ×. It is visible at low power with 25 cm. At high power it appears about 1′ diameter and has a conspicuous stellar nucleus. Photographs show eg I 775 (V = 13.3) 14′ NW.

eg 4268 dimen. 1.′6 × 0.′6 V = 12.5 sfc. br. 12.4
The faintest of three galaxies including eg 4273 and eg 4281, *q.v.*, this object is only marginally visible in 15 cm. It is best viewed at moderate powers in 25 cm. The lenticular 1′ × 0.′5 halo is elongated in pa 50°, growing brighter to a poorly defined oval core and conspicuous stellar nucleus. At 250 × in 30 cm the thinly tapered tips of the halo reach to an overall length of 1.′2. The core is about 40″ × 10″ and contains a faint sparkly nucleus. A mag. 14.5 star is visible 50″ NW: it appears fainter than the nucleus.

eg 4269 dimen. 1.′5 × 1.′0 B = 13.7 sfc. br. 14.1
eg I3155 dimen. 1.′1 × 0.′6 B = 14.9 sfc. br. 14.2
Both of these galaxies lie close to a mag. 7.5 star. eg 4269 lies 1.′6 SE of the star, and in 25 cm is visible at high power as a 20″ spot with a sharply brighter substellar center. A mag. 14 star lies 2.′2 SE. 30 cm shows the halo 30″ across; the stellar nucleus is comparable in brightness to the faint star SE. eg I3155 is located 2.′1 in pa 195° from the mag. 7.5 star, or 1.′1 SW of eg 4269. At 200 × in 25 cm it appears completely stellar, like a fainter version of the center of −69. High power in 30 cm shows an unconcentrated enucleate glow 30″ across.

eg 4270 dimen. 2.′2 × 1.′0 V = 12.2 sfc. br. 12.9
A member of the eg 4281 group, this galaxy is visible in 15 cm 5.′4 SSE of a mag. 8.5 star (ADS 8510: 8.5,12.3; 1.″4; 274°). In 25 cm it is a small lenticular object extending to only 40″ × 30″ in pa 110°. The central regions grow moderately brighter to a faint nearly stellar nucleus. With 30 cm the halo grows to 1.′2 × 0.′7. The core is broadly concentrated and contains a stellar nucleus that occasionally seems multiple. To the E and W, each at about 3.′5, are

Chart XXIV. Virgo South.

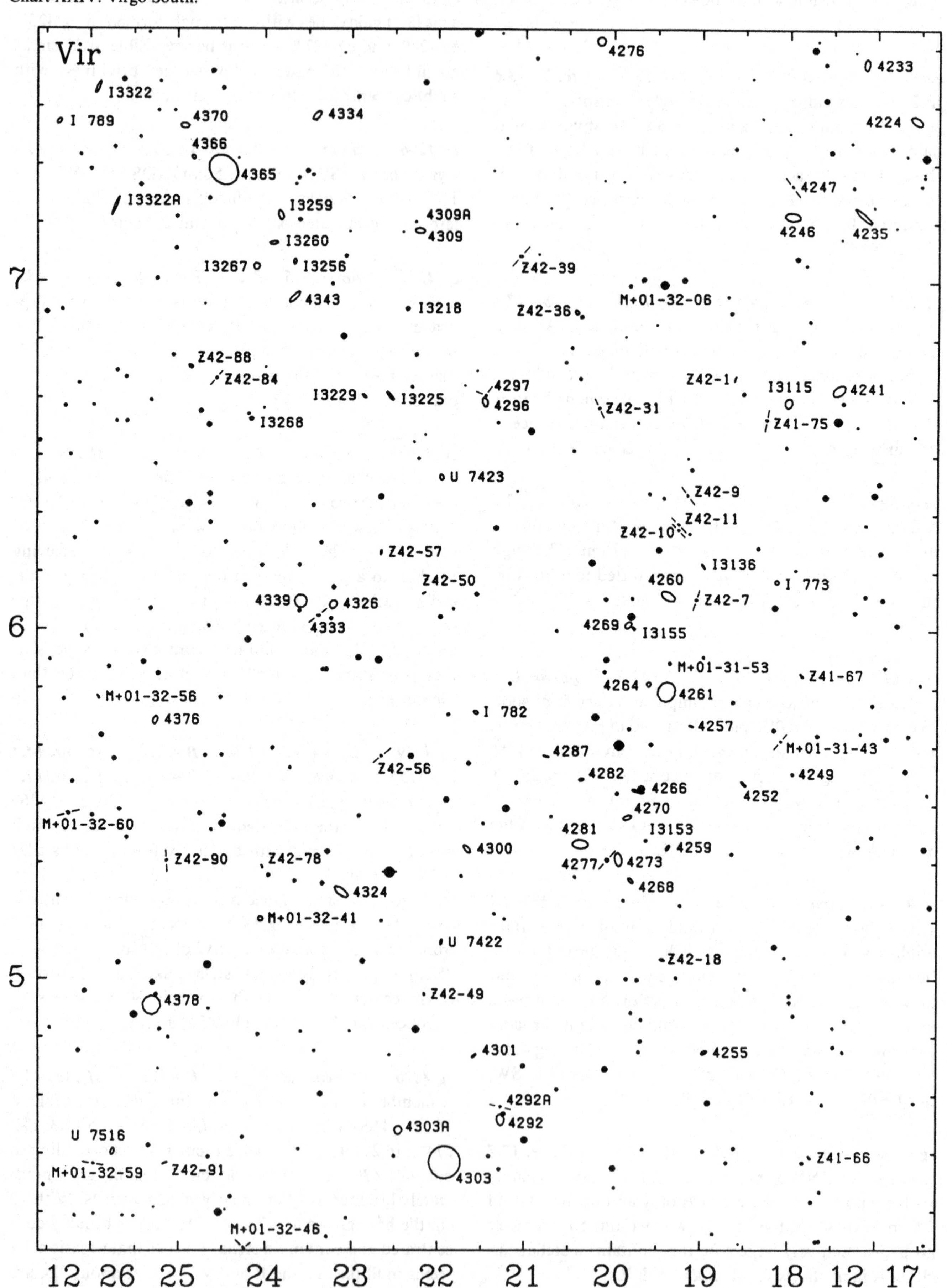

two mag. 13.5–14 stars. Just S of the pair to the NNW is eg 4266, *q.v.*

eg 4273 dimen. 2′.3 × 1′.5 V=11.9 sfc. br. 13.1
eg 4277 dimen. 0′.9 × 0′.8 V=13.5 sfc. br. 13.1
These galaxies, lying 1′.9 apart, are members of the eg 4281 group. eg 4273 is a small but definite spot in 15 cm, less concentrated but of similar size to eg 4281, *cf.* With 25 cm the halo is about 1′.5 × 1′, elongated nearly N-S. At higher powers there occasionally seems to be a brighter bar along the major axis. In 30 cm the broad, bright core is 1′.2 across, fading rapidly at its edges to the faint halo that reaches to perhaps 1′.5 diameter, slightly elongated in pa 10°. No distinct nucleus is visible at any magnification. Lying 2′.2 SE is a mag. 14 star; just 1′ N of this is eg 4277, which is barely visible in 25 cm as a 20″ spot at high power. In 30 cm it is a faint unconcentrated patch about 40″ diameter at 250 ×, though it is visible at 125 ×. Here the galaxy lies in the eastern side of an equilateral triangle formed by the mag. 14 star S, eg 4273, and a threshold star 1′ N of eg 4277.

eg 4281 dimen. 3′.1 × 1′.5 V=11.3 sfc. br. 12.9
This galaxy is the brightest of a group including eg 4273, eg 4270, and others, *q.v.* In 15 cm it is by far the easiest of the group to see. At 100 × it is a very small, fairly bright spot with a conspicuous stellar nucleus. 25 cm shows the stellar nucleus embedded in an ill-defined oval core and a 2′ × 0′.4 halo elongated in pa 90°. A pretty pair (12,12.5; 13″; 115°) is 5′.5 S, easily resolvable at 125 ×. With 30 cm the halo is 1′.4 × 0′.8, and the nucleus is not stellar.

eg 4292 dimen. 2′.1 × 1′.4 B=13.5 sfc. br. 14.5
In 25 cm this galaxy is a moderately faint, diffuse spot about 1′.2 SSE of a mag. 10.5 star. With 30 cm the halo is 1′.25 × 1′, elongated in pa 5°. There is an moderate even concentration to a faint nucleus that occasionally appears stellar. The nearby star interferes somewhat with viewing. Messier 61 (eg 4303, *q.v.*) lies 12′ SE.

eg 4294 dimen. 3′.1 × 1′.3 V=12.1 sfc. br. 13.5
25 cm shows this galaxy as a moderately low surface brightness object elongated in pa 155°, 1′.75 × 0′.5. The elongated core is only slightly brighter than the halo. eg 4299, *q.v.*, lies 5′.6 ESE.

eg 4299 dimen. 1′.7 × 1′.6 V=12.5 sfc. br. 13.4
With 25 cm this galaxy is a faint, circular spot less than 1′ diameter with a stellar center. Photographs show eg 4313 (B=12.7) 23′ NE, eg 4330 (B=13.1) 25′ SE, and eg 4352 (B=13.6) 40′ SE.

eg 4300 dimen. 1′.6 × 0′.7 B=13.8 sfc. br. 13.7
While not visible in 15 cm, this galaxy is a moderately low surface brightness 45″ × 30″ oval with strong central concentration in 25 cm. At 125 × there is no distinct nucleus though the center contains a few faint stellarings. With 30 cm the galaxy is best viewed at medium powers. The smoothly textured halo extends to 1′.25 × 0′.6 in pa 40°, but shows almost no central brightening other than a very faint but distinct stellar nucleus.

eg 4301 dimen. 1′.4 × 0′.5 B=14.8 sfc. br. 14.2
Viewed in 25 cm this galaxy is a very faint spot about 45″ diameter with slight brightening toward the center. Messier 61 (eg 4303, *q.v.*) is 19′ S.

eg 4303 dimen. 6′.0 × 5′.5 V=9.7 sfc. br. 13.4
Messier 61 is visible in 6 cm as a faint circular spot with a bright center. It appears about 3′ diameter in 15 cm, the core showing a stellar nucleus at high power. The galaxy is very bright in 25 cm, the mottled oval halo extending to 2′ × 1′.75, elongated NE-SW. There is hardly any central brightening except for the mag. 13 stellar nucleus. On the W edge 1′.1 from center is a mag. 14 star; another is visible outside the halo 2′.5 SW. 30 cm shows spiral structure weakly; toward the N side is a brighter patch, while just E of the center a dark region is visible. Overall the halo reaches to about 4′ × 3′.5, while the bright core is 30″ across. eg 4303A, *q.v.*, lies 9′.7 NE; eg 4292, *q.v.*, lies 12′ NW.

eg 4303A dimen. 1′.7 × 1′.4 V=13.0 sfc. br. 13.8
This galaxy is a marginal object for 25 cm, located 2′.4 E of a mag. 13 star that has a faint companion 30″ WSW. The galaxy is of moderately low surface brightness but is not difficult to see in 30 cm. The 1′ halo shows a slight, even concentration to a faint occasionally visible stellar nucleus. A mag. 15 star lies 45″ WNW of center.

eg 4307 dimen. 3′.7 × 0′.9 B=12.8 sfc. br. 14.0
This galaxy is a faint but well-defined object in 25 cm. The narrow spindle is elongated in pa 25°, 1′.5 × 0′.3, and only a little brighter toward the center. Photographs show eg I3211 (B=14.8) 3′.2 S, and eg 4316 (B=13.7) 19′ NNE.

eg 4324 dimen. 2′.5 × 1′.2 B=12.6 sfc. br. 13.6
Lying 9′ ESE of 17 Virginis (ADS 8531: V=6.5,9.4; 21″; 337°), 15 cm shows this object as a faint, nearly circular spot 45″ diameter. With 25 cm the lenticular 1′ × 0′.5 halo is elongated in pa 55°, showing a small core and a bright stellar nucleus at high power. In 30 cm the halo is 2′ × 0′.8 overall. The round, broadly concentrated core is sharply defined on the NW; a sharp nucleus is occasionally visible within.

eg 4334 ***dimen. 2.′4 × 1.′1*** ***B = 13.9*** ***sfc. br. 14.9***
A faint object for 25 cm, this galaxy appears as a circular 30″ glow containing a faint stellar nucleus. A mag. 12.5 star lies about 40″ S.

eg 4339 ***dimen. 2.′3 × 2.′3*** ***V = 11.4*** ***sfc. br. 13.3***
Located 1.′5 N of a mag. 11 star, this galaxy is only faintly visible in 15 cm. In 25 cm it is moderately bright, circular, and about 1′ diameter. The brightness rises smoothly and moderately toward the center: no core or nucleus is distinguishable. The halo remains circular in 30 cm, growing to 1.′5 diameter. Photographs show eg 4333 (B = 14.3) 4.′0 SW.

eg 4343 ***dimen. 2.′8 × 0.′9*** ***V = 12.3*** ***sfc. br. 13.1***
This galaxy is visible at 100 × in 25 cm, though it is moderately faint and fairly diffuse. The 45″ × 30″ oval is elongated SE-NW and becomes a bit brighter across the center, where a substellar nucleus is visible. With 30 cm the halo extends to 1.′2 × 0.′8. A conspicuous stellar nucleus shows in the center, but no other brightening is visible.

eg I3256 ***dimen. 1.′4 × 0.′7*** ***V = 12.6*** ***sfc. br. 12.4***
This galaxy looks like a faint star in 15 cm. 25 cm confirms its high surface brightness relative to eg 4343, *cf.* 6′ S. The tiny 30″ × 10″ oval halo is elongated SSE-NNW and contains a very bright stellar nucleus. It is an easy object for 30 cm, but remains quite small, showing the same features as in 25 cm. eg I3260, *q.v.*, lies 4.′6 NE.

eg I3258 ***dimen. 1.′6 × 1.′4*** ***B = 13.8*** ***sfc. br. 14.5***
A difficult object for 25 cm, this galaxy is best viewed at medium powers. At 100 × a bit of haziness is visible with a sharply concentrated center.

eg I3259 ***dimen. 1.′8 × 1.′1*** ***V = 13.6*** ***sfc. br. 14.1***
eg I3260 ***dimen. 1.′9 × 0.′7*** ***V = 13.3*** ***sfc. br. 13.4***
eg I3267 ***dimen. 1.′1 × 1.′1*** ***V = 13.5*** ***sfc. br. 13.5***
Faint for 25 cm, eg I3260 has a surface brightness between that of eg 4343 and eg I3256, *cf.*, and seems more diffuse than these neighbors to the southwest. Overall it is about 45″ across, slightly elongated E-W. 30 cm shows it about 2′ × 1′ in pa 100°. The core is only slightly brighter, but a stellar nucleus is occasionally visible. eg I3259, 5′ NNW, is at the limit of detection in 25 cm, having extremely low surface brightness. eg I3267, 5′ SE, is very faint and nearly stellar.

eg 4348 ***dimen. 3.′5 × 1.′0*** ***B = 12.9*** ***sfc. br. 14.1***
In 15 cm this galaxy is a tiny streak elongated in pa 35°. In 25 cm the halo is 1.′5 × 0.′5, concentrated to an elongated core and nonstellar nucleus. A mag. 12.5 star lies 1.′1 W of center.

eg 4351 ***dimen. 2.′0 × 1.′4*** ***V = 12.4*** ***sfc. br. 13.4***
This galaxy is best viewed at medium power in 25 cm. The small, low surface brightness haze is not conspicuous, though it lies in a blank field. The halo is about 50″ × 35″, elongated ENE-WSW, and is very broadly brighter across the center.

eg I3268 ***dimen. 0.′8 × 0.′8*** ***B = 14.2*** ***sfc. br. 13.6***
This object is located on the NW side of a mag. 12 star that is the southeastern vertex of a nearly right triangle of stars 2.′6 long. In 25 cm the galaxy is a fairly faint circular patch about 20″ diameter.

eg 4365 ***dimen. 6.′2 × 4.′6*** ***B = 10.6*** ***sfc. br. 14.2***
eg 4366 ***dimen.*** ***B = 14.7*** ***sfc. br.***
eg 4365 is a fairly faint object for 15 cm, showing a prominent core within a 1′ × 0.′75 halo. The galaxy is large and bright in 25 cm, the extensive low surface brightness halo reaching to 2.′5 × 1.′5, elongated NE-SW. The inner regions have a smooth texture, and grow evenly brighter toward a poorly defined 30″ core and weak substellar nucleus. A similar view is given by 30 cm. eg 4366, 5′ NE, is visible at 125 × in 25 cm, a faint spot with a faint stellar center. eg 4370, *q.v.*, lies 9.′9 NE. Two large, very narrow spindles, eg I3322 (B = 14.1) and eg I3322A (B = 13.8), are located 25′ NE and 19′ ESE, respectively, from eg 4365.

eg 4370 ***dimen. 1.′6 × 0.′9*** ***B = 13.7*** ***sfc. br. 14.0***
Located 9.′9 NE of eg 4365, *q.v.*, this galaxy appears in 25 cm as a fairly diffuse fat oval with broad central brightening.

eg 4371 ***dimen. 3.′9 × 2.′5*** ***V = 10.9*** ***sfc. br. 13.2***
In 15 cm this galaxy appears as a 1′ circular glow containing a stellar nucleus. It is fairly bright for 25 cm, in which it is conspicuous at low power. The well-defined circular 1.′5 halo is smoothly concentrated to a bright stellar nucleus.

eg 4374 ***dimen. 5.′0 × 4.′4*** ***V = 9.3*** ***sfc. br. 12.6***
Messier 84 is visible in 6 cm, appearing similar to nearby Messier 86 (*cf.* eg 4406), but a little brighter. 15 cm shows a bright oval haze with an intense, broadly concentrated core. In 25 cm it is elongated ESE-WNW, reaching to about 2.′5 × 2′, but fading so smoothly to the sky that the boundary is indefinite. The innermost regions, however, rise sharply in brightness toward a substellar nucleus. With 30 cm the 4′ × 3′ halo has a strong, very even con-

Chart XXV. Virgo North.

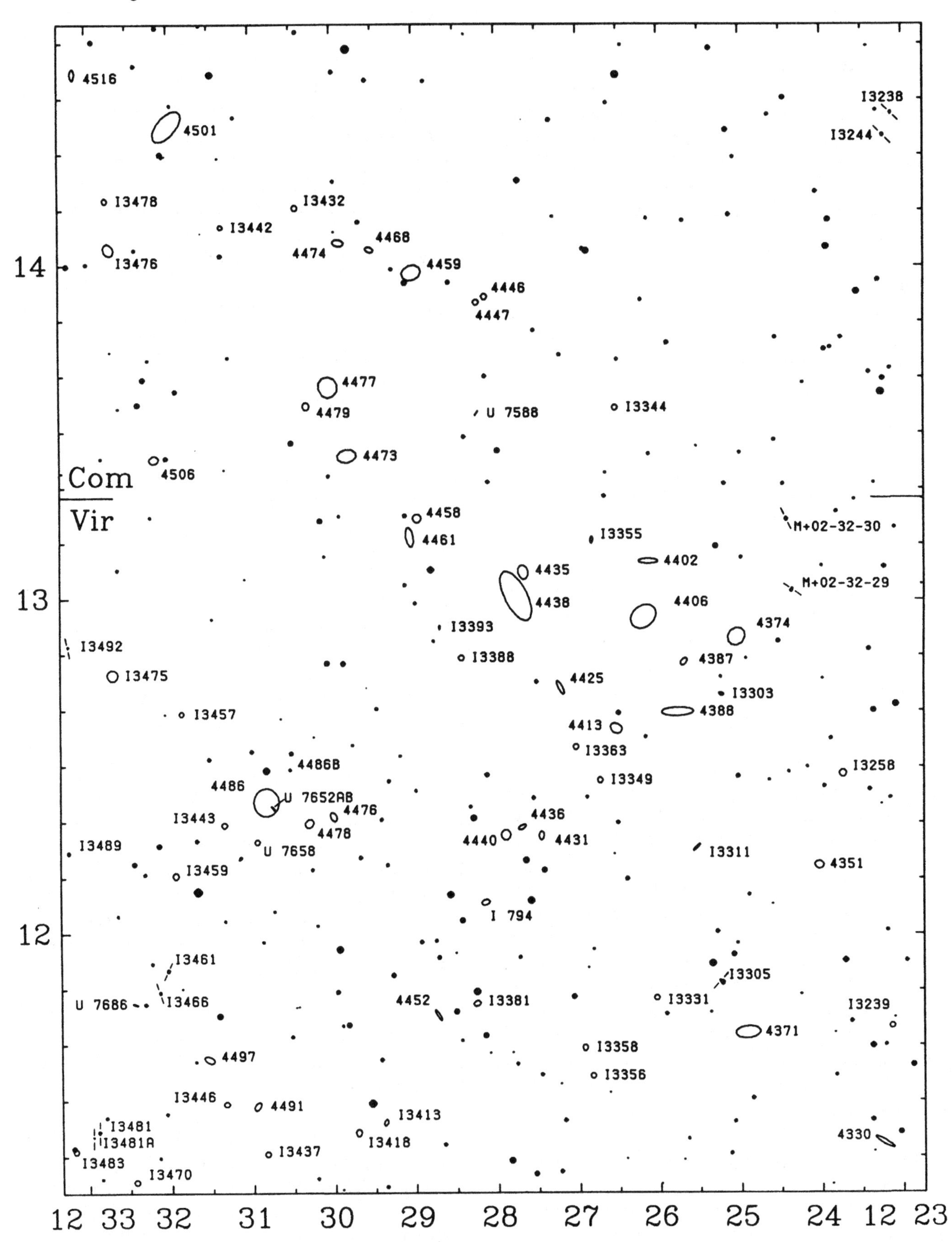

centration, and the nucleus is clearly nonstellar. A mag. 14 star is involved 1.25 WSW of center.

eg 4378 dimen. 3.3 × 3.1 B = 12.3 sfc. br. 14.7
This galaxy can be seen faintly in 15 cm forming an obtuse triangle with two mag. 9 stars 3.9 N and 3.4 SE. It is moderately bright in 25 cm, the 45″ × 30″ oval rising to a well-defined core and a faint stellar nucleus. In 30 cm it appears elongated in pa 160°, 1.5 × 0.8, with a uniformly bright core and an extensive halo. The SW flank is sharply defined.

eg 4380 dimen. 3.7 × 2.2 B = 12.4 sfc. br. 14.5
This galaxy appears moderately bright in 25 cm, though its surface brightness is low. The halo is elongated SSE-NNW, 2′ × 0.75, and will show some faint mottling with averted vision. A mag. 13 star lies 2.4 S. Photographs show eg I3328 (B = 14.4) 8.9 E, and eg 4390 (B = 13.3) 27′ NNE.

eg 4385 dimen. 2.3 × 1.5 V = 12.4 sfc. br. 13.5
25 cm shows this fairly faint galaxy as a circular 1.5 glow with a sharply brighter core.

eg 4387 dimen. 1.9 × 1.1 V = 12.0 sfc. br. 12.9
This galaxy is quite faint in 15 cm. 25 cm shows it as a circular 30″ patch with smooth, moderate concentration to a sharp center. A mag. 13 star is visible 1.5 NNW. 30 cm shows it much larger, 1.8 × 0.75, elongated SE-NW. The halo has a moderate, broad concentration to a small, faint nucleus.

eg 4388 dimen. 5.1 × 1.4 V = 11.2 sfc. br. 13.1
This galaxy forms a triangle with the brighter galaxies Messier 84 and 86 (eg 4374 and eg 4406, *q.v.*), which lie in the same medium-power field to the N. An easy object for 15 cm, the 3′ × 0.5 spindle is elongated E-W and shows a slight central brightening. With 25 cm the small core lies W of center; stellarings and a stellar nucleus are visible in it at high power. The E side of the core is squared off abruptly by a dark patch that seems at times to detach the extended eastern extremity of the lenticular 3.5 × 0.75 halo. A mag. 14.5 star lies 1.3 NE. The northern flank of the 4.5 × 1′ halo fades abruptly in 30 cm. The light is evenly concentrated to a less elongated core and very faint stellar nucleus.

eg 4402 dimen. 4.1 × 1.3 V = 11.3 sfc. br. 13.0
Much more difficult than nearby Messier objects, this galaxy is visible in 15 cm at 100 ×. It is clearly elongated, but less distinctly so here than in larger apertures. With 25 cm the 3′ × 0.75 halo is elongated in pa 90°, and completely unconcentrated, though a little mottling is discernable at 100 ×.

eg 4406 dimen. 7.4 × 5.5 V = 9.2 sfc. br. 13.2
Messier 86 is visible with 6 cm in the same low-power field with Messier 84 (eg 4374, *q.v.*) and appears as a slightly elongated spot with a bright center. In 15 cm the 2′ × 1.5 halo grows much brighter toward the center, but is otherwise featureless. The galaxy appears similar to M84 in 25 cm, fading very smoothly to the sky background. However, the 3.5 × 3′ halo, elongated ESE-WNW, is distinctly larger than that of M84, and seems more diffuse. The light is broadly brighter except at the very center, where a sharp nucleus is visible. With 30 cm it is more elongated and strongly concentrated than M84, though their brightnesses are similar. The 5′ × 3′ halo contains a condensed 10″ core that occasionally shows a stellar nucleus.

eg 4412 dimen. 1.5 × 1.4 B = 13.1 sfc. br. 13.7
This galaxy is moderately faint in 25 cm, appearing as a circular 1′ spot without much central concentration.

eg 4413 dimen. 2.5 × 1.7 B = 13.0 sfc. br. 14.3
This galaxy is located 2.9 S of a mag. 10.5 star, and 1.3 S of a mag. 11.5 star. In 15 cm it is faintly visible at low power as an unconcentrated, low surface brightness patch elongated roughly NE-SW. 25 cm shows a 1.5 × 1′ halo of moderately low surface brightness that is broadly brighter and faintly mottled across the center. The galaxy reaches 1.75 × 1′ in 30 cm, elongated in pa 60°, showing broad concentration and a granular texture. About 2′ S is a widely separated pair of mag. 12.5 stars (50″; 138°).

eg 4417 dimen. 3.6 × 1.4 V = 11.1 sfc. br. 12.8
This galaxy appears as a circular 1′ patch in 15 cm, showing a stellar nucleus at 100 ×. It is fairly bright in 25 cm, the halo extending to 1.5 × 0.5 in pa 50°. Except for a slight central bulge, the galaxy is very thin. The bright, granular core is 20″ across and contains a stellar nucleus. eg 4424, *q.v.*, lies 11′ SE.

eg 4420 dimen. 2.2 × 1.2 B = 12.7 sfc. br. 13.6
Though this galaxy is faintly visible in 15 cm, 25 cm shows it 1.5 × 0.75 in pa 10°. There is no distinct core, but it has a sharp, nearly stellar nucleus.

eg 4424 dimen. 3.7 × 1.9 V = 11.6 sfc. br. 13.6
Lying 11′ SE of eg 4417, *q.v.*, this galaxy appears as a diffuse, roundish spot about 1.25 diameter in 15 cm. A mag. 11.5 star is visible 2.3 N. With 25 cm the halo reaches 1.75 × 0.75 in pa 95°, showing rounded tips and a slight concentration toward the center. The brightest

part is off-center to the W. Photographs show a faint group of galaxies including eg 4410 and eg 4411 about 30′ SSW; eg 4445 (V = 12.7) lies 16′ E; eg 4451 (V = 12.5) lies 24′ ESE.

eg 4425 dimen. 3′.4 × 1′.2 V = 11.8 sfc. br. 13.2
In 15 cm this galaxy appears as a slight brightening without structure about 1′ diameter. 25 cm shows a sharply defined 1′.25 × 0′.75 oval elongated in pa 30°. The core is moderately brighter, much more elongated than the halo, and contains a very faint stellar nucleus. A mag. 13 star is visible 1′.2 W, a mag. 14 star 1′.8 NW, and a mag. 15 star 1′.3 SE.

eg 4428 dimen. 1′.9 × 0′.9 B = 13.3 sfc. br. 13.8
eg 4433 dimen. 2′.3 × 1′.1 B = 13.0 sfc. br. 13.9
These galaxies, which lie 7′ apart in pa 160°, are both faint objects for 25 cm. eg 4428 is elongated in pa 80°, 1′ × 0′.5, with little central brightening. eg 4433 appears a bit longer than −28, reaching 1′.25 × 0′.5 in pa 5°, and somewhat brighter due to a faint stellar nucleus. In 30 cm −33 grows to 2′.5 × 1′, with weak, even concentration to an unconcentrated 1′.5 × 0′.75 core. Two closely associated stars—mag. 13.5 on the N tip 50″ from center, and mag. 13 just W of the S tip 55″ from center—make further details difficult to discern. eg 4428 is 2′ × 1′ and has a concentration similar to −33, but here a very faint nucleus is occasionally visible at 250×.

eg 4429 dimen. 5′.5 × 2′.6 V = 10.2 sfc. br. 12.9
A bright object for 15 cm, this galaxy lies 1′.9 SW of a mag. 8.5 star. The 2′ × 1′ haze has a small bright core. 25 cm shows it 3′ × 0′.75 in pa 100°. The core is conspicuous and contains a small nucleus.

eg 4431 dimen. 2′.0 × 1′.3 V = 13.2 sfc. br. 14.1
eg 4436 dimen. 1′.9 × 0′.9 V = 13.5 sfc. br. 13.9
eg 4440 dimen. 2′.0 × 1′.7 V = 11.8 sfc. br. 12.9
eg 4440 is a moderately bright galaxy for 25 cm. The circular 45″ halo contains a sharply brighter 10″ × 5″ core elongated SE-NW. eg 4436, 2′ WNW, is much fainter than −40, a low surface brightness patch about 20″ across without central brightening. eg 4431 lies 6′.5 due W, and in 25 cm is more conspicuous than −36. It is also about 20″ diameter but has a bright center.

eg 4435 dimen. 3′.0 × 1′.9 V = 10.8 sfc. br. 12.6
eg 4438 dimen. 9′.3 × 3′.9 V = 10.0 sfc. br. 13.8
These interacting galaxies lie 4′.3 apart in pa 165°. In 15 cm, eg 4435 is about 1′ diameter with a bright stellar nucleus. 25 cm shows it about 1′.5 × 1′ in pa 15°, rising sharply from a smooth halo to the stellar nucleus. With averted vision, 15 cm will show eg 4438 to 3′ × 1′ in pa 20° with a small bright core. In 25 cm it is as large as 4′ × 2′ with a smooth broad concentration to a distinct mag. 14 stellar nucleus. The surface brightness is much lower here than for −35.

eg 4442 dimen. 4′.6 × 1′.9 V = 10.4 sfc. br. 12.6
This is a bright galaxy in 15 cm, extending to about 2′ × 1′ in pa 90°. With 25 cm the distinct 30″ × 15″ oval core contains a stellar nucleus. Compared to the core, the halo is much fainter and is of similar size to the 15 cm view.

eg 4452 dimen. 2′.4 × 0′.6 V = 12.3 sfc. br. 12.5
25 cm shows this galaxy as a small 1′.25 × 0′.2 spindle elongated in pa 35°. The high surface brightness halo has a slightly mottled texture. The core appears as a short line in the center, but there is no distinct nucleus. eg I3381 (B = 14.3) lies 7′.1 NW.

eg 4454 dimen. 2′.2 × 1′.9 V = 12.1 sfc. br. 13.5
This galaxy appears as a fairly faint circular patch about 1′ diameter in 25 cm. No brighter core is evident, but a stellar nucleus lies at the center. Two mag. 10 stars lie 4′.8 WSW and 4′.0 NNW.

eg 4457 dimen. 3′.0 × 2′.5 V = 10.8 sfc. br. 12.9
In 15 cm this galaxy appears as a moderately bright circular spot about 30″ diameter with a sharp center. The halo has a broad lenticular form in 25 cm, extending to 1′.25 × 1′ in pa 70°. A stellar nucleus is visible within a concentrated circular core.

eg 4458 dimen. 1′.9 × 1′.8 V = 12.1 sfc. br. 13.4
eg 4461 dimen. 3′.7 × 1′.5 V = 11.2 sfc. br. 12.9
These galaxies lie 3′.7 apart in pa 160°. eg 4461 is clearly the brighter object, easily visible in 15 cm as a concentrated 1′.5 × 0′.5 haze elongated nearly N-S. The halo is about the same size with 25 cm and brightens within to a circular 15″ core and stellar nucleus. The companion to the NNW is difficult to view in 15 cm, and is much fainter and smaller than −61 with 25 cm. The 45″ × 30″ halo is elongated nearly parallel to that of −61. The moderate, sharp central concentration contains a very faint substellar nucleus. A faint pair of stars (24″; 345°) 3′.6 WSW of eg 4458 can appear nebulous at low power.

eg 4469 dimen. 3′.9 × 1′.5 B = 12.3 sfc. br. 14.0
A faint object for 15 cm, this galaxy appears as an oval glow elongated in pa 90°, with a brighter circular core. In 25 cm the sharply tipped halo grows to 2′ × 1′, but is only 30″ wide except at the central bulge. The bright oval core contains some very faint stellarings along its major axis.

eg 4472 ***dimen. 8.′9 × 7.′4*** ***V = 8.4*** ***sfc. br. 12.9***
Messier 49 is visible in 6 cm as a bright circular spot, appearing much like an unresolved globular cluster. It is fairly bright in 15 cm, about 4′ diameter with a bright, distinct core and still brighter nucleus. A mag. 12.5 star is visible in the E side about 45″ from the center. The bright circular galaxy has a 45″ core and 2′ inner halo in 25 cm. The outer parts fade smoothly into the background to an overall diameter of more than 4′. Photographs show eg 4467 (V = 14.5) 4.′2 W, eg 4470 (B = 13.0) 11′ S, and eg 4464 (V = 12.6) 12′ NW.

eg 4476 ***dimen. 1.′9 × 1.′3*** ***V = 12.3*** ***sfc. br. 13.2***
eg 4478 ***dimen. 2.′0 × 1.′8*** ***V = 11.2*** ***sfc. br. 12.6***
Located about 10′ WSW of Messier 87 (eg 4486, *q.v.*), these galaxies lie 4.′5 apart. eg 4478 is a 40″ haze in 15 cm, appearing concentrated and with well-defined edges. It is little larger in 25 cm and shows an evenly concentrated halo that becomes suddenly brighter to a faint stellar nucleus. eg 4476 is much fainter than −78 and is difficult to view in 15 cm. In 25 cm the 55″ × 35″ halo is elongated NE-SW and has a brightness profile similar to −78. The sharp central condensation contains a stellar nucleus.

eg 4483 ***dimen. 1.′8 × 1.′1*** ***B = 13.4*** ***sfc. br. 14.0***
This galaxy is barely visible in 15 cm, and it appears small and fairly faint in 25 cm. The circular glow is 30″ diameter with a relatively bright core and a faint stellar nucleus.

eg 4486 ***dimen. 7.′2 × 6.′8*** ***V = 8.6*** ***sfc. br. 12.9***
Messier 87 is one of the brightest galaxies in the Virgo cluster, and 6 cm will show it as a bright and well-concentrated circular spot 6′ S of a mag. 8 star. In 15 cm it appears similar to Messier 94 (*cf.* eg 4472 in Canes Venatici). The galaxy is up to 4′ diameter in 25 cm. The 45″ core is of very high surface brightness; there is no distinct nucleus, but a central stellaring may be discerned with difficulty at high power.

eg 4487 ***dimen. 4.′1 × 3.′0*** ***B = 11.7*** ***sfc. br. 14.2***
This galaxy is a large, diffuse patch in 15 cm, appearing about 3′ diameter and slightly elongated E-W. A mag. 12.5 star is visible in the N side 1.′0 from center, and a mag. 12.5 star lies at 2.′6 in pa 100°. The low surface brightness halo reaches 4′ × 2.′5 in pa 80° with 30 cm. The light is very poorly and broadly concentrated: no nucleus is visible, but the central half of the galaxy is definitely mottled. At 250 × the S side seems to fade more abruptly than the N, but the star on the N edge makes this uncertain.

eg 4491 ***dimen. 1.′9 × 1.′0*** ***V = 12.6*** ***sfc. br. 13.2***
This galaxy is not a bright object for 25 cm, appearing as a broadly concentrated 45″ glow of low surface brightness. A weak stellar nucleus is occasionally visible in the center. A mag. 13 star lies 2.′2 NE.

eg 4496A ***dimen. 3.′9 × 3.′1*** ***B = 11.7*** ***sfc. br. 14.3***
eg 4496B ***dimen. 1.′0 × 0.′9*** ***mag.*** $_{vv}$***15*** ***sfc. br.***
This pair of galaxies consists of a spiral and a tiny irregular companion in contact; they are difficult to view in 15 cm. 25 cm shows a simple 3′ × 1′ glow of low surface brightness elongated in pa 150°; the two components are inseparable. eg 4480 (V = 12.4) lies 26′ NW.

eg 4497 ***dimen. 2.′3 × 1.′1*** ***V = 12.5*** ***sfc. br. 13.4***
Appearing in the same low-power field as eg 4491, *q.v.*, this galaxy is a little brighter than its neighbor in 25 cm. The oval halo is 1′ × 0.′75 in pa 65°, and shows moderate, even concentration to an indefinite but strongly elongated core.

eg 4503 ***dimen. 3.′5 × 1.′8*** ***V = 11.1*** ***sfc. br. 12.9***
With 15 cm this galaxy is 1′ × 0.′3, elongated in pa 10°, and contains a small circular core. It is a high surface brightness object in a barren field for 25 cm. The spindle-shaped halo extends to 1.′25 × 0.′5, brightening to a circular 20″ core and a stellar nucleus.

eg 4504 ***dimen. 4.′0 × 2.′8*** ***B = 11.9*** ***sfc. br. 14.4***
At low power in 15 cm this galaxy appears similar to but fainter than nearby eg 4487, *cf.*, though no faint stars are visible in the field. 25 cm shows a 3′ × 1′ glow of low surface brightness elongated in pa 135°. The light is unconcentrated, but some stellarings are visible on the surface. In 30 cm the halo is 3.′5 × 2′, showing a very poor, broad concentration and no nucleus. A mag. 14 star is visible at 2.′0 in pa 105°.

eg 4517 ***dimen. 10′ × 1.′9*** ***V = 10.4*** ***sfc. br. 13.5***
Just visible in 15 cm, this galaxy lies on the SW side of a mag. 10 star. The galaxy appears faint and very elongated in 25 cm, with the center immediately SW of the star. The halo is elongated in pa 80°, 6′ × 1′, with a very small core and a few stellarings near the center. Photographs show eg 4517A (V = 12.2) 17′ NW.

eg 4519 ***dimen. 3.′1 × 2.′2*** ***V = 11.7*** ***sfc. br. 13.7***
This galaxy is just visible in 15 cm, and in 25 cm it appears brighter than eg 4522, *cf.* 32′ N. The broadly concentrated oval is about 2′ × 1′, elongated E-W, and the center is faintly mottled. The low surface brightness halo is unconcentrated in 30 cm except for an occasion-

ally visible stellar nucleus. Three faint stars are visible within 3′ in the northwest quadrant.

eg 4522 dimen. 3′.7 × 1′.1 B = 12.7 sfc. br. 14.1
This galaxy is a faint object for 25 cm. The diffuse unconcentrated halo is about 1′.5 diameter.

eg 4526 dimen. 7′.2 × 2′.3 V = 9.6 sfc. br. 12.6
Situated between two mag. 7 stars, this galaxy is easy to find in 15 cm, and has a mag. 12.5 star on the S. The halo is about 2′ × 1′, and seems brighter than eg 4535, *cf.* 30′ NNE, while the core seems off-center within it. The brightest part is clearly off-center in 25 cm. The well-defined core is 20″ across; the halo reaches 3′ × 1′ in pa 110°. With 30 cm the halo extends to 3′.5 × 1′.25, rising in brightness very evenly toward the center, where a stellar nucleus is occasionally visible at 250 ×. In this aperture the galaxy appears nearly symmetric, though the eastern half of the halo seems slightly more extensive. Halfway to the mag. 7 star 7′ W is a mag. 12 star.

eg 4527 dimen. 6′.3 × 2′.3 V = 10.4 sfc. br. 13.1
In 15 cm this galaxy is a fairly bright, narrow spindle at least 3′ × 0′.7 in pa 65°. The outer edge is sharply defined in 25 cm, which occasionally gives the impression of a contrasting dark area surrounding the object. The halo reaches to 4′ × 1′ around a mottled oval core that contains a stellar nucleus slightly off-centered to the E. Photographs show eg 4533 (B = 14.5) 20′ S; eg 4536, *q.v.*, is 28′ S.

eg 4532 dimen. 2′.9 × 1′.3 V = 11.9 sfc. br. 13.2
Barely visible in 15 cm, this galaxy appears as a faint patch about 5′ NNW of a mag. 8 star. With 25 cm the unconcentrated, low surface brightness halo extends to 2′ × 1′ in pa 160°. A mag. 13.5 star lies on the E edge, 33″ from center. 30 cm shows it 2′.5 × 1′ with only a slight, broad concentration and no distinct nucleus.

eg 4535 dimen. 6′.8 × 5′.0 V = 10.0 sfc. br. 13.6
This large galaxy is easily visible in 15 cm at low power. The unconcentrated patch is about 4′ diameter and has a mag. 13 star embedded in the N side. the 5′ × 4′ halo has a slight, very broad concentration in 25 cm, with a granular appearance due to several stars in and near the galaxy. The halo grows to 6′ × 4′ in 30 cm, slightly elongated N-S. The mag. 13 star lies nearly 1′ N of center; a mag. 14.5 star is visible 50″ SW. A faint stellar nucleus is embedded in an otherwise unconcentrated core about 50″ across. The halo outside the core is unevenly bright. Other nearby stars lie 2′.5 ESE (mag. 14.5), 2′.2 S (mag. 13.5), and 3′.2 SW (mag. 13). A beautiful and detailed galaxy!

eg 4536 dimen. 7′.4 × 3′.5 V = 10.4 sfc. br. 13.8
This galaxy appears fainter than eg 4527, *cf.* 28′ N, in 15 cm, though the halo is more diffuse and a bit larger. It has a fat lenticular outline in 25 cm, 6′ × 2′, elongated SE-NW. eg 4533 (B = 14.5) lies 8′.5 N.

eg 4546 dimen. 3′.5 × 1′.7 V = 10.3 sfc. br. 12.1
This galaxy appears oblong in 15 cm, about 1′.5 × 0′.3 in pa 80°. The conspicuous core holds a prominent stellar nucleus; overall the center has a similar brightness to a mag. 11 star 1′.9 SE. In 25 cm the lenticular halo reaches 2′ × 0′.5, with a 20″ circular core and stellar nucleus.

eg 4550 dimen. 3′.5 × 1′.1 V = 11.5 sfc. br. 12.8
Small and indistinct with 25 cm, this galaxy is 1′ × 0′.25, elongated N-S, and without details except for a brighter center. Photographs show eg 4551 (V = 11.9) 3′ NE, forming a triangle with −50 and a mag. 12 star to the S.

eg 4552 dimen. 4′.2 × 4′.2 V = 9.8 sfc. br. 12.9
Messier 89 is visible in 6 cm as a small, sharply concentrated circular spot. In 15 cm the halo is about 1′.5 diameter, and strongly concentrated to a stellar nucleus. The halo is very bright in 25 cm, rising to an even brighter core and stellar nucleus.

eg 4564 dimen. 3′.1 × 1′.4 V = 10.9 sfc. br. 12.5
15 cm shows this object as a 1′ diameter patch without concentration. It remains ill-defined in 25 cm, but is slightly elongated NE-SW.

eg 4567 dimen. 3′.0 × 2′.1 V = 11.3 sfc. br. 13.1
eg 4568 dimen. 4′.6 × 2′.1 V = 10.8 sfc. br. 13.1
These interacting galaxies lie 1′.2 apart. eg 4567 is on the N, and in 25 cm it appears as a small, faint patch with stellar character. eg 4568 is somewhat brighter, and 15 cm shows a large fat oval that grows gradually brighter toward the center.

eg 4569 dimen. 9′.5 × 4′.7 V = 9.5 sfc. br. 13.5
Messier 90 is one of the brightest objects in the Virgo cluster; 6 cm will show it as a bright haze elongated NNE-SSW. 15 cm reveals a stellar center, which photographs show to be a superposed mag. 12 star. The halo is about 6′ × 2′ in 25 cm, elongated in pa 25°. With averted vision the E flank is flattened, while the W flank has a bulge in the center. The superposed star lies just W of the bright nucleus. Photographs show eg I3583 (B = 13.9) 6′ N, which may be interacting with M90. eg 4584 (B = 13.7) lies 32′ E.

eg 4570 dimen. 4′.1 × 1′.3 V = 10.8 sfc. br. 12.5
Viewed in 15 cm, this galaxy appears as a dull even glow about 1′.5 × 1′ with a bright stellar nucleus. In 25 cm it

appears elongated in pa 160°, 2′ × 0′.75, with a well-defined core and stellar nucleus. The halo is about the same size in 30 cm, and shows strong, sharp concentration to an elongated core and a small but nonstellar nucleus.

eg 4578 dimen. 3′.6 × 2′.8 V = 11.4 sfc. br. 13.7
In 15 cm this galaxy is moderately faint and circular. The halo is 1′ diameter and has a brighter center. It grows to 1′.5 diameter in 25 cm. The small bright core holds a stellar nucleus. The galaxy is a nice object in a rich field with 30 cm. The halo appears about 1′ diameter, slightly elongated NE-SW, and rises evenly to a faint stellar nucleus.

eg 4579 dimen. 5′.4 × 4′.4 V = 9.8 sfc. br. 13.1
Lying 7′.6 E of a mag. 8 star, Messier 58 is visible in 6 cm as a faint circular haze about 3′ diameter with a well-defined center. In 15 cm the halo is 4′.5 × 3′.5 overall, slightly elongated E-W, with a bright core and stellar nucleus. With 25 cm the stellar nucleus is surrounded by a well-defined core, which is elongated in pa 60°, while the halo remains in pa 90°.

eg 4580 dimen. 2′.4 × 1′.9 B = 12.7 sfc. br. 14.2
Lying 3′.3 NW of a mag. 11.5 star, 15 cm shows this galaxy as a slightly elongated spot 1′ diameter. With 25 cm it is elongated in pa 160°, 1′.25 × 1′, and shows a weak central concentration.

eg 4586 dimen. 4′.4 × 1′.6 V = 11.6 sfc. br. 13.6
Just visible in 15 cm, this galaxy appears quite elongated but still fairly faint in 25 cm. The overall size is 2′ × 0′.5, elongated ESE-WNW, with a few stellarings involved, but no central concentration.

eg 4592 dimen. 4′.6 × 1′.5 B = 12.2 sfc. br. 14.1
This galaxy appears about 2′ × 1′ in pa 95° with 25 cm. The uniformly bright oval has a smooth texture and low surface brightness. In 30 cm the halo grows to 2′.5 × 1′ and shows a weak, broad concentration across the center. There is no distinct nucleus, but a faint stellaring is visible on the N edge of the core.

eg 4593 dimen. 4′.0 × 3′.1 B = 11.7 sfc. br. 14.3
A moderately faint object for 15 cm, this galaxy appears as an oval glow at 75 ×, concentrated to a nonstellar nucleus. In 25 cm the halo is about 1′.5 × 0′.5, elongated NE-SW, and contains a stellar nucleus. The low surface brightness halo reaches to nearly 3′ × 1′.25 in pa 55° with 30 cm. The light is sharply and moderately concentrated to a circular core and substellar nucleus.

eg 4594 dimen. 8′.9 × 4′.1 V = 8.3 sfc. br. 12.0
Messier 104 is visible in 6 cm as a well-concentrated elongated patch 4′ ENE of a mag. 10 star. A loose cluster of faint stars lies WNW. The dark lane can be seen easily in 15 cm cutting across the core on the S side. The overall size is as large as 6′ × 2′ in excellent conditions elongated in pa 90°. The size increases to nearly 8′ × 3′ in 25 cm, the dark lane cutting clear through the object. The inner core is 30″ across and bisected by the dark lane, with the smaller portion on the S side.

eg 4596 dimen. 3′.9 × 2′.8 V = 10.5 sfc. br. 12.9
In 15 cm this galaxy appears slightly brighter than eg 4608, *cf.* 19′ E. The halo is about 2′ × 1′, elongated approximately E-W. In 25 cm the halo has an irregularly circular outline and a few dark patches. With 30 cm stars associated with the halo give it a mottled appearance at low power. The overall size is about 2′.5 × 2′.2, elongated roughly N-S, but the brightest parts form a 2′ × 0′.8 bar in pa 75°. At high power the inner halo occasionally appears mottled, particularly on the NW side. A mag. 11.5 star lies 1′.1 SSE, just within the edge of the halo.

eg 4597 dimen. 3′.6 × 1′.9 B = 12.6 sfc. br. 14.5
This galaxy is barely visible in 15 cm as an unconcentrated blob about 2′ diameter. It is a faint object even in 30 cm. The very low surface brightness halo reaches to 4′ × 1′.5, elongated NE-SW. The light is completely unconcentrated; no nucleus is visible even at 250 ×. A 2′ × 1′ triangle of mag. 13 stars is visible about 3′ NNW.

eg 4602 dimen. 3′.6 × 1′.4 B = 12.3 sfc. br. 13.9
Visible in 15 cm at 75 ×, this galaxy is somewhat larger than nearby eg 4597, *cf.*, and elongated E-W. 25 cm shows some indefinite details across the 2′.5 × 0′.7 halo. In 30 cm the broad core is mottled and lies within a broadly and very weakly concentrated 2′.7 × 1′ halo elongated in pa 105°. A mag. 14 star lies 1′.4 from the center at the E end of the halo.

eg 4608 dimen. 3′.2 × 2′.6 V = 11.1 sfc. br. 13.3
Located near ρ Virginis (V = 4.9), this galaxy is faint in 15 cm, appearing as a circular glow about 1′ diameter. In 25 cm the halo is perhaps 1′.5 diameter and has a sharply brighter center. 30 cm shows a bright core and a stellar nucleus within a 1′ × 0′.8 halo that is elongated in pa 30°. With averted vision the galaxy is startlingly bright. Stars are visible 1′.9 ESE, and 1′.6 and 1′.9 WNW. eg 4596, *q.v.*, lies 19′ W.

eg 4612 dimen. 2′.2 × 1′.8 B = 12.2 sfc. br. 13.6
Located only 1′ W of a mag. 10.5 star, this galaxy is visible in 15 cm as a 1′ circular spot with a stellar nucleus.

The mag. 10.5 star is the southernmost of a 9′ string of four similarly bright stars ending in a pair (10.5,12; 20″; 73°). In 30 cm the galaxy is circular, about 1′.5 across, but the star E can give the impression of E-W elongation. The light is well and sharply concentrated to a faint stellar nucleus.

eg 4621 dimen. 5′.1 × 3′.4 V = 9.8 sfc. br. 12.8
Messier 59 is visible in 6 cm as a nearly circular patch forming a 6′ triangle with two stars to the N. In 15 cm it is about 2′ × 0′.75 with a more circular bright core. The galaxy is smaller and fainter than Messier 60 (*cf.* eg 4649). In 25 cm it is elongated SE-NW with a brighter center. Photographs show a pair of galaxies, eg 4606–07 (V = 11.9,12.9), 3′.9 apart, 20′ NW. eg I 809 (B = 14.2) lies 6′.5 NNE; eg I3665 (B = 14.8) is 10′ SW.

eg 4623 dimen. 2′.6 × 0′.9 B = 13.1 sfc. br. 13.9
Barely observable in 15 cm, this galaxy is fairly faint for 25 cm. The halo appears elongated in pa 175°, 1′ × 0′.3, with a small circular core and stellar nucleus at the threshold of visibility. 30 cm shows the halo to nearly 2′ × 1′. The slightly concentrated inner regions are elongated and a string of distinct stellarings is visible along the major axis. A mag. 14.5 star lies about 1′.2 in pa 70° from the center.

eg 4630 dimen. 1′.7 × 1′.3 B = 13.0 sfc. br. 13.6
This galaxy appears as a faint circular blob in 25 cm. The halo is about 1′.5 diameter with a few stellarings involved, but little overall central concentration. A mag. 11.5 star lies 2′.8 NNE that has a mag. 13 companion 15″ away in pa 5°.

eg 4632 dimen. 3′.2 × 1′.3 B = 12.2 sfc. br. 13.6
This object is fairly faint in 15 cm, elongated in pa 60°, 1′.25 × 0′.75. In 25 cm the halo extends to about 2′ × 1′ with an even, moderate concentration to a well-defined oval core.

eg 4636 dimen. 6′.2 × 5′.0 V = 9.6 sfc. br. 13.3
A bright object in 15 cm, this galaxy appears about 1′.5 diameter and has a faint stellar nucleus. In 25 cm the halo is well concentrated to a bright core and stellar nucleus.

eg 4638 dimen. 2′.8 × 1′.6 V = 11.2 sfc. br. 12.7
In 15 cm this galaxy appears as a circular glow about 45″ diameter. 25 cm reveals a stellar nucleus in an otherwise featureless haze. 30 cm shows the halo to 1′ × 0′.6 in pa 125°. The galaxy is well concentrated to an elongated 1′.2 core and an elongated nonstellar nucleus. eg 4637 (B = 14.8) lies at 1′.6 in pa 97°.

eg 4639 dimen. 2′.9 × 2′.1 V = 11.5 sfc. br. 13.3
Though small compared to eg 4654, *cf.* 18′ SE, in 15 cm, this galaxy is nevertheless easy to see because of its moderately high surface brightness. The 1′ halo has a stellar center and a mag. 13 star 1′ SE from center. 30 cm shows the halo elongated nearly toward the star, about 1′.5 × 1′ in extent. A faint stellar nucleus stands out at center and seems multiple at times.

eg 4643 dimen. 3′.4 × 2′.7 V = 10.6 sfc. br. 12.9
A bright object for 15 cm, this galaxy appears circular, about 1′ diameter, with a bright core and nearly stellar nucleus. In 25 cm the halo is elongated in pa 135°, 1′.5 × 1′.25. The bright core contains a stellar nucleus. A mag. 11 star lies 2′.3 NW of center.

eg 4647 dimen. 3′.0 × 2′.5 V = 11.3 sfc. br. 13.3
eg 4649 dimen. 7′.2 × 6′.2 V = 8.8 sfc. br. 12.9
Messier 60, the brighter of this pair, is easily visible in 6 cm as an evenly fading circular glow with a faint central pip. The two galaxies are only 2′.5 apart, but 15 cm shows no contact between them. M60 is about 2′ diameter with a broad, diffuse core. eg 4647 has a similar size, but is much fainter and without central concentration. In 25 cm the two look like a double egg yolk embedded in a common halo; the portion that is −47 appears a little smaller than M60. In 30 cm, M60 is bright and well concentrated. Close scrutiny shows a slight E-W elongation to the broadly concentrated 1′.8 × 1′.2 halo, but the proximity of the companion makes this indistinct. Here the halos do not appear to touch. eg 4647, on the NW side, is about 1′ diameter with low surface brightness and a very weak central brightening; no nucleus is visible.

eg 4653 dimen. 2′.6 × 2′.4 V = 12.3 sfc. br. 14.1
Located 2′.6 NW of a mag. 11 star, 25 cm reveals only a faint glow here. The galaxy has very low surface brightness in 30 cm, the irregularly round halo reaching to at least 1′ diameter. eg 4642 (mag.$_Z$13.8) lies 9′.5 SW. eg 4653 is in a group with eg 4666 and eg 4668, *q.v.*

eg 4654 dimen. 4′.7 × 3′.0 V = 10.5 sfc. br. 13.2
Set among several stars, this is a nice galaxy in 15 cm. The 3′ diameter halo is slightly elongated approximately E-W. A mag. 10.5 star lies 3′.2 WNW; mag. 12.5 stars are visible 2′.1 and 3′.5 N; a mag. 11 star is 5′.8 NE. The galaxy is large in 30 cm, the rounded halo extending to 3′.75 × 2′.25 in pa 120°. There is a broad, weak concentration toward a small line of stellarings that is occasionally visible near the center along the major axis; the nebula has a smooth texture except in these inner regions. At 250 × the halo fades more abruptly on the south flank.

eg 4639, *q.v.*, lies 18′ NW; eg 4659 is 24′ NNE in Coma Berenices, *q.v.*

eg 4658 dimen. 2.′2 × 1.′0 B = 12.7 sfc. br. 13.5
Lying 2.′4 E of a mag. 8.5 star, this galaxy is barely discernable in 15 cm. It appears elongated N-S in 25 cm, 1.′25 × 0.′5 with an indistinct circular core. A faint streak is occasionally visible along the major axis. A mag. 13.5 star is 35″ from the center in pa 340°, near the N tip. Photographs show eg 4663 ($mag._{VV}14$) 7′ SSE.

eg 4660 dimen. 2.′8 × 1.′9 V = 10.9 sfc. br. 12.8
This galaxy appears nearly star-like in 15 cm, the faint halo being difficult to see. Lying in a blank high-power field, the halo is less than 30″ diameter in 25 cm. The small size and stellar nucleus make the galaxy look like a planetary nebula. With 30 cm the halo grows to 2′ × 1′ in pa 100°. The inner regions are quite elongated, showing a strong, sharp concentration to a bright stellar nucleus.

eg 4665 dimen. 4.′2 × 3.′5 B = 11.4 sfc. br. 14.2
Situated 1.′6 NE of a mag. 10 star, this galaxy is bright in 15 cm. The circular 2′ glow has a stellar nucleus. The galaxy grows to 2.′5 diameter in 25 cm, with a 30″ core and stellar nucleus. A mag. 11.5 star is visible 3.′3 NE of center.

eg 4666 dimen. 4.′5 × 1.′5 V = 10.8 sfc. br. 12.7
Easily visible at 50 × in 15 cm, this galaxy appears elongated in pa 40°, 3′ × 0.′5. The core is also elongated and contains a fairly distinct stellar nucleus. A pair of stars is visible 5′ SE. 25 cm shows the halo to 4′ × 0.′5, with a thin core 1′ long, and a stellar nucleus. The halo is more sharply defined next to the core; the tips are indefinite. The pair S has a third member between them (11.5,13,12.5; 17″,37″; 145°,172°). 30 cm shows a faint, nearly stellar nucleus embedded in a 45″ × 5″ inner core that fades smoothly to a 4′ × 1′ halo. eg 4668, *q.v.*, lies 7.′3 SE of −66, 2.′7 E of the trio of stars.

eg 4668 dimen. 1.′4 × 0.′9 V = 13.1 sfc. br. 13.2
Located 7.′3 SE of eg 4666, *q.v.*, this galaxy appears very faint in 25 cm, a small slightly concentrated spot about 45″ diameter. 30 cm reveals a circular glow of fairly low surface brightness. Some markings are visible at high power, with an occasional stellaring at the center. A conspicuous triplet of stars lies 2.′8 W (*cf.* eg 4666).

eg I3742 dimen. 1.′9 × 1.′0 B = 13.9 sfc. br. 14.4
This faint galaxy lies 18′ SE of eg 4659, *q.v.* in Coma Berenices, between a mag. 8 and a mag. 10.5 star about 10′ apart. In 30 cm the spotty, low surface brightness halo is about 1.′3 × 0.′6, elongated NE-SW. The light is unconcentrated, but a faint stellaring is visible near the center.

eg 4684 dimen. 2.′9 × 1.′1 B = 12.3 sfc. br. 13.4
This galaxy is a small object in 15 cm, but not difficult to view. The nebula is fairly bright but little concentrated in 30 cm. A mag. 13.5 star is visible 35″ from center on the N end of the squarish halo, which is elongated in pa 20°. The middle parts seem unevenly bright, but the closely associated star makes this impression uncertain.

eg 4688 dimen. 3.′3 × 3.′1 B = 12.5 sfc. br. 14.8
This object is faintly visible in 25 cm as an indefinite patch with a faint stellar center.

eg 4691 dimen. 3.′2 × 2.′7 V = 11.2 sfc. br. 13.4
This galaxy is visible in 15 cm as a uniformly bright 1.′5 × 1′ oval. It is elongated E-W in 25 cm, about 2′ × 1.′ A stellar nucleus is visible. In 30 cm it appears similar to eg 4684, *q.v.*, but more elongated. The core is 45″ × 20″, elongated in pa 80°, while the halo is 1.′5 × 0.′8 in pa 60°.

eg 4694 dimen. 3.′6 × 1.′7 B = 12.2 sfc. br. 14.1
Moderately faint in 15 cm, this galaxy appears circular, about 1′ diameter, with a nonstellar center. The halo is about 1.′25 diameter with 25 cm, and contains a slightly brighter core and faint stellar nucleus. 30 cm shows it 3′ × 1.′5 in pa 145°, with a poor broad concentration overall. The core is also quite elongated, 2′ × 0.′75, and a faint stellar nucleus is occasionally visible at 250 × off-center to the NE. A mag. 14 star lies 1.′2 W of center.

eg 4697 dimen. 6.′0 × 3.′8 V = 9.3 sfc. br. 12.7
With 15 cm this galaxy is elongated in pa 65°, 1.′25 × 0.′75. The bright core contains an inconspicuous stellar nucleus. The galaxy appears similar in 25 cm, but the overall size increases to 2′ × 1.′ Photographs show a faint anonymous galaxy 5.′9 WNW interacting with eg 4697.

eg 4698 dimen. 4.′3 × 2.′5 V = 10.5 sfc. br. 12.9
Located 6.′4 ESE of a mag. 7.5 star, this galaxy is visible in 15 cm between two mag. 10.5 stars. The galaxy is 1′ diameter with a brighter core. 25 cm shows a quite bright circular patch 1.′25 diameter with a bright core and sharp but nonstellar nucleus.

eg 4699 dimen. 3.′5 × 2.′7 V = 9.6 sfc. br. 11.8
Appearing much brighter in 15 cm than eg 4697, *cf.*, this galaxy shows an extensive, evenly concentrated 3′ × 2′ halo elongated NE-SW. The halo is very well and evenly concentrated to a stellar nucleus in a distinct 1′ × 0.′25 inner core. At 250 × in 30 cm the inner core immediately around the bright stellar nucleus is elongated NNE-

SSW. The nearly circular 1′ core is distinctly mottled, particularly near the nucleus, and fades evenly to a nearly uniform but faint halo, which extends to 3′.5 × 2′ in pa 45°. A mag. 13.5 star is visible 2′.2 SSW.

eg 4700 dimen. 3′.0 × 0′.7 B = 12.6 sfc. br. 13.2
In 15 cm this galaxy appears as a 1′ patch with a faint stellar nucleus. 25 cm reveals a mag. 13 star 2′ W of center. The central brightening is overshadowed by the star, but seems diffuse. The overall size is about 1′ × 0′.3, elongated NE-SW.

eg 4701 dimen. 3′.0 × 2′.5 V = 12.4 sfc. br. 14.4
This galaxy forms a triangle with two mag. 12 stars nearby to the WNW and SSW. 25 cm reveals a diffuse and unconcentrated glow about 1′.5 diameter with a faint stellar nucleus. In 30 cm the galaxy appears as a barely concentrated 1′.75 × 1′.25 oval elongated in pa 40°. No nucleus is visible even at 250 ×. A pretty triangle of mag. 13 stars, 30″ on a side, lies 4′ N.

eg M-02-33-15 dimen. 4′.0 × 3′.2 B = 12.1 sfc. br. 14.7
Though not visible in 15 cm, 25 cm reveals this galaxy as a large faint smudge perhaps 3′ × 1′.5 elongated E-W. No central concentration is evident. eg 4682 (B = 12.9) lies 32′ W.

eg 4713 dimen. 2′.8 × 1′.9 V = 11.8 sfc. br. 13.5
15 cm shows this galaxy as a faint circular patch 2′.5 NW of two mag. 11 stars aligned E-W. It is moderately faint in 25 cm, elongated E-W, 1′.5 × 1′. The halo is diffuse and without central brightening.

eg 4731 dimen. 6′.5 × 3′.4 B = 11.6 sfc. br. 14.7
25 cm shows this galaxy as a diffuse and ill-defined glow about 1′.5 diameter. Some stellarings are visible, but no central brightening is evident. A 1′ × 0′.4 trapezium of stars 7′ N appears nebulous at low power. eg M-01-33-27 (mag.$_{vv}$14.5) lies 11′ SE.

eg 4733 dimen. 2′.3 × 2′.1 B = 12.6 sfc. br. 14.3
This galaxy is visible in 15 cm as an unconcentrated patch 1′ diameter. The oval halo is 1′.5 × 1′.25 in 30 cm, elongated approximately E-W. The light is barely concentrated, though a core 1′ across is fairly distinct, and a very faint sparkle is occasionally glimpsed within. A mag. 13.5 star is visible 50″ W of the center.

eg 4742 dimen. 2′.3 × 1′.5 V = 11.1 sfc. br. 12.5
Appearing small and fairly faint in 15 cm, this galaxy nevertheless seems a little brighter than a mag. 11.5 star 1′.3 SE. The nebula is about 30″ diameter and contains a stellar nucleus. A wide unequal pair of stars lies 9′ NW [ADS 8684: V = 6.4,9.7; 30″; 301° (1959)]. The 1′ × 0′.5 halo has a fairly high surface brightness in 25 cm and is elongated in pa 60°. The stellar nucleus is startlingly bright in 30 cm, even though the core is sharply concentrated. The primary of the pair NW appears orange; the secondary seems blue by contrast.

eg 4746 dimen. 2′.5 × 0′.7 B = 13.3 sfc. br. 13.7
This galaxy lies within a 5′ equilateral triangle of mag. 11–13.5 stars and is not visible in 15 cm. In 30 cm the barely concentrated halo extends to 1′.5 × 0′.5, elongated in pa 120°. Weakly brighter spots appear along the major axis; the northeast flank seems to fade more abruptly.

eg 4753 dimen. 5′.4 × 2′.9 V = 9.9 sfc. br. 12.8
This is a conspicuous object for 15 cm, visible at low power. The overall size is 2′ × 1′.5 in pa 90°, with an elongated starry center. In 25 cm the stellar nucleus is prominent within a circular core. The halo reaches to 2′.5 × 1′.5, with a weak dark patch cutting across the NE side. The halo is 2′.5 × 1′.7 in 30 cm, brightening to a 1′.5 circular core. A 20″ nucleus is just discernable over the core, but does not contain a stellar center. Two stars, mag. 13.5 and 13, lie 2′.4 and 3′.8 W; the closer one is not far outside the halo.

eg 4754 dimen. 4′.7 × 2′.6 V = 10.5 sfc. br. 13.1
Lying in the same field with eg 4762, *q.v.* 11′ SE, in 15 cm this galaxy forms the northeastern end of an evenly spaced string including two mag. 11 stars 3′.2 and 5′.8 SW. The large, diffuse blob is unconcentrated except for a conspicuous stellar nucleus that has a brightness similar to the stars SW. At high power in 25 cm the core seems to have horns that project approximately N-S into the halo. With 30 cm the galaxy is encircled by six mag. 13–14 stars. The circular 2′ main body is moderately well and sharply concentrated, but a halo of very low surface brightness extends to about 3′.25 × 2′, elongated NNE-SSW. The nucleus remains stellar at 250 × and appears fainter than the mag. 11 stars SW. The region immediately surrounding the nucleus is quite bright, making the nucleus sometimes seem double or off-center.

eg 4757 dimen. 1′.8 × 0′.5 mag.$_{vv}$15 sfc. br.
eg 4760 dimen. 1′.8 × 1′.8 B = 12.5 sfc. br. 13.7
eg 4760 lies between a mag. 8.5 star SW and a mag. 9.5 star NNE. In 15 cm it is small, but not difficult to view at 75 ×. The 45″ halo is slightly concentrated. It is about the same size in 25 cm; the outer edge of the halo seems well defined. The texture is nearly smooth, the brightness nearly uniform: the 10″ core barely stands out in the center. This centermost region contains several stellarings in 30 cm at 125 ×; the slightly larger, unevenly

bright core is poorly concentrated. A very faint, low surface brightness halo extends to at most 1′.75 × 1′.25, elongated N-S. Beginning at the mag. 9.5 star NNE, a string of three fairly bright stars reaches N and W of the galaxy; 3′ NW of the last star is eg 4757, 12′ NNW of −60. At 250× in 30 cm it is faint, about 30″ diameter, and of moderate surface brightness, but without concentration. Photographs show another faint galaxy, eg 4766 (mag.$_{vv}$15), 6′.9 due N of −60; it has a tiny companion (mag.$_{vv}$15.5) 1′.2 NW of it.

eg 4759 dimen. mag.$_{vv}$15 sfc. br.
eg 4761 dimen. mag.$_{vv}$14.5 sfc. br.
This pair of galaxies in contact is visible in 15 cm as a very faint smudge 1′.5 N of a mag. 9 star. eg 4761 is 1′.25 across in 25 cm, the moderately high surface brightness halo rising to a concentrated core. The companion galaxy is not conspicuous. In 30 cm, −61 is about 1′.5 across. The circular glow is evenly concentrated to a bright substellar nucleus. eg 4759 lies on the NW edge and shows a sharply concentrated substellar nucleus comparable to that of −61. eg 4764 (mag.$_{vv}$16) is located only 1′ E of the pair.

eg 4762 dimen. 8′.7 × 1′.6 V = 10.2 sfc. br. 13.0
15 cm shows this large galaxy quite clearly at 50×, accompanied on the E, S, and W by bright field stars. The halo is elongated in pa 35°, about 4′ long. The very thin inner regions are almost needlelike. In 25 cm the very elongated center has a silvery texture like Messier 82 (*cf.* eg 3034 in Ursa Major). The beautiful spindle reaches to as much as 7′.5 length in 30 cm, but is only 30″ wide, hardly bulging toward the middle. The ends taper to sharp points. The halo is well concentrated along the major axis to a thin, high surface brightness outer core, indistinct 20″ circular inner core, and a small substellar nucleus, which seems very circular within the otherwise strongly elongated bar. eg 4754, *q.v.*, is 11′ NW.

eg 4765 dimen. 1′.4 × 1′.1 B = 13.3 sfc. br. 13.6
While faintly visible in 15 cm, 25 cm will show this galaxy more clearly as a moderately faint circular glow about 45″ diameter. The middle parts grow moderately brighter to a faint stellar nucleus. The halo extends to 1′.5 × 1′ in 30 cm, elongated in pa 70°. At 250× a 30″ × 10″ core is distinct and has a mottled texture, though no nucleus is visible. A mag. 14 star lies 1′.8 NW.

eg 4771 dimen. 4′.0 × 1′.0 B = 12.7 sfc. br. 14.1
Located 2′.4 E of a mag. 9.5 star, this object is visible in 15 cm as a faint circular spot 1′ diameter. It is elongated in in pa 135° with 25 cm, reaching an overall size of 1′.5 × 0′.5. The light is only a little brighter toward the center.

eg 4772 dimen. 3′.3 × 1′.7 B = 12.4 sfc. br. 14.1
This galaxy is visible in 15 cm as a faint patch less than 1′ diameter with a stellar nucleus. Moderately bright in 25 cm, the 1′ halo rises gradually in brightness to a sharply brighter but nonstellar center. The overall diameter is still about 1′.

eg 4775 dimen. 2′.2 × 2′.1 B = 11.7 sfc. br. 13.3
In 25 cm this galaxy appears as a moderately low surface brightness patch lying in a barren field. The irregular 1′.5 halo is slightly elongated SE-NW and has a weak central brightening.

eg 4781 dimen. 3′.5 × 1′.8 B = 11.7 sfc. br. 13.5
This galaxy is not difficult to view in 15 cm. The 2′ × 1′ halo is elongated roughly E-W and is poorly concentrated but of moderate surface brightness. Three mag. 12.5 stars curve with increasing separation in an arc from the western tip. 25 cm shows it 2′ × 0′.75, elongated in pa 115°. The middle is not brighter, but very faint stellarings are scattered over the halo. The smooth oval halo grows to 3′ × 1′.25 in 30 cm, again barely concentrated but of high surface brightness. There is no distinct nucleus. The closest mag. 12.5 star, 55″ W, is embedded in the nebula. The galaxy is the brightest of a group including eg 4742, eg 4760, eg 4784, and eg 4790, all *q.v.*

eg 4784 dimen. 1′.8 × 0′.4 mag.$_{vv}$15 sfc. br.
Forming the obtuse southeastern apex of a triangle including two mag. 11 stars, this galaxy lies 5′.5 SE of eg 4781, *q.v.* In 30 cm it is very small, only 30″ across, but dominated by a relatively bright mag. 13.5 stellar nucleus.

eg 4786 dimen. 2′.0 × 1′.5 B = 12.6 sfc. br. 13.8
Faintly visible in 15 cm, in 25 cm this galaxy lies in a blank field. The moderately high surface brightness halo has an overall size of 1′ × 0′.3, elongated in pa 165°, and contains a stellar nucleus.

eg 4790 dimen. 1′.8 × 1′.3 B = 12.8 sfc. br. 13.6
Located 18′ NNE of eg 4781, *q.v.*, this galaxy is just visible in 15 cm as an unconcentrated 1′ patch 4′.5 SSE of a mag. 10 star. The 1′.5 × 0′.75 halo is faint and unconcentrated in 25 cm, elongated E-W. The poorly, broadly concentrated, and granular glow extends to 2′ × 1′ in 30 cm. At 250× a faint stellaring is occasionally visible near the center. A threshold magnitude star is visible 1′.0 N; another, slightly brighter, lies 1′.5 E.

eg 4795 dimen. 1′.7 × 1′.5 B = 13.2 sfc. br. 14.0
eg 4796 dimen. 0′.3 × 0′.3 B = 14.0 sfc. br. 10.3
eg 4795 is just visible in 15 cm, and in 25 cm it is still moderately faint. The halo is a bit elongated SE-NW,

about 1′.5 diameter. A mag. 13 starlike spot is visible 27″ ENE of center, which photographs show to be eg 4796. eg 4803 (B = 14.0) lies 13′ NE; eg 4791 (B = 14.6) is 4′.7 W, and the pair U8042/45 (B = 14.4,14.5) is about 9′ SSE.

eg 4808 dimen. 2′.7 × 1′.3 B = 12.6 sfc. br. 13.7
While this galaxy is only barely visible in 15 cm, 25 cm shows an elongated patch with smooth edges. The halo is 1′.5 × 0′.5 in pa 130° without much central brightening. With 30 cm the galaxy appears both brighter and larger than eg 4765, *cf.*, located in the same low-power field about 40′ WNW. The broadly concentrated 2′ × 0′.8 halo has fairly distinct edges. There is no condensation in the center, but the halo is unevenly bright at 250×. A mag. 14 star is visible 1′.9 WNW.

eg 4818 dimen. 4′.5 × 1′.7 B = 11.9 sfc. br. 14.0
Located about 5′ E of an arc of three mag. 9–12 stars, this galaxy is faintly visible in 15 cm as a faint patch elongated roughly N-S. 25 cm shows it 2′ × 0′.75 in pa 10° with slight brightening toward the center. A mag. 12.5 star is visible near the S tip; a mag. 13.5 star lies off the E flank 1′.4 from center. The southernmost star in the arc to the W is a pair (10,11; 20″; 335°).

eg 4825 dimen. 2′.0 × 1′.5 V = 12.1 sfc. br. 13.2
This galaxy is a small, difficult object in 15 cm, just visible at 75×. 25 cm reveals a 40″ glow slightly elongated SE-NW, with a mag. 13.5 star 1′.6 WNW. The halo is smooth in 30 cm, showing a moderately sharp concentration to a nonstellar nucleus.

eg 4845 dimen. 5′.0 × 1′.6 B = 12.2 sfc. br. 14.3
Lying 12′ NE of a mag. 7 star, this is a faint galaxy in 15 cm. The halo is 1′.5 × 1′, elongated in pa 75°. A mag. 10.5 star lies 2′.1 NE, and a faint star is visible 1′.4 SSE. It grows to 3′ × 1′ in 25 cm, with a more circular and poorly concentrated core.

eg 4856 dimen. 4′.6 × 1′.6 V = 10.4 sfc. br. 12.4
This galaxy is located 7′ NW of a mag. 10.5 star and is moderately bright in 15 cm. The circular 45″ core contains a stellar nucleus, while the halo extends NE-SW to 2′ × 1′. Some interesting details appear with 25 cm. The nucleus is stellar, and a mag. 13 star matches it toward the ESE edge of the halo 25″ from the nucleus. A narrow spike seems to extend N from the center. This and the star ESE give the impression of a dark lane on the E side of the halo. In 30 cm the core is 1′ across, but does not show a stellar nucleus. The 4′ × 1′ halo is elongated in pa 40°; its southwestern half seems brighter. eg 4877, *q.v.*, lies 22′ SE.

eg 4866 dimen. 6′.5 × 1′.5 V = 11.2 sfc. br. 13.6
This galaxy is an easy object for 15 cm. The large spindle is at least 5′ long, elongated in pa 90°. The sparkly halo is concentrated to a stellar nucleus; a mag. 13.5 star (comparable in brightness to the nucleus) lies in the W side 47″ from center. 30 cm shows this star on the N edge of the well-concentrated 6′.5 × 1′ halo. The long bar is sharply defined on both N and S flanks, and unevenly bright within, most evidently in the eastern half where the star does not interfere. The concentrated core is 1′.75 long and contains a distinct stellar nucleus that is a little fainter than the star W.

eg 4877 dimen. 2′.7 × 1′.3 mag.$_{vv}$13 sfc. br.
This faint galaxy is located 2′.8 SE of a mag. 9 star that is 10′ SE of the mag. 10.5 star near eg 4856, *cf.* In 30 cm the 1′.25 × 0′.75 halo is rectangular in outline, barely concentrated to a faint stellar nucleus.

eg 4880 dimen. 3′.3 × 2′.5 B = 12.6 sfc. br. 14.7
25 cm will show this galaxy as a diffuse low surface brightness glow. The unconcentrated 2′ × 1′ halo is elongated roughly SE-NW. In 30 cm the halo extends to 3′ × 1′.5 in pa 160°, showing a broad moderate concentration across the center. A pretty 1′.5 triangle of mag. 13 stars is visible 5′.5 SE; several other faint stars lie at about the same distance from SE through W.

eg 4887 dimen. 1′.6 × 0′.8 mag.$_{vv}$14 sfc. br.
Lying 10′ SW of eg 4902, *q.v.*, this galaxy is just visible in 30 cm. The 30″ patch is unconcentrated, but contains a definite though very faint stellar nucleus.

eg 4891 dimen. 2′.8 × 2′.5 B = 12.6 sfc. br. 14.5
In 25 cm this galaxy appears a little brighter than eg 4899, *cf.* The halo is about 45″ diameter and contains a threshold magnitude stellar nucleus. 30 cm shows it 1′.5 across. The unevenly bright halo is unconcentrated except for a 15″ core and faint stellar nucleus. A mag. 14.5 star is visible 2′.0 NW. Photographs show eg 4855 (mag.$_{vv}$14) 26′ NW, and eg 4838 (mag.$_{vv}$13) 49′ NW.

eg 4899 dimen. 2′.7 × 1′.6 B = 12.6 sfc. br. 14.1
Lying 8′ SE of a mag. 7.5 star, this galaxy appears quite faint in 25 cm. The halo is elongated in pa 15°, 2′ × 0′.75, with a broadly brighter granular core. A mag. 13 star is visible 2′ NE. 30 cm shows a 1′.5 × 1′ halo with a very weak, broad concentration across the center.

eg 4900 dimen. 2′.3 × 2′.2 V = 11.5 sfc. br. 13.1
15 cm shows this object as a small diffuse spot just NW of a mag. 10 star. 25 cm reveals a nearly smooth 1′.5 patch of moderate surface brightness. The mag. 10 star lies near

the edge, 45″ SE of center. At high power, a threshold magnitude star is visible on the N edge, which photographs show to be a bright HII region.

eg 4902 dimen. 3′.0 × 2′.8 V = 11.2 sfc. br. 13.4
Forming a 2′ nearly equilateral triangle with two mag. 10 stars to the W, this galaxy is barely visible in 15 cm as a faint unconcentrated spot. 25 cm shows that the mag. 10 stars have a number of faint companions, notably mag. 13.5 stars about 40″ away from each in opposite directions. The nebula is 2′ diameter and without central brightening except for a faint stellar nucleus. The halo shows a slight, even concentration in 30 cm, but it is unevenly bright. No nucleus is definitely visible, but the inner parts seem elongated in pa 70°. eg 4887, *q.v.*, lies 10′ SW.

eg 4904 dimen. 2′.3 × 1′.6 V = 12.1 sfc. br. 13.4
This galaxy is located 1′.8 SSE of a mag. 11 star. 25 cm shows a faint 1′.25 × 1′ oval halo elongated NNE-SSW. The core is more elongated and a little brighter toward the center.

eg 4915 dimen. 1′.7 × 1′.4 V = 11.9 sfc. br. 12.8
This galaxy looks like a faint star in 15 cm, and it remains star-like at low power in 25 cm. High power reveals a well-defined 40″ spot of fairly high surface brightness with a conspicuous stellar nucleus. eg 4890 (mag._{VV}14) lies 12′ W; eg 4918 (no mag. available) is 6′.1 NE.

eg 4928 dimen. 1′.3 × 1′.0 B = 13.4 sfc. br. 13.5
Forming the northwestern member of a line with two stars, this galaxy is barely visible in 15 cm. In 25 cm it appears 45″ diameter with a broad concentration across the center.

eg 4933A dimen. 2′.5 × 1′.5 B = 12.7 sfc. br. 14.1
eg 4933B dimen. 0′.8 × 0′.7 mag._{VV}15 sfc. br.
Lying 7′ NE of a mag. 9 star, these interacting galaxies are 46″ apart in pa 50°. The brighter A component can be seen faintly with 15 cm, appearing less than 1′ diameter, with a slightly brighter center. Both galaxies are clearly visible in 25 cm using averted vision at 200 ×. The A component is NE of B and is about 1′ diameter with a stellar nucleus. In 30 cm, − 33A is 1′.2 across, with a smooth texture and a slight, even concentration to a distinct but substellar nucleus. Component B is 20″ diameter in 25 cm and has a stellar nucleus almost as bright as that of A. The fainter component grows to 45″ in 30 cm, where it has a similar character to its brighter companion. A mag. 13.5 star is visible 1′.7 NW. Photographs show a very faint (mag._{VV}17) third galaxy 1′ NE of − 33A.

eg 4939 dimen. 5′.8 × 3′.2 B = 11.6 sfc. br. 14.6
This galaxy is visible in 15 cm at low power as a faint, low surface brightness patch. The 2′ halo is unconcentrated except for a very faint nonstellar nucleus occasionally visible in the center. Two wide pairs of stars about 12′ E are resolved in 25 cm. In 30 cm the 3′ × 2′.5 halo is elongated N-S, the elongation expressed most strongly in the inner regions. The light is smooth and very evenly concentrated toward a distinct nonstellar nucleus 5″ across. The western side of the halo seems to fade more abruptly. A mag. 15 star is visible 1′.4 NW.

eg 4941 dimen. 3′.7 × 2′.1 V = 11.1 sfc. br. 13.2
Lying 2′.6 N of a mag. 11 star, 15 cm shows this object as a faint circular spot 1′ diameter. It is about 2′ diameter in 25 cm. The texture is smooth with no rise in brightness toward the center except at the faint stellar nucleus.

eg 4951 dimen. 3′.3 × 1′.4 B = 12.6 sfc. br. 14.1
25 cm reveals this galaxy as a 2′.5 × 0′.75 haze elongated in pa 90°, with a slightly brighter center. The south flank is more sharply defined. Photographs show a star 8″ from the nucleus.

eg 4958 dimen. 4′.1 × 1′.4 V = 10.5 sfc. br. 12.2
A bright object, this galaxy is visible in 15 cm as a 1′ × 0′.5 patch elongated in pa 15°. The bright core contains a stellar nucleus, which seems off-center toward the N edge of the core. 25 cm shows the halo to 1′.75 × 0′.5. eg 4948 (mag._{VV}14) lies 13′ NW; eg 4948A (mag._{VV}15) is 13′ SW.

eg 4981 dimen. 2′.8 × 2′.2 B = 11.8 sfc. br. 13.6
Lying on the NNW side of a mag. 10 star, this galaxy is faintly visible in 15 cm. 25 cm reveals an oval 1′.5 × 0′.75 glow elongated in pa 150°. A very faint stellar nucleus is visible. The SSE tip of the halo reaches nearly to the star 1′.0 away.

eg 4984 dimen. 2′.8 × 2′.2 B = 11.8 sfc. br. 13.7
This galaxy appears circular in 15 cm, about 1′ diameter, with a stellar nucleus. Many faint stars are visible within 6′ to the E and NE. 25 cm gives a similar view, but the halo seems more extensive on the S side, making the nucleus appear off-center.

eg 4995 dimen. 2′.5 × 1′.7 V = 11.0 sfc. br. 12.4
15 cm shows this galaxy as a large diffuse spot 3′.2 SSE of a mag. 8.5 star. The halo appears about 2′ diameter in 25 cm, with a broad even brightening across the center.

eg 4999 dimen. 2′.6 × 2′.2 B = 12.6 sfc. br. 14.3
This galaxy is not visible in 15 cm, and it is only barely visible in 25 cm. The indefinite, low surface brightness

halo has a granular center, but no other details are discernable.

eg 5017 dimen. 1.′7 × 1.′4 B = 13.1 sfc. br. 14.1
25 cm will show this galaxy as a small, moderately faint glow situated between a wide mag. 11 pair about 6′ WNW and a mag. 10 star 8′ E. The galaxy is only 20″ diameter, and well concentrated to a stellar nucleus. In 30 cm the smoothly textured halo grows to 40″ diameter; a distinct stellar nucleus is still visible, but otherwise the light is broadly concentrated.

eg 5018 dimen. 2.′6 × 2.′1 V = 10.8 sfc. br. 12.6
In 30 cm this galaxy is bright and of high surface brightness. The smooth 1.′5 × 0.′75 halo is elongated E-W, growing evenly and strongly brighter toward a small circular core and relatively bright stellar nucleus. A faint star is visible 1.′9 NNE. eg 5022 ($mag._{VV}$13.5) lies 7′ ESE.

eg 5037 dimen. 2.′5 × 0.′9 B = 13.1 sfc. br. 13.8
In 15 cm this galaxy appears as a circular glow with a mag. 13 star on the NE tip. The halo is 1′ × 0.′3 in 25 cm, elongated in pa 45°. A stellar nucleus appears to lie off-center toward the star, which causes the halo to seem more extensive to the SW. In 30 cm the halo is 1.′5 × 0.′4, reaching just S of the star 50″ NE. The 15″ circular core is unconcentrated except for a faint, distinct stellar nucleus. eg 5035 ($mag._{VV}$14) lies 6.′2 N.

eg 5044 dimen. 2.′6 × 2.′6 V = 11.0 sfc. br. 13.1
The brightest of a group, this galaxy is moderately faint in 15 cm, a circular glow about 1′ diameter. 25 cm reveals a granular core and an occasionally stellar nucleus. In 30 cm the halo grows to about 1.′5 diameter, showing a strong, even concentration to the stellar nucleus. Nearby members of the group include eg 5046 ($mag._{VV}$15) 6′ NE, and eg 5047 9.′7 SE and eg 5049 8.′4 E, both *q.v.*

eg 5047 dimen. 3.′0 × 0.′8 $mag._{VV}$13.5 sfc. br.
Located 9.′7 SE of eg 5044, *q.v.*, this galaxy forms a small nearly equilateral triangle with two faint stars 50″ E and 1.′0 SSE. In 30 cm the halo is 1′ × 0.′5, elongated in pa 70°, growing slightly and evenly brighter to a faint stellar nucleus.

eg 5049 dimen. 2.′1 × 0.′7 V = 12.9 sfc. br. 13.2
This high surface brightness galaxy is located 8.′4 E of eg 5044. In 30 cm it is very small, the very faint halo reaching only to 30″ diameter. The distinct 15″ core contains a stellar nucleus.

eg 5054 dimen. 5.′0 × 3.′1 B = 11.5 sfc. br. 14.3
A faint unconcentrated glow in 15 cm, this galaxy has an overall size of about 2′, with a mag. 13.5 star off the NE edge 1.′3 from the center. In 25 cm the halo has a granular texture with many stellarings over the surface. A mag. 14.5 star is visible 2.′2 NW. At 250 × in 30 cm, the round middle part of the halo is distinctly mottled and contains a very faint stellar nucleus. The evenly concentrated halo is 2.′5 × 2′, elongated SE-NW. A second mag. 14.5 star is visible 3.′0 NW. This galaxy lies near, but is not a member of the eg 5044 group.

eg 5068 dimen. 6.′9 × 6.′3 B = 10.5 sfc. br. 14.5
In 15 cm this galaxy is very large, appearing as an unconcentrated, very low surface brightness glow about 5′ across. The halo is 5′ × 4′ in 30 cm, elongated NE-SW, with a very slight, very broad concentration. The indistinct core is 2′ × 1′, elongated in pa 155°, and does not have a bright nucleus. A mag. 13.5 star is visible within the halo 1.′7 NNE of center; another lies 2.′5 W.

eg 5072 dimen. $mag._{VV}$14.5 sfc. br.
At 250 ×, 30 cm shows a mag. 13.5 star superposed 13″ SW of the center of this galaxy, hardly distinguishable from a nucleus. The small, faint haze is only 50″ × 20″, elongated ENE-WSW. eg 5070 ($mag._{VV}$15) lies 3′ NNE.

eg 5076 dimen. 1.′6 × 0.′9 $mag._{VV}$13.5 sfc. br.
eg 5079 dimen. 1.′7 × 1.′0 $mag._{VV}$12 sfc. br.
These two galaxies lie within 5′ to the south of eg 5077, *q.v.* eg 5076 appears as a faint, low surface brightness glow in 30 cm. The light is slightly concentrated to a distinct but faint stellar nucleus. Two mag. 13.5 stars 29″ apart are visible about 2′ SW. Nearly 3′ SE of eg 5077 is −79, just visible in 30 cm as an unconcentrated patch 45″ diameter.

eg 5077 dimen. 2.′0 × 1.′6 V = 11.5 sfc. br. 12.8
Lying 8′ E of a mag. 7 star, this galaxy appears faint and unconcentrated in 15 cm, a 1′ glow with a stellar nucleus. 25 cm shows a roundish, diffuse patch about 1′ diameter. A stellar nucleus is visible as well as a threshold magnitude star 18″ S of center. In 30 cm the 15″ core and substellar nucleus are sharply concentrated within the smooth, round 1′ halo. A mag. 13.5 star is visible 55″ SE, near which photographs show a very faint galaxy. eg 5077 is the brightest of a group, including eg 5072, −76, −79, and −88, all *q.v.*

eg 5084 dimen. 4.′8 × 1.′3 B = 12.0 sfc. br. 13.9
In 15 cm this galaxy is characterized by a bright center and very faint halo: the 1.′5 × 0.′5 glow is elongated E-W; the nucleus is almost stellar. A mag. 13 star is visible 2.′2 ENE. The halo is 3.′5 × 0.′75 in 30 cm, elongated in pa 80° and bulges at the middle. The light is moderately well concentrated to a fairly distinct 1′ × 0.′3 core and nonstel-

Chart XXVI. NGC 5077 group.

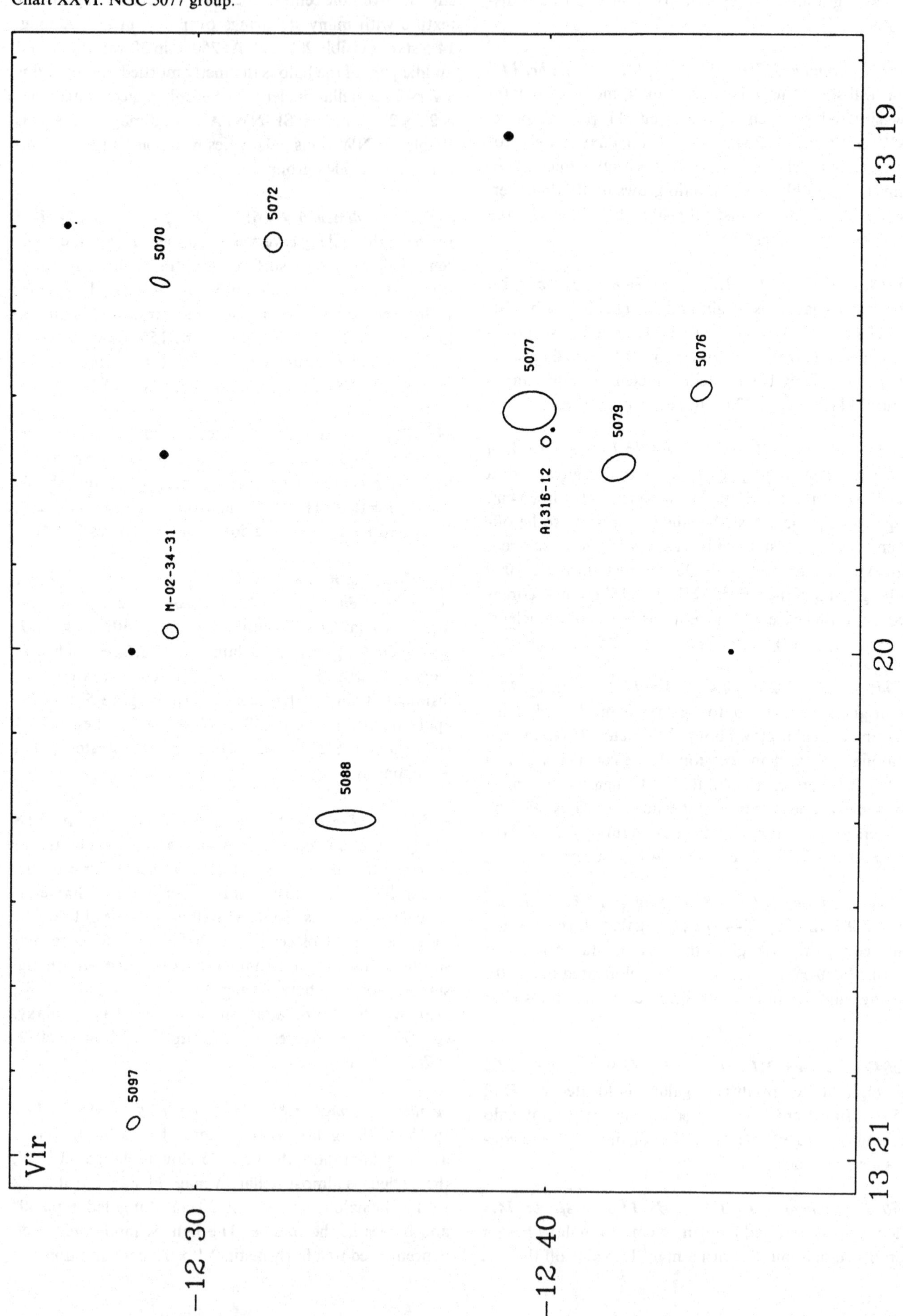

lar nucleus. Two faint stars are visible 1′.7 and 2′.4 ESE, the nearer lying just S of the halo.

eg 5087 dimen. 2′.3 × 1′.5 V = 11.0 sfc. br. 12.3
Visible at 50× in 15 cm, this galaxy lies about 7′ E of a curved group of five mag. 9–10 stars. The relatively bright and concentrated center of the nebula occasionally seems multiple, but not quite stellar. With 30 cm the halo is 1′.25 × 0′.8, elongated in pa 10°, and very faint in its outer reaches. The 50″ × 30″ core is distinctly brighter, but is itself unconcentrated. The nucleus is stellar and has no brightening immediately around it, making it look like the central star of a planetary nebula.

eg 5088 dimen. 2′.7 × 0′.9 B = 13.2 sfc. br. 14.0
Located about 13′ NE of eg 5077, *q.v.*, this galaxy is faint for 30 cm. The 45″ × 20″ halo is elongated N-S and is unconcentrated, except for a faint nonstellar nucleus.

eg I4237 dimen. 1′.9 × 1′.4 mag.$_{VV}$13 sfc. br.
This galaxy is located 11′ W of eg 5134, *q.v.* It is not bright in 30 cm: the low surface brightness 1′ halo is unconcentrated and shows no distinct nucleus. A mag. 14 star is visible 1′.4 E.

eg 5134 dimen. 2′.8 × 1′.7 B = 12.4 sfc. br. 13.9
In 30 cm this galaxy is 2′.5 × 1′.25, elongated in pa 155°. The halo has a poor, broad concentration and nearly circular inner parts. The stellar nucleus is sharply brighter, however, and is quite prominent even at 250×. eg I4237, *q.v.*, lies 11′ W.

eg 5147 dimen. 1′.8 × 1′.5 V = 11.8 sfc. br. 12.8
Viewed in 15 cm, this galaxy appears as a faint glow 1′ diameter with a stellar nucleus. 25 cm shows no distinct core, but a mag. 13.5 stellar nucleus lies in the center. The halo has a coarse-grained spotty texture. Photographs show that a star is superposed on the nucleus.

eg 5170 dimen. 8′.1 × 1′.3 B = 12.0 sfc. br. 14.3
This is quite a faint object in 15 cm, with an overall size of 2′ × 0′.5, elongated SE-NW. 30 cm shows a thin 5′ × 0′.4 spindle elongated in pa 125°. The light is poorly concentrated to an elongated 1′.5 core. A very faint stellar nucleus lies among several spots near it along the major axis.

eg 5230 dimen. 2′.2 × 2′.0 B = 12.7 sfc. br. 14.2
This galaxy has a moderately low surface brightness in 25 cm. The halo is about 1′.25 diameter and contains a brighter core that appears granular like the center of a globular cluster. Photographs show several faint galaxies to the NW: the brightest are eg 5221 (mag.$_Z$14.5) at 13′, and eg 5222AB (mag.$_Z$14.1), a double interacting system at 9′.3.

eg 5247 dimen. 5′.4 × 4′.7 V = 10.5 sfc. br. 13.8
Located 7′ SW of a mag. 9 star, this galaxy is faintly visible in 15 cm. It appears large and diffuse, about 2′ diameter, and without central concentration. 25 cm shows a smooth 2′.5 × 2′ halo elongated approximately N-S, with a small circular core 30″ across.

eg 5300 dimen. 3′.9 × 2′.7 B = 11.9 sfc. br. 14.3
This galaxy is faintly visible in 15 cm, where it appears as an unconcentrated 2′ × 1′ glow elongated SE-NW. A wide pair of mag. 8.5–9.5 stars lies 8′ SW. In 25 cm the halo is about 2′ diameter and without central concentration, though the surface has a fine-grained texture. The smooth, weakly and broadly concentrated halo grows to 3′ × 2′ in 30 cm, elongated in pa 150°. A mag. 13 star is visible 1′.7 SE near the edge of the halo.

eg 5324 dimen. 2′.4 × 2′.3 B = 12.4 sfc. br. 14.2
Visible in 15 cm, this galaxy appears as a faint diffuse spot about 1′.25 across. A mag. 12 pair (24″; 122°) is resolved 3′.8 NW. 25 cm shows the galaxy 1′.75 diameter, the core represented by a patchy region in the center. A mag. 14 star is visible 1′.4 SSE. In 30 cm the diffuse and unconcentrated halo extends to about 2′ diameter. Another faint star is visible within the halo, 26″ due E of the center.

eg 5334 dimen. 4′.4 × 3′.3 B = 11.9 sfc. br. 14.6
This galaxy is visible in 15 cm as a very low surface brightness patch 2′ across lying 3′.1 N of a mag. 12 star. In 25 cm the diffuse halo shows a little brightening toward the center. The galaxy grows to about 4′ × 2′.5 in 30 cm, elongated in pa 20°. The light is broadly and very weakly concentrated, and at times the impression of a narrower inner portion is strong. eg U8801 (mag.$_Z$15.0) lies 10′ SE; eg 5345 (mag.$_Z$13.8) is 28′ SE, near the mag. 5 star 90 Virginis.

eg 5363 dimen. 4′.2 × 2′.7 V = 10.2 sfc. br. 12.7
Visible at low power in 15 cm, this galaxy lies 3′.8 SW of a close pair of stars [ADS 9060: 8.4,8.9; 1″.0; 112° (1981)]. The circular 2′ halo is well concentrated and has a mag. 13.5 star 2′.1 SE of center. 25 cm shows a bright core and a stellar nucleus, which is visible without difficulty over it: photographs reveal a star (V = 12.1), however, superposed 5″ or 6″ SW of the true nucleus. In 30 cm the moderately and evenly concentrated halo grows to 2′.5 × 2′, elongated in pa 135°. The galaxy is the brightest of a group including eg 5364, *q.v.* 15′ S, eg 5360

(mag.$_Z$14.9), eg 5356 (mag.$_Z$14.1), eg 5348 (mag.$_Z$14.5), eg 5338 (mag.$_Z$14.3), and eg U8818 (mag.$_Z$15.2).

eg 5364 dimen. 7.′1 × 5.′0 V = 10.4 sfc. br. 14.2
Lying 15′ S of eg 5363, *q.v.*, this galaxy appears much fainter than its companion in 15 cm: the halo is larger but much less concentrated. In 25 cm, −64 is diffuse and of much lower surface brightness than −63. The surface has a granular texture and is without much central concentration. In 30 cm the large, weakly concentrated halo is 4.′5 × 2.′5, elongated in pa 30°. The unevenly bright core is granular, and elongated in a somewhat larger position angle than the halo. Two mag. 14 stars are visible 1.′6 from center in the NW quadrant. eg 5360 (mag.$_Z$14.9), one of a group of objects near eg 5363/64, lies about 9′ WSW.

eg 5426 dimen. 2.′9 × 1.′6 V = 12.2 sfc. br. 13.7
eg 5427 dimen. 2.′5 × 2.′3 V = 11.4 sfc. br. 13.2
Lying 2.′3 apart in pa 7°, both of these interacting galaxies are visible in 15 cm. Both have small, fairly sharp cores, but −27, the northern member, has a stellar nucleus and a larger surrounding halo. In 25 cm, −26 is about 1′ across, with the core elongated toward −27. The halo is smooth and diffuse, but contains a stellar nucleus. 30 cm shows it 1′ × 0.′75, elongated N-S. The halo is only slightly concentrated, but the inner portions are clearly elongated and have a granular texture. A mag. 14 star is visible 50″ NNE of center, E of a line connecting the two galaxies. eg 5427 is small and concentrated in 25 cm. It is distinctly larger than −26 in 30 cm, reaching 2′ diameter, slightly elongated in pa 75°. A dark patch lies immediately E and NE of center, while the western side of the halo appears more extended.

eg 5468 dimen. 2.′5 × 2.′5 B = 12.2 sfc. br. 14.1
eg 5472 dimen. 1.′4 × 0.′4 mag.$_{VV}$15 sfc. br.
eg 5468 lies about 4′ NW of a mag. 8 star. In 30 cm the unevenly bright, low surface brightness halo is 2.′5 diameter. At high power a faint knot is visible about 30″ N of center. eg 5472 lies 5′ E within a triangle of mag. 13–14 stars, near the easternmost and brightest star. The galaxy is 50″ across and poorly concentrated.

eg 5493 dimen. 2.′0 × 1.′4 V = 11.5 sfc. br. 12.4
A small object in 30 cm, this galaxy is nevertheless bright, lying in the same low-power field with a mag. 8.5 star 16′ SE. The halo is 1.′5 × 0.′75, elongated in pa 120°. The inner regions show a strong, even concentration, and become progressively more elongated toward a 20″ × 5″ core. A substellar nucleus is visible at 250 ×.

eg 5496 dimen. 4.′4 × 1.′0 B = 12.6 sfc. br. 14.1
This is a large galaxy for 30 cm, reaching 5′ × 1′, elongated in pa 175°. The low surface brightness bar is not tapered toward the ends and is unconcentrated except for an exceedingly weak brightening to an elongated 2′ core. A mag. 14.5 star is visible in the N half on the eastern edge of the bar about 50″ from the center. The high-power field around the object is empty of stars.

eg 5534 dimen. 1.′4 × 0.′8 B = 13.5 sfc. br. 13.4
This galaxy forms a link in an 11′ chain including five stars NW of a mag. 6.5 star. In 25 cm it is elongated E-W, about 40″ × 15″, with a well-defined lenticular outline. The center is sharply concentrated but nonstellar. The halo reaches 2′ × 1′ in pa 80° with 30 cm. The light is barely concentrated to a moderately faint stellar nucleus, which is off-center to the W side. The nearest star in the chain is 1.′5 WSW.

eg 5560 dimen. 3.′9 × 0.′9 V = 12.4 sfc. br. 13.6
eg 5566 dimen. 6.′5 × 2.′4 V = 10.5 sfc. br. 13.3
In 15 cm eg 5566 appears moderately bright, lying 1.′5 W of a mag. 12 star. The light is well concentrated to a stellar nucleus. It is 2′ × 0.′5 in 25 cm, elongated in pa 30°. The brighter elongated core contains a stellar nucleus. 30 cm shows the halo to 2.′5 × 1.′25, and a mag. 14 star 1.′2 SW of center. The central 10″ × 5″ is elongated more nearly N-S, in about pa 10°; this nucleus is composed of three or four bright knots. A dark area lies immediately SW of the nucleus. About 5′ NW is eg 5560, which is faintly visible in 25 cm. In 30 cm it is about 2.′5 × 1′, elongated in pa 105°. The relatively large, elongated core is about 1′ long, and poorly defined against the halo; there is no sharp nucleus. A mag. 13.5 star is visible 38″ NW of center. eg 5569 (mag.$_Z$14.9), a third member of this interacting group, lies 4.′2 NE of −66.

eg 5574 dimen. 1.′6 × 1.′0 V = 12.4 sfc. br. 12.7
eg 5576 dimen. 3.′2 × 2.′2 V = 10.9 sfc. br. 13.0
eg 5577 dimen. 3.′4 × 1.′1 mag.$_Z$13.6 sfc. br.
eg 5576 is the brightest of this group. In 30 cm it is 2′ × 1.′2, elongated E-W. The well-concentrated halo contains a circular core about 1′ across and a substellar nucleus that is comparable in brightness to a star 1.′2 NW. Nearly 3′ SW is eg 5574. 30 cm shows it 1.′25 × 0.′8 in pa 65°. The light is moderately concentrated to an occasionally visible stellar nucleus. A mag. 14 star is visible 1.′9 NW. About 10′ NNE of −76 is eg 5577, which in 30 cm is the largest but faintest of the three galaxies. The diffuse and poorly concentrated 3′ × 1′ halo is elongated in pa 60°. The central third of the object is unevenly bright. A mag. 14 star is visible 2.′0 NW; a fainter star lies at 52″ in pa 75°, on the S side of the NE tip.

eg 5584 dimen. 3.′3 × 2.′6 B = 11.9 sfc. br. 14.1
This galaxy lies 3.′4 SW of a mag. 10.5 star. The halo is faint for 25 cm, about 2′ × 1′ in extent, elongated in pa 145°, with a weakly brighter center. In 30 cm the broadly oval, very low surface brightness halo reaches 3.′5 × 2′, and has a very slight, broad concentration. Several other stars are visible near the galaxy: the brightest, mag. 12, lies 2.′1 NNE; another is visible 2.′8 NNW just off the end of the halo; a third is 1.′9 SE.

gc 5634 diam. 4.′9 V = 9.4 class IV
This isolated globular cluster is set among three mag. 8.5–11 stars, the brightest located only 1.′3 ESE of the center. The cluster is visible in 15 cm at 75 × as a broadly concentrated circular spot. In 25 cm it is about 2.′5 across; there is little outer envelope and only a broad brightening across the center, where there is a faint stellar nucleus. The cluster remains unresolved in 30 cm, though the core appears granular at 250 ×, and several very faint stars are visible within about 3′ radius. The broadly concentrated halo is 3′ diameter.

eg 5636 dimen. 1.′9 × 1.′4 mag.$_z$14.6 sfc. br.
eg 5638 dimen. 2.′6 × 2.′3 B = 12.2 sfc. br. 14.1
The brighter of these two galaxies is visible in 15 cm as a faint spot with a broadly brighter core. It is 1.′25 diameter in 25 cm, with a 15″ core containing a stellar nucleus. The halo reaches 1.′75 diameter in 30 cm. The light is moderately well concentrated to a distinct nucleus, which occasionally appears stellar. eg 5636 is visible 2′ N as a small spot that sometimes makes −38 appear elongated. The low surface brightness halo is 45″ × 30″, elongated NE-SW, and is unconcentrated except for a very faint stellar nucleus.

eg 5645 dimen. 2.′6 × 1.′7 V = 12.3 sfc. br. 13.7
25 cm shows this galaxy 6.′2 W of a mag. 9 star. The halo is 1.′5 × 1′, elongated ENE-WSW, with a broadly brighter core. In 30 cm it appears larger than eg 5665, *cf.* 55′ NE in Bootes. The halo is 2′ × 1.′5 in pa 75° and seems more sharply defined on the SE flank. Stars are visible NNE, NW, and SE, all within about 3′ radius.

eg 5668 dimen. 3.′3 × 3.′1 V = 11.5 sfc. br. 13.8
This galaxy is faintly visible in 15 cm as an unconcentrated 1′ glow with a very faint star on its NE edge 40″ from center. A wide pair of mag. 9 stars lies 5.′5 NW. In 25 cm the 1.′5 halo contains a grainy core. 30 cm still shows no overall concentration, but the surface is distinctly mottled, and a very faint nucleus is visible. The halo extends to 2′ diameter, encompassing the star NE.

eg 5690 dimen. 3.′5 × 1.′2 B = 12.5 sfc. br. 13.9
This galaxy lies 3.′2 ENE of a mag. 6.5 star [ADS 9323: 6.6,10.6; 1.″6; 135°(1964)]. In 25 cm the galaxy is circular, about 2′ diameter, with a moderately brighter core. 30 cm shows a 2.′5 × 1′ halo elongated SE-NW. A threshold magnitude star is visible 45″ SE of center near the major axis.

eg 5691 dimen. 2.′0 × 1.′7 B = 13.0 sfc. br. 14.1
Barely visible in 15 cm, this galaxy is still moderately faint for 25 cm. The halo is circular, about 1′ diameter, and has a small core. In 30 cm it is 1′ × 0.′75, elongated in pa 110°. The faint halo is unconcentrated except for a faint but distinct stellar nucleus.

eg 5701 dimen. 4.′7 × 4.′5 B = 11.8 sfc. br. 14.9
Located between two mag. 11 stars, this galaxy is easily visible to 15 cm at 75 ×, though it is fairly small. The circular halo is moderately well concentrated to a faint stellar nucleus. 25 cm shows a 1.′5 × 1.′25 halo elongated in pa 0°. A very faint and inconspicuous stellar nucleus resides in a bright, elongated core. In 30 cm the 2′ × 1.′75 halo has a moderately strong, even concentration to a substellar nucleus. A curving arc of three stars leads off from the NE side, starting with a mag. 14.5 star 1.′3 from center and ending with the mag. 11 star 3.′7 NE.

eg 5713 dimen. 2.′8 × 2.′5 V = 11.4 sfc. br. 13.4
eg 5719 dimen. 3.′4 × 1.′3 mag.$_z$13.8 sfc. br.
15 cm will show eg 5713 as a 1′ spot. 25 cm reveals a fat 1.′5 × 1′ lens elongated E-W with a slightly brighter center. In 30 cm the weakly concentrated, irregular halo is 2′ × 1.′75. The indistinct, irregular core has an abrupt edge on its N and NNE sides, and a stellar nucleus is occasionally visible near this edge. eg 5719 lies 11′ E and a bit S, between two widely separated mag. 11 stars. It is weakly visible in 25 cm. 30 cm shows a long, thin glow reaching 1.′75 × 0.′5 in pa 95°. There is some concentration to elongated inner parts, but no sharp nucleus is present.

eg 5740 dimen. 3.′1 × 1.′7 V = 11.9 sfc. br. 13.6
Lying 19′ SSW of eg 5746, *q.v.*, this object has low surface brightness in 25 cm. The galaxy is about 1.′5 diameter and has a faint stellar nucleus at the threshold of visibility. 30 cm shows it 2′ diameter with a brighter 45″ core. Photographs show eg 5738 (mag.$_z$14.7) 8.′2 SW.

eg 5746 dimen. 7.′9 × 1.′7 V = 10.5 sfc. br. 13.2
This galaxy lies 20′ WNW of the naked-eye star 109 Virginis (V = 3.7). 25 cm shows a bright object elongated nearly N-S. The brighter part is about 2′ × 0.′8, but very faint extensions are visible to 6′ length; the S end seems a

little longer than the N. In 30 cm the halo is 4′ × 0′.6 in pa 170°, with faint extensions reaching to perhaps 8′ length. The slightly brighter core is 2′ long and mottled, but without central concentration. A faint star is visible in the halo 2′ S of center along the major axis.

eg 5750 dimen. 2′.9 × 1′.7 V = 11.6 sfc. br. 13.2
This galaxy appears as a moderately faint spot 1′ diameter in 15 cm. The overall size is similar in 25 cm, while 30 cm shows it 2′ × 1′.5 in pa 65°. The halo has a slight, even concentration to a distinct substellar nucleus. A faint star is visible 1′.1 N.

eg I1066 dimen. 1′.5 × 0′.9 mag.$_Z$14.2 sfc. br.
eg I1067 dimen. 2′.3 × 2′.0 mag.$_Z$13.6 sfc. br.
These galaxies lie 2′.2 apart just W of a three-star arc of mag. 12–13 stars 3′.5 long. eg I1067 is both larger and brighter in 30 cm. The 1′.75 × 0′.75 halo is elongated in pa 75° and is only slightly concentrated with a few faint stellarings along the major axis. Photographs show two faint mag. 15 stars close by WSW of the galaxy that may cause the observed elongation. eg I1066 lies 1′.2 WNW of the southern star of the arc. In 30 cm at 250× it is circular, about 40″ across, with hardly any concentration and no distinct nucleus.

eg 5774 dimen. 3′.2 × 2′.7 V = 12.2 sfc. br. 14.4
eg 5775 dimen. 4′.3 × 1′.2 V = 11.4 sfc. br. 13.1
eg 5775 is clearly visible in 15 cm as a poorly concentrated 3′ × 1′ glow elongated in pa 145°. It grows to 4′.5 × 1′.25 in 30 cm, with a weak, broad concentration to an elongated 2′ × 0′.25 inner core. The SW side of the bar is more sharply defined, and a mag. 13.5 star is visible 53″ NE of the center. eg 5774 is visible 4′.5 NW as a very weakly concentrated 1′.1 × 0′.8 patch elongated roughly E-W. A stellaring is occasionally visible near the center. eg I1066/67, *q.v.*, lie 19′ SW; eg 5776, *q.v.*, is 36′ SSE.

eg 5776 dimen. 1′.4 × 1′.1 mag.$_Z$14.7 sfc. br.
This galaxy is located 36′ SSE of eg 5775, and about 4′ NNE of a mag. 9 star. In 30 cm it is small, about 45″ × 30″, elongated E-W, and unconcentrated except for a faint stellar nucleus.

eg 5806 dimen. 3′.1 × 1′.7 V = 11.6 sfc. br. 13.2
A member of the eg 5846 group, this galaxy is faintly visible in 15 cm. In 25 cm the oval halo is elongated in pa 170°, 1′.5 × 0′.75, and shows some central concentration. It seems much fainter than eg 5813, *cf.*, in 30 cm. The overall size is 2′ × 0′.9, with some brightening to a 50″ core and a faint, occasionally stellar nucleus.

eg 5813 dimen. 3′.6 × 2′.8 V = 10.7 sfc. br. 13.2
Visible in 15 cm, this galaxy lies in a 3′.5 triangle of mag. 12 stars. It is elongated SE-NW, 1′ × 0′.5, with a faint stellar nucleus. 25 cm shows a small concentrated core within a bright, well-defined 1′.5 × 0′.75 halo. In 30 cm four nearby stars mark the cardinal points (three form the triangle noted with 15 cm), all at about 3′.5; the stars E and W are the brightest. The galaxy is about 1′.3 diameter and has a brighter nonstellar nucleus. This galaxy is a member of the eg 5846 group; eg 5814 (mag.$_Z$14.7) lies 4′.6 SSE.

eg 5831 dimen. 2′.2 × 2′.0 V = 12.5 sfc. br. 13.0
This member of the eg 5846 group is moderately faint in 15 cm, less than 1′ diameter, and has a small, nonstellar nucleus. 30 cm shows it 50″ diameter, moderately concentrated to a nucleus that is only occasionally visible. A mag. 13 star is located 1′.5 NNE of center.

eg 5838 dimen. 4′.2 × 1′.6 V = 10.8 sfc. br. 12.7
15 cm and 25 cm show this galaxy as a small concentrated spot about 1′ diameter with a faint stellar nucleus. 30 cm shows a circular 50″ core surrounded by an elongated halo extending up to 2′.2 × 0′.8 in pa 40°. A mag. 14 star lies 2′.1 SW. The galaxy is a member of the eg 5846 group.

eg 5839 dimen. 1′.4 × 1′.4 B = 13.6 sfc. br. 14.3
eg 5846 dimen. 3′.4 × 3′.2 V = 10.2 sfc. br. 12.8
eg 5846A dimen. 0′.5 × 0′.5 V = 13.2 sfc. br. 11.7
eg 5846 is the brightest of a group. In 15 cm it appears about 1′.5 diameter and has a stellar nucleus. 25 cm shows the stellar nucleus in a bright core, the halo extending to at least 1′ diameter. It is fairly bright in 30 cm, with a strong, even concentration to the nucleus. The overall diameter is about 1′.8, with a slight elongation in pa 160° occasionally discernable. On the S side is a mag. 13.5 star-like spot, eg 5846A, a compact galaxy 40″ from the center. 30 cm also shows eg 5839, 15′ W, as a faint spot 30″ diameter and well concentrated to a faint nearly stellar nucleus. eg 5850, *q.v.*, lies 10′ SE of eg 5846; eg 5845 (V = 12.3) is 7′.3 E.

eg 5850 dimen. 4′.3 × 3′.9 V = 11.0 sfc. br. 13.8
A member of the eg 5846 group, this object is just visible in 15 cm 10′ SE of eg 5846, *q.v.* 25 cm shows a 2′.25 × 0′.5 halo elongated in pa 110°. The circular core is much brighter than the halo and contains a stellar nucleus. In 30 cm it is about 1′ diameter, slightly elongated in the outer parts. Overall the galaxy seems poorly concentrated, but a stellar nucleus is occasionally visible.

eg 5854 dimen. 2′.7 × 0′.8 V = 11.8 sfc. br. 12.5
This member of the eg 5846 group appears as a star-like spot in 15 cm. The halo is 40″ diameter in 25 cm, with a much brighter core. In 30 cm the elongated halo is 1′.8 × 0′.5 in pa 55°, with moderate concentration to more circular inner parts; no distinct nucleus is visible. A mag. 13 star lies 1′.8 ESE.

eg 5864 dimen. 2′.8 × 1′.0 B = 12.7 sfc. br. 13.7
While only just visible in 15 cm as a small diffuse spot, in 25 cm this galaxy is a well defined 1′ × 0′.3 lenticular glow elongated in pa 70°. A mag. 14 star lies 29″ ESE of the center. In 30 cm the halo reaches 1′.8 × 0′.5 and is moderately concentrated to an occasionally stellar nucleus. A second mag. 14 star is visible at 1′.1 in pa 200°. This galaxy is a member of the eg 5846 group.

Vulpecula

Embedded in the Milky Way, this inconspicuous constellation contains rich star fields and several open clusters. The center culminates at midnight about 25 July.

oc 6793 diam. 7′ mag. stars 15

This cluster is located in a rich star field and consequently does not stand out well; it is nevertheless observable in smaller apertures at low powers. 15 cm shows about 30 stars in an area 30′ across. A mag. 8 star stands out on the W side, and near the center is a 1′ triangle of mag. 10–11 stars. The northern star in this triangle is a resolved pair in 25 cm (ADS 12385: 10.5,11.5; 8″.3; 45°). About 45 stars are visible in the 30′ area, including a 45″ quadrilateral of fainter stars 2′ SSW of the triangle.

oc 6802 diam. 5′ V=8.8 stars 201

This cluster lies at the E end of the naked-eye "coat hanger" asterism called Brocchi's cluster (oc Collinder 399, V=3.6) and is enclosed within a box of four stars that includes two wide pairs. It is dimly visible in 6 cm at 25× as a hazy patch. At 50× in 15 cm it is oval, elongated N-S, and only a few stars are resolved. 25 cm shows partial resolution at 200×. 30 cm will show about fifteen stars in a 3′ × 2′ area with a faint background haze. A small trapezium of mag. 13 stars appears in the northern section of the group.

oc Stock 1 diam. 80′ V=5.3: stars 157

A widespread cluster of bright stars, this object forms a conspicuous group in binoculars and finderscopes. The cluster breaks into two sections in larger instruments. In 15 cm the western group is larger and contains about 20 stars; on the E is a small group of half a dozen stars including a nice pair (ADS 12669: 9.0,10.0; 9″.2; 101°). 25 cm gives a similar view but shows twice as many stars. A fainter, wider pair is visible on the S side of the western group.

oc 6823 diam. 7′ V=7.1 stars 79

This cluster is involved with an extended region of emission nebulosity that is generally invisible in amateur instruments. The cluster is conspicuous in 6 cm at 25×. Many of the stars are well concentrated at the center, but other stars are scattered widely over the field. 15 cm shows about 15 stars, including a tiny 20″ × 10″ diamond-shaped asterism at the center composed of four mag. 9–11 stars. 25 cm and 30 cm will reveal about three dozen stars within the central 5′ diameter.

oc 6830 diam. 10′ V=7.9 stars 82

Easily visible in 6 cm at low power, this cluster appears as a small, diffuse spot of high surface brightness that is granular to partially resolved at 25×. In 15 cm the cluster is 5′ diameter and contains about 15 stars, including a bright mag. 9 star off the E side. The stars are grouped like a four-legged spider. It is 10′ diameter in 25 cm and shows marked central concentration. About 30 stars are resolved, including a clump of faint stars on the NW. 30 cm shows about 30 stars within a 6′ circle. The brightest stars are on the SE, including a triangle of five stars, and to the NW is a string.

pn 6842 dimen. 53″ × 48″ nebula 13.1v star V=16.0:

30 cm will show this planetary as a diffuse annular glow with low surface brightness. No central star is evident.

pn 6853 dimen. 8′.0 × 5′.7 nebula 7.3v star V=13.8

Messier 27, the Dumbbell Nebula, is easily visible in 6 cm at 25× as a rounded rectangular haze elongated in pa 25°. The two bright lobes that form the dumbbell are barely distinguishable. A large halo surrounding the nebula is very faintly visible, elongated in pa 120°. A mag. 9 star lies on the W edge, and at high power a faint star is visible near the center of the northeastern lobe. 15 cm shows that the southwestern lobe is brighter and that the center of the nebula is slightly darker. In 30 cm the faint outer halo is 7′.5 × 6′: the brighter parts of the nebula are elongated almost perpendicular to this. The central star is visible here, and the southwestern lobe has two knots on its outer edge.

oc 6882 diam. 10′ V=8.1: stars 34
oc 6885 diam. 22′ V=5.9 stars 30

These clusters lie near 20 Vulpeculae (V=5.9) but for observers they are not clearly distinguishable. In 6 cm at low power, it is possible to imagine two objects: a small one (oc 6882) with a few faint stars resolved on the NW side of 20 Vul, and a larger one (oc 6885) of about 15 brighter stars scattered around and S of the bright star. 15 cm shows a garland of stars curving away E and S of oc 6882, passing N and E of 20 Vul. About 30 stars total

Chart XXVII. NGC 6802. A 40-minute exposure on IIa-D plus GG14 filter with the USNO 1.55 meter telescope taken by C. Luginbuhl on 24 July 1982. Magnitude sequence from: Hoag *et al.* 1961, Pubs. of the U.S. Naval Observatory, second series, volume 17, part 7. U.S. Naval Observatory photograph.

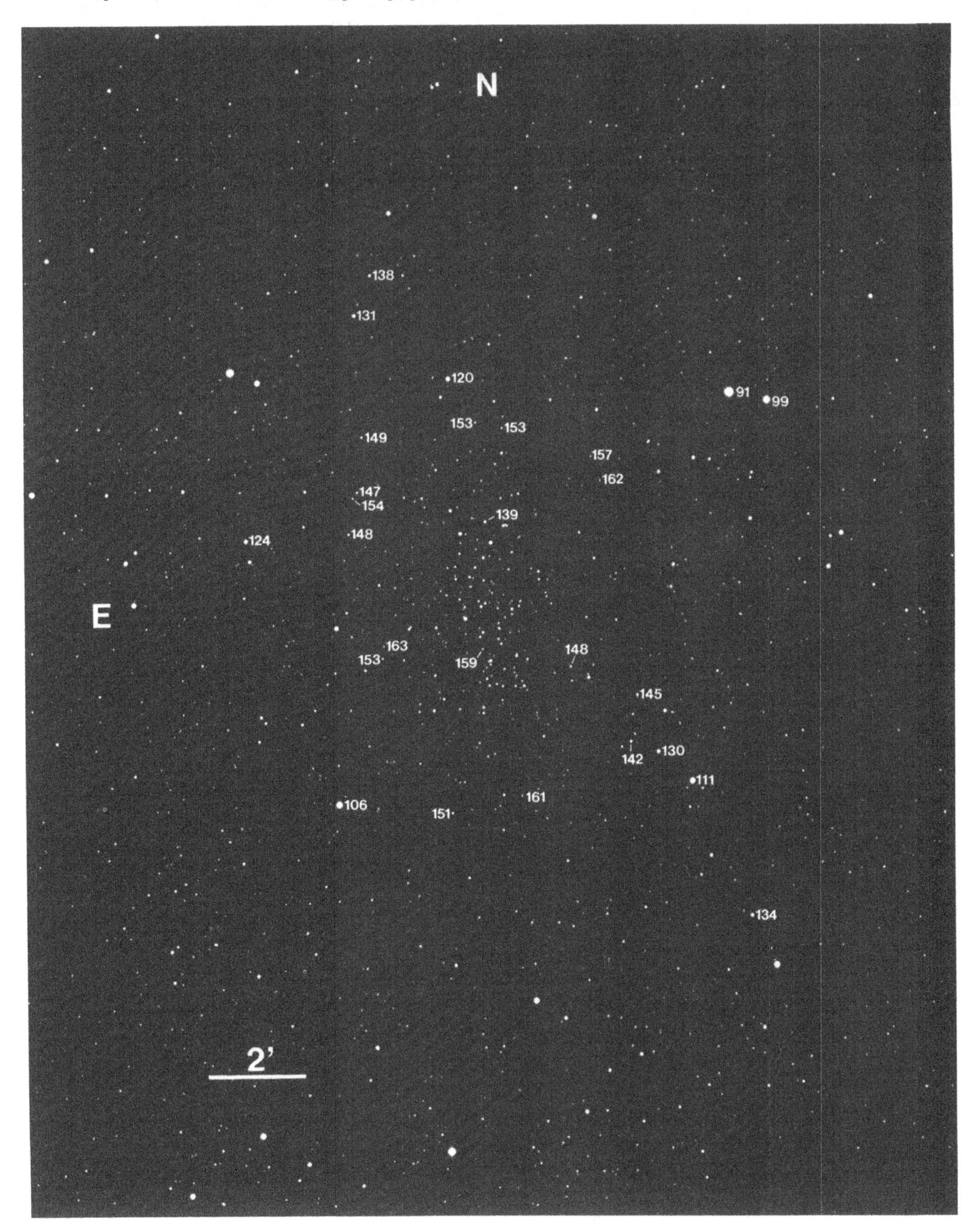

are visible in the region. As many as 75 stars are shown by 30 cm in the area, most on the W side of 20 Vul.

oc 6940 ***diam. 25′*** ***V=6.3*** ***stars 170***

Viewed in 6 cm, this cluster appears as a bright and fairly well concentrated haze about 20′ across. The brightest star is on the NE; about a dozen others are visible scattered unevenly across the haze, which seems composed of a mass of stars at the threshold of visibility. About 75 stars are resolved in 15 cm within 25′ diameter. Counting outliers, with 20 cm about 100 stars fill a 40′ area that does not stand out well from the rich stellar background. 25 cm shows 100 stars in a 30′ circle, giving a grand view at 100 × . 30 cm shows about 120 stars in a 23′ low-power field, mostly mag. 11–12. A brighter reddish star (FG Vulpeculae) lies in the center. A band without stars runs E-W just N of the center, while the corresponding part S of center is particularly rich. Three brighter stars lie on the NE edge.

Notes on references for deep-sky observers

Useful information concerning deep-sky objects and deep-sky observing is scattered throughout the astronomical literature. Many of the available compilations are now quite out of date or lack what we feel are the most useful data. To help the aspiring deep-sky observer, we describe below the major books and articles that provide general background for observing as well as information concerning individual deep-sky objects. For our work we were fortunate to have available the resources of two observatory libraries, which was particularly helpful for articles published in the specialized periodical literature. Many general references will be available in larger public libraries (perhaps via interlibrary loan), but most of the specialized material will require the resources of a college or university library. Some will probably be found only in observatory libraries.

We invariably suggest that the amateur's first task is to learn the constellations and that his first instrument should be a common pair of binoculars. Basic observing handbooks, intended for the beginner with his or her first telescope, are always in abundance. New twists on the basic theme appear monthly, it seems, and here we merely list a few that we have recommended to others starting out in the hobby. To help pick out the constellations, one can hardly do better than H. A. Rey's *The Stars: A New Way to See Them* ("enlarged worldwide edition," Houghton Mifflin, 1976). To show what can be seen with small instruments, try Muirden's *Astronomy with Binoculars* (Prentice-Hall, 1985) or his *Astronomy with a Small Telescope* (Prentice-Hall, 1986); also there is Menzel and Pasachoff's *A Field Guide to the Stars and Planets* (second edition, Houghton Mifflin, 1983), one of the famous Peterson Field Guide series, and the very similar *Universe Guide to the Stars and Planets* by Ridpath and Tirion (Universe Books, 1984). The latter two books include star charts drawn by the Dutch draftsman Wil Tirion. For slightly more advanced readers, there is Brown's venerable, information-packed *All About Telescopes* (Edmund Scientific, 1967), which contains instructions on making homebuilt telescopes, and Howard's *Telescope Handbook and Star Atlas* (third edition, Crowell, 1975). Good references for deep-sky observing are Jones' *Messier's Nebulae and Star Clusters* (American Elsevier, 1969) and Hartung's *Astronomical Objects for Southern Telescopes* (second edition, Cambridge, 1985).

Any of the works listed above provide, in addition to viewing tips, some general background on the science of astronomy. The good deep-sky observer is at least passingly familiar with basic concepts of astronomy, as he is often called upon by the lay public as a local "expert" during eclipses, the passage of comets, and so on. *The Milky Way* by Bok and Bok (fifth edition, Harvard, 1982), Ferris' coffee-table book *Galaxies* (Stewart, Tabori, & Chang, 1982), Roach and Gordon's *The Light of the Night Sky* (Reidel, 1973), Payne-Gaposchkin's *Stars and Clusters* (Harvard, 1979), and Malin and Murdin's *Colours of the Stars* (Cambridge, 1984) all provide enrichment in both the scientific and aesthetic aspects of astronomy on an introductory or semitechnical level.

On the question of dark adaptation and the use of red lighting to preserve it, we found the following articles helpful: "The Sky and Eye," by Roach and Jamnick, *Sky and Telescope* 17:165 (February 1956); "Time of Dark Adaptation after Stimulation by Various Brightnesses and Colors," Hulburt, *Journal of the Optical Society of America* (*JOSA*) 41:402 (1951); "Effects of Exposure to Various Red Lights upon Subsequent Dark Adaptation Measured by the Method of Constant Stimuli," Smith *et al.*, *JOSA* 45:502 (1955); "Sensitivity of the Eye to Spectral Radiation at Scotopic and Mesopic Intensity Levels," Kinney, *JOSA* 45:507 (1955); "Seasonal Changes in Scotopic Sensitivity," Sweeney *et al.*, *JOSA* 50:237 (1960).

Though they are of limited use while actually observing, photographic atlases provide a means of checking estimates made at the telescope and verifying details in your notes. As you collect notes on objects, particularly those not described in this book, you may want to compare your descriptions to an objective photograph, perhaps to help keep your imagination within bounds. In many cases, the small-scale (3′.9/mm) *Lick Observatory Sky Atlas* (no author, Lick Observatory, 1965) and its southern extension, the *Canterbury Sky Atlas* by Doughty, Shane, and Wood (University of Canterbury, New Zealand, 1972), will suffice to check positions of field stars and so on near an object, though finding copies of these atlases will be difficult as few libraries have them. These atlases were taken with the same 12.5 cm f/7 lens and show stars as faint as photographic-blue magnitude 16. If you have access to an observatory library, you may be able to use the much larger scale (1′.1/mm) *National Geographic Society—Palomar Observatory Sky Survey* (POSS) prints (California Institute of Technology, 1958). This atlas was taken with the 1.2 meter Palomar Schmidt telescope in both blue and red light, and so can give a better idea of the visual appearance of objects. Several other works contain beautiful and useful images of large numbers of objects, such as: *The Hubble Atlas of Galaxies* by Sandage (Carnegie, 1961); Sandage and Tammann's *Revised Shapley-Ames Catalogue of Bright Galaxies* (Carnegie, 1981); Arp's *Atlas of Peculiar Galaxies* (California Institute of Technology, 1966); the study by Hoag *et al.*, "Photometry of Stars in Galactic Cluster Fields" (*Publications of the U. S. Naval Observatory*, second series, volume 17, part 7, 1961); *Atlas of Deep-Sky Splendors* by Vehrenberg (fourth edition, Cambridge, 1983); *The Messier Album* by Mallas and Kreimer (Sky Publishing, 1978).

For the special cases of M31 and M33, references useful to the visual observer include: *Atlas of the Andromeda Galaxy* by Hodge (University of Washington, 1981) for identification of objects and detailed charts [don't miss Hodge's "Corrections to the Globular Cluster Identifications in the *Atlas of the Andromeda Galaxy*" in the *Publications of the Astronomical Society of the Pacific* (*PASP*) 95:20 (1983)]; "The M31 Globular Cluster System" by Crampton *et al.* [*Astrophysical Journal* (*Ap. J.*) 288:494 (1985)] for sizes and magnitudes of the globulars. Photometry of brighter stars ($V \lesssim 13$) superposed on the galaxy is contained in "S Andromedae 1885: A Centennial Review" by de Vaucouleurs and Corwin [*Ap. J.* 295:287 (1985)], and of fainter stars in "Novae in the Andromeda Nebula" by Arp [*Astronomical Journal* (*A. J.*) 61:15 (1956)]. A detailed "observing guide" for M33 is found in "On the Stellar Content and Structure of the Spiral Galaxy M33" by Humphreys and Sandage [*Ap. J. Supplements* 44:319 (1980)], which uses measures of stars in the field found in "A Preliminary Photoelectric Sequence in the Galaxy M33 of the Local Group" by Sandage and Johnson [*Ap. J.* 191:63 (1974)].

Among primary data sources, the one we most often referred to was the *Second Reference Catalogue of Bright Galaxies* by de Vaucouleurs, de Vaucouleurs, and Corwin (University of Texas, 1976). This contains a complete summary of data (positions, types, dimensions, magnitudes, radial velocities, etc.) and bibliographic references for over 4300 galaxies generally brighter than magnitude 15. Several other more comprehensive catalogues based on the POSS are also useful for identifications and other basic data. Nilson's *Uppsala General Catalogue of Galaxies* (Uppsala Astronomical Observatory, 1973) provides dimensions and descriptive information for 13,000 galaxies north of $-2°30'$ Declination brighter than magnitude 14.5 selected from Zwicky's six-volume *Catalogue of Galaxies and Clusters of Galaxies* (California Institute of Technology, 1960–1968). The CGCG covers the same area and gives positions, magnitudes, and charts including over 30,000 galaxies brighter than photographic-blue magnitude 15.7. Another descriptive survey of galaxies was completed by Vorontsov-Veljaminov, the five-volume *Morphological Catalogue of Galaxies* (Sternberg Institute, 1962–1974). This work includes dimensions and descriptive information for over 30,000 objects north of $-45°$ Declination. For objects south of $-17°.5$ Declination, we preferred to quote information from *The ESO/Uppsala Survey of the ESO (B) Atlas* by Lauberts (European Southern Observatory, 1982) and the *Southern Galaxy Catalogue* by Corwin, de Vaucouleurs, and de Vaucouleurs (University of Texas, 1985). The ESO catalogue is based on an inspection of a blue-light atlas completed with the 1 meter ESO Schmidt camera and provides positions, dimensions, and brief descriptions for 18,500 objects, about 16,000 of which are galaxies. The SGC, based on a UK Schmidt blue-light survey, gives morphological types, dimensions, and detailed descriptions for 5500 of the larger, brighter galaxies. Corwin also clears up many NGC/IC identification problems here.

The third edition of Lyngå's *Catalogue of Open Cluster Data* (Lund Observatory, 1983) is a good source for open cluster data. This is a condensed tabular listing (available only on microfiche and magnetic tape) of 1200 open clusters reported up to 1983. For detailed information on specific clusters, we referred to the periodical literature, which contains studies of over 400 objects. A thorough list of these papers is cited in a review by Janes and Adler, "Open Clusters and Galactic Structure" [*Ap. J. Supplements* 49:425 (1982)]. The integrated magnitudes derived by Skiff are included in the Lyngå catalogue and in Hirshfeld and Sinnott's *Sky Catalogue 2000.0*, volume 2 (Sky Publishing, 1985), a companion volume to Tirion's *Sky Atlas 2000.0* (Sky Publishing, 1981).

The diameters of globular clusters are poorly known: recently published values for many clusters vary by huge amounts, not uncommonly 50% or more. We used those given by Alcaino in "Basic Data for Galactic Globular Clusters" [*PASP* 89:491 (1977)]. Integrated magnitudes are more precisely known and are given in a summary of globular cluster data by Webbink ("Structural Parameters of Galactic Globular Clusters," in *Dynamics of Star Clusters*, Reidel, 1985). This source also includes data for the Fornax dwarf galaxy and its globulars. The best positions are in Webbink and "Accurate Positions for the Centers of Galactic Globular Clusters" [*A. J.* 91:312 (1985)] by Shawl and White. As with open clusters, magnitudes of stars are available from photometric studies of individual clusters, a list of which is contained in the Webbink paper.

Perek and Kohoutek's *Catalogue of Galactic Planetary Nebulae* (Academia Publishing House of the Czechoslovak Academy of Sciences, 1967) is the fundamental source of planetary nebulae positions and dimensions. Accurate visual magnitudes for planetaries have not been published. We used visual magnitudes by Marling and central star magnitudes measured by Kaler and his colleagues. A sample of the latter work can be found in Shaw and Kaler's "Apparent Magnitudes of Luminous Planetary Nebula Nuclei. I. Method and Application" [*Ap. J.* 295:537 (1985)]. The measurements of accurate positions by Milne are published in *A. J.* 78:239 (1973) and *A. J.* 81:753 (1976); those of Higgs are in the *Journal of the Royal Astronomical Society of Canada* 63:200 (1969).

Lynds' catalogues of northern dark and bright nebulae are published in *Ap. J. Supplements* 7:1 (1962) and *Ap. J. Supplements* 12:163 (1965), respectively. These lists are again based on the POSS, and thus contain large numbers of objects, most of which are visible only on such photographs. A useful list of objects for the visual observer is contained in the *Sky Catalogue 2000.0*, volume 2. The references there also include sources of photometry of stars involved in nebulae.

While up-to-date information on the astrometric properties of double and multiple stars is contained in the *Washington Visual Double Star Catalog* (U.S. Naval Observatory, 1984), accurate positions, magnitudes, and colors must be sought elsewhere. One of the larger compilations of photometry is Wallenquist's *Catalogue of Photoelectric Magnitudes and Colours of Visual Double and Multiple Systems* (Almqvist & Wiksell, 1981).

Accurate positions, magnitudes, and colors for stars brighter than magnitude 9 are available in several general star catalogues. The SAO *Star Catalogue* (Smithsonian Institution, 1966) and the third *Astronomische Gesellschaft Katalog* (AGK3) (Hamburger Sternwarte, 1975) are large catalogues of precise star positions and include approximate visual and/or photographic-blue magnitudes. The *Sky Catalogue 2000.0*, volume 1, by Hirshfeld and Sinnott (Sky Publishing, 1982) lists 45,000 stars to V magnitude 8.0. The magnitudes used in this catalogue were computed from the visual magnitudes in the SAO *Catalogue*, systematically corrected as a function of spectral type to the standard UBV system. Since the spectral types are only approximately known for the majority of these stars, the derived magnitudes, though usefully accurate, are not precise. The best available information for more than 10,000 stars brighter than V = 7.1 can be found in *The Bright Star Catalogue* by Hoffleit and Jaschek (Yale University Observatory, 1982), and its *Supplement* by Hoffleit, Saladyga, and Wlasuk (Yale University Observatory, 1983).

Catalogue

The following catalogue contains data on 2828 deep-sky objects. All objects with descriptions or references in the text and all objects plotted on the charts are included. The list is complimented by a number of objects not described in this book above Declination −45°. The object names are sorted in alphanumeric order, except that NGC objects come first.

Preceding the catalogue designation in the first column is a two-letter abbreviation that indicates the type of object: eg for galaxies, oc for open clusters, gc for globular clusters, pn for planetary nebulae, gn for diffuse nebulae.

The second and third columns contain the equinox 1950.0 Right Ascension and Declination; columns four and five are the Right Ascension minutes and Declination arcminutes for equinox 2000.0. If the hour or degree of the coordinate has also changed, it is not so indicated: such cases should be evident from comparison with the adjacent entries in the catalogue.

The sixth column is the morphological type, as explained in the chapter on Deep-Sky Data Sources.

The size of the object appears in the seventh column. The default unit is the arcminute; when arcseconds are given they are so indicated.

The eighth column is the V magnitude (indicated by regular numerals: 10.7), or the visual magnitude for planetaries (italic numerals: *10.7*).

The ninth column is the three-letter abbreviation for the constellation in which the object lies, while the tenth column lists the page number on which the description (indicated by regular numerals: 123) or reference (italic numerals: *123*) appears.

The last column indicates Messier objects, popular names, group or cluster membership of galaxies, mildly interacting [P(b)] or strongly interacting [P(c)] galaxies (following de Vaucouleurs), misidentifications in older catalogues, names of stars illuminating diffuse nebulae and other information.

Figures 4 through 7 show histograms based on the objects included in the catalogue. Figure 4a shows the distribution of galaxies as a function of integrated V magnitude; the dashed line is for all galaxies included in the catalogue, while the solid line indicates those for which descriptions can be found in the body of the book. Of 2230 galaxies included in the catalogue, only 1012 have published V magnitudes and are included in the histogram. Among the brighter galaxies ($V < 11$), our sample is essentially complete for the northern hemisphere. Fainter than this the selection rapidly becomes incomplete. In addition to the obvious limitation of not being able to personally view every thirteenth magnitude galaxy, this reflects the limitations of basic astronomical knowledge of galaxies. Because photometry of faint galaxies is not complete, we cannot say presently how many galaxies are brighter than, say, $V = 13$. The major portion of the galaxies observed in this book were taken from the Shapley-Ames catalogue of 1930, the contents

Figure 4. The magnitude and size distribution of galaxies in this book (see text).

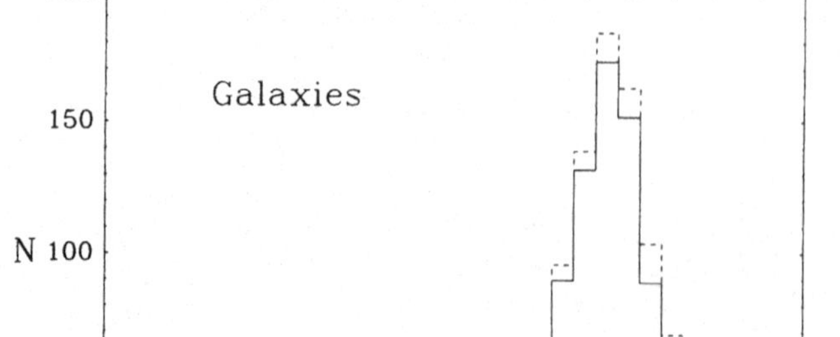
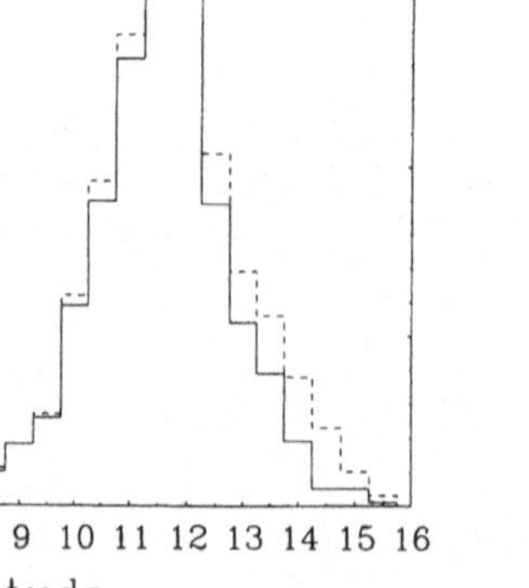

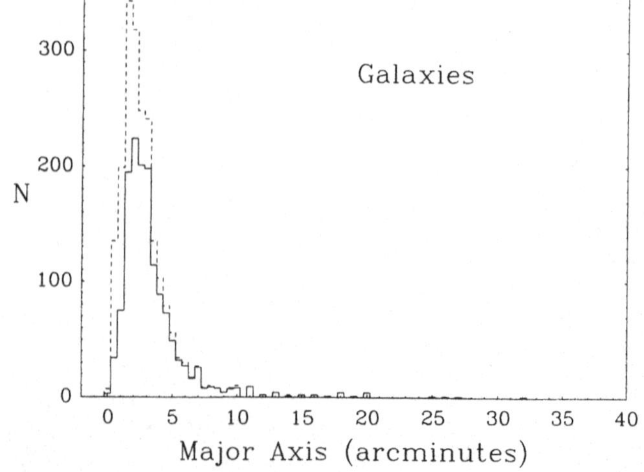

of which is plotted on the popular *Skalnate Pleso* and *Sky Atlas 2000.0* atlases. The Shapley-Ames catalogue, containing 1250 objects, was intended to be complete to photographic-blue magnitude 13.0, but de Vaucouleurs has shown that it is only 50 percent complete even at mag. 12.5 (corresponding to about V = 12.0 in Figure 4a). These factors together account for the rapid fall-off in numbers of galaxies after V = 11.5. The declining portion of the graph is comprised largely of faint companions near brighter galaxies, and members of clusters of galaxies such as those described under Coma Berenices and Perseus.

Figure 4b shows the distribution of the lengths of the major axes among galaxies, with the solid and dashed lines indicating the described and catalogued samples as before. Here the data are more complete: sizes are available for 2058 of the 2230 galaxies listed. The peak for the observed galaxies, at 2'.0, corresponds closely to the typical size of those at the peak of the magnitude distribution at V = 11.5. Note that the saw-tooth appearance of this histogram above 10 arcminutes is an artifact due to the bin size of the plot combined with the precision to which the data are expressed in the catalogue: sizes smaller than ten arcminutes are expresses to tenths of an arcminute; those larger are given only to whole arcminutes. This shows also in the other plots, though for the open clusters sizes larger than five arcminutes are given to whole arcminutes, those smaller to tenths.

Figures 5, 6, and 7 show the magnitude and size distributions for open clusters, globular clusters, and planetary nebulae, respectively. (No histogram of the sizes of planetary nebulae is plotted because of the lack of any consistent set of sizes for these objects: see the discussion under Deep-Sky Data Sources.) All the magnitude histograms show broader distributions than that of the galaxies, and brighter peak magnitudes. This results in part because the sample available for viewing from Earth is restricted by interstellar absorption in the Milky Way galaxy. Open star clusters and planetary nebulae are blocked from view beyond a relatively close distance,

Figure 5. The magnitude and size distribution of open clusters in this book (see text).

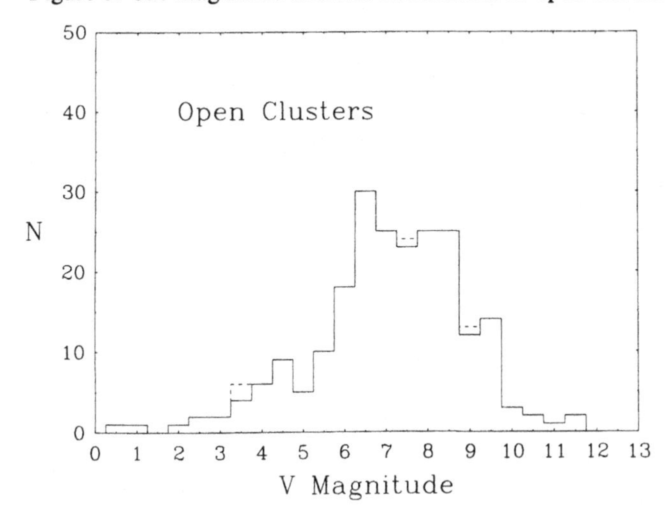

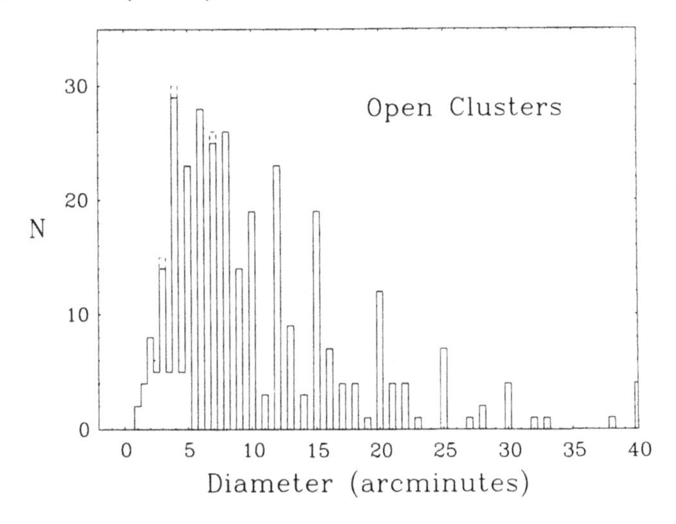

Figure 6. The magnitude and size distribution of globular clusters in this book (see text).

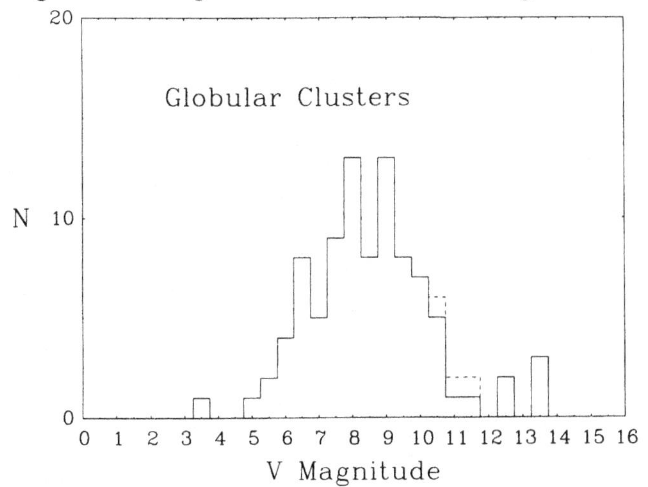

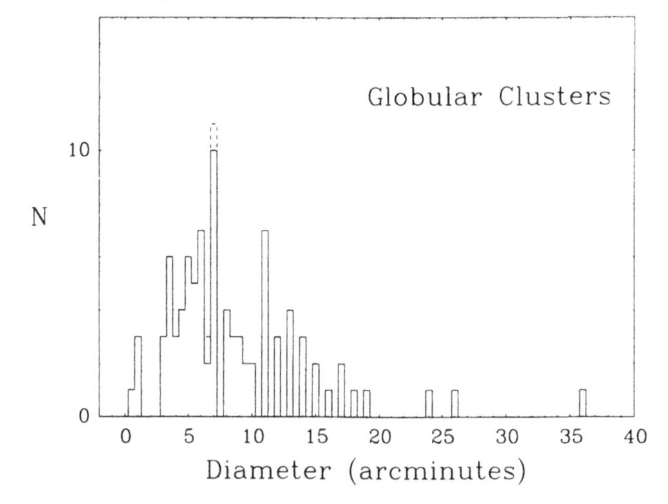

whereas galaxies, comparatively unaffected by absorption in most of the sky, fade gradually with distance. For these Milky Way objects, then, we see only the bright end of the total sample in the Galaxy.

The distribution of globular clusters with magnitude reflects more accurately the true distribution, as they show less correlation between distance and amount of obscuration. In addition, the sample of globular clusters is more nearly complete than that of the open clusters or planetary nebulae. The total number of open clusters in the Galaxy probably exceeds several thousand, of which only 351 are included in the catalogue (and only about 1200 are known); if the disk of the Milky Way could be removed, perhaps 40 or 50 new globular clusters would be found in addition to the 150 or so known today, 90 of which are included in the catalogue (excluding the four listed in the Fornax Dwarf, eg E356-G04).

Figure 7. The magnitude distribution of planetary nebulae in this book (see text).

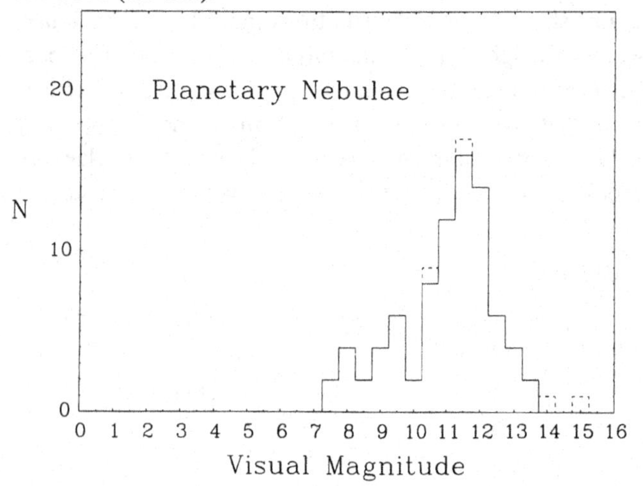

Finally, the described objects are displayed on an all-sky map on the endpapers of the book. Galaxies are plotted as filled circles, open clusters as stars, globular clusters as crossed circles, planetary nebulae as empty circles, and galactic nebulae as empty squares. Several features can be readily picked out. The Milky Way stands out prominently as a curving band of open clusters, diffuse and planetary nebulae. The southernmost portion, between 9 and 16 hours Right Ascension, is cut off by the horizon from northern latitudes. Another straight band composed entirely of galaxies stretches nearly north-south from Ursa Major to Centaurus at about 12 hours. This is the equatorial plane of the "Local Supercluster" of galaxies dominated by the Virgo cluster, centered near 12 hours and +10° Declination. The Coma cluster shows as a dense clump at about 13 hours and +30°. Other galaxy clusters are also evident: Perseus I at about 3 hours and +42°; Fornax at 4 hours and −35°. One can also see that absorption in the Milky Way causes a wide "zone of avoidance" on either side of it: few galaxies appear within 30° of the galactic plane.

	ID	RA (1950)	DEC	(2000)		TYPE	SIZE	V	CON	PG	NOTES C 01
eg	1	00 04.7	27 26	07.3	43	SA(s)b:	1.9x1.5	12.9	Peg	181	
eg	2	00 04.7	27 24	07.3	41	Sa	1.4x0.9	14.0	Peg	*181*	
eg	16	00 06.5	27 27	09.1	44	SAB0⁻sp	2.1x1.2	12.0	Peg	181	
eg	23	00 07.3	25 39	09.9	56	SB(s)a	2.3x1.6	12.0	Peg	181	
eg	24	00 07.4	−25 15	09.9	58	SA(s)c	5.5x1.6	11.5	Scl	221	in Sculptor group?
pn	40	00 10.3	72 15	13.0	32		38"x35"	*12.4*	Cep	69	
eg	45	00 11.5	−23 28	14.0	11	SA(s)dm	8.1x5.8	10.4	Cet	72	in Sculptor group?
eg	55	00 12.4	−39 28	14.9	11	SB(s)m:sp	32x6.5		Scl	221	brightest in Sculptor group; includes I1537
eg	95	00 19.7	10 13	22.3	30	SAB(rs)cP	1.9x1.5	12.6	Psc	197	
oc	103	00 22.5	61 04	25.3	21	II 1 m	5	9.8	Cas	62	
eg	125	00 26.3	02 34	28.9	51	SA0/a:	2.0x1.8	12.3	Psc	197	in 128 group
eg	126	00 26.6	02 32	29.2	49	SB0°:	1.4x0.8		Psc	*197*	in 128 group
eg	127	00 26.6	02 36	29.2	53	SA0°:	1.0x0.8	14.0	Psc	*197*	in 128 group
eg	128	00 26.7	02 35	29.3	52	S0Psp	3.4x1.0	11.6	Psc	197	brightest of group
oc	129	00 27.1	59 57	29.9	14	III 2 m	21	6.5	Cas	62	
eg	130	00 26.7	02 36	29.3	53	SA0⁻:	0.9x0.6	14.3	Psc	*197*	in 128 group
eg	131	00 27.2	−33 32	29.7	15	SB(s)b:sp	2.1x0.6		Scl	221	
oc	133	00 28.4	63 05	31.2	22	IV 1 p	7		Cas	62	doubtful cluster
eg	134	00 27.9	−33 31	30.4	14	SAB(s)bc	8.1x2.6	10.1	Scl	221	
oc	136	00 28.7	61 15	31.5	32	II 1 p	1.2		Cas	62	
oc	146	00 30.2	63 01	33.1	18	II 2 p	7	8.4	Cas	62	
eg	147	00 30.5	48 14	33.2	31	E5P	13x8.1	9.5	Cas	62	in Local Group
eg	148	00 31.8	−32 04	34.3	47	S0⁺:sp	2.4x1.2	12.1	Scl	221	
eg	150	00 31.8	−28 05	34.3	48	SB(rs)b:	4.2x2.3	11.1	Scl	221	
eg	151	00 31.5	−09 59	34.0	42	SB(r)bc	3.7x1.9	11.5	Cet	72	=153
eg	157	00 32.2	−08 40	34.7	23	SAB(rs)bc	4.3x2.9	10.4	Cet	72	
eg	175	00 34.9	−20 13	37.4	57	SB(r)ab	2.6x2.4	12.1	Cet	72	
eg	178	00 36.6	−14 27	39.1	11	SB(s)m	2.0x1.2	12.6	Cet	72	=I39; in 210 group
eg	185	00 36.2	48 04	39.0	20	E3P	11x9.8	9.2	Cas	62	in Local Group
oc	188	00 39.4	85 04	44.4	20	I 2 r	14	8.1	Cep	69	
oc	189	00 36.7	60 48	39.6	04	III 1 p	3.7	8.8	Cas	62	
eg	205	00 37.6	41 25	40.3	41	E5P	17x9.8	8.0	And	16	=M110; in Local Group; P(b) w/ 224
eg	210	00 38.1	−14 09	40.6	53	SAB(s)b	5.4x3.7	10.9	Cet	72	brightest of group
eg	214	00 38.8	25 14	41.5	30	SAB(r)c	2.1x1.6	12.2	And	16	
eg	221	00 40.0	40 36	42.7	52	cE2	7.6x5.8	8.2	And	16	=M32; in Local Group
eg	224	00 40.0	41 00	42.7	16	SA(s)b	180x63	3.5	And	16	=M31; brightest in Local Group
oc	225	00 40.5	61 31	43.4	47	III 1 p	12	7.0	Cas	63	
eg	227	00 40.1	−01 48	42.7	32	E3?	2.1x1.7		Cet	72	
eg	237	00 40.9	−00 24	43.5	08	SAB(rs)cd	1.8x1.2		Cet	72	
eg	245	00 43.5	−02 00	46.1	44	SA(rs)bP?	1.4x1.3		Cet	72	
pn	246	00 44.5	−12 09	47.0	53		4.6x4.1	*10.9*	Cet	72	
eg	247	00 44.7	−21 02	47.2	46	SAB(s)d	20x7.4	8.8	Cet	73	in Sculptor group
eg	253	00 45.1	−25 34	47.6	18	SAB(s)c	25x7.4	7.1	Scl	221	in Sculptor group
eg	254	00 45.0	−31 42	47.4	26	SAB(r)0/a?	2.1x1.1	11.8	Scl	221	
eg	255	00 45.3	−11 45	47.8	29	SAB(rs)bc	3.1x2.8	11.8	Cet	73	
eg	268	00 47.6	−05 28	50.1	12	SAB(rs)b?	1.7x1.3		Cet	73	
eg	273	00 48.3	−07 09	50.8	53	S0sp	2.6x0.9		Cet	73	in group w/ 274–5
eg	274	00 48.5	−07 20	51.0	04	SAB(r)0⁻P	1.7x1.6		Cet	73	P(b) w/ 275
eg	275	00 48.6	−07 20	51.1	04	SB(rs)cdP	1.5x1.2	12.5	Cet	73	P(b) w/ 274
eg	278	00 49.3	47 17	52.1	33	SAB(rs)b	2.2x2.1	10.9	Cas	63	

ID		RA (1950) DEC (2000)				TYPE	SIZE	V	CON	PG	NOTES C 02
oc	281	00 49.9	56 21	52.8	37		4.0		Cas	63	assoc. w/ gn
gn	281	00 50.4	56 19	53.3	35	e	35x30		Cas	63	assoc. w/ oc
gc	288	00 50.4	−26 52	52.8	36	X	14	8.1	Scl	222	
eg	289	00 50.3	−31 29	52.7	13	SB(rs)bc	3.7x2.7		Scl	222	
eg	300	00 52.5	−37 57	54.9	41	SA(s)d	20x15		Scl	222	in Sculptor group
eg	309	00 54.2	−10 11	56.7	55	SAB(r)c	3.1x2.7	11.9	Cet	73	
eg	337	00 57.3	−07 51	59.8	35	SB(s)d	2.8x2.0	11.6	Cet	73	
eg	337A	00 59.0	−07 51	01.5	35	SAB(s)dm	5.8x4.6		Cet	73	
eg	355	01 00.6	−06 36	03.1	20	S0sp	1.4x0.3		Cet		
eg	357	01 00.8	−06 36	03.3	20	SB(r)0/a:	2.6x1.9	11.8	Cet	73	brightest of group
oc	366	01 03.3	61 58	06.4	14	II 3 m	3.0		Cas	63	
oc	381	01 05.2	61 19	08.4	35	III 1 m	6		Cas	63	
eg	404	01 06.7	35 27	09.5	43	SA(s)0⁻:	4.4x4.2	10.1	And	18	in Local Group?
eg	428	01 10.4	00 43	13.0	59	SAB(s)m	4.1x3.2	11.4	Cet	73	
oc	433	01 12.1	59 52	15.3	08	III 2 p	4.0		Cas	*63*	
oc	436	01 12.5	58 33	15.6	49	I 2 m	6	8.8	Cas	63	
eg	439	01 11.5	−32 01	13.8	45	SA(r)0°?	1.8x1.1		Scl	222	
eg	441	01 11.5	−32 03	13.8	47	SB0/a	1.1x0.8		Scl	222	
eg	442	01 12.1	−01 17	14.7	01	S0/asp	1.2x0.8		Cet	*74*	
eg	450	01 13.0	−01 08	15.6	52	SAB(s)cd:	3.2x2.6	12.1	Cet	73	
oc	457	01 15.9	58 04	19.1	20	II 3 r	12	6.4	Cas	63	
eg	467	01 16.6	03 02	19.2	18	SA(s)0°P?	2.4x2.3	11.9	Psc	197	in group w/ 470/74
eg	470	01 17.2	03 09	19.8	25	SA(rs)b	3.0x2.0	11.9	Psc	197	in group w/ 467/74
eg	473	01 17.3	16 17	20.0	33	SAB(r)0/a:	2.2x1.4		Psc	197	
eg	474	01 17.5	03 09	20.1	25	(R')SA(s)0°	7.9x7.2	11.1	Psc	197	in group w/ 467/70
eg	488	01 19.2	05 00	21.8	16	SA(r)b	5.2x4.1	10.3	Psc	197	
eg	491	01 19.0	−34 20	21.3	04	SB(r)b?	1.5x1.3		Scl	222	brightest of group
eg	514	01 21.4	12 40	24.0	56	SAB(rs)c	3.5x2.9	11.9	Psc	197	
eg	516	01 21.5	09 18	24.1	34	SBb?sp	1.6x0.6		Psc	197	
eg	520	01 22.0	03 32	24.6	48	P	4.8x2.1	11.2	Psc	197	
eg	521	01 22.0	01 28	24.6	44	SB(r)bc	3.4x3.2	11.7	Cet	74	
eg	524	01 22.2	09 17	24.8	33	SA(rs)0⁺	3.2x3.2	10.6	Psc	198	
eg	532	01 22.7	09 00	25.3	16	Sb:sp	2.8x1.0		Psc	198	
eg	533	01 23.0	01 30	25.6	46	E3:	3.7x2.6		Cet	74	
oc	559	01 26.1	63 03	29.5	18	I 1 m	4.4	9.5	Cas	64	
eg	578	01 28.1	−22 55	30.5	40	SAB(rs)c	4.8x3.2	10.9	Cet	74	
oc	581	01 29.9	60 27	33.2	42	II 2 m	6	7.4	Cas	64	=M103
eg	584	01 28.8	−07 08	31.3	53	E4	3.8x2.4	10.4	Cet	74	=I1712
eg	586	01 29.1	−07 09	31.6	54	SA(s)a:?	1.6x0.8	13.2	Cet	*74*	
eg	596	01 30.4	−07 17	32.9	02	E2P	3.5x2.2	10.9	Cet	74	
eg	598	01 31.1	30 24	33.9	39	SA(s)cd	62x39	5.7	Tri	236	=M33; in Local Group
eg	613	01 32.0	−29 41	34.3	26	SB(rs)bc	5.8x4.6	10.0	Scl	222	
eg	615	01 32.6	−07 36	35.1	21	SA(rs)b	4.0x1.7	11.5	Cet	74	
eg	625	01 32.9	−41 42	35.1	27	SB(s)m?sp	3.0x1.3		Phe	196	
eg	628	01 34.0	15 32	36.7	47	SA(s)c	10x9.6	9.2	Psc	198	=M74
eg	636	01 36.6	−07 46	39.1	31	E3	2.3x1.9	11.3	Cet	74	
oc	637	01 39.4	63 45	42.9	00	I 2 m	3.5	8.2	Cas	64	
pn	650	01 39.2	51 19	42.4	34		2.7x1.8	*10.1*	Per	188	=M76
oc	654	01 40.6	61 38	44.1	53	II 2 r	8	6.5	Cas	64	
oc	659	01 40.8	60 27	44.2	42	I 2 m	5	7.9	Cas	64	

	ID	RA (1950)	DEC	(2000)		TYPE	SIZE	V	CON	PG	NOTES C 03
eg	660	01 40.4	13 23	43.1	38	SB(s)aP	9.1x4.1	10.7	Psc	198	P(b) w/ U 1195
eg	661	01 41.4	28 27	44.2	42	SA0:	2.2x1.8		Tri	236	
oc	663	01 42.6	61 00	46.0	15	II 3 r	8	7.1	Cas	64	
eg	669	01 44.3	35 18	47.2	33	Sab	3.4x0.8		Tri	236	
eg	670	01 44.6	27 38	47.4	53	SA0	2.5x1.1	12.1	Tri	238	
eg	672	01 45.1	27 11	47.9	26	SB(s)cd	6.6x2.7	10.8	Tri	238	P(b?) w/ I1727
eg	676	01 46.3	05 40	48.9	55	S0/a:sp	4.3x1.5		Psc	198	
eg	678	01 46.7	21 45	49.5	00	SB(s)b?sp	5.0x1.1		Ari	32	
eg	680	01 47.0	21 43	49.8	58	SA(s)0$^-$:P	2.9x2.5		Ari	32	in 697 group
eg	681	01 46.7	−10 41	49.2	26	SAB(s)absp	2.8x1.8	11.8	Cet	74	
eg	684	01 47.4	27 24	50.2	39	Sbsp	3.9x0.9		Tri	238	=I165
eg	691	01 47.9	21 31	50.7	46	SA(rs)bc	3.5x2.7		Ari	32	
eg	693	01 47.9	05 54	50.5	09	S0/a?	2.7x1.4		Psc	198	
eg	694	01 48.2	21 45	51.0	00	S0?P	0.8x0.5		Ari	32	
eg	697	01 48.5	22 07	51.3	22	SAB(r)bc:	4.7x1.8		Ari	32	brightest of group
eg	701	01 48.6	−09 57	51.1	42	SB(rs)c	2.5x1.3	12.2	Cet	74	
eg	706	01 49.2	06 03	51.8	18	Sb:	2.2x1.7		Psc	198	
eg	718	01 50.6	03 57	53.2	12	SAB(s)a	2.8x2.5	11.7	Psc	198	
eg	720	01 50.6	−13 59	53.0	44	E5	4.4x2.8	10.2	Cet	74	
eg	735	01 53.7	33 56	56.6	11	Sb	2.0x1.0		Tri	238	
eg	736	01 53.8	32 48	56.7	03	(R')SA0$^-$:	2.0x1.9	12.2	Tri	238	
eg	740	01 54.0	32 46	56.9	01	SBb?	1.7x0.5		Tri		
eg	741	01 53.7	05 23	56.3	38	E0:	3.2x3.2	11.3	Psc	198	brightest of group
eg	742	01 53.9	05 23	56.5	38	cE0:	0.2x0.2		Psc	198	
oc	743	01 55.2	59 56	58.7	11	IV 1 p	5		Cas	64	
oc	744	01 55.1	55 14	58.4	29	III 1 p	5	7.9	Per	188	
eg	750	01 54.6	32 58	57.5	13	EP	1.6x1.3	12.2	Tri	238	P(c) w/ 751
eg	751	01 54.6	32 58	57.5	13	EP	1.3x1.3	12.5	Tri	238	P(c) w/ 750
oc	752	01 54.8	37 26	57.8	41	III 1 m	45	5.7	And	18	
eg	753	01 54.8	35 40	57.8	55	SAB(rs)bc	2.9x2.1	12.4	And	18	
eg	761	01 54.9	33 08	57.8	23	SBab?	1.6x0.6		Tri	238	
eg	770	01 56.5	18 43	59.2	58	E3:	1.3x1.0		Ari	32	P(b)? w/ 772
eg	772	01 56.6	18 46	59.3	01	SA(s)b	7.1x4.5	10.3	Ari	32	P(b)? w/ 770
eg	777	01 57.4	31 11	00.3	26	E1	3.0x2.5		Tri	238	
eg	778	01 57.4	31 03	00.3	18	S0:	1.4x0.6		Tri	238	
eg	779	01 57.2	−06 12	59.7	57	SAB(r)b	4.1x1.4	11.0	Cet	74	
eg	783	01 58.2	31 38	01.1	52	SAB(s)bc	1.8x1.6		Tri	238	=I1765
eg	784	01 58.4	28 36	01.3	50	SBd:sp	6.2x1.7	11.8	Tri	239	
eg	785	01 58.8	31 35	01.7	49	E/S0	1.8x1.1		Tri	239	=I1766
eg	788	01 58.6	−07 03	01.1	49	SA(s)0/a:	1.8x1.5	12.1	Cet	75	
eg	821	02 05.7	10 46	08.4	00	E6?	3.5x2.2	10.8	Ari	32	
eg	864	02 12.8	05 46	15.4	00	SAB(rs)c	4.6x3.5	11.0	Cet	75	
oc	869	02 15.5	56 55	19.0	09	I 3 r	18	3.5	Per	188	=h Persei
eg	870	02 14.4	14 19	17.1	33	cE	0.1x0.1		Ari	32	
eg	871	02 14.5	14 19	17.2	33	SB(s)c:	1.3x0.5	13.5	Ari	32	
eg	876	02 15.2	14 17	17.9	31	SAc:sp	2.1x0.5		Ari	32	
eg	877	02 15.3	14 19	18.0	33	SAB(rs)bc	2.3x1.8	11.8	Ari	32	
oc	884	02 18.9	55 53	22.4	07	I 3 r	18	3.6	Per	188	=chi Persei
eg	890	02 19.0	33 02	22.0	16	SAB(r)0$^-$?	2.9x1.9	11.5	Tri	239	
eg	891	02 19.4	42 07	22.5	21	SA(s)b?sp	13x2.8	10.0	And	18	in 1023 group

	ID	RA (1950)	DEC	(2000)		TYPE	SIZE	V	CON	PG	NOTES C 04
eg	895	02 19.1	−05 45	21.6	31	SA(s)cd	3.6x2.8	11.8	Cet	75	
eg	908	02 20.8	−21 28	23.1	14	SA(s)c	5.5x2.8	10.2	Cet	75	
eg	922	02 22.8	−25 01	25.1	47	SB(s)cdP	1.9x1.8	12.2	For	116	colliding pair?
eg	925	02 24.3	33 21	27.3	34	SAB(s)d	9.8x6.0	10.0	Tri	239	in 1023 group
eg	936	02 25.1	−01 23	27.6	10	SB(rs)0^+	5.2x4.4	10.1	Cet	75	
eg	941	02 25.9	−01 23	28.4	10	SAB(rs)c	2.8x2.1	12.5	Cet	75	
eg	949	02 27.7	36 55	30.8	08	SA(rs)b:?	2.8x1.7	11.9	Tri	240	
eg	955	02 28.0	−01 20	30.5	07	Sab:sp	3.0x0.9	12.0	Cet	75	
oc	957	02 30.0	57 19	33.6	32	III 2 m	10	7.6	Per	188	
eg	958	02 28.2	−03 10	30.7	57	SB(s)c:	2.8x1.1	12.2	Cet	75	
eg	959	02 29.3	35 17	32.3	30	Sm:	2.2x1.4	12.4	Tri	240	
eg	972	02 31.3	29 06	34.2	19	I0	3.6x2.0	11.3	Ari	32	
eg	976	02 31.2	20 45	34.0	58	SA(rs)c:	1.7x1.5	12.4	Ari	32	
eg	986	02 31.6	−39 15	33.6	02	SB(rs)ab	3.7x2.8	11.0	For	116	
eg	991	02 33.1	−07 22	35.6	09	SAB(rs)c	2.7x2.5		Cet	75	in 1052 group
eg	1003	02 36.1	40 39	39.3	52	SA(s)cd	5.4x2.1	11.5	Per	188	in 1023 group
eg	1022	02 36.1	−06 53	38.6	40	(R')SB(s)a	2.5x2.1	11.4	Cet	75	in 1052 group
eg	1023	02 37.3	38 51	40.4	04	SB(rs)0^-	8.7x3.3	9.2	Per	188	brightest of group
oc	1027	02 38.8	61 20	42.7	33	II 3 m	20	6.7	Cas	65	
eg	1035	02 37.0	−08 21	39.5	08	SA(s)c?	2.2x0.9		Cet	75	in 1052 group
oc	1039	02 38.8	42 34	42.0	47	II 3 r	25	5.2	Per	189	=M34
eg	1042	02 37.9	−08 39	40.4	26	SAB(rs)cd	4.7x3.9	10.8	Cet	76	≠1048; in 1052 group
gc	1049	02 37.7	−34 28	39.8	15	V	0.8	12.6	For	116	
eg	1052	02 38.6	−08 28	41.1	15	E4	2.9x2.0	10.5	Cet	76	brightest of group
eg	1055	02 39.2	00 14	41.8	27	SBb:sp	7.6x3.0	10.6	Cet	76	in 1068 group
eg	1058	02 40.4	37 08	43.5	21	SA(rs)c	3.0x2.9	11.5	Per	189	in 1023 group
eg	1068	02 40.1	−00 14	42.7	01	(R)SA(rs)b	6.9x5.9	8.8	Cet	76	=M77; brightest of group; Seyfert
eg	1073	02 41.1	01 10	43.7	23	SB(rs)c	4.9x4.6	11.0	Cet	76	in 1068 group
eg	1079	02 41.6	−29 13	43.8	00	(R)SAB(rs)0/aP	3.1x1.9	11.4	For	116	
eg	1084	02 43.5	−07 47	46.0	34	SA(s)c	2.9x1.5	10.6	Eri	110	
eg	1087	02 43.9	−00 43	46.5	30	SAB(rs)c	3.5x2.3	11.0	Cet	76	in 1068 group
eg	1090	02 44.0	−00 27	46.6	14	SB(rs)bc	3.8x1.8	11.9	Cet	76	in 1068 group
eg	1097	02 44.2	−30 29	46.3	16	SB(s)b	9.3x6.6	9.3	For	116	
eg	1097A	02 44.0	−30 26	46.1	13	E5P:	0.7x0.4	12.9	For	*116*	
eg	1140	02 52.1	−10 14	54.5	02	Im	1.4x0.9	12.5	Eri	110	
eg	1156	02 56.8	25 02	59.7	14	IB(s)m	3.1x2.3	11.7	Ari	32	
eg	1160	02 57.9	44 46	01.2	58	Scd:	1.6x0.8		Per	189	
eg	1161	02 57.9	44 43	01.2	55	S0	3.1x2.2		Per	189	
eg	1169	03 00.2	46 11	03.6	23	SAB(r)b	4.4x3.0	11.7	Per	189	
eg	1172	02 59.3	−15 02	01.7	50	E1:	2.2x2.0	12.0	Eri	110	
eg	1175	03 01.3	42 09	04.6	21	SA(r)0^+	2.5x0.8	12.8	Per	189	
eg	1179	03 00.4	−19 06	02.7	54	SAB(r)cd	4.6x3.9		Eri	110	
eg	1186	03 02.2	42 39	05.5	51	SB(r)bc:	3.3x1.4		Per	189	
eg	1187	03 00.4	−23 04	02.6	52	SB(r)c	5.0x4.1		Eri	110	
oc	1193	03 02.5	44 11	05.8	23	I 2 m	3.0		Per	189	
eg	1199	03 01.3	−15 49	03.6	37	E3:	2.2x1.9	11.5	Eri	110	in 1209 group
eg	1201	03 02.0	−26 16	04.2	04	SA(r)0°:	4.4x2.8	10.6	For	116	
eg	1209	03 03.7	−15 48	06.0	36	E6:	2.6x1.5	11.4	Eri	110	brightest of group
oc	1220	03 08.0	53 09	11.7	20	I 1 p	2.0		Per	189	
eg	1224	03 08.0	41 11	11.3	22	$S0^-$:	1.7x1.5		Per	190	in Perseus cluster

ID		RA (1950) DEC (2000)				TYPE	SIZE	V	CON	PG	NOTES C 05
eg	1232	03 07.5	−20 46	09.8	35	SAB(rs)c	7.8x6.9	9.9	Eri	110	
eg	1232A	03 07.8	−20 47	10.1	36	SB(s)m	1.0x0.9		Eri	*110*	
eg	1241	03 08.8	−09 07	11.2	56	SB(rs)b	3.0x1.9		Eri	110	
eg	1242	03 08.9	−09 06	11.3	55	SB(rs)c:	1.3x0.7		Eri	110	
oc	1245	03 11.2	47 04	14.7	15	II 2 r	10	8.4	Per	190	
eg	1250	03 12.1	41 10	15.4	21	S0$^+$:sp	2.7x1.1		Per	190	in Perseus cluster; chart XX
eg	1255	03 11.4	−25 55	13.6	44	SAB(rs)bc	4.1x2.8	11.1	For	116	
eg	1259	03 14.0	41 12	17.3	23	E0	1.1x1.1	14.0	Per	190	chart XX
eg	1260	03 14.2	41 13	17.5	24	S0$^+$?sp	1.7x0.8	13.2	Per	190	in Perseus cluster; chart XX
eg	1264	03 14.7	41 20	18.0	31	SBab	1.3x1.2		Per		=M+07-07-50; chart XX
eg	1265	03 15.0	41 41	18.3	52	SA0$^-$	2.1x1.9		Per		in Perseus cluster; chart XX
eg	1267	03 15.4	41 17	18.7	28	SA0$^-$:	1.4x1.1	13.0	Per	190	in Perseus cluster; chart XX
eg	1268	03 15.5	41 18	18.8	29	SA(rs)b:	1.2x0.9	14.2	Per	190	in Perseus cluster; chart XX
eg	1270	03 15.7	41 17	19.0	28	E3:	1.1x0.9	12.9	Per	190	in Perseus cluster; chart XX
eg	1271	03 15.9	41 10	19.2	21	SB0?		13.9	Per	190	in Perseus cluster; chart XX
eg	1272	03 16.0	41 19	19.3	30	SA(s)0$^-$	2.5x2.3	12.5	Per	190	in Perseus cluster; chart XX
eg	1273	03 16.1	41 22	19.4	33	SA(r)0°?	1.4x1.4	12.9	Per	190	in Perseus cluster; chart XX
eg	1274	03 16.4	41 22	19.7	33	S0sp	0.6x0.4	14.0	Per	190	in Perseus cluster; chart XX
eg	1275	03 16.5	41 20	19.8	31	P	2.6x1.9	11.6	Per	191	≠1270; Seyfert; brightest in Perseus cl.; chart XX
eg	1277	03 16.6	41 24	19.9	35	S0$^+$:P	0.8x0.3	13.5	Per	191	in Perseus cluster; chart XX
eg	1278	03 16.6	41 23	19.9	34	E2P:	1.7x1.4	12.6	Per	191	in Perseus cluster; chart XX
eg	1279	03 16.7	41 18	20.0	29	S0	1.0x0.8		Per		chart XX
eg	1281	03 16.8	41 27	20.1	38	E5	0.7x0.5	13.5	Per	191	in Perseus cluster; chart XX
eg	1282	03 16.9	41 11	20.2	22	E1:	1.7x1.3	12.9	Per	191	in Perseus cluster; chart XX
eg	1283	03 17.0	41 13	20.3	24	E1:	1.3x0.8	13.6	Per	191	in Perseus cluster; chart XX
eg	1288	03 15.2	−32 46	17.2	35	SA(s)bc	2.3x2.2	12.1	For	117	
eg	1291	03 15.5	−41 17	17.3	06	(R)SB(s)0/a	10x9.1	8.5	Eri		=1269
eg	1292	03 16.1	−27 48	18.2	37	SA(s)c	3.2x1.7		For	117	
eg	1293	03 18.3	41 13	21.6	24	E0	1.1x1.1	13.3	Per	191	in Perseus cluster; chart XX
eg	1294	03 18.4	41 11	21.7	22	SA0$^-$:	1.8x1.6	13.1	Per	191	in Perseus cluster; chart XX
eg	1297	03 17.0	−19 17	19.3	06	SA0$^-$:	2.3x2.0		Eri	111	
eg	1300	03 17.4	−19 36	19.7	25	SB(rs)bc	6.5x4.3	10.4	Eri	111	
eg	1302	03 17.7	−26 14	19.8	03	(R)SB(s)0/a	4.4x4.2		For	117	
eg	1309	03 19.8	−15 35	22.1	24	SA(s)bc:	2.3x2.2	11.6	Eri	111	
eg	1315	03 20.9	−21 33	23.1	22	SB(rs)0$^+$?	1.8x1.8		Eri		
eg	1316	03 20.8	−37 23	22.7	12	(R')SAB(s)0°P	7.1x5.5	8.8	For	117	in Fornax I cluster
eg	1317	03 20.8	−37 17	22.7	06	(R')SAB(rs)a	3.2x2.8	11.0	For	117	=1318; in Fornax I cluster
eg	1319	03 21.7	−21 42	23.9	31	S0sp	1.3x0.7		Eri		
eg	1325	03 22.2	−21 43	24.4	32	SA(s)bc	4.6x1.8	11.6	Eri	111	
eg	1325A	03 22.6	−21 31	24.8	20	SB(r)b?	2.6x2.4	12.8	Eri		
eg	1326	03 22.0	−36 38	23.9	27	(R)SB(r)0$^+$	4.0x3.0	10.5	For	117	in Fornax I cluster
eg	1331	03 24.3	−21 32	26.5	22	E2:	1.1x1.0		Eri	111	=I 324
eg	1332	03 24.1	−21 31	26.3	21	S(s)0$^-$:sp	4.6x1.7	10.3	Eri	111	
gn	1333	03 26.1	31 12	29.2	22	c	9x7		Per	191	
eg	1337	03 25.7	−08 34	28.1	24	SA(s)cd	6.8x2.0	11.6	Eri	111	
eg	1339	03 26.1	−32 28	28.1	18	E4	2.3x1.9		For	117	
oc	1342	03 28.4	37 10	31.6	20	III 2 m	17	6.7	Per	191	
eg	1344	03 26.3	−31 14	28.3	04	(R')SAB(rs)0°	3.9x2.3	10.3	For	117	=1340
eg	1350	03 29.2	−33 48	31.2	38	(R')SB(r)ab	4.3x2.4	10.5	For	117	in Fornax I cluster
eg	1351	03 28.6	−35 01	30.5	51	SA0$^-$:	1.8x1.2	11.7	For	117	in Fornax I cluster; chart X

	ID	RA (1950)	DEC	(2000)		TYPE	SIZE	V	CON	PG	NOTES C 06
eg	1353	03 29.8	−20 59	32.0	49	SB(rs)b:	3.4x1.5	11.4	Eri	111	
eg	1355	03 30.9	−05 10	33.4	00	S0sp	1.8x0.5	13.1	Eri	111	
eg	1357	03 30.9	−14 00	33.2	50	SA(s)ab	2.5x1.8	11.7	Eri	111	
eg	1358	03 31.2	−05 16	33.7	06	SAB(r)0/a	2.8x2.1	12.1	Eri	111	
eg	1359	03 31.5	−19 40	33.7	30	SB(s)m?P	1.9x1.7	12.2	Eri	111	
pn	1360	03 31.1	−26 02	33.2	52		9x5	9.4	For	117	
eg	1365	03 31.7	−36 18	33.6	08	(R')SB(s)b	9.8x5.5	9.5	For	117	in Fornax I cluster?; chart X
eg	1366	03 31.9	−31 22	33.9	12	S0$^+$sp	2.7x1.2		For	117	
eg	1369	03 34.9	−36 25	36.8	15	SB(rs)0$^+$	1.4x1.3		Eri	112	=E358−G34; =M−06−09−4; chart X
eg	1371	03 32.9	−25 06	35.0	56	SAB(rs)a	5.4x4.0		For	117	=1367
eg	1373	03 33.1	−35 20	35.0	10	S0$^-$:	0.9x0.7		For	118	chart X
eg	1374	03 33.4	−35 24	35.3	14	SA(rs)0$^-$:	1.8x1.8	11.2	For	118	in Fornax I cluster; chart X
eg	1375	03 33.4	−35 26	35.3	16	SAB0$^-$:sp	1.9x0.6	12.2	For	118	in Fornax I cluster; chart X
eg	1376	03 34.6	−05 13	37.1	03	SA(s)cd	2.0x1.9		Eri	111	
eg	1379	03 34.1	−35 36	36.0	26	E0	2.0x1.9	11.1	For	118	in Fornax I cluster; chart X
eg	1380	03 34.5	−35 08	36.4	58	SB(s)0$^-$	4.9x1.9	10.2	For	118	in Fornax I cluster; chart X
eg	1380A	03 34.9	−34 54	36.8	44	S0°:sp	2.8x0.7		For	118	in Fornax I cluster; chart X
eg	1381	03 34.6	−35 28	36.5	18	SA0sp	2.9x0.8	11.6	For	118	in Fornax I cluster; chart X
eg	1382	03 35.2	−35 22	37.1	12	SAB(s)0$^-$	0.8x0.8	12.9	For	118	=1380B; chart X
eg	1385	03 35.3	−24 40	37.4	30	SB(s)cd	3.0x2.0	11.2	For	118	
eg	1386	03 34.9	−36 10	36.8	00	S0°	3.5x1.5		Eri	112	in Fornax I cluster; chart X
eg	1387	03 35.0	−35 40	36.9	30	SAB(s)0$^-$	2.4x2.2		For	118	in Fornax I cluster; chart X
eg	1389	03 35.3	−35 55	37.2	45	SAB(s)0$^-$	2.1x1.4	11.5	Eri	112	in Fornax I cluster; chart X
eg	1395	03 36.3	−23 11	38.5	01	E2P	3.2x2.5		Eri	112	
eg	1396	03 36.1	−35 37	38.0	27	SAB0$^-$:			For		chart X
eg	1398	03 36.8	−26 30	38.9	20	(R')SB(r)ab	6.6x5.2	9.7	For	118	
eg	1399	03 36.6	−35 37	38.5	27	E+1	3.2x3.1	9.8	For	118	in Fornax I cluster; chart X
eg	1400	03 37.3	−18 51	39.6	41	SA0$^-$	1.9x1.7	11.1	Eri	112	
eg	1401	03 37.2	−22 53	39.4	43	SB(s)0°:sp	2.8x0.7		Eri	112	
eg	1404	03 37.0	−35 45	38.9	35	E1	2.5x2.3	10.1	Eri	112	in Fornax I cluster; chart X
eg	1406	03 37.4	−31 29	39.4	19	SB(s)bc?sp	3.9x1.0		For	120	
eg	1407	03 38.0	−18 45	40.3	35	E$^+$0	2.5x2.5	9.8	Eri	112	brightest of group
eg	1411	03 37.1	−44 16	38.8	06	S(r)0$^-$:	2.8x2.3	11.3	Hor		=I1943
eg	1415	03 38.8	−22 43	41.0	33	(R)SAB(s)0/a	3.6x2.1		Eri	112	=I1983
eg	1417	03 39.5	−04 52	42.0	42	SAB(rs)b	2.8x1.9	12.0	Eri	112	
eg	1418	03 39.8	−04 54	42.3	44	SB(s)b:	1.5x1.2		Eri	112	
eg	1421	03 40.2	−13 39	42.5	30	SAB(rs)bc:	3.6x1.0	11.2	Eri	112	
eg	1425	03 40.2	−30 03	42.2	54	SA(s)b	5.4x2.7	11.0	For	120	in Fornax I cluster?
eg	1426	03 40.6	−22 16	42.8	07	(R')SA(rs)0$^-$:	2.1x1.5	11.4	Eri	112	
eg	1427	03 40.4	−35 33	42.3	24	SA0$^-$:	2.8x2.0	11.0	For	120	in Fornax I cluster; chart X
eg	1427A	03 38.3	−35 47	40.2	37	IB(s)m	2.2x1.5		Eri		chart X
eg	1428	03 40.5	−35 19	42.4	10	SAB0$^-$P:	1.8x0.8		For		in Fornax I cluster; chart X
eg	1433	03 40.5	−47 23	42.1	14	SB(r)a	6.8x6.0	10.0	Hor		
gn	1435	03 43.2	23 36	46.2	45	c	30x30		Tau	233	=Merope; assoc. w/ oc Pleiades
eg	1437	03 41.7	−36 01	43.6	52	SB(rs)a	2.9x2.1		Eri	112	=1436; in Fornax I cluster
eg	1439	03 42.7	−22 05	44.9	56	E1	2.3x2.2		Eri	112	
eg	1440	03 42.8	−18 25	45.1	16	SB(rs)0°?	2.3x1.9		Eri	113	=1442
oc	1444	03 45.6	52 31	49.4	40	IV 1 p	4.0	6.6	Per	191	
eg	1448	03 42.9	−44 48	44.5	39	SAcd:sp	8.1x1.8		Hor		=1457
eg	1452	03 43.1	−18 47	45.3	38	(R')SB(r)0/a	1.7x1.5		Eri	113	=1455

ID		RA (1950) DEC (2000)				TYPE	SIZE	V	CON	PG	NOTES C 07
eg	1453	03 44.0	−04 08	46.5	59	E2.5	2.1x1.6	11.6	Eri	113	brightest of group
eg	1461	03 46.2	−16 33	48.5	24	SA(r)0°	3.3x1.1	11.7	Eri	113	
eg	1487	03 54.1	−42 31	55.8	22	Im:P	2.0x1.4		Eri		P(c?) at 0.4
gn	1491	03 59.5	51 10	03.3	18	e	9x6		Per	193	
eg	1493	03 55.9	−46 21	57.5	12	SB(r)cd	2.6x2.3		Hor		
oc	1496	04 00.6	52 29	04.4	37	III 2 p	3.0		Per	193	
gn	1499	04 00.1	36 17	03.4	25	e	160x40		Per	193	
pn	1501	04 02.7	60 47	07.0	55		56"x48"	*11.5*	Cam	40	
oc	1502	04 03.3	62 12	07.7	20	I 3 m	8	5.7	Cam	40	
eg	1507	04 01.9	−02 20	04.4	12	SB(s)mP?	3.4x1.0	12.2	Eri	113	
eg	1510	04 01.9	−43 32	03.5	24	SA0°?P	1.0x0.9	13.0	Hor	*128*	P(b) w/ 1512
eg	1512	04 02.3	−43 29	03.9	21	SB(r)a	4.0x3.2	10.6	Hor	128	P(b) w/ 1510
oc	1513	04 06.3	49 23	10.0	31	II 1 m	12	8.4	Per	193	
pn	1514	04 06.1	30 39	09.2	47		2.3x2.0	*10.9*	Tau	233	
eg	1518	04 04.7	−21 19	06.9	11	SB(s)dm	3.0x1.4	11.8	Eri	113	
eg	1521	04 06.1	−21 11	08.3	03	SAB0°	2.9x1.9	11.4	Eri	113	
oc	1528	04 11.6	51 07	15.4	14	II 2 m	18	6.4	Per	193	
eg	1530	04 17.1	75 11	23.5	18	SB(rs)b	4.9x2.9		Cam	41	
eg	1531	04 10.1	−32 59	12.0	51	S0°P	1.3x0.8	12.1	Eri	113	P(b) w/ 1532
eg	1532	04 10.2	−33 00	12.1	52	SB(s)ab?sp	5.6x1.8		Eri	113	P(b) w/ 1531
pn	1535	04 12.0	−12 52	14.3	45		48"x42"	*9.6*	Eri	113	
eg	1537	04 11.7	−31 46	13.6	38	SAB0°P:	2.4x1.6	10.7	Eri	113	
oc	1545	04 17.1	50 08	20.9	15	IV 2 p	12	6.2	Per	194	
eg	1569	04 26.1	64 44	30.8	51	IBm	2.9x1.5	11.2	Cam	41	
gn	1579	04 26.9	35 10	30.2	16	c	12x8		Per	194	
eg	1587	04 28.1	00 33	30.7	39	E1P	2.0x1.9	11.6	Tau	233	P(b) w/ 1588
eg	1588	04 28.2	00 33	30.8	39	E1P:	1.8x1.0	13.0	Tau	233	P(b) w/ 1587
eg	1589	04 28.2	00 45	30.8	51	Sabsp	3.1x1.2		Tau	234	
eg	1590	04 28.5	07 31	31.2	37	P	1.3x1.2		Tau	234	
eg	1600	04 29.2	−05 12	31.7	06	E3	2.5x1.8	10.9	Eri	114	brightest of group
eg	1601	04 29.4	−05 12	31.9	06	S0:sp	1.0x0.7	13.8	Eri	114	in 1600 group
eg	1603	04 29.4	−05 12	31.9	06	(R)SA0			Eri	*114*	
eg	1615	04 33.1	19 51	36.0	57	SA0⁻:	1.6x0.9		Tau	234	
eg	1618	04 33.6	−03 15	36.1	09	SB(r)b:	2.8x1.2	12.6	Eri	114	
eg	1622	04 34.1	−03 17	36.6	11	SAB(r)ab:	3.9x1.1	12.3	Eri	114	
gn	1624	04 36.6	50 21	40.4	27	e	5x5		Per	195	
oc	1624	04 36.6	50 21	40.4	27	II 1 p	4.0	10.4	Per	195	
eg	1625	04 34.6	−03 24	37.1	18	SB(rs)b:	2.7x0.8	12.7	Eri	114	
eg	1637	04 39.0	−02 57	41.5	51	SAB(rs)c	3.3x2.9	11.0	Eri	114	
eg	1638	04 39.1	−01 54	41.6	48	SAB(rs)0°?	2.5x1.9	12.1	Eri	114	
eg	1640	04 40.1	−20 32	42.3	26	SB(r)b	2.8x2.0	11.7	Eri	114	
eg	1642	04 40.3	00 32	42.9	38	SA(rs)c:	2.0x1.8		Tau	234	
oc	1647	04 43.1	18 59	46.0	04	II 2 r	40	6.4	Tau	234	
eg	1659	04 44.0	−04 53	46.5	48	SA(r)bcP	1.7x1.3	12.5	Eri	114	
oc	1662	04 45.7	10 51	48.5	56	II 3 m	12	6.4	Ori	176	
oc	1664	04 47.5	43 37	51.1	42	III 1 p	13	7.6	Aur	33	
eg	1666	04 46.1	−06 39	48.5	34	SB(r)0⁺	1.3x1.2		Eri	*114*	
eg	1667	04 46.2	−06 24	48.6	19	SAB(r)c	1.5x1.3	12.1	Eri	114	
eg	1699	04 54.5	−04 50	57.0	45	SA(rs)b	1.1x0.7		Eri	*115*	
eg	1700	04 54.5	−04 57	57.0	52	E4	2.9x2.0	11.0	Eri	114	

ID		RA (1950)	DEC	(2000)		TYPE	SIZE	V	CON	PG	NOTES C 08
eg	1720	04 56.9	−07 56	59.3	52	SB(s)ab	1.8x1.2		Eri	*115*	
eg	1726	04 57.3	−07 50	59.7	46	SA(s)0°:	1.4x1.1	11.7	Eri	115	
eg	1729	04 57.8	−03 25	00.3	21	SA(rs)bc	1.6x1.4		Ori	176	
eg	1740	04 59.4	−03 22	01.9	18	E			Ori	176	
eg	1741AB	04 59.1	−04 20	01.6	16	see note		13.4	Eri	115	P(c); SB(s)mP, SB(s)m:sp; combined V; P(b) w/ I 399
eg	1744	04 57.9	−26 06	59.9	02	SB(s)d	6.8x4.1	11.2	Lep	153	
oc	1746	05 00.6	23 45	03.6	49	III 2 p	40		Tau	234	
oc	1778	05 04.7	36 59	08.1	03	III 2 p	10	7.7	Aur	33	
eg	1784	05 03.1	−11 56	05.4	52	SB(r)c	4.2x2.8	11.8	Lep	153	
gn	1788	05 04.5	−03 24	07.0	20	c	8x5		Ori	176	
eg	1792	05 03.5	−38 03	05.2	59	SA(rs)bc	4.0x2.1	10.2	Col	77	
oc	1798	05 08.1	47 34	11.9	38	I 1 m	6		Aur	33	
eg	1800	05 04.5	−32 01	06.4	57	IB(s)m	1.6x0.9	12.6	Col	77	
oc	1807	05 07.8	16 28	10.7	32	II 2 p	12	7.0	Tau	234	doubtful cluster
eg	1808	05 06.0	−37 35	07.7	31	(R)SAB(s)a	7.2x4.1	9.9	Col	77	
oc	1817	05 09.2	16 38	12.1	42	IV 2 r	16	7.7	Tau	235	
eg	1832	05 09.8	−15 45	12.1	41	SB(r)bc	2.8x1.9	11.4	Lep	153	
gc	1851	05 12.5	−40 06	14.1	03	II	11	7.2	Col	77	
oc	1857	05 16.7	39 18	20.2	21	I 3 m	6	7.0	Aur	33	
eg	1875	05 19.1	06 38	21.8	41	SA0⁻:			Ori	176	
oc	1883	05 22.2	46 30	25.9	33	II 1 m	2.5		Aur	33	
oc	1893	05 19.4	33 21	22.7	24	II 3 r	12	7.5	Aur	33	asso. w/ gn I 405
gc	1904	05 22.1	−24 34	24.2	31	V	8.7	7.8	Lep	153	=M79
oc	1907	05 24.7	35 17	28.0	19	I 1 m	7	8.2	Aur	35	
oc	1912	05 25.3	35 48	28.7	50	II 2 r	21	6.4	Aur	35	=M38
gn	1931	05 28.1	34 13	31.4	15	ce	4x4		Aur	35	asso. w/ oc
oc	1931	05 28.1	34 13	31.4	15	I 3 p	1.0	11.3	Aur	35	asso. w/ gn
gn	1952	05 31.5	21 59	34.5	01	e(SNR)	6x4		Tau	235	=M1; Crab; supernova AD 1054
eg	1954	05 30.5	−14 06	32.8	04	SA(rs)c	4.1x2.3		Lep	153	
eg	1957	05 30.7	−14 09	33.0	07	E0			Lep	153	
oc	1960	05 32.8	34 06	36.1	08	I 3 r	12	6.0	Aur	35	=M36
eg	1961	05 36.6	69 21	42.1	23	SAB(rs)c	4.3x3.0	11.2	Cam	41	=I2133
eg	1964	05 31.3	−21 59	33.4	57	SAB(s)b	6.2x2.5	10.7	Lep	153	
gn	1976	05 32.9	−05 25	35.4	23	e	90x60		Ori	176	=M42; asso. w/ oc
oc	1976	05 32.9	−05 25	35.4	23		45	3.7	Ori	176	incl. Trapezium; =M42; asso. w/ gn
oc	1977	05 32.8	−04 50	35.3	48		20	4.2	Ori	177	asso. w/ gn
gn	1977	05 33.0	−04 54	35.5	52	ce	40x25		Ori	177	asso. w/ oc
gn	1980	05 33.0	−05 56	35.4	54	e	14x14		Ori	177	asso. w/ oc
oc	1980	05 32.9	−05 58	35.3	56	III 3 m	20	2.5	Ori	177	asso. w/ gn
oc	1981	05 32.7	−04 28	35.2	26	III 3 p	28	4.6	Ori	177	
gn	1982	05 33.1	−05 18	35.6	16	ce	20x15		Ori	178	=M43
gn	1999	05 34.1	−06 45	36.5	43	c	2x2		Ori	178	
oc	2017	05 37.1	−17 53	39.3	51				Lep	153	
pn	2022	05 39.4	09 04	42.1	05		28"x27"	*11.9*	Ori	178	
gn	2023	05 39.2	−02 15	41.7	14	ce	10x8		Ori	178	
gn	2024	05 39.4	−01 52	41.9	51	e	30x30		Ori	178	
gn	2068	05 44.2	00 02	46.8	03	c	8x6		Ori	179	=M78
gn	2071	05 44.6	00 17	47.2	18	c	7x5		Ori	179	
eg	2090	05 45.2	−34 16	47.0	15	SA(rs)c	6.5x3.0		Col	77	
oc	2099	05 49.1	32 32	52.4	33	I 2 r	20	5.6	Aur	35	=M37

ID		RA (1950)	DEC (1950)	RA (2000)	DEC (2000)	TYPE	SIZE	V	CON	PG	NOTES C 09
oc	2112	05 51.3	00 23	53.9	24	II 2 m	18		Ori	179	
oc	2126	05 59.1	49 54	03.0	54	III 2 m	6		Aur	35	
oc	2129	05 58.0	23 18	01.0	18	I 3 m	7	6.7	Gem	121	
eg	2139	05 59.1	−23 40	01.2	40	SAB(rs)cd	2.2x1.9	11.7	Lep	154	=I2154
oc	2141	06 00.3	10 26	03.1	26	I 2 r	10	9.4	Ori	179	
eg	2146	06 10.8	78 23	18.8	22	SB(s)abP	6.0x3.8	10.5	Cam	41	P(b?) w/ 2146A
eg	2146A	06 15.9	78 33	23.9	32	SAB(s)c:	3.2x1.4		Cam	*41*	P(b?) w/ 2146
oc	2158	06 04.4	24 06	07.5	06	II 3 r	5	8.6	Gem	121	
oc	2168	06 05.8	24 21	08.9	20	III 3 r	28	5.1	Gem	121	=M35
oc	2169	06 05.6	13 58	08.4	57	III 3 m	6	5.9	Ori	179	
gn	2170	06 05.2	−06 23	07.6	23	c	2x2		Mon	167	
gn	2175	06 06.7	20 31	09.7	30	e	40x30		Ori	179	assoc. w/ oc
oc	2175	06 06.8	20 20	09.8	19	III 2 r	22	6.8	Ori	179	assoc. w/ gn
eg	2179	06 05.9	−21 44	08.0	45	SA(s)0/a	1.5x1.0	12.5	Lep	154	
gn	2183	06 08.4	−06 12	10.8	13	c	2x2		Mon	167	
gn	2185	06 08.7	−06 12	11.1	13	c	2x2		Mon	167	
oc	2186	06 09.5	05 28	12.2	27	II 2 m	5	8.7	Ori	179	
eg	2188	06 08.4	−34 06	10.2	07	SB(s)msp	3.7x1.1	11.8	Col		
oc	2192	06 11.7	39 52	15.2	51	II 2 m	6		Aur	36	
oc	2194	06 11.0	12 49	13.8	48	II 2 r	8	8.5	Ori	180	
eg	2196	06 10.1	−21 48	12.2	49	(R')SA(s)a	2.8x2.2	11.2	Lep	154	
oc	2204	06 13.5	−18 38	15.7	39	II 2 r	13	8.6	CMa	55	
eg	2207	06 14.2	−21 21	16.3	22	SAB(rs)bcP	4.3x2.9	10.7	CMa	55	P(b) w/ I2163
oc	2215	06 18.6	−07 16	21.0	17	II 2 m	10	8.4	Mon	167	
eg	2217	06 19.7	−27 13	21.7	15	(R)SB(rs)0+	4.8x4.4	10.4	CMa	55	
eg	2223	06 22.5	−22 49	24.6	51	SAB(r)b	3.3x3.0	11.4	CMa	55	
oc	2232	06 24.1	−04 43	26.6	45	III 2 p	45	3.9	Mon	167	
oc	2236	06 27.0	06 52	29.7	50	II 2 m	9	8.5	Mon	167	
gn	2237	06 29.6	04 40	32.3	38	e	90x90		Mon	167	=Rosetta
oc	2243	06 27.9	−31 15	29.8	17	I 2 r	7	9.4	CMa	55	
oc	2244	06 29.7	04 54	32.4	52	II 3 r	27	4.8	Mon	167	asso. w/ gn 2237
gn	2245	06 29.9	10 12	32.7	10	ce	2x2		Mon	167	
gn	2247	06 30.3	10 23	33.1	21	ce	2x2		Mon	167	
oc	2251	06 32.0	08 24	34.7	22	III 2 m	7	7.3	Mon	169	
oc	2252	06 32.3	05 25	35.0	23	III 2 m	20		Mon	169	
oc	2254	06 33.3	07 43	36.0	40	I 1 m	4.0	9.1	Mon	169	
oc	2259	06 35.8	10 56	38.6	53	II 1 p	4.5		Mon	169	
gn	2261	06 36.4	08 46	39.1	43	ce	var		Mon	169	=Hubble's variable; asso. w/ R Mon
eg	2263	06 36.4	−24 48	38.5	51	(R)SB(r)ab	2.8x1.8		CMa	55	
oc	2264	06 38.3	09 56	41.1	53	III 3 m	20	3.9	Mon	169	asso. w/ gn
gn	2264	06 38.2	09 57	41.0	54	e	10x7		Mon	169	=Cone
oc	2266	06 40.1	27 01	43.2	58	II 2 m	7		Gem	121	
eg	2268	07 00.8	84 28	14.3	23	SAB(r)bc	3.4x2.2	11.5	Cam	41	
oc	2269	06 41.3	04 37	44.0	34	II 1 p	3.0	10.0	Mon	169	
eg	2276	07 10.5	85 51	27.3	45	SAB(rs)c	2.6x2.5	11.4	Cep	69	P(b) w/ 2300
eg	2280	06 42.8	−27 35	44.8	38	SA(s)cd	5.6x3.2		CMa	55	
oc	2281	06 45.8	41 07	49.3	04	I 3 m	15	5.4	Aur	36	
gn	2282	06 44.3	01 23	46.9	20	c	3x3		Mon	169	
eg	2283	06 43.7	−18 09	45.9	12	SB(s)cd	3.4x2.6		CMa	57	
oc	2286	06 45.1	−03 07	47.6	10	III 2 m	15	7.5	Mon	169	

ID		RA (1950)	DEC	(2000)		TYPE	SIZE	V	CON	PG	NOTES C 10
oc	2287	06 44.9	−20 41	47.0	44	I 3 r	38	4.5	CMa	57	=M41
eg	2292	06 45.7	−26 41	47.7	44	SAB0°P	3.0x2.8		CMa	57	P(b) w/ 2293/5; in common envelope w/ 2293
eg	2293	06 45.7	−26 42	47.7	45	SAB(s)0+P	3.0x2.8		CMa	57	P(b) w/ 2292/5; in common envelope w/ 2292
eg	2295	06 45.4	−26 41	47.4	44	Sab:sp	1.5x0.4		CMa	57	P(b) w/ 2292/3
gc	2298	06 47.2	−35 57	49.0	00	VI	6.8	9.2	Pup	201	
eg	2300	07 15.8	85 49	32.4	43	SA0°	3.1x2.6	10.8	Cep	69	P(b) w/ 2276
oc	2301	06 49.2	00 32	51.8	28	I 3 r	15	6.0	Mon	169	
oc	2302	06 49.5	−07 00	51.9	04	III 2 m	5	8.9	Mon	170	
oc	2304	06 52.1	18 05	55.0	01	II 1 m	5		Gem	121	
oc	2309	06 53.8	−07 08	56.2	12	I 2 m	3.0		Mon	170	
eg	2310	06 52.3	−40 48	53.9	52	S0sp	5.0x1.2		Pup		
oc	2311	06 55.3	−04 31	57.8	35	III 2 m	7		Mon	170	
eg	2314	07 03.9	75 24	10.5	19	E3	2.1x1.8	12.1	Cam	41	
oc	2323	07 00.8	−08 16	03.2	20	II 3 r	16	5.9	Mon	170	=M50
oc	2324	07 01.6	01 08	04.2	03	II 2 r	8	8.4	Mon	170	
eg	2325	07 00.7	−28 38	02.7	42	SAB(s)0−:	2.3x1.5	11.2	CMa	58	
gn	2327	07 01.9	−11 14	04.2	19				CMa	58	
oc	2335	07 04.2	−10 00	06.6	05	III 2 m	12	7.2	Mon	170	
eg	2336	07 18.5	80 17	27.1	11	SAB(r)bc	6.9x4.0	10.3	Cam	41	
eg	2339	07 05.4	18 52	08.3	47	SAB(rs)bc	2.8x2.1	11.6	Gem	121	
eg	2341	07 06.3	20 40	09.3	35	SP	1.1x1.0		Gem	121	
eg	2342	07 06.4	20 43	09.4	38	SAB(rs)bcP	1.6x1.5		Gem	121	
oc	2343	07 05.9	−10 34	08.3	39	II 2 p	13	6.7	Mon	171	
oc	2345	07 06.0	−13 05	08.3	10	II 3 r	12	7.7	CMa	58	
pn	2346	07 06.8	−00 43	09.3	48		60"x50"	*11.8*	Mon	171	
eg	2347	07 11.3	64 48	16.1	43	(R')SA(r)b:	2.0x1.5	12.5	Cam	41	
eg	2350	07 10.4	12 21	13.2	16	S0/a	1.6x0.9		CMi	60	
oc	2353	07 12.2	−10 13	14.6	18	III 3 p	20	7.1	Mon	171	
oc	2354	07 12.2	−25 39	14.3	44	III 2 r	20	6.5	CMa	58	
oc	2355	07 14.1	13 52	16.9	47	II 2 m	9		Gem	122	
gn	2359	07 15.4	−13 07	17.7	12	e	10x5		CMa	58	includes gn I 468
oc	2360	07 15.5	−15 32	17.8	37	I 3 r	13	7.2	CMa	59	
oc	2362	07 16.7	−24 51	18.8	57	I 3 r	8	4.1	CMa	59	
eg	2366	07 23.6	69 19	28.9	13	IB(s)m	7.6x3.5	10.7	Cam	41	
oc	2367	07 18.0	−21 50	20.1	56	II 3 m	3.5	7.9	CMa	59	
oc	2368	07 18.6	−10 17	21.0	23	IV 1 p	5		Mon	171	
pn	2371−2	07 22.4	29 35	25.5	29		2.2x0.9	*11.3*	Gem	122	
oc	2374	07 21.7	−13 10	24.0	16	IV 2 p	19	8.0	CMa	59	
eg	2377	07 22.6	−09 34	25.0	40	SA(s)bc:	1.8x1.6		Mon	171	
eg	2380	07 21.9	−27 26	23.9	32	SAB0°:	1.2x1.2		CMa	59	=2382
oc	2383	07 22.6	−20 50	24.8	56	II 3 m	6	8.4	CMa	59	
oc	2384	07 22.9	−20 56	25.1	02	IV 3 p	2.5	7.4	CMa	59	
pn	2392	07 26.2	21 01	29.2	55		47"x43"	*9.2*	Gem	123	
oc	2394	07 25.9	07 08	28.6	02				CMi	60	doubtful cluster
oc	2395	07 24.3	13 41	27.1	35	IV 2 m	12	8.0	Gem	123	
oc	2396	07 25.8	−11 38	28.1	44	IV 1 m	10		CMa	59	
oc	2401	07 27.1	−13 52	29.4	58	I 1 p	2.0		Pup	201	
eg	2403	07 32.1	65 43	36.9	36	SAB(s)cd	18x11	8.4	Cam	41	in 3031 group
oc	2414	07 31.0	−15 20	33.3	27	I 3 m	4.0	7.9	Pup	201	
gc	2419	07 34.8	39 00	38.2	53		4.1	10.3	Lyn	159	

	ID	RA (1950)	DEC	(2000)		TYPE	SIZE	V	CON	PG	NOTES
oc	2420	07 35.5	21 41	38.5	34	I 1 r	10	8.3	Gem	123	
oc	2421	07 34.1	−20 30	36.3	37	I 2 r	8	8.3	Pup	201	
oc	2422	07 34.3	−14 23	36.6	30	I 3 m	25	4.4	Pup	201	=M47
oc	2423	07 34.8	−13 45	37.1	52	II 2 m	12	6.7	Pup	201	
oc	2425	07 36.0	−14 15	38.3	22	II 1 m	5		Pup	202	
oc	2432	07 38.7	−18 58	40.9	05	II 2 m	7		Pup	202	
oc	2437	07 39.5	−14 42	41.8	49	II 2 r	20	6.1	Pup	202	=M46
pn	2438	07 39.5	−14 37	41.8	44		73"x68"	*11.0*	Pup	202	
oc	2439	07 38.9	−31 32	40.8	39	II 3 r	9	6.9	Pup	202	
pn	2440	07 39.7	−18 05	41.9	12		74"x42"	*9.4*	Pup	202	
eg	2441	07 46.3	73 09	52.1	01	SAB(r)b:	2.2x2.0	12.2	Cam	42	
oc	2447	07 42.5	−23 45	44.6	52	I 3 r	10	6.2	Pup	203	=M93
oc	2451	07 43.6	−37 51	45.4	58	II 2 m	50	2.8	Pup	203	
pn	2452	07 45.4	−27 13	47.4	20		31"x24"	*12.2*	Pup	203	
oc	2453	07 45.7	−27 07	47.8	15	I 3 m	4.8	8.3	Pup	203	
oc	2455	07 46.8	−21 10	49.0	18	III 2 m	15		Pup	203	
eg	2460	07 52.6	60 29	56.9	21	SA(s)a	2.9x2.2	11.7	Cam	42	
gn	2467	07 50.3	−26 19	52.4	27	e	8x7		Pup	204	asso. w/ oc
oc	2467	07 50.3	−26 19	52.4	27	I 3 m	15		Pup	204	asso. w/ gn
eg	2474	07 54.1	53 00	58.0	52	E0	0.6x0.6		Lyn	*159*	
eg	2475	07 54.2	53 00	58.0	52	E1	0.8x0.8		Lyn	*159*	
oc	2477	07 50.5	−38 25	52.3	33	I 2 r	20	5.8	Pup	204	
oc	2479	07 52.8	−17 35	55.0	43	III 1 m	11		Pup	204	
oc	2482	07 52.8	−24 10	54.9	18	IV 1 m	10	7.3	Pup	204	
eg	2485	07 54.1	07 37	56.8	29	SAa	1.7x1.7		CMi	60	
oc	2489	07 54.2	−29 56	56.2	04	I 2 m	6	7.9	Pup	204	
eg	2500	07 58.1	50 53	01.8	45	SB(rs)d	2.9x2.7	11.6	Lyn	159	
oc	2506	07 57.8	−10 39	00.2	47	I 2 r	10	7.6	Mon	171	
eg	2507	07 58.8	15 51	01.6	43	S0/aP	2.5x2.1		Cnc	44	
oc	2509	07 58.5	−18 56	00.7	04	I 1 r	12		Pup	204	
eg	2514	08 00.0	15 57	02.8	49	SB(s)bc	1.4x1.4		Cnc	44	
eg	2517	08 00.4	−12 11	02.8	19	SAB(s)0:	0.8x0.6	12.3	Pup	205	
eg	2523	08 09.3	73 44	15.0	35	SB(r)bc	3.0x2.0	12.0	Cam	42	
eg	2523B	08 07.2	73 43	13.0	34	SA(s)b:sp	2.3x0.4		Cam	*42*	
eg	2525	08 03.3	−11 17	05.7	26	SB(s)c	2.9x2.0	11.6	Pup	205	
oc	2527	08 03.2	−28 01	05.3	10	II 2 m	16	6.5	Pup	205	
oc	2533	08 05.0	−29 45	07.0	54	II 2 r	10	7.6	Pup	205	
eg	2535	08 08.2	25 21	11.2	12	SA(r)cP	3.0x1.7	12.6	Cnc	44	P(b) w/ 2536
eg	2536	08 08.3	25 20	11.3	11	SB(rs)cP	0.9x0.6	14.1	Cnc	44	P(b) w/ 2535
eg	2537	08 09.7	46 09	13.2	00	IB(s)mP	1.7x1.5	11.7	Lyn	159	
eg	2537A	08 10.2	46 09	13.7	00	SB(rs)c	0.7x0.7		Lyn		
oc	2539	08 08.4	−12 41	10.7	50	III 2 m	15	6.5	Pup	205	
eg	2541	08 11.0	49 13	14.6	04	SA(s)cd	6.6x3.5	11.8	Lyn	159	
eg	2545	08 11.3	21 31	14.2	22	(R)SB(r)ab	2.2x1.3	12.4	Cnc	44	in Cancer cluster
oc	2546	08 10.6	−37 29	12.4	38	III 2 m	70	6.3	Pup	205	
oc	2547	08 09.2	−49 07	10.7	16	II 2 p			Vel	256	
oc	2548	08 11.3	−05 39	13.8	48	I 3 r	54	5.8	Hya	129	=M48
eg	2549	08 15.0	57 58	19.0	49	SA(r)0°sp	4.2x1.5	11.1	Lyn	159	
eg	2551	08 19.2	73 35	24.8	25	SA(s)0/a	1.9x1.3	12.0	Cam	42	
eg	2552	08 15.7	50 10	19.4	01	SA(s)m?	3.2x2.2	12.2	Lyn	159	

ID		RA (1950) DEC (2000)				TYPE	SIZE	V	CON	PG	NOTES C 12
eg	2557	08 16.3	21 36	19.2	27	SB0	1.4x1.3		Cnc	44	in Cancer cluster
eg	2560	08 17.0	21 08	19.9	59	S0/a	1.7x0.5		Cnc	44	
eg	2562	08 17.5	21 17	20.4	07	S0⁺:	1.4x1.0	12.9	Cnc	44	in Cancer cluster
eg	2563	08 17.7	21 14	20.6	04	S0°:	2.3x1.8	12.3	Cnc	44	in Cancer cluster
eg	2566	08 16.6	−25 20	18.7	29	SB(r)ab:	3.1x2.0		Pup	205	P(b?) w/ I2311
oc	2567	08 16.6	−30 29	18.6	38	II 2 m	11	7.4	Pup	205	
oc	2571	08 16.9	−29 35	18.9	44	II 3 m	7	7.0	Pup	206	
oc	2580	08 19.6	−30 09	21.6	19	II 2 m	8		Pup	206	
eg	2583	08 20.7	−04 51	23.2	01	E1	0.7x0.7	13.5	Hya	129	
eg	2584	08 20.8	−04 50	23.3	00	Sd:	1.3x0.7		Hya	129	
eg	2585	08 21.0	−04 46	23.5	56	Sb:	2.2x0.8		Hya	129	
oc	2587	08 21.4	−29 20	23.5	30	III 2 m	10		Pup	206	
eg	2608	08 32.3	28 39	35.3	29	SB(s)b:	2.5x1.6	12.1	Cnc	44	
pn	2610	08 31.1	−15 59	33.4	09		50"x47"	12.8	Hya	129	
eg	2612	08 31.4	−12 59	33.8	09	SAab	3.2x0.7		Hya	129	
eg	2613	08 31.2	−22 48	33.4	58	SA(s)b	7.2x2.1	10.4	Pyx	207	
gn	2626	08 33.7	−40 30	35.5	40	c	5x5		Vel	256	
oc	2627	08 35.2	−29 46	37.3	57	II 2 r	9		Pyx	207	
oc	2632	08 37.2	20 10	40.1	59	II 3 m	95	3.1	Cnc	44	=M44; Praesepe
eg	2633	08 42.6	74 17	48.1	06	SB(s)b	2.6x1.7	11.9	Cam	42	brightest of group
eg	2634	08 42.9	74 09	48.4	58	E1:	2.2x2.1		Cam	42	
eg	2634A	08 43.2	74 07	48.7	56	SB(s)bc?sp	1.9x0.5		Cam	42	
oc	2635	08 36.5	−34 35	38.5	46	II 1 p	3.0		Pyx		
eg	2639	08 40.1	50 23	43.7	12	(R)SA(r)a:?	2.0x1.3	11.8	UMa	241	
eg	2642	08 38.2	−03 57	40.7	08	SB(r)bc	2.3x2.2		Hya	129	
eg	2646	08 45.0	73 39	50.4	28	SB(r)0°:	1.7x1.6	12.0	Cam	42	
eg	2654	08 45.2	60 24	49.2	13	SBab:sp	4.3x0.9	11.8	UMa	241	
eg	2655	08 49.1	78 25	55.6	14	SAB(s)0/a	5.1x4.4	10.1	Cam	42	
oc	2658	08 41.4	−32 28	43.4	39	I 2 r	10		Pyx	207	
oc	2659	08 40.9	−44 46	42.6	57	III 3 m		8.6	Vel	256	
oc	2671	08 44.4	−41 42	46.2	53	I 3 p			Vel	256	
eg	2672	08 46.5	19 16	49.3	05	E1.5	2.6x2.4	11.6	Cnc	45	brightest of group
eg	2673	08 46.6	19 16	49.4	05	E0P	1.4x1.4	12.9	Cnc	45	in 2672 group
eg	2681	08 50.0	51 30	53.6	19	(R')SAB(rs)0/a	3.8x3.5	10.3	UMa	241	
oc	2682	08 47.7	12 00	50.4	49	II 3 r	30	6.9	Cnc	45	=M67
eg	2683	08 49.6	33 37	52.7	26	SA(rs)b	9.3x2.5	9.7	Lyn	159	
eg	2685	08 51.7	58 56	55.6	45	(R)SB0P	5.2x3.0	11.1	UMa	241	
eg	2693	08 53.4	51 32	57.0	20	E3:	2.2x1.7	11.7	UMa	241	
eg	2701	08 55.4	53 58	59.1	46	SAB(rs)c:	2.1x1.4	12.4	UMa	241	
eg	2712	08 56.2	45 07	59.6	55	SB(r)b:	2.9x1.7	12.0	Lyn	159	
eg	2713	08 54.7	03 07	57.3	55	SB(rs)ab	3.9x1.7	11.7	Hya	129	
eg	2715	09 01.8	78 17	08.0	05	SAB(rs)c	5.0x1.9	11.4	Cam	42	
eg	2716	08 55.0	03 17	57.6	05	(R)SB(r)0⁺	1.8x1.4	12.4	Hya	129	
eg	2732	09 06.9	79 24	13.4	12	S0sp	2.3x0.9	11.9	Cam	42	
eg	2742	09 03.6	60 41	07.5	29	SA(s)c:	3.1x1.7	11.7	UMa	241	
eg	2744	09 01.8	18 40	04.6	28	SB(s)ab:P	1.8x1.2	13.4	Cnc	45	in 2749 group
eg	2748	09 08.0	76 41	13.7	29	SAbc	3.1x1.3	11.7	Cam	43	
eg	2749	09 02.5	18 31	05.3	19	E3	2.0x1.7	12.0	Cnc	45	brightest of group
eg	2751	09 02.7	18 27	05.5	15	(R)SB(r)0°	1.1x0.7		Cnc	45	in 2749 group
eg	2752	09 02.9	18 32	05.7	20	SBb?sp	2.0x0.5		Cnc	45	in 2749 group

ID		RA (1950)	DEC	(2000)		TYPE	SIZE	V	CON	PG	NOTES C 13
eg	2763	09 04.5	−15 18	06.8	30	SB(r)cdP	2.3x2.1	12.1	Hya	129	
eg	2764	09 05.4	21 39	08.3	27	S0:	1.7x1.0	12.7	Cnc	45	
eg	2768	09 07.8	60 15	11.7	03	E6:	6.3x2.8	10.0	UMa	241	
eg	2770	09 06.5	33 20	09.6	08	SA(s)c:	3.7x1.3		Lyn	161	
eg	2773	09 07.1	07 23	09.8	11	S0sp	1.0x0.5		Cnc	*45*	
eg	2775	09 07.7	07 15	10.4	03	SA(r)ab	4.5x3.5	10.3	Cnc	45	
eg	2776	09 08.9	45 10	12.2	58	SAB(rs)c	2.9x2.7	11.6	Lyn	161	
eg	2777	09 08.0	07 25	10.7	13	Sab?	1.0x0.7		Cnc	*45*	
eg	2781	09 09.1	−14 37	11.5	49	SAB(r)0+	3.9x1.9	11.5	Hya	130	
eg	2782	09 10.9	40 19	14.1	07	SAB(rs)aP	3.8x2.9	11.5	Lyn	161	
eg	2784	09 10.1	−23 58	12.3	10	SA(s)0°:	5.1x2.3	10.1	Hya		
eg	2787	09 14.8	69 25	19.3	12	SB(r)0+	3.4x2.3	10.8	UMa	242	
pn	2792	09 10.6	−42 13	12.5	25		13"x13"	*11.7*	Vel	256	
eg	2793	09 13.7	34 38	16.8	25	SB(s)mP	1.3x1.1		Lyn	161	
eg	2798	09 14.2	42 13	17.4	00	SB(s)aP	2.8x1.1	12.3	Lyn	161	P(b) w/ 2799
eg	2799	09 14.3	42 12	17.5	59	SB(s)m?	1.9x0.6		Lyn	161	P(b) w/ 2798
eg	2805	09 16.3	64 19	20.4	06	SAB(rs)d	6.3x5.0	11.3	UMa	242	P(b) w/ 2814, 2820, I2458
eg	2811	09 13.8	−16 06	16.1	19	SB(rs)a	2.7x1.0	11.3	Hya	130	
eg	2814	09 17.1	64 28	21.2	15	I0?	1.4x0.4	13.8	UMa	242	P(b) w/ 2805, 2820, I2548
eg	2815	09 14.1	−23 26	16.3	39	SB(r)b:	3.5x1.3		Hya		
oc	2818	09 14.2	−36 25	16.2	38	III 1 m	8	8.2	Pyx	207	
pn	2818A	09 14.0	−36 25	16.0	38		35"x35"	*11.9*	Pyx	207	
eg	2820	09 17.7	64 28	21.8	15	SB(s)cPsp	4.3x0.7	12.8	UMa	242	P(b) w/ 2805, 2814, I2458
eg	2823	09 16.2	34 13	19.3	00	SBa	1.0x0.6		Lyn		in Abell 779; chart XVI
eg	2825	09 16.3	33 57	19.3	44	Sa:sp	1.1x0.5		Lyn	161	chart XVI
eg	2826	09 16.4	33 50	19.4	37	S0:sp	1.8x0.4		Lyn		chart XVI
eg	2827	09 16.3	34 07	19.4	54	SA(r)a?			Lyn		chart XVI
eg	2828	09 16.4	34 07	19.5	54	S0			Lyn		chart XVI
eg	2830	09 16.7	33 57	19.7	44	SB0/a:sp	1.3x0.3		Lyn		in Abell 779; chart XVI
eg	2831	09 16.7	33 57	19.7	44	E0	0.8x0.8	13.6	Lyn	*163*	in Abell 779; chart XVI
eg	2832	09 16.8	33 58	19.8	45	E+2:	3.3x2.2	11.5	Lyn	161	in Abell 779; chart XVI
eg	2833	09 16.9	34 09	19.9	56	Sa?			Lyn		chart XVI
eg	2834	09 17.0	33 55	20.0	42	E0			Lyn		chart XVI
eg	2835	09 15.6	−22 09	17.9	22	SB(rs)c	6.3x4.4		Hya		
eg	2839	09 17.5	33 52	20.5	39	E			Lyn		chart XVI
eg	2841	09 18.6	51 11	22.0	58	SA(r)b:	8.1x3.8	9.3	UMa	242	
eg	2844	09 18.6	40 22	21.8	09	SA(r)a:	1.9x1.0	12.9	Lyn	163	brightest of group
eg	2848	09 17.8	−16 19	20.1	32	SAB(s)c:	2.7x1.8	12.1	Hya	130	
oc	2849	09 17.5	−40 19	19.4	32				Vel	256	
eg	2851	09 18.2	−16 16	20.5	29	SA(r)0-	1.7x0.7		Hya	*130*	
eg	2852	09 20.1	40 23	23.3	10	SAB(r)a?	1.2x1.2		Lyn	*163*	in 2844 group
eg	2853	09 20.1	40 25	23.3	12	SB0:	2.0x1.0		Lyn	*163*	in 2844 group
eg	2855	09 19.0	−11 42	21.4	55	(R)SA(rs)0/a	2.7x2.4	11.6	Hya	130	
eg	2859	09 21.3	34 44	24.3	31	(R)SB(r)0+	4.8x4.2	10.7	LMi	150	
eg	2865	09 21.3	−22 57	23.6	10	(R')SAB(rs)0°P	2.0x1.5	11.4	Hya		
eg	2872	09 23.0	11 39	25.7	26	E2	2.1x1.9		Leo	137	
eg	2873	09 23.1	11 40	25.8	27	S(r)c:	0.8x0.2		Leo	137	
eg	2874	09 23.1	11 39	25.8	26	SB(r)bc	2.5x0.9		Leo	137	
eg	2880	09 25.7	62 43	29.6	30	SB0-	2.6x1.6	11.6	UMa	242	
eg	2881AB	09 23.6	−11 47	26.0	00	P, P	1.5x1.2		Hya	*130*	interacting pair

ID		RA (1950)	DEC	(2000)		TYPE	SIZE	V	CON	PG	NOTES C 14
eg	2884	09 24.0	−11 20	26.4	33	S0/a?	2.6x1.3		Hya	*130*	
eg	2888	09 24.1	−27 49	26.3	02	SA0°:	0.8x0.7	12.5	Pyx	207	
eg	2889	09 24.8	−11 25	27.2	38	SAB(rs)c	2.0x1.8	11.8	Hya	130	
eg	2902	09 28.5	−14 31	30.9	44	SA(s)0°:	1.3x1.2		Hya	130	
eg	2903	09 29.3	21 43	32.1	30	SAB(rs)bc	13x6.6	8.9	Leo	137	
eg	2907	09 29.3	−16 31	31.7	44	SA(s)a?sp	1.9x1.3		Hya	130	
eg	2911	09 31.1	10 23	33.8	10	SA(s)0:P	4.3x3.2	11.6	Leo	137	
eg	2914	09 31.4	10 20	34.1	07	SB(s)ab	1.2x0.8	13.1	Leo	*137*	
eg	2916	09 32.1	21 56	34.9	43	SA(rs)b?	2.6x1.9	12.0	Leo	137	
eg	2919	09 32.1	10 31	34.8	18	SAB(r)b:	1.9x0.8		Leo	*137*	
eg	2921	09 32.2	−20 42	34.5	55	SAB(r)a	2.5x0.9		Hya	*130*	
eg	2924	09 32.8	−16 11	35.2	24	E0	1.6x1.5		Hya	130	
eg	2935	09 34.4	−20 54	36.7	07	(R')SAB(s)b	3.5x3.0		Hya	130	
eg	2942	09 36.1	34 14	39.1	00	SA(s)c:	2.2x1.8		LMi	150	
eg	2950	09 39.0	59 05	42.6	51	(R)SB(r)0°	3.2x2.1	11.0	UMa	243	
eg	2955	09 38.2	36 07	41.2	53	(R')SA(r)b	1.8x1.0	12.7	LMi	150	
eg	2962	09 38.3	05 24	40.9	10	(R)SAB(rs)0⁺	3.3x2.4	11.7	Hya	130	
eg	2964	09 39.9	32 05	42.9	51	SAB(r)bc:	3.0x1.7	11.3	Leo	137	brightest of group
eg	2967	09 39.5	00 34	42.1	20	SA(s)c	3.0x2.9	11.6	Sex	231	
eg	2968	09 40.2	32 10	43.2	56	I0	2.2x1.5	11.8	Leo	137	in 2964 group
eg	2970	09 40.6	32 13	43.6	59	E1:	1.0x0.9		Leo	*137*	in 2964 group
eg	2974	09 40.0	−03 28	42.5	42	E4	3.4x2.1	10.8	Sex	231	
eg	2976	09 43.2	68 09	47.3	55	SAcP	4.9x2.5	10.2	UMa	243	in 3031 group
eg	2983	09 41.4	−20 15	43.7	29	SB(rs)0⁺	2.6x1.8	11.7	Hya	130	
eg	2985	09 45.9	72 31	50.4	17	(R')SA(rs)ab	4.3x3.4	10.5	UMa	243	
eg	2986	09 42.0	−21 03	44.3	17	S0⁻:	2.5x2.3	10.9	Hya	130	
eg	2989	09 43.1	−18 09	45.5	23	SA(s)bc?	1.4x1.1		Hya		
eg	2990	09 43.7	05 56	46.3	42	Sc:	1.3x0.8	12.8	Sex	231	
eg	2992	09 43.3	−14 06	45.7	20	SaP	4.1x1.4	11.9	Hya	131	P(b) w/ 2993
eg	2993	09 43.4	−14 08	45.8	22	SaP	1.6x1.3	12.6	Hya	131	P(b) w/ 2992
eg	2997	09 43.5	−30 58	45.7	12	SAB(rs)c	8.1x6.5		Ant	21	
eg	2998	09 45.6	44 19	48.8	05	SAB(rs)c	3.0x1.5		UMa	243	brightest of group
eg	3001	09 44.1	−30 12	46.3	26	SAB(rs)bc:	3.1x2.1		Ant	21	
eg	3003	09 45.6	33 39	48.6	25	Sbc?	5.9x1.7	11.7	LMi	150	
eg	3016	09 47.2	12 56	49.9	42	SA(s)b	1.3x1.1		Leo	137	
eg	3019	09 47.4	12 58	50.1	44	SAB(s)b	1.0x0.7		Leo	137	
eg	3020	09 47.4	13 03	50.1	49	SB(r)cd:	3.2x1.8		Leo	137	
eg	3021	09 48.0	33 47	51.0	33	SA(rs)bc:	1.7x1.0		LMi	150	
eg	3024	09 47.8	13 00	50.5	46	SAd:sp	2.2x0.6		Leo	138	
eg	3027	09 51.3	72 26	55.7	12	SB(rs)d:	4.7x2.3		UMa	*243*	in 3031 group?
eg	3031	09 51.5	69 18	55.6	04	SA(s)ab	26x14	6.8	UMa	243	=M81; brightest of group
eg	3032	09 49.2	29 28	52.1	14	SAB(r)0°	2.5x2.1	11.9	Leo	138	
eg	3034	09 51.7	69 55	55.8	41	I0sp	11x4.6	8.4	UMa	243	=M82; in 3031 group
eg	3038	09 49.1	−32 31	51.3	45	SA(s)a	2.6x1.7		Ant	21	brightest of group
eg	3041	09 50.4	16 55	53.1	41	SAB(rs)c	3.7x2.5	11.5	Leo	138	
eg	3043	09 52.7	59 33	56.2	19	Sb:sp	2.0x0.7		UMa	244	
eg	3044	09 51.1	01 49	53.7	35	SB(s)c?sp	4.8x0.9	12.0	Sex	231	
eg	3045	09 50.9	−18 25	53.3	39	Sb:	1.0x0.5		Hya	*131*	
eg	3052	09 52.1	−18 24	54.5	38	SAB(r)c:	2.1x1.5		Hya	131	
eg	3054	09 52.2	−25 28	54.5	42	SAB(r)b	3.9x2.6		Hya		

ID		RA (1950)	DEC	(2000)		TYPE	SIZE	V	CON	PG	NOTES
eg	3055	09 52.7	04 30	55.3	16	SAB(s)c	2.2x1.4	12.1	Sex	231	
eg	3056	09 52.3	−28 04	54.5	18	SAB(rs)0^{+}:	2.0x1.4		Ant	21	
eg	3065	09 57.6	72 25	01.9	11	SA(r)0°	2.0x1.9	12.0	UMa	244	P(b) w/ 3066
eg	3066	09 57.9	72 22	02.2	08	(R')SAB(s)bcP	1.2x1.1	12.9	UMa	244	P(b) w/ 3065
eg	3067	09 55.4	32 37	58.3	23	SAB(s)ab?	2.5x1.0	11.7	Leo	138	
eg	3073	09 57.5	55 52	00.9	38	SAB0^{-}	1.5x1.4		UMa	*244*	
eg	3077	09 59.4	68 59	03.4	44	I0P	4.6x3.6	9.9	UMa	244	in 3031 group
eg	3078	09 56.1	−26 41	58.4	55	E2.5	1.9x1.5	11.1	Hya		
eg	3079	09 58.6	55 55	02.0	41	SB(s)csp	7.6x1.7	10.6	UMa	244	
eg	3081	09 57.2	−22 35	59.5	49	(R)SAB(r)0^{+}	2.2x1.8		Hya		=I2529
eg	3087	09 57.0	−33 59	59.2	13	SA(rs)0^{-}:	1.9x1.3		Ant	21	
eg	3089	09 57.4	−28 05	59.7	19	SAB(rs)b	1.9x1.3		Ant	21	
eg	3091	09 57.9	−19 24	00.3	38	SA(r)0^{-}:	2.2x1.7		Hya	131	brightest of group
eg	3095	09 57.9	−31 19	00.1	33	SAB(rs)c	3.2x2.2		Ant	21	
eg	3096	09 58.2	−19 25	00.6	39	SB(rs)0°	0.7x0.6		Hya	*131*	in 3091 group
eg	3098	09 59.5	24 57	02.3	43	S0sp	2.6x0.8	12.0	Leo	138	
eg	3100	09 58.5	−31 25	00.7	39	SAB(s)0°P	2.8x1.7		Ant	21	=3103
eg	3108	10 00.3	−31 26	02.5	41	SA(s)0^{+}	2.7x1.8		Ant	22	
eg	3109	10 00.8	−25 55	03.1	10	SB(s)msp	14x3.5		Hya		
eg	3115	10 02.7	−07 29	05.2	44	S0^{-}sp	8.3x3.2	9.1	Sex	231	
eg	3124	10 04.3	−18 59	06.7	14	SAB(rs)bc	3.2x2.7		Hya	131	
eg	3125	10 04.3	−29 42	06.6	57	I0?	1.5x0.8		Ant	22	
pn	3132	10 04.9	−40 11	07.0	26		84"x53"	*9.4*	Vel	256	
eg	3145	10 07.7	−12 11	10.1	26	SB(rs)bc	3.3x1.7		Hya		
eg	3147	10 12.7	73 39	16.9	24	SA(rs)bc	4.0x3.5	10.7	Dra	105	
eg	3153	10 10.2	12 55	12.9	40	Scd:	2.3x1.1		Leo	138	
eg	3156	10 10.1	03 23	12.7	08	S0:	2.1x1.2		Sex	232	in 3166 group
eg	3158	10 10.9	39 01	13.9	46	E3:	2.3x2.1	11.8	LMi	150	brightest of group
eg	3162	10 10.8	22 59	13.6	44	SAB(rs)bc	3.1x2.7	11.6	Leo	138	
eg	3163	10 11.2	38 54	14.2	39	SA0^{-}:	1.4x1.4	13.1	LMi	150	in 3158 group
eg	3165	10 10.9	03 38	13.5	23	SA(s)dm:	1.6x0.9		Sex	232	in 3166 group
eg	3166	10 11.2	03 41	13.8	26	SAB(rs)0/a	5.2x2.7	10.6	Sex	232	brightest of group; P(b) w/ 3169
eg	3169	10 11.6	03 43	14.2	28	SA(s)aP	4.8x3.2	10.5	Sex	232	in 3166 group; P(b) w/ 3166
eg	3175	10 12.4	−28 37	14.7	52	SAB(s)a?	4.8x1.7	11.3	Ant	22	
eg	3177	10 13.8	21 22	16.6	07	SA(rs)b	1.7x1.3	12.3	Leo	138	
eg	3184	10 15.3	41 40	18.3	25	SAB(rs)cd	6.9x6.8	9.8	UMa	244	
eg	3185	10 14.9	21 56	17.7	41	(R)SB(r)a	2.3x1.6	12.2	Leo	138	in 3190 group
eg	3187	10 15.0	22 08	17.8	53	SB(s)cP	3.3x1.5	13.1	Leo	138	in 3190 group; P(b) w/ 3190
eg	3190	10 15.4	22 05	18.2	50	SA(s)aPsp	4.6x1.8	11.0	Leo	138	brightest of group; P(b) w/ 3187
eg	3193	10 15.7	22 09	18.5	54	E2	2.8x2.6	10.9	Leo	138	in 3190 group
eg	3198	10 16.9	45 48	20.0	33	SB(rs)c	8.3x3.7	10.4	UMa	244	
eg	3200	10 16.2	−17 44	18.6	59	SA(s)ab:?	4.8x1.8		Hya	131	
gc	3201	10 15.6	−46 10	17.7	25	X	18	6.7	Vel	256	
eg	3203	10 17.3	−26 27	19.6	42	SA(r)0^{+}?sp	3.0x0.7		Hya		
eg	3213	10 18.6	19 54	21.3	39	Sc:	1.2x1.0		Leo	*139*	
eg	3221	10 19.6	21 49	22.3	34	SB(s)c:sp	3.3x0.9		Leo	138	
eg	3222	10 19.8	20 08	22.5	53	SB0:	1.3x1.1	12.8	Leo	139	
eg	3223	10 19.4	−34 01	21.6	16	SA(s)b	4.1x2.6		Ant	22	=I2571
eg	3226	10 20.7	20 09	23.4	54	E2:P	2.8x2.5	11.4	Leo	139	P(b) w/ 3227
eg	3227	10 20.8	20 07	23.5	52	SAB(s)aP	5.6x4.0	10.8	Leo	139	Seyfert; P(b) w/ 3226

ID		RA (1950) DEC (2000)				TYPE	SIZE	V	CON	PG	NOTES C 16
eg	3241	10 22.0	−32 14	24.3	29	SA(r)ab:	1.5x1.1	12.7	Ant	22	
pn	3242	10 22.4	−18 23	24.8	38		45"x36"	*7.8*	Hya	131	
eg	3244	10 23.3	−39 34	25.5	49	SA(s)d	1.9x1.7		Ant	22	
eg	3245	10 24.5	28 46	27.3	31	SA(r)0°:?	3.2x1.9	10.8	LMi	150	
eg	3249	10 24.1	−34 43	26.4	58	SB(r)c:	1.4x1.1		Ant		chart II
eg	3250	10 24.4	−39 41	26.6	56	E4	3.2x2.1	11.0	Ant	22	
eg	3254	10 26.5	29 45	29.3	30	SA(s)bc	5.1x1.9	11.5	LMi	150	
eg	3256	10 25.7	−43 39	27.8	54	P	3.5x2.0	11.3	Vel		P(c) at 0.45
eg	3257	10 26.5	−35 24	28.8	39	SAB(s)0$^-$:?	1.1x1.1		Ant	22	chart II
eg	3258	10 26.7	−35 21	29.0	36	SA0$^-$	1.8x1.7	11.7	Ant	22	chart II
eg	3258A	10 26.1	−35 12	28.4	27	SAB0°:	1.8x0.5		Ant		chart II
eg	3258B	10 28.1	−35 18	30.4	33	SA0/a:	1.6x0.7		Ant	*24*	chart II
eg	3258C	10 29.2	−34 58	31.5	13	SB(r)a	1.5x1.1		Ant		chart II
eg	3258D	10 29.7	−35 09	32.0	24	SB(s)b	1.9x1.1		Ant		chart II
eg	3258E	10 30.2	−34 45	32.5	00	Sb?sp	1.8x0.6		Ant		chart II
eg	3259	10 29.1	65 18	32.6	03	SAB(rs)bc:	2.3x1.4		UMa	244	
eg	3260	10 26.9	−35 20	29.2	35	SA(r)0$^-$?P	1.3x1.1	13.2	Ant	22	chart II
eg	3261	10 26.9	−44 24	29.0	39	SB(rs)b	4.1x3.2		Vel		
eg	3266	10 29.8	65 01	33.3	46	SAB0°?	1.7x1.4		UMa	*244*	
eg	3267	10 27.6	−35 04	29.9	19	SAB(r)0°	2.4x1.4		Ant	22	chart II
eg	3268	10 27.8	−35 04	30.1	19	SA(rs)0$^-$	2.0x1.7	11.8	Ant	22	chart II
eg	3269	10 27.7	−34 58	30.0	13	SA(r)0$^+$	3.1x1.5		Ant	24	chart II
eg	3271	10 28.2	−35 06	30.5	21	SB(r)0°	2.3x1.1	11.7	Ant	24	=I2585; chart II
eg	3273	10 28.2	−35 21	30.5	36	SA(r)0°	2.3x1.1	12.5	Ant	24	chart II
eg	3274	10 29.5	27 56	32.3	41	SABcd:	2.2x1.1	12.8	Leo	139	
eg	3275	10 28.6	−36 29	30.8	44	SB(r)ab	2.8x2.3		Ant	24	
eg	3277	10 30.1	28 46	32.9	31	SA(r)ab	2.0x1.9	11.7	LMi	150	
eg	3281	10 29.6	−34 36	31.9	51	SA(s)b?	3.3x1.8	11.7	Ant	24	chart II
eg	3281B	10 29.6	−34 57	31.9	12	SB0$^-$?			Ant		chart II
eg	3281C	10 30.7	−34 38	33.0	53	S0:sp	1.9x0.5		Ant	24	chart II
eg	3285	10 31.3	−27 12	33.6	28	SB(s)aP	2.6x1.7		Hya	132	
eg	3285B	10 32.3	−27 24	34.6	40	SAB(r)bc	1.3x1.0		Hya		chart XI
eg	3287	10 32.1	21 55	34.8	39	SB(s)d	2.2x1.1		Leo	139	
eg	3289	10 31.9	−35 04	34.2	20	SB(rs)0$^+$:sp	2.2x0.5		Ant		P(b) w/ E375–G64; chart II
eg	3294	10 33.4	37 35	36.3	19	SA(s)c	3.3x1.8	11.7	LMi	150	
eg	3300	10 34.0	14 26	36.7	10	SAB(r)0°:?	2.1x1.1		Leo	139	
eg	3301	10 34.2	22 08	36.9	52	(R')SB(rs)0/a	3.6x1.2	11.4	Leo	139	
eg	3305	10 33.9	−26 54	36.3	10	E0	0.6x0.6	12.8	Hya		chart XI
eg	3307	10 33.9	−27 16	36.3	32	SB(r)0$^+$:	1.0x0.5	14.5	Hya		in Hydra cluster; chart XI
eg	3308	10 34.0	−27 11	36.4	27	SAB(s)0$^-$:	2.0x1.4	12.2	Hya	132	in Hydra cluster; chart XI
eg	3309	10 34.3	−27 16	36.7	32	E3	1.9x1.7	11.9	Hya	132	brightest in Hydra cluster; chart XI
eg	3310	10 35.7	53 46	38.8	30	SAB(r)bcP	3.6x3.0	11.1	UMa	244	
eg	3311	10 34.4	−27 16	36.8	32	SA(r)0°?	2.5x2.2	11.6	Hya	132	chart XI
eg	3312	10 34.7	−27 18	37.1	34	SA(s)bP?	3.6x1.5		Hya	132	=I 629; in Hydra cluster?; chart XI
eg	3314AB	10 34.9	−27 26	37.3	42	SAB:sp, SA(s)c:	2.0x1.2		Hya		interacting pair?; in Hydra cluster; chart XI
eg	3315	10 35.0	−26 56	37.4	12	S0$^-$?	0.7x0.6		Hya		=M–04–25–42; =E501–G48; chart XI
eg	3316	10 35.3	−27 20	37.7	36	SB(rs)0°	1.2x0.9		Hya		chart XI
eg	3318	10 35.1	−41 22	37.3	38	SAB(rs)b	2.6x1.5		Vel		
eg	3319	10 36.2	41 57	39.1	41	SB(rs)cd	6.8x3.9	11.3	UMa	245	
eg	3320	10 36.6	47 39	39.6	23	Scd:	2.2x1.2		UMa	245	

ID	RA (1950) DEC (2000)				TYPE	SIZE	V	CON	PG	NOTES C 17
eg 3329	10 40.5	77 04	44.7	48	(R)SA(r)b:	2.1x1.3		Dra	105	=3397; brightest of group
eg 3338	10 39.5	14 01	42.2	45	SA(s)c	5.5x3.7	10.8	Leo	140	in 3368 group
eg 3344	10 40.8	25 11	43.5	55	(R)SAB(r)bc	6.9x6.5	10.0	LMi	151	
eg 3346	10 41.0	15 08	43.7	52	SB(rs)cd	2.8x2.5		Leo	140	
eg 3347	10 40.5	−36 06	42.8	22	SB(rs)b	4.4x2.6		Ant	24	P(b) w/ 3354
eg 3348	10 43.5	73 06	47.2	50	E0	2.2x2.2	11.2	UMa	245	
eg 3351	10 41.3	11 58	43.9	42	SB(r)b	7.4x5.1	9.7	Leo	140	=M95; in 3368 group
eg 3353	10 42.3	56 13	45.4	57	Sb?P	1.5x1.1	12.7	UMa	245	P(b) w/ 3347
eg 3354	10 40.8	−36 06	43.1	22	S:P	0.6x0.6	13.2	Ant	24	
eg 3358	10 41.3	−36 09	43.6	25	(R)SAB(s)0/a	3.8x2.3		Ant	24	
eg 3359	10 43.4	63 29	46.7	13	SB(rs)c	6.8x4.3	10.5	UMa	245	
eg 3364	10 44.8	72 41	48.5	25	SAB(rs)bc	1.9x1.8		UMa	245	
eg 3367	10 43.9	14 01	46.5	45	SB(rs)c	2.3x2.1	11.5	Leo	140	
eg 3368	10 44.1	12 05	46.7	49	SAB(rs)ab	7.1x5.1	9.2	Leo	140	=M96; brightest of group
eg 3370	10 44.4	17 32	47.1	16	SA(s)c	3.1x1.9		Leo	140	
eg 3377	10 45.1	14 15	47.8	59	E5.5	4.4x2.7	10.2	Leo	140	in 3368 group
eg 3377A	10 44.7	14 20	47.3	04	SAB(s)m	2.0x1.9		Leo	*140*	
eg 3379	10 45.2	12 51	47.8	35	E1	4.5x4.0	9.3	Leo	140	=M105; in 3368 group
eg 3384	10 45.6	12 54	48.2	38	SB(s)0⁻:	5.9x2.6	10.0	Leo	140	in 3368 group
eg 3389	10 45.8	12 48	48.4	32	SA(s)c	2.7x1.5	11.8	Leo	141	
eg 3390	10 45.7	−31 16	48.0	32	Sbsp	4.0x0.7	12.2	Hya	132	
eg 3395	10 47.0	33 15	49.8	59	SAB(rs)cdP:	1.9x1.2	12.1	LMi	151	P(b) w/ 3396
eg 3396	10 47.1	33 15	49.9	59	IBmP	2.8x1.2	12.2	LMi	151	P(b) w/ 3395
eg 3403	10 50.2	73 57	53.9	41	SAbc:	3.1x1.3		Dra	105	
eg 3412	10 48.3	13 41	50.9	25	SB(s)0°	3.6x2.0	10.6	Leo	141	in Leo I cluster
eg 3413	10 48.6	33 02	51.4	46	S0sp	2.4x1.0		LMi	*152*	
eg 3414	10 48.5	28 15	51.2	59	S0P	3.6x2.7	10.8	LMi	152	
eg 3415	10 48.8	43 59	51.7	43	SA0⁺:	2.4x1.5		UMa	245	
eg 3418	10 48.7	28 03	51.4	47	SAB(s)a?	1.4x1.1		LMi	*152*	
eg 3423	10 48.6	06 06	51.2	50	SA(s)cd	3.9x3.5	11.2	Sex	232	
eg 3424	10 49.0	33 10	51.8	54	SB(s)b:?	3.0x0.9		LMi	*152*	
eg 3430	10 49.4	33 13	52.2	57	SAB(rs)c	3.9x2.3	11.5	LMi	152	=I2613
eg 3432	10 49.7	36 53	52.5	37	SB(s)msp	6.2x1.5	11.3	LMi	152	
eg 3433	10 49.4	10 25	52.0	09	SA(s)c	3.5x3.2		Leo	141	
eg 3437	10 49.9	23 12	52.6	56	SAB(rs)c:	2.6x0.9		Leo	141	
eg 3440	10 50.8	57 23	53.9	07	SBb?sp	2.3x0.6		UMa	*245*	
eg 3443	10 50.3	17 50	53.0	34	SAd	2.6x1.3		Leo	141	
eg 3445	10 51.6	57 15	54.7	59	SAB(s)m	1.6x1.5	12.4	UMa	245	
eg 3447	10 50.7	17 02	53.4	46	SAB(s)dmP	3.8x2.3		Leo	141	
eg 3447A	10 50.9	17 03	53.6	47	IB(s)mP	1.7x1.0		Leo	141	
eg 3448	10 51.7	54 35	54.7	19	I0	5.4x1.9	11.7	UMa	245	
eg 3449	10 50.5	−32 40	52.8	56	SAB(r)0/a:	2.6x1.2		Ant	24	
eg 3454	10 51.8	17 37	54.5	21	SB(s)c?sp	2.2x0.5		Leo	141	
eg 3455	10 51.9	17 33	54.6	17	(R')SAB(rs)b	2.8x1.8		Leo	141	
eg 3457	10 52.1	17 53	54.8	37	E:P	1.3x1.3		Leo	141	
eg 3458	10 53.0	57 23	56.0	07	SAB0:	1.7x1.1		UMa	246	
eg 3464	10 52.2	−20 48	54.6	04	SAB(rs)bc	2.8x2.0		Hya		
eg 3478	10 56.6	46 23	59.5	07	SB(rs)bc	2.8x1.4		UMa	246	
eg 3485	10 57.4	15 07	00.0	51	SB(r)b:	2.5x2.2		Leo	141	
eg 3486	10 57.7	29 15	00.4	59	SAB(r)c	6.9x5.4	10.3	LMi	152	

ID		RA (1950)	DEC	(2000)		TYPE	SIZE	V	CON	PG	NOTES C 18
eg	3489	10 57.7	14 10	00.3	54	SAB(rs)0^{+}	3.7x2.1	10.3	Leo	141	in Leo I cluster
eg	3495	10 58.7	03 54	01.3	38	Sd:	4.6x1.3		Leo	141	
eg	3501	11 00.1	18 16	02.8	00	Scsp	3.7x0.6		Leo	142	
eg	3504	11 00.5	28 15	03.2	59	(R)SAB(s)ab	2.7x2.2	11.1	LMi	152	
eg	3506	11 00.6	11 21	03.2	05	Sc:	1.3x1.3		Leo	142	
eg	3507	11 00.8	18 24	03.5	08	SB(s)b	3.5x3.0		Leo	142	
eg	3510	11 01.0	29 09	03.7	53	SB(s)msp	3.8x0.9	12.9	LMi	152	
eg	3511	11 01.0	−22 49	03.4	05	SA(s)c	5.4x2.2		Crt	95	
eg	3512	11 01.3	28 18	04.0	02	SAB(rs)c	1.7x1.5	12.4	LMi	152	
eg	3513	11 01.3	−22 59	03.7	15	SB(rs)c	2.8x2.3		Crt	95	
eg	3515	11 01.9	28 30	04.6	14	SAc?	1.1x0.9		LMi	*152*	
eg	3516	11 03.4	72 50	06.8	34	(R)SB(s)0°:	2.3x1.8	11.7	UMa	246	Seyfert
eg	3521	11 03.3	00 14	05.9	02	SAB(rs)bc	9.5x5.0	8.9	Leo	142	
eg	3547	11 07.3	11 00	09.9	44	Sb:	2.2x1.1	12.8	Leo	142	
eg	3549	11 08.1	53 40	11.0	24	SA(s)c:	3.2x1.3		UMa	246	
eg	3556	11 08.6	55 57	11.5	41	SB(s)cdsp	8.3x2.5	10.0	UMa	246	=M108
eg	3557	11 07.6	−37 16	10.0	32	E3	4.0x2.7	10.4	Cen		brightest of group
eg	3571	11 09.0	−18 01	11.5	17	SB(rs)a	3.3x1.3		Crt	95	=3544
eg	3577	11 10.9	48 33	13.7	17	SB(r)a	1.7x1.7		UMa	*246*	
eg	3583	11 11.4	48 36	14.2	20	SB(s)b	2.8x2.0		UMa	246	
eg	3585	11 10.8	−26 29	13.2	45	SAB(s)0^{-}:P	2.9x1.6	10.0	Hya	132	
pn	3587	11 11.9	55 18	14.8	02		3.4x3.3	*9.9*	UMa	246	=M97; Owl
eg	3592	11 11.8	17 32	14.4	16	Sbc?sp	2.0x0.6		Leo	*142*	
eg	3593	11 12.0	13 05	14.6	49	SA(s)0/a:	5.8x2.5	11.0	Leo	142	in 3627 group
eg	3596	11 12.5	15 04	15.1	48	SAB(rs)c	4.2x4.1		Leo	142	
eg	3598	11 12.6	17 32	15.2	16	E/S0	1.9x1.4		Leo	142	
eg	3599	11 12.8	18 23	15.4	07	SA0:	2.8x2.8	11.9	Leo	142	in 3607 group
eg	3605	11 14.1	18 18	16.7	02	E4.5	1.7x1.0	12.3	Leo	142	in 3607 group
eg	3607	11 14.3	18 20	16.9	04	SA(s)0°:	3.7x3.2	10.0	Leo	142	brightest of group
eg	3608	11 14.4	18 26	17.0	10	E2	3.0x2.5	11.0	Leo	142	in 3607 group
eg	3610	11 15.5	59 04	18.4	48	E5:	3.2x2.5	10.8	UMa	246	
eg	3611	11 14.9	04 50	17.5	34	SA(s)aP	2.4x2.0	12.2	Leo	142	
eg	3613	11 15.7	58 16	18.6	00	E6	3.6x2.0		UMa	246	
eg	3614	11 15.6	46 01	18.4	45	SAB(r)c	4.6x2.9		UMa	247	
eg	3619	11 16.5	58 02	19.4	46	(R)SA(s)0^{+}:	3.1x2.6		UMa	247	
eg	3621	11 15.9	−32 32	18.3	48	SA(s)d	10x6.5		Hya		
eg	3623	11 16.3	13 22	18.9	06	SAB(rs)a	10x3.3	9.3	Leo	142	=M65; in 3627 group
eg	3625	11 17.6	58 03	20.5	47	SAB(s)b:	2.2x0.8		UMa	247	
eg	3626	11 17.4	18 38	20.0	22	(R)SA(rs)0^{+}	3.1x2.2	10.9	Leo	143	=3622
eg	3627	11 17.6	13 16	20.2	00	SAB(s)b	8.7x4.4	9.0	Leo	143	=M66; brightest of group
eg	3628	11 17.7	13 52	20.3	36	SbPsp	15x3.6	9.4	Leo	143	in 3627 group
eg	3629	11 17.9	27 14	20.6	58	SA(s)cd:	2.2x1.7		Leo	143	
eg	3630	11 17.7	03 14	20.3	58	S0sp	2.3x0.9		Leo	143	
eg	3631	11 18.2	53 27	21.0	11	SA(s)c	4.6x4.1	10.4	UMa	247	
eg	3636	11 17.9	−10 01	20.4	17	E0	1.1x1.1		Crt	95	
eg	3637	11 18.1	−09 59	20.6	15	(R)SB(r)0°	1.7x1.6		Crt	95	
eg	3640	11 18.5	03 31	21.1	15	E3	4.1x3.4	10.3	Leo	143	
eg	3641	11 18.6	03 28	21.2	12	cE1	1.1x1.1		Leo	143	
eg	3642	11 19.4	59 21	22.3	05	SA(r)bc:	5.8x4.9	11.1	UMa	247	
eg	3646	11 19.1	20 27	21.7	11	P(ring)	3.9x2.6	11.2	Leo	143	ring galaxy

ID		RA (1950)	DEC	(2000)		TYPE	SIZE	V	CON	PG	NOTES C 19
eg	3649	11 19.6	20 29	22.2	13	SB(s)a	1.6x0.8		Leo	*143*	
eg	3655	11 20.3	16 52	22.9	36	SA(s)c:	1.6x1.1	11.6	Leo	143	
eg	3658	11 21.3	38 50	24.0	34	SA(r)0°:	2.1x1.9		UMa	*247*	
eg	3659	11 21.1	18 06	23.7	50	SB(s)m?	2.1x1.2		Leo	143	
eg	3664	11 21.9	03 36	24.5	20	SB(s)mP	2.0x1.9		Leo	143	P(b) w/ 3664A
eg	3664A	11 21.9	03 30	24.5	14	SBmP?	1.1x1.1		Leo		P(b) w/ 3664
eg	3665	11 22.0	39 02	24.7	46	SA(s)0°	3.2x2.6	10.8	UMa	247	
eg	3666	11 21.8	11 37	24.4	21	SA(rs)c:	4.2x1.4		Leo	143	
eg	3672	11 22.5	−09 31	25.0	48	SA(s)c	4.1x2.1		Crt	95	
eg	3673	11 22.7	−26 28	25.2	45	SB(rs)b	3.5x2.5		Hya	132	
eg	3674	11 23.6	57 20	26.4	04	S0sp	2.0x0.7		UMa	*247*	
eg	3675	11 23.4	43 52	26.1	36	SA(s)b	5.9x3.2		UMa	247	
oc	3680	11 23.3	−42 58	25.7	15	I 2 m	7	7.6	Cen		
eg	3681	11 23.9	17 08	26.5	52	SAB(r)bc	2.5x2.4	11.7	Leo	143	
eg	3683	11 24.7	57 09	27.5	52	SB(s)c?	2.0x0.9		UMa	247	
eg	3683A	11 26.4	57 25	29.2	08	SB(rs)c	2.5x1.8		UMa	*247*	
eg	3684	11 24.6	17 18	27.2	01	SA(rs)bc	3.2x2.3	11.7	Leo	144	
eg	3686	11 25.1	17 30	27.7	13	SB(s)bc	3.3x2.6	11.4	Leo	144	
eg	3687	11 25.3	29 47	28.0	30	(R')SAB(r)bc?	2.0x2.0		UMa	247	
eg	3689	11 25.6	25 56	28.2	39	SAB(rs)c	1.6x1.2	12.3	Leo	144	
eg	3690	11 25.7	58 50	28.5	33	IBmP	2.4x1.9		UMa	247	P(c) w/ I 694
eg	3691	11 25.6	17 12	28.2	55	SBb?	1.3x1.0		Leo	144	
eg	3692	11 25.8	09 41	28.4	24	Sb	3.3x0.9		Leo	*144*	
eg	3705	11 27.5	09 33	30.1	16	SAB(r)ab	5.0x2.3		Leo	144	
eg	3706	11 27.3	−36 07	29.8	24	SA(rs)0$^-$:	2.9x1.9		Cen		
eg	3717	11 29.1	−30 02	31.6	19	SAb:sp	5.8x1.4		Hya		
eg	3718	11 29.8	53 21	32.5	04	SB(s)aP	8.7x4.5	10.5	UMa	248	P(b) w/ 3729
eg	3719	11 29.7	01 06	32.3	49	SA(rs)b	2.0x1.5		Leo	*144*	
eg	3720	11 29.8	01 05	32.4	48	Sab?	1.1x1.0	13.3	Leo	144	
eg	3722	11 31.9	−09 25	34.4	42	S0:P	0.5x0.5		Crt	*95*	
eg	3724	11 32.0	−09 24	34.5	41	S0$^+$:sp	0.5x0.4		Crt	*95*	
eg	3726	11 30.6	47 18	33.3	01	SAB(r)c	6.0x4.5	10.4	UMa	248	
eg	3729	11 31.1	53 24	33.8	07	SB(r)aP	3.1x2.1	11.4	UMa	248	P(b) w/ 3718
eg	3732	11 31.7	−09 34	34.2	51	SAB(s)0/a:	1.3x1.2		Crt	95	brightest of group
eg	3733	11 32.3	55 08	35.0	51	SAB(s)c:	4.8x2.3		UMa	*248*	
eg	3735	11 33.1	70 49	36.0	32	SAc:sp	4.2x1.0		Dra	105	
eg	3738	11 33.1	54 48	35.8	31	Im	2.6x2.0	11.7	UMa	248	
eg	3745	11 35.1	22 18	37.7	01	SB(s)0$^-$:	0.4x0.2		Leo	144	in Copeland Septet; chart XII
eg	3746	11 35.1	22 17	37.7	00	SB(r)b	1.3x0.7		Leo	144	in Copeland Septet; chart XII
eg	3748	11 35.2	22 18	37.8	01	SB0°?sp	0.8x0.3		Leo	144	in Copeland Septet; chart XII
eg	3750	11 35.3	22 15	37.9	58	SAB0$^-$?	0.9x0.7		Leo	144	in Copeland Septet; chart XII
eg	3751	11 35.3	22 13	37.9	56	S0$^-$:P	0.4x0.3		Leo	144	in Copeland Septet; chart XII
eg	3753	11 35.3	22 16	37.9	59	Sab?spP	2.0x0.6		Leo	144	in Copeland Septet; chart XII
eg	3754	11 35.3	22 16	37.9	59	SBb?P	0.4x0.4		Leo	144	in Copeland Septet; chart XII
eg	3756	11 34.1	54 34	36.8	17	SAB(rs)bc	4.4x2.4	11.5	UMa	248	
eg	3769	11 35.0	48 10	37.7	53	SB(r)b:	3.2x1.1		UMa	248	P(b) w/ 3769A
eg	3769A	11 35.1	48 10	37.8	53	SBmP:	1.0x0.4		UMa	*248*	P(b) w/ 3769
eg	3773	11 35.6	12 23	38.2	06	SA0:	1.6x1.4		Leo	144	
eg	3780	11 36.7	56 33	39.4	16	SA(s)c:	3.1x2.6		UMa	248	
eg	3782	11 36.7	46 47	39.4	30	SAB(s)cd:	1.7x1.2		UMa	248	

C 20

ID		RA (1950)		DEC (1950)		RA (2000)	DEC (2000)	TYPE	SIZE	V	CON	PG	NOTES
eg	3783	11	36.6	−37	28	39.1	45	SB(r)a	1.9x1.5		Cen		Seyfert
eg	3790	11	37.2	17	59	39.8	42	S0/a	1.2x0.3		Leo	*144*	
eg	3801	11	37.7	18	00	40.3	43	S0?P	3.2x1.9	12.1	Leo	144	
eg	3802	11	37.7	18	03	40.3	46	Scd:sp	1.4x0.4	13.6	Leo	144	
eg	3804	11	38.2	56	29	40.9	12	SAB(s)cd	2.4x1.7		UMa	*248*	=3794
eg	3805	11	38.1	20	37	40.7	20	S0	1.7x1.3		Leo		chart XIII
eg	3806	11	38.2	18	04	40.8	47	SABb	1.6x1.5		Leo	*144*	=3807
eg	3808	11	38.1	22	42	40.7	25	SAB(rs)c:P	1.8x1.0		Leo	144	P(b) w/ 3808A
eg	3808A	11	38.1	22	44	40.7	27	I0?P	0.7x0.4		Leo	144	P(b) w/ 3808
eg	3810	11	38.4	11	45	41.0	28	SA(rs)c	4.3x3.1	10.7	Leo	144	
eg	3813	11	38.7	36	50	41.3	33	SA(rs)b:	2.3x1.2	11.7	UMa	248	
eg	3816	11	39.2	20	23	41.8	06	SA0P	2.2x1.3		Leo		chart XIII
eg	3818	11	39.4	−05	53	42.0	10	E5	2.1x1.4	11.8	Vir	257	
eg	3821	11	39.6	20	36	42.2	19	(R)SAB(s)ab	1.6x1.4		Leo	144	chart XIII
eg	3837	11	41.4	20	10	44.0	53	E	1.0x0.9		Leo	146	chart XIII
eg	3840	11	41.4	20	22	44.0	05	SA(r)b	1.2x0.8		Leo	146	chart XIII
eg	3841	11	41.4	20	15	44.0	58	SA0$^-$	0.9x0.4		Leo	146	chart XIII
eg	3842	11	41.5	20	14	44.1	57	E3	1.2x1.0		Leo	146	in Abell 1367; chart XIII
eg	3844	11	41.4	20	19	44.0	02	SA(r)0°sp	1.7x0.3		Leo	146	chart XIII
eg	3845	11	41.5	20	17	44.1	00	(R')SAB0°	1.3x0.5		Leo	*146*	chart XIII
eg	3851	11	41.7	20	16	44.3	59	SAB0°	0.3x0.3		Leo	146	chart XIII
eg	3857	11	42.2	19	49	44.8	32	SA0$^-$	1.4x0.7		Leo	146	chart XIII
eg	3859	11	42.3	19	44	44.9	27	SA(r)cP	1.1x0.3		Leo		chart XIII
eg	3860	11	42.2	20	04	44.8	47	SAB(r)ab	1.3x0.7		Leo	146	chart XIII
eg	3861	11	42.5	20	15	45.1	58	(R')SAB(r)b	2.4x1.5		Leo		chart XIII
eg	3862	11	42.5	19	53	45.1	36	E0	1.6x1.6	12.6	Leo	146	in Abell 1367; chart XIII
eg	3864	11	42.7	19	40	45.3	23	SAbc?	1.2x0.9		Leo		chart XIII
eg	3865	11	42.3	−08	57	44.8	14	SAB(s)b?	2.3x1.7		Crt	96	=3854
eg	3866	11	42.7	−09	04	45.2	21	S0/a:	1.4x1.0		Crt	96	
eg	3867	11	42.9	19	41	45.5	24	(R)SBa?	1.3x0.5		Leo		chart XIII
eg	3868	11	42.9	19	43	45.5	26	SA0:	0.8x0.4		Leo		chart XIII
eg	3872	11	43.2	14	03	45.8	46	E5	2.2x1.5	11.7	Leo	146	
eg	3873	11	43.2	20	03	45.8	46	E	1.1x1.0		Leo	146	chart XIII
eg	3875	11	43.2	20	03	45.8	46	SA0$^-$P	1.3x0.3		Leo	146	chart XIII
eg	3877	11	43.5	47	47	46.1	30	SA(s)c:	5.4x1.5		UMa	249	
eg	3884	11	43.6	20	40	46.2	23	SA(r)0/aP	1.9x1.5		Leo		chart XIII
eg	3885	11	44.3	−27	39	46.8	56	SA(s)0/a	1.7x0.9		Hya		
eg	3887	11	44.5	−16	35	47.0	52	SB(r)bc	3.3x2.7	11.0	Crt	96	
eg	3888	11	44.9	56	15	47.6	58	SAB(rs)c	1.8x1.4		UMa	249	in 3898 group
eg	3892	11	45.5	−10	41	48.0	58	SB(rs)0$^+$	2.8x2.2		Crt	96	
eg	3893	11	46.0	48	59	48.6	42	SAB(rs)c:	4.4x2.8		UMa	249	
eg	3896	11	46.3	48	57	48.9	40	SB0/a:P	1.7x1.3		UMa	249	
eg	3898	11	46.6	56	22	49.2	05	SA(s)ab	4.4x2.6	10.8	UMa	249	brightest of group
eg	3900	11	46.6	27	18	49.2	01	SA(r)0$^+$	3.5x1.9	11.4	Leo	149	
eg	3902	11	46.7	26	24	49.3	07	SAB(s)b:	1.8x1.5		Leo	149	
eg	3904	11	46.7	−29	00	49.2	17	SA0$^-$:	2.2x1.7	11.0	Hya		
eg	3906	11	47.1	48	42	49.7	25	SB(s)d	1.9x1.8		UMa	249	
eg	3911	11	46.8	25	13	49.4	56	(R')SAB(s)b:	1.3x1.0		Leo	149	
eg	3912	11	47.5	26	46	50.1	29	SAB(s)b?P	1.7x1.0		Leo	149	=3899
eg	3917	11	48.1	52	06	50.7	49	SAcd:	4.9x1.4		UMa	249	

	ID	RA (1950)	DEC	(2000)		TYPE	SIZE	V	CON	PG	NOTES
eg	3920	11 47.5	25 12	50.1	55	Sa?	1.4x1.3		Leo	149	
eg	3923	11 48.5	−28 32	51.0	49	SAB(s)0$^-$?P	2.9x1.9	10.1	Hya		
eg	3931	11 48.6	52 17	51.2	00	SA0$^-$:	1.4x1.1		UMa	249	
eg	3936	11 49.8	−26 38	52.3	55	SB(s)b?	4.0x0.8		Hya		
eg	3938	11 50.2	44 24	52.8	07	SA(s)c	5.4x4.9	10.4	UMa	249	
eg	3941	11 50.3	37 16	52.9	59	SB(s)0°	3.8x2.5		UMa	249	
eg	3944	11 50.5	26 29	53.1	12	E/S0	1.7x1.3		Leo	*149*	
eg	3945	11 50.6	60 57	53.2	40	SB(rs)0$^+$	5.5x3.6	10.6	UMa	249	
eg	3949	11 51.1	48 08	53.7	51	SA(s)bc:	3.0x1.8	11.0	UMa	249	
eg	3952	11 51.1	−03 43	53.7	00	I0P?	1.5x0.7		Vir		
eg	3953	11 51.2	52 37	53.8	20	SB(r)bc	6.6x3.6	10.1	UMa	250	
eg	3955	11 51.4	−22 53	53.9	10	S0/aP	3.2x1.3	11.9	Crt	96	
eg	3956	11 51.5	−20 17	54.0	34	SA(s)c:	3.5x1.2		Crt	96	
eg	3957	11 51.5	−19 17	54.0	34	SA0$^+$:sp	3.5x0.8		Crt	96	=I2965
eg	3958	11 52.0	58 39	54.6	22	SB(s)a	1.6x0.8		UMa	250	
eg	3962	11 52.1	−13 42	54.7	59	E1	2.9x2.6	10.6	Crt	96	
eg	3963	11 52.4	58 46	55.0	29	SAB(rs)bc	2.8x2.6		UMa	250	
eg	3972	11 53.2	55 36	55.8	19	SA(s)bc	4.0x1.3		UMa	*250*	
eg	3976	11 53.4	07 02	56.0	45	SAB(s)b	3.9x1.4		Vir	257	
eg	3977	11 53.5	55 40	56.1	23	(R)SA(rs)ab:	1.7x1.6		UMa	250	=3980
eg	3981	11 53.6	−19 37	56.2	54	SA(rs)bc	3.9x1.5		Crt	96	
eg	3982	11 53.9	55 24	56.5	07	SAB(r)b:	2.5x2.2		UMa	250	
eg	3985	11 54.1	48 37	56.7	20	SB(s)m:	1.2x0.8		UMa	250	
eg	3987	11 54.8	25 28	57.4	11	Sbsp	2.5x0.5		Leo	149	chart XIV
eg	3989	11 54.9	25 31	57.5	14	Sa?	0.8x0.4		Leo	*149*	chart XIV
eg	3990	11 55.0	55 44	57.6	27	S0$^-$:sp	1.7x1.0	12.6	UMa	250	
eg	3991	11 54.9	32 37	57.5	20	ImPsp	1.4x0.5	13.2	UMa	250	P(b) w/ 3994, 3995
eg	3992	11 55.0	53 39	57.6	22	SB(rs)bc	7.6x4.9	9.8	UMa	250	=M109
eg	3993	11 55.1	25 31	57.7	14	Sb:sp	1.9x0.6		Leo	149	chart XIV
eg	3994	11 55.0	32 33	57.6	16	SA(r)cP?	1.1x0.7	12.7	UMa	250	P(b) w/ 3991, 3995
eg	3995	11 55.2	32 34	57.8	17	SAmP	2.8x1.1	12.6	UMa	250	P(b) w/ 3991, 3994
eg	3997	11 55.2	25 33	57.8	16	SBbP	1.8x1.0		Leo	149	chart XIV
eg	3998	11 55.3	55 44	57.9	27	SA(r)0°?	3.1x2.5	10.6	UMa	250	
eg	3999	11 55.4	25 21	58.0	04	S0?			Leo		chart XIV
eg	4000	11 55.4	25 25	58.0	08	S0sp	1.4x0.3		Leo	149	chart XIV
eg	4004	11 55.5	28 09	58.1	52	SP	2.2x0.8		Leo	*149*	
eg	4005	11 55.6	25 24	58.2	07	SA(r)ab:	1.2x0.7		Leo	149	chart XIV
eg	4008	11 55.7	28 28	58.3	11	E5	2.5x1.5	12.0	Leo	149	
eg	4009	11 55.9	25 30	58.5	13	S			Com		chart XIV
eg	4011	11 55.8	25 23	58.4	06	Sab?			Leo		chart XIV
eg	4013	11 56.0	44 14	58.6	57	Sbsp	5.2x1.3		UMa	251	
eg	4015A	11 56.1	25 19	58.7	02	E	1.4x1.1		Com	78	chart XIV
eg	4015B	11 56.2	25 19	58.8	02	Sd?sp	0.9x0.3		Com	*78*	chart XIV
eg	4016	11 55.8	27 48	58.4	31	SBdm:	1.6x0.9		Leo	*149*	=U 6954
eg	4017	11 56.2	27 44	58.8	27	SAB(s)bc	1.8x1.6		Leo	*149*	=U 6967
eg	4018	11 56.1	25 36	58.7	19	Sabsp	1.8x0.4		Com		chart XIV
eg	4021	11 56.4	25 22	59.0	05	cE	0.9x0.7		Com	*78*	chart XIV
eg	4022	11 56.5	25 30	59.1	13	SAB0:	1.6x1.5		Com	78	chart XIV
eg	4023	11 56.5	25 16	59.1	59	Sab?	1.2x0.8		Com	78	chart XIV
eg	4024	11 56.0	−18 04	58.6	21	SB(s)0°:	2.3x1.7		Crv	93	

ID		RA (1950) DEC (2000)				TYPE	SIZE	V	CON	PG	NOTES C 22
eg	4026	11 56.8	51 14	59.4	57	S0sp	5.1x1.4		UMa	251	
eg	4027	11 57.0	−18 59	59.6	16	SB(s)dm	3.0x2.3	11.1	Crv	93	
eg	4030	11 57.8	−00 49	00.4	06	SA(s)bc	4.3x3.2		Vir	257	
eg	4032	11 58.0	20 21	00.6	04	Im	2.1x2.0		Com	78	
eg	4033	11 58.0	−17 34	00.6	51	SA0$^-$?	2.5x1.1		Crv	93	
eg	4036	11 58.9	62 11	01.5	54	S0$^-$	4.5x2.0	10.5	UMa	251	
eg	4037	11 58.8	13 41	01.4	24	SB(rs)b:	2.7x2.3		Com		
eg	4038	11 59.3	−18 35	01.9	52	SB(s)mP	2.6x1.8	10.7	Crv	93	P(c) w/ 4039
eg	4039	11 59.3	−18 36	01.9	53	SA(s)mP	3.2x2.2		Crv	93	P(c) w/ 4038
eg	4041	11 59.7	62 25	02.3	08	SA(rs)bc:	2.8x2.7	11.2	UMa	251	
eg	4045	12 00.2	02 16	02.8	59	SAB(r)a	2.8x2.0	11.8	Vir	257	
eg	4045A	12 00.2	02 14	02.8	57	SB0$^+$:	0.7x0.3		Vir	257	
eg	4047	12 00.3	48 55	02.9	38	(R)SA(rs)b:	1.5x1.3		UMa	251	
eg	4050	12 00.3	−16 06	02.9	23	SB(r)ab	3.1x2.2		Crv	93	
eg	4051	12 00.6	44 49	03.2	32	SAB(rs)bc	5.0x4.0	10.3	UMa	251	Seyfert
eg	4062	12 01.5	32 11	04.1	54	SA(s)c	4.3x2.0	11.2	UMa	251	
eg	4064	12 01.6	18 43	04.2	26	SB(s)a:P	4.5x1.9	11.4	Com	78	
eg	4068	12 01.5	52 52	04.0	35	Im	3.2x1.8		UMa	252	
eg	4073	12 01.9	02 11	04.5	54	SA0$^-$:	2.5x1.9		Vir	257	
eg	4085	12 02.8	50 38	05.3	21	SAB(s)c:?	2.8x0.9	12.3	UMa	252	
eg	4088	12 03.1	50 49	05.6	32	SAB(rs)bc	5.8x2.5	10.5	UMa	252	
eg	4094	12 03.3	−14 15	05.9	32	SA(rs)c:	4.2x1.8		Crv	93	
eg	4096	12 03.5	47 45	06.0	28	SAB(rs)c	6.5x2.0	10.6	UMa	252	
eg	4100	12 03.6	49 52	06.1	35	(R')SA(rs)bc	5.2x1.9		UMa	252	
eg	4102	12 03.9	52 59	06.4	42	SAB(s)b?	3.2x1.9		UMa	252	
eg	4105	12 04.1	−29 29	06.7	46	E1:P	2.4x1.9		Hya		P(b) w/ 4106
eg	4106	12 04.2	−29 29	06.8	46	SAB(s)0°:P	1.9x1.5	11.4	Hya		P(b) w/ 4105
eg	4109	12 04.3	43 16	06.8	59	Sa?	0.8x0.7		CVn	46	
eg	4111	12 04.5	43 21	07.0	04	SA(r)0$^+$:sp	4.8x1.1	10.6	CVn	46	
eg	4114	12 04.6	−13 54	07.2	11	(R')SAB(s)a:	2.3x1.1		Crv	93	
eg	4116	12 05.1	02 58	07.7	41	SB(rs)dm	3.8x2.4	11.9	Vir	257	
eg	4117	12 05.2	43 24	07.7	07	S0$^-$:	2.8x1.1		CVn	46	
eg	4118	12 05.4	43 24	07.9	07	S0$^+$?	0.9x0.5		CVn	46	
eg	4121	12 05.4	65 24	07.9	07	E2	0.6x0.5		Dra	105	
eg	4123	12 05.6	03 09	08.2	52	SB(r)c	4.5x3.5	11.2	Vir	257	
eg	4124	12 05.6	10 40	08.2	23	SA(r)0$^+$	4.6x1.7		Vir	257	
eg	4125	12 05.6	65 27	08.1	10	E6P	5.1x3.2	9.7	Dra	105	
eg	4128	12 06.1	69 03	08.6	46	SA0:sp	2.8x1.0		Dra	105	
eg	4129	12 06.3	−08 46	08.9	03	SB(s)ab:sp	2.6x0.8	12.6	Vir	257	
eg	4136	12 06.8	30 12	09.3	55	SAB(r)c	4.1x3.9		Com	78	
eg	4138	12 07.0	43 58	09.5	41	SA(r)0$^+$	2.9x1.9		CVn	46	
eg	4143	12 07.1	42 49	09.6	32	SAB(s)0°	2.9x1.8		CVn	46	
eg	4144	12 07.5	46 44	10.0	27	SAB(s)cd?sp	5.9x1.5		UMa	252	
eg	4145	12 07.5	40 10	10.0	53	SAB(rs)d	5.8x4.4	11.0	CVn	46	
gc	4147	12 07.6	18 49	10.1	32	IX	4.0	10.2	Com	78	
eg	4150	12 08.0	30 41	10.5	24	SA(r)0°?	2.5x1.8	11.7	Com	78	
eg	4151	12 08.0	39 41	10.5	24	(R')SAB(rs)ab:	5.9x4.4	10.4	CVn	46	Seyfert
eg	4152	12 08.1	16 19	10.6	02	SAB(rs)c	2.3x1.9	12.0	Com	78	
eg	4156	12 08.3	39 45	10.8	28	SB(rs)b	1.5x1.3	13.0	CVn	46	
eg	4157	12 08.6	50 46	11.1	29	SAB(s)b?sp	6.9x1.7		UMa	252	

ID	RA (1950) DEC (2000)				TYPE	SIZE	V	CON	PG	NOTES C 23
eg 4158	12 08.6	20 27	11.1	10	SA(r)b:	2.0x1.8		Com	78	
eg 4162	12 09.3	24 24	11.8	07	(R)SA(rs)bc	2.5x1.6	11.5	Com	79	
eg 4164	12 09.6	13 29	12.1	12	E3	0.3x0.3		Vir	257	
eg 4165	12 09.7	13 32	12.2	15	SAB(r)a:?	1.5x1.1	13.2	Vir	257	
eg 4168	12 09.7	13 29	12.2	12	E2	2.8x2.6	11.3	Vir	257	brightest of group
eg 4178	12 10.2	11 09	12.8	52	SB(rs)dm	5.0x2.0	11.4	Vir		=I3042
eg 4179	12 10.3	01 35	12.9	18	S0sp	4.2x1.2	10.9	Vir	257	
eg 4183	12 10.8	43 59	13.3	42	SA(s)cd?sp	5.0x0.9		CVn	46	
eg 4186	12 11.6	15 00	14.1	43	SA(s)ab:	1.4x1.1		Com	79	
eg 4189	12 11.2	13 42	13.7	25	SAB(rs)cd?	2.5x2.1	11.7	Com	79	=I3050
eg 4190	12 11.2	36 55	13.7	38	ImP	1.7x1.6		CVn	46	
eg 4192	12 11.3	15 11	13.8	54	SAB(s)ab	9.5x3.2	10.1	Com	79	=M98
eg 4193	12 11.4	13 27	13.9	10	SAB(s)c:?	2.3x1.2	12.4	Vir	258	
eg 4203	12 12.6	33 29	15.1	12	SAB0$^-$:	3.6x3.3	10.7	Com	79	
eg 4206	12 12.7	13 18	15.2	01	SA(s)bc:	5.2x1.2	12.1	Vir	258	
eg 4212	12 13.1	14 11	15.6	54	SAbc?	3.0x2.1	11.2	Com	79	=4208
eg 4214	12 13.1	36 37	15.6	20	IAB(s)m	7.9x6.3	9.7	CVn	47	=4228
eg 4215	12 13.4	06 41	16.0	24	SA(r)0$^+$:sp	1.9x0.8		Vir	258	
eg 4216	12 13.4	13 25	15.9	08	SAB(s)b:	8.3x2.2	9.9	Vir	258	
eg 4217	12 13.4	47 22	15.9	05	Sbsp	5.5x1.8		CVn	47	
eg 4218	12 13.3	48 25	15.8	08	Sa?	1.2x0.8		CVn	47	
eg 4219	12 13.8	−43 03	16.4	20	SA(s)bc	4.5x1.7		Cen		
eg 4220	12 13.7	48 10	16.2	53	SA(r)0$^+$	4.1x1.5		CVn	47	
eg 4222	12 13.8	13 35	16.3	18	Sd:sp	3.3x0.6		Com	79	
eg 4224	12 14.0	07 44	16.6	27	SA(s)a:sp	2.4x1.0	11.8	Vir	258	chart XXIV
eg 4226	12 14.0	47 18	16.5	01	SaP?	1.3x0.7		CVn	47	
eg 4233	12 14.6	07 54	17.2	37	S0$^+$	2.3x1.1	12.0	Vir	258	chart XXIV
eg 4234	12 14.6	03 58	17.2	41	SB(s)dm	1.3x1.2	12.9	Vir	258	
eg 4235	12 14.6	07 28	17.2	11	SA(s)asp	4.3x1.1	11.6	Vir	258	=I3098; chart XXIV
eg 4236	12 14.3	69 45	16.7	28	SB(s)dm	19x6.9	9.6	Dra	105	
eg 4237	12 14.7	15 36	17.2	19	SAB(rs)bc	2.3x1.6	11.7	Com	79	
eg 4239	12 14.7	16 48	17.2	31	S0	1.9x1.2		Com	79	
eg 4241	12 14.9	06 58	17.5	41	SA(s)0/a:	2.5x1.4	12.0	Vir	258	chart XXIV
eg 4242	12 15.0	45 54	17.5	37	SAB(s)dm	4.8x3.8	11.1	CVn	47	
eg 4244	12 15.0	38 05	17.5	48	SA(s)cd:sp	16x2.5	10.1	CVn	47	
eg 4245	12 15.1	29 53	17.6	36	SB(r)0/a:	3.3x2.6	11.4	Com	80	
eg 4246	12 15.4	07 28	18.0	11	SA(s)c	2.5x1.5	12.7	Vir	258	=I3113; chart XXIV
eg 4247	12 15.4	07 33	18.0	16	(R)SAB(s)abP?	0.7x0.6		Vir	259	chart XXIV
eg 4248	12 15.4	47 41	17.9	24	I0?sp	3.0x1.2	12.6	CVn	47	
eg 4249	12 15.4	05 53	18.0	36	S0			Vir		chart XXIV
eg 4251	12 15.6	28 27	18.1	10	SB0?sp	4.2x1.9		Com	80	
eg 4252	12 16.0	05 50	18.6	33	Sb?sp	1.5x0.4		Vir		chart XXIV
eg 4253	12 15.9	30 05	18.4	48	(R')SB(s)a?	1.1x1.0		Com	*80*	
eg 4254	12 16.3	14 42	18.8	25	SA(s)c	5.4x4.8	9.8	Com	80	=M99
eg 4255	12 16.4	05 04	19.0	47	SB(r)0°	1.5x0.8		Vir	259	chart XXIV
eg 4256	12 16.4	66 11	18.8	54	SA(s)b:sp	4.6x1.0		Dra	105	
eg 4257	12 16.6	06 00	19.2	43	Sab:sp	1.2x0.4		Vir	*259*	chart XXIV
eg 4258	12 16.5	47 35	19.0	18	SAB(s)bc	18x7.9	8.3	CVn	47	=M106
eg 4259	12 16.8	05 39	19.4	22	S0sp	1.1x0.5	13.6	Vir	259	chart XXIV
eg 4260	12 16.8	06 23	19.4	06	SB(s)a	2.6x1.4	11.7	Vir	259	chart XXIV

ID		RA (1950) DEC (2000)				TYPE	SIZE	V	CON	PG	NOTES C 24
eg	4261	12 16.8	06 06	19.4	49	E2.5	3.9x3.2	10.3	Vir	259	chart XXIV
eg	4262	12 17.0	15 09	19.5	52	SB(s)0^{-}?	2.2x2.0	11.5	Com	80	
eg	4264	12 17.0	06 08	19.6	51	SB(rs)0^{+}	1.1x0.9	12.9	Vir	259	chart XXIV
eg	4266	12 17.2	05 49	19.8	32	SB(s)a?sp	2.1x0.5		Vir	259	in 4281 group; chart XXIV
eg	4267	12 17.2	13 05	19.7	48	SB(s)0^{-}?	3.5x3.2	10.9	Vir	259	
eg	4268	12 17.2	05 34	19.8	17	SB0/a?sp	1.6x0.6	12.5	Vir	259	in 4281 group; chart XXIV
eg	4269	12 17.3	06 18	19.9	01	S0	1.5x1.0		Vir	259	chart XXIV
eg	4270	12 17.3	05 45	19.9	28	S0	2.2x1.0	12.2	Vir	259	in 4281 group; chart XXIV
eg	4273	12 17.4	05 37	20.0	20	SB(s)c	2.3x1.5	11.9	Vir	261	in 4281 group; chart XXIV
eg	4274	12 17.3	29 53	19.8	36	(R)SB(r)ab	6.9x2.8	10.3	Com	80	
eg	4276	12 17.6	07 58	20.1	41	SA(s)c	1.7x1.7		Vir		chart XXIV
eg	4277	12 17.5	05 37	20.1	20	SAB(rs)0/a:	0.9x0.8	13.5	Vir	261	in 4281 group; chart XXIV
eg	4278	12 17.6	29 34	20.1	17	E1.5	3.6x3.5	10.2	Com	80	
eg	4281	12 17.8	05 40	20.4	23	S0^{+}:sp	3.1x1.5	11.3	Vir	261	brightest of group; chart XXIV
eg	4282	12 17.9	05 51	20.5	34	S0sp	0.8x0.5		Vir		chart XXIV
eg	4283	12 17.8	29 35	20.3	18	E0	1.4x1.4	12.0	Com	80	
eg	4284	12 17.8	58 22	20.2	05	Sbc	2.9x1.4		UMa	252	
eg	4286	12 18.2	29 37	20.7	20	SA(r)0/a:	1.9x1.2		Com	*81*	=I3181
eg	4287	12 18.3	05 55	20.9	38	Ssp	1.4x0.2		Vir		chart XXIV
eg	4290	12 18.4	58 22	20.8	05	SB(rs)ab:	2.5x1.9		UMa	252	
eg	4291	12 18.1	75 39	20.3	22	E2.5	2.2x1.9		Dra	105	
eg	4292	12 18.7	04 52	21.3	35	SB0°:	2.1x1.4		Vir	261	chart XXIV
eg	4292A	12 18.7	04 54	21.3	37	SAB(r)c:	0.4x0.4		Vir		chart XXIV
eg	4293	12 18.7	18 40	21.2	23	(R)SB(s)0/a	6.0x3.0		Com	81	
eg	4294	12 18.8	11 47	21.3	30	SB(s)cd	3.1x1.3	12.1	Vir	261	
eg	4296	12 18.9	06 56	21.5	39	SB0P	1.8x1.1		Vir		chart XXIV
eg	4297	12 18.9	06 57	21.5	40	S0spP	0.7x0.3		Vir		chart XXIV
eg	4298	12 19.0	14 53	21.5	36	SA(rs)c	3.2x1.9	11.4	Com	81	
eg	4299	12 19.1	11 47	21.6	30	SAB(s)dm:	1.7x1.6	12.5	Vir	261	
eg	4300	12 19.1	05 40	21.7	23	(R)SA:	1.6x0.7		Vir	261	chart XXIV
eg	4301	12 19.0	05 03	21.6	46	SAa?	1.4x0.5		Vir	261	chart XXIV
eg	4302	12 19.2	14 53	21.7	36	Sc:sp	5.2x1.1	11.6	Com	81	
eg	4303	12 19.4	04 45	22.0	28	SAB(rs)bc	6.0x5.5	9.7	Vir	261	=M61; chart XXIV
eg	4303A	12 19.9	04 51	22.5	34	SB(s)cd	1.7x1.4	13.0	Vir	261	chart XXIV
eg	4304	12 19.6	−33 12	22.2	29	SB(s)b	2.4x2.4	12.0	Hya		
eg	4307	12 19.6	09 19	22.1	02	Sbsp	3.7x0.9		Vir	261	
eg	4309	12 19.7	07 25	22.2	08	SAB(r)0^{+}	2.0x1.2		Vir		chart XXIV
eg	4309A	12 19.7	07 27	22.2	10	E0			Vir		chart XXIV
eg	4312	12 20.0	15 49	22.5	32	SA(rs)ab:sp	4.7x1.3	11.8	Com	82	
eg	4313	12 20.1	12 05	22.6	48	SA(rs)ab?sp	3.9x1.1		Vir	*261*	
eg	4314	12 20.1	30 10	22.6	53	SB(rs)a	4.8x4.3	10.5	Com	82	
eg	4316	12 20.2	09 37	22.7	20	SAbcsp	2.7x0.6		Vir	*261*	
eg	4319	12 19.6	75 36	21.8	19	SB(r)ab	3.1x2.5		Dra	105	
eg	4321	12 20.4	16 06	22.9	49	SAB(s)bc	6.9x6.2	9.4	Com	82	=M100
eg	4322	12 20.5	16 11	23.0	54	SB(r)0°:	1.3x1.1	13.9	Com	82	=4323
eg	4324	12 20.6	05 32	23.2	15	SA(r)0^{+}	2.5x1.2		Vir	261	chart XXIV
eg	4326	12 20.7	06 21	23.3	04	SAB(r)ab:	1.7x1.4		Vir		chart XXIV
eg	4328	12 20.8	16 06	23.3	49	SA0^{-}:	1.5x1.4	13.5	Com	82	
eg	4330	12 20.8	11 39	23.3	22	Scd:sp	4.3x1.0		Vir	*261*	chart XXVI
eg	4333	12 20.8	06 19	23.4	02	SB(s)ab	1.1x0.9		Vir	*262*	chart XXIV

ID		RA (1950) DEC (2000)				TYPE	SIZE	V	CON	PG	NOTES C 25
eg	4334	12 20.9	07 45	23.4	28	SB(s)ab	2.4x1.1		Vir	262	chart XXIV
eg	4339	12 21.0	06 22	23.5	05	E0	2.3x2.3	11.4	Vir	262	chart XXIV
eg	4340	12 21.1	17 00	23.6	43	SB(r)0+	4.1x3.2	11.0	Com	82	
eg	4343	12 21.1	07 14	23.6	57	SA(rs)b:	2.8x0.9	12.3	Vir	262	chart XXIV
eg	4344	12 21.1	17 49	23.6	32	SB0:	1.9x1.8		Com	82	
eg	4346	12 21.0	47 16	23.4	59	S0sp	3.5x1.4		CVn	47	
eg	4348	12 21.3	−03 10	23.9	27	Sb:sp	3.5x1.0		Vir	262	
eg	4350	12 21.4	16 58	23.9	41	SA0sp	3.2x1.1	11.0	Com	82	
eg	4351	12 21.5	12 29	24.0	12	SB(rs)abP:	2.0x1.4	12.4	Vir	262	=4354; chart XXVI
eg	4352	12 21.6	11 30	24.1	13	SA0:sp	1.9x0.9	12.7	Vir	*261*	
eg	4359	12 21.7	31 48	24.2	31	SB(rs)c?sp	3.5x1.0		Com	*83*	
pn	4361	12 21.9	−18 31	24.5	48		1.9x1.9	*10.9*	Crv	93	
eg	4365	12 21.9	07 36	24.4	19	E3	6.2x4.6		Vir	262	chart XXIV
eg	4366	12 22.2	07 38	24.7	21	dE6			Vir	262	chart XXIV
eg	4369	12 22.1	39 40	24.6	23	(R)SA(rs)a	2.5x2.4		CVn	47	
eg	4370	12 22.4	07 43	24.9	26	Sasp	1.6x0.9		Vir	262	chart XXIV
eg	4371	12 22.4	11 59	24.9	42	SB(r)0+	3.9x2.5	10.9	Vir	262	chart XXVI
eg	4373	12 22.7	−39 29	25.4	46	SAB(s)0−:	3.2x2.3	11.1	Cen		
eg	4373A	12 23.0	−39 03	25.7	20	SA0°:sp	2.6x0.8		Cen		
eg	4374	12 22.5	13 10	25.0	53	E1	5.0x4.4	9.3	Vir	262	≡M84; chart XXVI
eg	4375	12 22.5	28 50	25.0	33	SB(r)abP?	1.6x1.4		Com	82	
eg	4376	12 22.8	06 01	25.4	44	SA(s)cd	1.7x1.0		Vir		chart XXIV
eg	4377	12 22.7	15 02	25.2	45	SA0−	1.8x1.5	11.8	Com	83	
eg	4378	12 22.8	05 12	25.4	55	(R)SA(s)a	3.3x3.1		Vir	264	chart XXIV
eg	4379	12 22.7	15 53	25.2	36	S0−P?	2.1x1.8	11.5	Com	83	
eg	4380	12 22.8	10 18	25.3	01	SA(rs)b:?	3.7x2.2		Vir	264	
eg	4382	12 22.9	18 28	25.4	11	SA(s)0+P	7.1x5.2	9.2	Com	83	≡M85
eg	4383	12 22.9	16 45	25.4	28	Sa?P	2.2x1.2		Com	83	
eg	4385	12 23.2	00 51	25.8	34	SB(rs)0+:	2.3x1.5	12.4	Vir	264	
eg	4386	12 22.4	75 48	24.5	31	SAB0−?	3.0x1.7		Dra	106	
eg	4387	12 23.2	13 05	25.7	48	E5	1.9x1.1	12.0	Vir	264	chart XXVI
eg	4388	12 23.2	12 56	25.7	39	SA(s)b:sp	5.1x1.4	11.2	Vir	264	chart XXVI
eg	4389	12 23.1	45 58	25.5	41	SB(rs)bcP:	2.7x1.5		CVn	48	
eg	4390	12 23.3	10 44	25.8	27	SAB(s)bc:	1.8x1.4		Vir	*264*	
eg	4394	12 23.4	18 29	25.9	12	(R)SB(r)b	3.9x3.5	10.9	Com	83	
eg	4395	12 23.3	33 50	25.8	33	SA(s)m:	13x11	10.2	CVn	48	
eg	4396	12 23.5	15 57	26.0	40	SAd:sp	3.5x1.2		Com	83	
eg	4402	12 23.6	13 23	26.1	06	Sbsp	4.1x1.3	11.3	Vir	264	chart XXVI
eg	4405	12 23.6	16 28	26.1	11	SA(rs)0/a:	2.0x1.4		Com	83	
eg	4406	12 23.7	13 13	26.2	56	E3	7.4x5.5	9.2	Vir	264	≡M86; chart XXVI
eg	4410AB	12 23.9	09 18	26.4	01	Sab?P, S0?P	1.3x0.8		Vir		P(b) at 0.3 in common envelope
eg	4411A	12 23.9	09 09	26.4	52	SB(rs)dm	2.2x2.0	12.8	Vir		
eg	4411B	12 24.3	09 10	26.8	53	SAB(rs)dm	2.7x2.7	12.4	Vir		
eg	4412	12 24.1	04 14	26.7	57	SB(r)b?P	1.5x1.4		Vir	264	
eg	4413	12 24.0	12 53	26.5	36	(R')SB(rs)b:	2.5x1.7		Vir	264	chart XXVI
eg	4414	12 24.0	31 30	26.5	13	SA(rs)c?	3.6x2.2	10.3	Com	83	
eg	4417	12 24.3	09 52	26.8	35	SB0:sp	3.6x1.4	11.1	Vir	264	
eg	4419	12 24.4	15 19	26.9	02	SB(s)asp	3.4x1.3	11.0	Com	83	
eg	4420	12 24.4	02 46	27.0	29	SB(r)bc:	2.2x1.2		Vir	264	=4409
eg	4421	12 24.5	15 44	27.0	27	SB(s)0/a	2.7x2.2	11.6	Com	84	

ID		RA (1950)	DEC (1950)	RA (2000)	DEC (2000)	TYPE	SIZE	V	CON	PG	NOTES C 26
eg	4424	12 24.7	09 42	27.2	25	SB(s)a:	3.7x1.9	11.6	Vir	264	
eg	4425	12 24.7	13 01	27.2	44	SB0:sp	3.4x1.2	11.8	Vir	265	chart XXVI
eg	4428	12 24.9	−07 54	27.5	11	SAB(rs)c	1.9x0.9		Vir	265	
eg	4429	12 24.9	11 23	27.4	06	SA(r)0$^+$	5.5x2.6	10.2	Vir	265	
eg	4431	12 24.9	12 34	27.4	17	SA(r)0	2.0x1.3	13.2	Vir	265	chart XXVI
eg	4433	12 25.1	−08 00	27.7	17	SAB(s)ab	2.3x1.1		Vir	265	
eg	4435	12 25.1	13 21	27.6	04	SB(s)0°	3.0x1.9	10.8	Vir	265	P(b) w/ 4438; chart XXVI
eg	4436	12 25.2	12 36	27.7	19	S0	1.9x0.9	13.5	Vir	265	chart XXVI
eg	4438	12 25.2	13 17	27.7	00	SA(s)0/aP:	9.3x3.9	10.0	Vir	265	P(b) w/ 4435; chart XXVI
eg	4440	12 25.4	12 34	27.9	17	SB(rs)a	2.0x1.7	11.8	Vir	265	chart XXVI
eg	4442	12 25.5	10 05	28.0	48	SB(s)0°	4.6x1.9	10.4	Vir	265	
eg	4445	12 25.7	09 43	28.2	26	Sab:sp	2.8x0.6	12.7	Vir	265	
eg	4446	12 25.6	14 11	28.1	54	SA(s)c	1.2x1.2		Com		chart XXVI
eg	4447	12 25.7	14 11	28.2	54	(R)SB0	1.1x0.9		Com		chart XXVI
eg	4448	12 25.8	28 54	28.3	37	SB(r)ab	4.0x1.6	11.1	Com	84	
eg	4449	12 25.8	44 22	28.2	05	IBm	5.1x3.7	9.4	CVn	48	
eg	4450	12 26.0	17 22	28.5	05	SA(s)ab	4.8x3.5	10.1	Com	84	
eg	4451	12 26.1	09 32	28.6	15	SA0:	1.5x1.0	12.5	Vir	265	
eg	4452	12 26.2	12 02	28.7	45	Sa?sp	2.4x0.6	12.3	Vir	265	chart XXVI
eg	4454	12 26.3	−01 40	28.9	57	(R)SB(r)0/a	2.2x1.9	12.1	Vir	265	
eg	4455	12 26.2	23 06	28.7	49	SB(s)d?sp	2.8x1.0		Com	84	
eg	4457	12 26.4	03 51	29.0	34	(R)SAB(s)0/a	3.0x2.5	10.8	Vir	265	
eg	4458	12 26.4	13 31	28.9	14	E0.5	1.9x1.8	12.1	Vir	265	chart XXVI
eg	4459	12 26.5	14 15	29.0	58	SA(r)0$^+$	3.8x2.8	10.4	Com	84	chart XXVI
eg	4460	12 26.3	45 08	28.7	51	SB(s)0$^+$?sp	4.4x1.4		CVn	48	
eg	4461	12 26.5	13 28	29.0	11	SB(s)0$^+$:	3.7x1.5	11.2	Vir	265	chart XXVI
eg	4462	12 26.7	−22 53	29.3	10	SB(r)ab	3.7x1.6		Crv	93	
eg	4464	12 26.8	08 26	29.3	09	S0/a?	1.1x0.9	12.6	Vir	266	
eg	4467	12 27.0	08 16	29.5	59	E2	0.7x0.6	14.5	Vir	266	
eg	4468	12 27.0	14 20	29.5	03	SA0:	1.5x1.1	13.0	Com	84	chart XXVI
eg	4469	12 26.9	09 02	29.4	45	SB(s)0/a?sp	3.9x1.5		Vir	265	
eg	4470	12 27.1	08 06	29.6	49	Sa?	1.5x1.1		Vir	266	
eg	4472	12 27.2	08 17	29.7	00	E2	8.9x7.4	8.4	Vir	266	=M49
eg	4473	12 27.3	13 42	29.8	25	E5	4.5x2.6	10.2	Com	84	chart XXVI
eg	4474	12 27.4	14 21	29.9	04	S0P:	2.3x1.2	11.7	Com	85	chart XXVI
eg	4476	12 27.5	12 38	30.0	21	SA(r)0$^-$:	1.9x1.3	12.3	Vir	266	chart XXVI
eg	4477	12 27.5	13 55	30.0	38	SB(s)0:?	4.0x3.5	10.4	Com	85	chart XXVI
eg	4478	12 27.8	12 36	30.3	19	E2	2.0x1.8	11.2	Vir	266	chart XXVI
eg	4479	12 27.8	13 51	30.3	34	SB(s)0°:?	1.8x1.5	12.5	Com	85	chart XXVI
eg	4480	12 27.9	04 31	30.5	14	SAB(s)c	2.6x1.4	12.4	Vir	266	
eg	4483	12 28.1	09 18	30.6	01	SB(s)0/a:	1.8x1.1		Vir	266	
eg	4485	12 28.1	41 59	30.5	42	IB(s)mP	2.4x1.7	12.0	CVn	48	P(b) w/ 4490
eg	4486	12 28.3	12 40	30.8	23	E$^+$0.5P	7.2x6.8	8.6	Vir	266	=M87; chart XXVI
eg	4486B	12 28.0	12 46	30.5	29	cE0	0.5x0.4	13.3	Vir		chart XXVI
eg	4487	12 28.5	−07 47	31.1	04	SAB(rs)cd	4.1x3.0		Vir	266	
eg	4489	12 28.3	17 02	30.8	45	E1	2.2x2.1		Com	85	
eg	4490	12 28.2	41 55	30.6	38	SB(s)dP	5.9x3.1	9.8	CVn	48	P(b) w/ 4485
eg	4491	12 28.4	11 46	30.9	29	SB(s)a:	1.9x1.0	12.6	Vir	266	chart XXVI
eg	4494	12 28.9	26 03	31.4	46	E1.5	4.8x3.8	9.9	Com	85	
eg	4496A	12 29.1	04 13	31.7	56	SB(rs)m	3.9x3.1		Vir	266	=4505; P(c?) w/ 4496B

ID		RA (1950)	DEC (2000)			TYPE	SIZE	V	CON	PG	NOTES C 27
eg	4496B	12 29.1	04 12	31.7	55	IB(s)m:	1.0x0.9		Vir	266	=4505; P(c?) w/ 4496A
eg	4497	12 29.0	11 54	31.5	37	SAB(s)0/a:	2.3x1.1	12.5	Vir	266	chart XXVI
eg	4498	12 29.1	17 08	31.6	51	SAB(s)cd	3.2x1.9		Com	85	
eg	4501	12 29.5	14 42	32.0	25	SA(rs)b	6.9x3.9	9.5	Com	85	=M88; chart XXVI
eg	4502	12 29.5	16 58	32.0	41	SB(s)c	1.4x0.7		Com	85	
eg	4503	12 29.6	11 27	32.1	10	SB0⁻:	3.5x1.8	11.1	Vir	266	
eg	4504	12 29.7	−07 17	32.3	34	SA(s)cd	4.0x2.8		Vir	266	
eg	4506	12 29.7	13 42	32.2	25	SaP?	1.6x1.3		Com	85	chart XXVI
eg	4507	12 32.9	−39 38	35.6	55	SAB(r)0⁺	2.3x2.0		Cen		
eg	4515	12 30.6	16 33	33.1	16	S0⁻:	1.6x1.3		Com	85	
eg	4516	12 30.6	14 51	33.1	34	SB(rs)ab?	1.9x1.1		Com		chart XXVI
eg	4517	12 30.2	00 23	32.8	06	SA(s)cd:sp	10x1.9	10.4	Vir	266	=4437
eg	4517A	12 29.9	00 40	32.5	23	SB(rs)dm:	4.2x3.0	12.2	Vir	266	
eg	4519	12 31.0	08 56	33.5	39	SB(rs)d	3.1x2.2	11.7	Vir	266	
eg	4522	12 31.1	09 27	33.6	10	SB(s)cd:sp	3.7x1.1		Vir	267	
eg	4526	12 31.5	07 59	34.0	42	SAB(s)0°:	7.2x2.3	9.6	Vir	267	
eg	4527	12 31.6	02 56	34.2	39	SAB(s)bc	6.3x2.3	10.4	Vir	267	
eg	4532	12 31.8	06 45	34.3	28	IBm	2.9x1.3	11.9	Vir	267	
eg	4533	12 31.8	02 36	34.4	19	Sd:sp	2.0x0.4		Vir	267	
eg	4535	12 31.8	08 29	34.3	12	SAB(s)c	6.8x5.0	10.0	Vir	267	
eg	4536	12 31.9	02 28	34.5	11	SAB(rs)bc	7.4x3.5	10.4	Vir	267	
eg	4540	12 32.3	15 50	34.8	33	SAB(rs)cd	2.0x1.6		Com		
eg	4546	12 32.9	−03 31	35.5	48	SB(s)0⁻:	3.5x1.7	10.3	Vir	267	
eg	4548	12 32.9	14 46	35.4	29	SB(rs)b	5.4x4.4	10.2	Com	85	
eg	4550	12 33.0	12 30	35.5	13	SB0⁺?sp	3.5x1.1	11.5	Vir	267	
eg	4551	12 33.1	12 32	35.6	15	E3:	2.0x1.6	11.9	Vir	267	
eg	4552	12 33.1	12 50	35.6	33	E0	4.2x4.2	9.8	Vir	267	=M89
eg	4553	12 33.4	−39 10	36.1	27	(R)SAB(r)0⁺	2.0x1.0		Cen		
eg	4559	12 33.5	28 14	36.0	57	SAB(rs)cd	10x4.9	9.9	Com	85	
eg	4561	12 33.6	19 36	36.1	20	SB(rs)dm	1.5x1.3		Com		=I3569
eg	4562	12 33.1	26 08	35.6	51	SB(s)m?sp	2.5x0.9		Com	86	
eg	4564	12 33.9	11 43	36.4	27	S0⁻	3.1x1.4	10.9	Vir	267	
eg	4565	12 33.9	26 16	36.4	00	SA(s)b?sp	16x2.8	9.5	Com	85	
eg	4567	12 34.0	11 32	36.5	16	SA(rs)bc	3.0x2.1	11.3	Vir	267	P(b) w/ 4568
eg	4568	12 34.0	11 31	36.5	15	SA(rs)bc	4.6x2.1	10.8	Vir	267	P(b) w/ 4567
eg	4569	12 34.3	13 26	36.8	10	SAB(rs)ab	9.5x4.7	9.5	Vir	267	=M90; P(b?) w/ I3583
eg	4570	12 34.4	07 31	36.9	15	S0sp	4.1x1.3	10.8	Vir	267	
eg	4571	12 34.4	14 30	36.9	14	SA(r)d	3.8x3.4	11.3	Com	86	=I3588
eg	4572	12 33.9	74 31	35.9	15	SAB(s)	1.8x1.1		Dra	106	
eg	4575	12 35.2	−40 16	37.9	32	SB(s)bc:	2.2x1.8		Cen		
eg	4578	12 35.0	09 50	37.5	34	SA(r)0°?	3.6x2.8	11.4	Vir	268	
eg	4579	12 35.2	12 06	37.7	50	SAB(rs)b	5.4x4.4	9.8	Vir	268	=M58
eg	4580	12 35.3	05 39	37.8	23	SAB(rs)aP	2.4x1.9		Vir	268	
eg	4584	12 35.8	13 23	38.3	07	SAB(s)a?	1.5x1.2		Vir	267	
eg	4586	12 35.9	04 36	38.4	20	SA(s)a:sp	4.4x1.6	11.6	Vir	268	
eg	4589	12 35.5	74 28	37.4	12	E2	3.0x2.7		Dra	106	
gc	4590	12 36.8	−26 28	39.5	44	X	12	7.7	Hya	132	=M68
eg	4592	12 36.7	−00 15	39.3	31	SA(s)dm:	4.6x1.5		Vir	268	
eg	4593	12 37.1	−05 04	39.7	20	(R)SB(rs)b	4.0x3.1		Vir	268	
eg	4594	12 37.4	−11 21	40.0	37	SA(s)asp	8.9x4.1	8.3	Vir	268	=M104; Sombrero

ID		RA (1950)	DEC (2000)			TYPE	SIZE	V	CON	PG	NOTES C 28
eg	4595	12 37.4	15 34	39.9	18	SAB(rs)b?	1.8x1.2		Com		
eg	4596	12 37.4	10 27	39.9	11	SB(r)0$^+$	3.9x2.8	10.5	Vir	268	
eg	4597	12 37.6	−05 32	40.2	48	SB(rs)m	3.6x1.9		Vir	268	
eg	4601	12 38.1	−40 37	40.8	53	SAB(r)0$^+$	2.2x0.6		Cen		
eg	4602	12 38.0	−04 52	40.6	08	SAB(rs)bc	3.6x1.4		Vir	268	
eg	4603	12 38.2	−40 42	40.9	58	SAB(rs)c	3.8x2.5		Cen		
eg	4605	12 37.8	61 53	40.0	37	SB(s)cP	5.5x2.3		UMa	252	
eg	4606	12 38.4	12 11	40.9	55	SB(s)a:	2.8x1.5	11.9	Vir	*269*	
eg	4607	12 38.7	12 10	41.2	54	SBb?sp	3.2x0.8	12.9	Vir	*269*	
eg	4608	12 38.7	10 26	41.2	10	SB(r)0°	3.2x2.6	11.1	Vir	268	
eg	4612	12 39.0	07 35	41.5	19	(R)SAB0°	2.2x1.8		Vir	268	
eg	4616	12 39.6	−40 22	42.3	38	SA(s)0$^-$:	1.6x1.6		Cen		
eg	4618	12 39.1	41 26	41.5	10	SB(rs)m	4.4x3.8	10.8	CVn	48	P(b?) w/ 4625
eg	4621	12 39.5	11 55	42.0	39	E5	5.1x3.4	9.8	Vir	269	=M59
eg	4622	12 39.9	−40 28	42.6	44	(R')SA(r)a	2.1x2.0		Cen		
eg	4623	12 39.6	07 57	42.1	41	SB0$^+$?sp	2.6x0.9		Vir	269	
eg	4625	12 39.5	41 33	41.9	17	SAB(rs)mP	2.4x2.0	12.3	CVn	48	=I3675; P(b?) w/ 4618
eg	4627	12 39.6	32 51	42.0	35	E4P	2.7x2.0	12.3	CVn	48	P(b) w/ 4631
eg	4630	12 40.0	04 14	42.5	58	IB(s)m:	1.7x1.3		Vir	269	
eg	4631	12 39.7	32 49	42.1	33	SB(s)dsp	15x3.3	9.2	CVn	48	P(b) w/ 4627
eg	4632	12 40.0	00 11	42.6	05	SAc	3.2x1.3		Vir	269	
eg	4635	12 40.2	20 13	42.7	57	SAB(s)d	2.0x1.5		Com		
eg	4636	12 40.3	02 58	42.9	42	E0.5	6.2x5.0	9.6	Vir	269	
eg	4637	12 40.4	11 43	42.9	27	S0?	1.5x0.8		Vir	*269*	
eg	4638	12 40.3	11 43	42.8	27	S0$^-$sp	2.8x1.6	11.2	Vir	269	
eg	4639	12 40.4	13 32	42.9	16	SAB(rs)bc	2.9x2.1	11.5	Vir	269	
eg	4642	12 40.7	−00 22	43.3	38	Sbsp	2.0x0.7		Vir	*269*	
eg	4643	12 40.8	02 15	43.4	59	SB(rs)0/a	3.4x2.7	10.6	Vir	269	
eg	4645	12 41.4	−41 29	44.1	45	SA0$^-$	2.2x1.5		Cen		
eg	4647	12 41.0	11 51	43.5	35	SAB(rs)c	3.0x2.5	11.3	Vir	269	
eg	4648	12 39.9	74 42	41.7	26	E3	2.2x1.7		Dra	106	
eg	4649	12 41.2	11 50	43.7	34	E2	7.2x6.2	8.8	Vir	269	=M60
eg	4651	12 41.2	16 40	43.7	24	SA(rs)cd	3.8x2.7	10.7	Com	86	
eg	4653	12 41.3	−00 17	43.9	33	SAB(rs)d	2.6x2.4	12.3	Vir	269	
eg	4654	12 41.4	13 24	43.9	08	SAB(rs)cd	4.7x3.0	10.5	Vir	269	=I3708
eg	4656	12 41.5	32 27	43.9	11	SB(s)mP	14x3.3	10.3	CVn	48	includes eg 4657
eg	4658	12 42.0	−09 49	44.6	05	SB(s)bc	2.2x1.0		Vir	270	
eg	4659	12 42.0	13 46	44.5	30	S0/a	1.8x1.3		Com	86	
eg	4660	12 42.0	11 28	44.5	12	E5.5:	2.8x1.9	10.9	Vir	270	
eg	4663	12 42.2	−09 56	44.8	12	SB(s)0°?	1.4x1.1		Vir	*270*	=I 811
eg	4665	12 42.6	03 20	45.2	04	SB(s)0/a	4.2x3.5		Vir	270	=4664
eg	4666	12 42.6	−00 11	45.2	27	SABc:	4.5x1.5	10.8	Vir	270	
eg	4668	12 43.0	−00 16	45.6	32	SB(s)d:	1.4x0.9	13.1	Vir	270	
eg	4670	12 42.8	27 24	45.3	08	SB(s)0/aP:	1.8x1.4	12.7	Com	86	
eg	4673	12 43.1	27 20	45.6	04	E1.5	1.2x1.0		Com	86	
eg	4677	12 44.2	−41 19	47.0	35	SB(s)0$^+$:	2.9x1.0	12.7	Cen		
eg	4679	12 44.8	−39 18	47.5	34	SA(s)bc?	2.3x1.2		Cen		
eg	4681	12 44.7	−43 04	47.5	20	Sab	1.1x1.0		Cen		
eg	4682	12 44.7	−09 47	47.3	03	SAB(rs)c	2.8x1.6		Vir	*271*	
eg	4684	12 44.7	−02 27	47.3	43	SB(r)0$^+$	2.9x1.1		Vir	270	

ID	RA (1950)	DEC (2000)			TYPE	SIZE	V	CON	PG	NOTES C 29
eg 4688	12 45.2	04 37	47.7	21	SB(s)cd	3.3x3.1		Vir	270	
eg 4689	12 45.3	14 02	47.8	46	SA(rs)bc	4.0x3.5	10.9	Com	86	
eg 4691	12 45.7	−03 04	48.3	20	(R)SB(s)0/aP	3.2x2.7	11.2	Vir	270	
eg 4694	12 45.7	11 15	48.2	59	SB0P	3.6x1.7		Vir		
eg 4696	12 46.1	−41 02	48.9	18	E$^+$1P	3.5x3.2	10.7	Cen		brightest in Centaurus I cluster
eg 4697	12 46.0	−05 32	48.6	48	E6	6.0x3.8	9.3	Vir	270	P(b) w/ A1245–05
eg 4698	12 45.9	08 46	48.4	30	SA(s)ab	4.3x2.5	10.5	Vir	270	
eg 4699	12 46.4	−08 24	49.0	40	SAB(rs)b	3.5x2.7	9.6	Vir	270	
eg 4700	12 46.5	−11 08	49.1	24	SB(s)c?sp	3.0x0.7		Vir	271	
eg 4701	12 46.7	03 40	49.2	24	SA(s)cd	3.0x2.5	12.4	Vir	271	
eg 4710	12 47.2	15 26	49.7	10	SA(r)0$^+$?sp	5.1x1.4	11.0	Com	86	
eg 4712	12 47.1	25 45	49.5	29	SA(s)bc	2.6x1.3	13.0	Com	86	
eg 4713	12 47.4	05 35	49.9	19	SAB(rs)d	2.8x1.9	11.8	Vir	271	
eg 4725	12 48.0	25 47	50.4	31	SAB(r)abP	11x7.9	9.2	Com	86	
eg 4731	12 48.4	−06 07	51.0	23	SB(s)cd	6.5x3.4		Vir	271	
eg 4733	12 48.6	11 11	51.1	55	SA0$^-$:	2.3x2.1		Vir	271	
eg 4736	12 48.5	41 24	50.9	08	(R)SA(r)ab	11x9.1	8.1	CVn	48	=M94
eg 4742	12 49.2	−10 11	51.8	27	E4:	2.3x1.5	11.1	Vir	271	
eg 4746	12 49.4	12 21	51.9	05	Sb:sp	2.5x0.7		Vir	271	
eg 4747	12 49.3	26 03	51.7	47	SBcd?spP	3.6x1.4	12.4	Com	86	
eg 4750	12 48.3	73 09	50.1	53	(R)SA(rs)ab	2.3x2.1		Dra	106	
eg 4753	12 49.8	−00 56	52.4	12	I0	5.4x2.9	9.9	Vir	271	
eg 4754	12 49.8	11 35	52.3	19	SB(r)0$^-$:	4.7x2.6	10.5	Vir	271	
eg 4756	12 50.3	−15 09	52.9	25	S0/a?	2.0x1.7		Crv	93	
eg 4757	12 50.2	−10 02	52.8	18	S0sp	1.8x0.5		Vir	271	
eg 4759	12 50.5	−08 55	53.1	11	S0P			Vir	272	
eg 4760	12 50.5	−10 13	53.1	29	E0?	1.8x1.8		Vir	271	
eg 4761	12 50.5	−08 55	53.1	11	S0P			Vir	272	
eg 4762	12 50.4	11 30	52.9	14	SB(r)0°?sp	8.7x1.6	10.2	Vir	272	
eg 4763	12 50.8	−16 44	53.4	00	Sa:	1.7x1.3		Crv	94	
eg 4764	12 50.6	−08 55	53.2	11	(R)SA0°:	0.7x0.3		Vir	272	
eg 4765	12 50.7	04 44	53.2	28	S0/a?	1.4x1.1		Vir	272	
eg 4766	12 50.5	−10 06	53.1	22	S0sp	1.6x0.4		Vir		pair w/ Anon. 1.2 NW
eg 4767	12 51.1	−39 27	53.9	43	SA0$^-$	2.6x1.4		Cen		
eg 4771	12 50.8	01 32	53.4	16	SAd?sp	4.0x1.0		Vir	272	
eg 4772	12 50.9	02 26	53.5	10	SA(s)a	3.3x1.7		Vir	272	
eg 4775	12 51.2	−06 21	53.8	37	SA(s)d	2.2x2.1		Vir	272	
eg 4781	12 51.8	−10 16	54.4	32	SB(rs)d	3.5x1.8		Vir	272	
eg 4782	12 52.0	−12 18	54.6	34	E0P	1.5x1.5	11.7	Crv	94	P(b) w/ 4783
eg 4783	12 52.0	−12 17	54.6	33	E0P	1.7x1.6	11.8	Crv	94	P(b) w/ 4782
eg 4784	12 52.0	−10 21	54.6	37	S0sp	1.8x0.4		Vir	272	
eg 4786	12 52.0	−06 35	54.6	51	E3P	2.0x1.5		Vir	272	
eg 4790	12 52.3	−09 59	54.9	15	SB(rs)c:?	1.8x1.3		Vir	272	
eg 4791	12 52.2	08 19	54.7	03	(R)SA0	1.6x1.1		Vir	273	
eg 4792	12 52.5	−12 14	55.1	30	S0sp	1.0x0.7		Crv	94	
eg 4793	12 52.2	29 13	54.6	57	SAB(rs)c	2.9x1.7	11.7	Com	86	
eg 4794	12 52.6	−12 20	55.2	36	SB(rs)a	2.3x1.1		Crv	94	
eg 4795	12 52.5	08 20	55.0	04	(R')SB(r)aP?	1.7x1.5		Vir	272	
eg 4796	12 52.5	08 20	55.0	04	cE	0.3x0.3		Vir	272	
eg 4800	12 52.3	46 48	54.6	32	SA(rs)b	1.8x1.4		CVn	49	

ID		RA (1950)	DEC	(2000)		TYPE	SIZE	V	CON	PG	NOTES C 30
eg	4802	12 53.2	−11 47	55.8	03	SA0?			Crv	94	=4804
eg	4803	12 53.1	08 30	55.6	14	Sab?	0.6x0.4		Vir	273	
eg	4808	12 53.3	04 34	55.8	18	SA(s)cd:	2.7x1.3		Vir	273	
eg	4814	12 53.2	58 37	55.3	21	SA(s)b	3.2x2.5		UMa	253	
eg	4818	12 54.2	−08 15	56.8	31	SAB(s)ab	4.5x1.7		Vir	273	
eg	4825	12 54.6	−13 24	57.2	40	SA0$^-$	2.0x1.5	12.1	Vir	273	
eg	4826	12 54.3	21 57	56.8	41	(R)SA(rs)ab	9.3x5.4	8.5	Com	86	=M64; Blackeye
eg	4835	12 55.3	−46 00	58.1	16	SAB(rs)bc:	3.4x1.0		Cen		
eg	4838	12 55.4	−12 48	58.0	04	SB0/a?	2.0x1.7		Vir	273	
eg	4839	12 55.0	27 46	57.4	30	SA0$^-$	4.2x2.1	12.4	Com	87	
eg	4840	12 55.1	27 53	57.5	37	E1:	0.6x0.6	13.5	Com		chart VIII
eg	4841A	12 55.1	28 45	57.5	29	SA0P	1.7x1.7	12.7	Com	87	
eg	4841B	12 55.2	28 45	57.6	29	E0P	1.3x1.3	13.3	Com	87	
eg	4842A	12 55.2	27 46	57.6	30	E0	0.6x0.4	13.9	Com	87	in Coma cluster
eg	4842B	12 55.2	27 45	57.6	29	E3:	0.4x0.3	15.0	Com	87	in Coma cluster
eg	4845	12 55.5	01 51	58.1	35	SA(s)absp	5.0x1.6		Vir	273	
eg	4848	12 55.7	28 31	58.1	15	Sa:sp	1.8x0.6	13.6	Com	87	chart VIII
eg	4850	12 56.0	28 14	58.4	58	SA0	1.6x1.6	14.1	Com		chart VIII
eg	4851	12 55.9	28 25	58.3	09	SAB0		14.6	Com		chart VIII
eg	4853	12 56.2	27 52	58.6	36	(R')SA0$^-$?	1.2x1.0	13.1	Com		in Coma cluster; chart VIII
eg	4854	12 56.4	27 57	58.8	41	SB0		13.9	Com		in Coma cluster; chart VIII
eg	4855	12 56.7	−12 57	59.3	13	E			Vir	273	
eg	4856	12 56.7	−14 46	59.3	02	SB(s)0/a	4.6x1.6	10.4	Vir	273	
eg	4858	12 56.6	28 23	59.0	07	SBb	0.4x0.3	15.0	Com	87	in Coma cluster; chart VIII
eg	4860	12 56.7	28 24	59.1	08	E2:	1.0x0.9	13.5	Com	87	in Coma cluster; chart VIII
eg	4861	12 56.7	35 08	59.1	52	SB(s)m:	4.1x1.6	12.2	CVn	49	
eg	4864	12 56.8	28 15	59.2	59	E2	0.6x0.4	13.6	Com	87	in Coma cluster; chart VIII
eg	4865	12 56.9	28 21	59.3	05	SA0°sp	1.4x0.8	13.3	Com	87	in Coma cluster; chart VIII
eg	4866	12 57.0	14 27	59.5	11	SA(r)0$^+$:sp	6.5x1.5	11.2	Vir	273	
eg	4867	12 56.8	28 14	59.2	58	S0$^-$	1.1x1.1	14.2	Com		in Coma cluster; chart VIII
eg	4868	12 56.8	37 35	59.1	19	SAb?	1.7x1.6		CVn	49	
eg	4869	12 57.0	28 11	59.4	55	E3	1.1x1.1	13.5	Com	87	in Coma cluster; chart VIII
eg	4871	12 57.1	28 14	59.5	58	SA0	0.5x0.4	14.1	Com	87	in Coma cluster; chart VIII
eg	4872	12 57.2	28 13	59.6	57	SB0	1.0x1.0	13.7	Com	87	in Coma cluster; chart VIII
eg	4873	12 57.1	28 15	59.5	59	SA0	0.8x0.6	14.2	Com	87	in Coma cluster; chart VIII
eg	4874	12 57.2	28 14	59.6	58	E$^+$0	2.7x2.7	11.9	Com	87	in Coma cluster; chart VIII
eg	4875	12 57.2	28 11	59.6	55	SA0		14.7	Com		in Coma cluster; chart VIII
eg	4876	12 57.3	28 11	59.7	55	E5	0.4x0.4	14.2	Com		in Coma cluster; chart VIII
eg	4877	12 57.8	−15 01	00.4	17	SA(s)ab:	2.7x1.3		Vir	273	
eg	4880	12 57.7	12 45	00.2	29	SAB(r)0/a:	3.3x2.5		Vir	273	
eg	4881	12 57.6	28 31	00.0	15	SA0$^-$	1.0x1.0	13.5	Com	87	in Coma cluster; chart VIII
eg	4883	12 57.5	28 18	59.9	02	SB0	0.4x0.3	14.3	Com	89	in Coma cluster; chart VIII
eg	4886	12 57.7	28 15	00.1	59	E0	0.8x0.8	13.9	Com	89	=4882; in Coma cluster; chart VIII
eg	4887	12 58.0	−14 24	00.6	40	S0$^+$P?	1.6x0.8		Vir	273	
eg	4889	12 57.7	28 15	00.1	59	E$^+$4	3.0x2.1	11.4	Com	89	=4884; brightest in Coma cluster; chart VIII
eg	4890	12 58.0	−04 18	00.6	34	SP	1.1x0.8		Vir	274	
eg	4891	12 58.3	−13 10	00.9	26	SB(r)bc:	2.8x2.5		Vir	273	
eg	4894	12 57.9	28 14	00.3	58	SA0sp		13.9	Com		in Coma cluster; chart VIII
eg	4895	12 57.9	28 28	00.3	12	SA0Psp	2.3x0.9	12.8	Com	89	in Coma cluster; chart VIII
eg	4895A	12 57.8	28 26	00.2	10	S0$^-$			Com		in Coma cluster; chart VIII

ID		RA (1950)	DEC (2000)			TYPE	SIZE	V	CON	PG	NOTES C 31
eg	4896	12 58.1	28 37	00.5	21	S0$^-$P:	1.3x0.7	13.7	Com		in Coma cluster; chart VIII
eg	4898AB	12 57.9	28 14	00.3	58	EP, EP		13.6	Com	89	in Coma cluster; double system; chart VIII
eg	4899	12 58.3	−13 41	00.9	57	SAB(rs)c:	2.7x1.6		Vir	273	
eg	4900	12 58.1	02 46	00.6	30	SB(rs)c	2.3x2.2	11.5	Vir	273	
eg	4902	12 58.4	−14 15	01.0	31	SB(r)b	3.0x2.8	11.2	Vir	274	
eg	4904	12 58.4	00 15	01.0	01	SB(s)cd	2.3x1.6	12.1	Vir	274	
eg	4906	12 58.3	28 12	00.7	56	E3	0.6x0.5	14.2	Com		in Coma cluster; chart VIII
eg	4907	12 58.4	28 26	00.8	10	SB(r)b	1.5x1.3	13.4	Com	89	chart VIII
eg	4908	12 58.5	28 19	00.9	03	E5	1.1x1.0	13.5	Com		in Coma cluster; chart VIII
eg	4911	12 58.5	28 04	00.9	48	SAB(r)bc	1.3x1.2	12.8	Com	89	in Coma cluster; chart VIII
eg	4914	12 58.4	37 35	00.7	19	SA0$^-$	3.6x2.2		CVn	49	
eg	4915	12 58.9	−04 17	01.5	33	E0	1.7x1.4	11.9	Vir	274	
eg	4918	12 59.2	−04 14	01.8	30	Sb:			Vir	274	
eg	4919	12 58.9	28 05	01.3	49	(R')SA(r)0°?	1.4x0.8	13.7	Com	89	chart VIII
eg	4921	12 59.0	28 09	01.4	53	SB(rs)ab	2.7x2.4	12.1	Com	89	in Coma cluster; chart VIII
eg	4923	12 59.1	28 07	01.5	51	(R')SA(r)0$^-$?	1.3x1.2	13.6	Com	89	chart VIII
eg	4926	12 59.5	27 54	01.9	38	SA0$^-$	1.4x1.3	12.9	Com		chart VIII
eg	4926A	12 59.7	27 55	02.1	39	S0P?	0.8x0.6	14.4	Com		in Coma cluster; chart VIII
eg	4927	12 59.6	28 16	02.0	00	SA0$^-$:		13.6	Com		chart VIII
eg	4928	13 00.4	−07 49	03.0	05	SA(s)bcP	1.3x1.0		Vir	274	
eg	4933A	13 01.3	−11 14	03.9	30	S0/aP	2.5x1.5		Vir	274	P(b) w/ 4933B
eg	4933B	13 01.3	−11 14	03.9	30	EP	0.8x0.7		Vir	274	P(b) w/ 4933A
eg	4936	13 01.5	−30 15	04.2	31	SA0$^-$?	1.9x1.9	11.3	Cen		
eg	4939	13 01.6	−10 04	04.2	20	SA(s)bc	5.8x3.2		Vir	274	
eg	4941	13 01.6	−05 17	04.2	33	(R)SAB(r)ab:	3.7x2.1	11.1	Vir	274	
eg	4945	13 02.5	−49 12	05.4	28	SB(s)cd:sp	20x4.4		Cen	68	
eg	4947	13 02.6	−35 04	05.4	20	SB(r)ab	2.8x1.7		Cen		=I3974
eg	4948	13 02.3	−07 41	04.9	57	SB(s)d?sp	1.8x0.7		Vir	274	
eg	4948A	13 02.5	−07 54	05.1	10	SB(s)d	1.4x1.2	13.4	Vir	274	
eg	4951	13 02.5	−06 14	05.1	30	SAB(rs)cd:	3.3x1.4		Vir	274	
eg	4957	13 02.8	27 50	05.2	34	E3	1.3x1.1	12.9	Com	89	
eg	4958	13 03.2	−07 45	05.8	01	SB(r)0?sp	4.1x1.4	10.5	Vir	274	
eg	4961	13 03.4	28 00	05.8	44	SB(s)cd	1.7x1.3	13.5	Com	89	
eg	4981	13 06.2	−06 31	08.8	47	SAB(r)bc	2.8x2.2		Vir	274	
eg	4984	13 06.3	−15 15	08.9	31	(R)SAB(rs)0$^+$	2.8x2.2		Vir	274	
eg	4995	13 07.1	−07 34	09.7	50	SAB(rs)b	2.5x1.7	11.0	Vir	274	
eg	4999	13 07.0	01 56	09.6	40	SB(r)b	2.6x2.2		Vir	274	
eg	5005	13 08.6	37 19	10.9	03	SAB(rs)bc	5.4x2.7	9.8	CVn	49	
eg	5011	13 10.0	−42 50	12.9	06	E1.5	2.0x2.0		Cen	68	
eg	5012	13 09.2	23 11	11.6	55	SAB(rs)c	2.9x1.8		Com		
eg	5016	13 09.7	24 22	12.1	06	SAB(rs)c	1.9x1.4		Com		
eg	5017	13 10.3	−16 30	13.0	46	E2:	1.7x1.4		Vir	275	
eg	5018	13 10.3	−19 15	13.0	31	(R')SAB(rs)0°:P	2.6x2.1	10.8	Vir	275	brightest of group
eg	5022	13 10.8	−19 17	13.5	33	Sb:Psp	2.5x0.4		Vir	275	
gc	5024	13 10.5	18 26	12.9	10	V	13	7.5	Com	89	=M53
eg	5033	13 11.1	36 52	13.4	36	SA(s)c	10x5.6	10.1	CVn	49	
eg	5035	13 12.2	−16 14	14.9	30	SAB(r)0$^+$	1.6x1.2		Vir	275	in 5044 group
eg	5037	13 12.3	−16 20	15.0	36	SA(s)a:	2.5x0.9		Vir		in 5044 group
eg	5044	13 12.7	−16 07	15.4	23	E0	2.6x2.6	11.0	Vir	275	brightest of group
eg	5046	13 13.1	−16 04	15.8	20	E2?	1.2x1.0		Vir	275	in 5044 group

ID		RA (1950)	DEC	(2000)		TYPE	SIZE	V	CON	PG	NOTES C 32
eg	5047	13 13.1	−16 15	15.8	31	S0sp	3.0x0.8		Vir	275	in 5044 group
eg	5049	13 13.3	−16 08	16.0	24	S0sp	2.1x0.7	12.9	Vir	275	in 5044 group
gc	5053	13 14.0	17 58	16.4	42	XI	11	9.9	Com	89	
eg	5054	13 14.3	−16 22	17.0	38	SA(s)bc	5.0x3.1		Vir	275	
eg	5055	13 13.6	42 18	15.8	02	SA(rs)bc	12x7.6	8.6	CVn	49	=M63
eg	5061	13 15.3	−26 34	18.0	50	SA0−	2.6x2.3		Hya		
eg	5068	13 16.2	−20 47	18.9	03	SAB(rs)cd	6.9x6.3		Vir	275	
eg	5070	13 16.6	−12 13	19.2	29	Sc?	0.7x0.2		Vir	275	chart XVII
eg	5072	13 16.6	−12 16	19.2	32	E			Vir	275	chart XVII
eg	5074	13 16.1	31 44	18.4	28	SAbP?	1.0x1.0	14.0	CVn	49	
eg	5076	13 16.8	−12 29	19.4	45	(R')SB0+:	1.6x0.9		Vir	275	chart XVII
eg	5077	13 16.9	−12 24	19.5	40	E3.5	2.0x1.6	11.5	Vir	275	brightest of group; chart XVII
eg	5079	13 17.0	−12 26	19.6	42	SB(rs)bc:P	1.7x1.0		Vir	275	chart XVII
eg	5082	13 17.8	−43 26	20.7	42	SB(s)0+	3.1x1.0		Cen		
eg	5084	13 17.6	−21 34	20.3	50	SAb?Psp	4.8x1.3		Vir	275	
eg	5085	13 17.6	−24 11	20.3	27	SA(s)c	3.4x3.0		Hya	132	
eg	5087	13 17.7	−20 21	20.4	37	SAB0−?	2.3x1.5	11.0	Vir	277	
eg	5088	13 17.7	−12 19	20.3	35	SA(s)b:?	2.7x0.9		Vir	277	in 5077 group; chart XVII
eg	5090	13 18.3	−43 27	21.2	43	E2	2.6x2.5		Cen		
eg	5091	13 18.4	−43 28	21.3	44	Sb:sp	2.2x0.4		Cen		
eg	5097	13 18.3	−12 13	20.9	29	S0?			Vir		chart XVII
eg	5101	13 19.0	−27 10	21.8	26	(R)SB(rs)0/a	5.5x4.9		Hya		
eg	5102	13 19.1	−36 22	21.9	38	SA(s)0−P:	9.3x3.5	9.7	Cen		
eg	5107	13 19.2	38 48	21.5	32	SB(s)cd?sp	1.9x0.7		CVn	*49*	
eg	5112	13 19.7	39 00	22.0	44	SB(rs)cd	3.9x2.9		CVn	49	
eg	5116	13 20.6	27 15	23.0	59	SB(s)c:	2.2x0.9		Com	90	
eg	5121	13 21.9	−37 25	24.8	41	(R')SA(s)a	2.3x2.0		Cen		
eg	5128	13 22.6	−42 45	25.5	01	S0P	18x14	7.0	Cen	68	=Centaurus A
eg	5134	13 22.6	−20 52	25.3	08	SA(s)b?	2.8x1.7		Vir	277	
eg	5135	13 23.0	−29 34	25.8	50	SB(s)ab	2.4x1.0		Hya		
gc	5139	13 23.8	−47 13	26.8	29	VIII	36	3.5	Cen	68	=omega Centauri
eg	5147	13 23.8	02 22	26.3	06	SB(s)dm	1.8x1.5	11.8	Vir	277	
eg	5161	13 26.4	−32 55	29.2	10	SA(s)c:	5.4x2.3		Cen		
eg	5169	13 26.1	46 56	28.2	41	SB(rs)b:	2.4x1.0		CVn	*50*	
eg	5170	13 27.1	−17 43	29.8	58	SA(s)c:sp	8.1x1.3		Vir	277	
eg	5172	13 26.9	17 19	29.3	04	SAB(rs)bc:	3.3x1.9	11.9	Com	90	
eg	5173	13 26.3	46 51	28.4	36	E0:	1.3x1.3		CVn	*50*	
eg	5180	13 27.0	17 05	29.4	50	S0?	1.8x1.2		Com	90	
eg	5188	13 28.6	−34 32	31.5	47	SB(rs)b:P	3.2x1.3		Cen		
eg	5193	13 29.1	−32 59	31.9	14	E0P:	1.8x1.6		Cen		P(b?) w/ 5193A
eg	5193A	13 29.0	−32 59	31.8	14	S0°:sp	0.8x0.3		Cen		P(b?) w/ 5193
eg	5194	13 27.8	47 27	29.9	12	SA(s)bcP	11x7.8	8.4	CVn	49	=M51; P(b) w/ 5195
eg	5195	13 27.9	47 32	30.0	17	I0P	5.4x4.3	9.6	CVn	49	=M51; P(b) w/ 5194
eg	5198	13 28.1	46 56	30.2	41	E1.5:	2.1x1.9		CVn	50	
eg	5204	13 27.7	58 41	29.6	26	SA(s)m	4.8x3.0	11.3	UMa	253	in 5457 group
eg	5221	13 32.5	14 05	35.0	50	(R)SBb:P	2.9x0.8		Vir	*277*	P(b) w/ 5222AB
eg	5222A	13 32.5	14 00	35.0	45	E	1.6x1.2		Vir	*277*	P(c) w/ 5222B; P(b) w/ 5221
eg	5222B	13 32.5	14 00	35.0	45	SP	0.7x0.4		Vir	*277*	P(c) w/ 5222A; P(b) w/ 5221
eg	5230	13 33.1	13 56	35.6	41	SA(s)c	2.2x2.0		Vir	277	
eg	5236	13 34.2	−29 37	37.0	52	SAB(s)c	11x10		Hya	132	=M83

ID		RA (1950)	DEC	(2000)		TYPE	SIZE	V	CON	PG	NOTES C 33
eg	5247	13 35.4	−17 38	38.1	53	SA(s)bc	5.4x4.7	10.5	Vir	277	
eg	5248	13 35.1	09 09	37.6	54	SAB(rs)bc	6.5x4.9	10.2	Boo	37	
eg	5253	13 37.1	−31 24	39.9	39	Im:P	4.0x1.7	10.6	Cen		
eg	5263	13 37.6	28 39	39.9	24	Sab:	1.7x0.4		CVn	*50*	
gc	5272	13 39.9	28 38	42.2	23	VI	16	5.9	CVn	50	=M3
eg	5273	13 39.9	35 55	42.1	40	SA(s)0°	3.1x2.7	11.6	CVn	50	
eg	5276	13 40.2	35 53	42.4	38	SAB(s)b	1.1x0.6		CVn	*50*	
eg	5296	13 44.2	44 06	46.3	51	S0+:	1.2x0.7		CVn	*50*	P(b?) w/ 5297
eg	5297	13 44.3	44 07	46.4	52	SAB(s)bc:sp	5.6x1.4		CVn	50	P(b?) w/ 5296
eg	5300	13 45.7	04 12	48.2	57	SAB(r)c	3.9x2.7		Vir	277	
eg	5301	13 44.4	46 21	46.4	06	SA(s)b:sp	4.4x1.1		CVn	50	
eg	5302	13 46.0	−30 16	48.9	31	SB(s)0+:	1.7x1.3	12.2	Cen		
eg	5308	13 45.4	61 13	47.1	58	S0-sp	3.5x0.8	11.3	UMa	253	
eg	5311	13 46.8	40 14	48.9	59	S0/a	2.2x1.8		CVn	50	chart V
eg	5313	13 47.6	40 14	49.7	59	Sb?	1.9x1.2		CVn	50	chart V
eg	5322	13 47.6	60 26	49.3	11	E3.5	5.5x3.9	10.0	UMa	253	
eg	5324	13 49.5	−05 49	52.1	04	SA(rs)c:	2.4x2.3		Vir	277	
eg	5326	13 48.7	39 49	50.8	34	SAa:	2.5x1.3		CVn	50	chart V
eg	5328	13 50.1	−28 15	52.9	30	E1.5:	1.7x1.4	11.8	Hya	132	
eg	5330	13 50.2	−28 13	53.0	28	cE0	0.3x0.3		Hya	*134*	
eg	5334	13 50.3	−00 52	52.9	07	SB(rs)c:	4.4x3.3		Vir	277	=I4338
eg	5337	13 50.3	39 56	52.4	41	S0/a:	1.8x0.9		CVn		chart V
eg	5338	13 50.9	05 27	53.4	12	SB:0	2.6x1.5		Vir	*278*	
eg	5341	13 50.4	38 04	52.6	49	Sb:	1.5x0.7		CVn	*50*	
eg	5345	13 51.7	−01 11	54.3	26	(R')SAab:P	1.6x1.6		Vir	*277*	
eg	5346	13 51.0	39 49	53.1	34	SA(s)c	2.4x1.0		CVn		chart V
eg	5347	13 51.1	33 44	53.3	29	(R')SB(s)ab	1.9x1.5	12.6	CVn	50	
eg	5348	13 51.7	05 28	54.2	13	SBbc:sp	3.6x0.6		Vir	*278*	
eg	5349	13 51.0	38 08	53.2	53	SBb	1.8x0.6		CVn	*50*	
eg	5350	13 51.3	40 37	53.4	22	SB(r)b	3.2x2.6	11.4	CVn	52	chart V
eg	5351	13 51.3	38 10	53.5	55	SA(r)b:	3.1x1.8	12.1	CVn	50	
eg	5353	13 51.4	40 32	53.5	17	S0sp	2.8x1.5	11.1	CVn	52	chart V
eg	5354	13 51.4	40 33	53.5	18	S0sp	2.3x2.0	11.5	CVn	52	chart V
eg	5355	13 51.7	40 35	53.8	20	S0?	1.5x0.9		CVn	52	chart V
eg	5356	13 52.5	05 35	55.0	20	SABbc:sp	3.2x1.0		Vir	*278*	
eg	5357	13 53.1	−30 06	56.0	21	E1	1.1x1.0		Cen		in I4329 group
eg	5358	13 51.9	40 31	54.0	16	S0/a	1.4x0.4		CVn	52	chart V
eg	5360	13 53.1	05 14	55.6	59	I0	2.0x0.9		Vir		
eg	5362	13 52.8	41 34	54.9	19	Sb?P	2.4x1.1		CVn	52	
eg	5363	13 53.6	05 30	56.1	15	I0?	4.2x2.7	10.2	Vir	277	
eg	5364	13 53.7	05 16	56.2	01	SA(rs)bcP	7.1x5.0	10.4	Vir	278	
eg	5365	13 54.8	−43 41	57.9	56	(R)SB(s)0-	3.1x2.4		Cen		
eg	5371	13 53.6	40 42	55.7	27	SAB(rs)bc	4.4x3.6	10.8	CVn	52	=5390; chart V
eg	5376	13 53.6	59 45	55.3	30	SAB(r)b?	2.1x1.4		UMa	253	
eg	5377	13 54.3	47 29	56.3	14	(R)SB(s)a	4.6x2.7	11.2	CVn	52	
eg	5378	13 54.7	38 03	56.8	48	(R')SB(r)a	2.7x2.3		CVn	*54*	
eg	5379	13 53.9	59 59	55.5	44	SAB(r)b?	2.2x1.1		UMa	*253*	
eg	5380	13 54.8	37 51	56.9	36	SA0-	2.1x2.1		CVn	52	
eg	5383	13 55.0	42 06	57.1	51	SB(rs)b:P	3.5x3.1	11.4	CVn	54	
eg	5389	13 54.5	59 59	56.1	44	SAB(r)0/a:?	4.1x1.3		UMa	*253*	

ID		RA (1950)	DEC	(2000)		TYPE	SIZE	V	CON	PG	NOTES C 34
eg	5394	13 56.4	37 42	58.5	27	SB(s)bP	1.9x1.1	13.0	CVn	54	P(b) w/ 5395
eg	5395	13 56.5	37 40	58.6	25	SA(s)bP	3.1x1.7	11.6	CVn	54	P(b) w/ 5394
eg	5398	13 58.4	−32 49	01.3	03	(R')SBdm?P	2.9x1.4		Cen		
eg	5406	13 58.2	39 09	00.3	55	SAB(rs)bc	2.1x1.6		CVn	54	
eg	5408	14 00.3	−41 08	03.4	22	IB(s)m	1.8x1.2		Cen		
eg	5419	14 00.7	−33 44	03.6	58	E4	2.4x2.1		Cen		brightest in cluster
eg	5422	13 58.9	55 24	00.7	10	S0sp	3.9x0.9		UMa	253	
eg	5426	14 00.8	−05 50	03.4	04	SA(s)cP	2.9x1.6	12.2	Vir	278	P(b) w/ 5427
eg	5427	14 00.8	−05 48	03.4	02	SA(s)cP	2.5x2.3	11.4	Vir	278	P(b) w/ 5426
eg	5430	13 59.1	59 34	00.7	20	SB(s)b	2.4x1.5		UMa	253	
eg	5440	14 00.8	35 00	03.0	46	SAa	3.3x1.4		CVn	54	
eg	5443	14 00.5	56 03	02.2	49	SB(s)b?	2.8x1.2		UMa	253	
eg	5444	14 01.2	35 22	03.4	08	SA0⁻:	2.7x2.3		CVn	54	
eg	5445	14 01.4	35 16	03.6	02	S0?	2.0x0.9		CVn	54	
eg	5448	14 00.9	49 25	02.8	11	(R)SAB(r)a	4.2x2.0		UMa	253	
eg	5457	14 01.5	54 36	03.3	22	SAB(rs)cd	27x26	7.7	UMa	253	=M101; brightest of group
gc	5466	14 03.2	28 46	05.4	32	XII	11	9.0	Boo	37	
eg	5468	14 04.0	−05 13	06.6	27	SAB(rs)cd	2.5x2.5		Vir	278	
eg	5472	14 04.3	−05 13	06.9	27	SA(r)ab?sp	1.4x0.4		Vir	278	
eg	5473	14 03.0	55 08	04.7	54	SAB(s)0⁻:	2.6x1.8	11.4	UMa	254	
eg	5474	14 03.3	53 54	05.1	40	SA(s)cdP	4.5x4.2	10.9	UMa	254	in 5457 group
eg	5480	14 04.5	50 58	06.3	44	SA(s)c:	1.8x1.3		UMa	254	
eg	5481	14 04.8	50 58	06.6	44	SA0⁻	1.7x1.4		UMa	254	
eg	5483	14 07.3	−43 05	10.4	19	SA(s)c	3.1x2.8		Cen		
eg	5484	14 05.1	55 16	06.8	02	E2			UMa	254	
eg	5485	14 05.5	55 14	07.2	00	SA0P	2.6x2.1	11.5	UMa	254	
eg	5486	14 05.7	55 20	07.4	06	SA(s)m:	1.7x1.1		UMa	254	
eg	5493	14 08.9	−04 49	11.5	03	S0Psp	2.0x1.4	11.5	Vir	278	
eg	5494	14 09.5	−30 25	12.4	39	SA(s)c	2.2x2.0		Cen		
eg	5496	14 09.1	−00 55	11.7	09	Sd:sp	4.4x1.0		Vir	278	
eg	5523	14 12.6	25 33	14.9	19	SA(s)cd:	4.5x1.4		Boo	37	
eg	5530	14 15.3	−43 10	18.5	24	SA(rs)bc	4.1x2.2		Lup	158	
eg	5533	14 14.0	35 35	16.1	21	SA(rs)ab	3.2x2.1	11.8	Boo	37	
eg	5534	14 15.0	−07 11	17.6	25	(R')SAB(s)abP:	1.4x0.8		Vir	278	
eg	5548	14 15.7	25 22	18.0	08	(R')SA(s)0/a	1.9x1.7	12.5	Boo	37	Seyfert
eg	5556	14 17.7	−29 01	20.6	15	SAB(rs)d	3.1x2.7		Hya	134	
eg	5557	14 16.3	36 43	18.4	29	E1	2.4x2.2	11.1	Boo	37	
eg	5560	14 17.6	04 13	20.1	59	SB(s)bP	3.9x0.9	12.4	Vir	278	P(b) w/ 5566, 5569
eg	5566	14 17.8	04 10	20.3	56	SB(r)ab	6.5x2.4	10.5	Vir	278	P(b) w/ 5560, 5569
eg	5569	14 18.0	04 13	20.5	59	SAB(rs)cd:	1.9x1.7		Vir	278	P(b) w/ 5560, 5566
eg	5574	14 18.4	03 28	20.9	14	SB0⁻:?sp	1.6x1.0	12.4	Vir	278	
eg	5576	14 18.5	03 30	21.0	16	E3	3.2x2.2	10.9	Vir	278	
eg	5577	14 18.7	03 40	21.2	26	SA(rs)bc:	3.4x1.1		Vir	278	
eg	5584	14 19.8	−00 10	22.4	24	SAB(rs)cd	3.3x2.6		Vir	279	
eg	5585	14 18.2	56 58	19.8	44	SAB(s)d	5.5x3.7	10.9	UMa	254	in 5457 group
eg	5595	14 21.5	−16 30	24.3	44	SAB(rs)c	2.0x1.2		Lib	155	
eg	5597	14 21.7	−16 32	24.5	46	SAB(s)cd	2.0x1.8	12.1	Lib	155	
eg	5600	14 21.4	14 52	23.8	38	ScP	1.4x1.4	12.7	Boo	37	
eg	5605	14 22.4	−12 56	25.1	10	(R')SAB(rs)c:	1.8x1.5		Lib	155	
eg	5613	14 22.0	35 07	24.1	53	(R)SAB(r)0⁺	1.1x1.0		Boo	37	P(b) w/ 5614

ID	RA (1950) DEC (2000)				TYPE	SIZE	V	CON	PG	NOTES C 35
eg 5614	14 22.0	35 05	24.1	51	SA(r)abP	2.7x2.3	11.7	Boo	37	P(b) w/ 5613; includes eg 5615
eg 5631	14 25.0	56 48	26.6	35	SA(s)0°	2.2x2.1		UMa	254	
eg 5633	14 25.6	46 22	27.5	09	(R)SA(rs)b	2.3x1.4	12.3	Boo	37	
gc 5634	14 27.0	-05 45	29.6	58	IV	4.9	9.4	Vir	279	
eg 5636	14 27.1	03 29	29.6	16	SAB(r)0^+	1.9x1.4		Vir	279	
eg 5638	14 27.2	03 27	29.7	14	E1	2.6x2.3	11.3	Vir	279	
eg 5641	14 27.1	29 03	29.3	50	(R')SAB(r)ab	2.7x1.6		Boo	37	
eg 5643	14 29.5	-43 57	32.7	10	SAB(rs)c	4.6x4.1		Lup		
eg 5645	14 28.2	07 30	30.7	17	SB(s)d	2.6x1.7	12.3	Vir	279	
eg 5653	14 28.0	31 26	30.2	13	(R')SA(rs)b	1.8x1.5	12.2	Boo	38	
eg 5660	14 28.1	49 51	29.9	38	SAB(rs)c	2.8x2.6	11.8	Boo	38	
eg 5665	14 30.0	08 18	32.5	05	SAB(rs)cP?	2.1x1.5		Boo	38	P(b) w/ 5665A
eg 5665A	14 30.0	08 18	32.5	05	cE0P:			Boo		P(b) w/ 5665
eg 5668	14 30.9	04 40	33.4	27	SA(s)d	3.3x3.1	11.5	Vir	279	
eg 5669	14 30.3	10 07	32.7	54	SAB(rs)cd	4.1x3.2		Boo	38	
eg 5673	14 29.8	50 11	31.5	58	SBc?sp	2.6x0.8		Boo	38	
eg 5676	14 31.0	49 41	32.8	28	SA(rs)bc	3.9x2.0	10.9	Boo	38	
eg 5678	14 30.6	58 08	32.1	55	SAB(rs)b	3.2x1.7		Dra	106	
eg 5687	14 33.3	54 42	34.9	29	S0^-?	2.6x1.9	11.8	Boo	38	
eg 5689	14 33.7	48 58	35.5	45	SB(s)0/a:	3.7x1.2	11.9	Boo	38	brightest of group
eg 5690	14 35.2	02 30	37.7	17	Sc?sp	3.5x1.2		Vir	279	
eg 5691	14 35.3	-00 11	37.9	24	SAB(s)a:P	2.0x1.7		Vir	279	
eg 5693	14 34.4	48 48	36.2	35	SB(rs)d	1.9x1.6		Boo	39	
gc 5694	14 36.7	-26 19	39.6	32	VIII	3.6	9.2	Hya	134	
eg 5701	14 36.7	05 35	39.2	22	(R)SB(rs)0/a	4.7x4.5		Vir	279	
eg 5705	14 37.3	-00 30	39.9	43	SB(rs)d	2.8x1.9		Vir		
eg 5713	14 37.6	-00 05	40.2	18	SAB(rs)bcP	2.8x2.5	11.4	Vir	279	P(b) w/ 5719
eg 5716	14 38.3	-17 16	41.1	29	SB(rs)c?	1.9x1.5		Lib	155	
eg 5719	14 38.4	-00 06	41.0	19	SAB(s)abP	3.4x1.3		Vir	279	P(b) w/ 5713
eg 5728	14 39.6	-17 02	42.4	15	SAB(r)a:	2.8x1.6	11.3	Lib	155	
eg 5738	14 41.4	01 49	43.9	36	S0	1.4x0.4		Vir	279	
eg 5739	14 40.6	42 03	42.5	50	SAB(r)0^+:	2.2x2.1		Boo	39	
eg 5740	14 41.9	01 53	44.4	40	SAB(rs)b	3.1x1.7	11.9	Vir	279	
eg 5746	14 42.4	02 10	44.9	57	SAB(rs)b?sp	7.9x1.7	10.5	Vir	279	
eg 5750	14 43.6	-00 01	46.2	14	SB(r)0/a	2.9x1.7	11.6	Vir	280	
eg 5756	14 44.8	-14 39	47.6	52	SAB(s)b?sp	2.0x1.0		Lib	155	
eg 5757	14 45.0	-18 52	47.8	04	(R)SB(r)b	2.1x1.9		Lib	155	
eg 5768	14 49.6	-02 20	52.2	32	SA(rs)c:	2.0x1.6		Lib	155	
eg 5774	14 51.2	03 47	53.7	35	SAB(rs)d	3.2x2.7	12.2	Vir	280	
eg 5775	14 51.5	03 45	54.0	33	SBc?sp	4.3x1.2	11.4	Vir	280	
eg 5776	14 52.0	03 10	54.5	58	S0^-	1.4x1.1		Vir	280	
eg 5791	14 55.9	-19 04	58.7	16	S0^-:sp	2.4x1.4		Lib	155	brightest of group
eg 5792	14 55.8	-00 53	58.4	05	SB(rs)b	7.2x2.1		Lib	156	
eg 5793	14 56.6	-16 30	59.4	42	Sb:sp	1.8x0.6	13.2	Lib	156	
eg 5796	14 56.6	-16 26	59.4	38	E0.5	1.9x1.7		Lib	156	
eg 5806	14 57.5	02 05	00.0	53	SAB(s)b	3.1x1.7	11.6	Vir	280	in 5846 group
eg 5812	14 58.3	-07 16	01.0	28	E0	2.4x2.2	11.2	Lib	156	
eg 5813	14 58.7	01 54	01.2	42	E1.5	3.6x2.8	10.7	Vir	280	in 5846 group
eg 5814	14 58.8	01 49	01.3	37	Sc?	1.2x0.7		Vir	280	
eg 5820	14 57.2	54 05	58.7	53	S0sp	2.5x2.3	11.9	Boo	39	

ID		RA (1950)	DEC	(2000)		TYPE	SIZE	V	CON	PG	NOTES C 36
eg	5821	14 57.5	54 07	59.0	55	S	1.7x1.0		Boo	*39*	
gc	5824	15 00.9	−32 52	04.0	04	I	6.2	7.8	Lup	158	
eg	5831	15 01.6	01 25	04.1	13	E3	2.2x2.0	11.5	Vir	280	
eg	5838	15 02.9	02 18	05.4	06	SA0$^-$	4.2x1.6	10.8	Vir	280	in 5846 group
eg	5839	15 02.9	01 50	05.4	38	SA0°?	1.4x1.4		Vir	280	in 5846 group
eg	5845	15 03.5	01 50	06.0	38	E3	0.9x0.5	12.3	Vir	*280*	in 5846 group
eg	5846	15 03.9	01 48	06.4	36	E0.5	3.4x3.2	10.2	Vir	280	brightest of group
eg	5846A	15 03.9	01 47	06.4	35	cE2.5	0.5x0.5	13.2	Vir	280	
eg	5850	15 04.6	01 44	07.1	33	SB(r)b	4.3x3.9	11.0	Vir	280	in 5846 group
eg	5854	15 05.3	02 46	07.8	35	SB(s)0$^+$sp	2.7x0.8	11.8	Vir	281	
eg	5858	15 06.1	−11 01	08.8	12	E6:	1.4x0.6		Lib	156	
eg	5861	15 06.6	−11 08	09.3	19	SAB(rs)c	3.0x1.8		Lib	156	
eg	5864	15 07.0	03 15	09.5	04	SB(s)0°?sp	2.8x1.0		Vir	281	
eg	5866	15 05.1	55 57	06.5	46	SA0$^+$sp	5.2x2.3	10.0	Dra	106	≠M102
pn	5873	15 09.6	−37 56	12.8	07		3"x3"	*11.2*	Lup	158	
eg	5874	15 06.5	54 57	07.9	46	SAB(rs)bc	2.5x1.8		Boo	39	
eg	5876	15 08.1	54 42	09.5	31	SB(r)ab:	2.6x1.4		Boo	39	
eg	5878	15 11.0	−14 05	13.8	16	SA(s)b	3.5x1.7	11.5	Lib	156	
eg	5879	15 08.5	57 11	09.8	00	SA(rs)bc:?	4.4x1.7	11.5	Dra	106	
eg	5885	15 12.4	−09 54	15.1	05	SAB(r)c	3.5x3.2	11.7	Lib	156	
eg	5893	15 11.8	42 09	13.6	58	SB(r)b	1.4x1.3		Boo	*39*	
eg	5895	15 12.0	42 11	13.8	00	Sb:sp	0.9x0.2		Boo	*39*	P(b) w/ 5896
eg	5896	15 12.0	42 11	13.8	00	Sc?	0.2x0.2		Boo	*39*	P(b) w/ 5895
gc	5897	15 14.5	−20 50	17.4	01	XI	13	8.6	Lib	157	
eg	5898	15 15.3	−23 55	18.2	06	SAB0$^-$	1.7x1.7	11.5	Lib	157	
eg	5899	15 13.2	42 14	15.0	03	SAB(rs)c	3.0x1.3	11.8	Boo	39	
eg	5900	15 13.3	42 24	15.1	13	Sb:sp	1.6x0.6		Boo	*39*	
eg	5903	15 15.7	−23 53	18.6	04	SA(s)0$^-$:	2.0x1.7	11.5	Lib	157	
gc	5904	15 16.0	02 16	18.5	05	V	17	5.7	Ser	228	=M5
eg	5905	15 14.1	55 42	15.4	31	SB(r)b	4.2x3.3		Dra	106	
eg	5907	15 14.6	56 30	15.9	19	SA(s)c:sp	12x1.8	10.2	Dra	106	
eg	5908	15 15.4	55 36	16.7	25	SA(s)b:sp	3.2x1.3	11.9	Dra	107	
eg	5915	15 18.8	−12 55	21.6	06	SB(s)abP	1.6x1.2		Lib	157	P(b) w/ 5916, 5916A
eg	5916	15 18.9	−13 00	21.7	11	SB(rs)aP	2.9x1.0		Lib	157	P(b) w/ 5915, 5916A
eg	5916A	15 18.5	−12 56	21.3	07	SB(s)cP	1.3x0.5		Lib	*157*	P(b) w/ 5915, 5916
eg	5921	15 19.5	05 15	22.0	04	SB(r)bc	4.9x4.2	10.8	Ser	228	
eg	5936	15 27.7	13 10	30.1	00	SB(rs)b	1.5x1.4	12.4	Ser	228	
eg	5949	15 27.3	64 56	28.0	46	SA(r)bc?	2.4x1.2		Dra	107	
eg	5957	15 33.0	12 13	35.4	03	(R')SAB(r)b	3.0x2.9		Ser	228	
eg	5958	15 32.8	28 49	34.9	39	S:	1.2x1.1		CrB	92	
eg	5961	15 33.2	31 02	35.2	52	S	1.0x0.4		CrB	92	
eg	5962	15 34.2	16 46	36.5	36	SA(r)c	2.8x2.1	11.4	Ser	228	
eg	5970	15 36.1	12 21	38.5	11	SB(r)c	3.0x2.1	11.4	Ser	228	
eg	5981	15 36.9	59 33	37.9	23	Sc?sp	2.8x0.6	13.0	Dra	107	
eg	5982	15 37.6	59 31	38.6	21	E3	2.9x2.2	11.1	Dra	107	
eg	5984	15 40.6	14 23	42.9	14	SB(rs)d:	3.0x0.9		Ser	228	
eg	5985	15 38.6	59 30	39.6	20	SAB(r)b	5.5x3.2	11.0	Dra	107	
gc	5986	15 42.8	−37 38	46.1	47	VII	9.8	7.5	Lup	158	
eg	5989	15 40.6	59 55	41.6	45	Sc:	1.1x1.0		Dra	*107*	
eg	6015	15 50.7	62 28	51.5	19	SA(s)cd	5.4x2.3	11.1	Dra	107	

	ID	RA (1950)	DEC	(2000)		TYPE	SIZE	V	CON	PG	NOTES C 37
pn	6026	15 58.1	−34 24	01.3	32		54"x36"	*12.9*	Lup	158	
eg	6052AB	16 03.0	20 41	05.2	33	Sc, Sc	1.0x0.7	13.0	Her	125	=6064; P(c)
pn	6058	16 02.7	40 49	04.4	41		25"x20"	*13.0*	Her	125	
eg	6070	16 07.4	00 50	09.9	42	SA(s)cd	3.6x2.1	11.7	Ser	228	
pn	6072	16 09.7	−36 06	13.0	14		70"x70"	*11.8*	Sco	216	
gc	6093	16 14.1	−22 51	17.1	58	II	8.9	7.3	Sco	216	=M80
eg	6106	16 16.4	07 32	18.8	25	SA(s)c	2.6x1.5	12.2	Her	125	
eg	6118	16 19.2	−02 10	21.8	17	SA(s)cd	4.7x2.3		Ser	228	
eg	6120	16 18.0	37 54	19.8	47	P	0.6x0.5		CrB		in Abell 2199
gc	6121	16 20.5	−26 25	23.6	32	IX	26	5.8	Sco	216	=M4
oc	6124	16 22.2	−40 33	25.6	40	I 3 r	40	5.8	Sco	216	
gc	6139	16 24.3	−38 44	27.7	51	II	5.5	8.9	Sco	216	
gc	6144	16 24.2	−25 55	27.3	02	XI	9.3	9.0	Sco	216	
eg	6145	16 23.4	41 04	25.1	57	Sb:	1.0x0.4		Her	*125*	
eg	6146	16 23.5	41 00	25.2	53	E	1.6x1.2		Her	125	
pn	6153	16 28.1	−40 09	31.5	15		28"x21"	*10.9*	Sco	216	
eg	6158	16 26.0	39 30	27.7	23	E			Her	125	in Abell 2199
eg	6160	16 26.0	41 02	27.7	55	E2	2.1x1.7		Her	125	in Abell 2197
eg	6166	16 26.9	39 40	28.6	33	E+2P	2.4x1.8	12.0	Her	125	brightest in Abell 2199
gc	6171	16 29.7	−12 57	32.5	03	X	10	8.1	Oph	172	=M107
eg	6173	16 28.1	40 55	29.8	49	E3	2.2x1.7		Her	125	in Abell 2197
eg	6175AB	16 28.3	40 45	30.0	39	S, E:	1.3x0.8		Her	*125*	double system
oc	6178	16 32.1	−45 32	35.7	38	III 3 p	5	7.2	Sco	216	
eg	6181	16 30.2	19 56	32.4	50	SAB(rs)c	2.6x1.3	11.9	Her	125	
oc	6192	16 36.8	−43 16	40.3	22	I 2 r	9		Sco	216	
oc	6193	16 37.5	−48 40	41.3	46	II 3 p	16	5.2	Ara	31	
eg	6194	16 34.8	36 18	36.6	12	E1:			Her	125	
eg	6196	16 36.1	36 10	37.9	04	SAB0⁻:P	1.7x1.2	12.8	Her	126	=I4615; brightest of group
eg	6197	16 36.3	36 05	38.1	59	Sab:	0.8x0.3		Her	126	=I4616
oc	6204	16 42.8	−46 56	46.5	01	I 3 m	6	8.2	Ara	31	
gc	6205	16 39.9	36 33	41.7	27	V	17	5.7	Her	126	=M13
eg	6207	16 41.3	36 56	43.1	50	SA(s)c	3.0x1.4	11.6	Her	126	
pn	6210	16 42.4	23 53	44.5	48		48"x8"	*8.8*	Her	126	
eg	6217	16 35.1	78 18	32.7	12	(R)SB(rs)bc	3.1x2.7	11.2	UMi	255	
gc	6218	16 44.6	−01 52	47.2	57	IX	15	6.8	Oph	172	=M12
oc	6222	16 45.8	−44 39	49.4	44		4.0		Sco	217	
gc	6229	16 45.6	47 37	47.0	32	VII	4.5	9.4	Her	126	
oc	6231	16 50.5	−41 43	54.0	48	I 3 p	15	2.6	Sco	217	
gc	6235	16 50.4	−22 06	53.4	11	X	5.0	10.0	Oph	172	
eg	6239	16 48.5	42 49	50.1	44	SB(s)bP?	2.8x1.3	12.3	Her	127	
oc	6242	16 52.2	−39 25	55.6	30	I 3 m	9	6.4	Sco	217	
oc	6249	16 54.0	−44 42	57.6	47	II 2 m	6	8.2	Sco	217	
gc	6254	16 54.5	−04 01	57.1	06	VII	15	6.6	Oph	172	=M10
eg	6255	16 53.0	36 35	54.8	30	SBcd:	3.5x1.5		Her	127	
gc	6256	16 56.2	−37 03	59.6	07		6.6	11.3	Sco		
oc	6259	16 57.1	−44 36	00.7	40	II 2 r	15	8.0	Sco	217	
gc	6266	16 58.0	−30 02	01.2	06	IV	14	6.7	Oph	172	=M62
oc	6268	16 58.9	−39 40	02.4	44	II 2 p	6		Sco	217	
gc	6273	16 59.5	−26 12	02.6	16	VIII	14	6.7	Oph	173	=M19
oc	6281	17 01.4	−37 50	04.8	54	II 2 p	8	5.4	Sco	217	

ID		RA (1950)	DEC	(2000)		TYPE	SIZE	V	CON	PG	NOTES C 38
gc	6284	17 01.4	−24 42	04.5	46	IX	5.6	8.9	Oph	173	
gc	6287	17 02.1	−22 38	05.1	42	VII	5.1	9.3	Oph	173	
gc	6293	17 07.1	−26 31	10.2	35	IV	7.9	8.2	Oph	173	
gn	6302	17 10.4	−37 03	13.8	06	e	83"x24"	9.6	Sco	217	=Bug
gc	6304	17 11.4	−29 24	14.6	27	VI	6.8	8.4	Oph	173	
pn	6309	17 11.2	−12 51	14.0	54		52"x52"	*11.5*	Oph	173	
gc	6316	17 13.5	−28 05	16.6	08	III	4.9	8.8	Oph	173	
oc	6318	17 14.3	−39 24	17.8	27	III 2 p	4.0		Sco	218	
oc	6322	17 14.9	−42 54	18.5	57	I 3 m	9	6.0	Sco	218	
gc	6325	17 14.9	−23 43	17.9	46	IV	4.3	10.6	Oph	173	
gc	6333	17 16.3	−18 28	19.2	31	VIII	9.3	7.6	Oph	173	=M9
pn	6337	17 18.8	−38 26	22.2	29		49"x45"	*12.3*	Sco	218	
eg	6340	17 11.3	72 22	10.4	18	SA(s)0/a	3.4x3.0	11.0	Dra	107	
gc	6341	17 15.6	43 11	17.1	08	IV	11	6.4	Her	127	=M92
gc	6342	17 18.2	−19 32	21.2	35	IV	3.0	9.8	Oph	174	
gc	6352	17 21.7	−48 23	25.5	26	XI	7.1	8.1	Ara	31	
gc	6355	17 20.9	−26 18	24.0	21		5.0	9.7	Oph	174	
gc	6356	17 20.7	−17 46	23.6	49	II	7.2	8.2	Oph	174	
gn	6357	17 21.4	−34 09	24.7	12	e	12x11		Sco	218	assoc. w/ oc Pismis 24
gc	6366	17 25.1	−05 02	27.8	04	IX	8.3	8.9	Oph	174	
pn	6369	17 26.3	−23 43	29.3	45		58"x34"	*11.4*	Oph	174	
gc	6380	17 31.9	−39 02	35.4	04		3.9	11.1	Sco		
oc	6383	17 31.5	−32 32	34.8	34	II 3 m	4.0	5.5	Sco	218	
eg	6384	17 30.0	07 06	32.4	04	SAB(r)bc	6.0x4.3	10.6	Oph	174	
gc	6388	17 32.6	−44 42	36.3	44	III	8.7	6.7	Sco	218	
oc	6396	17 34.8	−34 58	38.1	00	II 3 m	3.0	8.5	Sco	219	
oc	6400	17 37.4	−36 55	40.8	57	II 2 m	12		Sco	219	
gc	6401	17 35.6	−23 53	38.7	55		5.6	9.5	Oph	174	
gc	6402	17 35.0	−03 13	37.6	15	VIII	12	7.6	Oph	174	=M14
oc	6404	17 36.3	−33 13	39.6	15	III 2 m	6		Sco	219	
oc	6405	17 36.8	−32 11	40.1	13	II 3 r	33	4.2	Sco	219	=M6
eg	6412	17 31.4	75 44	29.6	42	SA(s)c	2.3x2.1	11.8	Dra	107	
oc	6416	17 41.1	−31 20	44.3	21	III 2 m	30	5.7	Sco	219	
oc	6425	17 43.7	−31 31	46.9	32	II 1 m	15	7.2	Sco	219	
gc	6426	17 42.4	03 11	44.9	10	IX	3.2	11.1	Oph	174	
gc	6440	17 45.9	−20 21	48.9	22	V	5.4	9.1	Sgr	209	
gc	6441	17 46.8	−37 02	50.2	03	III	7.8	7.2	Sco	219	
oc	6444	17 46.2	−34 48	49.5	49	IV 1 p	12		Sco	220	=Ruprecht 132
pn	6445	17 46.3	−20 00	49.3	01		3.1x0.9	*11.2*	Sgr	209	
oc	6451	17 47.5	−30 12	50.7	13	I 2 r	8		Sco	220	
gc	6453	17 47.5	−34 35	50.8	36	IV	3.5	9.8	Sco	220	
oc	6469	17 49.9	−22 20	52.9	21	IV 2 m	8		Sgr	209	
oc	6475	17 50.6	−34 48	53.9	49	I 3 r	80	3.3	Sco	220	=M7
eg	6482	17 49.7	23 05	51.8	04	E3:	2.3x2.0	11.2	Her	127	
oc	6494	17 53.9	−19 01	56.8	01	II 2 r	30	5.5	Sgr	209	=M23
gc	6496	17 55.4	−44 16	59.0	16	XII	6.9	8.5	CrA	91	
eg	6503	17 50.0	70 09	49.5	08	SA(s)cd	6.2x2.3	10.2	Dra	107	
oc	6506	17 56.8	−24 46	59.9	46	IV 1 p	6		Sgr	209	=Ruprecht 138
gn	6514	17 58.9	−23 02	01.9	02	ce	20x20		Sgr	209	=M20; Trifid
oc	6514	17 59.7	−22 58	02.7	58		13	6.3	Sgr	209	

ID		RA (1950) DEC (2000)				TYPE	SIZE	V	CON	PG	NOTES	C 39
gc	6517	17 59.1	−08 58	01.8	58	IV	4.3	10.3	Oph	175		
oc	6520	18 00.2	−27 54	03.4	54	I 2 r	5		Sgr	210		
gc	6522	18 00.4	−30 02	03.6	02	VI	5.6	8.4	Sgr	210		
gn	6523	18 01.6	−24 20	04.7	20	e	90x40		Sgr	210	≡M8; Lagoon; assoc. w/ oc 6530	
gc	6528	18 01.6	−30 04	04.8	04	V	3.7	9.5	Sgr	210		
oc	6530	18 01.7	−24 20	04.8	20	II 2 m	15	4.6	Sgr	210	assoc. w/ gn 6523	
oc	6531	18 01.6	−22 30	04.6	30	I 3 r	15	5.9	Sgr	210	≡M21	
gc	6535	18 01.3	−00 18	03.9	18	XI	3.6	10.5	Ser	230		
gc	6539	18 02.1	−07 35	04.8	35	X	6.9	9.8	Ser	230		
oc	6540	18 03.0	−27 46	06.1	46	I 1 p	1.5		Sgr	211		
gc	6541	18 04.4	−43 43	08.0	43	III	13	6.1	CrA	91		
pn	6543	17 58.6	66 38	58.6	38		23"x17"	*8.1*	Dra	107		
gc	6544	18 04.3	−25 00	07.4	00		8.9	8.1	Sgr	211		
oc	6546	18 04.2	−23 19	07.2	19	II 1 r	15	8.0	Sgr	211		
gc	6553	18 06.2	−25 55	09.3	54	XI	8.1	8.1	Sgr	211		
gc	6558	18 07.1	−31 46	10.4	45		3.7	9.8	Sgr	211		
pn	6563	18 08.7	−33 53	12.0	52		54"x41"	*11.0*	Sgr	211		
pn	6565	18 08.7	−28 11	11.9	10		10"x8"	*11.6*	Sgr	211		
pn	6567	18 10.8	−19 05	13.7	04		11"x7"	*11.0*	Sgr	211		
oc	6568	18 09.8	−21 37	12.8	36	IV 1 m	12		Sgr	211		
gc	6569	18 10.4	−31 50	13.7	49	VIII	5.8	8.7	Sgr	211		
pn	6572	18 09.7	06 51	12.1	52		16"x13"	*8.1*	Oph	175		
eg	6574	18 09.6	14 58	11.9	59	SAB(rs)bc:	1.4x1.1	12.0	Her	127		
oc	6583	18 12.8	−22 09	15.8	08	I 2 m	5		Sgr	212		
gn	6589	18 13.4	−19 49	16.4	48	c	5x3		Sgr	212		
gn	6590	18 13.5	−19 54	16.5	53	c	3x2		Sgr	212		
oc	6596	18 14.6	−16 41	17.5	40	II 2 m	10		Sgr	212		
oc	6603	18 15.5	−18 26	18.4	25	I 2 r	4.5		Sgr	212	≡M24	
oc	6604	18 15.3	−12 15	18.1	14	I 3 m	4.0	6.5	Ser	230		
oc	6611	18 16.0	−13 48	18.8	47	II 3 m	21	6.0	Ser	230	≡M16; assoc. w/ gn I4703	
oc	6613	18 17.0	−17 01	19.9	00	II 3 p	8	6.9	Sgr	212	≡M18	
oc	6618	18 17.9	−16 12	20.8	11	III 3 m	25	6.0	Sgr	212	assoc. w/ gn	
gn	6618	18 18.0	−16 12	20.9	11	e	40x30		Sgr	212	≡M17; Swan; assoc. w/ oc	
gc	6624	18 20.5	−30 23	23.7	21	VI	5.9	8.0	Sgr	212		
gc	6626	18 21.5	−24 54	24.6	52	IV	11	6.8	Sgr	212	≡M28	
pn	6629	18 22.7	−23 14	25.7	12		16"x14"	*11.3*	Sgr	212		
oc	6631	18 24.4	−12 04	27.2	02	II 1 m	7		Sct	224		
oc	6633	18 25.3	06 32	27.7	34	III 2 m	20	4.6	Oph	175		
gc	6637	18 28.1	−32 23	31.4	21	V	7.1	7.6	Sgr	213	≡M69	
gc	6638	18 27.8	−25 32	30.9	30	VI	5.0	9.1	Sgr	213		
gc	6642	18 28.9	−23 31	31.9	29		4.5	9.4	Sgr	213		
eg	6643	18 21.2	74 33	19.7	34	SA(rs)c	3.9x2.1	11.1	Dra	108		
oc	6645	18 29.7	−16 56	32.6	54	IV 1 m	15		Sgr	213		
oc	6649	18 30.7	−10 26	33.5	24	I 3 m	7	8.9	Sct	224		
gc	6652	18 32.5	−33 02	35.8	00	VI	3.5	8.8	Sgr	213		
gc	6656	18 33.4	−23 57	36.5	54	VII	24	5.1	Sgr	213	≡M22	
oc	6664	18 34.0	−08 16	36.7	13	III 2 m	12	7.8	Sct	224		
gc	6681	18 40.0	−32 21	43.3	18	V	7.8	8.0	Sgr	213	≡M70	
oc	6683	18 39.5	−06 20	42.2	17	II 1 p	3.0	9.4	Sct	224		
oc	6694	18 42.5	−09 27	45.2	24	II 3 m	8	8.0	Sct	226	≡M26	

	ID	RA (1950) DEC (2000)				TYPE	SIZE	V	CON	PG	NOTES	C 40
eg	6702	18 45.5	45 39	46.9	42	E3:	2.1x1.6	12.2	Lyr	164		
eg	6703	18 45.9	45 30	47.3	33	SA0⁻	2.6x2.5	11.4	Lyr	164		
oc	6704	18 48.2	−05 16	50.9	12	I 2 m	6	9.2	Sct	226		
oc	6705	18 48.4	−06 20	51.1	16	I 2 r	14	5.8	Sct	226	=M11	
oc	6709	18 49.1	10 17	51.5	21	IV 2 m	13	6.7	Aql	28		
eg	6710	18 48.6	26 47	50.6	51	SA0⁺:	2.0x1.2	12.8	Lyr	164		
gc	6712	18 50.3	−08 46	53.0	42	IX	7.2	8.2	Sct	226		
gc	6715	18 51.9	−30 33	55.1	29	III	9.1	7.6	Sgr	214	=M54	
oc	6716	18 51.6	−19 57	54.6	53	IV 1 p	10	6.9	Sgr	214		
gc	6717	18 52.1	−22 46	55.1	42	VIII	3.9	9.2	Sgr	214		
pn	6720	18 51.7	32 58	53.6	02		86"x62"	*8.8*	Lyr	164	=M57; Ring	
gc	6723	18 56.2	−36 42	59.6	38	VII	11	7.2	Sgr	214		
gn	6726–7	18 58.4	−36 58	01.8	54	c	8x8		CrA	91	TY CrA	
gn	6729	18 58.4	−37 02	01.8	58	ce	var		CrA	*91*	R CrA	
oc	6738	18 59.1	11 32	01.4	36	IV 2 p	15		Aql	28		
pn	6741	19 00.0	−00 31	02.6	27		9"x7"	*11.5*	Aql	28		
gc	6749	19 02.6	01 42	05.1	47		6.3	12.4	Aql	28		
pn	6751	19 03.3	−06 04	06.0	59		21"x21"	*11.9*	Aql	28		
oc	6755	19 05.3	04 09	07.8	14	II 2 r	15	7.5	Aql	28		
oc	6756	19 06.2	04 36	08.7	41	I 1 m	4.0		Aql	29		
gc	6760	19 08.7	00 57	11.2	02	XI	6.6	9.1	Aql	29		
pn	6772	19 11.9	−02 48	14.5	43		70"x56"	*12.7*	Aql	29		
pn	6778	19 15.8	−01 41	18.4	35		25"x19"	*12.3*	Aql	29		
gc	6779	19 14.6	30 06	16.5	11	X	7.1	8.3	Lyr	165	=M56	
pn	6781	19 16.0	06 27	18.4	33		1.9x1.8	*11.4*	Aql	30		
oc	6791	19 19.1	37 33	20.9	39	I 2 r	16	8.2	Lyr	165		
oc	6793	19 21.0	22 05	23.1	11	III 2 p	7		Vul	282		
oc	6802	19 28.4	20 10	30.6	16	I 1 m	5	8.8	Vul	282		
pn	6803	19 28.9	09 57	31.3	03		5"x5"	*11.4*	Aql	30		
pn	6804	19 29.2	09 07	31.6	13		62"x49"	*12.0*	Aql	30		
gc	6809	19 36.8	−31 05	40.0	58	XI	19	6.4	Sgr	214	=M55	
oc	6811	19 36.7	46 27	38.2	34	III 1 r	20	6.8	Cyg	97		
eg	6814	19 39.9	−10 27	42.6	20	SAB(rs)bc	3.2x3.0	11.2	Aql	30	Seyfert	
pn	6818	19 41.2	−14 16	44.0	09		22"x15"	*9.3*	Sgr	214		
oc	6819	19 39.6	40 04	41.3	11	I 1 r	10	7.3	Cyg	97		
eg	6822	19 42.1	−14 56	44.9	49	IB(s)m	10x9.5		Sgr	214	=I4895; in Local Group	
oc	6823	19 41.0	23 11	43.1	18	I 3 p	7	7.1	Vul	282		
pn	6826	19 43.5	50 24	44.8	31		27"x24"	*8.8*	Cyg	97		
oc	6830	19 48.9	22 56	51.0	04	II 2 p	10	7.9	Vul	282		
oc	6834	19 50.2	29 17	52.2	25	II 2 m	5	7.8	Cyg	97		
eg	6835	19 51.8	−12 42	54.6	34	SB(s)a?sp	2.7x0.7	12.5	Sgr	215		
eg	6836	19 51.9	−12 49	54.7	41	SABm	1.2x1.2	12.9	Sgr	*215*		
gc	6838	19 51.5	18 39	53.7	47		7.2	8.0	Sge	208	=M71	
pn	6842	19 53.0	29 09	55.0	17		53"x48"	*13.1*	Vul	282		
pn	6853	19 57.4	22 35	59.6	43		8.0x5.7	*7.3*	Vul	282	=M27; Dumbell	
gn	6857	19 59.9	33 23	01.8	31	e	1x1		Cyg	97		
gc	6864	20 03.1	−22 04	06.0	55	I	6.0	8.5	Sgr	215	=M75	
oc	6866	20 02.1	43 51	03.7	00	II 2 r	7	7.6	Cyg	97		
oc	6871	20 04.0	35 38	05.9	47	II 2 p	25	5.2	Cyg	97	assoc. w/ 27 Cyg	
pn	6881	20 09.0	37 16	10.8	25		5"x5"	*14.0*	Cyg			

C 41

ID		RA (1950) DEC (2000)				TYPE	SIZE	V	CON	PG	NOTES
oc	6882	20 09.6	26 24	11.7	33	II 2 p	10	8.1	Vul	282	
oc	6883	20 09.4	35 42	11.3	51	IV 2 m	15		Cyg	98	
oc	6885	20 09.9	26 20	12.0	29	III 2 p	22	5.9	Vul	282	
pn	6886	20 10.5	19 50	12.7	59		9"x9"	*11.4*	Sge	208	
gn	6888	20 10.7	38 16	12.5	25	e(SNR)	20x10		Cyg	98	
eg	6890	20 14.8	−44 58	18.3	49	SA(rs)b	1.5x1.2	12.5	Sgr		
pn	6891	20 12.8	12 33	15.2	42		74"x62"	*10.5*	Del	103	
pn	6894	20 14.4	30 25	16.4	34		44"x39"	*12.3*	Cyg	98	
eg	6902	20 21.0	−43 49	24.4	39	(R')SB(r)a	2.2x1.8		Sgr		
pn	6905	20 20.1	19 57	22.3	07		42"x35"	*11.1*	Del	103	
eg	6907	20 22.1	−24 58	25.1	48	SB(s)bc	3.4x3.0	11.3	Cap	61	=6908
oc	6910	20 21.3	40 37	23.1	47	I 3 m	8	6.6	Cyg	98	
oc	6913	20 22.1	38 22	23.9	32	II 3 m	7	6.6	Cyg	98	=M29
eg	6923	20 28.6	−31 00	31.7	50	SB(s)b:	2.5x1.4	12.1	Mic	166	
eg	6925	20 31.2	−32 09	34.3	59	SA(s)bc	4.1x1.6	11.3	Mic	166	
eg	6927	20 30.2	09 45	32.6	55	S0:sp	0.6x0.2	14.5	Del	*103*	in 6928 group
eg	6928	20 30.4	09 46	32.8	56	SB(s)ab	2.2x0.8	12.6	Del	103	brightest of group
eg	6930	20 30.6	09 42	33.0	52	SB(s)ab?sp	1.4x0.6	13.1	Del	103	in 6928 group
gc	6934	20 31.7	07 14	34.1	24	VIII	5.9	8.7	Del	104	
oc	6939	20 30.4	60 28	31.4	38	II 1 r	8	7.8	Cep	69	
oc	6940	20 32.5	28 08	34.6	18	III 2 m	25	6.3	Vul	284	
eg	6944	20 35.9	06 49	38.4	00	S0⁻:	1.7x0.8	13.3	Del	104	
eg	6944A	20 35.7	06 44	38.2	55	SB(rs)dP:	1.2x0.9		Del	104	
eg	6946	20 33.8	59 59	34.8	09	SAB(rs)cd	11x9.8	8.8	Cyg	99	
eg	6951	20 36.6	65 56	37.2	07	SAB(rs)bc	3.8x3.3	11.1	Cep	69	=6952
eg	6954	20 41.5	03 02	44.0	13	Sab?	1.1x0.6	13.2	Del	104	
eg	6956	20 41.5	12 20	43.9	31	SBb	2.1x2.0		Del	104	
eg	6958	20 45.5	−38 11	48.7	00	SA(rs)0°	2.4x2.2	11.4	Mic	166	
gn	6960	20 43.6	30 32	45.7	43	e(SNR)			Cyg	99	=Veil
gc	6981	20 50.7	−12 44	53.4	33	IX	5.9	9.3	Aqr	25	=M72
gn	6992	20 54.3	31 30	56.4	42	e(SNR)			Cyg	99	=Veil
oc	6994	20 56.3	−12 50	59.0	38	IV 1 p	1.5		Aqr	25	=M73; doubtful cluster
oc	6996	20 54.7	44 26	56.5	38	III 2 m	7		Cyg	99	
gn	7000	20 57.0	44 08	58.8	20	e	100x60		Cyg	99	=North America
gc	7006	20 59.2	15 59	01.5	11	I	2.8	10.5	Del	104	
pn	7008	20 59.1	54 21	00.6	33		98"x75"	*10.7*	Cyg	99	
pn	7009	21 01.5	−11 34	04.2	22		44"x23"	*8.0*	Aqr	25	=Saturn
eg	7015	21 03.2	11 13	05.6	25	Sbc	2.0x1.8		Equ	109	
pn	7026	21 04.6	47 39	06.3	51		27"x11"	*10.9*	Cyg	99	
pn	7027	21 05.2	42 02	07.1	14		18"x10"	*8.5*	Cyg	99	
oc	7039	21 09.4	45 27	11.2	39	IV 2 m	16	7.6	Cyg	99	
oc	7044	21 11.1	42 17	13.0	29	I 1 r	3.5		Cyg	99	
eg	7046	21 12.4	02 38	14.9	50	SB(r)cd	2.0x1.5		Equ	109	
pn	7048	21 12.4	46 04	14.2	16		60"x60"	*12.1*	Cyg	99	
oc	7062	21 21.4	46 10	23.2	23	II 2 m	7	8.3	Cyg	100	
oc	7063	21 22.4	36 17	24.4	30	III 1 p	8	7.0	Cyg	100	
eg	7070	21 27.2	−43 18	30.4	05	SA(s)cd	2.1x1.9	12.3	Gru		
eg	7072	21 27.4	−43 22	30.6	09	SAB(s)d:?	0.8x0.7	13.9	Gru		
gc	7078	21 27.6	11 57	30.0	10	IV	12	6.0	Peg	181	=M15
eg	7079	21 29.4	−44 17	32.6	04	SB(s)0°	2.8x1.7	11.6	Gru		

ID		RA (1950) DEC (2000)				TYPE	SIZE	V	CON	PG	NOTES C 42
oc	7086	21 28.8	51 22	30.5	35	II 2 m	9	8.4	Cyg	100	
gc	7089	21 30.9	-01 03	33.5	50	II	13	6.4	Aqr	25	=M2
oc	7092	21 30.4	48 13	32.2	26	III 2 m	32	4.6	Cyg	100	=M39
eg	7097	21 37.1	-42 46	40.3	32	E5	2.5x1.5		Gru		=7095; brightest of group
gc	7099	21 37.5	-23 24	40.3	10	V	11	7.3	Cap	61	=M30
eg	7107	21 39.3	-45 01	42.5	47	SB(s)dm	1.9x1.8		Gru		
oc	7127	21 42.2	54 23	43.9	37	IV 1 p	2.8		Cyg	100	
oc	7128	21 42.3	53 29	44.0	43	I 3 m	3.1	9.7	Cyg	100	
gn	7129	21 42.0	65 52	43.2	06	c	7x7		Cep	70	assoc. w/ oc
oc	7129	21 40.2	65 52	41.3	06	IV 2 p	2.7		Cep	70	assoc. w/ gn
gn	7133	21 43.4	65 56	44.6	10		3x3		Cep	70	
eg	7135	21 46.8	-35 07	49.8	53	SA0-P	2.8x2.0	11.7	PsA	200	
eg	7137	21 45.9	21 56	48.2	10	SAB(rs)c	1.5x1.4	12.4	Peg	181	
pn	7139	21 44.9	63 33	46.2	47		86"x70"	*13.3*	Cep	70	
oc	7142	21 44.7	65 34	45.9	48	I 2 r	9	9.3	Cep	70	
oc	7160	21 52.3	62 22	53.7	36	I 3 p	7	6.1	Cep	70	
eg	7162	21 56.6	-43 33	59.7	19	SA(s)c	2.8x1.1		Gru		
eg	7163	21 56.4	-32 07	59.3	53	(R)SABab:	1.8x1.2	13.4	PsA	*200*	
eg	7166	21 57.5	-43 38	00.6	24	SA0-	2.4x0.9	11.8	Gru		
eg	7171	21 58.3	-13 31	01.0	17	SB(rs)b	2.8x1.7	12.3	Aqr	25	
eg	7172	21 59.1	-32 07	02.0	53	SabP?sp	2.2x1.3	11.9	PsA	200	
eg	7173	21 59.2	-32 13	02.1	59	E2:	1.3x1.1	12.1	PsA	200	P(b) w/ 7174/6
eg	7174	21 59.2	-32 14	02.1	00	SbP?sp	1.3x0.7	12.6	PsA	200	P(c) w/ 7176; P(b) w/ 7173
eg	7176	21 59.2	-32 14	02.1	00	S0-?P	1.3x1.3	11.9	PsA	200	P(c) w/ 7174; P(b) w/ 7173
eg	7177	21 58.3	17 30	00.7	44	SAB(r)b	3.3x2.3	11.2	Peg	181	
eg	7184	21 59.9	-21 03	02.7	48	SB(r)c	5.8x1.8	11.2	Aqr	25	
oc	7209	22 03.2	46 15	05.2	30	III 1 m	25	7.7	Lac	135	
eg	7217	22 05.6	31 07	07.8	22	(R)SA(r)ab	3.7x3.2	10.2	Peg	181	
eg	7218	22 07.5	-16 55	10.2	40	SB(rs)cd	2.5x1.3	12.1	Aqr	25	
oc	7226	22 08.7	55 10	10.5	25	I 3 m	4.0	9.6	Cep	70	
oc	7235	22 10.8	57 02	12.6	17	II 3 m	4.0	7.7	Cep	70	
eg	7240	22 13.2	37 02	15.4	17	S0-:	0.8x0.8	13.8	Lac	135	
eg	7242	22 13.5	37 03	15.7	18	SA(r)0°:	2.6x2.0	12.0	Lac	135	brightest of group
oc	7243	22 13.3	49 38	15.3	53	II 2 m	21	6.4	Lac	135	
oc	7245	22 13.4	54 05	15.3	20	II 2 m	7	9.2	Lac	136	
eg	7248	22 14.7	40 15	16.8	30	SA0-:	2.1x1.1		Lac	136	
eg	7250	22 16.1	40 19	18.3	34	Sbc:	1.4x0.7		Lac	136	
eg	7252	22 18.0	-24 56	20.8	41	(R)SA(r)0°:P	2.2x1.8	12.1	Aqr	25	colliding system
oc	7261	22 18.6	57 50	20.4	05	II 3 m	6	8.4	Cep	70	
oc	7281	22 20.9	57 35	22.7	50	IV 2 p	12		Cep	71	
pn	7293	22 26.9	-21 06	29.6	51		12x10	7.3	Aqr	26	=Helix
oc	7296	22 26.2	52 02	28.2	17	II 2 p	4.0		Lac	136	
eg	7298	22 28.2	-14 27	30.9	12	SA(s)c:	1.5x1.3		Aqr	*26*	
eg	7300	22 28.3	-14 16	31.0	01	SAB(rs)b:	2.2x1.2	12.9	Aqr	26	
eg	7302	22 29.7	-14 23	32.4	08	SA(s)0-:	1.9x1.3	12.1	Aqr	26	=I5228
eg	7307	22 31.0	-41 12	33.9	57	SAB(s)c:P	4.2x1.0		Gru		
eg	7309	22 31.7	-10 37	34.3	21	SAB(rs)c	2.1x2.0	12.5	Aqr	26	
eg	7314	22 33.0	-26 19	35.8	03	SAB(rs)bc	4.6x2.3	10.9	PsA	200	
eg	7315	22 33.2	34 33	35.5	49	S0	1.9x1.9		Peg	181	chart XVIII
eg	7317	22 33.6	33 41	35.9	57	E2	1.0x0.8	13.6	Peg	181	in Stephan's Quintet; chart XVIII

ID		RA (1950)	DEC	(2000)		TYPE	SIZE	V	CON	PG	NOTES
eg	7318A	22 33.7	33 42	36.0	58	E2P	1.0x1.0	13.3	Peg	181	in Stephan's Quintet; P(c) w/ 7318B; chart XVIII
eg	7318B	22 33.7	33 42	36.0	58	SB(s)bcP	1.9x1.3	13.1	Peg	181	in Stephan's Quintet; P(c) w/ 7318A; chart XVIII
eg	7319	22 33.8	33 43	36.1	59	SB(s)bcP	1.7x1.3	13.1	Peg	181	in Stephan's Quintet; chart XVIII
eg	7320	22 33.8	33 41	36.1	57	SA(s)d	2.2x1.2	12.7	Peg	181	in Stephan's Quintet?; chart XVIII
eg	7320A	22 34.3	33 33	36.6	49	Sbc:sp			Peg		chart XVIII
eg	7320B	22 35.2	33 40	37.5	56	SABa:sp			Peg		chart XVIII
eg	7320C	22 34.1	33 44	36.4	00	(R)SAB(s)0/a	0.8x0.8	15.5	Peg		in Stephan's Quintet?; chart XVIII
eg	7325	22 34.2	34 14	36.5	30	S0^{-}?			Peg		chart XVIII
eg	7326	22 34.1	34 17	36.4	33	SAB(r)ab			Peg		chart XVIII
eg	7331	22 34.8	34 10	37.1	26	SA(s)bc	11x4.0	9.5	Peg	183	brightest of group; chart XVIII
eg	7332	22 35.0	23 32	37.4	48	S0Psp	4.2x1.3	11.0	Peg	183	
eg	7335	22 35.0	34 11	37.3	27	SA(rs)0^{+}	1.7x0.8		Peg	183	chart XVIII
eg	7336	22 35.1	34 13	37.4	29	(R)SB0^{+}P	1.0x0.4		Peg		chart XVIII
eg	7337	22 35.2	34 07	37.5	23	SB(rs)b	1.3x1.0		Peg	183	chart XVIII
eg	7339	22 35.4	23 32	37.8	48	SAB(s)bc:?	3.0x0.9	12.1	Peg	183	
eg	7340	22 35.4	34 09	37.7	25	cE	1.1x0.8		Peg	183	chart XVIII
eg	7343	22 36.3	33 49	38.6	05	SB(s)bc	1.2x1.0	13.4	Peg		chart XVIII
pn	7354	22 38.5	61 01	40.4	17		22"x18"	*12.2*	Cep	71	
eg	7361	22 39.5	−30 19	42.3	03	S(r)c:?	3.5x1.0	12.5	PsA	200	=I5237
eg	7371	22 43.4	−11 16	46.0	00	(R)SA(r)0/a:	2.1x2.0	12.1	Aqr	26	
eg	7377	22 45.1	−22 35	47.8	19	(R')SAB(rs)0^{+}?	2.2x1.8	11.6	Aqr	26	
oc	7380	22 45.0	57 50	47.0	06	III 2 m	12	7.2	Cep	71	assoc. w/ gn
eg	7392	22 49.1	−20 52	51.8	36	SA(s)bc	2.0x1.3	11.9	Aqr	26	
eg	7410	22 52.2	−39 56	55.0	40	SB(s)a	5.5x2.0	10.4	Gru	124	
eg	7412	22 52.9	−42 55	55.8	39	SB(s)b	4.0x3.1	11.4	Gru		
eg	7418	22 53.8	−37 18	56.6	02	SAB(rs)cd	3.3x2.8	11.4	Gru	124	
oc	7419	22 52.3	60 34	54.3	50	I 2 m	2.0		Cep	71	
eg	7421	22 54.1	−37 37	56.9	21	SB(rs)ab	2.2x2.1	12.0	Gru	124	
eg	7424	22 54.5	−41 20	57.3	04	SAB(rs)cd	7.6x6.8		Gru		
eg	7442	22 57.0	15 17	59.5	33	SAc:	1.2x1.2		Peg		
eg	7448	22 57.6	15 43	00.1	59	SA(rs)bc	2.7x1.3	11.7	Peg	183	
eg	7454	22 58.6	16 07	01.1	23	E4	2.0x1.6		Peg	*183*	
eg	7456	22 59.4	−39 50	02.2	34	SA(s)cd:	5.9x1.8		Gru		
eg	7457	22 58.6	29 53	01.0	09	SA(rs)0^{-}?	4.4x2.5	10.6	Peg	183	
eg	7462	23 00.0	−41 06	02.8	50	SB(s)c?sp	3.7x0.6		Gru		
eg	7469	23 00.7	08 36	03.2	52	(R')SAB(rs)a	1.8x1.3	11.9	Peg	183	Seyfert
eg	7479	23 02.4	12 03	04.9	19	SB(s)c	4.1x3.2	11.0	Peg	183	
gc	7492	23 05.8	−15 53	08.4	37	XII	6.2	11.4	Aqr	26	
eg	7496	23 07.0	−43 42	09.8	26	SB(s)b	3.5x2.8	11.1	Gru		
eg	7507	23 09.4	−28 49	12.1	33	E0	2.6x2.6	10.4	Scl	222	
oc	7510	23 09.4	60 18	11.5	34	II 3 r	4.0	7.9	Cep	71	
eg	7513	23 10.5	−28 38	13.2	22	SB(s)bP	3.2x2.6	11.8	Scl	*222*	
eg	7531	23 12.0	−43 52	14.8	36	SA(r)bc	3.5x1.5	11.3	Gru		
eg	7537	23 12.0	04 14	14.5	30	SAbc:	2.3x0.7	13.2	Psc	198	
gn	7538	23 11.5	61 14	13.6	30		8x7		Cep	71	
eg	7541	23 12.2	04 16	14.7	32	SB(rs)bc:P	3.5x1.4	11.7	Psc	198	
eg	7552	23 13.4	−42 51	16.2	35	(R')SB(s)ab	3.5x2.5	10.7	Gru	124	=I5294
eg	7576	23 14.8	−05 00	17.4	44	SA(r)0^{+}	1.5x1.2	13.0	Aqr	*26*	
eg	7582	23 15.6	−42 39	18.4	23	(R')SB(s)ab	4.6x2.2	10.6	Gru	124	brightest of group
eg	7585	23 15.5	−04 55	18.1	39	(R')SA(s)0^{+}P	2.3x1.9	11.7	Aqr	26	

ID		RA (1950)	DEC	(2000)		TYPE	SIZE	V	CON	PG	NOTES C 44
eg	7586	23 15.2	07 59	17.7	15	SBbc:P			Peg		chart XIX
eg	7590	23 16.2	−42 31	18.9	15	SA(rs)bc:	2.7x1.1	11.5	Gru	124	in 7582 group
eg	7592AB	23 15.8	−04 41	18.4	25	Sab:P	1.4x1.1		Aqr	*26*	colliding pair at 0.25
eg	7599	23 16.6	−42 32	19.3	16	SA(s)c	4.4x1.5	11.4	Gru	124	in 7582 group
eg	7600	23 16.3	−07 51	18.9	35	S0$^-$sp	2.4x1.1	11.8	Aqr	26	
eg	7606	23 16.5	−08 46	19.1	30	SA(s)b	5.8x2.6	10.8	Aqr	26	
eg	7608	23 16.7	08 05	19.2	21	SAab:	1.6x0.5		Peg		chart XIX
eg	7611	23 17.1	07 47	19.6	03	SB0/a?	1.5x0.7	12.6	Psc	199	in Pegasus I cluster; chart XIX
eg	7612	23 17.2	08 18	19.7	34	SAB0	1.8x0.9		Peg		chart XIX
eg	7615	23 17.4	08 08	19.9	24	Sb?	1.2x0.7		Peg	184	in Pegasus I cluster; chart XIX
eg	7617	23 17.6	07 54	20.1	10	SA0°:	1.0x0.8	13.8	Psc	199	chart XIX
eg	7619	23 17.7	07 56	20.2	12	E2	2.9x2.6	11.1	Peg	184	brightest in Pegasus I cluster; chart XIX
eg	7621	23 17.9	08 06	20.4	22	S0sp	0.8x0.2		Peg		chart XIX
eg	7623	23 18.0	08 07	20.5	23	SA0$^+$:	1.9x1.3	12.4	Peg	184	in Pegasus I cluster; chart XIX
eg	7625	23 18.0	16 57	20.5	13	SA(rs)aP	1.8x1.7	12.1	Peg	184	
eg	7626	23 18.2	07 57	20.7	13	E1P:	2.5x2.0	11.2	Peg	184	in Pegasus I cluster; chart XIX
eg	7631	23 18.9	07 57	21.4	13	SA(r)b?	1.9x1.0		Peg	184	in Pegasus I cluster; chart XIX
eg	7632	23 19.3	−42 45	22.0	29	SB(s)0$^-$:	1.5x0.8		Gru	*124*	=I5313
eg	7634	23 19.1	08 36	21.6	52	SB0	1.5x1.1		Peg	184	
gn	7635	23 18.5	60 54	20.7	10	e	15x8		Cas	65	
eg	7640	23 19.7	40 34	22.1	50	SB(s)c	11x2.5	10.8	And	19	
eg	7648	23 21.3	09 23	23.8	39	S0	1.9x1.3	12.9	Peg	186	=I1486
oc	7654	23 22.0	61 19	24.2	35	II 2 r	13	6.9	Cas	65	=M52
pn	7662	23 23.5	42 16	25.9	33		32"x28"	*8.3*	And	20	
eg	7673	23 25.2	23 19	27.7	36	(R')SAc?P	1.7x1.6	12.7	Peg	186	
eg	7674	23 25.4	08 30	27.9	47	SA(r)bcP	1.2x1.1	13.3	Peg	186	
eg	7675	23 25.5	08 29	28.0	46	SAB(s)0$^-$:	0.7x0.4	13.9	Peg	186	
eg	7677	23 25.6	23 15	28.1	32	SAB(r)bc	1.9x1.2		Peg	186	
eg	7678	23 26.0	22 09	28.5	26	SAB(rs)c	2.3x1.8	12.2	Peg	186	
eg	7679	23 26.2	03 14	28.8	31	SB0P:	1.9x1.3	12.7	Psc	199	P(b?) w/ 7682
eg	7682	23 26.5	03 16	29.1	33	SB(r)ab	1.2x1.0	13.4	Psc	199	P(b?) w/ 7679
oc	7686	23 27.8	48 51	30.2	08	IV 1 p	13	5.6	And	20	
eg	7713	23 33.6	−38 13	36.3	56	SB(r)d:	4.3x2.0		Scl	222	
eg	7713A	23 34.5	−38 00	37.2	43	SAB(r)cd:	2.0x1.7		Scl	*222*	
eg	7716	23 34.0	00 01	36.6	18	SAB(r)b:	2.3x1.9	12.2	Psc	199	
eg	7721	23 36.2	−06 48	38.8	31	SA(s)c	3.4x1.5	11.8	Aqr	26	
eg	7723	23 36.4	−13 14	39.0	57	SB(r)b	3.6x2.6	11.1	Aqr	26	
eg	7724	23 36.5	−12 30	39.1	13	SB(s)ab	1.9x1.4		Aqr	*27*	
eg	7727	23 37.3	−12 34	39.9	17	SAB(s)aP	4.2x3.4	10.7	Aqr	27	
eg	7741	23 41.4	25 48	43.9	05	SB(s)cd	4.0x2.8	11.4	Peg	186	
eg	7742	23 41.7	10 29	44.2	46	SA(r)b	2.0x2.0	11.5	Peg	186	
eg	7743	23 41.8	09 39	44.3	56	(R)SB(s)0$^+$	3.1x2.6	11.2	Peg	186	
eg	7744	23 42.4	−43 11	45.0	54	SAB(s)0°	2.3x1.8		Phe	196	=I5348
eg	7755	23 45.3	−30 48	47.9	31	SAB(rs)c	3.7x3.0		Scl	222	
oc	7762	23 47.4	67 45	49.8	02	II 2 m	11		Cep	71	
eg	7764	23 48.3	−41 01	50.9	44	IB(s)m	1.5x1.0	12.3	Phe	196	
eg	7769	23 48.5	19 52	51.0	09	(R)SA(rs)b	1.8x1.8	12.1	Peg	186	
eg	7770	23 48.8	19 49	51.3	06	S0/a?	1.0x0.9		Peg	186	
eg	7771	23 48.9	19 50	51.4	07	SB(s)a	2.7x1.3	12.3	Peg	186	
eg	7778	23 50.8	07 36	53.4	53	E	1.4x1.3		Psc	199	

ID	RA (1950)	DEC	(2000)		TYPE	SIZE	V	CON	PG	NOTES C 45
eg 7779	23 50.9	07 36	53.5	53	(R')SA0/a:	1.6x1.3		Psc	199	in 7782 group
eg 7780	23 51.0	07 50	53.6	07	Sab	1.2x0.7		Psc		in 7782 group
eg 7781	23 51.2	07 35	53.8	52	S0/a	0.9x0.3		Psc	*199*	
eg 7782	23 51.3	07 42	53.9	59	SA(s)b	2.4x1.4		Psc	199	brightest of group
eg 7785	23 52.8	05 38	55.4	55	E5.5	2.3x1.4	11.6	Psc	199	
oc 7788	23 54.2	61 07	56.7	24	I 2 p	9		Cas	67	
oc 7789	23 54.5	56 27	57.0	44	II 2 r	16	6.7	Cas	67	
oc 7790	23 55.9	60 58	58.4	15	II 2 m	17	8.5	Cas	67	
eg 7793	23 55.3	−32 52	57.9	35	SA(s)dm	9.1x6.6	9.1	Scl	223	
eg 7814	00 00.7	15 52	03.3	09	SA(s)ab:sp	6.3x2.6	10.5	Peg	187	
eg A0313+40	03 13.7	40 33	17.0	44	Sb?			Per		chart XX
eg A0329–35	03 29.8	−35 13	31.7	03	SAB(rs)0°:			For		chart X
eg A0337–35	03 37.3	−35 32	39.2	22	SAB(s)0°:			For		chart X
eg A0530–14	05 30.8	−14 07	33.1	05	cE			Lep	153	
oc A0612+12	06 12.0	12 56	14.8	55		2.0		Ori	180	
gn A0702–12	07 02.1	−12 16	04.4	21	c			CMa	58	
eg A1026–35	10 26.2	−35 27	28.4	42	S0			Ant	22	chart II
eg A1027–34	10 27.6	−34 55	29.9	10	S0			Ant		chart II
eg A1027–35A	10 27.6	−35 10	29.9	25	S			Ant		chart II
eg A1027–35B	10 27.6	−35 07	29.9	22	SAB(s)0⁻:			Ant	22	chart II
eg A1033–27	10 33.7	−27 15	36.1	31	S0⁻P?			Hya		in Hydra cluster; chart XI
eg A1034–27A	10 34.1	−27 04	36.5	20	S0sp			Hya		chart XI
eg A1034–27B	10 34.3	−27 18	36.7	34	S0sp			Hya		chart XI
eg A1034–27C	10 34.7	−27 09	37.1	25	S0			Hya		chart XI
eg A1131–09	11 31.6	−09 12	34.1	29	SAB?			Crt	95	=3721?
eg A1135+22	11 35.5	22 07	38.1	50	S0?			Leo		chart XII
eg A1245–05	12 45.5	−05 30	48.1	46	SAB(s)cP			Vir	*270*	P(b) w/ 4697
eg A1254+27	12 55.0	27 46	57.4	30	E	0.3x0.2	14.5	Com	*87*	
eg A1255+28A	12 55.1	28 27	57.5	11	S0:	0.7x0.4		Com		chart VIII
eg A1255+28B	12 55.4	28 09	57.8	53	E3?	0.6x0.6		Com		chart VIII
eg A1256+27	12 56.7	27 55	59.1	39	SBa	0.6x0.5	14.7	Com		in Coma cluster; chart VIII
eg A1257+28	12 57.7	28 15	00.1	59	SA0sp		14.9	Com	89	chart VIII
eg A1259+27A	12 59.4	27 52	01.8	36	S0sp			Com		chart VIII
eg A1259+27B	12 59.6	27 55	02.0	39	S0			Com		chart VIII
eg A1316–12	13 16.8	−12 24	19.4	40	cE			Vir	*275*	chart XVII
eg A1350+40	13 50.1	40 24	52.2	09	S0?			CVn		chart V
eg A2236+33	22 36.3	33 59	38.6	15	S0			Peg		chart XVIII
eg A2315+08	23 15.4	08 19	17.9	35	E			Peg		chart XIX
eg A2317+08A	23 17.1	08 20	19.6	36	cE			Peg		chart XIX
eg A2317+08B	23 17.1	08 04	19.6	20	S0?			Peg		chart XIX
eg A2317+08C	23 17.9	08 04	20.4	20	cE			Peg		chart XIX
eg A2318+08	23 18.1	08 07	20.6	23	S0?	0.7x0.6		Peg		chart XIX
pn BD+30°3639	19 32.8	30 24	34.8	31		13"x10"	*11.4*	Cyg	97	
pn Baade 1	03 50.7	19 21	53.6	30		40"x40"	*15.1*	Tau		
oc Berkeley 43	19 13.2	11 08	15.6	13	II 1 m	6		Aql	29	
oc Berkeley 45	19 16.9	15 37	19.2	43	II 1 p	4.0		Aql	30	
oc Berkeley 58	23 57.6	60 41	00.2	58	II 1 m	8	9.7	Cas	67	
oc Berkeley 70	05 22.2	41 51	25.7	54	III 1 m	12		Aur	33	
oc Berkeley 79	18 42.6	−01 16	45.2	13	II 1 r	7		Aql	28	
oc Berkeley 80	18 51.9	−01 19	54.5	15	II 1 p	4.0		Aql	28	

ID	RA (1950)	DEC	(2000)		TYPE	SIZE	V	CON	PG	NOTES C 46
oc Berkeley 82	19 09.1	12 59	11.4	04	III 1 p	4.0		Aql	29	
oc Biurakan 9	06 55.1	03 17	57.7	13	II 2 m	3.0		Mon	170	
oc Biurakan 10	06 49.6	03 00	52.2	56	I 1 p	4.0	10.4	Mon	170	
gn Ced 62	06 04.8	18 42	07.7	42	c	2x1		Ori	179	
oc Cr 21	01 47.3	26 55	50.1	10	III 3 p	6		Tri	238	doubtful cluster
oc Cr 115	06 43.9	01 49	46.5	46	IV 2 m	10		Mon	169	
oc Cr 333	17 28.0	−34 03	31.3	05	II 2 m	8		Sco	218	
oc Cr 394	18 50.5	−20 27	53.5	23	IV 2 m	22		Sgr	214	
oc Cr 399	19 23.2	20 05	25.4	11	III 3 m	60	3.6	Vul	*282*	=Brocchi's cluster
oc Cr 428	21 01.4	44 23	03.2	35	IV 1 p	14		Cyg	99	
oc Cr 465	07 04.8	−10 32	07.2	37	IV 2 p	9		Mon	171	
oc Cr 466	07 04.9	−10 44	07.3	49	IV 1 p	4.0		Mon	171	
eg E285–G007	20 20.5	−44 10	23.9	00	SAB(rs)a:	2.5x1.0		Sgr		=M–07–42–1
eg E352–G041	01 16.8	−34 22	19.1	06	SAB(rs)0°?	1.5x0.9		Scl	*222*	=M–06–04–5; in 491 group
eg E356–G004	02 37.9	−34 40	40.0	27	dE2	20x14	8.4	For	116	=M–06–07–1; Fornax Dwarf; in Local Group
gc E356–SC01	02 36.7	−35 02	38.8	49	VIII	1.2	13.5	For	116	
gc E356–SC05	02 38.1	−34 45	40.2	32	IV	0.6	13.6	For	116	
gc E356–SC08	02 40.3	−34 19	42.4	06	III	0.8	13.4	For	116	
eg E358–G014	03 31.0	−35 54	32.9	44	dE5:	0.7x0.5		For		chart X
eg E358–G015	03 31.2	−34 59	33.1	49	SBm:P	1.1x0.6		For		chart X
eg E358–G016	03 31.2	−35 53	33.1	43	SBc?	1.0x0.3		For		chart X
eg E358–G040	03 35.9	−36 19	37.8	09		0.7x0.5		Eri		chart X
eg E358–G042	03 36.2	−34 41	38.1	31	SB0°:	1.0x0.4		For		chart X
eg E358–G051	03 39.6	−35 03	41.5	53	S0/a:	1.0x0.4		For		=M–06–09–19; chart X
eg E358–G054	03 41.2	−36 26	43.1	17	SAB(s)dm	1.6x1.6		Eri		chart X
eg E358–IG31	03 34.7	−35 05	36.6	55	E, S	0.5x0.1		For		interacting pair; chart X
eg E375–G028	10 25.1	−35 01	27.4	16	SAB0°:	0.9x0.3		Ant		chart II
eg E375–G033	10 26.2	−34 13	28.5	28	SA(rs)bc	0.9x0.7		Ant		chart II
eg E375–G034	10 26.2	−35 16	28.5	31	E	0.3x0.3		Ant		chart II
eg E375–G041	10 27.3	−35 00	29.6	15	S0sp	0.7x0.1		Ant	22	=M–06–23–35; chart II
eg E375–G057	10 29.7	−34 44	32.0	59	S0	0.7x0.4		Ant		chart II
eg E375–G059	10 29.7	−34 56	32.0	11	SA(rs)0°:	0.9x0.8		Ant		chart II
eg E375–G064	10 31.8	−35 02	34.1	18	SAB(rs)aP	0.9x0.9		Ant		P(b) w/ 3289; chart II
eg E437–G011	10 34.5	−27 40	36.8	56	S0	0.8x0.4		Hya		=M–05–25–22; chart XI
eg E437–G013	10 34.6	−27 40	37.0	56	S0	0.6x0.2		Hya		=M–05–25–24; chart XI
eg E437–IG03	10 32.8	−27 44	35.1	00	E?, E?	0.6x0.2		Hya		chart XI
eg E479–G004	02 24.1	−24 31	26.4	18	SBd	2.7x1.6		For		=M–04–06–41
eg E501–G020	10 32.5	−26 57	34.9	13	S0	0.6x0.5		Hya		=M–04–25–23; chart XI
eg E501–G021	10 33.0	−27 06	35.4	22	S0/asp	0.8x0.2		Hya		=M–04–25–25; chart XI
eg E501–G022	10 33.0	−27 26	35.3	42	Sbc?sp	0.6x0.1		Hya		chart XI
eg E501–G026	10 33.1	−27 13	35.5	29	S0	0.7x0.2		Hya		=M–04–25–26; chart XI
eg E501–G027	10 33.6	−27 04	36.0	20	S0	0.6x0.3		Hya		chart XI
eg E501–G035	10 34.1	−26 44	36.5	00	SB(r)0°:	1.0x0.4		Hya		=M–04–25–33; chart XI
eg E501–G041	10 34.5	−26 48	36.9	04	SAbc	0.6x0.4		Hya		=M–04–25–37; chart XI
eg E501–G047	10 34.9	−27 13	37.3	29	SB0°:	0.8x0.3	13.5	Hya		=M–04–25–43; chart XI
eg E501–G049	10 35.0	−27 18	37.4	34	S0	0.6x0.2		Hya		chart XI
eg E501–G052	10 35.3	−27 08	37.7	24	S0/a	0.7x0.2		Hya		=M–04–25–47; chart XI
eg E501–G059	10 35.5	−26 52	37.9	08	S0:	0.7x0.6		Hya		=M–04–25–50; chart XI
oc Harvard 13	17 01.6	−48 07	05.4	11	IV 1 p	15		Ara	31	
oc Harvard 16	17 28.0	−36 49	31.4	51	III 2 r	15		Sco	218	

ID	RA (1950)	DEC	(2000)		TYPE	SIZE	V	CON	PG	NOTES C 47
oc Harvard 20	19 50.9	18 12	53.1	20	IV 2 p	9	7.7	Sge	208	
oc Harvard 21	23 51.6	61 29	54.1	46	IV 1 p	4.0		Cas	65	
oc Haffner 3	07 01.6	−06 03	04.0	08	III 2 p	5		Mon	170	
oc Haffner 4	07 03.9	−14 54	06.2	59	III 1 m	2.4		CMa	58	
oc Haffner 8	07 21.1	−12 14	23.4	20	II 2 m	4.2	9.1	CMa	59	
oc Haffner16	07 48.2	−25 19	50.3	27	I 2 m	5	10.0	Pup	204	
oc Haffner18AB	07 50.6	−26 15	52.7	23	I 2 p	2.0	9.3	Pup	204	
oc Haffner19	07 50.7	−26 09	52.8	17	I 1 m	3.2	9.4	Pup	204	
oc Hogg 22	16 43.0	−47 01	46.7	06	IV 3 p	1.5	6.7	Ara	31	
pn Humason 2–1	18 47.7	20 47	49.8	51		3"x3"	*11.5*	Her	127	
oc Hyades	04 24.0	15 45	26.9	52	II 3 m	2100	0.5	Tau	233	
eg I 10	00 17.7	59 01	20.4	18	IBm?	5.1x4.3	10.3	Cas	62	in Local Group
eg I 167	01 48.4	21 40	51.2	55	SAB(s)c	3.0x1.9		Ari	*32*	
eg I 184	01 57.4	−07 05	59.9	50	Sab:	1.5x0.7	13.8	Cet	75	
eg I 239	02 33.3	38 45	36.4	58	SAB(rs)cd	4.6x4.3	11.2	Per	*189*	in 1023 group
eg I 284	03 02.9	42 11	06.2	23	SAdm	4.0x2.2	11.8	Per	189	
pn I 289	03 06.3	61 08	10.3	19		42"x28"	*13.3*	Cas	65	
eg I 309	03 12.9	40 37	16.2	48	SA(s)0°	1.0x1.0		Per		in Perseus cluster; chart XX
eg I 310	03 13.4	41 09	16.7	20	SA(r)0°:	1.6x1.6	12.7	Per	190	in Perseus cluster; chart XX
eg I 312	03 14.8	41 34	18.1	45	S0sp	1.5x0.8		Per		in Perseus cluster; chart XX
eg I 313	03 17.7	41 43	21.0	54	SA0⁻:	1.6x1.1	13.5	Per	191	in Perseus cluster; chart XX
eg I 316AB	03 18.0	41 45	21.3	56	S0?, S0?	1.8x1.0	14.8	Per	191	double system; chart XX
eg I 335	03 33.6	−34 37	35.5	27	S0sp	1.7x0.4		For	118	=I1963; chart X
eg I 342	03 42.0	67 56	46.8	05	SAB(rs)cd	18x17		Cam	40	
eg I 344	03 38.9	−04 50	41.4	40	SAB(s)b:	1.3x0.6		Eri	*112*	
pn I 351	03 44.3	34 54	47.5	03		8"x6"	*12.0*	Per	191	
eg I 356	04 02.6	69 41	07.8	49	SA(s)abP	5.2x4.1	10.5	Cam	40	
oc I 361	04 14.8	58 11	19.0	18	II 2 r	8	11.7	Cam	40	
eg I 381	04 37.9	75 33	44.5	39	SAB(rs)bc	2.8x1.7		Cam	41	
eg I 399	04 59.3	−04 22	01.8	18	IAB(s)mP	0.6x0.5	14.5	Eri	115	P(b) w/ 1741AB
gn I 405	05 13.0	34 16	16.3	19	e	50x30		Aur		AE Aur
gn I 410	05 19.3	33 28	22.6	31	e	40x30		Aur	*33*	
gn I 417	05 20.0	34 12	23.3	15	e	13x10		Aur	*33*	
pn I 418	05 25.2	−12 44	27.5	42		14"x11"	*9.3*	Lep	153	
eg I 455	07 19.0	85 38	34.9	32	S0	1.4x0.9		Cep	69	
eg I 467	07 21.9	79 59	30.3	53	SAB(s)c:	3.4x1.5		Cam	*41*	
eg I 469	07 41.6	85 17	56.0	09	SAB(rs)ab:	2.3x1.3		Cep	69	
eg I 520	08 48.4	73 41	53.7	30	SAB(rs)ab?	2.3x1.9		Cam	*42*	
eg I 546	09 32.5	−16 10	34.9	23	SB0⁺:	0.8x0.5		Hya	*130*	
eg I 694	11 25.7	58 50	28.5	33	SBm?P	1.2x1.0		UMa	247	P(c) w/ 3690
eg I 732A	11 43.4	20 43	46.0	26	S0P			Leo		chart XIII
eg I 732B	11 43.4	20 43	46.0	26	SAB0/a:P			Leo		chart XIII
eg I 749	11 56.0	43 01	58.6	44	SB(rs)c	2.5x2.1	12.2	UMa	251	
eg I 750	11 56.3	43 00	58.9	43	Sab:sp	2.9x1.4	11.8	UMa	251	
eg I 764	12 07.7	−29 28	10.3	45	SA(s)c?	4.8x1.8		Hya		
eg I 773	12 15.6	06 25	18.2	08	SB(s)b	1.8x1.0		Vir		chart XXIV
eg I 775	12 16.4	13 12	18.9	55	SB0⁻P:	1.4x1.0	13.3	Vir	*259*	
eg I 782	12 19.1	06 03	21.7	46	SB(r)0/a:	0.7x0.6		Vir		chart XXIV
eg I 783	12 19.1	16 01	21.6	44	SAB(rs)0/a?	1.5x1.4	13.6	Com	82	
eg I 789	12 23.8	07 44	26.3	27	SB0/a	1.1x0.6		Vir		chart XXIV

ID	RA (1950)	DEC	(2000)		TYPE	SIZE	V	CON	PG	NOTES C 48
eg I 794	12 25.6	12 22	28.1	05	SB0⁻:	1.7x1.2		Vir		chart XXVI
eg I 809	12 39.6	12 02	42.1	46	dE1	1.4x1.4		Vir	*269*	=I3672
eg I 839	12 56.0	28 25	58.4	09	SA0		15.3	Com		chart VIII
eg I 844	13 00.6	−30 15	03.3	31	S0⁻sp	1.8x0.6		Cen		
eg I1029	14 30.7	50 08	32.4	55	SAb?sp	3.0x0.7		Boo	*38*	
eg I1066	14 50.5	03 30	53.0	18	S	1.5x0.9		Vir	280	
eg I1067	14 50.6	03 32	53.1	20	SB(s)b	2.3x1.9		Vir	280	
eg I1077	14 54.5	−19 01	57.3	13	(R')SB(s)b:P	0.7x0.6		Lib	155	in 5791 group
eg I1081	14 56.1	−19 03	58.9	15	SAB(s)a:	0.9x0.4		Lib	156	
eg I1084	14 58.6	−07 15	01.3	27	S0?			Lib	*156*	
eg I1091	15 05.5	−10 57	08.2	08	SB(s)b?	1.3x0.9		Lib	156	
eg I1131	15 36.5	12 15	38.9	05	E1:	1.0x0.7		Ser	*228*	
gc I1276	18 08.0	−07 14	10.7	13	XII	7.1	10.3	Oph		
gn I1287	18 27.6	−10 50	30.4	48	c	20x10		Sct		
pn I1295	18 51.9	−08 54	54.6	50		1.7x1.4	*12.7*	Sct	227	
eg I1296	18 51.4	33 00	53.3	04	SB(s)c	1.3x1.0		Lyr	*165*	
pn I1297	19 14.0	−39 42	17.4	37		8"x6"	*10.7*	CrA	91	
oc I1310	20 08.5	34 47	10.4	56	II 1 p	4.0		Cyg	98	
oc I1311	20 08.6	41 04	10.3	13	I 1 r	9		Cyg	98	
gn I1318A	20 14.7	41 39	16.4	48	e	45x20		Cyg	98	
oc I1369	21 10.4	47 32	12.1	44	II 2 m	5	8.8	Cyg	*99*	
oc I1396	21 37.5	57 16	39.1	30	IV 3 m	90	3.5	Cep		assoc. w/ gn
gn I1396	21 37.5	57 14	39.1	28	e	170x130		Cep		assoc. w/ oc
eg I1417	21 57.7	−13 23	00.4	09	SB?sp	1.4x0.4		Aqr	*25*	
oc I1434	22 08.6	52 35	10.5	50	III 2 m	8		Lac	135	
eg I1441	22 13.1	37 03	15.3	18	SBa?	0.9x0.4		Lac	*135*	
oc I1442	22 14.6	53 48	16.5	03	III 1 p	3.5	9.1	Lac	136	
eg I1459	22 54.4	−36 44	57.2	28	E3.5	3.4x2.5	10.0	Gru	124	=I5265; brightest of group
gn I1470	23 03.2	59 59	05.3	15	e	70"x45"		Cep	71	
eg I1602	00 53.4	−10 15	55.9	59	S0P	1.3x1.1		Cet	*73*	
eg I1727	01 44.7	27 05	47.5	20	SB(s)m	6.2x2.9	11.6	Tri	238	P(b?) w/ 672
eg I1731	01 47.4	26 57	50.2	12	SAB(s)c	1.7x1.1		Tri	238	
eg I1738	01 48.7	−10 02	51.2	47	SAB(r)b	1.2x1.0		Cet	74	
pn I1747	01 54.0	63 05	57.6	20		13"x13"	*12.1*	Cas	64	
eg I1784	02 13.3	32 25	16.3	39	SA(rs)bcP:	1.6x1.0	13.2	Tri	239	
oc I1805	02 28.9	61 14	32.7	27	II 3 m	22	6.5	Cas	65	assoc. w/ gn
oc I1848	02 47.3	60 14	51.2	26	I 3 p	12	6.5	Cas	65	assoc. w/ gn
gn I1848	02 47.2	60 13	51.1	25	e	40x10		Cas	65	assoc. w/ oc
eg I1953	03 31.5	−21 39	33.7	29	SB(rs)d	2.8x2.3		Eri	111	
pn I2003	03 53.2	33 43	56.4	52		7"x6"	*11.5*	Per	193	
eg I2006	03 52.6	−36 07	54.5	58	(R)SA0⁻	2.3x2.1		Eri	113	in Fornax I cluster?
eg I2035	04 07.5	−45 39	09.1	31	S0:	0.7x0.6		Hor		
eg I2132	05 30.2	−13 58	32.5	56	SABb:	1.8x0.8		Lep	153	
pn I2149	05 52.7	46 06	56.4	06		15"x10"	*10.7*	Aur	35	
oc I2157	06 01.9	24 00	05.0	00	II 1 p	7	8.4	Gem	121	
eg I2163	06 14.3	−21 21	16.4	22	SB(rs)cP	2.7x1.2		CMa	55	P(b) w/ 2207
pn I2165	06 19.4	−12 58	21.7	00		9"x7"	*10.6*	CMa	55	
gn I2167	06 28.3	10 29	31.1	27		2x1		Mon	167	
oc I2169	06 28.4	10 03	31.2	01				Mon	167	assoc. w/ gn
gn I2169	06 28.4	10 03	31.2	01	c	22x17		Mon	167	assoc. w/ oc

ID	RA (1950)	DEC (2000)			TYPE	SIZE	V	CON	PG	NOTES
eg I2174	07 02.4	75 26	09.1	21	(R')SB(r)0/a:	1.1x1.0		Cam	41	
gn I2177	07 03.1	-10 29	05.5	34	e	20x20		CMa	58	
eg I2209	07 52.0	60 26	56.3	18	SB(rs)b:	1.1x1.0	13.8	Cam	42	
eg I2233	08 10.5	45 54	14.0	45	SB(s)d:sp	4.7x0.6	13.0	Lyn	159	
eg I2293	08 16.7	21 32	19.6	23	SB(rs)ab	1.0x0.8		Cnc	44	
eg I2311	08 16.6	-25 13	18.7	22	E0:	1.3x1.3		Pup	205	P(b?) w/ 2566
eg I2389	08 42.6	73 43	48.0	32	SB(s)b?	1.7x0.4	13.4	Cam	42	
eg I2458	09 17.5	64 27	21.6	14	I0P:	0.5x0.3		UMa	242	P(b) w/ 2805, 2814, 2820
eg I2482	09 24.6	-11 53	27.0	06	E	3.0x1.9		Hya	130	
eg I2522	09 53.0	-32 54	55.2	08	SB(s)cP:	2.8x1.7		Ant	21	
eg I2523	09 53.0	-32 58	55.2	12	SB(s)bcP?	1.4x0.8		Ant	21	
eg I2532	09 57.9	-33 59	00.1	13	SB(r)a	1.1x1.1		Ant	21	
eg I2533	09 58.3	-31 00	00.5	14	SA0⁻:	1.5x1.2		Ant	21	
eg I2537	10 01.6	-27 20	03.9	35	SAB(rs)c	2.7x1.9	12.2	Ant	21	
eg I2584	10 27.6	-34 39	29.9	54	S0:sp	4.9x2.3		Ant		chart II
eg I2587	10 28.7	-34 18	31.0	33	(R')SB(rs)0°	2.5x2.3	12.6	Ant		chart II
eg I2597	10 35.4	-26 49	37.8	05	E⁺4	1.3x1.1		Hya		chart XI
eg I2627	11 07.4	-23 27	09.9	43	SA(s)bc:	2.7x2.6	12.0	Crt	95	
eg I2951	11 40.8	20 02	43.4	45	SAB(rs)ab:	1.5x0.8		Leo		chart XIII
eg I2955	11 42.5	19 54	45.1	37	SAB0°?	0.4x0.3		Leo	146	chart XIII
eg I2969	11 49.9	-03 34	52.5	51	SAB(rs)bc:	1.5x0.8		Vir		
eg I2995	12 03.2	-27 40	05.8	57	SB(s)c:	3.1x1.2		Hya		
eg I3044	12 10.3	14 15	12.8	58	SB(s)cdP:	2.1x1.0		Com	79	
eg I3061	12 12.5	14 19	15.0	02	SBc?sp	2.3x0.6	13.6	Com	79	
eg I3115	12 15.5	06 56	18.1	39	SB(s)c	1.7x1.4	13.1	Vir		chart XXIV
eg I3136	12 16.4	06 28	19.0	11	SB(s)c?	1.1x0.4		Vir		chart XXIV
eg I3153	12 17.1	05 41	19.7	24	SA(r)c			Vir	259	chart XXIV
eg I3155	12 17.2	06 17	19.8	00	S0sp	1.1x0.6		Vir	259	chart XXIV
eg I3211	12 19.6	09 16	22.1	59	SA(r)c	1.2x1.0		Vir	261	
eg I3218	12 19.8	07 12	22.3	55	dE0	0.8x0.8		Vir		chart XXIV
eg I3225	12 20.1	06 57	22.6	40	SAcd	1.8x0.8		Vir		chart XXIV
eg I3229	12 20.3	06 57	22.8	40	Sbc	1.2x0.4		Vir		chart XXIV
eg I3238	12 20.6	14 44	23.1	27	S0/a:	0.7x0.4		Com		chart XXVI
eg I3239	12 20.6	12 00	23.1	43	Sm			Vir		chart XXVI
eg I3244	12 20.7	14 40	23.2	23	SA(s)c	0.7x0.6		Com		chart XXVI
eg I3253	12 21.1	-34 21	23.7	38	SA(s)c	3.3x1.6		Cen		
eg I3256	12 21.1	07 20	23.6	03	S0⁻sp	1.4x0.7	12.6	Vir	262	=4342; chart XXIV
eg I3258	12 21.2	12 45	23.7	28	SB(s)mP:	1.6x1.4		Vir	262	chart XXVI
eg I3259	12 21.3	07 28	23.8	11	SAB(s)dm?	1.8x1.1	13.6	Vir	262	chart XXIV
eg I3260	12 21.4	07 23	23.9	06	SAB(s)0°:sp	1.9x0.7	13.3	Vir	262	=4341; chart XXIV
eg I3267	12 21.6	07 19	24.1	02	SA(s)cd	1.1x1.1	13.5	Vir	262	chart XXIV
eg I3268	12 21.6	06 53	24.1	36	SA(s)c	0.8x0.8		Vir	262	chart XXIV
eg I3290	12 22.5	-39 30	25.2	47	(R')SB(s)0/a	1.8x0.8		Cen		
eg I3292	12 22.3	18 28	24.8	11	S0			Com	83	
eg I3303	12 22.7	13 00	25.2	43	E1:	1.4x0.9		Vir		chart XXVI
eg I3305	12 22.7	12 08	25.2	51	dE7	1.4x0.5		Vir		chart XXVI
eg I3311	12 23.0	12 32	25.5	15	Scsp	2.1x0.4		Vir		chart XXVI
eg I3322	12 23.4	07 50	25.9	33	SAB(s)cd:sp	2.5x0.6		Vir	262	chart XXIV
eg I3322A	12 23.2	07 30	25.7	13	SB(s)cd:sp	3.5x0.5		Vir	262	chart XXIV
eg I3327	12 23.5	15 10	26.0	53	SBa			Com	83	

ID	RA (1950)	DEC	(2000)		TYPE	SIZE	V	CON	PG	NOTES C 50
eg I3328	12 23.4	10 20	25.9	03	dE1	0.8x0.8		Vir	*264*	
eg I3331	12 23.6	12 06	26.1	49	S0			Vir		chart XXVI
eg I3344	12 24.0	13 51	26.5	34	dE6	0.7x0.4		Com		chart XXVI
eg I3349	12 24.3	12 44	26.8	27	dE1			Vir		chart XXVI
eg I3355	12 24.3	13 27	26.8	10	Im	1.3x0.6	15.1	Vir		chart XXVI
eg I3356	12 24.4	11 50	26.9	33	Sm	1.7x1.2		Vir		chart XXVI
eg I3358	12 24.4	11 57	26.9	40	dE2:P	1.5x1.2		Vir		chart XXVI
eg I3363	12 24.5	12 50	27.0	33	dE7			Vir		chart XXVI
eg I3370	12 25.0	−39 04	27.7	21	SB0°?	2.8x2.4	11.1	Cen		
eg I3381	12 25.7	12 04	28.2	47	SA0⁻:	1.4x1.1		Vir	*265*	chart XXVI
eg I3388	12 25.9	13 06	28.4	49	dE5			Vir		chart XXVI
eg I3392	12 26.2	15 17	28.7	00	SAb:	2.3x1.1		Com	84	
eg I3393	12 26.2	13 12	28.7	55	S0⁺:	1.3x0.5		Vir		chart XXVI
eg I3413	12 26.9	11 43	29.4	26	E4?	1.6x1.0		Vir		chart XXVI
eg I3418	12 27.2	11 41	29.7	24	IBm:	1.5x1.0		Vir		chart XXVI
eg I3432	12 27.9	14 26	30.4	09	Sc	0.7x0.4		Com		chart XXVI
eg I3437	12 28.2	11 37	30.7	20	dE6			Vir		chart XXVI
eg I3442	12 28.8	14 24	31.3	07	E0:	1.5x1.5		Com		chart XXVI
eg I3443	12 28.7	12 37	31.2	20	dE0P			Vir		chart XXVI
eg I3446	12 28.9	11 46	31.4	29	Sm:			Vir		chart XXVI
eg I3457	12 29.3	12 56	31.8	39	E3:	1.6x1.2		Vir		chart XXVI
eg I3459	12 29.4	12 27	31.9	10	SB0	1.3x1.2		Vir		chart XXVI
eg I3461	12 29.5	12 10	32.0	53	dE2	0.8x0.7		Vir		chart XXVI
eg I3466	12 29.6	12 06	32.1	49	P	0.7x0.5		Vir		chart XXVI
eg I3470	12 29.9	11 32	32.4	15	dE0	0.8x0.8		Vir		chart XXVI
eg I3475	12 30.1	13 03	32.6	46	E3:	2.6x2.5	13.3	Vir		chart XXVI
eg I3476	12 30.2	14 20	32.7	03	IB(s)m:	2.2x1.9	12.8	Com	85	chart XXVI
eg I3478	12 30.2	14 28	32.7	11	SAB0:	1.3x1.1		Com		chart XXVI
eg I3481	12 30.4	11 41	32.9	24	SAB0⁻:P	0.7x0.5	13.6	Vir		P(b) w/ I3481A; chart XXVI
eg I3481A	12 30.4	11 40	32.9	23	E1:P	0.2x0.2		Vir		P(b) w/ I3481; chart XXVI
eg I3483	12 30.7	11 37	33.2	20	SAB(s)b:P	0.5x0.4	15.0	Vir		chart XXVI
eg I3489	12 30.7	12 32	33.2	15	SA(r)bc	0.7x0.6		Vir		chart XXVI
eg I3492	12 30.8	13 08	33.3	51	E3P	0.4x0.4		Vir		chart XXVI
pn I3568	12 31.6	82 50	32.9	33		18"x18"	*10.6*	Cam	43	
eg I3583	12 34.2	13 32	36.7	16	IBm	2.1x1.1		Vir	*267*	P(b?) w/ 4569
eg I3665	12 39.3	11 46	41.8	30	dE4P	1.3x0.9		Vir	*269*	
eg I3742	12 43.0	13 36	45.5	20	SB(s)c	1.9x1.0		Vir	270	
eg I3943	12 56.2	28 23	58.6	07	SAB0/a			Com		chart VIII
eg I3946	12 56.4	28 05	58.8	49	SA0sp	1.0x0.5	13.9	Com		in Coma cluster; chart VIII
eg I3947	12 56.5	28 03	58.9	47	SA0		14.6	Com		in Coma cluster; chart VIII
eg I3949	12 56.5	28 06	58.9	50	SA0Psp	1.3x0.2	14.1	Com		in Coma cluster; chart VIII
eg I3955	12 56.7	28 16	59.1	00	SA0		14.3	Com	87	in Coma cluster; chart VIII
eg I3957	12 56.7	28 02	59.1	46	E	0.4x0.4	14.5	Com		chart VIII
eg I3959	12 56.7	28 03	59.1	47	E3	0.6x0.6	13.9	Com		in Coma cluster; chart VIII
eg I3960	12 56.7	28 07	59.1	51	SB0	0.4x0.4	14.5	Com		in Coma cluster; chart VIII
eg I3963	12 56.8	28 03	59.2	47	SA0sp	0.6x0.3	13.5	Com		in Coma cluster; chart VIII
eg I3973	12 57.1	28 09	59.5	53	SB0		14.2	Com		in Coma cluster; chart VIII
eg I3976	12 57.1	28 07	59.5	51	SA0sp		14.4	Com		in Coma cluster; chart VIII
eg I3998	12 57.4	28 15	59.8	59	SB0	0.6x0.4	14.7	Com	87	in Coma cluster; chart VIII
eg I4011	12 57.7	28 16	00.1	00	E0	0.4x0.4	15.0	Com		in Coma cluster; chart VIII

ID	RA (1950) DEC (2000)				TYPE	SIZE	V	CON	PG	NOTES C 51
eg I4012	12 57.7	28 21	00.1	05	E3	0.4x0.3	14.8	Com		in Coma cluster; chart VIII
eg I4021	12 57.8	28 19	00.2	03	E0	0.3x0.3	14.9	Com		in Coma cluster; chart VIII
eg I4026	12 58.0	28 19	00.4	03	SB0	0.5x0.4	14.7	Com		in Coma cluster; chart VIII
eg I4040	12 58.2	28 20	00.6	04	Sdm:	1.1x0.6	14.7	Com		in Coma cluster; chart VIII
eg I4041	12 58.3	28 16	00.7	00	SA0	0.7x0.7	14.5	Com		in Coma cluster; chart VIII
eg I4042	12 58.3	28 14	00.7	58	SB0	0.6x0.5	14.3	Com		in Coma cluster; chart VIII
eg I4045	12 58.4	28 22	00.8	06	E4	0.7x0.5	13.9	Com		in Coma cluster; chart VIII
eg I4051	12 58.5	28 17	00.9	01	E0	1.3x1.1	13.4	Com		in Coma cluster; chart VIII
eg I4237	13 21.8	−20 53	24.5	09	SB(r)b?	1.9x1.4		Vir	277	
eg I4263	13 26.4	47 11	28.5	56	SB(s)d:sp	2.0x0.5		CVn	*50*	
eg I4296	13 33.8	−33 43	36.7	58	E0	2.7x2.7	10.6	Cen		brightest of group
eg I4299	13 33.9	−33 49	36.8	04	SAB(s)a:	1.6x0.7	12.7	Cen		in I4296 group
eg I4327	13 45.9	−29 58	48.8	13	SA(s)c?	1.0x0.7		Cen		
eg I4329	13 46.2	−30 03	49.1	18	SAB(s)0⁻	3.2x1.6	11.5	Cen		brightest of group
eg I4336	13 48.6	39 57	50.7	42	SBb	1.6x0.5		CVn		chart V
eg I4351	13 55.0	−29 04	57.9	19	SA(s)b:sp	5.6x1.2		Hya		
pn I4406	14 19.3	−43 55	22.5	09		1.7x0.6	*10.3*	Lup	158	
eg I4444	14 28.5	−43 12	31.7	25	SAB(rs)bc:	1.9x1.7		Lup		
eg I4538	15 18.3	−23 29	21.2	40	SAB(s)c	2.2x1.8		Lib	157	
pn I4593	16 09.4	12 12	11.7	04		12"x10"	*10.7*	Her	125	
eg I4614	16 36.0	36 13	37.8	07	SA(r)ab	1.0x0.9		Her	126	
gn I4628	16 53.3	−40 18	56.8	23	e	34x16		Sco	*217*	
pn I4634	16 58.6	−21 45	01.6	49		10"x8"	*10.9*	Oph	172	
oc I4651	17 20.8	−49 54	24.7	57	II 2 r	12	6.9	Ara	31	
oc I4665	17 43.8	05 44	46.3	43	III 2 m	70	4.2	Oph	175	
gn I4703	18 16.2	−13 48	19.0	47	e	120x25		Ser	230	=M16; assoc. w/ oc 6611
oc I4725	18 28.7	−19 17	31.6	15	I 3 m	30	4.6	Sgr	213	=M25
oc I4756	18 36.5	05 24	39.0	27	II 3 r	40	4.6	Ser	230	
pn I4776	18 42.6	−33 24	45.9	21		8"x8"	*10.4*	Sgr		
oc I4996	20 14.6	37 29	16.5	38	II 3 p	6	7.3	Cyg	98	
eg I5105	21 21.2	−40 45	24.4	32	SA(rs)0⁻	2.5x1.5	11.7	Mic	166	
eg I5105A	21 22.4	−40 29	25.6	16	SB(s)c	2.2x1.2		Mic	*166*	
eg I5105B	21 22.9	−41 03	26.1	50	SB(rs)b?	2.1x0.7		Mic	*166*	
pn I5117	21 30.6	44 22	32.5	35			*11.5*	Cyg		
eg I5131	21 44.4	−35 07	47.4	53	SAB0°:	1.6x1.4	12.4	PsA	*200*	
eg I5135	21 45.3	−35 11	48.3	57	SaP	1.3x1.2	12.3	PsA	*200*	=7130
gn I5146	21 51.3	47 02	53.2	16	c	10x10		Cyg	100	=Cocoon
oc I5146	21 51.5	47 02	53.4	16	III 2 p	8	7.2	Cyg	100	assoc. w/ gn
eg I5179	22 13.2	−37 06	16.1	51	SA(rs)bc	2.3x1.2	11.9	Gru		=I5183; =I5184
eg I5186	22 15.9	−37 03	18.8	48	Sbc:	2.1x1.0		Gru		
pn I5217	22 21.9	50 43	23.9	58		7"x6"	*11.3*	Lac	136	
eg I5240	22 38.9	−45 02	41.8	46	SB(r)a	3.2x2.5		Gru		
eg I5264	22 54.1	−36 49	56.9	33	Sab:Psp	2.2x0.4		Gru	*124*	
eg I5267	22 54.4	−43 40	57.3	24	SA(rs)0°:	5.0x4.1	10.5	Gru		
eg I5269B	22 53.8	−36 31	56.6	15	SB(rs)cd?	4.0x1.1		Gru	*124*	
eg I5271	22 55.3	−34 01	58.1	45	Sb:	2.3x0.8		PsA	200	
eg I5273	22 56.7	−37 58	59.5	42	SB(rs)cd:	2.9x2.1	11.5	Gru		
eg I5283	23 00.8	08 37	03.3	53	SA(r)cdP?	0.9x0.4	13.8	Peg	*183*	
eg I5309	23 16.6	07 50	19.1	06	SB(r)a:	1.6x0.7		Psc	198	chart XIX
eg I5325	23 26.0	−41 37	28.7	20	SAB(rs)bc	2.5x2.5		Phe	196	

ID	RA (1950) DEC (2000)				TYPE	SIZE	V	CON	PG	NOTES C 52
eg I5328	23 30.6	-45 18	33.3	01	SAB(s)0°	2.5x1.6		Phe		
eg I5328A	23 30.5	-45 18	33.2	01	SB(s)0+?	1.0x0.4		Phe		
eg I5328B	23 31.3	-45 29	34.0	12	SB(s)c:	1.6x0.5		Phe		
eg I5332	23 31.8	-36 23	34.5	06	SA(s)d	6.6x5.1	10.6	Scl	222	
pn J 320	05 02.8	10 38	05.6	42		22"x22"	*12.0*	Ori	176	
pn J 900	06 23.0	17 49	25.9	47		12"x10"	*11.8*	Gem	121	
oc King 5	03 11.0	52 32	14.6	43	I 2 m	6		Per	190	
oc King 8	05 46.1	33 37	49.4	38	II 2 m	8	11.2	Aur	35	
oc King 9	22 13.6	54 09	15.5	24	I 1 m	2.5		Lac	*136*	
oc King 12	23 50.5	61 41	53.0	58	II 1 p	2.0		Cas	65	
oc King 14	00 29.0	62 53	31.9	10	III 1 p	7	8.5	Cas	62	
oc King 17	05 05.0	39 01	08.4	05	II 2 m	1.5		Aur	33	
oc King 18	22 50.1	58 01	52.1	17	II 2 p	4.0		Cep	71	
oc King 21	23 47.4	62 26	49.9	43	I 2 p	2.5	9.6	Cas	65	
eg M-06-08-024	03 29.2	-36 29	31.1	19	SA0⁻:			For		chart X
eg M-06-08-025	03 29.5	-35 30	31.4	20	SB(s)0°			For		chart X
eg M-06-08-027	03 32.6	-35 43	34.5	33	SAB(rs)0°			For		chart X
eg M-06-09-008	03 34.9	-35 32	36.8	22	SA0⁻:			For		chart X
eg M-06-23-039	10 27.6	-34 51	29.9	06				Ant	*24*	
eg M-03-26-006	09 57.8	-19 23	00.2	37	cE0	0.4x0.4		Hya	131	
eg M-02-05-053	01 46.7	-10 18	49.2	03	SAB(s)cd	2.8x2.2		Cet	74	
eg M-02-33-015	12 46.8	-09 51	49.4	07	SB(s)m	4.0x3.2		Vir	271	
eg M-02-33-103	13 01.4	-11 14	04.0	30	SA(s)mP	0.4x0.4		Vir	*274*	
eg M-02-34-031	13 17.3	-12 14	19.9	30	SABab	0.7x0.7		Vir		chart XVII
eg M-02-38-013	14 45.0	-14 56	47.8	08	Sc?	1.2x0.9		Lib	*155*	
eg M-02-38-015	14 45.2	-14 05	47.9	17	SB(s)d	2.3x2.3		Lib	155	
eg M-02-39-007	15 11.0	-15 17	13.8	28	SA(s)d	3.2x2.7		Lib	156	
eg M-01-03-085	01 02.6	-06 29	05.1	13	SAB(rs)d	4.0x3.5		Cet	73	
eg M-01-06-023	01 57.8	-06 21	00.3	07	SBm:	0.6x0.6		Cet	*75*	
eg M-01-22-006	08 20.5	-04 47	23.0	57	S0			Hya	129	
eg M-01-30-003	11 31.8	-09 19	34.3	36	(R')SA(r)a:	1.5x1.2		Crt	95	=3730?
eg M-01-30-004	11 31.8	-09 20	34.3	37	SBd:sp	1.3x0.3		Crt	95	=3730?
eg M-01-30-008	11 32.2	-09 23	34.7	40	(R')SB(s)a:	0.7x0.3		Crt		
eg M-01-30-022	11 38.9	-06 13	41.5	30	IAm	1.4x1.2		Vir	*257*	
eg M-01-33-027	12 48.6	-06 17	51.2	33	SBm:	0.6x0.6		Vir	*271*	
eg M+01-31-043	12 15.6	05 58	18.2	41	S0/a:			Vir		chart XXIV
eg M+01-31-053	12 16.8	06 12	19.4	55	cI	0.5x0.4		Vir	*259*	chart XXIV
eg M+01-32-006	12 17.3	07 16	19.9	59	SB(s)c	0.6x0.4		Vir		chart XXIV
eg M+01-32-041	12 21.5	05 27	24.1	10	SBm?	1.1x1.1		Vir		chart XXIV
eg M+01-32-046	12 21.7	04 30	24.3	13	S0:	1.4x0.3		Vir		chart XXIV
eg M+01-32-056	12 23.3	06 05	25.8	48	cIP			Vir		chart XXIV
eg M+01-32-059	12 23.4	04 45	26.0	28	S0P?	0.3x0.3		Vir		chart XXIV
eg M+01-32-060	12 23.7	05 45	26.3	28	S0	0.5x0.4		Vir		chart XXIV
eg M+01-59-058	23 18.6	07 50	21.1	06	SA(r)b:	1.2x0.6		Psc		chart XIX
eg M+01-59-061	23 18.9	08 16	21.4	32	SBc?	0.6x0.4		Peg		chart XIX
eg M+01-59-081	23 25.4	08 30	27.9	47	S0P	0.3x0.2		Peg	186	
eg M+02-32-029	12 21.8	13 18	24.3	01	S0?	0.4x0.4		Vir		chart XXVI
eg M+02-32-030	12 21.9	13 31	24.4	14	SA(r)bc	1.1x0.8		Vir		chart XXVI
eg M+03-30-043	11 38.8	20 14	41.4	57	Sa:	0.7x0.7		Leo		chart XIII
eg M+03-30-048	11 39.7	20 23	42.3	06	Sa?	0.6x0.3		Leo		chart XIII

ID	RA (1950)	DEC	(2000)		TYPE	SIZE	V	CON	PG	NOTES C 53
eg M+03-30-051	11 39.8	20 24	42.4	07	SAB(rs)bc:P	1.4x0.7		Leo		chart XIII
eg M+03-30-055	11 40.2	20 19	42.8	02	SA(r)ab:	1.3x0.6		Leo		chart XIII
eg M+03-30-063	11 40.9	19 55	43.5	38	SA(r)bc:	1.1x0.9		Leo		chart XIII
eg M+03-30-064	11 41.0	19 53	43.6	36	E	0.2x0.2		Leo		chart XIII
eg M+03-30-067	11 41.4	20 04	44.0	47	S0	0.8x0.4		Leo		chart XIII
eg M+03-30-071	11 41.4	20 04	44.0	47	S0/a?	1.3x0.4		Leo		chart XIII
eg M+03-30-076	11 41.7	20 08	44.3	51	(R)SB(r)0/a:	0.6x0.5		Leo		chart XIII
eg M+03-30-078	11 41.8	20 06	44.4	49	(R)SB0	0.6x0.4		Leo		chart XIII
eg M+03-30-079	11 41.8	20 07	44.4	50	SA0	0.8x0.2		Leo		chart XIII
eg M+03-30-083	11 41.9	20 21	44.5	04	SA(r)0°sp	1.4x0.3		Leo		chart XIII
eg M+03-30-085	11 42.2	19 58	44.8	41	SBa?	0.6x0.5		Leo		chart XIII
eg M+03-30-086	11 42.2	19 59	44.8	42	Sbc?	0.5x0.5		Leo		chart XIII
eg M+03-30-087	11 42.2	20 03	44.8	46	SAc?P	0.5x0.5		Leo		chart XIII
eg M+03-30-092	11 42.3	20 03	44.9	46	SA(s)c:P	0.7x0.3		Leo		chart XIII
eg M+03-30-094	11 42.5	20 15	45.1	58	SB(r)bc:	0.6x0.2		Leo	146	chart XIII
eg M+03-30-098	11 42.7	20 07	45.3	50	SA0	0.4x0.3		Leo		chart XIII
eg M+03-30-107	11 43.4	19 48	46.0	31	Sab?	0.5x0.5		Leo		chart XIII
eg M+03-30-108	11 43.5	19 43	46.1	26	S0:	0.5x0.5		Leo		chart XIII
eg M+04-28-029	11 39.3	20 36	41.9	19	(R')SAB(s)b	0.6x0.4		Leo		chart XIII
eg M+04-28-043	11 41.7	20 30	44.3	13	SAa:?	0.7x0.4		Leo	146	chart XIII
eg M+04-28-044	11 41.7	20 30	44.3	13	S0⁻:	0.4x0.3		Leo	146	chart XIII
eg M+04-28-047	11 42.8	20 36	45.4	19	(R)SAB0:	0.5x0.4		Leo		chart XIII
eg M+05-31-046	12 56.1	28 17	58.5	01	S0	1.4x1.4	13.8	Com		chart VIII
eg M+05-31-063	12 56.8	28 21	59.2	05	E	0.4x0.4	14.4	Com	87	chart VIII
eg M+05-31-074	12 57.4	27 59	59.8	43	S0sp	1.3x0.4	13.9	Com		chart VIII
eg M+05-31-094	12 58.5	28 03	00.9	47	SA0	0.6x0.4	15.3	Com	89	chart VIII
eg M+05-31-095	12 58.5	28 38	00.9	22	S0:		13.8	Com		chart VIII
eg M+05-31-096	12 58.7	28 05	01.1	49	E		14.8	Com	89	chart VIII
eg M+06-21-004	09 15.3	34 06	18.4	53	SA0	0.7x0.7		Lyn		chart XVI
eg M+06-21-005	09 15.3	33 46	18.3	33	S0/a:	0.6x0.6		Lyn		chart XVI
eg M+06-21-006	09 15.5	34 06	18.6	53	SB0	0.5x0.5		Lyn		chart XVI
eg M+06-21-012	09 16.6	33 38	19.6	25	S0	0.8x0.8		Lyn		chart XVI
eg M+06-21-024	09 17.7	33 56	20.7	43	SAB0:			Lyn		chart XVI
eg M+07-07-048	03 14.2	41 12	17.5	23	S0			Per		chart XX
eg M+07-07-061	03 16.3	41 24	19.6	35	S0⁻		13.7	Per	190	chart XX
eg M+07-07-070	03 17.0	41 27	20.3	38	Sc?	0.6x0.3		Per		chart XX
eg M+07-07-078	03 18.9	40 32	22.2	43	SAB(r)a:	0.7x0.5		Per		chart XX
eg M+07-28-078	13 48.6	40 31	50.7	16	E0:	0.8x0.8		CVn		chart V
eg M+07-29-002	13 49.3	40 28	51.4	13	cE0	0.9x0.9		CVn		chart V
eg M+07-29-014	13 51.9	40 14	54.0	59	IBm?	0.6x0.6		CVn		chart V
eg M+08-26-038	14 27.8	49 50	29.6	37	SBm:	1.0x0.5		Boo	*38*	
eg M+09-17-009	09 57.8	55 57	01.2	43	S0	1.1x0.5		UMa	*244*	
eg M+09-19-014	11 12.0	55 14	14.9	58	Sc:	0.4x0.3		UMa	*246*	
eg M+10-17-002	11 25.8	58 51	28.6	34	E	0.4x0.4		UMa	*248*	
oc Markarian 6	02 25.9	60 26	29.6	39	III 1 p	4.5	7.1	Cas	65	
oc Markarian38	18 12.3	-19 01	15.2	00	I 1 p	2.0	6.9	Sgr	211	
pn Merrill 2-1	15 19.4	-23 27	22.3	38		6"x6"	*11.6*	Lib	157	=VV72
oc Melotte 71	07 35.2	-11 57	37.5	04	II 2 r	8	7.1	Pup	202	
oc Melotte 72	07 36.0	-10 34	38.4	41	III 1 m	9		Mon	171	
oc Melotte 111	12 22.6	26 23	25.1	06	III 3 r	275	1.8	Com	83	=Coma

ID	RA (1950) DEC (2000)				TYPE	SIZE	V	CON	PG	NOTES C 54
eg Mk205	12 19.5	75 35	21.7	18	QSO			Dra	105	
pn PK164+31°01	07 54.0	53 33	57.9	25		6.8x6.0	*12.1*	Lyn	159	
oc Pismis 24	17 21.4	-34 09	24.7	12	III 2 p	5	9.6	Sco	218	assoc. w/ gn 6357
oc Pleiades	03 44.0	23 58	47.0	07	I 3 r	120	1.2	Tau	233	=M45; assoc. w/ gn
oc Ruprecht 7	06 55.4	-13 09	57.7	13	II 2 m	4.0		CMa	57	
oc Ruprecht 8	06 59.4	-13 31	01.7	35	IV 1 p	4.0		CMa	58	
oc Ruprecht 11	07 05.2	-20 43	07.4	48	III 2 p	2.9		CMa	58	
oc Ruprecht 18	07 22.7	-26 07	24.8	13	III 1 m	4.0	9.4	CMa	59	
oc Ruprecht 24	07 29.6	-12 39	31.9	45	IV 1 p	8		Pup	201	
oc Ruprecht 26	07 34.9	-15 32	37.2	39	III 1 p	4.0		Pup	201	
oc Ruprecht136	17 56.2	-24 42	59.3	42	IV 1 m	3.0		Sgr	209	
oc Ruprecht141	18 25.5	-12 21	28.3	19	IV 1 p	6		Sct	224	
oc Ruprecht142	18 29.3	-12 17	32.1	15	IV 1 p	5		Sct	224	
oc Ruprecht143	18 29.8	-12 10	32.6	08	IV 1 m	6		Sct	224	
oc Ruprecht144	18 30.6	-11 28	33.4	26	IV 1 p	2.0		Sct	224	
oc Stock 1	19 33.7	25 06	35.8	13	IV 2 p	80	5.3	Vul	282	
oc Stock 2	02 11.4	59 02	15.0	16	I 2 m	60	4.4	Cas	65	
oc Stock 8	05 24.3	34 23	27.6	25	I 3 m	5		Aur	33	
oc Stock 10	05 35.6	37 54	39.0	56	IV 2 p	25		Aur	35	
oc Stock 23	03 12.3	59 11	16.3	22	II 3 p	15		Cam	40	
oc Stock 24	00 36.8	61 41	39.7	57	III 1 p	4.0	8.8	Cas	63	
oc Tombaugh 1	06 58.2	-20 24	00.4	28	III 1 m	6		CMa	57	
oc Tombaugh 2	07 01.2	-20 47	03.4	51	I 1 m	3.0		CMa	58	
oc Tombaugh 5	03 43.7	58 54	47.8	03	III 2 r	17	8.4	Cam	40	
oc Trumpler 1	01 32.3	61 02	35.7	17	II 2 p	4.5	8.1	Cas	64	
oc Trumpler 2	02 33.7	55 46	37.3	59	II 2 p	17	5.9	Per	188	
oc Trumpler 3	03 07.6	63 04	11.8	15	III 2 m	23		Cas	65	=Harvard 1
oc Trumpler 6	07 24.0	-24 12	26.1	18	III 2 p	6		CMa	59	
oc Trumpler 7	07 25.2	-23 56	27.3	02	II 3 m	5	7.9	CMa	59	
oc Trumpler 9	07 53.2	-25 48	55.3	56	II 2 p	6	8.7	Pup	204	=Harvard 2
oc Trumpler 24	16 53.5	-40 35	57.0	40	IV 2 p	60		Sco	217	=Harvard 12; assoc. w/ gn
oc Trumpler 25	17 21.3	-38 57	24.8	00	II 1 m	4.0		Sco	218	=Harvard 14
oc Trumpler 26	17 25.3	-29 27	28.5	29	II 1 m	7	9.5	Oph	174	=Harvard 15
oc Trumpler 27	17 32.9	-33 27	36.2	29	III 3 m	7	6.7	Sco	218	
oc Trumpler 28	17 33.5	-32 27	36.8	29	III 2 m	13	7.7	Sco	219	
oc Trumpler 29	17 38.1	-40 05	41.6	06	III 2 m	12		Sco	219	=Harvard 17
oc Trumpler 30	17 53.1	-35 19	56.5	19	IV 1 p	8		Sco	220	=Harvard 18
oc Trumpler 32	18 14.7	-13 22	17.5	21	I 2 m	6		Ser	230	=Harvard 19
oc Trumpler 34	18 37.1	-08 32	39.8	29	IV 2 m	5	8.6	Sct	224	
oc Trumpler 35	18 40.3	-04 11	42.9	08	I 2 m	6	9.2	Sct	226	
eg U 1195	01 39.7	13 43	42.4	58	P	3.4x1.2		Psc	*198*	=M+02-05-11; P(b) w/ 660
eg U 2598	03 10.8	41 06	14.1	17	SAB0:	1.8x1.1		Per		chart XX
eg U 2608	03 11.7	41 51	15.0	02	(R')SB(s)b	1.1x1.0		Per		=M+07-07-37; chart XX
eg U 2612	03 11.9	41 48	15.2	59	Sc	1.1x0.8		Per		=M+07-07-38; chart XX
eg U 2616	03 12.5	41 41	15.8	52	S0:	1.4x0.6		Per		chart XX
eg U 2617	03 12.7	40 42	16.0	53	SAB(s)d	2.9x1.1		Per		=M+07-07-41; chart XX
eg U 2618	03 12.7	41 53	16.0	04	SABab:	1.5x0.7		Per		=M+07-07-42; chart XX
eg U 2619	03 13.0	41 00	16.3	11	SB0	1.5x1.5		Per		chart XX
eg U 2621	03 13.1	41 21	16.4	32	S0/a:sp	1.7x0.3		Per		=M+07-07-44; chart XX
eg U 2626	03 13.7	41 10	17.0	21	Sa?sp	1.4x0.3		Per		chart XX

ID	RA (1950) DEC (2000)				TYPE	SIZE	V	CON	PG	NOTES C 55
eg U 2639	03 14.5	41 47	17.8	58	S0sp:	1.5x0.2		Per		chart XX
eg U 2642	03 14.7	40 45	18.0	56	SBb:	1.1x0.8		Per		chart XX
eg U 2656	03 15.5	40 53	18.8	04	S0	1.3x0.9		Per		chart XX
eg U 2664	03 16.2	40 34	19.5	45	SBm	1.5x1.1		Per		chart XX
eg U 2665	03 16.1	41 27	19.4	38	S0	1.4x0.6		Per		=M+07-07-60; chart XX
eg U 2672	03 16.8	40 44	20.1	55	Sa?sp	1.1x0.4		Per		chart XX
eg U 2673	03 16.7	41 04	20.0	15	S0	1.9x1.7		Per		=M+07-07-66; chart XX
eg U 2686	03 17.7	40 37	21.0	48	SABa	1.7x1.0		Per		chart XX
eg U 2689	03 18.2	40 38	21.5	49	SAB0?	1.8x1.2		Per		chart XX
eg U 2698	03 18.8	40 41	22.1	52	S0⁻:	1.3x1.1		Per		=M+07-07-77; chart XX
eg U 4832	09 08.5	79 25	15.0	13	Sc?P	1.6x0.5		Cam	*42*	=M+13-07-17
eg U 4847	09 09.7	79 29	16.2	17	Scd:	1.2x0.2		Cam	*43*	=M+13-07-20
eg U 4904	09 14.1	42 07	17.3	54	SBc?	1.0x0.7		Lyn	161	=M+07-19-54
eg U 5832	10 40.2	13 43	42.9	27	SB?	1.2x1.1		Leo	140	=M+02-27-42; in 3368 group
eg U 6035	10 52.8	17 24	55.5	08	IBm	1.4x1.3		Leo	*141*	=M+03-28-35
eg U 6296	11 14.2	18 04	16.8	48	SBc:	1.4x0.5		Leo	*142*	=M+03-29-21
eg U 6680	11 40.4	19 56	43.0	39	SA(r)b:	1.3x0.7		Leo		=M+03-30-58; chart XIII
eg U 6683	11 40.7	20 02	43.3	45	S0/a	1.1x0.3		Leo		=M+03-30-59; chart XIII
eg U 6697	11 41.2	20 15	43.8	58	SB(s)m:sp	1.7x0.4		Leo	144	=M+03-30-66; in Abell 1367; chart XIII
eg U 6719	11 42.2	20 24	44.8	07	SB(r)bP	1.3x0.9		Leo	146	=M+03-30-89; chart XIII
eg U 6725	11 42.5	20 43	45.1	26	(R)SAB0⁺:	1.8x1.5		Leo		=M+04-28-46; chart XIII
eg U 6791	11 46.7	27 01	49.3	44	Sc	1.9x0.4		Leo	149	=M+05-28-35
eg U 6806	11 47.7	26 15	50.3	58	SB(r)bc:P	2.2x0.8		Leo	149	=M+04-28-60
eg U 6930	11 54.7	49 33	57.3	16	SAB(s)d	4.3x3.0		UMa	250	=M+08-22-46
eg U 7000	11 58.6	−01 01	01.2	18	IB:m	1.2x1.0		Vir	257	=M+00-31-17
eg U 7422	12 19.5	05 23	22.1	06	SB(s)ab	1.1x0.4		Vir		=M+01-32-23; chart XXIV
eg U 7423	12 19.4	06 44	22.0	27	Sm:	1.4x0.8		Vir		=M+01-32-24; chart XXIV
eg U 7516	12 23.2	04 47	25.8	30	SA(s)c	1.4x0.8		Vir		=M+01-32-55; chart XXIV
eg U 7588	12 25.7	13 50	28.2	33	Sc	1.1x0.1		Com		chart XXVI
eg U 7652AB	12 28.2	12 39	30.7	22	P, P	1.2x0.3		Vir		interacting system; chart XXVI
eg U 7658	12 28.4	12 33	30.9	16	E2	1.3x1.1		Vir		=M+02-32-110; chart XXVI
eg U 7686	12 29.9	12 04	32.4	47	Sc	1.4x0.3		Vir		=M+02-32-121; chart XXVI
eg U 8041	12 52.7	00 23	55.3	07	SB(s)d	3.2x2.1		Vir		=M+00-33-21
eg U 8042	12 52.6	08 11	55.1	55	SBm:	1.5x1.0		Vir	*273*	
eg U 8045	12 52.9	08 11	55.4	55	IB:m	1.1x0.9		Vir	*273*	
eg U 8071	12 55.1	28 28	57.5	12	S0:	1.4x0.4		Com		chart VIII
eg U 8736	13 46.9	39 45	49.0	30	Sab?	1.5x0.7		CVn		=M+07-28-73; chart V
eg U 8801	13 51.0	−00 58	53.6	13	Sb:	1.4x0.4		Vir	*277*	=M+00-35-25
eg U 8818	13 51.6	05 36	54.1	21	Sbc	1.1x1.0		Vir	*278*	=M+01-35-50
eg U 8840	13 53.0	40 10	55.1	55	S0	1.4x0.5		CVn		=M+07-29-17; chart V
eg U 8841	13 53.0	40 24	55.1	09	SBb	1.7x1.1		CVn		=M+07-29-18; chart V
eg U 8851	13 53.6	39 57	55.7	42	Sdm	1.7x0.6		CVn		=M+07-29-19; chart V
eg U 9920	15 33.3	30 58	35.3	48	Sbsp	1.5x0.2		CrB	*92*	=M+05-37-6
eg U10473	16 35.1	36 31	36.9	25	SBa	1.7x0.5		Her	126	=M+06-36-55
eg U12121	22 35.4	34 35	37.7	51	S0	1.4x0.6		Peg		=M+06-49-51; chart XVIII
eg U12132	22 36.6	34 03	38.9	19	SB(r)b	1.1x0.6		Peg		chart XVIII
eg U12510	23 17.1	07 59	19.6	15	dE?	1.6x1.1		Peg		chart XIX
eg U12518	23 17.7	07 39	20.2	55	Sbsp	1.4x0.3		Psc		=M+01-59-53; chart XIX
eg U12522	23 17.8	07 44	20.3	00	Sm	1.7x1.7		Psc		=M+01-59-54; chart XIX
eg U12535	23 18.5	07 54	21.0	10	SBbc:	1.2x0.2		Psc		chart XIX

ID	RA (1950)	DEC	(2000)		TYPE	SIZE	V	CON	PG	NOTES C 56
eg VZw319	03 02.9	42 11	06.2	23	cE	0.2x0.2		Per	189	
eg Z041-066	12 15.2	04 45	17.8	28	Ssp			Vir		chart XXIV
eg Z041-067	12 15.4	06 09	18.0	52	SA(s)c			Vir		chart XXIV
eg Z041-075	12 15.7	06 53	18.3	36	S0			Vir		chart XXIV
eg Z042-001	12 16.1	06 59	18.7	42	Scsp			Vir		chart XXIV
eg Z042-007	12 16.5	06 23	19.1	06	SB(s)a			Vir		chart XXIV
eg Z042-009	12 16.7	06 39	19.3	22	S0sp			Vir		chart XXIV
eg Z042-010	12 16.7	06 34	19.3	17	P			Vir		chart XXIV
eg Z042-011	12 16.7	06 35	19.3	18	S0sp			Vir		chart XXIV
eg Z042-018	12 16.9	05 20	19.5	03	E7			Vir		chart XXIV
eg Z042-031	12 17.7	06 54	20.3	37	dE5?			Vir		chart XXIV
eg Z042-036	12 17.9	07 11	20.5	54	Sd:			Vir		chart XXIV
eg Z042-039	12 18.5	07 21	21.0	04	S0P			Vir		chart XXIV
eg Z042-049	12 19.6	05 14	22.2	57	Sa?			Vir		chart XXIV
eg Z042-050	12 19.6	06 23	22.1	06	Sab			Vir		chart XXIV
eg Z042-056	12 20.2	05 55	22.8	38	Sc			Vir		chart XXIV
eg Z042-057	12 20.1	06 30	22.6	13	Ssp			Vir		chart XXIV
eg Z042-078	12 21.5	05 35	24.1	18	Ssp			Vir		chart XXIV
eg Z042-084	12 22.0	06 59	24.5	42	SB(s)bc			Vir		chart XXIV
eg Z042-088	12 22.3	07 02	24.8	45	dE5			Vir		chart XXIV
eg Z042-090	12 22.5	05 36	25.0	19	S0sp			Vir		chart XXIV
eg Z042-091	12 22.6	04 45	25.2	28	Sc:sp			Vir		chart XXIV
eg Z069-113	12 13.3	13 30	15.8	13	S0			Vir	*258*	
eg Z097-062	11 39.6	20 16	42.2	59	S			Leo		chart XIII
eg Z097-063	11 39.7	20 20	42.3	03	S			Leo		chart XIII
eg Z097-073	11 40.3	20 15	42.9	58	SP			Leo		chart XIII
eg Z097-074	11 40.4	20 22	43.0	05	S0?			Leo		chart XIII
eg Z097-077	11 40.7	19 40	43.3	23	S0			Leo		chart XIII
eg Z097-079	11 40.6	20 17	43.2	00	I			Leo		chart XIII
eg Z097-086	11 41.2	20 19	43.8	02	SA0+:sp			Leo	146	chart XIII
eg Z097-090	11 41.4	20 13	44.0	56	SA0	0.2x0.2		Leo	146	chart XIII
eg Z097-092	11 41.4	20 28	44.0	11	SAB(r)0/a			Leo		chart XIII
eg Z097-094	11 41.5	20 05	44.1	48	E:			Leo		chart XIII
eg Z097-099	11 41.5	20 01	44.1	44	E			Leo		chart XIII
eg Z097-109	11 41.9	20 01	44.5	44	S0?			Leo		chart XIII
eg Z097-111	11 41.8	20 23	44.4	06	cE			Leo		chart XIII
eg Z097-113	11 42.2	20 02	44.8	45	S0?			Leo		chart XIII
eg Z097-115	11 42.2	20 09	44.8	52	S0sp			Leo		chart XIII
eg Z097-118	11 42.3	19 53	44.9	36	S0			Leo		chart XIII
eg Z097-123	11 42.3	19 46	44.9	29	S0?			Leo		chart XIII
eg Z097-124	11 42.4	20 01	45.0	44	E			Leo		chart XIII
eg Z097-133AB	11 42.7	20 18	45.3	01	P, P			Leo		P(b)?; chart XIII
eg Z097-138	11 43.2	20 19	45.8	02	SAaP?			Leo		chart XIII
eg Z097-142	11 43.5	19 55	46.1	38	S0P?sp			Leo		chart XIII
eg Z097-143	11 43.5	20 04	46.1	47	SB0			Leo		chart XIII
eg Z127-011	11 35.3	22 05	37.9	48	cE			Leo		chart XII
eg Z127-040	11 41.1	20 33	43.7	16	SA(r)0:sp			Leo		chart XIII
eg Z160-049	12 55.4	28 27	57.8	11	E			Com		chart VIII
eg Z160-057	12 55.8	28 24	58.2	08	S0			Com		chart VIII
eg Z160-086	12 58.2	27 54	00.6	38	S0			Com		chart VIII

ID	RA (1950) DEC (2000)				TYPE	SIZE	V	CON	PG	NOTES C 57
eg Z160-089	12 58.3	28 36	00.7	20	S0			Com		chart VIII
eg Z160-093	12 58.8	28 04	01.2	48	SA0			Com		chart VIII
eg Z160-100	12 59.4	28 10	01.8	54	E			Com		chart VIII
eg Z160-101	12 59.4	28 22	01.8	06	S0sp		14.2	Com		chart VIII
eg Z160-108	12 59.8	28 29	02.2	13	P			Com		chart VIII
eg Z160-214	12 56.6	28 30	59.0	14	SAB0		14.1	Com		chart VIII
eg Z160-251	12 58.2	28 25	00.6	09	SA0sp			Com		chart VIII
eg Z160-261	12 58.6	28 10	01.0	54	SA(r)0+:			Com		chart VIII
eg Z219-013	13 50.9	39 58	53.0	43	cI			CVn		chart V
eg Z219-021	13 51.9	39 57	54.0	42	SBb:			CVn		chart V
eg Z268-049A	11 29.8	53 13	32.5	56	SB0:P	0.6x0.4	14.3	UMa	*248*	in chain
eg Z268-049B	11 29.8	53 14	32.5	57	SA(s)0/a?P			UMa	*248*	in chain
eg Z268-049C	11 29.9	53 14	32.6	57	(R')S0/a?P			UMa	*248*	in chain
eg Z268-049D	11 29.9	53 14	32.6	57	Sa?Psp			UMa	*248*	in chain
eg Z268-049E	11 30.0	53 13	32.7	56	Scd:sp	1.1x0.2		UMa	*248*	in chain
eg Z312-012	09 25.7	62 46	29.6	33	cE	0.5x0.5		UMa	*243*	
eg Z406-069	23 17.4	08 17	19.9	33	IBm?	0.7x0.6		Peg		chart XIX
eg Z406-080	23 18.6	07 54	21.1	10	E	0.7x0.3		Psc		chart XIX
eg Z540-059	03 10.2	40 54	13.5	05	SAB0:sp			Per		chart XX
eg Z540-067	03 12.1	41 26	15.4	37	E			Per		chart XX
eg Z540-074	03 13.2	41 27	16.5	38	S0			Per		chart XX
eg Z540-077	03 13.5	40 31	16.8	42	S0			Per		chart XX
eg Z540-079	03 13.7	41 27	17.0	38	S0/a:			Per		chart XX
eg Z540-085	03 14.6	41 16	17.9	27	S0-:			Per		chart XX
eg Z540-087	03 15.1	41 14	18.4	25	S0		13.7	Per	190	chart XX
eg Z540-089	03 15.3	41 18	18.6	29	S0			Per		chart XX
eg Z540-113	03 18.0	41 18	21.3	29	S0sp			Per		chart XX
eg Z540-115	03 18.2	41 20	21.5	31	S0?			Per		chart XX

Appendix of double stars

Following are data on 152 catalogued double and multiple stars referred to in the descriptions, including 39 intended as aids for gauging seeing. The stars are sorted in alphanumeric order.

The first five columns give the name and position in the same format used in the catalogue. The sixth column lists V (V =) or visual magnitudes for the components. In the seventh column is the separation in arcseconds, followed in the eighth and ninth columns by the position angle and date. The date usually refers to the most recent published measure for the pair, but in a few cases where proper motions or orbits are accurately known the data have been predicted from these for a later or future date.

Column ten lists the constellation in which the star appears, and eleven and twelve show the page number and object under which the pair is described. Stars intended for use in estimating seeing have no page number or object listed.

The last column gives special names, spectral types when available, and other information. For pairs with known orbits, the separation and position angle for the year 2000 have been calculated and are listed. These, combined with the entries in columns seven and eight, allow approximate interpolation for other dates.

ID	RA (1950) DEC (2000)				MAGNITUDES	SEP	PA	DATE	CON	PG	OBJECT	NOTES A 1
ADS 460	00 30.7	63 01	33.6	18	10.0,10.0	6.8	223	1923	Cas	62	oc 146	
ADS 719AB	00 49.9	56 21	52.8	38	7.9, 9.9	1.5	82	1969	Cas	63	oc 281	AB:O5V
AC					, 8.9	3.9	134	1969	Cas	63	oc 281	C:O9Vn
AD					, 9.4	9.3	194	1969	Cas	63	oc 281	
AE					,12.1	16.8	333	1969	Cas	63	oc 281	
CD					8.9, 9.4	7.6	220	1969	Cas	63	oc 281	
ADS 1073	01 16.9	57 58	20.1	14	V= 5.0, 7.0	133.8	231	1956	Cas	63	oc 457	=phi-1,2 Cas; F0Ia+B5Ia
ADS 1081	01 17.2	-00 46	19.8	31	V= 6.3, 7.0	1.6	11	1974	Cet			=42 Cet; A2:V+G4:III
ADS 1254	01 33.4	07 23	36.0	39	7.3, 7.3	1.7	53	1975	Psc			AB:F6V
ADS 1342	01 39.5	63 47	43.0	02	10.0,11.2	8.7	359	1904	Cas	64	oc 637	
ADS 1381	01 42.6	60 58	46.0	13	10.7,11.2	7.1	38	1929	Cas	64	oc 663	
ADS 1384	01 42.6	60 51	46.1	06	9.4,10.9	9.3	105	1922	Cas	64	oc 663	B2
ADS 1390	01 43.1	61 01	46.6	16	9.7,10.9	7.6	69	1918	Cas	64	oc 663	B3
ADS 1487	01 49.4	10 34	52.0	49	7.8, 7.8	3.1	201	1968	Ari			F1V+Am
ADS 1534AB	01 53.2	37 00	56.1	15	V= 5.7, 5.9	200.2	298	2000	And	18	oc 752	K0III+gM0; optical
AP					V= ,11.9	18.4	79	1934	And	18	oc 752	common proper motion
ADS 1562	01 55.0	32 56	57.9	10	8.9, 9.2	1.5	229	1978	Tri	238	eg 750	AB:K0; widening
ADS 1579	01 56.5	24 35	59.3	50	8.0, 8.3	1.1	276	1965	Ari			A3
ADS 1624	02 03.4	24 52	06.2	07	8.0, 8.5	1.9	163	1971	Ari			K5
ADS 1937	02 29.9	57 19	33.5	32	V= 8.0, 9.9	23.4	19	1926	Per	188	oc 957	A:B2Ib
ADS 1953	02 31.6	-05 51	34.1	38	V= 7.8, 8.0	3.6	346	1961	Cet			K2; fixed since 1832
ADS 2042	02 38.3	18 35	41.1	48	V= 7.7, 8.0	3.4	118	1968	Ari			A:A0V
ADS 2052	02 39.0	42 29	42.2	42	V= 8.4, 9.1	1.4	55	1959	Per	189	oc 1039	B9
ADS 2257	02 56.3	21 09	59.2	20	V= 5.2, 5.5	1.4	207	1972	Ari			=epsilon Ari; A2Vs+A2Vs
ADS 2426AB	03 12.3	59 51	16.3	02	7.7, 8.0	7.2	142	1981	Cam	40	oc St 23	A0
AC					,10.5	26.1	43	1915	Cam	40	oc St 23	
ADS 2755CD	03 43.3	24 02	46.3	11	8.1,12.1	18.1	8	1915	Tau	233	Pleiades	K2
ADS 2766	03 44.5	24 30	47.4	40	8.5,11.0	6.8	330	1975	Tau	233	Pleiades	B8+A0; fixed since 1832
ADS 2767	03 44.5	23 46	47.5	55	V= 7.3,10.4	6.1	265	1952	Tau	233	Pleiades	B9; fixed since 1832
ADS 2783AB	03 45.7	52 30	49.5	39	V= 6.9, 9.1	8.6	253	1934	Per	191	oc 1444	A:B0.5III
AC					,12.0	12.1	39	1910	Per	191	oc 1444	
DE					10.3,10.7	2.6	231	1976	Per	191	oc 1444	
ADS 2795AB	03 47.3	23 42	50.3	51	V= 6.7, 9.8	3.2	238	1974	Tau	233	Pleiades	A:B9.5Vp
AC					V= , 9.0	10.2	236	1937	Tau	233	Pleiades	
ADS 2984AB	04 03.4	62 12	07.8	20	V= 7.0, 7.0	17.9	304	1967	Cam	40	oc 1502	B1IIn+B0V; B comp is eclipsing binary SZ Cam, though both may be variable
ADS 3136	04 17.1	50 16	20.9	23	7.9, 9.4	18.3	346	1926	Per	194	oc 1545	K2
ADS 3297	04 30.7	17 55	33.5	01	V= 7.0, 7.1	3.1	277	1968	Tau			AB:B9IVn; fixed since 1830
ADS 3734	05 06.9	37 14	10.3	18	V= 6.7, 7.0	1.6	221	1960	Aur			B2II+K3; fixed since 1828
ADS 3745	05 07.9	16 26	10.7	30	V= 9.6,11.0	9.9	267	1961	Tau	234	oc 1807	G5
ADS 3928	05 18.1	33 20	21.4	23	9.5, 9.7	14.6	210	1918	Aur	33	oc 1893	B8
ADS 4182	05 32.6	-06 02	35.0	00	V= 4.8, 5.7	36.2	223	1962	Ori	177	oc 1980	B0.5V+B1Vv
ADS 4192	05 33.0	-04 24	35.5	22	V= 6.6, 8.4	4.2	60	1940	Ori	177	oc 1981	B2.5IV
ADS 4194	05 33.1	34 06	36.4	08	V= 9.1, 9.4	11.0	305	1962	Aur	35	oc 1960	B2V+B2V
ADS 4254AB	05 37.1	-17 53	39.3	51	V= 6.4, 7.8	0.6	155	1981	Lep	153	oc 2017	B8V
CD	05 37.1	-17 51	39.3	50	8.5, 9.2	1.5	357	1977	Lep	153	oc 2017	C:A2V
ADS 4374AC	05 44.2	00 04	46.7	05	V=10.2,10.6	50.5	202	1923	Ori	179	gn 2068	
ADS 4490	05 51.4	18 54	54.4	54	8.0, 8.0	3.1	148	1954	Ori			A0; fixed since 1831
ADS 4728AB	06 05.7	13 59	08.5	58	V= 7.4, 8.0	2.5	109	1970	Ori	179	oc 2169	B1V+B2V
ADS 4744	06 06.2	24 26	09.3	26	V= 7.3, 9.1	31.0	188	1973	Gem	121	oc 2168	G0

ID	RA (1950)	DEC	(2000)		MAGNITUDES	SEP	PA	DATE	CON	PG	OBJECT	NOTES A 2
ADS 4991	06 19.7	17 35	22.6	34	V= 7.3, 8.3	2.3	20	1953	Gem			AB:A0Vn; fixed since 1831
ADS 5482AB	06 46.5	41 03	50.0	00	V= 9.0,10.6	7.7	49	1904	Aur	36	oc 2281	A0
AC					,10.0	46.1	210	1904	Aur	36	oc 2281	
CD					V=10.1,10.6	7.7	108	1903	Aur	36	oc 2281	
ADS 5669	06 57.4	75 18	04.1	14	V= 7.1, 8.2	12.6	31	1971	Cam	41	eg 2314	G0+G0; common proper motion
ADS 5692	06 58.4	−20 34	00.6	38	V= 6.8,10.8	5.1	143	1968	CMa	57	oc To 1	A3
ADS 5740	07 01.1	−08 24	03.5	28	9.8, 9.9	5.0	38	1940	Mon	170	oc 2323	A0
ADS 5761	07 02.3	−11 27	04.6	31	V= 7.4, 9.4	5.3	150	1938	CMa	58	gn 2327	B5
ADS 5817	07 05.9	−10 33	08.3	37	V= 8.5,10.7	10.9	302	1931	Mon	171	oc 2343	G5
ADS 5871	07 09.7	27 19	12.8	14	V= 7.2, 7.2	1.2	317	1990	Gem			AB:F8V; orbit: [1.1;313(2000)]
ADS 5977AB	07 16.6	−24 29	18.7	34	V= 4.4,10.5	8.5	89	1960	CMa	59	oc 2362	=tau CMa; O9Ib
AC					V= ,11.2	14.8	79	1960	CMa	59	oc 2362	
AD					V= , 8.2	85.0	75	1960	CMa	59	oc 2362	
ADS 5997AB	07 18.0	−21 48	20.1	54	V= 9.4, 9.7	4.5	275	1937	CMa	59	oc 2367	B3
ADS 6062	07 22.9	−20 55	25.1	01	V= 9.1, 9.6	4.5	347	1964	CMa	59	oc 2384	B5
ADS 6104AC	07 25.5	−11 27	27.9	33	V= 5.8, 8.5	20.0	313	1958	CMa	59	oc 2396	G8Ib+B8V
ADS 6180	07 31.7	12 25	34.5	18	V= 8.4, 9.1	1.8	100	1969	Gem			B8
ADS 6208AC	07 33.8	−14 23	36.1	30	V= 5.7, 9.7	19.8	41	1974	Pup	201	oc 2422	B4III:n(shell)+A0V
ADS 6216	07 34.3	−14 22	36.6	29	V= 7.0, 7.3	7.4	305	1952	Pup	201	oc 2422	B6V+B6V
ADS 6223	07 34.8	−13 46	37.1	53	9.6,10.2	0.5	159	1943	Pup	201	oc 2423	A0
ADS 6309AB	07 40.2	−18 09	42.4	16	9.4, 9.9	1.8	291	1965	Pup	202	pn 2440	A
AC					,10.3	24.9	2	1942	Pup	202	pn 2440	
AE					,11.1	22.4	334	1942	Pup	202	pn 2440	
ADS 6425	07 50.1	03 31	52.7	23	7.0, 7.5	1.0	20	1966	CMi			A2; slowly closing
ADS 7093	08 53.0	−07 47	55.5	58	V= 6.7, 6.9	4.1	2	1968	Hya			=17 Hya; A2m+A7m
ADS 7303	09 17.3	51 29	20.7	16	V= 6.1,10.2	5.7	133	1973	UMa	242	eg 2841	=37 Lyn; A:F5V
ADS 7565	09 51.0	69 11	55.1	56	10.9,10.9	9.0	271	1967	UMa	243	eg 3031	G+G
ADS 7566	09 52.9	69 08	57.0	54	9.5, 9.5	2.1	112	1981	UMa	243	eg 3031	F5
ADS 7641	10 04.3	−19 04	06.6	18	8.8,10.0	9.5	146	1959	Hya	131	eg 3124	G5; optical
ADS 7704	10 13.5	17 59	16.3	44	V= 7.2, 7.5	1.4	181	1980	Leo			AB:A9IV; orbit: [1.5;179(2000)]
ADS 7836	10 31.7	21 51	34.4	36	7.6, 9.0	11.0	259	1931	Leo	139	eg 3287	K0+F8
ADS 7936	10 46.8	−03 46	49.3	01	V= 7.0, 7.8	2.2	10	1982	Sex			=40 Sex; AB:A2IV; slowly widening
ADS 8043	11 01.4	03 55	04.0	38	7.5, 7.6	1.3	117	1973	Leo			AB:Am; slowly widening
ADS 8236	11 33.9	56 25	36.6	08	7.9, 8.4	6.0	166	1977	UMa	248	eg 3780	G5
ADS 8446	12 08.3	40 10	10.8	53	V= 7.3, 8.0	0.4	265	1980	CVn	46	eg 4145	AB:A8III; orbit: [0.2;134(2000)]
ADS 8510	12 17.1	05 49	19.7	33	8.5,12.3	1.4	274	1939	Vir	259	eg 4266	G5; see also eg 4270
ADS 8531	12 20.0	05 35	22.5	18	V= 6.5, 9.4	20.6	337	1964	Vir	261	eg 4324	=17 Vir; F8V+dK5
ADS 8561	12 25.7	45 04	28.1	48	V= 7.5, 8.1	9.9	158	1981	CVn	48	eg 4460	A:F9V; common proper motion
ADS 8575	12 28.0	10 00	30.6	43	V= 8.1, 8.4	1.4	237	1966	Vir			F2; slowly widening
ADS 8684	12 48.8	−10 04	51.4	20	V= 6.4, 9.7	30.2	301	1959	Vir	271	eg 4742	A:G8III; optical
ADS 8708	12 53.9	−00 41	56.4	57	7.2, 7.6	0.9	89	1966	Vir			AB:F7V; slowly widening
ADS 8972	13 35.0	−07 37	37.6	52	7.5, 7.5	2.8	41	1957	Vir			=81 Vir; K2
ADS 9053	13 52.3	−07 49	55.0	04	V= 6.6, 7.5	3.4	94	1981	Vir			F8V+G0; orbit: [3.5;97(2000)]
ADS 9060	13 53.8	05 32	56.3	17	8.4, 8.9	1.0	112	1981	Vir	277	eg 5363	F5
ADS 9323	14 34.9	02 30	37.5	17	6.6,10.6	1.6	135	1964	Vir	279	eg 5690	A:F8V
ADS 9425	14 51.0	15 54	53.4	42	V= 6.9, 7.5	1.2	171	1980	Boo			F4IV+G1IV; orbit: [0.8;159(2000)]
ADS 9474	14 58.1	54 03	59.6	52	V= 6.8, 7.4	40.4	342	1956	Boo	39	eg 5820	F1V+F1V; fixed since 1823
ADS 9666	15 26.5	65 03	27.2	52	9.6,11.1	19.3	45	1915	Dra	107	eg 5949	K0
ADS 9959	16 09.8	12 02	12.1	55	V= 8.5, 9.7	7.3	147	1962	Her	125	pn I4593	A3; fixed since 1830
ADS 9969	16 11.0	13 39	13.3	32	V= 7.4, 7.5	4.1	348	1971	Her			=49 Ser; G9V+G9V

ID	RA (1950)	DEC	(2000)		MAGNITUDES	SEP	PA	DATE	CON	PG	OBJECT	NOTES A 3
ADS 10075	16 26.7	18 31	28.9	25	V= 7.7, 7.9	1.4	135	1980	Her			K3V+K3V; orbit: [2.0;124(2000)]
ADS 10650	17 33.3	01 02	35.8	00	7.5, 7.5	2.9	81	1953	Oph			AB:B8IV; fixed since 1831
ADS 10905	17 54.2	18 20	56.4	20	7.0, 7.0	2.6	293	1960	Her			A0IV+F5:III; fixed since 1829
ADS 10991AB	17 59.3	−23 02	02.4	02	V= 7.6,10.4	6.0	22	1934	Sgr	209	oc 6514	A:O8V
AC					V= , 8.7	10.8	212	1956	Sgr	209	oc 6514	
CD					V= 8.7,10.5	2.3	281	1934	Sgr	209	oc 6514	
ADS 11193BR	18 12.3	−19 00	15.3	59	V=11.3,12.5	5.0	324	1914	Sgr	211	oc Ma 38	
ADS 11441	18 30.7	−10 30	33.5	28	V= 9.7,11.4	4.1	90	1913	Sct	224	oc 6649	
ADS 11483	18 33.7	16 56	35.9	59	V= 6.9, 7.1	1.6	161	1980	Her			AB:F8V; orbit: [1.3;147(2000)]
ADS 11640	18 43.0	05 27	45.5	30	V= 6.4, 6.7	2.5	116	1974	Ser			"Tweedledee & Tweedledum"; A1V+A1V; slowly
												widening; each component is an extremely
												close pair (0.2").
ADS 12385	19 21.1	22 03	23.2	09	10.5,11.5	8.3	45	1910	Vul	282	oc 6793	
ADS 12669	19 34.4	24 53	36.5	00	9.0,10.0	9.2	101	1951	Vul	282	oc St 1	B9
ADS 12962	19 46.3	11 41	48.7	49	V= 6.3, 6.8	1.5	108	1967	Aql			=pi Aql; A3V+F9III
ADS 13374AF	20 04.1	35 39	06.0	47	V= 6.8, 7.4	36.0	28	1965	Cyg	97	oc 6871	A comp=V1676 Cyg: WN5+O9.5I/III; F:B1Ia
ADS 13465	20 08.8	34 43	10.7	52	9.2, 9.4	4.1	172	1965	Cyg	98	oc I1310	G5
ADS 13486AD	20 09.4	35 41	11.3	50	10.4,10.5	17.2	52	1926	Cyg	98	oc 6883	B2+B2
ADS 13490	20 09.5	35 43	11.4	52	10.2,11.5	8.5	359	1933	Cyg	98	oc 6883	
ADS 13506	20 10.0	00 43	12.6	52	V= 6.9, 7.2	3.0	208	1958	Aql			A:B9p; fixed since 1830
ADS 13515	20 10.3	38 12	12.1	21	V= 7.2,10.5	13.8	60	1929	Cyg	98	gn 6888	A:G8IIICN+1
ADS 14259	20 43.6	30 32	45.7	43	V= 4.2, 9.4	6.2	67	1964	Cyg	99	gn 6960	=52 Cyg; A:G9III
ADS 14556	20 59.7	06 59	02.2	11	7.1, 7.1	2.6	219	1964	Equ			=2 Equ; F8
ADS 14715	21 08.6	09 21	11.0	33	7.8, 8.0	2.7	81	1965	Equ			AB:A3IV; slowly closing
ADS 15434	21 52.4	62 23	53.8	37	V= 7.0, 7.9	62.5	145	1970	Cep	70	oc 7160	A comp is eclipsing binary EM Cep: B0.5+B1Ve;
												B:B1III/V
ADS 15562	21 59.7	−17 12	02.4	58	V= 7.1, 7.2	3.7	244	1973	Aqr			=29 Aqr; A2V+K0III; A comp is eclipsing binary
												DX Aqr
ADS 15639	22 04.6	00 20	07.1	34	7.6, 8.0	2.5	98	1970	Aqr			G0; slowly widening
ADS 15785	22 13.2	49 38	15.2	53	V= 9.3, 9.6	9.4	11	1967	Lac	135	oc 7243	B8V+B9V
ADS 15934	22 23.9	−17 00	26.6	45	V= 6.2, 6.4	3.1	334	1976	Aqr			=53 Aqr; G0V+G0V
ADS 16260	22 44.1	57 49	46.1	04	V= 7.6, 8.6	30.9	117	1924	Cep	71	oc 7380	F8:G0; common proper motion; fixed since 1845
ADS 16334	22 50.8	60 39	52.8	55	7.8, 9.8	8.3	138	1916	Cep	71	oc 7419	A:A0:V+G0:V
ADS 16579	23 09.4	−12 12	12.0	56	7.2, 7.2	3.6	278	1979	Aqr			K0
ADS 16777	23 25.9	03 17	28.5	34	10.0,10.1	1.3	48	1978	Psc	199	eg 7679	G5
ADS 17081	23 50.7	61 41	53.2	58	V=10.4,10.7	6.6	341	1977	Cas	65	oc K 12	
Ali 516	03 28.0	37 13	31.2	23	V=10.4,11.2	13.7	234	1930	Per	191	oc 1342	
Barton 2120	06 32.1	08 22	34.8	20	10.6,11.8	4.8	81	1978	Mon	169	oc 2251	
Barton 2926	07 50.7	−26 09	52.8	17	V=10.8,12.6	4.4	106	1911	Pup	204	oc Ha 19	
Brisbane 14	18 57.7	−37 08	01.1	04	V= 6.5, 6.7	12.8	281	1967	CrA	91	gn 6726	B8IV−V+B9V
Leonard 55	23 55.7	60 56	58.2	13	10.9,11.1	2.4	264	1970	Cas	67	oc 7790	=CEab Cas; double Cepheid variable
Melbourne 8AB	16 37.6	−48 40	41.3	46	V= 5.7, 9.0	1.6	14	1938	Ara	31	oc 6193	O5IIIn
OΣΣ 82	07 01.7	01 34	04.3	29	V= 6.6, 7.4	90.4	318	1925	Mon	170	oc 2324	A:A0V
South 698	18 01.2	−22 30	04.2	30	V= 7.3, 8.8	29.6	316	1900	Sgr	210	oc 6531	B0+B8
Stein 185	01 05.2	61 20	08.4	36	10.8,12.5	8.1	129	1903	Cas	63	oc 381	
Stein 1223	23 47.6	62 26	50.1	43	V=11.4,12.4	9.8	72	1908	Cas	65	oc K 21	
Stein 1560	01 16.0	58 05	19.2	21	9.1, 9.9	13.5	319	1911	Cas	63	oc 457	B2
Stein 1741	01 55.4	55 12	58.7	27	V=11.4,11.4	14.4	235	1912	Per	188	oc 744	
Stein 2630	22 08.8	56 10	10.6	25	11.7,12.7	10.6	197	1917	Cep	70	oc 7226	
Winnecke 4	12 19.8	58 22	22.2	05	9.6, 9.9	51.3	80	1958	UMa	252	eg 4284	=Messier 40; G0; AGK3 measure

ID	RA (1950) DEC (2000)				MAGNITUDES	SEP	PA	DATE	CON	PG	OBJECT	NOTES A 4
Wirtz 12	15 14.3	55 38	15.6	27	11.1,11.5	9.1	17	1917	Dra	106	eg 5905	=Stein 2320
h 367AB	05 28.1	34 15	31.4	17	11.1,12.1	8.0	239	1902	Aur	35	oc 1931	
AC					,12.5	10.2	310	1902	Aur	35	oc 1931	
AE					,13.8	14.8	17	1899	Aur	35	oc 1931	
h 684AB	04 45.7	10 51	48.5	56	V= 8.3,10.0	24.2	264	1910	Ori	176	oc 1662	G5
AC					V= , 9.6	46.5	211	1910	Ori	176	oc 1662	
AE					V= , 9.5	77.7	239	1910	Ori	176	oc 1662	E:A2
CD					V= 9.6,11.3	10.5	307	1910	Ori	176	oc 1662	
h 699	05 25.1	35 21	28.4	23	V=10.4,11.6	9.7	216	1932	Aur	35	oc 1907	
h 1043	00 36.7	60 48	39.6	04	11.6,12.7	11.3	173	1908	Cas	62	oc 189	
h 1098	01 55.4	59 56	58.9	11	10.3,12.7	16.5	343	1911	Cas	64	oc 743	
h 1606	20 59.1	54 20	00.6	32	9.3,10.2	18.4	186	1921	Cyg	99	pn 7008	K7
h 2167	02 57.8	44 41	01.1	53	9.4, 9.9	37.2	34	1909	Per	189	eg 1161	F5+G0
h 2301	06 09.5	05 28	12.2	27	V=10.9,11.8	9.6	352	1911	Ori	179	oc 2186	
h 2491	09 13.7	34 44	16.8	31	10.2,10.3	14.7	200	1912	Lyn	161	eg 2793	
h 2493	09 16.8	33 57	19.8	44	10.1,11.7	10.0	157	1906	Lyn	161	eg 2825	
h 2596	12 04.8	43 23	07.3	06	8.2,10.7	34.1	239	1916	CVn	46	eg 4111	K0
h 2810	17 46.7	−19 59	49.7	00	7.7,10.5	40.8	189	1918	Sgr	209	pn 6445	A2
h 2827	18 14.2	−19 53	17.2	52	10.0,10.1	19.9	254	1930	Sgr	212	gn 6589	
h 3265AB	05 04.7	36 59	08.0	03	V=10.2,10.2	15.0	137	1932	Aur	33	oc 1778	B8+B8
AC					V= ,13.0	15.1	22	1895	Aur	33	oc 1778	
h 3930	07 06.0	−13 05	08.3	10	V= 9.9,10.9	15.4	72	1910	CMa	58	oc 2345	
h 4962AB	17 31.4	−32 33	34.7	35	V= 5.6,10.9	5.4	102	1933	Sco	218	oc 6383	A:O6V+O6V
AC					V= ,10.2	13	83	1936	Sco	218	oc 6383	
h 4966AB	17 34.3	−34 59	37.6	01	10.0,10.3	1.0	117	1936	Sco	219	oc 6396	=Innes 1007AB
AC					V= 9.8,10.8	12.3	273	1936	Sco	219	oc 6396	
CD					11.7,12.0	2.5	245	1936	Sco	219	oc 6396	=Innes 1007CD
h 4990	17 50.2	−22 20	53.2	21	9.5,11.0	23.4	300	1920	Sgr	209	oc 6469	B8
h 5501	18 42.1	−01 03	44.7	00	9.5,10.8	23.4	7	1908	Aql	28	oc Be 79	A0
Σ 172	01 47.2	26 51	50.0	06	10.5,10.7	17.6	194	1916	Tri	238	oc Cr 21	F2
Σ 263AB	02 22.9	60 26	26.6	40	V= 8.4,11.2	14.9	103	1962	Cas	65	oc Ma 6	B2
Σ 264CD	02 23.0	60 27	26.7	41	V= 9.9,10.9	16.6	226	1962	Cas	65	oc Ma 6	B2
Σ 745AB	05 32.4	−06 02	34.8	00	V= 8.4, 9.3	28.7	347	1918	Ori	177	oc 1980	A+A
Σ 1052	07 12.2	−10 12	14.6	18	9.1, 9.3	20.0	21	1918	Mon	171	oc 2353	B+B8

Printed in the United States
By Bookmasters